FUNDAMENTALS AND APPLICATIONS OF ROCK MECHANICS

SECOND EDITION

DEBASIS DEB, Ph.D.
HAG Professor
Department of Mining Engineering,
Indian Institute of Technology, Kharagpur

ABHIRAM KUMAR VERMA, Ph.D.
Professor
Department of Mining Engineering
Indian Institute of Technology, Kharagpur

PHI Learning Private Limited
Delhi-110092
2026

In fond memory of ***Shri Asoke K. Ghosh*** *(October 1942 – February 2024), Founder Chairman and Managing Director of PHI Learning, whose vision endlessly inspires.*

The Legacy Continues....

Published by Pushpita Ghosh, PHI Learning Private Limited, Rimjhim House, 111, Patparganj Industrial Estate, Delhi-110092 and Gopsons Printers Pvt. Ltd., A-14, Sector-60, Noida, Gautambudh Nagar, Noida, Uttar Pradesh-201301.

₹ 1295.00

FUNDAMENTALS AND APPLICATIONS OF ROCK MECHANICS, Second Edition
Debasis Deb and Abhiram Kumar Verma

ISBN-978-93-5443-846-2 (Print Book)
ISBN-978-93-5443-936-0 (e-Book)

The export rights of this book are vested solely with the publisher.

To
My Parents

Late Kumud Chandra Deb, and **Smt. Chaya Deb**

—Debasis Deb

To
In Loving Memory of My Wife

Shobhita Verma

—Abhiram Kumar Verma

Contents

Preface

It has been more than eight years since the first edition of this book was published in 2016. The book was presented to the readers with an aim to provide a lucid approach to understand the theory and applications of rock mechanics. The text, in the **first edition** covered the pertinent topics for undergraduate and graduate level courses. Each topic was provided with examples and exercise problems. The wide reception of the book in multiple disciplines compels us for preparing a new edition comprising the latest developments in rock mechanics and rock engineering.

In this edition, several new examples and exercise problems are added along with the answers to the exercise problems. Each chapter is reviewed thoroughly. Feedback and suggestions received from the students and our peers are incorporated by providing in-depth explanation of the concepts. New topics are also added almost in every chapter by augmenting figures and tables, wherever necessary. The instrumentation concepts given in Chapter 6 on *Measurement of in situ stresses and deformation* are expanded further and their applications are included in Chapter 13 on slope stability. Chapter 14 is now more focused on application of numerical methods, AI and ML techniques in rock engineering.

This edition will not only benefit the students of mining, civil and geotechnical engineering but also to the practicing engineers working in mines, tunnels, dams and slopes. The readers are encouraged to contact the authors, Prof. Debasis Deb at deb@mining.iitkgp.ac.in and Prof. Abhiram Kumar Verma at akverma@mining.iitkgp.ac.in, or the publisher at phi@phindia.com for any feedback to improve this textbook in the future.

Debasis Deb
Abhiram Kumar Verma

Preface to the First Edition

The title *Fundamentals and Applications of Rock Mechanics* deals with the foundation of mechanics applied to rock engineering. It is a discipline firmly engaged with the basic and applied researches and activities comprising mining, civil constructions and foundations, dams, vehicular and hydroelectric tunnels, caverns, slope stability, underground storage facilities for foods, fuels, and waste material, underground disposal of municipal and radioactive waste, petroleum wellbore and others. The term 'rock mechanics' signifies the basic sciences of mechanics applied to rock and rock masses. It is also a discipline which analyses the stability of structures made of/on the rock masses. The main factor distinguishes the rock mechanics from other engineering mechanics discipline is the application of mechanics on a large scale pre-stressed naturally occurring material with the properties governed by discontinuities. For the last several decades, principles of rock mechanics have evolved through theoretical, laboratory and field experimentations and have imbedded firmly and popularly into the analysis of rock engineering problems. The theoretical development of rock mass failure criteria and subsequent understanding of non-linear behaviour of jointed rock masses have made it possible to analyse complex interactions of structures made of/on the rock mass. Rock mechanics also deal with the fundamental sciences related to rock-support interactions to improve stability and safety of structures. Today, rock mechanics is the first subject to be taught in mining and geotechnical engineering curricula.

Theoretical and applied rock mechanics literatures are very large to list all publications or even to discuss all possible concepts. This text is intended to address the fundamental concepts of rock mechanics and their applications in the area of rock engineering problems. The material presented in this book is the elaboration of a course of lectures originally prepared for the undergraduate mining engineering students at Indian Institute of Technology (IIT), Kharagpur. The authors believe that the contents of this book are sufficient for more than two semester course for undergraduate and postgraduate students of civil, mining, geotechnical and petroleum engineering. The professional engineers involving with rock mechanics applications and design may also use this book profitably as a reference or learn about state-of-the-art techniques for the applications of the rock mechanics principles in their specialisation. This book will be equally useful as a reference material for the practicing engineers and rock mechanics professionals working in design and applications aspect of rock mechanics.

Chapter 1 introduces rock as a natural material, its formation and comparison with soil. The geological material is like a rock mass which incorporates the intact rock and geological discontinuity like fractures, joints, fold, faults, bedding planes. Chapter 2 elaborates on three phases of soil and rock matrices that is air, water and solid. The discussion is limited to estimate various physical properties of soil and rock matrices. Concept of porosity and permeability is also provided with explanation of weight-volume relationships. Chapter 3 defines the characteristics of rock mass and discusses the rock mass classification system like RQD, RMR, Q, GSI and CMRR. Finally, these classification systems are used to determine the rock mass strength, modulus of deformation, standup time and support requirements for different rock masses. Chapter 4 introduces stress and strain analysis that has been enhanced by introducing stress and strain in polar coordinate system and by providing the concept of Airy stress function. It highlights the basic concepts of transforming stress and strain tensors and displacement vector from one coordinate system to another. The fundamentals of principal stress and strain, Mohr's circle of stress and strain and Airy stress function are explained in detail with worked-out example wherever appropriate.

Chapter 5 deals with the constitutive relations of isotropic and anisotropic elastic rocks. Detailed description of plane stress, plane strain and axisymmetric conditions is given with examples. The concept of elastic-plastic properties of rocks is also presented in this chapter. Chapter 6 introduces the concept of *in situ* and induced stress and their measurement techniques. The chapter also introduces the instruments used for the measurement of displacement, load and stresses in the rock structures. Chapter 7 discusses the induced stresses around circular and elliptical opening. The concept of zone of influence, stress field due to multiple opening, stability of discontinuity around circular opening with or without discontinuity and zone of failure around opening are discussed. In Chapter 8, different types of pillars (flat and inclined) are introduced with their design principles. The effect of size and shape on strength of pillars are discussed along with the major empirical strength relationships. Pillar stress during development (tributary area method) and depillaring operations (advanced pillar design) are also discussed with examples. Pillar stabilisation by bolting and backfilling/stowing are introduced. Some of the specialised methods based on a practical approach like yield pillar technique are also discussed. In Chapter 9, fundamentals of rock mass yield/failure criteria are explained. This chapter also introduces the concept of deviatoric plane followed by derivation of yield criteria using invariants of deviatoric stresses. Analytical solution of stress field around circular opening in plastic condition is also discussed.

Chapter 10 introduces rock joint in rock mass. Rock joint properties like shear strength models are introduced with the parameters like JRC, concept of joint compliance and their deformability. In Chapter 11, mechanics of different types of reinforcing supports like bolts, shotcrete/gunite, roof/side stitching, roof truss, fore poling, wire nets, W-strap and roof supports like prop, chock (or cogs/chockmate), roof bars, safety support (goal post), steel arches, and backfilling/stowing are discussed. The numerical formulation of rock bolts is discussed in detail. Chapter 12 introduces the concept of subsidence with the method of subsidence prediction and the concept of influence function, whereas Chapter 13 introduces the slope stability issues in rock mass as well as loose material like soil. The mechanics behind the failure along with stabilising techniques are discussed in the chapter. Plane failure, wedge failure, toppling failure and circular failure are explained with the help of worked-out examples. In Chapter 14, various numerical methods applied in rock engineering problems are introduced giving possible applications, their advantages and disadvantages. It is not the intent here to describe each method mathematically, but the main idea is to expose the readers with various numerical tools that are available for solving complex rock engineering problems.

The prime motivation for writing this book came from the students, academics, researchers and field practitioners at various institutes in general and students at IIT, Kharagpur in particular.

Over the last few years, authors have realised the need for a textbook which provides detail concepts of rock mechanics with worked-out examples and elaborates discussions for better comprehension of the subject. This book is intended to reach the students by introducing the fundamentals of mechanics for rock material and extend their theoretical knowledge in applied areas of rock mechanics. The fundamentals are introduced with illustrations, solved problems and exercises in a comprehensive manner. With this background, the authors have been motivated to pen down the above chapters so that the students can apprehend the concepts and at the end, able to solve the problems on rock engineering. Any suggestions for improvement of the book will thankfully acknowledge. Please contact us at deb@iitkgp.ac.in, deb.kgp@gmail.com, verma@iitkgp.ac.in, or abhiram.verma@gmail.com. You may also send your feedback at phi@phindia.com.

Debasis Deb
Abhiram Kumar Verma

Acknowledgements

The authors are thankful to their teachers who inculcated the knowledge of rock mechanics and motivated them to disseminate the knowledge to the students, academicians, researchers, field practitioners and finally the humanity at large through classroom teaching and research.

The authors are thankful to their numerous students, especially Dr. Rakesh Kumar, Dr. I. Sreenivas Rao, Dr. Kamal C. Das, Dr. Ranjan Pramanik, Dr. Sumant Mohato, Dr. Shruti Narwal, Mr. Sonu Kumar Pasi, Mr. Prakash Chandra Jat, Mr. Babulal Rohidas, Dr. Sujeet Bharti, Mr. Ritam Roy, Mr. Asif Ahmed and many more, who have helped them directly or indirectly in solving exercise problems, critical discussions, proofreading the manuscript and incorporating figures and exercises over the years.

The authors express their deep gratitude to all their colleagues for their criticisms spanned from various institutions, being a part in their carrier, especially the Department of Mining Engineering, IIT, Kharagpur for their support during preparation of this book.

Authors are also thankful to readers' comments and suggestions on the first edition.

This book could not be completed without constant encouragement and loving support from the authors' family members. Finally, the authors express their deepest gratitude to their brothers, sister and in-laws for their encouragement and support.

Finally, the contribution of Dr. Deb would not have possible without loving cooperation from his wife, Ms. Samhita Deb; daughter, Didhiti; and son, Dibyabha for their unbounded understanding and freedom given to him for timely completion of the book.

Dr. Verma's deepest gratitude goes to his wife, Late Ms. Shobhita Verma, who encouraged him with unfailing support since the idea of this book was conceived. Lovely daughter, Aashvi Verma keeps him moving with her innocent comments with cheerful behaviour; while the unconditional support, blessings, and unwavering presence of his parents, Sri. Arun Kumar Verma and Smt. Manisha Verma in every walk of life keeps him stronger.

Special thanks are due to the editorial and production teams at PHI for their efforts and making necessary measures for publishing the second edition of this book.

Debasis Deb
Abhiram Kumar Verma

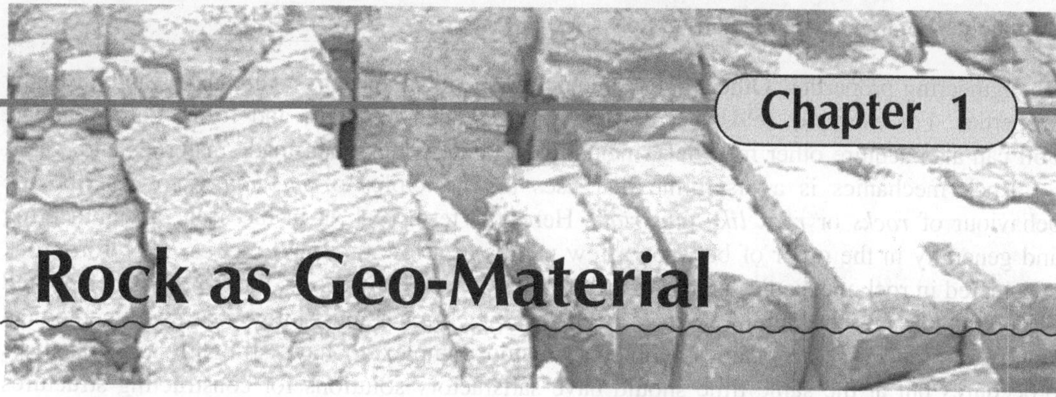

Chapter 1

Rock as Geo-Material

1.1 INTRODUCTION

This is the basic chapter introducing the rock mass as a natural material incorporated with geological discontinuities. These geological discontinuities make the rock different from the engineering material like steel. Rock geological cycle includes the transition from one form to the other over millions of year and billions of year to complete the cycle while passing through the various phases from rock to soil through weathering process, conversion of rock and soil from magma. The chapter also briefly discusses the rock types like igneous, sedimentary and metamorphic rock; type of discontinuities; types of faults; folds and bedding planes. At last the concept of stereonet projection are introduced with worked-out examples.

1.2 ROCK AS NATURAL MATERIAL

Rock is an aggregate of minerals which are naturally occurring substances having a fixed chemical composition and molecular structure. As the basic constituents of rocks, minerals control much of rock behaviour. Mineral is defined as a naturally occurring inorganic substance having fixed chemical composition and molecular structure. Some minerals are very strong and resistant to deterioration and they produce rocks with similar properties. On the other hand, some other minerals are softer in nature and hence produce weaker rock. More than 2500 different minerals are present in the earth's crust and are classified as Silicate, Sulphide, Sulfate, Carbonate, Oxide, Phosphate, Halide and Limonite minerals. The most common minerals include feldspar ($KAlSiO_8$–$NaAlSiO_8$–$CaAlSi_2O_8$), quartz (SiO_2), ferromagnesian minerals, iron oxide (Fe_2O_3), calcite ($CaCO_3$), dolomite [$CaMg(CO_3)_2$], mica and gypsum ($CaSO_4.2H_2O$). A mineral may be a single element also known as native element, such as Silver (Ag), Platinum (Pt), Iron (Fe), copper (Cu) or gold (Au) or it may be a compound made up of a number of elements. The upper part of earth crust includes 60% Feldspar, 17% Amphiboles and Pyrozine, 12% Quartz and 3.8% Mica.

Rock may break down due to weathering process into soils and many of these minerals remain in their original form. For example, many sand grains are made of quartz and thus reflect its engineering properties. Other minerals undergo chemical and physical changes and take new properties. For example, feldspar often experiences such changes, and forms clay minerals. Soil can also acquire other materials, including organic matter, man-made materials and water.

Rock mechanics is a discipline that uses the principles of *mechanics* to describe the behaviour of *rocks* or *rock like materials*. Here, the term rock is in the scale of engineering and generally in the order of between a few metres to a few thousand metres. Therefore, rock considered in rock mechanics is in fact the rock mass, which composes intact rock materials and discontinuities present therein. A rock mechanics practitioner must know that many geological engineering and rock mechanics problems are too complex to have straightforward analysis procedures but at the same time should have satisfactory solutions for constructing structures in rock mass tunnel, slope, surface and underground mines, etc.

For normal construction materials or engineering or man-made, e.g., steel and concrete, the mechanical behaviours are well defined because these materials are generally continuous, homogeneous, isotropic, and linearly elastic (CHILE) and free from any defects. In general, the properties of these man-made materials are known and hence their properties can be controlled. Some properties of rocks are unknown, unpredictable and often cannot be controlled. The basic theory and understanding of rock mechanics is primarily based on assuming that the rock mass is a CHILE material, but in reality it is a DIANE (discontinuous, inhomogeneous (heterogeneous), anisotropy, non-elastic) material, the behaviours of which have been understood over the years but with reservation. There are lots of scope to understand the mechanical behaviour of rock mass with discontinuities. The existence of discontinuity depends on the scale. The discontinuous nature and scale dependence feature is not common in other man-made materials.

Rock mass as a structural material is used in three ways, first as a construction material, second as structures made inside rock mass, finally as foundation material such as structures made on top of the rock mass such as bridges, high rise buildings or skyscraper, etc. Rock mechanics principles are applied to various engineering disciplines like mining, civil and petroleum, in their own ways. In civil engineering, it involves designing foundation, tunnels for hydropower and metro railways, underground civil utility, nuclear repository and underground storage for civil or military purposes and slope. In mining engineering, it involves designing mine slopes, surface opening, underground opening either horizontally or vertically, waste dump and shafts. In petroleum engineering, the stability of deep wellbore, reservoir mechanics during production phase, basin subsidence are important areas of applications.

As rock is a natural material, there is little scope to enhance its inherent physico-mechanical properties but for the application of artificial support or reinforcements, these properties can be upgraded based on the application in hand. Hence, the objectives of rock mechanics are to study:

(i) Physico-mechanical properties of intact rock, rock mass, joints and other form of discontinuities
(ii) Behaviour of structures made in rock mass under various loading conditions such as *in-situ* stress, blasting, earthquake and others
(iii) Mechanism of rock mass failure
(iv) Stability of rock structure, foundation design for surface and underground excavation

Based on the above study, design of rock structure and foundation in a given rock mass is performed for safe extraction of minerals or hydroelectric tunnels.

1.3 INTACT ROCK VERSUS ROCK MASS

Intact rock is defined in engineering terms as rock containing no significant fractures. However, on a small scale it is composed of grains in the form of microstructure being governed by the basic rock forming processes. Geological events over the years may affect its mechanical properties and its susceptibility to water penetration and weathering effect. This type of rock sample is normally tested in laboratory for estimation of strength, elastic modulus, and other engineering parameters.

Rock mass includes joints, fractures, cracks and other geological discontinuities in rock matrix. In other words, rock strata in abundance are termed as rock mass. It is obvious that the engineering properties obtained from laboratory testing of intact rock would be higher values than those of rock masses. It is also difficult to estimate rock mass properties directly in the field or in the laboratory. Several rock mass classification systems are developed over the years to quantify the condition of rock mass in terms of local geology, strength of intact rock matrix and water conditions. Detail descriptions of rock mass classification systems are given in Chapter 3.

Intact rock and rock mass are defined by antonyms CHILE and DIANE respectively as mentioned earlier. In short, intact rock is called CHILE material whereas rock mass (or real rock) is called DIANE material. In reality, the rock mass behaves in between the CHILE and DIANE materials defined below.

Rock without any defects are continuous whereas with defects are discontinuous. It is also true for displacement and strain fields. Homogeneity represents uniform composition throughout and hence exhibit similar properties at every location in rock. It is obvious that properties would vary based on the location in rock mass due to non-uniform composition and generally termed as inhomogeneous property. It is because rock masses are composed of various rock types, inter-bedding and intrusion of foreign material. If material properties are independent of direction, they are termed as *isotropic properties.* An anisotropic rock has different properties in different directions due to regular jointing, bedding or oriented fabric. Some sedimentary rocks, e.g., shale, have noticeable anisotropic characteristics. Other sedimentary rocks may not have clear anisotropy due to discontinuities within. However, under the influence of rock formation process and pressure, small degree of anisotropy is possible. Rock with most obvious anisotropy is slate. Phyllite and schist are the other foliated metamorphic rocks that exhibit anisotropy. Other variations are transversely isotropic and will be discussed in Chapter 5. A rock is called linear elastic if it follows similar loading and unloading load-deformation (stress-strain) curves, and the sample returns to its original strain-free state when the loads are removed. In other words, upon removal of deforming forces, the size and shape of a body return precisely to their original state. On the contrary, in non-linear elasticity, load varies non-linearly with deformation and rock sample returns to its original state after unloading. However, it may take a different path than loading curve.

1.4 ROCK AS GEOLOGICAL MATERIAL

In geological term, rock can be considered as particulate substance such as sand/clay particles or in large form such as mountains, tectonic plates, planetary cores, planets, etc. Figure 1.1

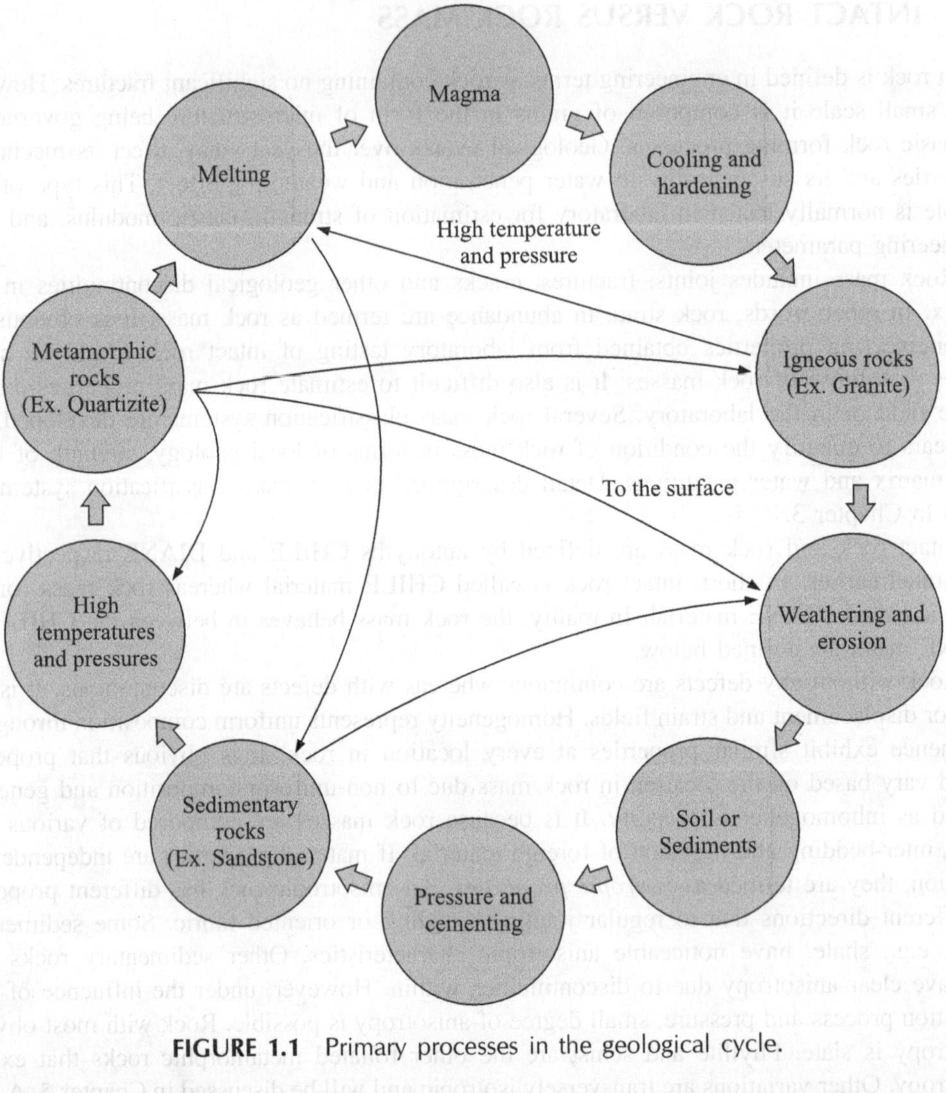

FIGURE 1.1 Primary processes in the geological cycle.

shows the geological process involved in the formations of various rocks. It should be noted that the processes are dynamic and continuous. The process begins with molten magma (from within the earth) forms the rock, and then continues with the rocks being broken down into soil. Again, soil can convert back into rock. Rocks are classified according to their place in the geologic cycle. The three major categories of rock are classified based on their origins: (i) igneous rocks from magma, (ii) sedimentary rock from sediments lithification and (iii) metamorphic rocks through metamorphism due to high pressure and temperature.

1.4.1 Igneous Rocks

Igneous rocks are formed when molten rock (magma) cools and solidifies, with or without crystallisation. There are two primary types of igneous rocks: *intrusive* (also plutonic) rocks form below the ground surface, where they cool slowly, whereas *extrusive* (also volcanic) rocks arrive at the ground surface in a molten state, such as through a volcano, and then cool very rapidly. Intrusive rocks include both large bodies, known as *plutons*, and smaller sheet like bodies, known as *sills* and *dikes*, those fill cracks inside other rocks. Extrusive rocks generally have finer grains and smoother surfaces.

Magma from which minerals crystallise is rich in silicon, oxygen, aluminium, sodium, potassium, calcium, iron, and magnesium elements. These elements combine to form a silicate mineral, which account for over 90% of all igneous rocks. Igneous rocks make up approximately 95% of the upper part of the earth's crust, but they are buried on the earth's surface by a relatively thin layer of sedimentary and metamorphic rocks.

The magma can be derived either from the earth's mantle or pre-existing rocks molten by extreme temperature and pressure changes. Figure 1.1 shows the origin of magma and igneous rock through the geologic cycle. Common igneous rocks are granite, basalt, diorite, andesite, rhyolite, gabbro, kimberlite, peridotite, serpentinite, and others.

1.4.2 Sedimentary Rocks

Rocks exposed to the atmosphere are subjected to physical, chemical, and biological breakdown through weathering. The process of weathering results in disintegration of rock into boulders, cobbles, pebbles, gravels, granule, sand or even soil. All rocks disintegrate slowly as a result of mechanical weathering and chemical weathering. *Mechanical weathering* is the breakdown of rock into particles without producing changes in the chemical composition of the minerals in the rock. *Chemical weathering* is the breakdown of rock by chemical reaction and changes in chemical composition.

The weathered material transformed back into rock through the hardening process called *induration* or *lithification*, thus forming the second major category of rocks called sedimentary rock. Sedimentary rocks cover 75% of the earth's surface, but count for only 5% of the rock in the earth's crust. There are three basic types of sedimentary rocks: (i) *clastic sedimentary rocks* such as breccia, conglomerate, sandstone and shale, out of which shale and sandstone are most common, formed from mechanical weathering; (ii) *chemical sedimentary rocks* such as dolomite, rock salt, some limestones, iron ores (like hematite, magnetite, goethite, limonite, siderite, etc.) that form when dissolved material precipitate from solution; and (iii) *organic sedimentary rocks* such as coal, oil shale (or kerogen shale) and some limestones which form from the accumulation of plant or animal debris. Second and third category rocks are also known as *carbonate sedimentary rock.*

Clastic rocks form when soil or sediment deposits become hardened as a result of pressure from overlying strata and cementation through precipitation of water-soluble minerals such as calcium carbonate or iron oxide. Because of their mode of deposition, many clastic rocks are layered or stratified, which makes them quite different from massive formations. The interface between these layers are called *bedding planes*.

1.4.3 Metamorphic Rocks

At times, igneous, sedimentary or older metamorphic rocks are subjected to intense heat and extreme pressure while deep in the earth's crust. Due to this reason, they undergo profound physical and/or chemical change, thus forming the third type of rock called *metamorphic rock.* The metamorphic processes generally improve the engineering behaviour of these rocks by increasing their hardness and strength. Above 200°C, heat causes minerals to re-crystallise. Pressure forces some crystals to re-orient. The combined effects of re-crystallisation and re-orientation usually lead to *foliation*, which is a unique feature of metamorphic rocks. It occurs when a strong compressive force is applied from one direction to a re-crystallising rock. This causes the platy or elongated crystals of minerals, such as mica and chlorite, to grow with their long axes perpendicular to the direction of the force. This results in a banded or foliated rock. These foliations are important because the shear strength is less for stresses acting parallel to the foliations. Common metamorphic rocks which are foliated are slate, schist, gneiss and non-foliated rocks are quartzite and marble. Gneiss is formed from ignious rock granite whereas, slate, marble, quartzite are formed from sedimentry rocks shale, limestone, and sandstone respectively. Schist is formed by the transformation of another metamorphic rock, i.e., slate, whereas slate is formed after metamorphosis of shale.

1.5 ROCK VERSUS SOIL

Rock and soil are the two broad categories of geo-material. Soil is formed by the disintegration of rock by weathering activities of nature. So, the composition of both rock and soil are same, but physical and mechanical behaviours differ significantly. The properties may vary because of chemical weathering. In a nutshell, weathering can take place together or independently in the form of physical weathering (disintegrating), chemical weathering (decomposition) and chemical transformation. Because soil is originated from rock, rock can be considered as mother of soil.

Goodman (1990) mentioned the differences between rock and soil as given below:

(i) Rocks are generally cemented; soils are rarely cemented

(ii) Rocks usually have much lower porosity than soils

(iii) Rocks can be found in stated of decay with greatly altered properties and attributes; effects of weathering on soils are more subtle and generally less variable

(iv) Rock masses are often discontinuous, and jointed; soil usually can be represented as continuous

(v) Rocks have more complex, and generally unknowable stress histories. In rock masses, the minor principal stress is vertical; in most soils the major principal stress is vertical.

1.6 DISCONTINUITIES IN ROCKS

The term discontinuity refers collectively to joints, faults, bedding planes, schistosity, folds and other similar defects in rock. A rock mass consists of rock matrix and combination of any of such discontinuities. In the following, introduction of a few discontinuities are given. Interested readers may learn more about these features in any structural geology textbook.

1.6.1 Fractures and Joints

Fractures are planar surfaces along which rocks have been broken causing two separate blocks. The surface may not have cohesion or may contain foreign material which causes cohesion.

Fractures are formed due to extensions and shear mode of failure. Understanding of physical properties of fractures are of great interest for determining permeability of rock to fluid flow, migration of oil and gas, stability of structures made in rock mass. In general, many fractures exist in the area having similar orientation and extension with very little displacement parallel to a fracture surface. A group of these fractures is termed as *joint set*. Two or more joint sets affecting the same volume of rock is considered as *joint system*. Figure 1.2 shows a joint system having three joint sets. Chapter 10 enumerates properties of joints and their analysis in great detail.

FIGURE 1.2 A typical joint system in iron ore at Barbil, Odisha, India.

Joints can be formed due to erosion of the overlying strata exposed at the surface. The removal of overlying rock results in change of stresses, and hence leads to the fracturing of underlying rock. Joints can also be caused by cooling of hot rock masses, which form cooling joints. Columnar jointing or columnar basalts are typical joint features by cooling. Joints are also formed by tectonic movement. The joints are often found in sets. A joint set is a group of parallel joints. Typically, a rock mass can have between one to a few joint sets. Joints are the most common type of rock discontinuities. They are generally considered as part of the rock mass, as the spacing of joints usually is between a few centimetres and a few metres. At times, joints persist over thousands of square kilometre having gradual variation in orientation. These joints appear on the surface as leaner fractures and normally termed as *lineaments*.

1.6.2 Faults

Geologic faults are planar rock surfaces along which one side has moved relative to other side. Most of the faults are occurred due to brittle shear fracture or zones (Twiss and Moores, 2007). Generally, faults are termed to those shear fractures or zones which extend over metres of distance or more. The largest examples are at tectonic plate boundaries, but many faults occur far from active plate boundaries. Since faults usually do not consist of a single, clean fracture, the term fault zone is used when referring to the zone of complex deformation associated with the fault plane as shown in Figure 1.3(a). Figure 1.3(b) shows a portion of San Andreas Fault in California, USA.

(a) A typical normal fault at Singrauli, Madhya Pradesh, India

(b) A portion of San Andres Fault

FIGURE 1.3 Fault zone.

A *shear zone* is a wide zone of distributed shearing in rock. Typically this is a type of fault, but it may be difficult to place a distinct fault plane into the shear zone. Shear zones can be only centimetre wide, or up to several kilometres wide. Fault zone and shear zone are large scale geological features. They are often dealt separately from the rock mass. Small scale single faults often have the similar effects as a joint. The behaviour of large scale fault and shear zones require specific investigation and analysis, if a project is to be constructed over or close to such zones.

The faults are classified as (Figure 1.4):

1. *Dip-slip fault*: It is of two types—(a) normal, and (b) thrust
2. *Strike-slip fault*: It is of two types—(a) right-lateral or dextral, and (b) left-lateral or sinistral
3. *Diagonal (oblique)-slip fault*: It is of two types—(a) sinistral-normal, and (b) sinistral-reverse
4. *Rotational fault:* Rotational movement is exhibited between the blocks.

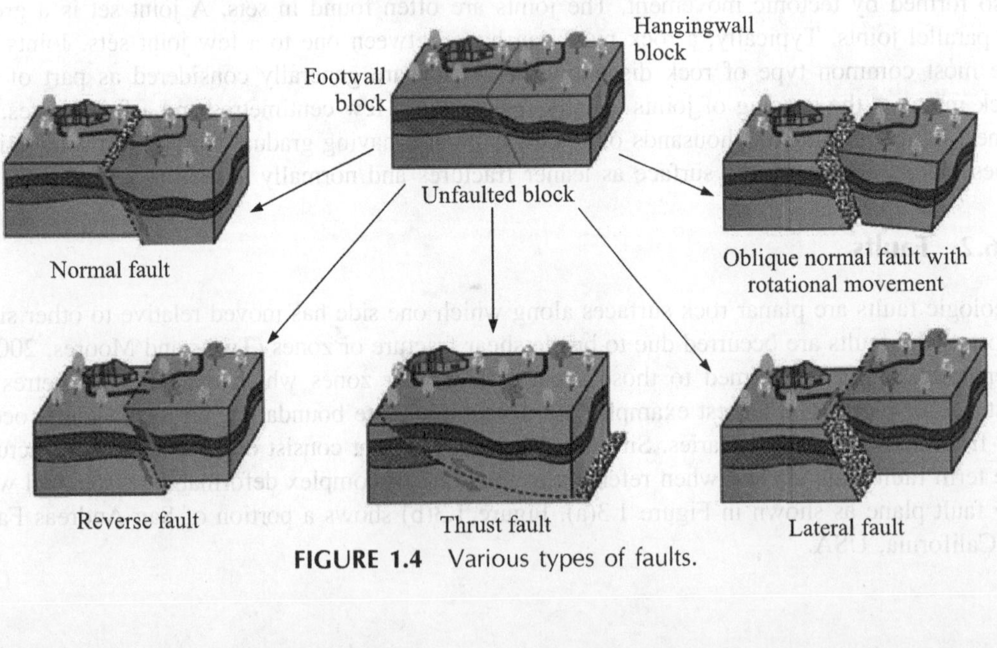

FIGURE 1.4 Various types of faults.

1.6.3 Folds

Folds are commonly observed in sedimentary formation. A fold is recognised by wavy form of crustal rock mass as shown in Figure 1.5. Existence of folds proves that ductile deformation had occurred in the earth's crust. Based on appearance, twenty different folds are defined such as anticline, syncline, inclined, vertical, recumbent, isoclined, overturned, chevron, box, monocline (or homocline), structural tress, anticlinorium, synclinorium, geanticline, geosyncline, basin, dome, drag, fan and pitching. The folds those are oriented concave downward they are called *anticline* and those are concave upward they are called *syncline*.

FIGURE 1.5 Folds.

Folds, particularly intense folds, are often associated with high degrees of fracturing in relatively weak and soft rocks. Although the folding feature may not be directly taken into account of rock mass, but the results of folding are often reflected in the rock mass consideration. Sometimes these folds are gradual; other times they are very abrupt.

1.6.4 Bedding Planes

Bedding plane is defined as a plane of separation between layers and stratification plane. Therefore, bedding plane is a discontinuity separating different rocks (Figure 1.6) formed in horizontal or near-horizontal layers over geological time. Bedding plane often can be fully closed and cemented. In metamorphic rock similar plane of weakness is called *schistosity*. However, some bedding planes could also become potential weathered zones and pocket of groundwater. For example, an interface between porous sandstone and limestone may lead to extensive weathering of the limestone, which leads to cavities along the interface.

FIGURE 1.6 Bedding planes.

1.7 STRIKE, DIP AND DIP DIRECTION OF DISCONTINUITY PLANE

Geological structures such as discontinuity planes are characterised by their orientation in space by strike, dip direction and dip as shown in Figure 1.7(a). The strike is the compass direction of the intersection of the discontinuity plane and the horizontal, and is expressed as a bearing from the true north. For example, if a joint has a strike of N40°W (also represented by bearing as 320°), then the traces of the line of intersection of the joint plane with a horizontal plane is oriented 40° west of the true north. The dip is the angle between the geologic surface and the horizontal surface, and is measured in a vertical plane oriented perpendicular to the strike. The dip also needs a direction. For example, a joint with strike N40°W may have a dip direction of N50°E or S50°W since strike and dip direction is 90° apart. When expressed together, this data is called an attitude, and may be written in condensed form as N40°W; N50°E, 60° (dip) or 320°/60°. Attitudes are usually measured in field using a Brunton compass, as shown in Figure 1.7(b).

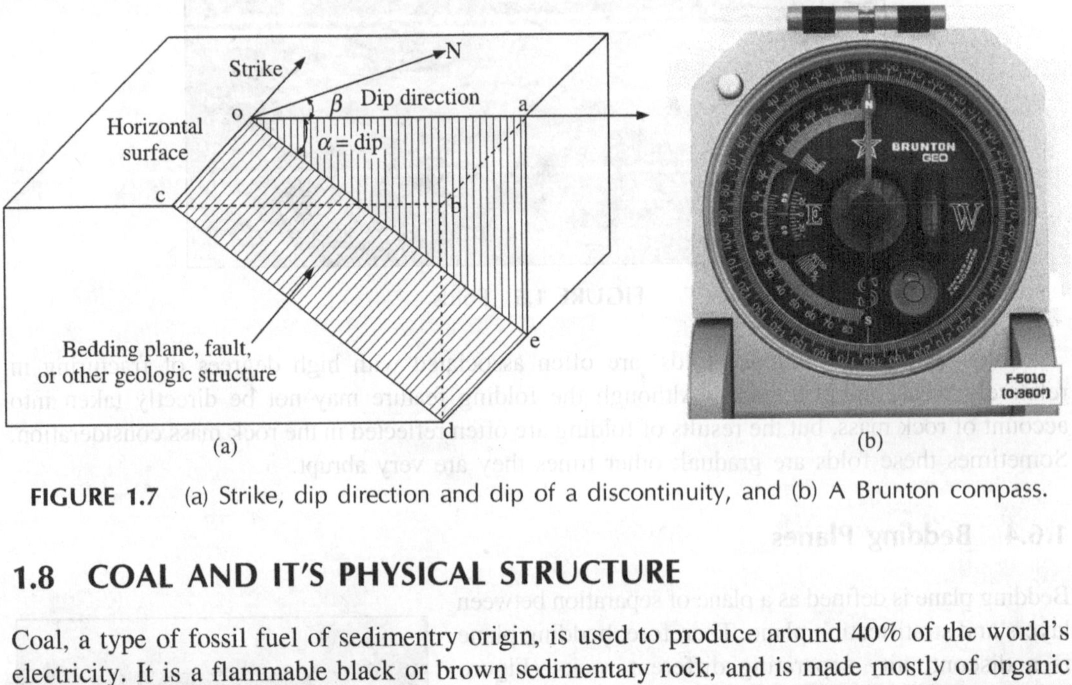

FIGURE 1.7 (a) Strike, dip direction and dip of a discontinuity, and (b) A Brunton compass.

1.8 COAL AND IT'S PHYSICAL STRUCTURE

Coal, a type of fossil fuel of sedimentry origin, is used to produce around 40% of the world's electricity. It is a flammable black or brown sedimentary rock, and is made mostly of organic carbon with variable quantities of other elements, chiefly hydrogen, sulphur, oxygen, and nitrogen. There are two main types of coal: (i) *thermal coal*, mostly used for power generation, and (ii) *metallurgical coal*, mostly used for steel production. Thermal coal is more abundant, has lower carbon content and is higher in moisture content than metallurgical coal.

Geological evidences indicate that coal is produced from plant remains slowly reducing hydrogen and oxygen content of molecules and converting it to peat, lignite, sub-bituminous, bituminous, anthracite and ultimately to graphite (Mazumder, 2012). Anthracite coal is dense having low porosity and fixed carbon of 92–98%. Bituminous coal has volatile matter of 15–20% with fixed carbon of 65–85%, and it is porous or soft in nature. Lignite or brown coal contains moisture of 45–55% with a fixed carbon of about 40–45%.

Mining methods and strata control techniques to extract coal deposit depend upon the morphology of the coal seams. Coal seams can be (i) regular tabular sheet-like geometry [Figure 1.8(a)], (ii) irregular geometry with abrupt change in geometry, (iii) composite geometry having coal and intervening clastic rocks, normally termed as *dirt coal seams* [Figure 1.8(b)], and (iv) split or fingered geometry [Figure 1.8(c)].

Coal seams are also jointed having sub-parallel fractures orienting nearly perpendicular to the bedding plane. These joints are commonly termed as *cleats* or *cleat surfaces*. The cleat directions, especially in bituminous coals, are important in establishing preferred directions of mine development (McCulloch, et al., 1974). The coal tends to break along its natural fracture systems called *face cleat*, also known as main or master cleat which are planar and more or less continuous throughout the coal seam. The *butt cleats*, also known as cross or board cleats, are discontinuous, non-planar and commonly ends at intersection of face cleats. Figure 1.8(d) shows a portion of a coal block identifying face and butt cleat. Therefore, study of cleat system is not only important for finding direction of mine development, but also for understanding methane liberation through CO_2 sequestration from industrial plants in coal seams of abandoned coal mines or deep coal seams.

(a) Regular sheet-like seam (b) Dirty coal seam

(c) Spliting coal seam (Wanless, 1964) (d) Cleat system

FIGURE 1.8 Coal seams and cleats.

1.9 STEREONETS

A discontinuity plane or a line is generally plotted in 3D. As mentioned in Section 1.7, a discontinuity plane is generally represented by strike, dip direction and a dip. If two planes intersect each other, then finding orientation (plunge and trend) of the intersection line becomes cumbersome. For this reason, stereographic projection system is developed to allow 3D orientation data to be represented and analysed in 2D. This projection consists of a reference sphere and a 3D discontinuity plane is projected and analysed in an equatorial horizontal plane. The major idea is to project a 3D incline surface into a horizontal plane by an arc of a circle whose diameter is the strike of the plane (see Figure 1.9).

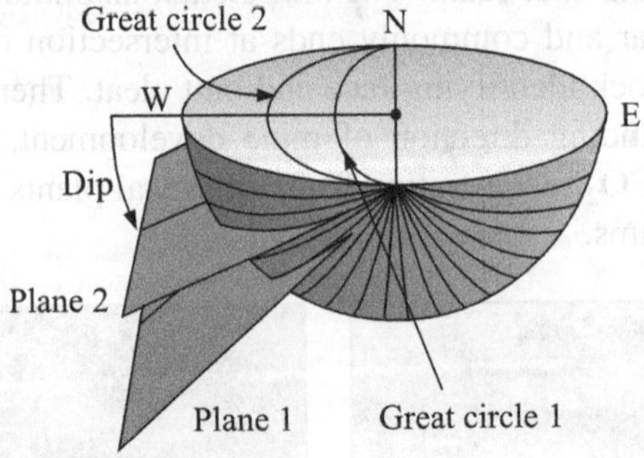

FIGURE 1.9 Schematic diagram of discontinuity planes and their great circles.

Consider a discontinuity plane having a strike N*x*°E having certain dip as shown Figure 1.10(a). Also assume that an inclined line OM having a trend at Ny°E plunge at a certain angle with the horizontal. The discontinuity or the dipping plane is plotted in a reference sphere dipping at the same angle. The centre of the sphere is point O. The strike of the dipping plane is a diameter of the horizontal circle which cuts the reference sphere into two halves as shown in Figure 1.10(b). The dip direction of the plane is a perpendicular line to the strike directed towards the dip of that plane. The dipping plane intersects the sphere producing spherical projection of the dipping plane. Each point on the spherical projection can be produced to the zenith or focus (the top most vertical point on the surface of the sphere above the point O) of the sphere. The point F refers to as the focus for lower hemisphere projection. These lines cut the horizontal plane at various points. The loci of these points make the great circle of the dipping plane on the horizontal plane. The centre of the great circle is the centre of the reference sphere. A line perpendicular to the dipping plane touches the sphere at two locations called *poles* [Figure 1.10(b)]. The pole at lower hemisphere is produced to the focus and the line intersects the horizontal plane at point P. This point is termed as *stereographic projection* of pole or simply pole of the great circle.

Figure 1.10(c) shows an equatorial projection plane (horizontal plane) having great circles at every 10°. The strike of these great circles is along the NS direction. The great circle of the discontinuity plane is plotted showing strike and dip of the plane. The pole of this great circle or normal of the plane is 90° apart from the circle in the opposite side of the dip direction. It is noteworthy that a 3D plane is conveniently represented in 2D without loosening any geometrical information. Similarly, an inclined line can also be plotted by a point as shown in the figure.

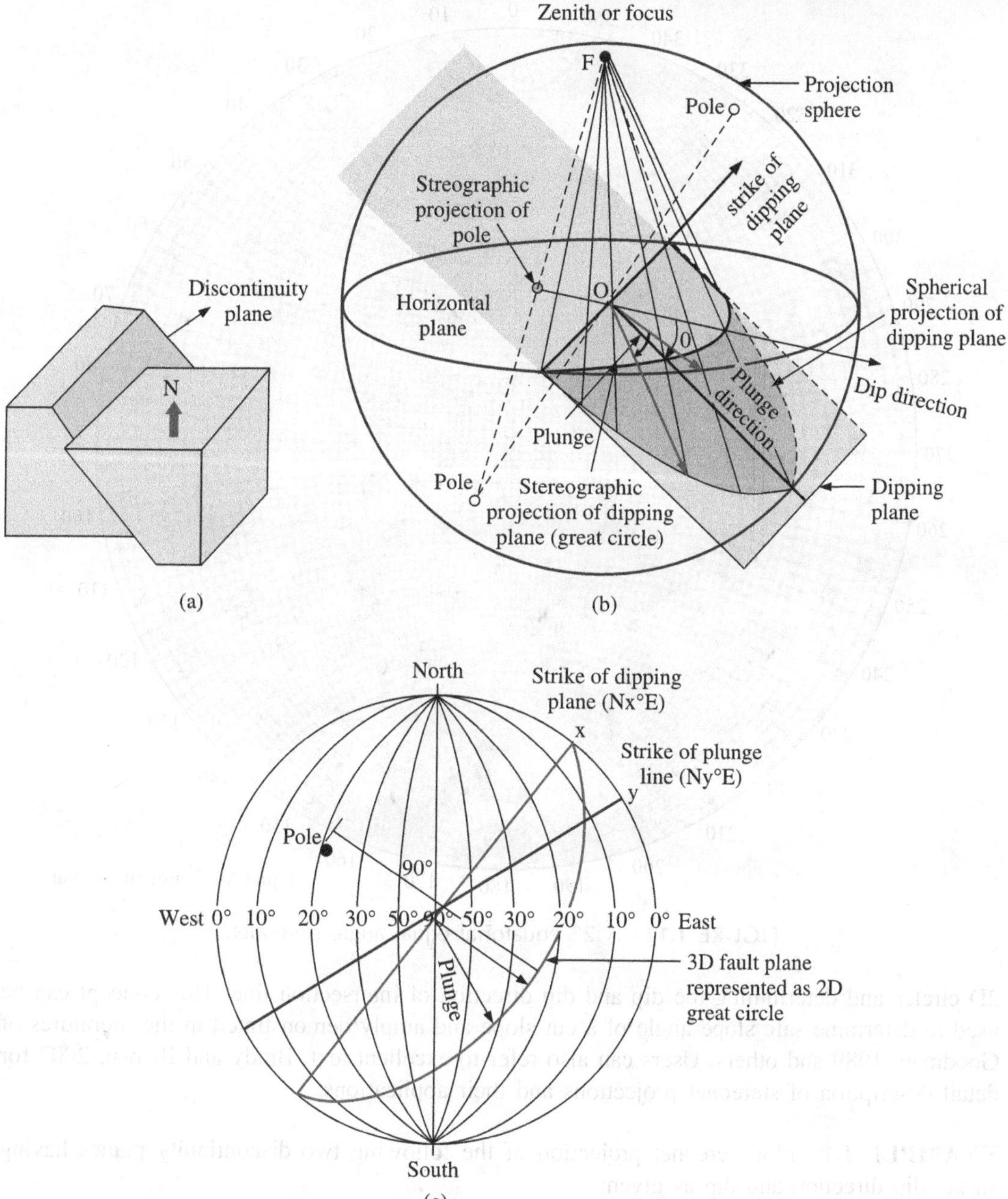

FIGURE 1.10 Stereonet projects of discontinuity plane and plunge line.

Figure 1.11 depicts an equatorial stereonet presenting series of great circles of planes having a dipping increments of 2°. A stereonet projection of a discontinuity plane is plotted using this equal angle equatorial net. There are several software available for projecting 3D planes into

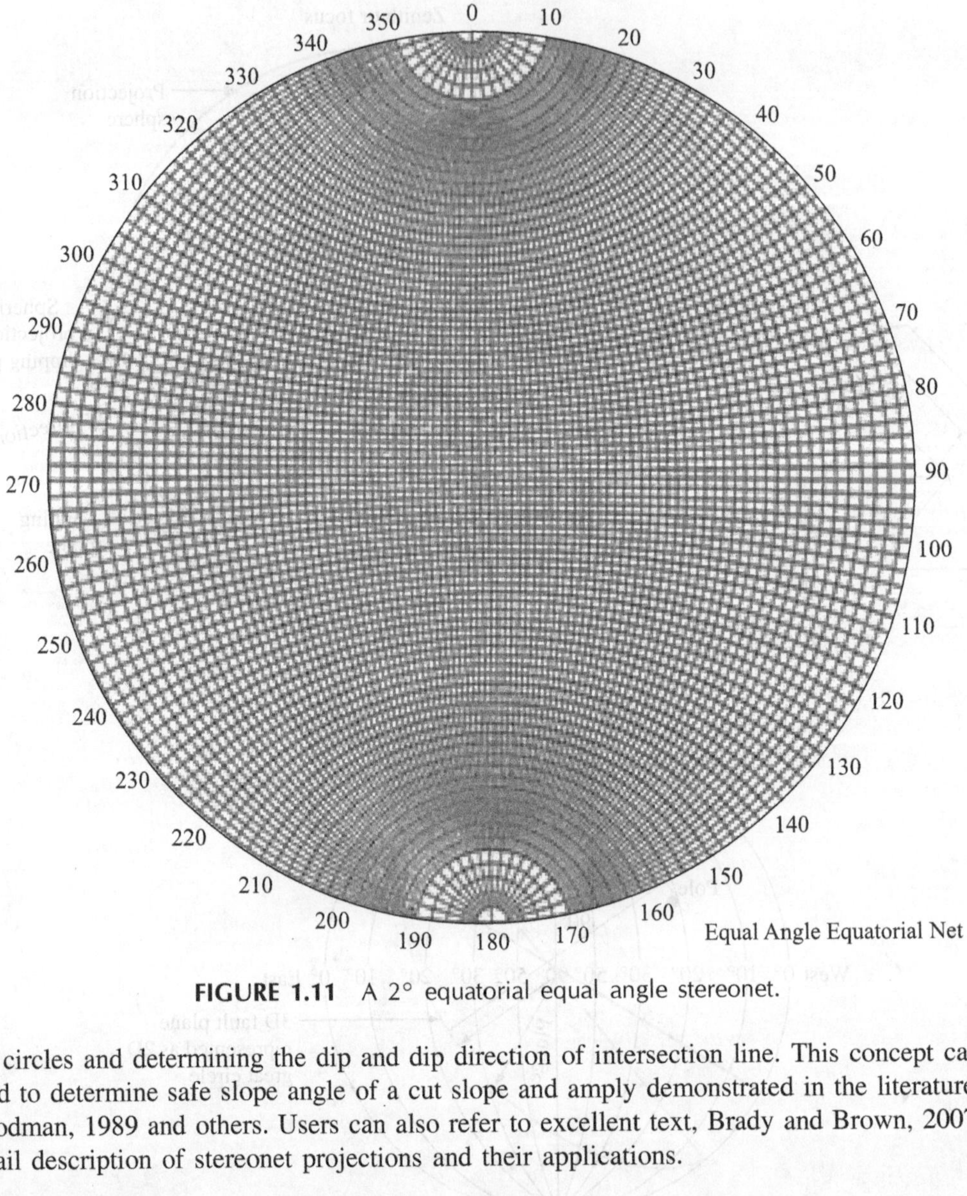

FIGURE 1.11 A 2° equatorial equal angle stereonet.

2D circles and determining the dip and dip direction of intersection line. This concept can be used to determine safe slope angle of a cut slope and amply demonstrated in the literatures of Goodman, 1989 and others. Users can also refer to excellent text, Brady and Brown, 2007 for detail description of stereonet projections and their applications.

EXAMPLE 1.1 Plot stereonet projection of the following two discontinuity planes having strike, dip direction and dip as given:

Discontinuity plane 1: Strike N50°E; Dip direction: S40°E; Dip: 60°
Discontinuity plane 2: Strike N50°W; Dip direction: S40°W; Dip: 40°
Determine the dip and dip direction of the line of intersection.

Solution: These can be achieved manually by plotting stereonet projections in a tracing paper using 2° equatorial equal angle stereonet. The following figure shows stereonet projections of discontinuity planes. Dip direction and dip of the intersection line are also shown in the figure.

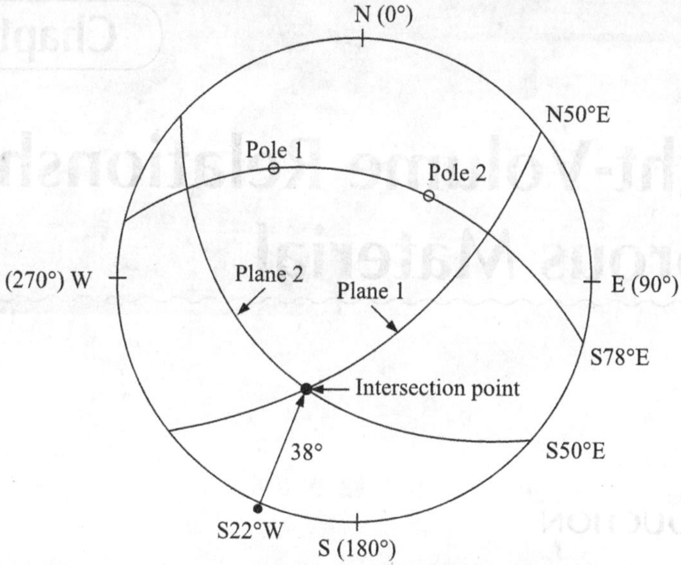

EXERCISES

1. What are the differences between natural and engineering materials?
2. How intact rock and rock mass are different with respect to structural integrity and material properties?
3. Discuss the geological cycle of rock formation. State the rock classification based on formation with an example.
4. What are the common rock-forming minerals?
5. Differentiate between rock and soil material. Discuss rock discontinuities.
6. What structures, such as cleavage, schistosity, and foliation, are found in metamorphic rocks?
7. How joints are formed in a rock mass?
8. What are faults, and how are they formed?
9. What is the purpose of stereographic projection? For stereographic projection, what geological information is required to represent the plane and line of a geological discontinuity? What do you mean by pitch?
10. Plot a point on a lower hemisphere equal angle projection that represents the orientation of a line of trend 245° and plunge 35°.

Chapter 2

Weight-Volume Relationships of Porous Material

2.1 INTRODUCTION

The term "porous material" is often associated with loose or unconsolidated soil or sand available on the earth's crust. Soil is generally formed by disintegration of rocks for over million years and found naturally on the earth's crust [Figure 2.1(a)]. The particle size of soil type material varies between 0.002 mm (clay) and 80 mm (gravel). Particle size of sand generally varies between 0.05 mm and 4 mm. Rock dumps, on the other hand, are formed by man-made activities to store broken waste rocks outside a mine pit. After blasting, overburden rocks are transported out of an open pit mine and are dumped in a place having multiple lifts or stages as shown in Figure 2.1(b). The particle size of a dump may vary from more than 1 m (boulders) to less than 1 mm (fines). A dump then contains "soil-rock mixture" (SRM). Several intact rocks such as coarse grained sandstone, sandyshale, limestone and others are also porous in their natural forms.

(a) (b)

FIGURE 2.1 (a) Soil on earth's crust, and (b) Dump slope in open pit mine.

Pore spaces are generally filled with air in dry condition, however, water may get into some of the pore spaces and can alter physical as well as chemical properties of soil or rock. This chapter elaborates on three phases of soil and rock matrix, i.e., air, water and solid. The discussion is limited to estimate various physical properties of soil and rock matrices. Concept of porosity and permeability is also provided with explanation of weight-volume relationships.

2.2 PHASE DIAGRAM

Porous material constitutes a complex system comprising with solid particles, air and water. In natural state, pore spaces are filled with both air and water as shown in Figure 2.2(a). A phase diagram of soil or rock represents a block diagram by distinctly showing the homogeneous part of solid particles and void spaces which may be filled with air and water or both. Figure 2.2(b) depicts such a three-phase diagram representing all three constituents or phases distinctly.

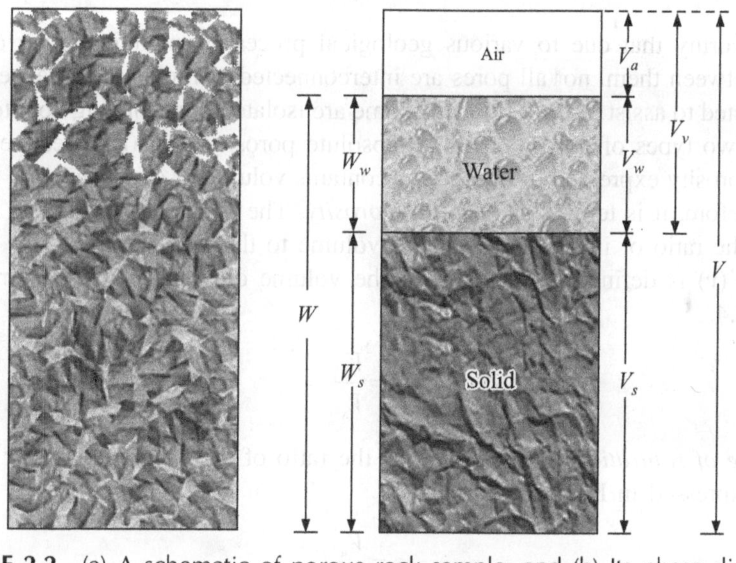

FIGURE 2.2 (a) A schematic of porous rock sample, and (b) Its phase diagram.

A dry soil or rock mass does not have water phase and is termed as two-phase system. Similarly, another two-phase system is a saturated soil or rock mass which contains only water in its void spaces.

2.3 WEIGHT-VOLUME RELATIONSHIPS

A porous material is composed of solid and void spaces. Void spaces may be fully or partially filled with water. A typical rock sample and its general three-phase diagrams are shown Figure 2.2. The volume and weight components of air, water and solid are marked in Figure 2.2(b). It is clear that the total volume, V of the sample can be expressed as:

$$V = V_s + V_v = V_s + V_w + V_a \tag{2.1}$$

where, V_s = volume of rock solids, V_v = volume of voids, V_w = volume of water in the voids and V_a = volume of air in the voids. Neglecting weight of air in void spaces, the total weight of the sample can be expressed as:

$$W = W_s + W_w \tag{2.2}$$

where, W_s = weight of solids, and W_w = weight of water in void spaces.

Based on the above volume relationships, porosity, void ratio and degree of saturation are defined as follows.

Porosity (p) is defined as the ratio of the volume of voids to the total volume as given in Eq. 2.3.

$$p = \frac{V_v}{V} \tag{2.3}$$

It is noteworthy that due to various geological processes including grain deposition and cementation between them, not all pores are interconnected in rock strata. Hence, certain pores are interconnected to assist in fluid flow and some are isolated pores having no interconnectivity. This leads to two types of porosity, viz. (i) absolute porosity, and (ii) effective porosity. The definition of porosity expressed in Eq. 2.3, V_v contains volume of connected and non-connected pores and therefore, it is termed as *absolute porosity*. The *effective porosity* on the other hand is defined as the ratio of interconnected pore volume to the total volume.

Void ratio (v) is defined as the ratio of the volume of voids to the volume of solids as given in Eq. 2.4.

$$v = \frac{V_v}{V_s} \tag{2.4}$$

The *degree of saturation* (s) is defined as the ratio of the volume of water to the volume of voids, as expressed in Eq. 2.5.

$$s = \frac{V_w}{V_v} \tag{2.5}$$

Porosity and degree of saturation are commonly expressed as a percentage and void ratio as a decimal number. From Eqs. 2.3 and 2.4, the relationship between porosity and void ratio can be obtained as given in Eq. 2.6.

$$p = \frac{V_v}{V} = \frac{V_v}{V_s + V_v} = \frac{(V_v / V_s)}{1 + (V_v / V_s)} = \frac{v}{1 + v} \tag{2.6}$$

Water content (w) in a porous material is defined as the ratio of the weight of water to the weight of solids as given in Eq. 2.7. It is also termed as moisture content.

$$w = \frac{W_w}{W_s} \tag{2.7}$$

Bulk density (ρ) is defined as the mass of the porous material (M) per unit volume (V) with unit kg/m^3 or g/cm^3. It is expressed as:

$$\rho = \frac{M}{V} \tag{2.8}$$

where, M represents the moist mass of the sample expressed in kg or g. Grain density (ρ_s) is expressed as the mass of solids (M_s) per unit volume of solids (V_s). In mathematical term, it is written as:

$$\rho_s = \frac{M_s}{V_s} \tag{2.9}$$

Specific gravity of solid (G_s) is expressed as the ratio of grain density to density of water. The density of water is 1000 kg/m^3 or 1 g/cm^3 at standard temperature and pressure. Unit weight (γ) is the weight expressed in N/m^3 or dyne/cm^{3}of material per unit volume or can be defined as the product of bulk density and acceleration due to gravity (g). This can be expressed as:

$$\gamma = \frac{W}{V} = \rho g \tag{2.10}$$

EXAMPLE 2.1 Show that $w = \gamma (V/W_s) - 1$

Solution: $w = \dfrac{W_w}{W_s} = \dfrac{W - W_s}{W_s} = \dfrac{W}{W_s} - 1 = \gamma(V/W_s) - 1$

Unit weight expressed in Eq. 2.10 is also referred to as moist unit weight, since W may contain W_w. If the soil or rock sample is completely dry, $W_w = 0$, then Eq. 2.10 can be written as:

$$\gamma_d = \frac{W_s}{V} \tag{2.11}$$

Here, γ_d is called dry unit weight. Similarly, dry bulk density of a sample can be estimated as:

$$\rho_d = \frac{M_s}{V} \tag{2.12}$$

EXAMPLE 2.2 Show that $\gamma_d = (1 - w)\gamma$

Solution: $\gamma_d = \dfrac{W_s}{V} = \dfrac{W_s}{W/\gamma} = \dfrac{\gamma}{(W_s + W_w)/W_s} = \dfrac{\gamma}{1+w}$

For all practical purpose, $w < 1$. Now, taking Tailor series expansion of $\dfrac{1}{1+w}$ $= 1 - w + w^2 - w^3 + w^4 - \ldots$ and truncated up to first order term, we find, $\gamma_d = (1 - w)\gamma$.

2.4 DETERMINATION OF POROSITY AND GRAIN DENSITY OF POROUS MATERIAL

Strength and stiffness of material are directly related to its bulk density. Bulk and grain densities are good measures of grain packing and porosity. The magnitude of bulk density depends on the maximum and minimum possible void ratios for the material while still maintaining inter-granular contact. On the basis of IS 13030: 1991, specific gravity of all samples is determined using specific gravity bottle generally termed as *pycnometer* (Figure 2.3).

In this test, small cylindrical sample of a porous material is weighted and its volume is determined to estimate the dry unit weight as given below.

$$\gamma_d = \frac{W_s}{V} = \frac{4W_s}{\pi D^2 L} \tag{2.13}$$

where, W_s is the weight of the sample, and D and L are the average diameter and length of the sample. The sample is crushed until grains are visible. After measuring the weight (W_1) of an empty pycnometer, the crushed samples are poured into the bottle and combined weight (W_2) is taken. After that distilled water is poured into the bottle to fill in the empty spaces and weight (W_3) is taken. Then bottle is washed thoroughly and is filled with only distilled water to take the final weight (W_4). Figure 2.3(a) to 2.3(d) depicts the four stages of weight measurements.

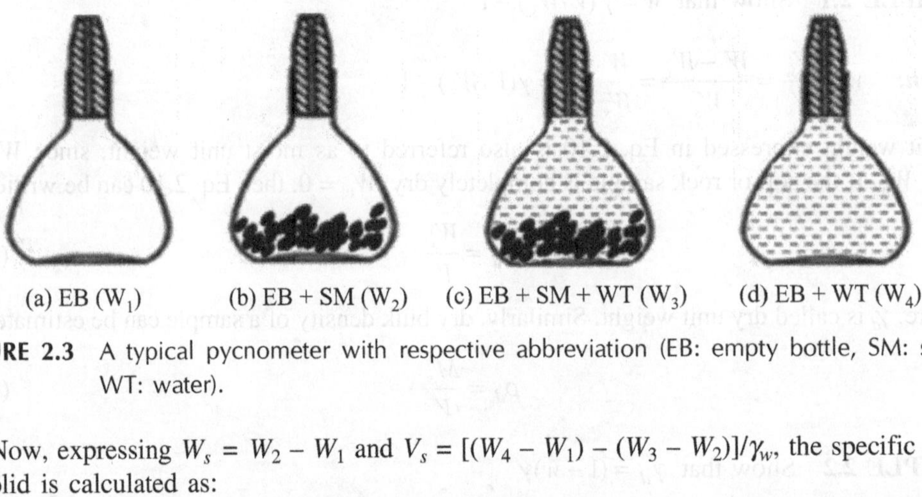

FIGURE 2.3 A typical pycnometer with respective abbreviation (EB: empty bottle, SM: sample, WT: water).

Now, expressing $W_s = W_2 - W_1$ and $V_s = [(W_4 - W_1) - (W_3 - W_2)]/\gamma_w$, the specific gravity of solid is calculated as:

$$G_s = \frac{W_s}{\gamma_w V_s} = \frac{(W_2 - W_1)}{(W_4 - W_1) - (W_3 - W_2)} \tag{2.14}$$

Table 2.1 lists specific gravity of some common minerals. In the above equation, if all weights are expressed in grams then unit weight of water is replaced by density of water as 1.0 g/cm^3. Porosity of the sample is estimated as:

$$p = 1 - \frac{V_s}{V} = 1 - \frac{(W_s/\gamma_w G_s)}{(W_s/\gamma_d)} = 1 - \frac{\gamma_d}{G_s \gamma_w} \tag{2.15}$$

TABLE 2.1 Specific Gravity of Some Common Minerals

Mineral (non-clay)	G_s	*Mineral (clay)*	G_s
Calcite	2.7	Chlorite	2.6–3.0
Dolomite	2.8	Illite	2.6–2.9
Feldspar	2.5–2.7	Kaolinite	2.6–2.7
Gypsum	2.3	Montmorillonite	2.7–2.8
Hematite	4.9–5.3		
Limonite	3.6–4.0		
Mica	2.8–3.2		
Pyrite	4.9–5.1		
Quartz	2.6		
Talc	2.7–2.8		

EXAMPLE 2.3 A dry sandstone sample of diameter 22 mm and length 35 mm weighs 30 g. The sample is crushed and a pyconometer is used to estimate the grain density having the following data: W_3 = 91 g and W_4 = 72 g. Determine the grain density of sandstone and porosity of the sample.

Solution: Volume of the soil sample is 13.304 cm^3. Using Eq. 2.13,

$$\gamma_d = \frac{30}{13.304} = 2.255 \text{ g/cc}$$

$$W_s = W_2 - W_1 = 30\,\text{g}$$

and
$$V_s = [(W_4 - W_1) - (W_3 - W_2)] = \frac{[72 - W_1 - (91 - 30 - W_1)]\,\text{g}}{1.0\,\dfrac{\text{g}}{\text{cc}}} = 11 \text{ cc.}$$

Hence, the specific gravity of soil particle, G_s = 30/12 = 2.727.

$$\text{Porosity, } p = \left(1 - \frac{2.255}{2.727}\right) \times 100 = 17.32\%$$

2.5 RELATIONSHIP BETWEEN WEIGHT AND VOLUME PARAMETERS

From the above definitions and discussions, relationships between weight and volume parameters can be derived for easy estimation of a parameter from a few given parameters. For example, weight of water in a moist porous sample can be estimated as follows:

$$W_w = wW_s = w\gamma_w G_s V_s = w\gamma_w (1-p) G_s V \tag{2.16}$$

Similarly, it can be shown that moist unit weight of porous material is equal to

$$\gamma = \frac{W}{V} = \frac{W_s + W_w}{V} = \frac{G_s\gamma_w V_s + wG_s\gamma_w V_s}{V_s(1+v)} = \frac{(1+w)\,G_s\gamma_w}{1+v} = (1-p)(1+w)G_s\gamma_w \tag{2.17}$$

Hence, dry density can be obtained by putting $w = 0$ in the above equation and then void ratio can be obtained as:

$$v = \frac{G_s \gamma_w}{\gamma_d} - 1 \tag{2.18}$$

EXAMPLE 2.4 A rock sample having 20% moisture content and volume of 75 cm^3 weighs 150 g. If the specific gravity of grain is 2.6, determine (i) dry unit weight, and (ii) porosity.

Solution: Using Eq. 2.17, moist or bulk density of the sample is estimated as:

$$\gamma = \frac{W}{V} = \frac{0.150 \times 9.81}{75 \times 100^{-3}} = 19.62 \text{ kN/m}^3$$

The porosity is then estimated as:

$$19.62 = (1 - p)(1.2) \times 2.6 \times 9.81$$

or,

$$p = 35.8\%$$

Hence, dry unit weight, $\gamma_d = (1 - 0.358)(1.0) \times 2.6 \times 9.81 = 16.33$ kN/m^3.

Now, similar relationships can be derived by considering the volume of water. We know that the volume occupied by water is

$$V_w = \frac{W_w}{\gamma_w} = \frac{wG_s\gamma_w V_s}{\gamma_w} = wG_sV_s = w(1-p)G_sV \tag{2.19}$$

Hence, relationship between degree of saturation and moisture content can be derived as:

$$s = \frac{V_w}{V_v} = \frac{wG_sV_s}{vV_s} = \frac{wG_s}{v} \tag{2.20}$$

Equation 2.20 states that if the porous material is fully saturated, i.e., $s = 1$, void ratio can be equated to

$$v = wG_s \tag{2.21}$$

Substituting Eq. 2.21 into Eq. 2.17, we find the expression of unit weight of saturated material as:

$$\gamma_{sat} = \frac{(G_s + v)\gamma_w}{1 + v} \tag{2.22}$$

It can be seen that moist density (ρ) and saturated density (ρ_{sat}) of porous material can be obtained using Eq. 2.22 by replacing γ_w (9.81 kN/m^3) by ρ_w of 1000 kg/m^3.

EXAMPLE 2.5 Using data given in Example 2.4, determine the degree of saturation of the moist rock.

Solution: Using the relationship given in Eq. 2.20, the void ratio is obtained as 0.56. Hence, degree of saturation is (Eq. 2.22)

$$s = \frac{0.2 \times 2.6}{0.56} \times 100 = 92.85\%$$

2.6 HYDRAULIC CONDUCTIVITY AND PERMEABILITY

Hydraulic conductivity is the measure of the capability of porous rock or soil to transmit water through network of interconnected pore spaces under certain driving pressure or force. In a geologic formation, the driving pressure originates by the differential ground pressure normally termed as *hydraulic gradient*. The hydraulic head is normally measured by combining pressure head or static head or static pressure head, velocity head and static heads. A schematic diagram of water flow from point A to point B through a porous rock stratum is shown in Figure 2.4.

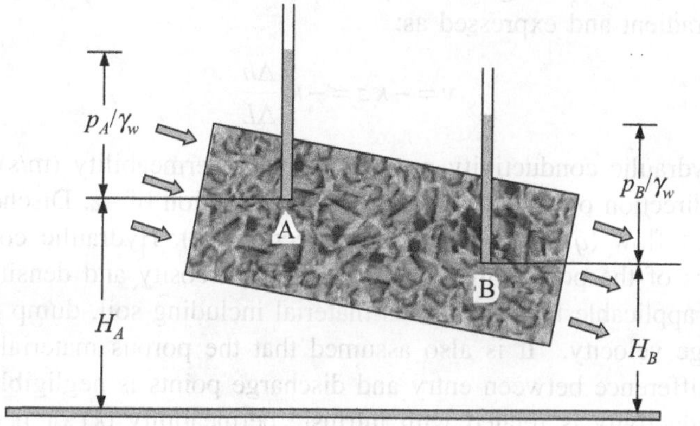

FIGURE 2.4 A schematic diagram of water flow through porous media.

The total hydraulic head at points A and B are h_A and h_B respectively and are expressed by Bernauli's equation as::

$$h_A = \frac{p_A}{\gamma_w} + \frac{v_A^2}{2g} + H_A \tag{2.23a}$$

$$h_B = \frac{p_B}{\gamma_w} + \frac{v_B^2}{2g} + H_B \tag{2.23b}$$

where, p, v and H denote pressure at a point, fluid flow velocity at a point and elevation of the point above a datum or reference plane respectively. The assumption behind Bernoulli's equation are (i) the fluid is ideal, i.e., viscosity is zero, (ii) the flow is steady, (iii) the flow is incompressible, and (iv) the flow is not rotational. It may be noted that static pressure head measured in m or mm is obtained as the measured pressure (height of water column multiplied by unit weight of water) at a point divided by the unit weight of water. The velocity head is obtained by dividing kinetic energy per unit volume of water $(1/2)\rho_w v^2$ with unit weight of water. Now, the hydraulic gradient (z) between two points is defined as difference in pressure (Δh) between the given two points divided by the horizontal distance (L) between them and is given as:

$$z = \frac{\Delta h}{\Delta L} = \frac{h_A - h_B}{L_A - L_B} \tag{2.24}$$

It may be seen that for linear flow, the value of z is negative since, $h_A > h_B$ and $L_A < L_B$. In general, velocity through porous media is very low since velocity of seepage water is mostly considered. However, it is found that for laminar flow, velocity of water in between two points is directly proportional to the hydraulic gradient between them (Das, 2002).

2.6.1 Darcy's Law

Observing the water flow through clean saturated sands, Darcy (1856) proposed that for laminar and steady state flow, the discharge velocity, v from the end (discharge) point is proportional to the hydraulic gradient and expressed as:

$$v = -\kappa z = -\kappa \frac{\Delta h}{\Delta L} \tag{2.25}$$

where, κ is the hydraulic conductivity or coefficient of permeability (m/s) and negative sign signifies that the direction of velocity is in the same direction of $-z$. Discharge velocity is the ratio of quantity of flow (q) and area of cross-section (A). Hydraulic conductivity depends upon the properties of the porous material along with viscosity and density of the fluid. The above equation is applicable for any porous material including soil, dump material and others having low seepage velocity. It is also assumed that the porous material is 100% saturated and temperature difference between entry and discharge points is negligible.

Hydraulic conductivity is related with intrinsic permeability (k) or permeability in terms of dynamic viscosity (μ) and unit weight (γ) of the fluid. Permeability is the property of the porous material and depends on its pore size, surface area, and others. The relationship is:

$$k = \frac{\kappa\mu}{\gamma} \tag{2.26}$$

The unit of permeability is m^2 or darcy and that of dynamic viscosity is $Nm^{-2}s$ or Pa-s or Poise. Kinematic viscosity is the ratio of dynamic viscosity and density of the fluid. Kinematic viscosity is expressed as m^2/s and in CGS unit as stroke, where 1 stroke is 10^{-4} m^2/s. Note that viscosity of fluid is dependent on temperature so as the permeability. Permeability is generally expressed in md (millidarcy) which equals to 1.01325×10^{-12} m^2. Similarly, 1 Poise = 0.1 $Nm^{-2}s$. The coefficient of viscosity of a fluid is 1 poise if a tangential force of 1 dyne is required to maintain a unit velocity gradient between two adjacent layers of that fluid each of area 1 cm^2.

2.6.2 Linear Flow

Considering water flow through a differential pipe element dx (Figure 2.5) due to a differential head dh along the direction of flow and the quantity of discharge water per unit time, $q = vA$ for a uniform cross-sectional area A, Eq. 2.25 can be rewritten as:

$$q = -\kappa A \frac{dh}{dx} \tag{2.27}$$

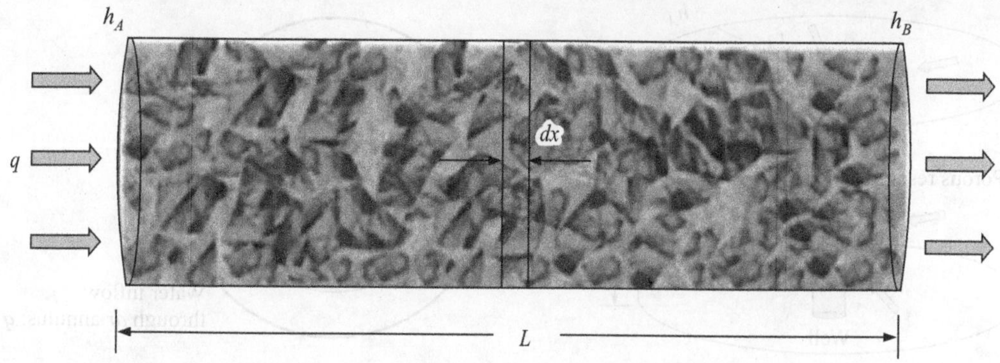

FIGURE 2.5 Hydraulic conductivity in linear flow.

Now, assuming steady state flow, i.e., quantity of water flowing through the pipe does not vary with time. Hence, at the entry point A and at the discharge B, quantities are the same as q, the above equation can be rewritten as:

$$q\int_{L_A}^{L_B} dx = -\kappa A\int_{h_A}^{h_B} dh \tag{2.28}$$

where,

$$\kappa = \frac{q}{A}\left(\frac{L_B - L_A}{h_A - h_B}\right)$$

EXAMPLE 2.6 An aquifer having a cross-sectional area of 100×25 m^2 has a pressure gradient of 25 cm between two points located 50 m apart. If the hydraulic conductivity of the aquifer is 2 m per day, determine the quantity of water flowing in a day between these two points.

Solution: Using Eq. 2.28, we find that

$$q = \kappa A\left(\frac{h_A - h_B}{L_B - L_A}\right) = \frac{2 \times 2500 \times 25 \times 10^{-2}}{50} = 25 \text{ m}^3/\text{day}$$

2.6.3 Radial Flow

Figure 2.6 shows a schematic diagram of radial flow of water to a well due to pressure difference between $r = r_A$ and $r = r_B$, where r_A denotes the radius of the reservoir or porous rock strata and r_B is the radius of the well. Considering a differential volume element having thickness of annulus as dr and length of rock strata to be L, the gradient of pressure head is dh/dr acting along the direction of the flow. Hence, Darcy's equation can be written considering positive sign as:

$$q = \kappa A\frac{dh}{dr} \tag{2.29}$$

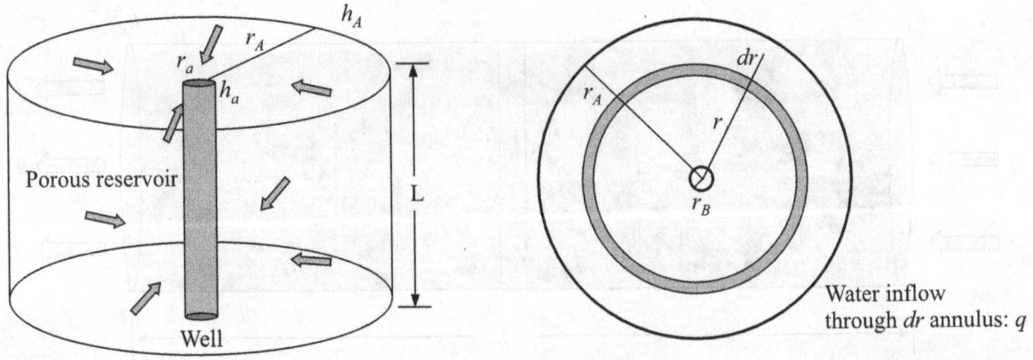

FIGURE 2.6 (a) A cylindrical block of porous media with water flow towards well, (b) cross-sectional view.

Expressing cross-sectional area, $A = 2\pi rL$, the above equation can be rewritten as:

$$q\int_{r_B}^{r_A} \frac{dr}{r} = 2\pi L\kappa \int_{h_B}^{h_A} dh \tag{2.30}$$

where,
$$\kappa = \frac{q}{2\pi L}\left[\frac{\ln(r_A/r_B)}{h_A - h_B}\right]$$

EXAMPLE 2.7 Consider a cylindrical oil reservoir of radius 1.2 km and height 20 m having *in situ* pressure of 45 MPa. A well of 200 mm is drilled in the middle of the reservoir to extract oil. Estimate the quantity of oil that may flow per day to the well. Assume viscosity of oil is 75 cp, specific gravity 0.85 and permeability 50 md *in situ* condition.

Solution: From the above data, hydraulic conductivity of oil can be estimated as:

$$\kappa = \left(\frac{50\times 1.01325\times 10^{-12}\times 0.85\times 1000\times 9.81}{75\times 10^{-2}\times 0.1}\right) = 5.63\times 10^{-6}\ \text{m/s}$$

Using Eq. 2.30, we find quantity as:

$$q = 2\pi L\kappa\left(\frac{h_A - h_B}{\ln(r_A/r_B)}\right)$$

$$= 2\pi\times 20\times 5.63\times 10^{-6}\ \frac{45\times 10^6/(850\times 9.81)}{\ln(1200/0.2)} = 0.44\ \text{m}^3/\text{s}$$

Hence, quantity of oil flows per day is 37937.3 m^3.

2.6.4 Measurement of Hydraulic Conductivity in Laboratory

In the laboratory, hydraulic conductivity or permeability of soil type material are determined using two methods: (i) the constant head method, and (ii) the falling head method. The constant head method is preferred for material having a hydraulic conductivity value over 10^{-4} cm/s.

On the other hand, falling head testing method is used if the hydraulic conductivity is less than 10^{-4} cm/s. For more detail on standard procedure for hydraulic conductivity using testing constant head method is given in reference ASTM D2434-22.

In order to determine the hydraulic conductivity of porous rock material, constant head procedure is adopted with slightly different apparatus. In this section, procedure for determining hydraulic conductivity using constant head method is elaborated giving an example.

Constant Head Method

Hydraulic conductivity or permeability refers to the capability of water to flow through a soil or porous material. This property is required for the calculation of ground water flow, seepage rate through earth dams or mine dumps, ponds, landfill and others. Hydraulic conductivity of porous rock provides rate of water flow through rock matrix, joints and fractures.

The instruments required for this test are Permeameter, tamper, scoop, 1000 mL graduated cylinders, watch (or stopwatch), thermometer, and filter paper.

At first, solid density or unit weight of the porous material is determined using pycnometer as mentioned above. Then, average inside diameter (D) of the chamber of the permeameter is measured and one porous rock or pumic stone pieces are placed on the inner support ring in the base of the chamber as shown in Figure 2.7. A filter paper is placed on top of the porous rock and moist soil type material is poured using a scoop into the chamber upto a pre-defined length (L). If required, samples are rammed for compaction. Water pipe is connected at a level above the top of the sample as shown in Figure 2.7. Water inflow rate is controlled in such a way that steady state condition is achieved. It means that water head remains constant throughout the test although discharge occurs from the other end of the chamber. Water head (h) is measured from the top level of porous stone to the height of the water inlet. Temperature of water is also measured using a thermometer.

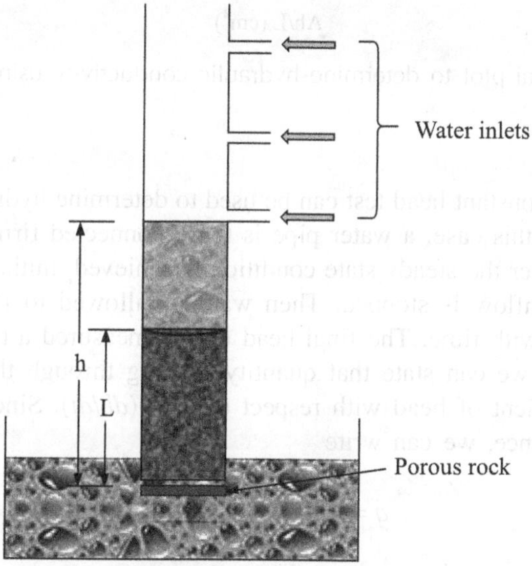

FIGURE 2.7 Schematic diagram of a constant head permeability setup.

Water discharged (Q) through the outlet is collected in a graduated cylinder for a pre-determined time (t). The experiment is repeated for varying length of sample. In some cases, water head is varied for a given sample length.

According to Darcy's law, we can write (Eq. 2.28) as:

$$q = \left(\frac{Q}{t}\right) = \kappa\left(\frac{hA}{L}\right) \tag{2.31}$$

where, $A = \pi D^2/4$

Data are collected for at least three different sample lengths or by varying water heads for same sample length. Figure 2.8 shows a typical plot of q versus hA/L for hematite ore samples. It is obvious that the slope of this line is the hydraulic conductivity of the sample. Once the hydraulic conductivity is known, permeability can be obtained by knowing the viscosity of water for the measured temperature and dry density of the sample.

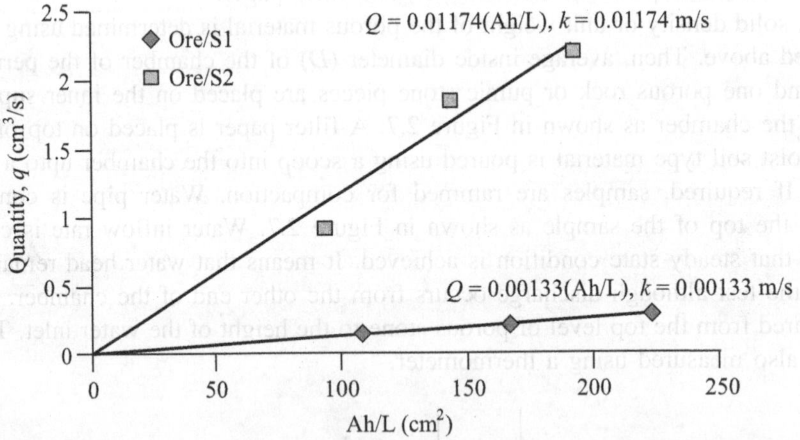

FIGURE 2.8 A typical plot to determine hydraulic conductivity using constant head test.

Falling Head Method

The apparatus used for constant head test can be used to determine hydraulic conductivity using falling head method. In this case, a water pipe is to be connected firmly with the chamber as shown in Figure 2.9. After the steady state condition is achieved, initial head (h_1) at time $t = t_1$ is recorded and water inflow is stopped. Then water is allowed to flow through the sample such that head reduces with time. The final head (h_2) is measured after time $t = t_2$.

Using Darcy's law, we can state that quantity flowing through the sample length at any time t is related to gradient of head with respect to time (dh/dt). Since water head is falling, dh/dt is negative and hence, we can write

$$q = \kappa\left(\frac{hA}{L}\right) = -A_p\frac{dh}{dt} \tag{2.32}$$

where, A_p is the cross-sectional area of the standpipe. The above equation can be rearranged as follows:

$$\int_{t_1}^{t_2} dt = -\frac{A_p L}{A\kappa}\int_{h_1}^{h_2}\frac{dh}{h} \tag{2.33}$$

$$\kappa = \frac{A_p L}{A(t_2 - t_1)}\ln\left(\frac{h_1}{h_2}\right)$$

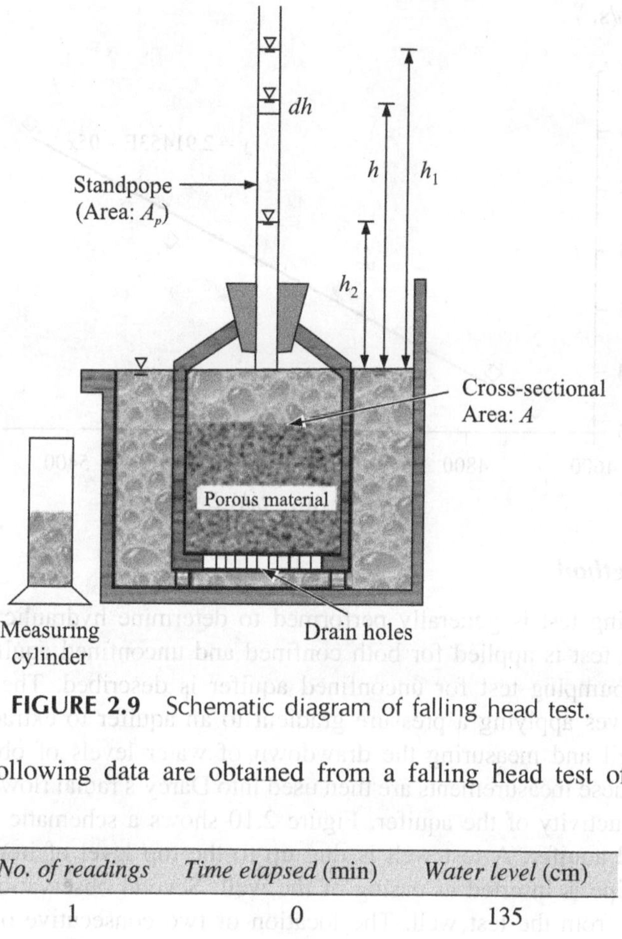

FIGURE 2.9 Schematic diagram of falling head test.

EXAMPLE 2.8 Following data are obtained from a falling head test of a soil sample of length 13.4 cm.

No. of readings	*Time elapsed* (min)	*Water level* (cm)
1	0	135
2	409	115
3	765	100
4	1158	86

The diameter of standpipe and that of sample are 4.4 cm and 7.6 cm respectively. Determine the hydraulic conductivity of the soil sample.

Solution: From the above data, following parameters can be obtained by manipulating Eq. 2.33. Here, subscript $j = i + 1$.

$x = A(t_j - t_i)/A_p L$	$y = \ln(h_i/h_j)$
5463.759	0.16034
4755.742	0.13976
5250.018	0.15082

Using above data, the following plot is constructed and hydraulic conductivity is found to be 2.914×10^{-5} cm/s.

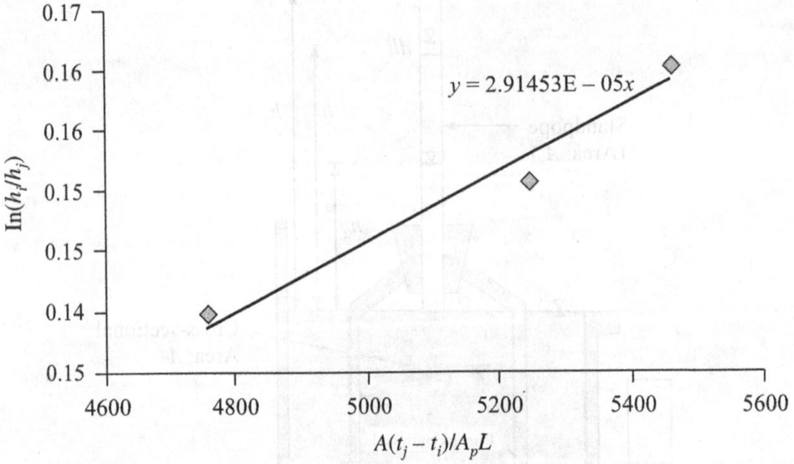

Field Pumping Method

In the field, pumping test is generally performed to determine hydraulic conductivity of soil or an aquifer. This test is applied for both confined and unconfined aquifers. In this text, the principle of field pumping test for unconfined aquifer is described. The principle of a field pumping test involves applying a pressure gradient to an aquifer to extract groundwater from a test pumping well and measuring the drawdown of water levels of observation wells as a function of time. These measurements are then used into Darcy's radial flow equation to calculate the hydraulic conductivity of the aquifer. Figure 2.10 shows a schematic diagram of pumping test for unconfined aquifer. A test well is dug up to the top level of impermeable rock layer and a perforated pipe is inserted as casing of the well. Several observation wells are also dug at least 5 m away from the test well. The location of two consecutive observation wells and the test well should be such that they lay on a line. In general, radial distance of observation well is taken in logarithmic scale and are measured from the test well.

Water is pumped out at a constant rate during the test and water levels or drawdown depths of observation wells are measured with time. Pumping rate of water is monitored using a flow meter and drawdown depth can be measured using mechanical or electrical level detection sensors. A steady state condition is achieved when water level of observation well and that of test well are constant. In other word, the pumping or discharge rate and rate of flow of groundwater through aquifer is the same.

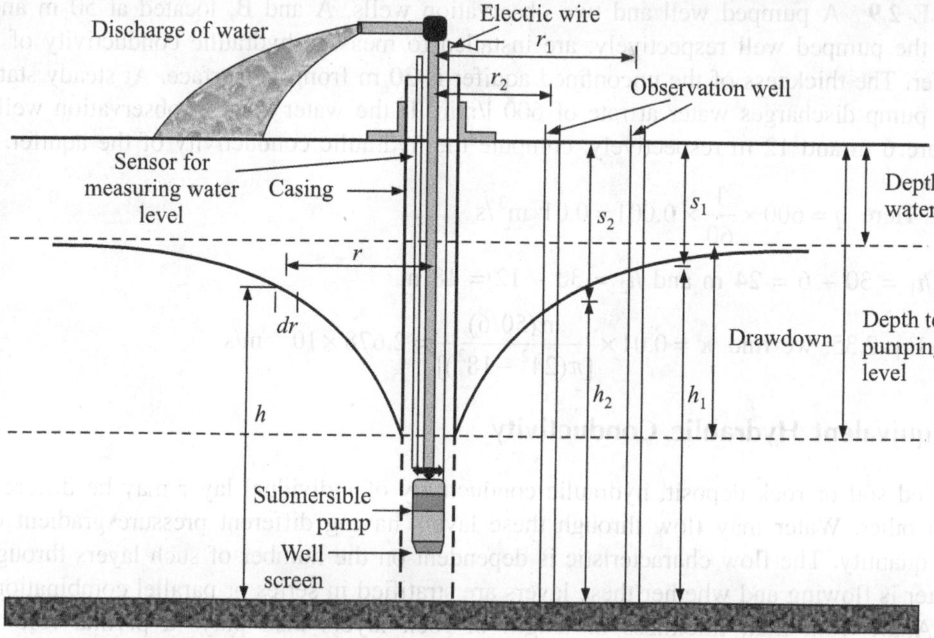

FIGURE 2.10 Schematic diagram of field pump test for unconfined aquifer.

Based on this assumption, quantity of water pumped out can be expressed using Darcy's radial flow equation as:

$$q = \kappa(2\pi rh)\left(\frac{dh}{dr}\right) \tag{2.34}$$

Figure 2.10 depicts two observation wells at r_1 and r_2 distance from the centre of the test well. Considering pressure heads as h_1 and h_2 respectively, Eq. 2.34 can be rewritten as:

$$\int_{r_1}^{r_2} \frac{dr}{r} = \left(\frac{2\pi\kappa}{q}\right)\int_{h_1}^{h_2} hdh \tag{2.35}$$

$$\kappa = \frac{q\ln\left(\dfrac{r_1}{r_2}\right)}{\pi(h_1^2 - h_2^2)}$$

In the above equation, pressure heads can be replaced by drawdown depth as $s_i = H - h_i$ ($i = 1, 2$) where H is the height of the surface from the impermeable layer. As mentioned before Eq. 2.35 is valid for unconfined aquifer only. For confined aquifer, transient flow model needs to be adopted and the description of that is beyond the scope of this book. Interested readers may also refer literatures to other methods for *in situ* hydraulic conductivity measurements such as Boutwell permeameter, constant head borehole permeameter and others.

EXAMPLE 2.9 A pumped well and two observation wells, A and B, located at 50 m and 6 m from the pumped well respectively, are installed to measure hydraulic conductivity of a sand aquifer. The thickness of the unconfined aquifer is 30 m from the surface. At steady state condition, pump discharges water at rate of 600 l/min. If the water level at observation wells A and B are 6 m and 12 m respectively, compute the hydraulic conductivity of the aquifer.

Solution: Here $q = 600 \times \frac{1}{60} \times 0.001 = 0.01 \text{ m}^3/\text{s}$

Now, $h_1 = 30 - 6 = 24$ m and $h_2 = 30 - 12 = 18$ m

Using Eq. 2.35, we find $\kappa = 0.01 \times \frac{\ln(50/6)}{[\pi(24^2 - 18^2)]} = 2.678 \times 10^{-5}$ m/s

2.6.5 Equivalent Hydraulic Conductivity

For stratified soil or rock deposit, hydraulic conductivity of individual layer may be different from each other. Water may flow through these layers having different pressure gradient or discharge quantity. The flow characteristic is dependent on the number of such layers through which water is flowing and whether these layers are stratified in series or parallel combinations or both. Apart from that, thickness or length of rock layers also plays a pivotal role in determining the effective conductivity of the system. In the following, effective or equivalent hydraulic conductivity of series and parallel combinations are derived for both linear and radial flow regimes.

Series Combination: Linear Flow

Let us assume that water is flowing through n number of rock or soil layers having length L_i and hydraulic conductivity κ_i for the ith layer. Water enters through layer 1 and discharges out from layer n as shown in Figure 2.11. We also assume that cross-sectional area through which water is flowing is the same for each layer and there is no loss of water through the plane of interaction between two consecutive layers. From the figure, pressure gradient across ith layer can be expressed based on Eq. 2.28 as:

$$\Delta h_i = \frac{q}{A}\left(\frac{L_i}{\kappa_i}\right) \tag{2.36}$$

Then the total pressure gradient for n layers considering same quantity flowing from layer 1 to layer n will be

$$h_n - h_0 = \sum_{i=1}^{n} \Delta h_i = \frac{q}{A}\sum_{i=1}^{n}\left(\frac{L_i}{\kappa_i}\right) \tag{2.37}$$

Now, if n layers are replaced by a single equivalent rock or soil layer of conductivity κ_{eq} having length, $L = \sum_{i=1}^{n} L_i$, the pressure gradient will be

$$h_n - h_0 = \frac{qL}{\kappa_{eq} A} \tag{2.38}$$

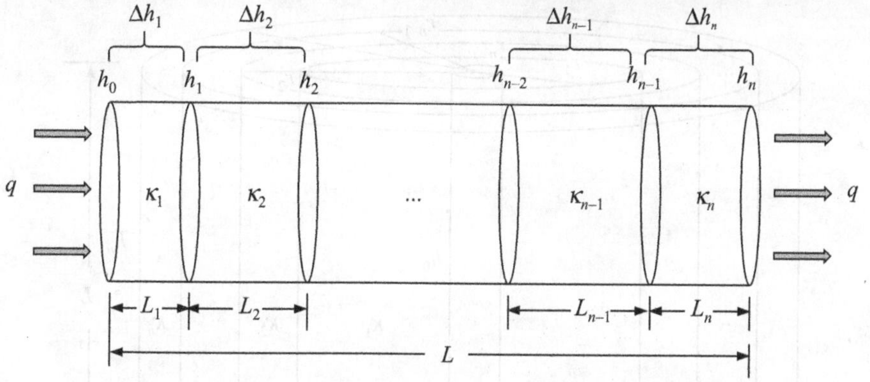

FIGURE 2.11 Series combination of layers: linear flow.

Comparing Eqs. 2.37 and 2.38, we find the equivalent hydraulic conductivity of series combination for linear flow as:

$$\kappa_{eq} = \frac{\sum_{i=1}^{n} L_i}{\sum_{i=1}^{n} (L_i/\kappa_i)} \tag{2.39}$$

Series Combination: Radial Flow

For radial flow, similar concept is applied to derive the equivalent hydraulic conductivity. In this case, we assume that ith layer is confined in between radius r_{i-1} and r_i as shown in Figure 2.12. Then, pressure gradient at ith layer is written as:

$$\Delta h_i = \frac{q}{2\pi L}\left(\frac{\ln(r_i / r_{i-1})}{\kappa_i}\right) \tag{2.40}$$

Similarly, the total pressure gradient from layer n to the well is

$$h_n - h_0 = \frac{q}{2\pi L}\left(\frac{\ln(r_n/r_0)}{\kappa_{eq}}\right) \tag{2.41}$$

Hence, the equivalent hydraulic conductivity is obtained as:

$$\kappa_{eq} = \frac{\ln(r_n/r_0)}{\sum_{i=1}^{n}\left(\frac{\ln(r_i/r_{i-1})}{\kappa_i}\right)} \tag{2.42}$$

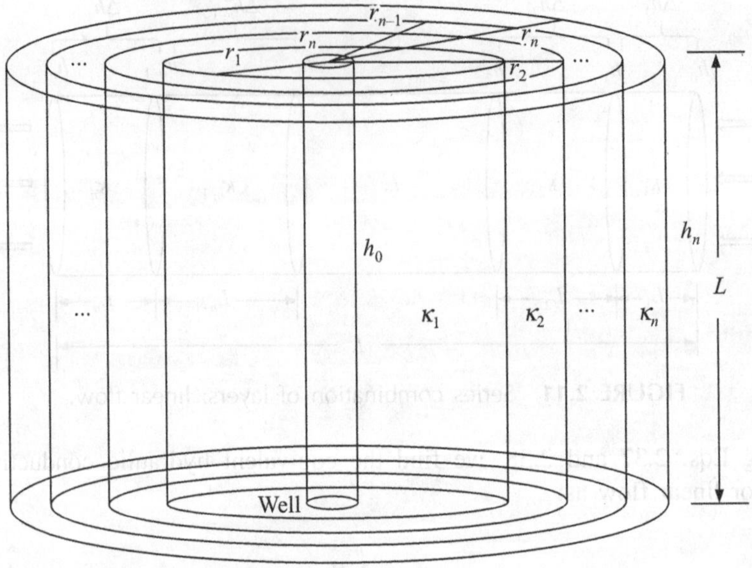

FIGURE 2.12 Series combination of layers: radial flow.

EXAMPLE 2.10 Find the equivalent hydraulic conductivity of 4 beds in series, having equal formation length for a radial system if the radius of the penetrating well bore is 150 mm.

Bed	*Thickness of bed*, m	κ_i, cm/s
1	15	2.5
2	40	1.6
3	35	3.6
4	20	1.5

Solution: Here, $r_0 = 0.15$ m, $r_1 = 15.15$ m, $r_2 = 55.15$ m, $r_3 = 90.15$ m, $r_4 = 110.15$ m. Now, using Eq. 2.42, we find the equivalent hydraulic conductivity is

$$\kappa_{eq} = \frac{\ln(110.15/0.15)}{\dfrac{\ln(110.15/90.15)}{1.5} + \dfrac{\ln(90.15/55.15)}{3.6} + \dfrac{\ln(55.15/15.15)}{1.6} + \dfrac{\ln(15.15/0.15)}{2.5}} = 2.257 \text{ cm/s}$$

Parallel Combination: Linear Flow

Figure 2.13 depicts a set of parallel beds of porous media having thickness T_i and hydraulic conductivity κ_i for the ith layer. Water enters through all layers from one side and discharges out from the other side. We also assume that length of each bed is the same and there is no loss of water through the plane of interaction between two consecutive layers. From the figure, quantity flowing through ith layer due to pressure difference of Δh can be expressed as:

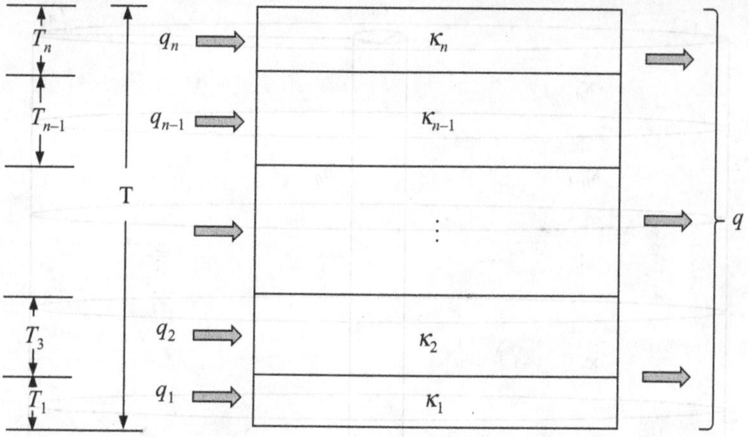

FIGURE 2.13 Parallel combination of layers: linear flow.

$$q_i = \frac{\kappa_i T_i W \Delta h}{L} \tag{2.43}$$

where, W represents the width of each bed. Then the total quantity flowing through all the beds are sum of all individual quantity and expressed as:

$$q = \sum_{i=1}^{n} q_i = \left(\frac{W\Delta h}{L}\right)\sum_{i=1}^{n}\kappa_i T_i \tag{2.44}$$

Now, if the n layers are replaced by a single equivalent rock or soil layer of conductivity κ_{eq} having thickness $T = \sum_{i=1}^{n} T_i$, the pressure gradient will be

$$q = \left(\frac{W\Delta h}{L}\right)\kappa_{eq}\sum_{i=1}^{n} T_i \tag{2.45}$$

Comparing Eqs. 2.44 and 2.45, we find the equivalent hydraulic conductivity of series combination for linear flow as:

$$\kappa_{eq} = \frac{\sum_{i=1}^{n}(\kappa_i T_i)}{\sum_{i=1}^{n} T_i} \tag{2.46}$$

Parallel Combination: Radial Flow

The similar concept can be adopted to derive the equivalent hydraulic conductivity for radial flow cases (Figure 2.14). The quantity q_i can be obtained from the following equation as:

$$q_i = \left(\frac{2\pi\,\Delta h}{\ln(r_n/r_0)}\right)\kappa_i L_i \tag{2.47}$$

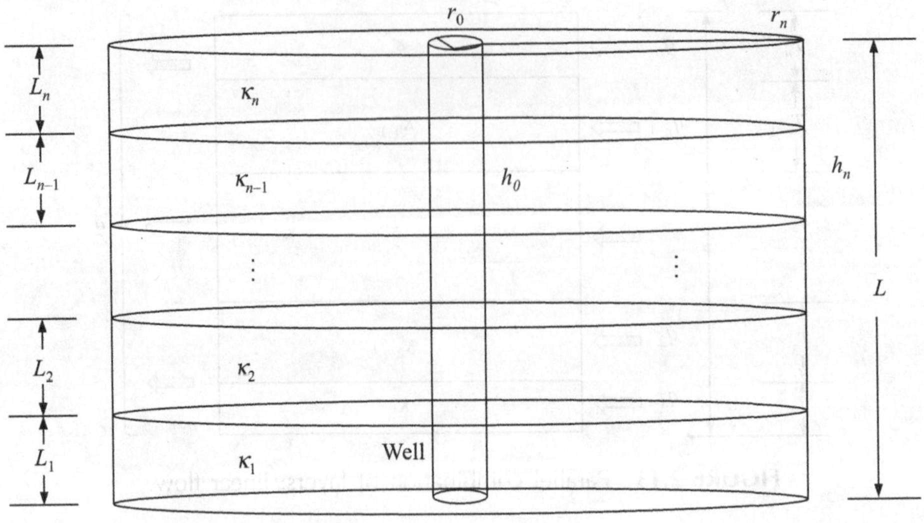

FIGURE 2.14 Parallel combination of layers: radial flow.

Hence, the total quantity is as follows:

$$q = \sum_{i=1}^{n} q_i = \left(\frac{2\pi \Delta h}{\ln (r_n / r_0)} \right) \sum_{i=1}^{n} \kappa_i L_i \tag{2.48}$$

From the above, equivalent hydraulic conductivity can be obtained as:

$$\kappa_{eq} = \frac{\sum_{i=1}^{n} (\kappa_i L_i)}{\sum_{i=1}^{n} L_i} \tag{2.49}$$

EXERCISES

1. A shale composed of 60% illite, 20% chlorite, and 20% pyrite with respective specific weights (G_s) are 2.75; 2.84;and 4.9 respectively. The porosity values at different depths are as follows:

   0 to 180 m: 33.5%; 180 m to 750 m: 25.4%; 750 m to 1015 m: 21.1%, and 1015 m to 1800 m: 9.6%.

   Estimate the vertical stress at 1800 m deep in the shale formation assuming that shale continues from surface to depth 1800 m and saturated with water.
2. A sandstone core composed of quartz and feldspar grains with calcite cement is 82 mm in diameter and 169 mm in length. On saturation in water, its wet weight is 21.42 N; after oven drying its weight is 20.31 N. Calculate its wet unit weight, dry unit weight, and its porosity.

3. Another core specimen from the same formation as the rock of Exercise 2, displays large voids. Its wet unit weight is 20.11 kN/m^3. Assuming its specific gravity is the same as for the rock in Exercise 2, estimate its porosity.
4. Water is flowing horizontally through 3 rock beds. If κ_i denotes the hydraulic conductivity of ith layer in cm/s, determine the equivalent hydraulic conductivity in terms of κ.
5. A sandstone specimen weight 320 g has a diameter of 54 mm and a height of 60 mm whose moisture content is 20% and has a specific gravity of 2.65. Find the porosity of sandstone sample.
6. A waste dump sample (granular) has a void ratio of 0.9. If the void ratio is reduced to 0.7 by compaction, find the percentage volume loss due to compaction.
7. A soil sample whose water content is 20% has a bulk density of 21.6 kN/m^3. The sample undergoes air drying with an insignificant change in the void ratio. What is the water content of this sample when its bulk density is reduced to 20 kN/m^3?
8. A laboratory test was conducted on a sand sample with water to find the coefficient of permeability (κ) by constant head method. Where A = area of cross section of permeameter; Diameter of permeameter is 44 cm; L = flow length of the sand sample (40 cm), h = hydraulic head. Three tests are conducted on a hydraulic head of 20, 40, and 60 cm and are plotted as follows. Find (a) κ; (b) for flow rate, q = 0.025 cm^3/s; find the corresponding hydraulic head.

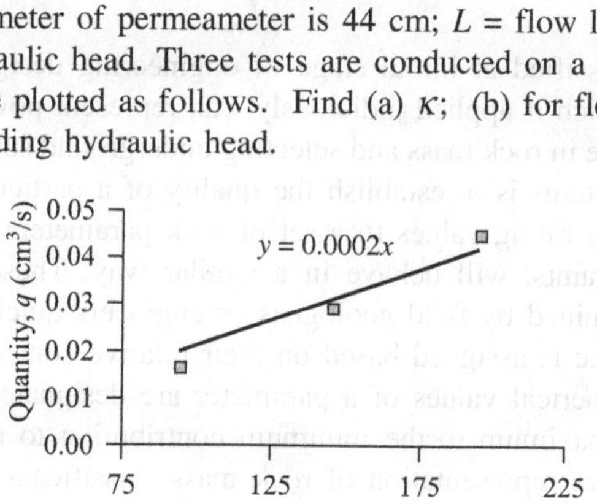

9. A one-dimensional gas percolation test was performed on dry rock, and the following results were obtained. Find the hydraulic conductivity in the same rock when it is in a saturated condition.
   - Flow rate of gas : $q = 2.0 \times 10^{-3}$ cm^3/s
   - Injection pressure : p_i = 395.525 kPa
   - Outflow-side pressure : p_o = 101.325 kPa
   - Diameter of the sample : D = 5 cm
   - Length of the sample : L = 10 cm
   - Test temperature : T = 20°C
10. The mass of moist soil having volume 0.005 m^3 is 10 kg. The moisture content and the specific gravity of soil solids were determined to be 13% and 2.65 respectively. Determine the followings:
    (a) Porosity (b) Void ratio
    (c) Dry density (d) Degree of saturation

Chapter 3

Rock Mass Classification Systems

3.1 INTRODUCTION

Rock mass is widely classified at initial stage of engineering design. There are varieties of classification systems, which if applied judiciously, can represent powerful tools for estimating stability of structures made in rock mass and selecting underground support system. The purpose of these classification systems is to establish the quality of a particular rock mass (or part of a rock mass) by assigning rating values to a set of rock parameters which, under a particular set of engineering constraints, will behave in a similar way. These parameters are selected such that it can be determined by field geologists or engineers quickly and easily. For a rock parameter, numerical value is assigned based on their relative contribution to the overall rock mass properties. The numerical values of a parameter are designated within upper and lower limits to quantify their maximum to the minimum contribution to rock mass properties. The quantitative and qualitative representation of rock mass classification systems are determined by mathematical operations like addition, multiplication and/or division of these parameters. Finally, based on numerical measures, the quality of rock mass is classified into a group. These classification systems are not intended to replace analytical studies, field observations, measurements or engineering judgment, but to categorise the quality of rock mass based on few measurable parameters without in site investigation procedures.

The objective of these classification systems are to: (i) identify the most significant parameters influencing the rock mass behaviour, (ii) classify groups of similar behaviour, (iii) provide the characteristics of each group, (iv) relate the behaviour of rock mass at one site to the other, (v) derive quantitative data and guidelines for engineering design, and (vi) provide common basis for communication between engineers and geologists.

The classification systems are simple and effective way of representing rock mass quality, but one cannot use them for different objectives or in significantly new circumstances. This chapter discusses the five most popular rock mass classification systems like rock quality designation (RQD), rock mass rating (RMR), tunnelling quality index (Q), geological strength index (GSI) and coal mine roof rating (CMRR). Finally, these classification systems are used to determine the rock mass strength, modulus of deformation, stand-up time and support requirements for different rock mass.

3.2 DEFINITION AND PROPERTIES OF ROCK MASS

In rock engineering, properties of rock mass are defined by a general term 'discontinuities'. *Discontinuity* is a collective term for most types of joints, weak bedding planes, weak schistosity planes, weakness or shear zone and faults. Joints are the most prevalent type of natural discontinuity. As mentioned in Chapter 1, a joint set is a break of geological origin in the continuity of a body of rock along which there has been no visible displacement. A group of parallel joints is called a *set* and several such joint sets intersect to form a *joint system*. Joints may be open or filled. They frequently form parallel to bedding planes, foliation and cleavage and may be termed as bedding joint, foliation joints and cleavage joints accordingly.

Discontinuities generally influence the engineering responses of rock masses in a variety of ways. A discontinuity plane

(i) lowers shear strength along joint plane which may slip.
(ii) reduces the overall shear and tensile strengths of the rock mass.
(iii) renders the overall mechanical response of the rock mass discontinuous in the sense that individual blocks may be free to rotate or to translate with associated slip and/or separation at block interfaces.
(iv) introduces a wide range of potential failure mechanisms (e.g., unravelling, toppling, slip or the gravity fall of blocks or wedges) not present in a continuum and requiring associated reinforcement measures.
(v) influences the stress distribution within the rock mass because of their low stiffness and strengths.
(vi) controls to a large extent the fragmentation achieved by excavation/blasting processes.
(vii) increases permeability in orders of magnitude higher than those of intact rocks, providing the major conduits for the flow of fluids through the rock masses.

Impact of discontinuities on rock engineering projects is evaluated only when discontinuities in the rock mass are characterised and their properties are established. "Suggested methods for the quantitative description of discontinuities in rock masses" of the International Society of Rock Mechanics (1978) have documented the ten most important parameters as discussed below and shown in Figure 3.1.

(i) Orientation: It is defined as an attitude of a discontinuity in space. It is described by the dip direction (azimuth) and dip angle of the line of steepest declination in the plane of the discontinuity. Example: strike, dip direction/dip angle (see Figure 1.7).

(ii) Spacing: It is a perpendicular distance between adjacent discontinuities. Normally refers to the mean or modal spacing of a set of joints, expressed as perpendicular distance between discontinuities or frequency (number of discontinuities per metre). Table 3.1 shows the description for spacing. Figure 3.1 shows the measured length (d_2) for joint set 2 by tape and α_2 is the smallest angle between the measuring tape and the observed joint set then the spacing between joints are expressed as:

$$S = d_2 \sin \alpha_2 \tag{3.1}$$

TABLE 3.1 Description for Spacing (ISRM 1978)

Description	*Spacing* (mm)
Extremely close spacing	< 20
Very close spacing	20–60
Close spacing	60–200
Moderate spacing	200–600
Wide spacing	600–2000
Very wide spacing	2000–6000
Extremely wide spacing	> 6000

(iii) Persistence: It is trace length of discontinuity as observed in an exposure. Persistence gives crude measure of the areal extent or penetration length of a discontinuity. Termination in solid rock or against other discontinuities reduces the persistence. Description for persistence is shown in Table 3.2.

TABLE 3.2 Description for Persistence (ISRM 1978)

Description	*Trace length* (m)
Very low persistence	< 1
Low persistence	1–3
Medium persistence	3–10
High persistence	10–20
Very high persistence	> 20

(iv) Roughness: Inherent surface roughness and waviness are relative to the mean plane of a discontinuity. Both roughness and waviness contribute to the shear strength. Large scale waviness may also alter the dip locally.

Barton (1973) quantified roughness by joint roughness coefficient (JRC) ranges from 0 to 20 as shown in Figure 10.9 of Chapter 10.

(v) Wall strength: It is an equivalent compressive strength of the adjacent rock walls of a discontinuity which may be lower than rock block strength due to weathering or alteration of the walls. It is an important component of shear strength, if rock walls are in contact. Barton (1973) defined wall strength by a term called joint wall compressive strength (JCS). Franklin and Dusseault (1989) determined the relationship between JCS and the Schmidt hammer rebound number (R), which is as follows:

$$\log_{10} \text{JCS} = 0.00088\gamma R + 1.01 \tag{3.2}$$

where, γ is the unit weight of rock (kN/m^3) and JCS in MPa.

(vi) Aperture: It is a perpendicular distance between adjacent rock walls of a discontinuity, in which the intervening space is air or water or soild filled. Table 3.3 shows the description for aperture.

TABLE 3.3 Description Associated with Aperture (ISRM 1978)

Description		*Aperture* (mm)
Closed feature	Very tight	< 0.1
	Tight	0.1–0.25
	Partly open	0.25–0.5
Gapped feature	Open	0.5–2.5
	Moderately wide	2.5–10
	Wide	> 10
Open feature	Very wide	10–100
	Extremely wide	100–1000
	Cavernous	> 1000

(vii) Filling: It is a geological material that separates the adjacent rock walls of a discontinuity and that is usually weaker than the parent rock. Typical filling materials are sand, silt, clay, breccia, gouge and mylonite. Also includes thin mineral coatings and healed discontinuities, e.g., quartz and calcite veins.

(viii) Seepage: Water flow and free moisture are visible in individual discontinuities or in the rock mass as a whole.

(ix) Number of sets: The number of joint sets comprising joint system. The rock mass may be further divided into individual discontinuities. Figure 3.1 shows three joint sets s_1, s_2, and s_3. Table 3.4 shows the description for joint sets.

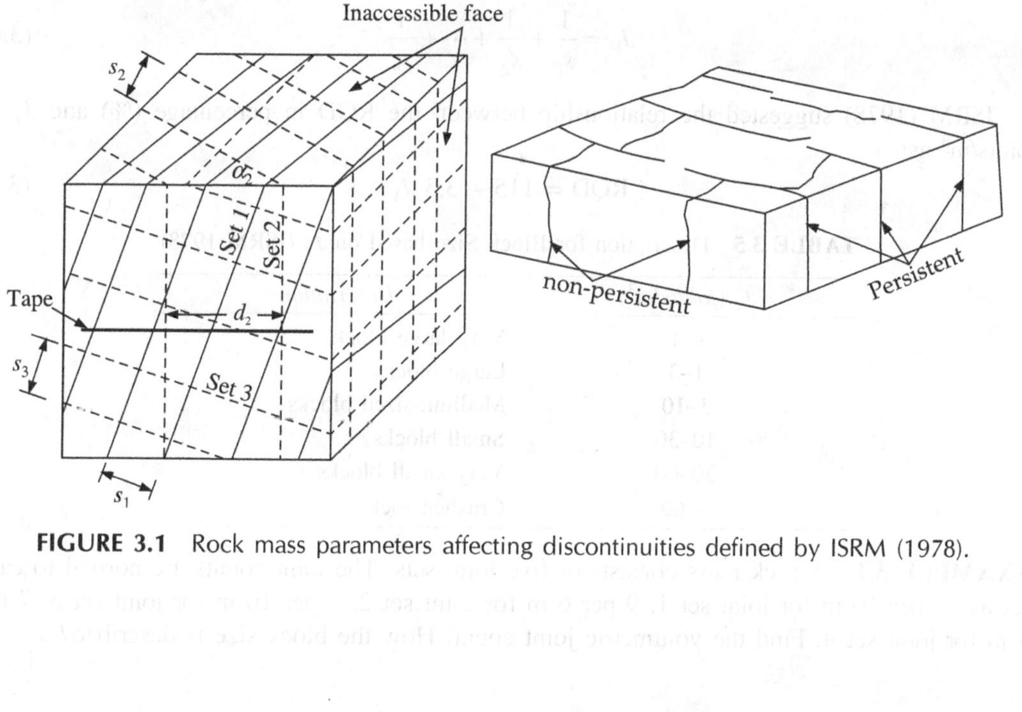

FIGURE 3.1 Rock mass parameters affecting discontinuities defined by ISRM (1978).

TABLE 3.4 Description for Joint Sets (ISRM 1978)

Description	*Joint sets*
I	Massive, occasional random joints
II	One joint set
III	One joint set plus random
IV	Two joint sets
V	Two joint sets plus random
VI	Three joint sets
VII	Three joint sets plus random
VIII	Four or more joint sets
IX	Crushed rock, similar to soils

(x) Block size: Rock block dimensions resulting from the mutual orientation of intersecting joint sets, and from the spacing of the individual sets. Individual discontinuities may further influence the block size and shape. Block size is typically estimated as one of three indices. These are RQD, volumetric joint count (J_V), and block size index (I_b). The block size index is estimated by selecting several typical block sizes and taking their average dimensions (ISRM, 1978) as shown in Table 3.5. In the special case of an orthogonal joint system of three joint sets (as shown in Figure 3.1), the index become

$$I_b = \frac{S_1 + S_2 + S_3}{3} \tag{3.3}$$

The *volumetric joint count* (J_V) is defined as the sum of number of joints per metre for each joint set present, and is measured along the joint set perpendicular (ISRM, 1978).

$$J_V = \frac{1}{S_1} + \frac{1}{S_2} + \cdots + \frac{1}{S_n} \tag{3.4}$$

ISRM (1978) suggested the relationship between the RQD in percentage (%) and J_V in joints/m^3 as:

$$\text{RQD} = 115 - 3.3\, J_V \tag{3.5}$$

TABLE 3.5 Description for Block Size based on J_V (ISRM 1978)

J_V *(joints/m^3)*	*Description*
< 1	Very large blocks
1–3	Large blocks
3–10	Medium-sized blocks
10–30	Small blocks
30–60	Very small blocks
> 60	Crushed rock

EXAMPLE 3.1 A rock mass consists of five joint sets. The joint counts are normal to each set as 12 per 10 m for joint set 1, 9 per 6 m for joint set 2, 8 per 10 m for joint set 3, 7 per 5 m for joint set 4. Find the volumetric joint count. How the block size is described?

Solution: The joint spacing are $s_1 = 10/12$; $s_2 = 6/9$; $s_3 = 10/8$; and $s_4 = 5/7$. Now, using Eq. 3.4, we get

$$J_V = \frac{12}{10} + \frac{9}{6} + \frac{8}{10} + \frac{7}{5} = 4.9 \text{ joints/m}^3$$

From Eq. 3.5, we have

$$\text{RQD} = 115 - 3.3 \times 4.9 = 98.83\%$$

3.3 ROCK MASS CLASSIFICATION SYSTEMS

A number of different rock mass classification systems have been developed as shown in Table 3.6. All these classification systems use three categories of parameters like joint parameters, material parameters and boundary conditions. These systems are developed by defining their (i) objective, (ii) parameters, (iii) calibration of parameters, and (iv) algebra. The chapter discusses five important and most commonly used classification systems as:

(i) Rock quality designation (RQD)
(ii) Rock mass rating (RMR)
(iii) Q-system or Norwegian Geotechnical Institute (NGI) classification
(iv) Geological strength index (GSI)
(v) Coal mine roof rating (CMRR)

RQD is itself a classification system, but it is being used as a parameter in almost all rock mass classification systems. RMR and Q–system are widely used for rock support system design and the GSI system is used for estimating design parameters. The GSI system is the only rock mass classification system that is directly linked to engineering parameters such as Hoek–Brown strength parameters (m_b and s) or rock mass modulus (E). RMR, Q and GSI systems are based on the modification of the existing ones to suit specific application. For example, the RMR system was modified by Laubscher (1990) for mine design and by Kendorski et al. (1983) for drift support design in caving mines. The Q–system (Potvin et al., 1989) and GSI system (Sonmez and Ulusay, 1999) is modified for stope design. GSI system is further modified to consider poor and very poor rock mass into the classification (Osgoui et. al., 2010).

3.3.1 Rock Quality Designation

The rock quality designation (RQD) index is developed by Deere, Hendron, Patton and Cording (1967) to provide a quantitative estimation of rock mass quality from drill core logs. RQD is defined as the percentage of intact drillcore pieces longer than 10 cm in length (Deere et al., 1967; Barton, 2002, ASTM D6032/D6032M-17) as shown in Eq. 3.6. RQD is influenced by the drilling technique and the size of core barrel. ISRM recommended RQD be computed from NX cores of 54 mm diameter double-tube core barrel. The definition of RQD in percentage (%) is

$$\text{RQD} = \frac{\sum \text{length of core pieces} > 10 \text{ cm}}{\text{Total length of core run}} \times 100 \tag{3.6}$$

TABLE 3.6 Rock Mass Classification System

Name of Classification	*Originator and Year*	*Applications*
Rock load theory	Terzaghi (1946)	Tunnels with steel support; Unsuitable for modern tunnelling
Stand–up time	Lauffer (1958)	Tunnelling; Conservative
New Austrian tunnelling method (NATM)	Rabcewicz (1964–65, 1972)	Tunnelling in incompetent (overstresses) ground; Unutilised in squeezing ground condition
Rock quality designation (RQD)	Deere, Hendron, Patton, Cording (1967)	Tunnelling; Sensitive to orientation effects
Rock structure rating (RSR) concept	Wickham, Tiedemann, Skinner (1972)	Tunnels with steel support; Not useful with steel fibre shotcrete
Rock mass rating (RMR)	Bieniawski (1973, 1989)	Tunnels, mines, slopes foundations
NGI/Q–system	Barton, Lien, Lunde (1974, 2002)	Tunnels and wide openings
Mining RMR	Laubscher (1975)	Mining
The typological classification	Matula, Holzer (1978)	For use in communication
Unified rock classification system (URCS)	Williamson (1980)	For use in communication
Basic geological description (BGD)	ISRM (1981)	For general use
Rock mass strength (RMS)	Stille, Groth, Fredriksson (1982)	Modified RMR
Modified basic RMR (MBR)	Kendorski, Cummings, Bieniawski, Skinner (1983)	Mining
Simplified RMR (SMR)	Brook, Dharmaratne (1985)	Mines and tunnels; Modified RMR and MRMR
Ramamurthy/Arora	Ramamurthy, Arora (1993)	For intact and jointed rocks; Modified Deere and Miller approach
Geological strength index (GSI)	Hoek, Kaiser, Bawden (1995)	Mines, tunnel
Rock mass index (RMi)	Palmstrom (1995)	Tunnels, mining openings and other openings in rock mass
Rock mass number (N)	Singh, Goel (1999)	Stress free Q–system

Table 3.7 shows the relationship between the numerical value of RQD and the engineering quality of the rock.

TABLE 3.7 Rock Quality Classification based on RQD

RQD (%)	< 25	25–50	50–75	75–90	90–100
Rock quality	Very poor	Poor	Fair	Good	Very good

RQD is simple and quick, if carried out in conjunction with the normal geological logging of a core, it adds very little to the cost of the site investigation. It depends on the orientation of the drill hole and considered to be a directional property. RQD does not take direct account of joint orientation/spacing/seepage and others which influences the behaviour of a rock mass around an underground and surface excavation.

EXAMPLE 3.2 A 200 cm borehole log of a rock is shown in Figure below. Find the RQD and the core recovery ratio.

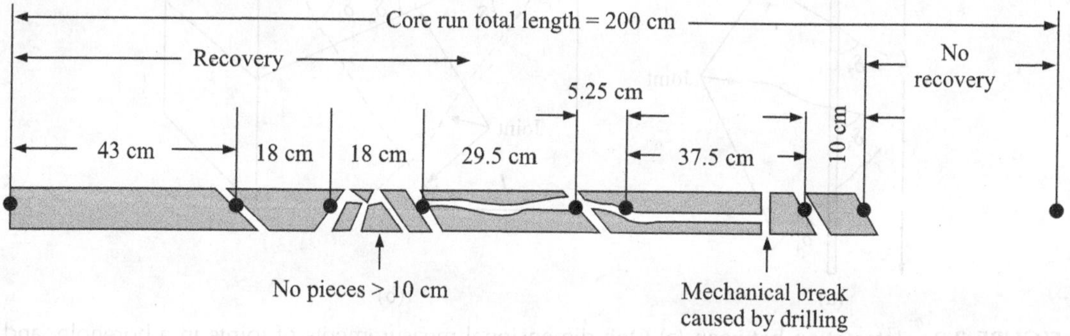

Solution: Use Eq. 3.6 to find the RQD as follows:

$$\text{RQD} = \frac{(43 + 18 + 29.5 + 37.5)}{200} \times 100 = \frac{128}{200} \times 100$$

$$\text{RQD} = 64\% \text{ (Rock Quality: FAIR)}$$

$$\text{Core recovery ratio} = \frac{\sum \text{length of core pieces recovered}}{\text{Total length of core run}} \times 100 = \frac{161.25}{200} \times 100$$

$$\text{Core recovery ratio} = 80.63\%$$

Indirectly, RQD in % can be obtained from joint spacing measurements made on an exposure by using Eq. 3.7 (Priest and Hudson 1976).

$$\text{RQD} = 100e^{-0.1L}(0.1L + 1) \tag{3.7}$$

where, L is number of discontinuities per metre of core run. RQD is a directional property and dependent on the borehole orientation. The interaction of a borehole with rock joints in 2D and 3D are shown in Figure 3.2. In principle, it is based on the measurement of the angle between each joint and the surface or the drill hole. The weighted joint density (WJD) is for measurements on rock surfaces:

$$\text{WJD} = \frac{1}{\sqrt{A}} \sum \frac{1}{\sin \delta_i} \tag{3.8}$$

and for measurements along a drill core or scan line:

$$\text{WJD} = \frac{1}{\sqrt{L}} \sum \frac{1}{\sin \delta_i} \tag{3.9}$$

where, δ_i is the intersection angle, i.e., the angle between the observed plane or drill hole and the individual joint; A is the size of the observed area in m^2; L is the length of measured section along the core or scan line.

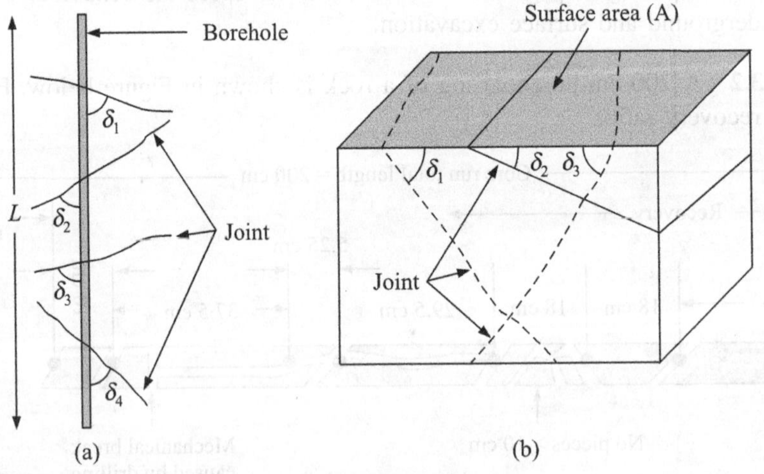

FIGURE 3.2 Interaction between (a) One dimensional measurements of joints in a borehole, and (b) Two dimensional measurement of joints on a surface.

3.3.2 Rock Mass Rating—RMR System

The RMR or geomechanics classification system was developed at South African Council for Scientific and Industrial Research (CSIR) by Prof. Z.T. Bieniawski in 1976 and later modified in the year 1989. The advantage of this system is that only few basic parameters relating to the geometry and mechanical properties of the rock mass are required.

Original RMR (1976) consists of eight parameters, i.e., rock quality designation (RQD), state of weathering, uniaxial compressive strength (UCS), spacing of joints and bedding, strike and dip orientation, separation of joints, continuity of joints, and ground water inflow. In the year 1989, Bieniawski proposed the modified system by eliminating the state of weathering since its effect is accounted for UCS, and clubbed two parameters like separation of joints and continuity of joints into condition of joints. In addition, the strike and dip orientations of joints are removed and their effects are adjusted into the rating after the basic parameters had been considered. Finally, RMR (1989) considers the following six parameters as also elaborated in Table 3.8 (Bieniawski, 1989).

(i) UCS
(ii) RQD
(iii) Spacing of discontinuities
(iv) Condition of discontinuities
(v) Groundwater conditions
(vi) Orientation of discontinuities

The overall rating of the classification system are given as:

$$\text{RMR} = \Sigma(\text{classification parameters}) + \text{discontinuity orientation adjustment} \quad (3.10)$$

Table 3.8A shows first five classification parameters and their ratings, the parameters are grouped into five classes, each one covers a range of values appropriate to that parameter.

TABLE 3.8A Classification Parameters and Their Ratings

Parameter			*Range of values*						
1	Strength of intact rock material	Point-load strength index (MPa)	>10	4–10	2–4	1–2	For this low range uniaxial compressive test is preferred		
		Uniaxial compressive strength (MPa)	>250	100–250	50–100	25–50	5–25	1–5	< 1
	Rating		15	12	7	4	2	1	0
2	Drill core quality RQD (%)		90–100	75–90	50–75	25–50	< 25		
	Rating		20	17	13	8	3		
3	Spacing of discontinuities (mm)		> 2000	600–2000	200–600	60–200	< 60		
	Rating		20	15	10	8	5		
4	Condition of discontinuities (See Table 3.8E)		Very rough surfaces; Not continuous; No separation; Unweathered wall rock	Slightly rough surfaces; Separation < 1 mm; Slightly weathered walls	Slightly rough surfaces; Separation < 1 mm; Highly weathered walls	Slicken sided surfaces; Gouge < 5 mm thick; Separation 1–5 mm; Continuous	Soft gouge >5 mm thick; Separation > 5 mm; Continuous		
	Rating		30	25	20	10	0		
5	Ground water	Inflow per 10 m tunnel length (L/m)	None	< 10	10–25	25–125	> 125		
		(Joint water pressure)/(Major principal stress, σ)	0	< 0.1	0.1–0.2	0.2 – 0.5	> 0.5		
		General conditions	Completely dry	Damp	Wet	Dripping	Flowing		
	Rating		15	10	7	4	0		

TABLE 3.8B Rating Adjustment for Discontinuity Orientations (see Table 3.8F)

Strike and dip orientations		*Very favourable*	*Favourable*	*Fair*	*Unfavourable*	*Very unfavourable*
Ratings	Tunnels and mines	0	–2	–5	–10	–12
	Foundations	0	2	–7	–15	–25
	Slopes	0	–5	–25	–50	

TABLE 3.8C Rock Quality Determined from Total Ratings

Rating	100–81	80–61	60–41	40–21	< 21
Class number	I	II	III	IV	V
Description	Very good	Good	Fair	Poor	Very poor

TABLE 3.8D Meaning of Rock Quality

Class number	*I*	*II*	*III*	*IV*	*V*
Average stand–up time	20 years for 15 m span	1 year for 10 m span	1 week for 5 m span	10 hours for 2.5 m span	30 minutes for 1 m span
Cohesion of rock mass (kPa)	> 400	300–400	200–300	100–200	< 100
Friction angle of rock mass (deg)	> 45	35–45	25–35	15–25	< 15

TABLE 3.8E Guidelines for Classification of Discontinuity Condition

Discontinuity length (persistence) (mm)	< 1	1–3	3–10	10–20	> 20
Rating	6	4	2	1	0
Aperture (mm)	None	< 0.1	0.1–10	1–5	> 5
Rating	6	5	4	1	0
Roughness	Very rough	Rough	Slightly rough	Smooth	Slickensided
Rating	6	5	3	1	0
Infilling (gouge)	None	Hard filling < 5 mm	Hard filling > 5 mm	Soft filling < 5 mm	Soft filling > 5 mm
Rating	6	4	2	2	0
Weathering	Unweathered	Slightly weathered	Moderately weathered	Highly weathered	Decomposed
Rating	6	5	3	1	0

TABLE 3.8F Effect of Discontinuity Strike and Dip Orientation in Tunnelling

Strike perpendicular to tunnel axis				*Strike parallel to tunnel axis*		
Drive with dip		*Drive against dip*		*Dip* 45–90°	*Dip* 20–45°	*Dip* 0–20° *irrespective of strike*
Dip 45–90°	*Dip* 20–45°	*Dip* 45–90°	*Dip* 20–45°			
Very favourable	Favourable	Fair	Unfavourable	Very unfavourable	Fair	Fair

When assessing a given rock mass, one establishes into that group in which the parameter lies, and then sums the resulting numerical ratings for the five parameters.

Apart from Table 3.8A, another Table 3.8E is given to quantify the condition of discontinuities by the addition of rating of five parameters like discontinuity length, aperture, roughness rating, infilling rating, and weathering.

Table 3.8B shows the adjustment of ratings with respect to orientation of rock structure like tunnel and mine, and foundation or slopes. The orientation of the discontinuities becomes progressively more important from tunnels and mines, through foundations, to slopes. Figure 3.3 shows the effect of joint orientation with respect to drive direction of excavation on their stability. Figure 3.3(a) is more stable at face and floor for dip angle less than 45° than Figure 3.3(b) and Figure 3.3(c), and hence considered as favorable, unfavourable and fair respectively as per Table 3.8F.

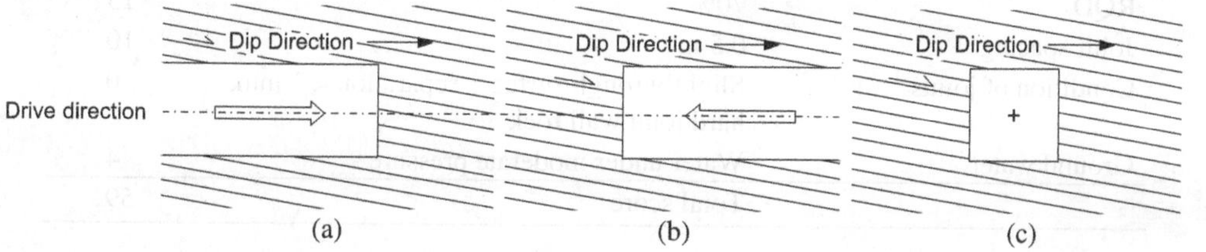

FIGURE 3.3 (a) Strike perpendicular to tunnel axis and drive with dip; (b) Strike perpendicular to tunnel axis and drive against dip; and (c) Strike parallel to tunnel axis.

In Table 3.8C, the rock mass classes are given with a description from 'very good' to 'very poor'. Based on rock mass class Table 3.8D proposes average stand–up time and the Mohr–Coulomb strength parameters (like cohesion and friction angle) for each class.

This classification scheme has been widely applied in mining and tunnelling applications for decades despite the simplicity of approach when dealing with complex rock masses. Although Q–system is more appropriate for hard and jointed rock mass, in coal mine applications RMR is still a prevailing rock mass classification system.

EXAMPLE 3.3 A joint set is found to be with slightly rough and highly weathered with a spacing less than 1 m. The joint surface is open, persistence is 2.8 m and aperture varies from 0.2 to 0.8 mm. Find the joint condition rating based on RMR classification system.

Solution: The condition of discontinuity is first determined from Table 3.8A. If five conditions of discontinuity (i.e., algebraic sum of the ratings for separation and filling/open, weathering, roughness, persistence, and aperture) do not match as defined in Table 3.8A, then the individual ratings of the parameters defining the condition of discontinuity shown in Table 3.8E are used to define the overall rating of the condition of discontinuity. The ratings for open joint without filling is 6; for highly weathered joint rating is 1; slightly rough joint rating is 3; for 2.8 m persistence, rating is 4 and finally for aperture of 0.2 mm to 0.8 mm, rating is 4.

So, the overall rating of joint condition is

$$\text{RMR joint condition rating} = 6 + 1 + 3 + 4 + 4 = 18$$

EXAMPLE 3.4 Consider the example of a granitic rock mass in which a tunnel is to be driven. It is given that the uniaxial compressive strength of rock is 150 MPa, RQD is 70%, Joint spacing is 0.5 m, joint is slightly rough and hard with a separation of less than 1 mm. If the presence of ground water is moderate, then classify the rock mass.

Solution: The tunnel has been oriented such that the strike of dominant joint set is perpendicular to the tunnel axis with a dip of 30° against the drive direction. From Table 3.8F, this situation is described as unfavourable for which a rating adjustment of –10 is to be applied. Thus, the final rock mass rating becomes 59 which places the rock mass at the upper end of Class III with a description of fair.

Classification parameter	*Value of description*	*Rating*
Strength of intact material	150 MPa	12
RQD	70%	13
Joint spacing	0.5 m	10
Condition of joints	Slightly rough surfaces separation < 1 mm, hard joint wall rock	20
Ground water	Water under moderate pressure	4
	Total score	59

EXAMPLE 3.5 A tunnel is to be driven through a slightly weathered granite with a dominant joint set dipping at 60° against the direction of the drive. Index testing and logging of diamond drilled core give typical point–load strength index value of 8 MPa and average RQD value of 70%. The slightly rough and slightly weathered joints with a separation of < 1 mm, are spaced at 300 mm. Tunnelling conditions are anticipated to be wet. Determine RMR of this rock mass.

Solution:

Classification parameter	*Value of description*	*Rating*
Point load index	8 MPa	12
RQD	70%	13
Spacing of discontinuities	Note 1	22
Ground water	Wet	7
Adjustment of joint orientation	Note 2	–5
	Total score	59

Note 1: For slightly rough and altered discontinuity surfaces with a separation of < 1 mm, Table 3.8E gives a rating of 25. When more detailed information is available, Table 3.8E can be used to obtain a more refined rating. Hence, in this case, the rating is the sum of: 4 (1–3 m discontinuity length), 4 (separation 0.1–1.0 mm), 3 (slightly rough), 6 (no infilling), and 5 (slightly weathered) = 22.

Note 2: Table 3.8F gives a description of 'Fair' for the conditions assumed where the tunnel is to be driven against the dip of a set of joints dipping at 60°. Using this description for 'Tunnels and Mines' in Table 3.8B gives an adjustment rating of –5.

3.3.3 Rock Tunnelling Quality Index (Q–System)

Barton, Lien and Lunde (1974, 2002) of the Norwegian Geotechnical Institute (NGI) proposed a tunnelling quality index (Q) for the determination of rock mass characteristics and tunnel support requirements on the basis of an evaluation of a large number of case histories of underground excavations. The numerical value of the index varies on a logarithmic scale from 0.001 to a maximum of 1,000 and is defined as:

$$Q = \frac{RQD}{J_n} \times \frac{J_r}{J_a} \times \frac{J_w}{SRF} \tag{3.11}$$

where, RQD is the rock quality designation; J_n is the joint set number; J_r is the joint roughness number; J_a is the joint alteration number; J_w is the joint water reduction factor; SRF is the stress reduction factor.

As evident from Eq. 3.11, Q–system consists of multiplication of three quotient parameters. The **first quotient** (RQD/J_n), representing the structure of the rock mass, is a crude measure of the block or particle size, with the two extreme values (100/0.5 or 200 and 10/20 or 0.5) differing by a factor of 400. If the quotient is interpreted in units of centimetres, the extreme 'particle sizes' of 200 cm to 0.5 cm are seen to be crude, but fairly realistic approximations. Probably the largest blocks should be several times this size and the smallest fragments are less than half the size excluding clay particles. The **second quotient** (J_r/J_a) represents the roughness and frictional characteristics of the joint walls or filling materials. This quotient is weighted in favour of rough, unaltered joints in direct contact. It is to be expected that such surfaces will be close to peak strength, that they will dilate strongly when sheared, and they will, therefore, be especially favourable to tunnel stability. When rock joints have thin clay mineral coatings and fillings, the strength is reduced significantly. Nevertheless, rock wall contact after small shear displacements may be an important factor for preserving the excavation from ultimate failure. If no rock wall contact exists, the conditions are extremely unfavourable to tunnel stability. The **third quotient** (J_w/SRF) consists of two stress parameters and describes the concept of 'active stress'. SRF is a measure of total stress parameter which includes: (i) loosening load in case of an excavation through shear zones and clay bearing rock, (ii) rock stress in competent rock, and (iii) squeezing loads in plastic incompetent rocks. The parameter (J_w) is a measure of water pressure, which has an adverse effect on the shear strength of joints due to reduction in effective normal stress. Water may, in addition, cause softening and possible out–wash of clay–filled joints. It has been proved that it is impossible to combine these two parameters in terms of inter–block effective stress, because paradoxically a high value of effective normal stress may sometimes signify less stable conditions than a low value, despite the higher shear strength. Based on the above discussions, it can be said that rock tunnelling quality, Q is considered to be a function of only three parameters which are crude measures of:

(i) Block size (RQD/J_n)
(ii) Inter–block shear strength (J_r/J_a)
(iii) Active stress (J_w/SRF)

Table 3.9 is reproduced from Barton et al., 1974, 2002 to show the full range of classes for the six parameters involved in the system. Finally, Table 3.10 shows the rock mass classification based on parameters used in Q-system.

TABLE 3.9A Joint Set Number

	Class	J_n
A	Massive, no. or few joints	0.5–1
B	One joint set	2
C	One joint set plus random joints	3
D	Two joint set	4
E	Two joint set plus random joints	6
F	Three joint set	9
G	Three joint set plus random joints	12
H	Four or more joint sets, random, heavily jointed, 'sugar–cube', etc.	15
J	Crushed rock, earth like	20

Note: (i) For tunnel intersections, use ($3.0 \times J_n$) (ii) For portals, use ($2.0 \times J_n$).

TABLE 3.9B Joint Roughness Number

	Class	J_r
(a) Rock-wall contact, and (b) rock-wall before 10 cm shear		
A	Discontinuous joints	4
B	Rough or irregular, undulating	3
C	Smooth, undulating	2
D	Slicken sided, undulating	1.5
E	Rough or irregular, planar	1.5
F	Smooth, planar	1.0
G	Slicken sided, planar	0.5
(b) No rock-wall contact when sheared		
H	Zone containing clay minerals thick enough to prevent rock-wall contact	1.0
J	Sandy, gravely or crushed zone thick enough to prevent rock-wall contact	1.0

Note: (i) Descriptions refer to small-scale features and intermediate scale features, in that order. (ii) Add 1.0 if the mean spacing of the relevant joint set is greater than 3 m. (iii) $J_r = 0.5$ can be used for planar joints having lineations, provided the lineations are oriented for minimum strength. (iv) J_r and J_a classification is applied to the joint set or discontinuity that is least favourable for stability, both from the point of view of orientation and shear resistance, τ (where $t \approx \sigma_n \tan^{-1} (J_r/J_a)$.

TABLE 3.9C Joint Alteration Number

	Class	$\varphi_r(°)$	J_a
(a) Rock-wall contact (no mineral fillings, only coatings)			
A	Tightly healed, hard, non-softening, impermeable filling, i.e., quartz or epidote	—	0.75
B	Unaltered joint walls, surface staining only	25–35	1.0
C	Slightly altered joint walls, non-softening mineral coatings, sandy particles, clay-free disintegrated rock, etc.	25–30	2.0
D	Silty- or sandy-clay coating, small clay fraction (non-softening)	20–25	3.0
E	Softening or low friction clay mineral coatings, i.e., kaolinite or mica. Also chlorite, talc, gypsum, graphite, etc., and small quantities of swelling clays	8–16	4.0

(Contd.)

Class		$\varphi_r(°)$	J_a
(b) Rock-wall contact 10 cm shear (thin mineral fillings)			
F	Sandy particles, clay-free disintegrated rock, etc.	25–30	4.0
G	Strongly over-consolidated non-softening clay mineral fillings (continuous, but < 5 mm thickness)	16–24	6.0
H	Medium or low over-consolidation, softening, clay mineral fillings (continuous, but < 5 mm thickness)	12–16	8.0
J	Swelling-clay fillings, i.e., montmorillonite (continuous, but < 5 mm thickness). Value of J_a depends on percent of swelling clay-size particles, and access to water, etc.	6–12	8–12
(c) No rock-wall contact when sheared (thick mineral fillings)			
KLM	Zones or bands of disintegrated or crushed rock and clay (see G, H, J for descriptions of clay conditions)	6–24	6, 8 or 8–12
N	Zones or bands of silty- or sandy-clay, small clay fraction (non-softening)		5.0
OPR	Thick, continuous zones or bands of clay (see G, H, J for description of clay condition)	6–24	10,13 or 13–20

TABLE 3.9D Joint Water Reduction Factor

Class		*Approx. water pressure* (kg/cm^2)	J_w
A	Dry excavations or minor inflow, i.e., < 5 l/min locally	< 1	1.0
B	Medium inflow or pressure, occasional outwash of joint filling	1–2.5	0.66
C	Large inflow or high pressure in competent rock with unfilled joints	2.5–10	0.50
D	Large inflow or high pressure in considerable outwash of joint fillings	2.5–10	0.33
E	Exceptionally high inflow or water pressure at blasting, decaying with time	> 10	0.2–0.1
F	Exceptionally high inflow or water pressure continuing without noticeable decay	> 10	0.1–0.05

Note: (i) Factors C to F are crude estimates. Increase J_w if drainage measures are installed. (ii) Special problems caused by ice formation are not considered. (iii) For general characterization of rock masses distant from excavation influences, the use of J_w = 1.0, 0.66, 0.5, 0.33, etc. as depth increases from say 0–5, 5–25, 25–250 to >250 m is recommended, assuming that RQD/J_n is low enough (e.g., 0.5–25) for good hydraulic connectivity. This will help to adjust Q for some of the effective stress and water softening effects, in combination with appropriate characterization value of SRF. Correlations with depth-dependent static deformation modulus and seismic velocity will then follow the practise used when these were developed.

TABLE 3.9E Stress Reduction Factor

Class		*SRF*
(a) Weakness zones intersecting excavation, which may cause loosening of rock mass when tunnel is excavated		
A	Multiple occurrences of weakness zones containing clay or chemically disintegrated rock, very loose surrounding rock (any depth)	10
B	Single weakness zones containing clay or chemically disintegrated rock (depth of excavation ≤ 50 m)	5

(Contd.)

TABLE 3.9E Stress Reduction Factor

	Class	*SRF*		
C	Single weakness zones containing clay or chemically disintegrated rock (depth of excavation > 50 m)	2.5		
D	Multiple shear zones in competent rock (clay-free), loose surrounding rock (any depth)	7.5		
E	Single shear zones in competent rock (clay-free), (depth of excavation ≤ 50 m)	5.0		
F	Single shear zones in competent rock (clay-free), (depth of excavation > 50 m)	2.5		
G	Loose, open joints, heavily jointed or 'sugar cube', etc. (any depth)	5.0		
(b)	**Competent rock, rock mass problems**	σ_c/σ_1	σ_θ/σ_c	SRF
H	Low stress, favourable stress condition	> 200	< 0.01	2.5
J	Medium stress, favourable stress condition	200–10	0.01–0.3	1
K	High stress, very tight pressure. Usually favourable to stability, may be unfavourable for wall stability	10–5	0.3–0.4	0.5–2
L	Moderate slabbing after > 1 h in massive rock	5–3	0.5–0.65	5–50
M	Slabbing and rock burst after a few minutes in massive rock	3–2	0.65–1	50–200
N	Heavy rock burst (strain–burst) and immediate dynamic deformation in massive rock	< 2	> 1	200–400
(c)	**Squeezing rock: plastic flow of incompetent rock under the influence of high rock pressure**	σ_θ/σ_c	SRF	
O	Mild squeezing rock pressure	1–5	5–10	
P	Heavy squeezing rock pressure	> 5	10–20	
(d)	**Squeezing rock: Chemical swelling activity depending on presence of water**	SRF		
R	Mild swelling rock pressure	5–10		
S	Heavy swelling rock pressure	10–15		

Note: (i) Reduce these values of SRF by 25–50% if the relevant shear zones only influence but do not intersect the excavation. This will also be relevant for characterization. (ii) For strongly anisotropic virgin stress field (if measured): when, $5 \le \sigma_1/\sigma_3 \le 10$, reduce σ_c to $0.75\ \sigma_c$. When $\sigma_1/\sigma_3 > 10$, reduce σ_c to $0.5\sigma_c$, where σ_c is the unconfined compressive strength, σ_1 and σ_3 are the major and minor principal stresses, and σ_θ the maximum tangential stress (estimated from elastic theory). (iii) Few case records available where depth below surface is less than span width, suggest as SRF increase from 2.5 to 5 for such case (see H). (iv) Cases L, M, and N are usually most relevant for support design of deep tunnel excavations in hard massive rock masses, with RQD/J_n ratios from about 50–200. (v) For general characterization of rock masses distant from excavation influences, the use of SRF = 5, 2.5, 1.0, and 0.5 is recommended as depth increases from say 0–5, 5–25, 25–250 or > 250 m. This will help to adjust Q for some of the effects, in combination with appropriate values of J_w. Correlations with depth-dependent static deformation modulus and seismic velocity will then follow the practice used when these were developed. (vi) Cases of squeezing rock may occur for depth $H > 350Q^{1/3}$ Singh (1993). Rock mass compression strength can be estimated from $\text{SIGMA}_{cm} \approx 5\gamma Q_c^{1/3}$ (MPa) where γ is the rock density in t/m^3, and $Q_c \approx Q \times \sigma_c/100$, Barton (2000).

TABLE 3.10 Rock Quality Classification for *Q*–system (Barton et al., 1974)

Q–value	*Class*	*Rock mass*
400–1000	A	Exceptionally good
100–400	A	Extremely good
40–100	A	Very good
10–40	B	Good
4–10	C	Fair
1–4	D	Poor
0.1–1.0	E	Very poor
0.01–0.1	F	Extremely poor
0.001–0.01	G	Exceptionally poor

Barton et al. (1974) defined a parameter called the *equivalent dimension* (D_e), of the excavation to relate the index Q to the stability and support requirements of underground excavations. The equivalent dimension is the ratio of the span, diameter or wall height of the excavation to the excavation support ratio (ESR). Hence,

$$D_e = \frac{\text{Excavation span, diameter or height (m)}}{\text{Excavation support ratio (ESR)}}$$

The value of ESR is related to the intended use of the excavation and to the degree of security which is demanded of the support system installed to maintain the stability of the excavation. The suggested values of ESR with excavation category are as given in Table 3.11.

TABLE 3.11 Values of ESR with Excavation Category

Type of Excavation	*ESR*
Temporary mine openings	3–5
Permanent mine openings, water tunnels for hydro power (excluding high pressure penstocks), pilot tunnels, drifts and headings for large excavations.	1.6
Storage rooms, water treatment plants, minor road and railway tunnels, surge chambers, access tunnels.	1.3
Power stations, major road and railway tunnels, civil defence chambers, portal intersections.	1.0
Underground nuclear power stations, railway stations, sports and public facilities, factories.	0.8

Figure 3.4 shows chart between the equivalent dimension and Q ratings, and is used to define a number of support categories.

Barton et al. (1980) defined the relationship between length of rockbolt (L) in m, excavation width (B) in m, and excavation support ratio (ESR) as:

$$L = \frac{2 + 0.15\,B}{\text{ESR}} \tag{3.12}$$

The maximum unsupported span in m, can be estimated as:

$$\text{Maximum unsupported span} = 2 \times \text{ESR} \times Q^{0.4} \tag{3.13}$$

RMR ≈ 9 In Q + 44 (Bieniawski, 1989) $Q \approx e^{\frac{[RMR-44]}{9}}$ [1]

RMR ≈ 15 log Q + 50 (Barton, 1995) $Q \approx 10^{\frac{[RMR-50]}{15}}$ [2]

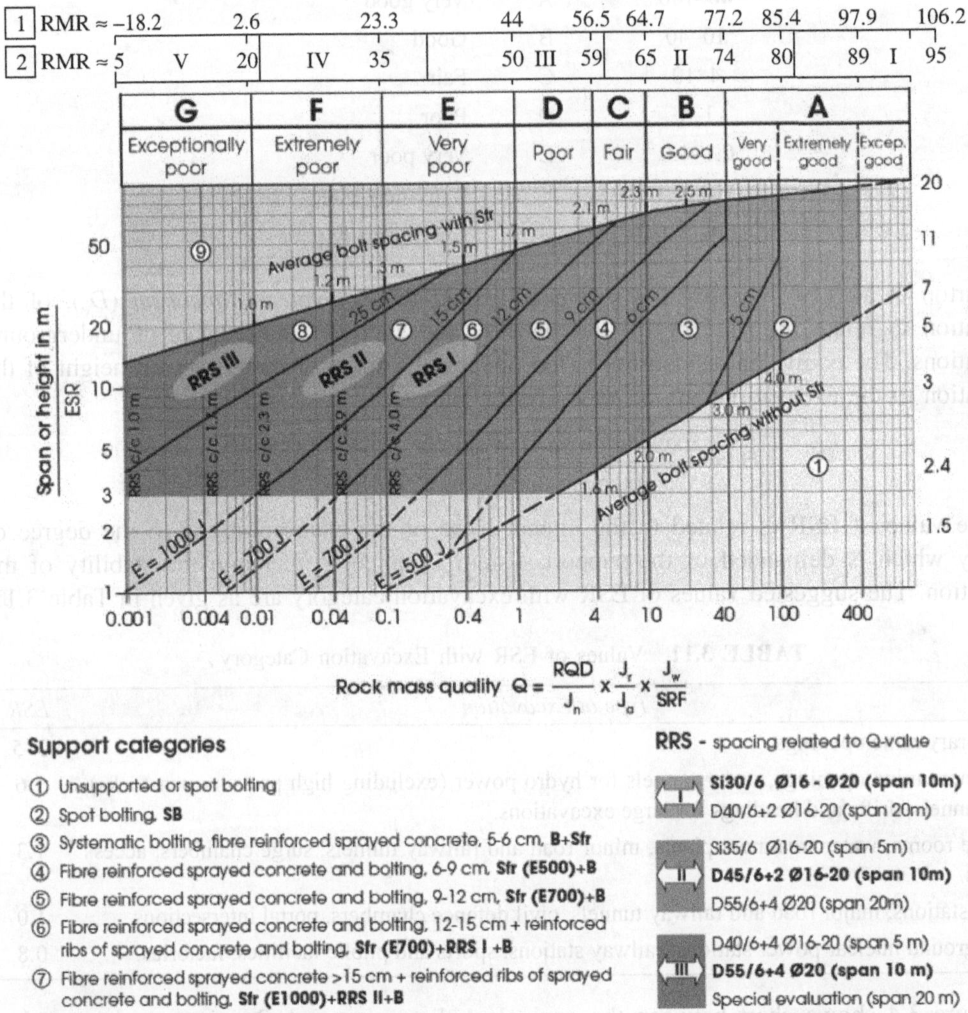

Support categories

① Unsupported or spot bolting
② Spot bolting, **SB**
③ Systematic bolting, fibre reinforced sprayed concrete, 5-6 cm, **B+Sfr**
④ Fibre reinforced sprayed concrete and bolting, 6-9 cm, **Sfr (E500)+B**
⑤ Fibre reinforced sprayed concrete and bolting, 9-12 cm, **Sfr (E700)+B**
⑥ Fibre reinforced sprayed concrete and bolting, 12-15 cm + reinforced ribs of sprayed concrete and bolting, **Sfr (E700)+RRS I +B**
⑦ Fibre reinforced sprayed concrete >15 cm + reinforced ribs of sprayed concrete and bolting, **Sfr (E1000)+RRS II+B**
⑧ Cast concrete lining, **CCA** or **Sfr (E1000)+RRS III+B**
⑨ Special evaluation

Bolts spacing is mainly based on Ø20 mm

E = Energy absorbtion in fibre reinforced sprayed concrete

ESR = Excavation Support Ratio

Areas with dashed lines have no empirical data

RRS - spacing related to Q-value

I: **Si30/6 Ø16 - Ø20 (span 10m)**
D40/6+2 Ø16-20 (span 20m)

II: Si35/6 Ø16-20 (span 5m)
D45/6+2 Ø16-20 (span 10m)
D55/6+4 Ø20 (span 20m)

III: D40/6+4 Ø16-20 (span 5 m)
D55/6+4 Ø20 (span 10 m)
Special evaluation (span 20 m)

Si30/6 = Single layer of 6 rebars, 30 cm thickness of sprayed concrete
D = Double layer of rebars
Ø16 = Rebar diameter is 16 mm
c/c = RSS spacing, centre - centre

FIGURE 3.4 Relation between Q and RMR system along with permanent support recommendations based on Q-values and span/ESR (NGI, 2022).

Grimstad and Barton (1993) suggested the relationship between the value of Q and the permanent roof support pressure (P_{roof}) in MPa as:

$$P_{roof} = \frac{2Q^{1/3}\sqrt{J_n}}{3J_r} \tag{3.14}$$

EXAMPLE 3.6 A 15 m span crusher chamber for an underground mine is to be excavated in a granitic rock at a depth of 2,100 m below the surface. The rock mass contains two sets of joints. These joints are undulating, rough and unweathered with very minor surface staining. RQD values range from 85% to 95% and laboratory tests on core samples of intact rock give an average uniaxial compressive strength of 170 MPa. The principal stress directions are approximately vertical and horizontal and the magnitude of the horizontal principal stress is approximately 1.5 times that of the vertical principal stress. The rock mass is locally damp but there is no evidence of flowing water. Discuss the support requirement of the above excavation.

Solution: The numerical value of RQD is used directly in the calculation of Q and for this rock mass, an average value of 90 will be used. Table 3.9A lists that for two joint sets, the value of J_n is 4. For rough or irregular joints which are undulating, Table 3.9B gives $J_r = 3$. Table 3.9C gives $J_a = 1.0$, for unaltered joint walls with surface staining only. Table 3.9D shows that, for an excavation with minor inflow, the joint water reduction factor, $J_w = 1.0$. For a depth of 2,100 m, the overburden stress will be approximately 57 MPa and in this case, the major principal stress $\sigma_1 = 85$ MPa. Since the UCS of the granite is approximately 170 MPa, this gives a ratio of $\sigma_c/\sigma_1 = 2$. Table 3.9E shows that, for competent rock with rock stress problems, this value of σ_c/σ_1 can be expected to produce heavy rock burst conditions and that value of SRF should lie between 10 and 20. A value of SRF = 1.5 will be assumed for this calculation. Using these values.

$$Q = \frac{90}{4} \times \frac{3}{1} \times \frac{1}{1.5} = 45$$

The crusher station discussed above falls into the category of permanent mine openings and is assigned an excavation support ratio ESR = 1.6. Hence, for an excavation span of 15 m, the equivalent dimension, $D_e = 15/1.6 = 9.4$.

For equivalent dimension ($D_e = 9.4$) and $Q = 45$, places this crusher excavation in category which requires a pattern of rock bolts (spaced at 3.0 m).

Løset (1992) suggests that, for rocks with $4 < Q < 30$, blasting damage will result in the creation of new 'joints' with a consequent local reduction in the value of Q for the rock surrounding the excavation. The blast damage can be accounted by reducing the RQD value for the blast damaged zone, say RQD drops to 50%, the resulting value of $Q = 2.9$. From Figure 3.4, this value of Q, for an equivalent dimension, D_e of 9.4, places the excavation just inside category (3) which requires systematically spaced steel rock bolt, at approximately 3.0 m spacing, and a 4 mm thick layer of steel fibre reinforced shotcrete and bolt length 3.13 m.

Comparison of RMR and Q–system

The common parameter used in both the system is RQD. RMR classification system includes joint spacing and orientation. Joint roughness and alteration are the strongest feature of Q–system and are not emphasised in the RMR system. In RMR system, rock stress is not used specially as a parameter when selecting support measures. In the Q–system, the ratio (σ_c/σ_1) is evaluated when treating rock stress problems. The Q–system also accounts for loosening caused by shear zone, faults and squeezing and swelling ground.

RMR system emphasises structural considerations and orientation and does not account for stress state. Q–system assesses the stress field but does not account for joints/fracture orientation. Both systems adequately assess discontinuity properties, i.e., roughness, infilling, and weathering.

3.3.4 Geological Strength Index (GSI) System

The geological strength index (GSI), introduced by Hoek (1994), Hoek et al., (1995) and Hoek and Brown (1998) provides a system for estimating the reduction in rock mass strength for different geological conditions as identified by field observations. It is a number ranging from about 5 for extremely poor rock masses to 100 for extremely strong unjointed rock masses. The rock mass characterisation is straightforward and it is based upon the visual impression of the rock structure, in terms of blockiness, and the surface condition of the discontinuities indicated by joint roughness and alteration. The combination of these two parameters provides a practical basis for describing a wide range of rock mass types, with diversified rock structure ranging from very tightly interlocked strong rock fragments to heavily crushed rock masses. Based on the rock mass description, the value of GSI is estimated from the contours given in Table 3.12. Table 3.13 shows the rock mass classification based on GSI values.

Hoek (1994) and Hoek and Brown (1997) reported that GSI is equal to RMR without considering the water and discontinuities orientation adjustments. However, GSI can be approximated from RMR as:

$$GSI \approx RMR_{76} = RMR_{89} - 5 \quad (3.15)$$

where, RMR_{89} and RMR_{76} are the values computed according to Bieniawski (1989) and Bieniawski (1976) RMR classification systems. The relationship is meaningless in the range of weak, very weak and heterogeneous rock masses.

3.3.5 Modified GSI for Poor or Very Poor Rock Mass

For very poor rock mass quality, RMR and Q–system cannot be applied because these are dependent on RQD, which in most of the poor rock masses is essentially zero. RQD plays a dominant role in rock mass classification by RMR and Q–system, but very subjective to apply in poor rock masses due to limitations and difficulty in the estimation of RQD (in most cases the value of RQD for such rock masses is lower than 10 and even zero). On the other hand, GSI is independent of RQD but dependent on structural fabric (i.e., degree of jointing and joint condition), have alleviated such constraints and difficulties, but there are still difficulties in measuring the degree of jointing of such a poor rock mass as denoted by N/A (not applicable) parts at the bottom of the existing GSI charts. However, the use of existing GSI chart is subjective and requires long-term experience. Generally, very poor rock masses are defined by a GSI value below 27 or 25. To improve the applicability of the GSI in poor or very poor rock masses and to effectively quantify the joint properties, four parameters are defined: (i) structural rating (SR) with its relation with J_v, (ii) broken structural domain (BSTR), (iii) intact core recovery (ICR), and (iv) joint surface. These properties are obtained from overall field observation and core-box surveying (Osgoui et al., 2010). Table 3.14 shows the modified GSI chart for poor and very poor rock masses.

TABLE 3.12 Characterisation of Rock Masses on the basis of Interlocking and Joint Alteration (Hoek et al., 1998)

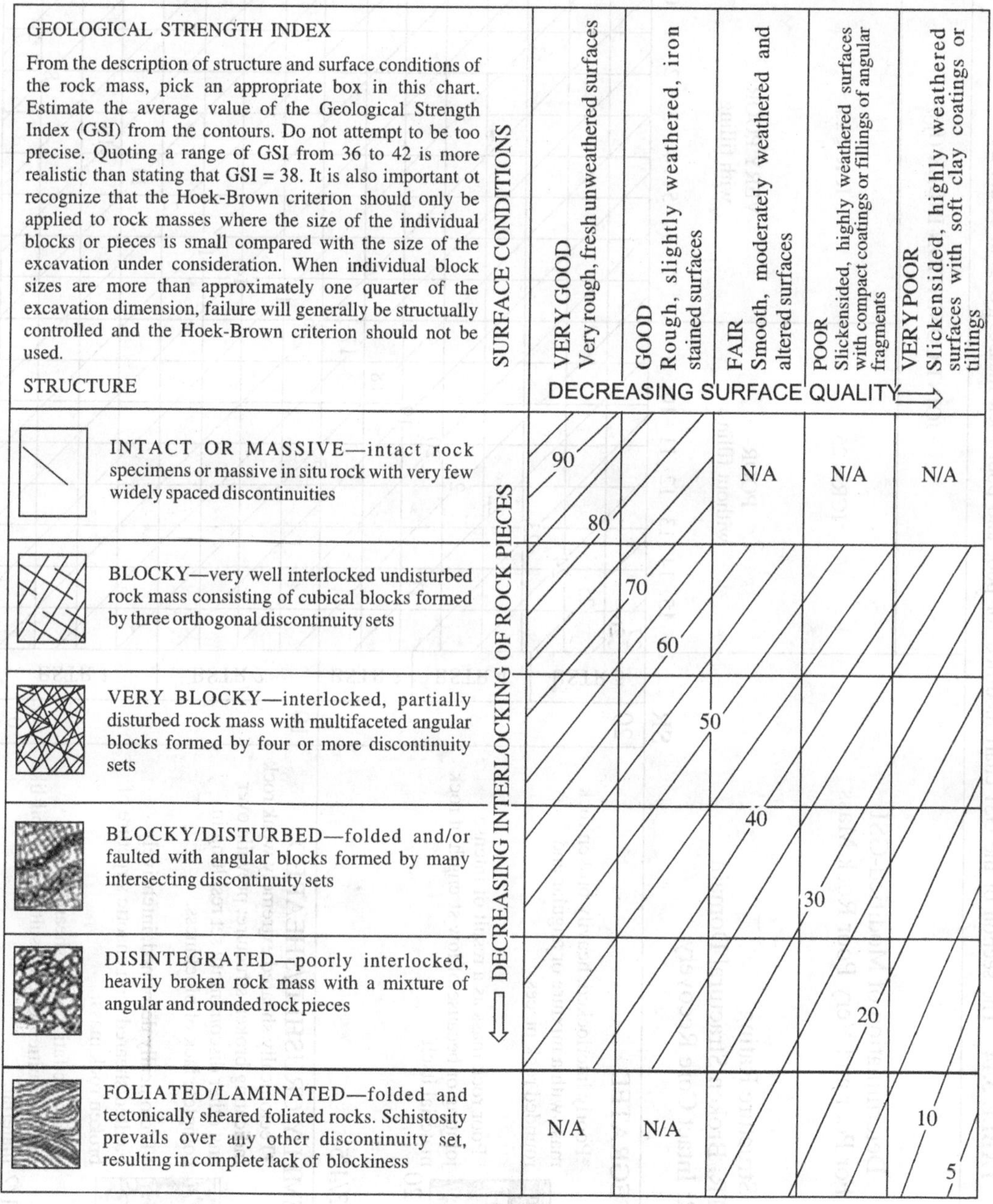

TABLE 3.13 GSI Rock Quality Classification

GSI value	95–76	75–56	55–36	35–21	< 20
Rock mass quality	Very good	Good	Fair	Poor	Very poor

TABLE 3.14 The section of the GSI chart modified for poor and poor rock masses (Osgoui et al., 2010)

Determination of Modified-GSI
For Poor and Very Poor Rock Mass

SR: Structure Rating
BSTR: Broken Stractural Domain
ICR: Intact Core Recovery

JOINT CONDITION INDEX (I_{JC})

ICR < 25	ICR < 25
POOR without filling	VERY POOR with filling

I_{JC}: 16 15 14 13 12 11 10 9 8 7 6 5 4 3 2 1 0

Rock mass	SR	BSTR
DISINTEGRATED *Poorly interlocked, heavily broken rock mass with a mixture or angular and rounded rock pieces. *Poor rock mass as a result of intense jointing or because of low strength of rock material itself. RQD < 20. BSTR 3/4/5	20	BSTR 5, BSTR 4, BSTR 3
EXTREMELY CRUSHED/SHEARED *Tectonically sheared extremely weak rock indicating broken structure; prevails over any other discontinuity set resulting in complete lack of blockiness. *Tectonically deformed, intensively Faulted/sheared with chaotic structure of broken rock mass. *Tectonic fatigue and sheared discontinuities, often resulting in a soil-like material. RQD=0. BSTR 1/2	10 … 0	BSTR 2, BSTR 1

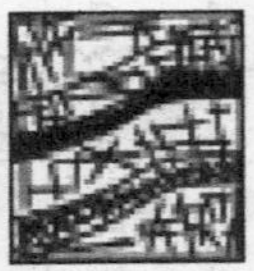

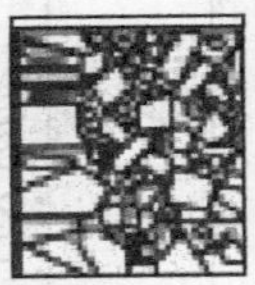

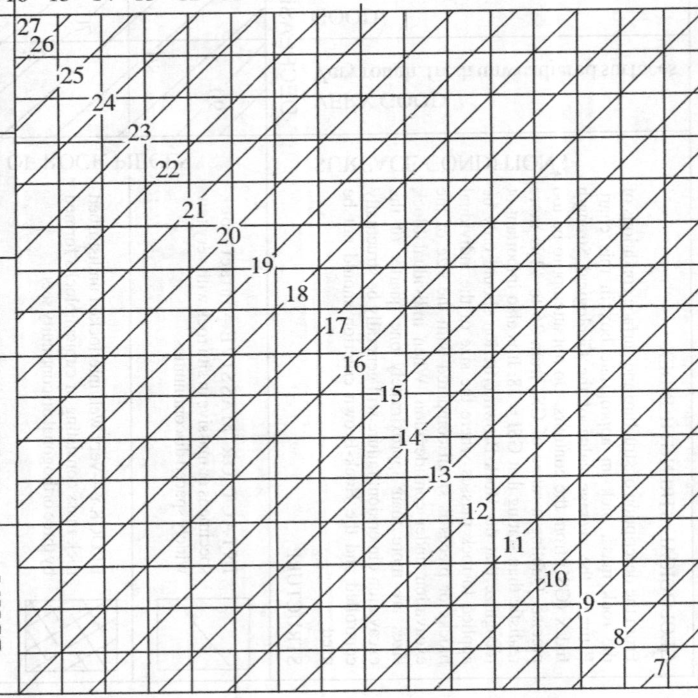

Broken Structure Domain (BSTR)

It is the drillcore zones recovered from a very weak rock mass having a length greater than 25 cm (Unal, 1996). Based on degree of size and composition, five groups of BSTR domain are defined. BSTR distinguishes the poor rock mass from very poor rock mass. The above reference suggests that the BSTR approach is based on overall observation of rock mass quality, but it is appropriate that BSTR be described through core-box logging rather than overall rock mass observation. It should be noted that the assignment of only one type of BSTR for a rock medium is not logical; rather, it is suggested to know the type of BSTR governing whole poor rock mass. Various types of BSTR are shown in Figure 3.5 and its classifications with descriptions are given in Table 3.15.

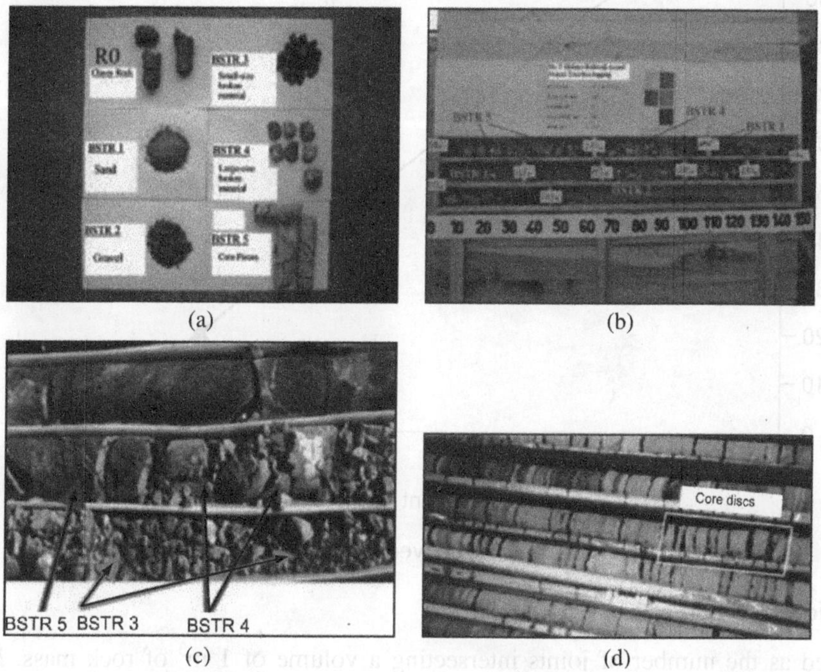

FIGURE 3.5 (a)–(c) Various types of BSTR, and (d) Disc types of broken rock cores (Osgoui, et al., 2010).

TABLE 3.15 BSTR Classification and Description

BSTR	*Description*	*Domain*
1/2	Soil material/extremely crushed/sheared rock mass	Very poor rock mass
3/4/5	Disintegrated decomposed rock mass	Poor rock mass
4	Disc type broken rock	

Structure Rating (SR)

The SR is defined qualitatively in terms of interlocking or blockiness of the rock pieces in original GSI chart and categorised quantitatively by Sonmez and Ulusay (2002) in five categories as massive/intact, blocky/very blocky, blocky/disturbed/seamy, disintegrated/decomposed and extremely crushed/sheared, based on the intervals of J_V. The correlation between SR limits and corresponding J_V values, encompassing the wide range of the rock quality as shown in Figure 3.6. The SR intervals are 100, 80, 60, 40, 20, 10 and 0 to adopt the BSTR types and the corresponding J_V values are 0.1, 0.3, 1, 3, 10, 30, 60 and 100.

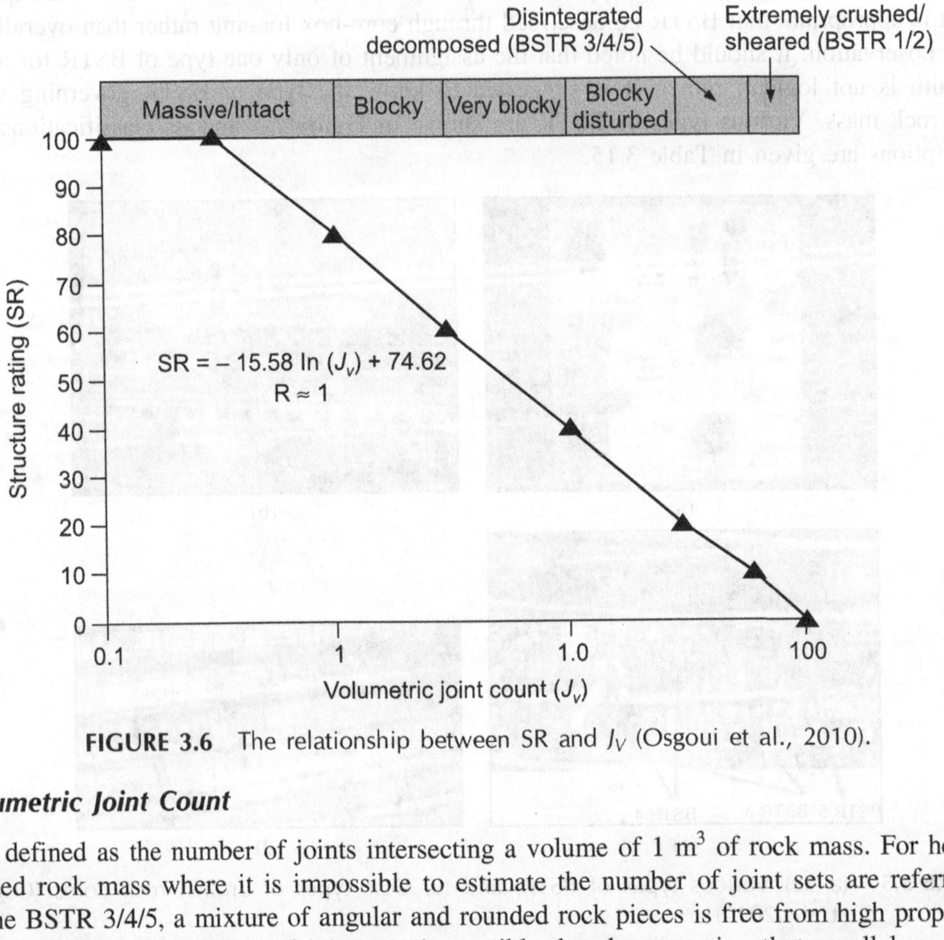

FIGURE 3.6 The relationship between SR and J_V (Osgoui et al., 2010).

Volumetric Joint Count

It is defined as the number of joints intersecting a volume of 1 m^3 of rock mass. For heavily jointed rock mass where it is impossible to estimate the number of joint sets are referred to as the BSTR 3/4/5, a mixture of angular and rounded rock pieces is free from high proportion of fines caused. If counting of joint sets is possible then by assuming that parallel or nearly parallel surfaces represent the same joint set, such parallel surfaces should be counted once. For example, in case of a rock piece from a rock mass including three joint sets approximately perpendicular to each other, prismatic blocks with six surfaces are formed. In this case, if the parallel surfaces are formed from a single joint set, the number of joint sets (J_n) is estimated as 3. While in the case of a tetrahedral rock piece of the surfaces which are not parallel to each other, the number of joint sets is considered as 4. Assuming that BSTR types of rock masses are homogeneous and isotropic, the following expression is suggested for the approximate estimation of J_V in conjuction with J_n (Sonmez and Ulusay, 2002).

$$J_V = J_n\left(\frac{1}{S_b}\right) \tag{3.16}$$

where, J_n is the estimated number of joint sets as mentioned above, and S_b is the average size of the block or rock pieces, which represents average spacing of joints and estimated from the selected piece of BSTR 3/4/5.

Joint Condition Indicators

Joint condition parameters like joint persistence and joint roughness contribute significantly to the existing classification systems. But those parameters are not useful in case of very poor rock mass with high degree of fracturing and decomposition. Considering BSTR in connection with ICR and weathering (W), joint condition index (I_{JC}) is defined (Osgoui et al., 2010). Here, ICR is defined as the total length of the cylindrical core pieces greater than 2 cm divided by the total length of the structural region or drillrun. The ICR for very poor rock masses is assumed to be less than 25 to satisfy the modified GSI requirements. The upper part of the modified GSI chart (Table 3.14) is divided into two categories, namely, poor and very poor. For ICR < 25%, the total rating of I_{JC} varies between 0 and 16. A simple way of determining I_{JC} by three different ways is shown in Table 3.16 (Osgoui et al., 2010).

TABLE 3.16 Joint Condition Index (I_{JC}) Ratings for Using in the Modified GSI (Osgoui et al., 2010)

			Joint Condition Index (I_{JC})		
Intact core recovery ICR < 25	Without filling	BSTR 1 bs = 0 BSTR 1 bs = 2 BSTR 1 bs = 4 BSTR 1 bs = 6 BSTR 1 bs = 8	I_{JC} = (bs) + (W/2) + 4		
	With filling	$t_f \geq 5$ mm	soft hard	0 4	Where, t_f: filling thickness
		1 mm $\leq t_f \geq 5$ mm	soft hard	I_{JC} = (bs/2) + (W/2) I_{JC} = (bs/2) + (W/2) + 4	
		$t_f \leq 1$ mm		I_{JC} = (bs) + (W/2) + 4	
Weathering "W"	*Unal* (1996)	*ISRM* (2007)	*Class*	*W*	*Rating*
	Unweathered	Fresh	1	< 1.1	Poor rock mass
	Slightly weathered	Slightly weathered	2	1.1 – 1.5	
	Moderately weathered	Moderately weathered	3	1.5 – 2.0	
	Highly weathered	Highly weathered	4	> 2.0	
	Very highly weathered	Completely weathered	5		
	Decomposed	Residual soil	6		
	$W = R_f/R_w$	Where R_f is the rebound number obtained from fresh rock surface and R_w is the rebound number gained from the weathered rock surface in Schmidt hammer test			
	If W is known, Rating = 10.7 – 2.7 W (Gokceoglu and Aksoy, 2000)				

3.3.6 Coal Mine Roof Rating (CMRR)

The coal mine roof rating (CMRR) is specifically designed for bedded coal measure rock by summing up various individual ratings of different parameters to obtain a value between 0 to 100 scale. The standard rock mass classification systems like RMR, Q and GSI are not readily applicable to coal mining because these classification systems have the following limitations (Mark and Molinda, 2005):

(i) They tend to focus on the joint properties, but in case of coal mine roof, bedding is generally the most significant discontinuity affecting roof stability.
(ii) They rate just one rock unit at a time, while coal mine roof often consists of several layers.
(iii) The dimensions and stability requirements of tunnels are different from coal mines.

The data required for the CMRR are collected from underground exposures such as roof, exploratory drill core etc. The parameters required for CMRR classification system are (Mark and Molinda, 2005, Molinda et al., 2000).

(i) Unconfined compressive strength (UCS)
(ii) Discontinuity intensity
(iii) Shear strength of discontinuities
(iv) Moisture sensitivity deductions
(v) Thickness weighted average roof rating
(vi) Strong bed adjustment
(vii) Number of layers
(viii) Groundwater
(ix) Surcharge

The CMRR is calculated in a two-step process illustrated in Figure 3.7.

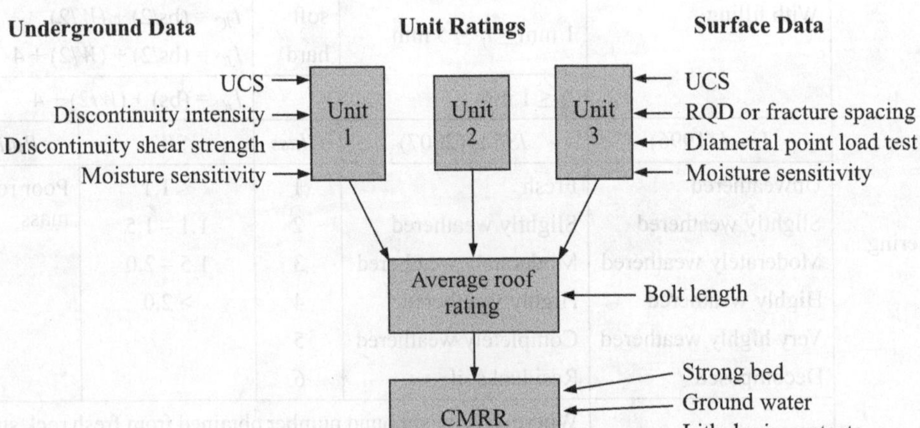

FIGURE 3.7 Flow chart for the CMRR (Mark and Molinda, 2005).

(i) ***First step:*** Mine roof is divided into structural units. Ratings are determined for each structural unit. A structural unit generally contains of one lithologic layer, but several rock layers may be lumped together if their engineering properties are similar.

(ii) ***Second step:*** The CMRR is determined by averaging all the unit ratings within the bolted interval (with the contribution of each unit weighted by its thickness) and applying appropriate adjustment factors. The second step is the same regardless of whether the unit ratings are from data collected from underground/surface or from drill core.

Calculation of the CMRR has been greatly simplified by the development of a CMRR computer program that can be obtained from the link www.cdc.gov.

Unconfined Compressive Strength (UCS)

CMRR recommends the point load test (PLT) for drill core to determine UCS of intact rock. Although the laboratory test is the standard method to determine UCS but because of expensive, time consuming and care for sample preparation/handling increase the variability in results. The UCS rating scale used in CMRR is shown in Table 3.17 (A and B). Approximately one-third of the CMRR rating is determined from UCS, which is approximately twice the parametric weight given to the UCS in the original RMR.

TABLE 3.17A Rating of UCS for Laboratory Test

UCS (MPa)	≤ 35.5	$35.5 < \text{UCS} < 149$	> 149
Rating	5 + (0.215 UCS)	7 + (0.157 UCS)	30

TABLE 3.17B Rating of UCS for Point Load Test

PLT	≤ 0.24	$0.24 < \text{PLT} < 1.2$	$1.2 < \text{PLT} \le 2.2$	> 2.2
Rating	25	209.9 PLT + 20	15.1 PLT + 27.5	60

EXAMPLE 3.7 A coal mine seam is 3 m to 3.2 m thick and developed by Bord and Pillar method. The galleries are 4.5 m wide and 3 m in height. The width of the pillars are 35 m × 35 m from center to center. The depth of cover is around 270 m. The parting between the above seam is around 10 m. The immediate roof is composed of medium to fine grained sandstone. In general, two sets of joints are prominent in the roof, some are random joints, and most of them are rough planar and unaltered slips. The average layer thickness in the sandstone is found to be 15 cm. In general, the mine roof has been observed to be moist in nature. The compressive strength of sandstone is 21.8 MPa. Find the contribution of UCS in unit rating.

Solution: For UCS equal to 21.8 MPa, the rating of UCS from Table 3.17A is 9.687.

Discontinuity Intensity

Intensity of discontinuities is determined by spacing and persistence. The more a discontinuity set is closely spaced, the greater weakening effect it has on the rock mass. Persistence is more important for discontinuities that are widely spaced. Like UCS, intensity accounts for about one–third of the total CMRR. Table 3.18 shows the bedding/discontinuity rating scale for underground data for each combination of spacing and persistence.

TABLE 3.18 Bedding/discontinuity intensity rating table for underground data

Persistence (m)	*Spacing* (mm) > 1800	600–1800	200–600	60–200	< 60
0–1	35	30	24	17	9
1–3	32	27	21	15	9
> 3	30	25	20	13	9

The discontinuity intensity is also measured by measuring fractured spacing (FS) and RQD. Fractured spacing (FS) is determined by counting the core breaks in a particular unit, and then dividing by the thickness of the unit. The RQD and L, i.e., fractured spacing are related as given in Eq. 3.7. As input, the CMRR uses either the RQD or the FS. When the fracture spacing is greater than about 300 mm, the RQD is not very sensitive, so the fracture spacing is used directly. At the other extreme, when the core is highly broken or lost, the RQD appears to be the better measure. The CMRR uses Eqs. 3.17a and 3.17b to calculate discontinuity spacing rating (DSR) of core from L and RQD respectively. DSR varies between 20 and 48.

$$\text{DSR} = 5.64 \ln(L) + 5.8 \quad (3.17a)$$

$$\text{DSR} = 10.5 \ln(\text{RQD}) - 11.6 \quad (3.17b)$$

where, L is number of discontinuities per metre representing fractured spacing (FS).

EXAMPLE 3.8 In continuation from previous example, find the DSR contribution in unit calculation if RQD of immediate roof, i.e., sandstone is 60%.

Solution: For 60% RQD, Eq. 3.17b is used to find the DSR as follows:

$$\text{DSR} = 29.34$$

Shear Strength of Discontinuities

Bedding plane shear strength is a critical parameter for coal mine ground control, because the most severe loading applied to coal mine roof is normally lateral, caused by horizontal stress (Mark and Barczak, 2000). Molinda and Mark (1996) found that the lateral strength of some shales are just one-sixth of their axial strength.

In underground, the cohesion of bedding surfaces is evaluated by using a 9 cm mason chisel and a hammer to split hand samples of rock. Weaker, less cohesive surfaces require fewer chisel to split (Table 3.19). Cohesion can also be estimated by observing the nature of fractured wall of a roof fall. Mark and Molinda (1996) commented that if the wall "stair steps" with most of the roof failures occurring along bedding, then the cohesion is probably low. On the other hand, if most of the failure surfaces cut across the bedding, then the strength of the bedding is most likely equal to or greater than that of the intact rock. Slicken sided surfaces are already planes of failure and receive the minimum rating.

The roughness along a discontinuity surface is the other component of the surface's shear strength. CMRR uses the estimation of surface roughness by visual inspection and classified into "jagged", "wavy", or "planar" as shown in Table 3.19 and ratings are listed in Table 3.20.

TABLE 3.19 Underground Data Sheet for the CMRR (Mark and Molinda, 2005)

DATE ________ MINE ________ LOCATION ________ PAGE ____ OF ____ CMRR ________

TYPE OF EXPOSURE ________ NAME ________

UNIT						UNIT DISCONTINUITIES					
Unit No.	Unit Thickness	Strip Log	Description	Strength	Moisture Sensitivity	Disco. I.D.	Description	Cohesion	Roughness	Spacing	Persistence Lateral/Vert
3						A.					
						B.					
						C.					
	CONTACT										
2						A.					
						B.					
						C.					
	CONTACT										
1						A.					
						B.					
						C.					

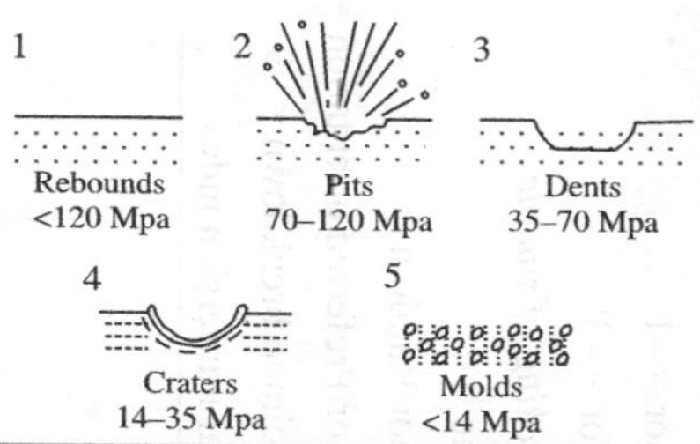

	Strength	Moisture Sensitivity		Cohesion	Roughness	Spacing	Persistence Lateral/Vert
1	Rebounds	Not Sensitive	1	Strong (>7)**	Jagged	>1.8 m	0–0.9 m
2	Pits	Slightly Sensitive	2	Moderate (4–7)	Wavy	0.6–1.8 m	0.9–3 m
3	Dents	Moderately Sensitive	3	Weak (1–3)	Planar	20–60 cm	3–9 m
4	Craters	Severely Sensitive	4	Slicken-sided (0)		6–20 cm	< 9 m
5	Molds		5			<6 cm	

Groundwater (inflow 10 m of entry length) (Circle one)

Dry 0	1	Heavy Drip 2.7 – 12.2	4
Damp 0–1.3	2	Flowing >13.2	5
Light Drip 1.3–2.7	3		

Describe condition in vicinity of fall (circle one)

1. Good 3. Heavy
2. Scaly 4. Failed

COMMENTS:

**Chisel blows necessary to split bedding.

TABLE 3.20 Bedding/Discontinuity Shear Strength Rating Table for Underground Data

Roughness	*Cohesion*			
	Strong	*Moderate*	*Weak*	*Slickensided*
Jagged	35	29	24	10
Wavy	35	27	20	10
Planar	35	25	16	10

EXAMPLE 3.9 In continuation from previous Examples 3.7 and 3.8, find the contribution of shear strength of discontinuity in unit rating calculation.

Solution: In general, two sets of joints are prominent in the roof, some are random joints, and most of them are rough planar and unaltered slips. Here, the rating depends on cohesion and surface roughness. The rough planar and unaltered slips means no displacement and highly cohesive. Hence, the rating of shear strength of discontinuity from Table 3.20 will be 35.

Moisture Sensitivity Deduction

Moisture sensitivity is determined by either immersion test data or by visual estimation in underground exposures. Maximum deduction for moisture sensitivity is 15 points. Immersion test data sheet is shown in Table 3.21. If immersion test results are not available, moisture sensitivity can sometimes be estimated visually in underground exposures (Table 3.22).

Table 3.22 indicates how the results from either test can be used for input to the CMRR.

TABLE 3.21 Data sheet for immersion test

Immersion		*Breakability*	
Observation	*Rating*	*Observation*	*Rating*
Appearance of water			
Clear = 0		No change = 0	
Misty = –1		Small change = 2	
Cloudy = –3		Large change = 6	
Talus formation		Breakability index	
None = 0			
Minor = –1			
Major = –3			
Cracking of sample			
None = 0			
Minor/Random = –1			
Major/Preferred orientation = –3			
Specimen breakdown = –9			
Total immersion index			

(Contd.)

TABLE 3.21 Data sheet for immersion test (*Contd.*)

Immersion		*Breakability*	
Observation	*Rating*	*Observation*	*Rating*

Procedure for immersion test

1. Select sample (s)–hand sized
2. Test for hand breakability
3. Rinse specimen (to remove surface dirt, dust, etc.)
4. Immerse in water for 1 hour
5. Observe and rate water appearance, talus formation, and cracking of sample
6. Sum rating for immersion index
7. Rates for hand breakability
8. Determine Breakability index
9. The final immersion test index is the greater of the Breakability index or the immersion index.

TABLE 3.22 Moisture Sensitivity Classes and Ratings from Both Immersion and Slake Durability Tests[a] (Mark and Malinda, 2005)

Moisture sensitivity class	*Rating adjustment*	*Immersion index*	*Slake durability index*
Not sensitive	0	0–1	100–98
Slightly sensitive	–3	2–4	98–92
Moderately sensitive	–7	5–9	92–80
Severely sensitive	–15	> 9	< 80

[a]Apply rating adjustment to unit rating only when unit forms the immediate roof or if water is leaking through the bolted interval.

EXAMPLE 3.10 Find the moisture rating of CMRR.

Solution: No immersion test data are available. The only information is that the roof is moist. Consider moist to be the average classification, hence, the rating will be 7.5, i.e., half of the maximum rating.

3.3.7 Calculation of Unit Rating

For underground data: The equation for calculating the unit rating is given as follows:

$$\begin{aligned}\text{Unit rating (UR)} = {} & \text{UCS rating} + \text{Discontinuity intensity rating} \\ & + \text{Discontinuity shear strength rating} \\ & + \text{Multiple discontinuity adjustment} \\ & + \text{Moisture sensitivity deduction}\end{aligned} \tag{3.18a}$$

For drill core data: The equation is

$$\text{Unit rating} = \text{UCS rating} + \text{Discontinuity rating} + \text{Moisture sensitivity deduction} \tag{3.18b}$$

EXAMPLE 3.11 Find the unit rating of the problem given in Example 3.7.

Solution: By using Eq. 3.18a, the overall rating

Unit rating = Ratings for [UCS (9.69) + Discontinuity intensity (29.34)
+ Discontinuity shear strength (35)
+ Multiple discontinuity adjustment (0)
– Moisture sensitivity deduction (7.5)]
= 74.03

Thickness-weighted Average Roof Rating

Let us consider a bolted roof consists of layers of sandstone (*ss*), shaly sandstone (*s–ss*) and shale (*s*) with thickness and unit ratings as (t_{ss}, UR_{ss}); (t_{s-ss}, UR_{s-ss}) and (t_s, UR_{ss}) respectively from top to bottom. If bolt length is (t_b) then thickness-weighted average (RRW) is

$$\text{RRW} = \frac{(t_s \times UR_s) + (t_{s-ss} \times UR_{s-ss}) + [\{t_d - (t_s + t_{s-ss})\} \times UR_{ss}]}{t_b} \tag{3.19}$$

Note that the thickness of uppermost layer of sandstone is only $\{t_d - (t_s + t_{s-ss})\}$ m and is in contact with bolt and so used in calculation.

EXAMPLE 3.12 Bolted mine roof consists of three units (from top down) 2 m thick sandstone layer followed by shaly sandstone of 0.8 m and 0.4 m of shale as an immediate roof. The unit rating of these layers are 60, 50 and 40 for sandstone, shaly sandstone and shale respectively. If the length of the roof bolt is 1.8 m, then find the thickness weighted average (RRW).

Solution: Use Eq. 3.19 to find RRW.

$$\text{RRW} = \frac{(0.4 \times 40) + (0.8 \times 50) + [\{1.8 - (0.4 + 0.8)\} \times 60]}{1.8} = 51.1$$

Strong Bed Adjustment (SB_A)

The strongest bed within the bolted interval often determines the performance of mine roof. The strongest bed's effect on the CMRR depends first upon how much stronger it is than the other roof layers. Second, the strong bed must be at least 0.3 m thick before it can be provide any additional support, and the amount of the adjustment is maximum when the bed is at least 1.2 m thick. Third, the roof bolts must obtain at least 0.3 m of anchorage in the strong bed for the adjustment to be considered. Finally, the higher into the roof that the strong bed is located, the less its positive effect will be SB_A is estimated as follows:

$$SB_A = [(0.72 \times \text{SBD} \times \text{THSB}) - 2.5] \times [1 - (0.33(\text{THWR} - 0.5))] \tag{3.20}$$

where,

- SBD is the Strong Bed Difference, the difference between the strong bed's unit rating and the thickness-weighted average of all the unit ratings within the bolted material,
- THSB is the thickness of the strong bed (m), and
- THWR is the thickness of the weak rock suspended from the strong bed (m).

Note that if the strong bed is at the top of the bolted interval, its full thickness is used in the calculated of the SB_A (up to a maximum of 1.2 m).

Number of Layers

A coal mine roof contains numerous lithologic contacts which are less competent than roof that consists of a single rock type, the value of CMRR is reduced considering a weakness in the system. The maximum deduction from the CMRR is 5 points if more than 4 weak contacts are present.

Groundwater

Groundwater is more prevalent in shallow mines, particularly beneath stream valleys, but it can also be introduced by leakage from pooled water in abandoned mines or fracturing of overlying aquifers during high extraction mining. The CMRR maintains the RMR system's rating scale, with a maximum deduction for flowing groundwater of 10 points.

Surcharge

The strength of rocks overlying the bolted interval is only considered when they are significantly weaker than the rocks within it. The CMRR accounts for the surcharge with a 3 point deduction.

3.3.8 Classification of Coal Mine Roof based on CMRR

Molinda and Mark (1994) suggest the following categorisation of roof competency based on CMRR classification system (Table 3.23).

Table 3.23 Roof Rock Quality based on CMRR Classification System

CMRR	< 45	45–65	> 65
Rock quality	Weak roof	Moderate roof	Strong roof

In this respect, a CMRR value of 30 would mean a poor roof which requires additional support immediately after mine opening and CMRR value of 75 would signify a competent roof which may sustain overburden load by itself.

Number of mine planning design tools are based on CMRR like analysis of longwall pillar stability (ALPS), analysis of longwall tailgate serviceability (ALTS), stability of extended cuts, roof bolt selection, roof fall evaluations and others. These softwares can be downloaded from the NIOSH website.

3.4 APPLICATION OF ROCK MASS CLASSIFICATION SYSTEM

3.4.1 Rock Mass Strength

The GSI classification of rock mass can be used to estimate its strength parameters. Hoek et al. (2002) presented a practical approach to estimate rock mass strength, equivalent cohesion and angle of friction angle of Mohr–Coulomb yield criterion of rock mass based on Hoek–Brown yield criterion as given in Section 9.8.3 of Chaper 9.

3.4.2 Deformation Modulus

Rock sample tested in laboratory are not representative for natural rock masses, the deformability modulus for a rock mass can be evaluated from the RMR. Bieniawski (1978) defined the following relation for hard rock when RMR > 50 and

$$E_m = 2\text{RMR} - 100 \tag{3.21}$$

where, E_m is modulus of deformation of rock mass in GPa.

Serafim and Pereira (1983) defined the following relationship for modulus of deformation in GPa for soft rock when RMR < 50.

$$E_m = 10^{(\text{RMR}-10)/40} \tag{3.22}$$

Figure 3.8 shows the relationship of *in situ* deformation modulus E_m from RMR and Q–system.

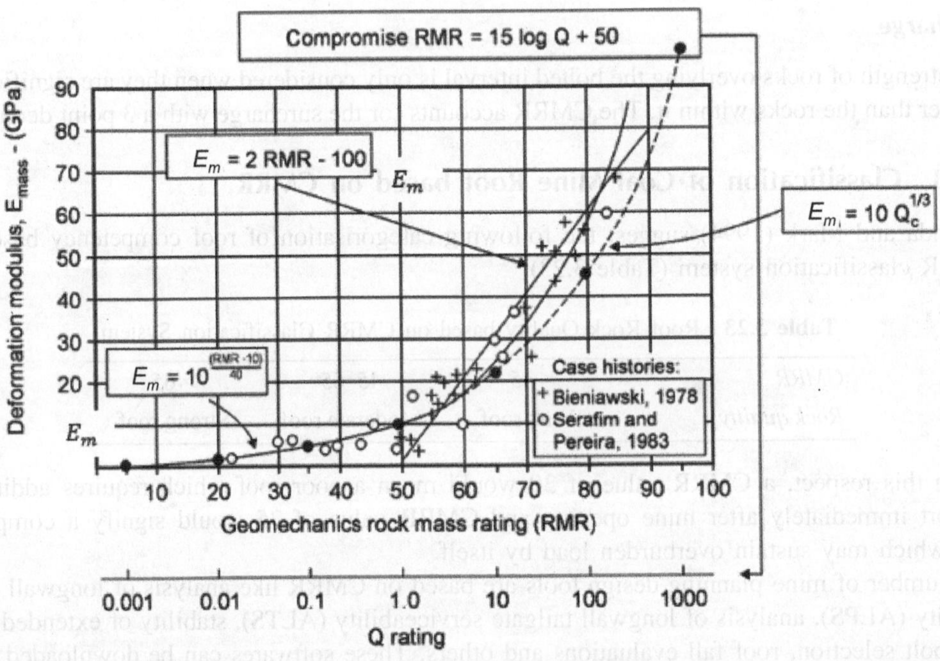

FIGURE 3.8 Static deformation modulus Emass; Q and RMR and some empirical inter-relationships (Barton, 2002).

The relationship given by Serafim and Pereira (1983) overpredict the values of Young's modulus for poor–quality rock masses. Hence, Hoek and Brown (1997) suggested the following modification of the Serafim and Pereira relationship when the intact unconfined compressive strength (σ_{ci}) is less than 100 MPa.

$$E_m = (\sigma_{ci}/100)^{0.5}\, 10^{[(\text{GSI}-10)/40]} \tag{3.23}$$

In this case the GSI has been substituted for the RMR. Barton et al. (1980) suggested that the Q–system could be used to estimate the *in situ* value of Young's Modulus by following relationship for $Q > 1$, where E_m is in GPa.

$$E_m = 25 \log_{10} Q \tag{3.24}$$

This relation gives good agreement with measured deformations when used in numerical analysis (Barton et al., 1992). However, rock mass moduli vary considerably and a range from $10 \log_{10} Q$ to $40 \log_{10} Q$ should be expected. Hoek et al. (2002) proposed the following relation to find the deformation modulus related to the GSI system.

$$E_m = [1 - (D/2)]\,(\sigma_{ci}/100)^{0.5}\,10^{(GSI-10)/40}\ \text{GPa} \tag{3.25}$$

Where D is a factor ranges from 0 to 1 and depends on the degree of disturbances due to blasting and stress relaxation (Hoek et al., 1997), also defined in Section 9.8 of Chapter 9.

According to Gokceoglu et al. (2003) relationship,

$$E_m = 0.001(E_i/\sigma_{ci})\,(\sigma_{ci}/100)^{0.5}\,10^{(GSI-10)/40}\ \text{GPa} \tag{3.26}$$

where E_i is elastic modulus of intact rock.

From Barton (2002) relationship of deformation modulus and Q,

$$E_m = 10(Q\sigma_{ci}/100)^{1/3} \tag{3.27}$$

3.4.3 Stand-up Time

The concept of stand-up time was originally conceived by Lauffer (1958) to indicate the time period within which an excavation will remain serviceable and after which significant instability and caving would be experienced. The stand-up time for an unsupported span is related to the quality of the rock mass in which the span is excavated.

Figure 3.9 shows the span versus stand-up time graph developed by Bieniawski (1989). Hutchinson and Diederichs (1996) have presented the maximum stable unsupported span as a function of Bieniawski's (1989) RMR value as shown in Figure 3.10. Barton et al. (1974) described the application of the Q–system for rock mass classification for the determination of no-support limits for various types of excavations as shown in Figure 3.4.

The stand-up time concept is a reduction in the time available for the installation of support when the tunnel span in increased. While a large tunnel span needs support to be stable, a small tunnel might be stable without support. Self-supporting characteristics of the rock mass are more for lesser span length.

3.4.4 Support Selection

Bieniawski (1989) developed a scheme to estimate the type of support systems used for different values of RMR as given in Table 3.24. Similarly, Barton's chart (Figure 3.4) can also be used to find the type of bolt to be used for roof rating in terms of RMR and Q.

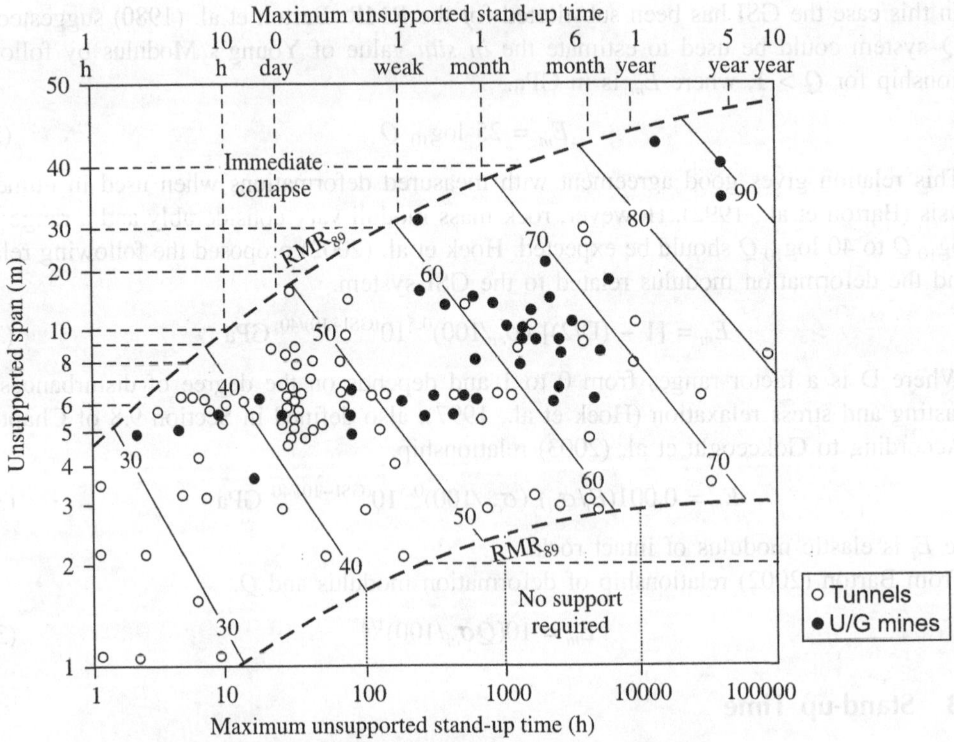

FIGURE 3.9 Unsupported tunnel limit (Bieniawski, 1989).

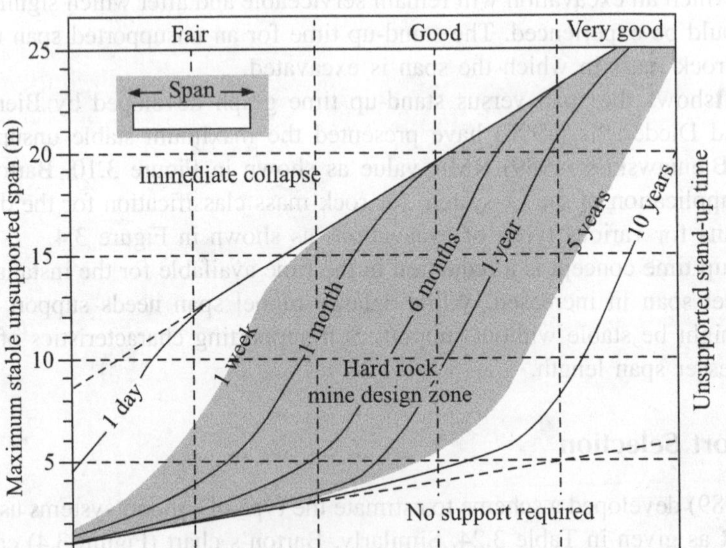

FIGURE 3.10 Alternate representation of RMR_{89} stand-up time guideline (Hutchinson and Diederichs, 1996).

TABLE 3.24 Guidelines for Excavation and Support of 10 m Span Rock Tunnels in Accordance with the RMR System (Bieniawski, 1989)

Rock mass class	*Excavation*	*Rock bolts (20 mm diameter, fully grouted)*	*Shotcrete*	*Steel sets*
I – Very good rock RMR: 81–100	Full face, 3 m advance	Generally no support required except spot bolting		
II – Good rock RMR: 61–80	Full face, 1–1.5 m advance, complete support 20 m from face	Locally, bolts in crown 3 m long, spaced 2.5 m with occasional wire mesh	50 mm in crown where required	None
III – Fair rock RMR: 41–60	Top heading and bench 1.5–3 m advance in top heading. Commence support after each blast. Complete support 10 m from face.	Systematic bolts 4 m long, spaced 1.5–2 m in crown and walls with wire mesh in crown	50–100 mm in crown and 30 mm in sides	None
IV – Poor rock RMR: 21–40	Top heading and bench 1.0–1.5 advance in top heading	Systematic bolts 4–5 m long, wire mesh. Bolt invert	100–150 mm in crown, 100 mm in sides	Light to medium ribs spaced 1.5 m where required
V – Very poor rock RMR: < 20	Multiple drifts 0.5–1.5 m advance in top heading. Install support concurrently with excavation. Shotcrete as soon as possible after blasting	Systematic bolts 5–6 m long, spaced 1–1.5 m in crown and walls with wire mesh. Bolt invert.	150–200 mm in crown, 150 mm in sides, and 50 mm on face	Medium to heavy ribs spaced 0.75 m with steel lagging and forepoling if required close invert

EXERCISES

1. A scanline survey is to be carried out on a rock mass which contains two sets of parallel discontinuities whose traces on the wall are mutually inclined at 55°. The trace of set A makes an angle of 30° with a horizontal scanline. Large numbers of measurements give the apparent mean spacing of sets A and B along the scanline as 0.70 m and 0.90 m respectively.
   (a) Calculate the mean Joint Spacing of each set.
   (b) What is the mean spacing of all discontinuities in the direction of the scanline?
   (c) Assuming that the combined discontinuity spacing follow a negative exponential distribution, estimate the RQD of the rock mass in the direction of the scanline.
2. A rock mass is interested by three sets of discontinuities as 010/35, 290/60 and 170/44. Draw a neat sketch of rock mass with the joint sets.
3. A rock mass consist of five joint sets with joint count of 8 per 10 m for joint set 1, 15 per 10 m for joint set 2, 12 per 10 m for joint set 3, 9 per 10 m for joint set 4, and

5 per 10 m for joint set 5. Find the volumetric joint count and describe the block size as ISRM classification. Also estimate the RQD based on empirical relationship.

4. Length of core obtained from 5 m of borehole long from an underground exploratory drilling in cm is 12, 2, 13, 24, 05, 10, 09, 24, 15, 06, 15, 9.75, 9.85, 10, 20, 15, 25, 5, 15, 6, 9, 2, 25, 24, 28, 12, 15, 18, 12, 15, 4, 5, 6, 7.8, 9, 12. Find for borehole (i) RQD, (ii) Core recovery %, and (iii) estimate volumetric joint count.

5. Determine the RQD in % of a total core length of 400 cm in sandstone rock mass as shown.

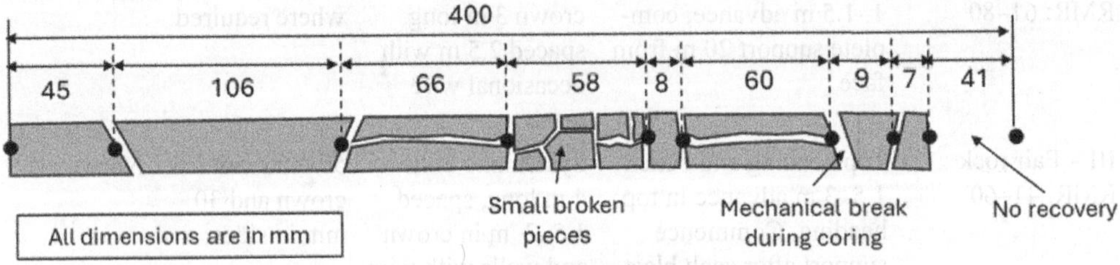

6. A tunnel is to be driven through a slightly weathered granite with a dominant joint set dipping at 60° against the direction of the drive. Index testing and logging of diamond drilled core give typical Point-load strength index values of 8 MPa and average RQD values of 70%. The joints, which are slightly rough and slightly weathered with a separation of < 1 mm, are spaced at 300 mm. Tunnelling conditions are anticipated to be wet. Find RMR?

7. A joint survey is conducted in an underground mine with host rock Quartz Chlorite Schist, revealing that the rock mass consists of 3 persistent joint sets. The average RQD of the rock mass is 68%, the average joint spacing is 0.15 m, and joint surfaces are generally undulating and smooth and opened with sign of weathering. The reported value of the average UCS of rock material is 60 MPa.
   (a) Calculate the RMR and state the class of rock mass.
   (b) Assume a cube of intact rock are formed then show the three joint sets.

8. A rock mass at a depth of 200 m contains three fracture sets. One set comprises bedding planes; which are highly weathered, slightly rough surfaces, and are continuous with an orientation of 180/10. Second set of joints are slightly weathered, slightly rough, and have an orientation of 185/75. Finally, the third set of joints are slightly weathered and slightly rough, and have an orientation of 090/80. (a) Draw a theoretical rock mass in the form of a cube. (b) If the average uniaxial compressive strength of the intact rock is 100 MPa; RQD is 70% and mean joint spacing is 0.5 m, then classify the rock mass based on RMR system. (c) If a tunnel of 6 m diameter is excavated from east to west, then assess its stability.

9. Use the Q-system to assess the stability of the rock mass as described in Exercise 8.

10. It is proposed to construct an underground tunnel 500 m below the ground. The drilled cores have an RQD of 85% and the number of joint sets is estimated to be 2. The joints are rough, undulating and unweathered with minor surface staining. The average uniaxial compressive strength of the cores is 190 MPa. The major principal stress acts

horizontally and is twice the vertical stress. The unit weight of the rock is approximately 30 kN/m^3. The excavation is relatively dry, with some dampness and negligible inflow. Estimate the Q-value and GSI. Classify rock mass rating based on RMR and Q system.

11. A coal mine seam is 4.0–4.2 m thick and developed by Bord and Pillar method. The galleries are 4.8 m wide and 4.0 m in height. Pillars are 35 m × 35 m from center to center. The depth of cover is around 300 m. The parting between the above seam is around 20 m. The immediate roof is composed of medium to fine grained sandstone with a slake durability index 95%. In general, two sets of joints are prominent in the roof, some are random joints, and most of them are rough planar and unaltered slips. The average layer thickness in the sandstone is found to be 15 cm. In general, the mine roof has been observed to be moist in nature. The compressive strength of immediate roof rock, i.e., sandstone is 20.8 MPa and 70%. Then find (i) the unit rating of sandstone, (ii) If the immediate roof rock consists of 1 m sandstone, followed by 0.3 m shale, 0.5 m shaly sandstone and 5 m of sandstone. The unit rating of shale and shaly sandstone rock is 40 and 55 and a bolt of 1.5 m length is used to stabilize the roof then find the CMRR of roof rock.

Chapter 4

Analysis of Stresses and Strains

4.1 INTRODUCTION

One of the major objectives of a mining, civil or a geotechnical engineer is to assess deformations and stresses in rock or rock mass. This understanding leads to the determination of failure characteristics of an underground excavation, failure of rock slopes or any structure made in rock. There are mainly two types of stresses exist in a geofield: (i) *in situ* stresses, and (ii) induced stresses. The pre-excavation or pre-mining stresses are termed as *in situ stresses* and can be classified into: (i) gravitational, (ii) tectonic, and (iii) locked-in stresses based on their origin. Once an excavation is made in a rock mass, deformations occur around the excavation boundaries and stresses are redistributed based on the geometry of the excavation and rock properties. The new equilibrium stress is termed as *induced stress*.

This book mainly deals with the computational aspect of induced deformations and stresses which are responsible for destabilisation or collapse of an excavation. In this chapter, analysis of stresses and strains are revisited for both Cartesian and polar coordinate systems. Transformation of displacements, forces, strains and stresses from one coordinate system to another are illustrated for two-dimensional (2D) and three-dimensional (3D) cases. Graphical representation of stresses and strains are also explained with the help of *Mohr's circles*. Concepts and computational aspects of principal stresses are enumerated in detail with solved examples. Strain compatibility conditions and stress equilibrium equations are the basis of solving solid mechanics problems. These equations are derived for a differential element in Cartesian and polar coordinate systems. In general, for 2D and 3D cases, stress equilibrium equations are established in the form of differential equations. The *Airy stress function* relates to stress tensors and converts stress equilibrium equations into a single differential equation. As a result, solution of Airy stress function for the given problem geometry, material properties and boundary conditions becomes possible. In this chapter, concept of Airy stress function is enumerated for Cartesian as well as polar coordinate systems. Note that in the entire text, bold character symbols are used for denoting vectors and matrices and italicised symbols are used for expressing equations.

4.2 STRESS, STRAIN AND STRAIN ENERGY

4.2.1 Definition of Stress

Let us assume a body, Ω in which traction force $\mathbf{F}_t$, point forces, $\mathbf{F}_i$ and body forces, $\mathbf{F}_b$ are applied as shown in Figure 4.1(a). The body is constrained at boundary B. An arbitrary plane P has bisected the body into two parts denoted by Ω^+ and Ω^-. Figure 4.1(b) shows the plane P and the resultant force $\mathbf{F}$, acting on an incremental area ΔA of the Ω^+ laying on the plane P. A Cartesian coordinate system is assumed to represent the force vector. Using this system, the force vector $\mathbf{F}$ can be resolved into x, y, and z directions as F_x, F_y, and F_z respectively and can be represented as:

$$\mathbf{F} = F_x\mathbf{i} + F_y\mathbf{j} + F_z\mathbf{k} \tag{4.1}$$

Again the force vector $\mathbf{F}$ can also be resolved along unit normal $\mathbf{n}$ and unit shear $\mathbf{s}$ directions on the plane P as shown in Figure 4.1(b). The resolved forces $\mathbf{F}_n$ and $\mathbf{F}_s$ are called normal (perpendicular) and shearing (tangential) forces respectively on the area ΔA. The magnitude $\mathbf{F}/\Delta A$, $\mathbf{F}_n/\Delta A$ and $\mathbf{F}_s/\Delta A$ are called *average stress*, *average normal stress* and *average shear stress* respectively acting on the area ΔA.

The concept of stress at a point is obtained by letting ΔA become an infinitesimal. The force vector $\mathbf{F}$, and components, $\mathbf{F}_n$ and $\mathbf{F}_s$ also approach to zero but the limit of $\mathbf{F}/\Delta A$, $\mathbf{F}_n/\Delta A$ and $\mathbf{F}_s/\Delta A$ when $\Delta A \to 0$ are finite. Hence, stress vector σ on a point is defined as:

$$\sigma = \lim_{\Delta A \to 0} \frac{\mathbf{F}}{\Delta A} \tag{4.2}$$

Similarly, normal stress σ_n and shear stress, τ on a point can be defined as:

$$\sigma_n = \lim_{\Delta A \to 0} \frac{\mathbf{F}_n}{\Delta A} \qquad \tau = \lim_{\Delta A \to 0} \frac{\mathbf{F}_s}{\Delta A} \tag{4.3}$$

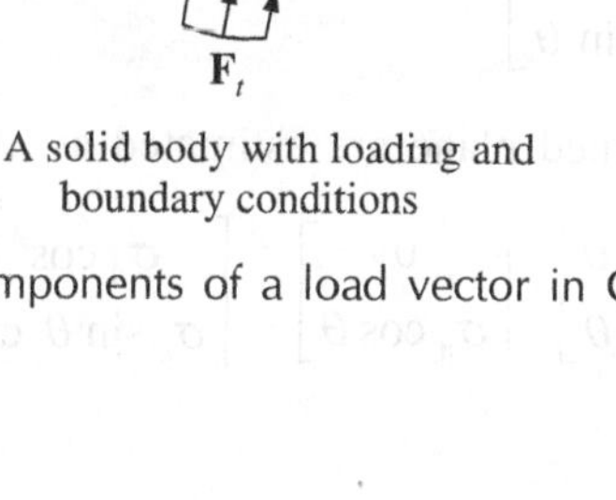

(a) A solid body with loading and boundary conditions

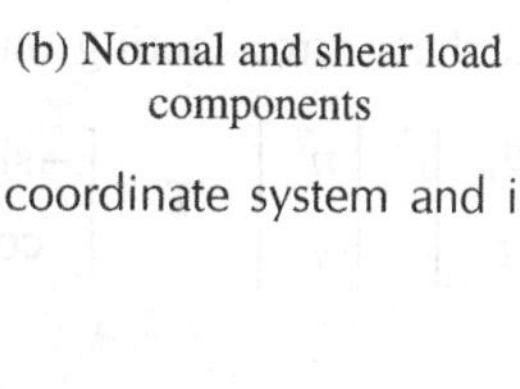

(b) Normal and shear load components

FIGURE 4.1 Components of a load vector in Cartesian coordinate system and in shear plane.

Vertical Stress at a point* h *m below the ground

Let us assume that average density of overburden rock h m below ground is ρ kg/m^3 as shown in Figure 4.2. Considering a horizontal plane P at h m below ground, the normal force on area ΔA m^2 will be $F_n = \rho g h \Delta A$ N. According to Eq. 4.3, normal (vertical) stress at a point h m below ground will be

$$\sigma_n = \lim_{\Delta A \to 0} \frac{F_n}{\Delta A} = \lim_{\Delta A \to 0} \frac{\rho g h \Delta A}{\Delta A} = \rho g h$$

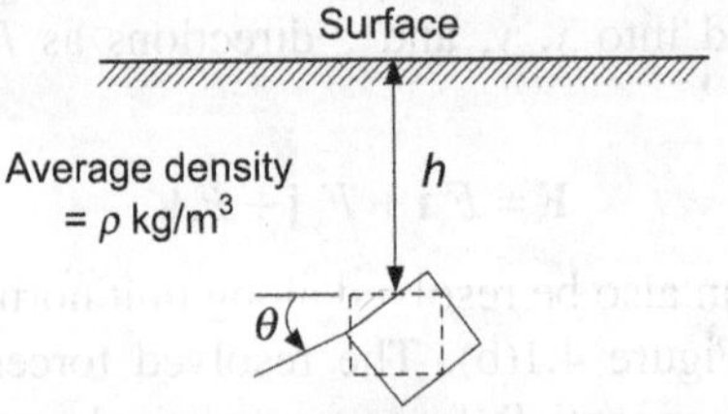

FIGURE 4.2 Normal and shear stresses in underground.

Since no horizontal and shear stress act on the dotted square, stress in matrix form is written as (refer Eq. 4.5a).

$$\boldsymbol{\sigma} = \begin{bmatrix} 0 & 0 \\ 0 & \sigma_n \end{bmatrix}$$

Now, let's assume that we have to determine normal and shear stresses on an inclined plane (shown by solid square element) directed at θ degree from the horizontal axis. From the above, we define a *traction vector* in horizontal and vertical directions are:

$$\mathbf{T} = \boldsymbol{\sigma} \cdot \boldsymbol{n} = \begin{bmatrix} 0 & 0 \\ 0 & \sigma_n \end{bmatrix} \begin{bmatrix} -\sin\theta \\ \cos\theta \end{bmatrix} = \begin{bmatrix} 0 \\ \sigma_n \cos\theta \end{bmatrix}$$

Here, unit vector is the outward normal vector of the inclined plane defined as

$$\boldsymbol{n} = \begin{bmatrix} -\sin\theta \\ \cos\theta \end{bmatrix}$$

Defining unit shear vector along the inclined plane as

$$\boldsymbol{s} = \begin{bmatrix} \cos\theta \\ \sin\theta \end{bmatrix}$$

Normal and shear stress acting on the inclined plane are estimated as

$$\begin{bmatrix} \sigma_n \\ \tau \end{bmatrix} = \begin{bmatrix} n^T \\ s^T \end{bmatrix} \cdot \mathbf{T} = \begin{bmatrix} -\sin\theta & \cos\theta \\ \cos\theta & \sin\theta \end{bmatrix} \cdot \begin{bmatrix} 0 \\ \sigma_n \cos\theta \end{bmatrix} = \begin{bmatrix} \sigma_n \cos^2\theta \\ \sigma_n \sin\theta \cos\theta \end{bmatrix}$$

Normal stress can be either compressive or tensile. In rock mechanics, sign convention of normal and shear stresses are elaborated as follows:

(i) A positive normal stress would signify a compressive stress; the outward normal of the plane points to the positive Cartesian direction and normal stress points to the negative direction, and conversely.

(ii) A negative normal stress would signify a tensile stress; the outward normal of the plane and normal stress point to the positive Cartesian direction, and conversely.

(iii) The sense of positive shear stress is obtained when shear stress is directed in a negative Cartesian direction while acting on a plane whose outward normal points in a positive direction, and conversely.

The positive sign convention of normal and shear stresses in rock mechanics is depicted in Figure 4.3. Gravitational (vertical) stress and horizontal stress beneath the earth's surface are one of the examples of normal stresses. *Shear strength* of intact rock, rock joint or plane of weakness is the shear resistance that can be obtained under given loading conditions and is of major concern for determination of overall strength of intact rock or rock mass against shearing. For joints, the magnitude of the shear strength depends on the nature of contact surfaces (wavy, zaggard, plane, slickensided), type of gauge material present, and the magnitude of the normal stress acting on the joint surface.

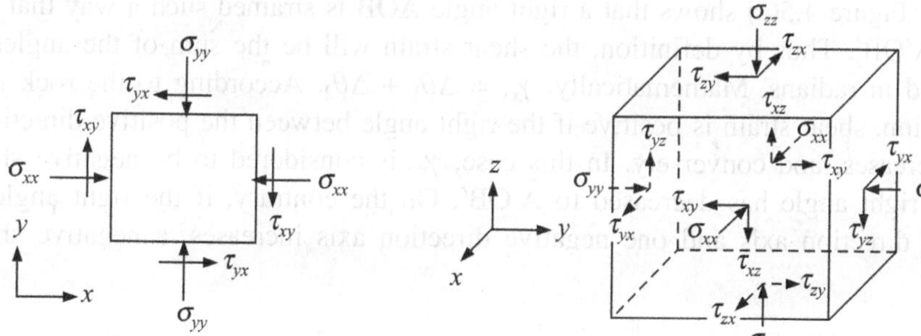

(a) Stress tensors in 2D (b) Stress tensors in 3D

FIGURE 4.3 Rock mechanics positive sign convention.

4.2.2 Definition of Strain

When a body is subjected to external loading, it undergoes deformation or movement. If the body is not constrained, unbalanced force creates rigid body motion or rotation. The rigid body rotation of an element is shown in Figure 4.4. However, rigid body motion does not contribute the strain field, and thus does not affect the stresses (Sadd, 2005). On the contrary, if the body is constrained at some part of the boundary, it deforms due to applied loading and thus, strains occur in the body. Figure 4.4 depicts horizontal and vertical contraction of an element.

As for example, during uniaxial compression loading of rock sample, load is applied axially from one end of the sample and the other end is constrained for any axial movement. As a result, axial deformation or strains occurs in the sample. Now, consider a small element of

FIGURE 4.4 Typical deformation of a rectangular element.

axial length Δl inside the sample before loading and δl be the change in length Δl after the load is applied. Then the *infinitesimal axial strain* in the sample is defined as:

$$\varepsilon = \lim_{\Delta l \to 0} \frac{\delta l}{\Delta l} \tag{4.4}$$

Here ε is termed as infinitesimal *normal strain* and it occurs when the body is deformed along a straight line. The length of the body, Δl is considered infinitesimal so that the variation of strain over the axial length can be ignored and the superposition principles of strains due to various stresses can also be applied (Obert and Duvall, 1967).

The *shear strain* occurs in a body when the angle between two original orthogonal straight line changes. Figure 4.5(a) shows that a right angle AOB is strained such a way that it's shape changes to A′OB′. Thus by definition, the shear strain will be the sum of the angles $\Delta\theta_1$ and $\Delta\theta_2$ measured in radians. Mathematically, $\gamma_{xy} = \Delta\theta_1 + \Delta\theta_2$. According to the rock mechanics sign convention, shear strain is positive if the right angle between the positive directions of the two axes increases, and conversely. In this case, γ_{xy} is considered to be negative shear strain because the right angle has decreased to A′OB′. On the contrary, if the right angle between one positive direction axis and one negative direction axis increases, a negative shear strain occurs.

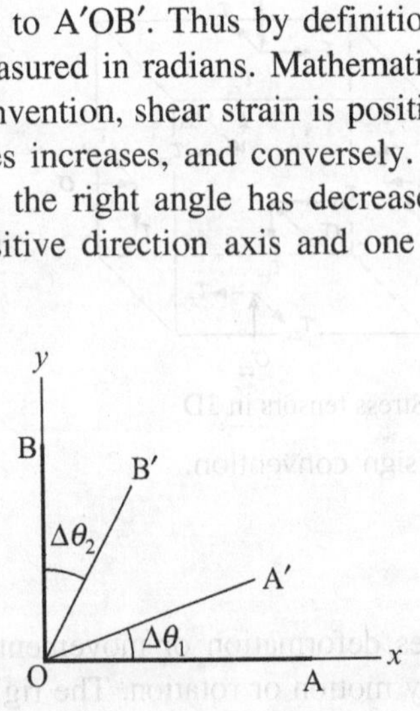

FIGURE 4.5(a) Definition of shear strain.

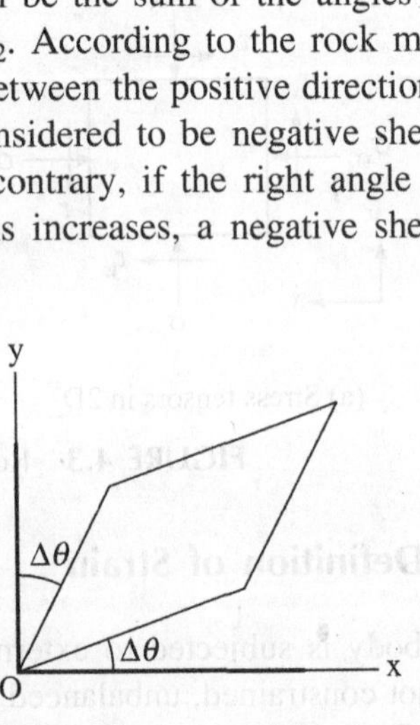

FIGURE 4.5(b) Shear strain due to pure shear.

Let us assume that an element is subjected to pure shear stress (τ_{xy}) and is strained as shown in Figure 4.5(b). Due to this, the shear strain γ_{xy} can be obtained as $\gamma_{xy} = 2\Delta\theta$ or $\Delta\theta = \gamma_{xy}/2$. If the angle $\Delta\theta$ is very small, it can be shown that change length in x-direction to the original length in y-direction and vice-versa is equal to $\Delta\theta$. In other words, distortion of the element in both planes is $\gamma_{xy}/2$ and it is caused by shear stress τ_{xy}.

4.2.3 Stress and Strain Tensor

A tensor is a physical quantity with magnitude, direction and a plane under consideration (Hudson et al., 1997) such as stress, strain, moment of inertia, etc. In general, a scalar, a vector and a matrix in Cartesian coordinate system are considered to be zero, first and second order tensor quantity respectively. Stress and strains are second order tensor quantities. The order of tensor quantity is determined based on the number of transformations required to transform the quantity from one coordinate system to another. Figures 4.3(a) and 4.3(b) show stress tensors in 2D and 3D respectively. Here, σ_{xx} signifies the normal stress acting on *yz* plane and directed along *x*-direction, while τ_{xy} defines the shear stress acting on the plane perpendicular to *x*-axis and is directed towards *y*-direction. Similar nomenclature is applied for other tensors. In matrix form, the stress tensors (**T**) are expressed as:

$$\mathbf{T} = \begin{bmatrix} \sigma_{xx} & \tau_{xy} \\ \tau_{yx} & \sigma_{yy} \end{bmatrix} \quad \text{for 2D} \tag{4.5a}$$

$$\mathbf{T} = \begin{bmatrix} \sigma_{xx} & \tau_{xy} & \tau_{xz} \\ \tau_{yx} & \sigma_{yy} & \tau_{yz} \\ \tau_{zx} & \tau_{zy} & \sigma_{zz} \end{bmatrix} \quad \text{for 3D} \tag{4.5b}$$

From the force equilibrium of the area or volume element, it can easily be shown that, $\tau_{yx} = \tau_{xy}$, $\tau_{zy} = \tau_{yz}$ and $\tau_{xz} = \tau_{zx}$ (Obert and Duvall, 1967). Thus, for two-dimensional stress analysis, σ_{xx}, σ_{yy} and τ_{xy} are in-plane stress tensors. In this text, the tensor σ_{zz} is considered to be the out of plane stress tensor in 2D. For three-dimensional problems, six stress tensors σ_{xx}, σ_{yy}, σ_{zz}, τ_{xy}, τ_{yz} and τ_{zx} are used to define a state of stress condition. From the above discussions, it can be noted that the tensor matrices given in Eqs. 4.5a and 4.5b are symmetric and can be rewritten as:

$$T = \begin{bmatrix} \sigma_{xx} & \tau_{xy} \\ \tau_{xy} & \sigma_{yy} \end{bmatrix} \quad \text{for 2D} \tag{4.6a}$$

$$T = \begin{bmatrix} \sigma_{xx} & \tau_{xy} & \tau_{zx} \\ \tau_{xy} & \sigma_{yy} & \tau_{yz} \\ \tau_{zx} & \tau_{yz} & \sigma_{zz} \end{bmatrix} \quad \text{for 3D} \tag{4.6b}$$

Strain tensors are obtained from the gradient of the displacements. Consider an original (undeformed) body Ω in two dimensions in Cartesian coordinate system *xy* as shown in Figure 4.6. When the body is deformed denoted by Ω^*, every particle in the body takes up a new position, which is described with new coordinate system x^*y^*. Let us consider a particle P which is located in the original body with coordinates *x* and *y*. After deformation, the particle is moved to location P* in the deformed body with coordinates x^* and y^*. The vector denoted by (OP*–OP) represents the displacement vector of the particle. Mathematically, coordinate x^* and y^* can be written, in terms of components of displacement vector as:

$$x^* = x + u \tag{4.7a}$$

$$y^* = y + v \tag{4.7b}$$

where u and v are the displacement fields in x and y directions respectively and are functions of both the coordinates x and y. Note that displacement fields are also function of coordinates x^* and y^*.

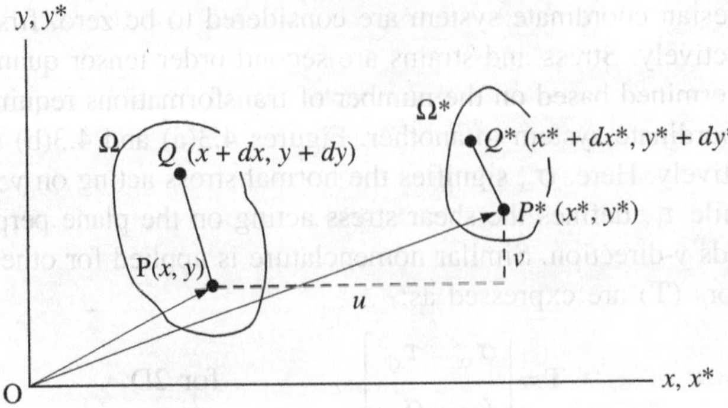

FIGURE 4.6 Displacement vector.

Now, consider an infinitesimal line element PQ in the original body Ω where the coordinates of a particle Q is defined by $x + dx$ and $y + dy$. The square of the length ds of the line element PQ, representing a square area, in its original configuration is given by

$$(ds)^2 = (dx)^2 + (dy)^2 \tag{4.8}$$

After the body is deformed, the square of the length ds^* of the deformed line element P*Q*, representing a square area in the deformed body, will be

$$(ds^*)^2 = (dx^*)^2 + (dy^*)^2 \tag{4.9}$$

From Eqs. 4.7a and 4.7b, differentials dx^* and dy^* are computed based on differentials dx and dy as

$$dx^* = \frac{\partial x^*}{\partial x}dx + \frac{\partial x^*}{\partial y}dy = \left(1 + \frac{\partial u}{\partial x}\right)dx + \frac{\partial u}{\partial y}dy \tag{4.10a}$$

$$dy^* = \frac{\partial y^*}{\partial x}dx + \frac{\partial y^*}{\partial y}dy = \frac{\partial v}{\partial x}dx + \left(1 + \frac{\partial v}{\partial y}\right)dy \tag{4.10b}$$

From Eqs. 4.8, 4.9 and 4.10, the change in square area is found as:

$$(ds^*)^2 - (ds)^2 = 2e_{xx}(dx)^2 + 2e_{yy}(dy)^2 + 2e_{xy}dxdy \tag{4.11}$$

where, the strain tensors e_{xx}, e_{yy} and e_{xy} are:

$$e_{xx} = \frac{1}{2}\left[\left(1 + \frac{\partial u}{\partial x}\right)^2 + \left(\frac{\partial v}{\partial x}\right)^2 - 1\right] = \frac{\partial u}{\partial x} + \frac{1}{2}\left[\left(\frac{\partial u}{\partial x}\right)^2 + \left(\frac{\partial v}{\partial x}\right)^2\right] \tag{4.12a}$$

$$e_{yy} = \frac{1}{2}\left[\left(\frac{\partial u}{\partial y}\right)^2 + \left(1 + \frac{\partial v}{\partial y}\right)^2 - 1\right] = \frac{\partial v}{\partial y} + \frac{1}{2}\left[\left(\frac{\partial u}{\partial y}\right)^2 + \left(\frac{\partial v}{\partial y}\right)^2\right] \tag{4.12b}$$

$$e_{xy} = \left(\frac{\partial v}{\partial x} + \frac{\partial u}{\partial y}\right) + \left(\frac{\partial u}{\partial x}\frac{\partial u}{\partial y} + \frac{\partial v}{\partial x}\frac{\partial v}{\partial y}\right) \tag{4.12c}$$

Then *Cauchy's infinitesimal strain tensors* in two dimensions are expressed by ignoring the quadratic terms in the Eqs. 4.12a, 4.12b and 4.12c as:

$$\varepsilon_{xx} = \frac{\partial u}{\partial x},\ \varepsilon_{yy} = \frac{\partial v}{\partial y},\ \gamma_{xy} = \frac{\partial v}{\partial x} + \frac{\partial u}{\partial y} \tag{4.13}$$

For three-dimensional displacement fields with $u = u(x, y, z)$, $v = v(x, y, z)$ and $w = w(x, y, z)$ in x, y and z directions respectively the additional strain tensors apart from that given in Eq. 4.13 will be

$$\varepsilon_{zz} = \frac{\partial w}{\partial z},\ \gamma_{yz} = \frac{\partial w}{\partial y} + \frac{\partial v}{\partial z},\ \gamma_{zx} = \frac{\partial u}{\partial z} + \frac{\partial w}{\partial x} \tag{4.14}$$

In this book, ε_{xx}, ε_{yy} and γ_{xy} are considered in-plane strain tensors for two-dimensional problems. The out-of-plane strain tensor is ε_{zz}. For three-dimensional problems, ε_{xx}, ε_{yy}, ε_{zz}, γ_{xy}, γ_{yz} and γ_{zx} represent six strain tensors. The strain tensor matrix is also symmetric and of similar form as given in Eq. 4.6 with the exception that stress notations are replaced by strain notations as shown in Eq. 4.16.

$$\begin{bmatrix} \varepsilon_{xx} & \gamma_{xy} \\ \tau_{yx} & \varepsilon_{yy} \end{bmatrix} \tag{4.15a}$$

$$\begin{bmatrix} \varepsilon_{xx} & \gamma_{xy} & \gamma_{xz} \\ \gamma_{yx} & \varepsilon_{yy} & \gamma_{yz} \\ \gamma_{zx} & \gamma_{xz} & \varepsilon_{zz} \end{bmatrix} \tag{4.15b}$$

EXAMPLE 4.1 Show that for a line element of length dx parallel to x-axis, the infinitesimal strain will be $\varepsilon_{xx} = \dfrac{dx^* - dx}{dx}$.

Solution: Since the line element is parallel to x-axis, $dy = dz = 0$. Hence, Eq. 4.11 becomes

$$(dx^*)^2 - (dx)^2 = 2\varepsilon_{xx}(dx)^2$$

$$dx^* - dx = \frac{2\varepsilon_{xx}(dx)^2}{dx^* + dx}$$

In this case dx* differs from dx by a small quantity of the second order if we assume displacement field u, v and w and the strain components to be infinitesimal. Hence, in the right hand side of the equation, we can assume $dx^* \approx dx$ and thus,

$$\varepsilon_{xx} = \frac{dx^* - dx}{dx}$$

EXAMPLE 4.2 Given displacement vector at a point in a body is $\mathbf{u} = \{(x^2y - xy^2)\ \mathbf{e}_x + (xy^2 - x^2y)\mathbf{e}_y\} \times 10^{-5}$ m where $\mathbf{e}_x$ and $\mathbf{e}_y$ are the unit vectors in x and y directions respectively. Determine the strains at (2 m, 2 m) point.

Solution: Here, $u = (x^2y - xy^2) \times 10^{-5}$ and $v = (xy^2 - x^2y) \times 10^{-5}$. Using Eq. 4.13, we find $\varepsilon_{xx} = y(2x - y) \times 10^{-5}$, $\varepsilon_{yy} = x(2y - x) \times 10^{-5}$ and $\gamma_{xy} = (x^2 + y^2 - 4xy) \times 10^{-5}$. At the point (2 m, 2 m), the values of strain are $\varepsilon_{xx} = 4.0 \times 10^{-5}$, $\varepsilon_{yy} = 4.0 \times 10^{-5}$ and $\gamma_{xy} = -8.0 \times 10^{-5}$.

4.2.4 Stresses and Strains in Polar Coordinate System

Many applications in rock mechanics are solved for curved boundaries, viz. stresses and displacements around circular openings (see Chapter 7). To formulate such problems in 2D, polar coordinate system is used. For 3D applications with curved boundaries cylindrical and/or spherical coordinate system are popularly used for problem formulation. In this section, definitions of displacement, strain and stress tensors in polar coordinate system are discussed.

The relationship between Cartesian coordinates (x, y) and polar coordinate variables (r, θ) can be expresses as:

$$x = r\cos\theta, \quad y = r\sin\theta \tag{4.16}$$

where, r = radial distance, and θ = angle measured counter-clockwise from positive x-axis to radial axis (r).

Figure 4.7 shows the positive sign convention of stresses in polar coordinate system. If the displacements in radial and tangential directions are u_r and u_θ respectively, then strain tensor ε_{rr}, $\varepsilon_{\theta\theta}$ and $\gamma_{r\theta}$ can be expressed as:

$$\varepsilon_{rr} = \frac{\partial u_r}{\partial r} \tag{4.17a}$$

$$\varepsilon_{\theta\theta} = \frac{1}{r}\left(u_r + \frac{\partial u_\theta}{\partial \theta}\right) \tag{4.17b}$$

$$\gamma_{r\theta} = \frac{1}{r}\frac{\partial u_r}{\partial \theta} + \frac{\partial u_\theta}{\partial r} - \frac{u_\theta}{r} \tag{4.17c}$$

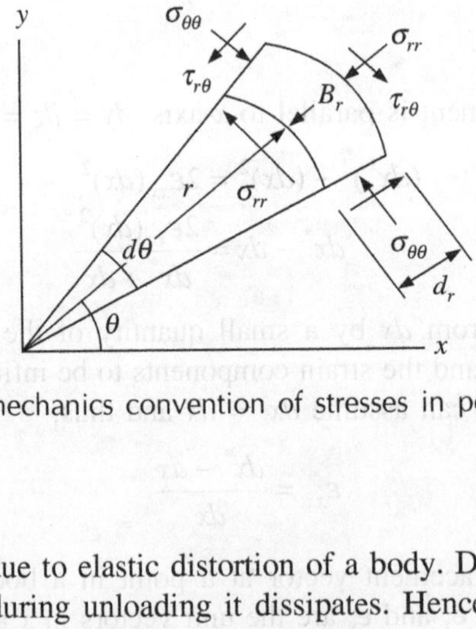

FIGURE 4.7 Rock mechanics convention of stresses in polar coordinate system.

4.2.5 Strain Energy

Strain energy is developed due to elastic distortion of a body. During loading, strain energy is contained in the body and during unloading it dissipates. Hence for a conservative body, the net strain energy of loading and unloading cycle is zero. *Strain energy density* is defined as strain energy per unit volume of a body and is expressed as:

$$U_0 = \frac{1}{2}\varepsilon^T\sigma = \frac{1}{2}[\varepsilon_{xx}\sigma_{xx} + \varepsilon_{yy}\sigma_{yy} + \varepsilon_{zz}\sigma_{zz} + \gamma_{xy}\tau_{xy} + \gamma_{yz}\tau_{yz} + \gamma_{zx}\tau_{zx}] \tag{4.18a}$$

Considering the volume of the body as V, the total strain energy will be

$$U = \int_V U_0\, dV \tag{4.18b}$$

4.3 TRANSFORMATION OF VECTORS AND TENSORS

In this section, techniques for transforming displacement and force vectors, strain and stress tensors from one coordinate system to another are reviewed. In this chapter, two-dimensional transformation methods are enumerated. The concepts of transformation in three dimensions are given in Appendix A.

4.3.1 Transformation Matrix for Displacement Vector

Let us assume $x' - y'$ represents a coordinate system where angle between x-axis and x'-axis is θ degrees as shown in Figure 4.8. Let us also assume that the displacement fields based on $x' - y'$ coordinate system are represented as u' and v' respectively. Now, u' and v' in terms of u and v can be represented using transformation technique as given below.

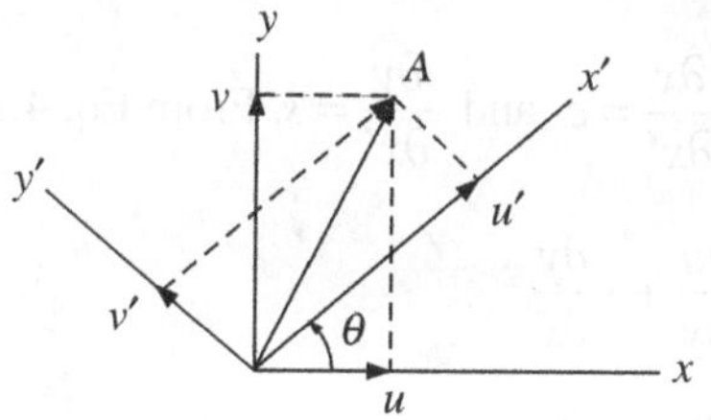

	x	y
x'	c	s
y'	$-s$	c

$c = \cos\theta$, $s = \sin\theta$

FIGURE 4.8 Transformation of displacement in 2D.

A displacement vector **A** can be represented by both the coordinate system as shown in Figure 4.8. Hence, displacements u' and v' in terms of u and v are obtained as:

$$u' = \mathbf{A}\cdot\mathbf{i}' = (u\mathbf{i} + v\mathbf{j})\cdot\mathbf{i}' = u\cos\theta + v\sin\theta \tag{4.19a}$$

$$v' = \mathbf{A}\cdot\mathbf{j}' = (u\mathbf{i} + v\mathbf{j})\cdot\mathbf{j}' = -u\sin\theta + v\cos\theta \tag{4.19b}$$

where, unit vector in $x' - y'$ co-ordinate system is: $\mathbf{i}' = \cos\theta\mathbf{i} + \sin\theta\mathbf{j}$, and $\mathbf{j}' = -\sin\theta\mathbf{i} + \cos\theta\mathbf{j}$.

Thus, the transformation matrix **R**, for transforming displacement from $x - y$ coordinate to $x' - y'$ coordinate system, can be written as:

$$\begin{Bmatrix} u' \\ v' \end{Bmatrix} = \underbrace{\begin{bmatrix} c & s \\ -s & c \end{bmatrix}}_{\mathbf{R}} \begin{Bmatrix} u \\ v \end{Bmatrix} \tag{4.20}$$

4.3.2 Transformation Matrix for Force Vector

Force vector can also be transformed from one Cartesian coordinate system to another. Let us assume, **F** and **F′** denote the force vectors in $x - y$ and $x' - y'$ Cartesian coordinate systems respectively as mentioned in the previous section. If the virtual displacement vectors in the two coordinate systems are $\delta\mathbf{q}$ and $\delta\mathbf{q}'$ respectively, the virtual work done by the real forces will be the same. Hence, it can be written as:

$$(\delta\mathbf{q})^T \mathbf{F} = (\delta\mathbf{q}')^T \mathbf{F}' = (\delta\mathbf{q})^T \mathbf{R}^T \mathbf{F}' \tag{4.21}$$

Since $\delta\mathbf{q} \neq 0$,

$$\mathbf{F} = \mathbf{R}^{\mathrm{T}}\mathbf{F}' \tag{4.22}$$

For above example, $\mathbf{R} = (\mathbf{R}^{\mathrm{T}})^{-1}$ and thus, $\mathbf{F}' = \mathbf{RF}$.

4.3.3 Transformation Matrix for Strain Tensors

Transformation of strain tensors from one coordinate system to another is not that straight forward as elaborated for displacement vector. In this case, strain vector $\boldsymbol{\varepsilon}'^{\mathrm{T}} = \{\varepsilon_{x'x'}\ \varepsilon_{y'y'}\ \gamma_{x'y'}\ \varepsilon_{zz}\}$ need to be represented in terms of strain vector $\boldsymbol{\varepsilon}^{\mathrm{T}} = \{\varepsilon_{xx}\ \varepsilon_{yy}\ \gamma_{xy}\ \varepsilon_{zz}\}$. Note that for two-dimensional problems the z-axis does not change during transformation. Considering $\varepsilon_{x'x'}$ as the gradient of u' with respect to x', and using chain rule, we find

$$\varepsilon_{x'x'} = \frac{\partial u'}{\partial x'} = \frac{\partial u'}{\partial x}\frac{\partial x}{\partial x'} + \frac{\partial u'}{\partial y}\frac{\partial y}{\partial x'} \tag{4.23}$$

The direction cosines are expressed as $\dfrac{\partial x}{\partial x'} = c$ and $\dfrac{\partial y}{\partial x'} = s$. From Eq. 4.18, we can write that

$$\frac{\partial u'}{\partial x} = c\frac{\partial u}{\partial x} + s\frac{\partial v}{\partial x} \tag{4.24a}$$

$$\frac{\partial u'}{\partial y} = c\frac{\partial u}{\partial y} + s\frac{\partial v}{\partial y} \tag{4.24b}$$

Combing Eqs., 4.22 and 4.23, we find

$$\varepsilon_{x'x'} = c^2\varepsilon_{xx} + s^2\varepsilon_{yy} + cs\gamma_{xy} \tag{4.25}$$

Similarly, other two strain tensors $\varepsilon_{y'y'}$ and $\gamma_{x'y'}$ expressed in $x' - y'$ coordinate system can be obtained using strain vector $\boldsymbol{\varepsilon}^T = \{\varepsilon_{xx}\ \varepsilon_{yy}\ \gamma_{xy}\ \varepsilon_{zz}\}$. After combining all results, we find the strain transformation matrix as

$$\mathbf{S} = \begin{bmatrix} c^2 & s^2 & cs & 0 \\ s^2 & c^2 & -cs & 0 \\ -2cs & 2cs & c^2 - s^2 & 0 \\ 0 & 0 & 0 & 1 \end{bmatrix} \tag{4.26}$$

$$\boldsymbol{\varepsilon}' = \mathbf{S}\boldsymbol{\varepsilon}$$

The last row and the last column are added to consider the strain transformation in z-direction. It is clear that strain in z-direction does not change due to transformation of coordinate system.

4.3.4 Transformation Matrix for Stress Tensors

Transformation of stress tensors from $x - y$ coordinate system to $x' - y'$ coordinate system is performed based on the transformation matrix given in Eq. 4.25. For a given loading condition, the strain energy of a body remains the same irrespective of the coordinate systems. Considering a virtual strain vector $\delta\varepsilon$, we can equate internal virtual strain energies estimated using the coordinate systems $x - y$ and $x' - y'$ as:

$$\delta\varepsilon^T\sigma = \delta\varepsilon'^T\sigma' = \delta\varepsilon^T\mathbf{S}^T\sigma' \tag{4.27}$$

Since for any virtual strain vector $\delta\varepsilon \neq 0$, we find that

$$\sigma' = (\mathbf{S}^T)^{-1}\sigma = \mathbf{T}\sigma \tag{4.28}$$

where stress transformation matrix $\mathbf{T}$ can be obtained from matrix $\mathbf{S}$ as:

$$\mathbf{T} = \begin{bmatrix} c^2 & s^2 & 2cs & 0 \\ s^2 & c^2 & -2cs & 0 \\ -cs & cs & c^2 - s^2 & 0 \\ 0 & 0 & 0 & 1 \end{bmatrix} \tag{4.29}$$

Using the matrix given in Eq. 4.28, normal and shear stresses in $x' - y'$ coordinate system can be explicitly written if the x'-axis is rotated θ degree counter-clockwise with respect to x-axis (Figure 4.9) as given below:

$$\sigma_{x'x'} = \left(\frac{\sigma_{xx} + \sigma_{yy}}{2}\right) + \left(\frac{\sigma_{xx} - \sigma_{yy}}{2}\right)\cos 2\theta + \tau_{xy}\sin 2\theta \tag{4.30a}$$

$$\sigma_{y'y'} = \left(\frac{\sigma_{xx} + \sigma_{yy}}{2}\right) - \left(\frac{\sigma_{xx} - \sigma_{yy}}{2}\right)\cos 2\theta - \tau_{xy}\sin 2\theta \tag{4.30b}$$

$$\tau_{x'y'} = -\left(\frac{\sigma_{xx} - \sigma_{yy}}{2}\right)\sin 2\theta + \tau_{xy}\cos 2\theta \tag{4.30c}$$

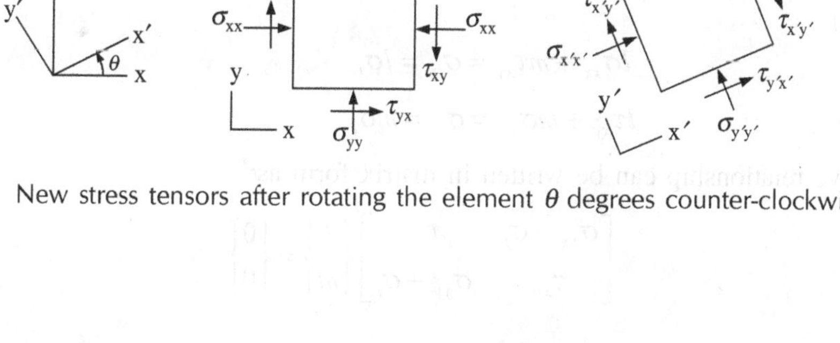

FIGURE 4.9 New stress tensors after rotating the element θ degrees counter-clockwise direction.

EXAMPLE 4.3 The vertical and horizontal stresses at the middle of a coal pillar are found to be 15 MPa and 3 MPa respectively. The shear stress is 2.5 MPa as shown in the figure below. Determine the stresses in $x' - y'$ coordinate system as shown in the figure.

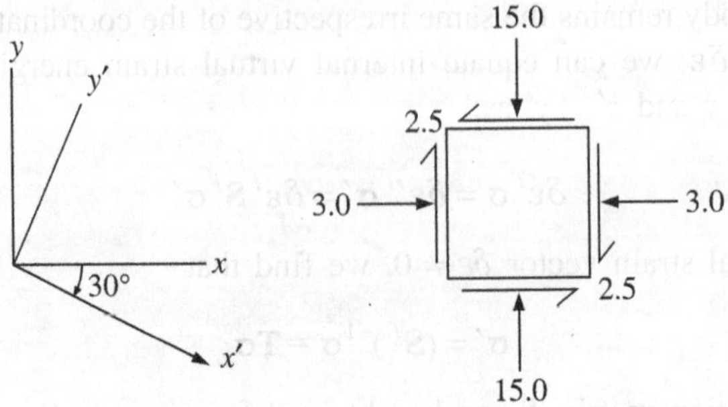

Solution: Here, $\theta = -30°$. Hence, $c = 0.866$ and $s = -0.5$. Using Eq. 4.29, we find

$$\begin{Bmatrix} \sigma_{x'x'} \\ \sigma_{y'y'} \\ \tau_{x'y'} \end{Bmatrix} = \begin{bmatrix} 0.75 & 0.25 & -0.866 \\ 0.25 & 0.75 & 0.866 \\ 0.433 & -0.433 & 0.5 \end{bmatrix} \begin{Bmatrix} 3 \\ 15 \\ 2.5 \end{Bmatrix} = \begin{Bmatrix} 3.84 \\ 14.16 \\ -3.95 \end{Bmatrix} \text{MPa}$$

4.4 STRESS ANALYSIS

4.4.1 Principal Stresses in Two Dimensions

The *principal stresses* are normal stresses that act in planes where shear stress does not exist. In two-dimensional problems, principal stresses also signify the maximum and the minimum normal stresses that can exist at a particular plane and generally termed as *major* and *minor principal stresses* respectively.

Consider that a principal stress σ_i acts on the oblique plane AB which has the normal vector **ON** as shown in Figure 4.10. The direction cosines of the normal vector are represented by $l = \cos\theta$ and $m = \sin\theta$. The components of σ_i in Cartesian x and y directions are σ_{ix} and σ_{iy} respectively. Hence we can write

$$\sigma_{ix} = l\sigma_i \text{ and } \sigma_{iy} = m\sigma_i \tag{4.31}$$

Applying force equilibrium relationships in x and y directions, it can be easily shown that (Figure 4.10)

$$l\sigma_{xx} + m\tau_{xy} = \sigma_{ix} = l\sigma_i \tag{4.32a}$$

$$l\tau_{xy} + m\sigma_{yy} = \sigma_{iy} = m\sigma_i \tag{4.32b}$$

The above relationship can be written in matrix form as:

$$\begin{bmatrix} \sigma_{xx} - \sigma_i & \tau_{xy} \\ \tau_{xy} & \sigma_{yy} - \sigma_i \end{bmatrix} \begin{Bmatrix} l \\ m \end{Bmatrix} = \begin{Bmatrix} 0 \\ 0 \end{Bmatrix} \tag{4.33}$$

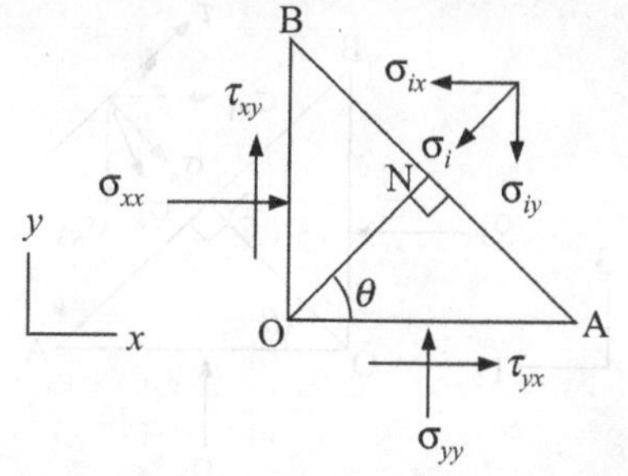

FIGURE 4.10 Estimation of principal stress.

If this system of equations has a non-zero solution then the determinant of the left hand side matrix must be zero. Thus, the following equation is obtained:

$$\sigma_i^2 - N_1\sigma_i - N_2 = 0 \tag{4.34}$$

where,

$$N_1 = \sigma_{xx} + \sigma_{yy} \tag{4.35a}$$

$$N_2 = -\sigma_{xx}\,\sigma_{yy} + \tau_{xy}^2 \tag{4.35b}$$

Here N_1 and N_2 are called *invariant of stress tensor* which means that these values are not dependent on the coordinate system. By solving Eq. 4.34, the magnitude of major (σ_1) and minor (σ_3) principal stresses can be obtained considering $\sigma_1 \geq \sigma_3$ as

$$\sigma_{1,3} = \left(\frac{\sigma_{xx} + \sigma_{yy}}{2}\right) \pm \sqrt{\frac{1}{4}(\sigma_{xx} - \sigma_{yy})^2 + \tau_{xy}^2} \tag{4.36}$$

In the above equation, we assume that normal stresses are compressive. Once the magnitude of principal stresses are known, Eq. 4.33 can be used to find the direction cosines using the additional relationship that states $l^2 + m^2 = 1$.

In solid and rock mechanics, principal stresses have a special significance since their magnitudes do not alter with the coordinate systems. Thus, it is convenient to represent a state of stress condition using principal stresses. Let us assume that the principal axes are defined by numerals 1 and 3 representing major and minor principal stress directions respectively. It is now of interest to find the normal and shear stresses as a function of principal stresses. Consider Figure 4.11 where an oblique plane AB makes an angle θ with 1 axis (major principal direction). The line ON represents the normal vector of the plane AB. The direction cosines of the normal vector are represented by $l = \cos\theta$ and $m = \sin\theta$ with respect to 1 and 3 axes respectively. Consider that the normal stress σ_n acts on the oblique plane parallel to the line ON and shear stress τ acts tangential to the plane AB. Considering area of the plane AB unity, the resultant force vector made by stresses of σ_n and τ is σ_R. Resolving σ_R into 1 and 3 directions produces σ_{R1} and σ_{R3} respectively. Hence from Figure 4.11, it can be written as

$$\sigma_{R1} = l\sigma_1 \text{ and } \sigma_{R3} = m\sigma_3 \tag{4.37}$$

The magnitude of σ_n can be obtained as:

$$\sigma_n = l\sigma_{R1} + m\sigma_{R3} = l^2\sigma_1 + m^2\sigma_3 \tag{4.38}$$

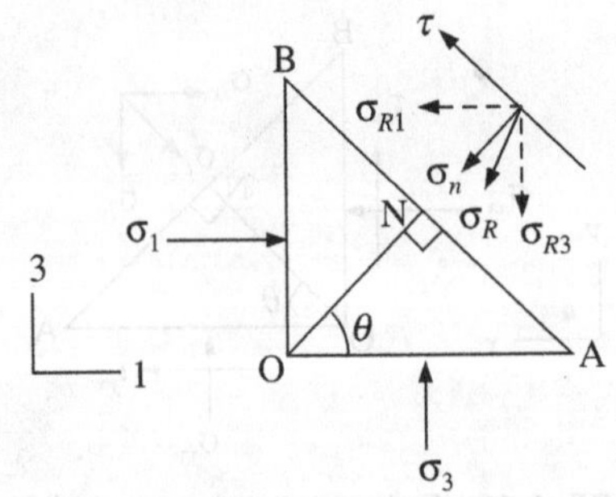

FIGURE 4.11 Estimation of normal and shear stress from principal stresses.

The magnitude of shear stress τ is then computed as:

$$\tau^2 = \sigma_R^2 - \sigma_n^2 = l^2\sigma_1^2 + m^2\sigma_3^2 - (l^2\sigma_1 + m^2\sigma_3)^2 = l^2m^2(\sigma_1 - \sigma_3)^2 \tag{4.39}$$

Therefore,

$$\tau = lm(\sigma_1 - \sigma_3) \tag{4.40}$$

4.4.2 Principal Stresses in Three Dimensions

Principal stresses in three dimensions can also be found in the similar fashion as given in two dimensional cases. As before, assume that a principal stress σ_i acts on the oblique plane ABC whose normal vector is designated by ON as shown in Figure 4.12. Consider that l, m and n represent direction cosines of the normal vector with respect to x, y and z axis respectively. It is obvious from Figure 4.12 that the components of σ_i in x, y and z directions are:

$$\sigma_{ix} = l\sigma_i, \quad \sigma_{iy} = m\sigma_i, \quad \sigma_{iz} = n\sigma_i \tag{4.41}$$

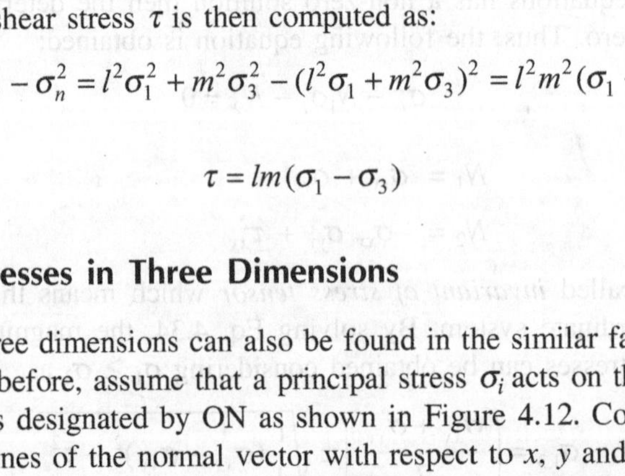

FIGURE 4.12 Concept of principal stress in 3D.

From Figure 4.12, it can be equated that area OBC = l, area OCA = m and area OAB = n if the oblique area ABC = 1. Then, equating force equilibrium conditions in three coordinate axes directions, we find

$$\begin{vmatrix} \sigma_{xx} - \sigma_i & \tau_{yx} & \tau_{zx} \\ \tau_{xy} & \sigma_{yy} - \sigma_i & \tau_{zy} \\ \tau_{xz} & \tau_{yz} & \sigma_{zz} - \sigma_i \end{vmatrix} = 0 \tag{4.42}$$

The above determinant gives to the following cubical equation:

$$\sigma_i^3 - N_1\sigma_i^2 - N_2\sigma_i - N_3 = 0 \tag{4.43}$$

where N_i (i = 1, 3) are the invariants of stress tensors and are expressed as:

$$N_1 = \sigma_{xx} + \sigma_{yy} + \sigma_{zz} \tag{4.44a}$$

$$N_2 = -(\sigma_{xx}\sigma_{yy} + \sigma_{yy}\sigma_{zz} + \sigma_{zz}\sigma_{xx}) + \tau_{xy}^2 + \tau_{yz}^2 + \tau_{zx}^2 \tag{4.44b}$$

$$N_3 = \sigma_{xx}\sigma_{yy}\sigma_{zz} + 2\tau_{xy}\tau_{yz}\tau_{zx} - \sigma_{xx}\tau_{yz}^2 - \sigma_{yy}\tau_{zx}^2 - \sigma_{zz}\tau_{xy}^2 \tag{4.44c}$$

The solution of cubic equation provides three principal stresses: (i) major principal stress, σ_1, (ii) intermediate principal stress, σ_2, and (iii) minor principal stress, σ_3. This is a typical eigenvalues (principal stresses) and eigenvectors (direction cosines) problem. Once the magnitude of a principal stress is known, Eq. 4.42 can be used to calculate the direction cosines. The magnitude of principal stresses are such that $\sigma_1 \geq \sigma_2 \geq \sigma_3$. The direction of these principal stresses can also be obtained in the similar fashion as described for 2D case. The best way to obtain the solution of Eq. 4.43 is by iterative method such as Newton-Raphson using a computer program or with a programmable calculator.

Once the magnitudes of three principal stresses are known, the invariants of stresses can be obtained based on principal stresses as follows:

$$N_1 = \sigma_1 + \sigma_2 + \sigma_3 \tag{4.45a}$$

$$N_2 = -(\sigma_1\sigma_2 + \sigma_2\sigma_3 + \sigma_3\sigma_1) \tag{4.45b}$$

$$N_3 = \sigma_1\sigma_2\sigma_3 \tag{4.45c}$$

As before, let us denote the principal axes by numerals 1, 2 and 3 to represent the major, intermediate and minor principal stress directions respectively as shown in Figure 4.13. Again let us consider an oblique plane ABC which has the normal vector ON having direction cosines l, m and n respectively with x, y and z axes. The normal stress σ_n acts parallel to the normal vector ON and shear stress τ acts tangential to the plane ABC. As shown for two-dimensional cases, let us assume that σ_R denotes the resultant of σ_n and τ. The resolved stresses of σ_R in 1, 2 and 3 directions are σ_{R1}, σ_{R2} and σ_{R3} respectively. Hence, we can write the following relationships:

$$\sigma_{R1} = l\sigma_1,\ \sigma_{R2} = m\sigma_2,\ \sigma_{R3} = n\sigma_3 \tag{4.46}$$

From Figure 4.13, it can be shown that the magnitude of normal stress will be

$$\sigma_n = l\sigma_1 + m\sigma_2 + n\sigma_3 = l^2\sigma_1 + m^2\sigma_2 + n^2\sigma_3 \tag{4.47}$$

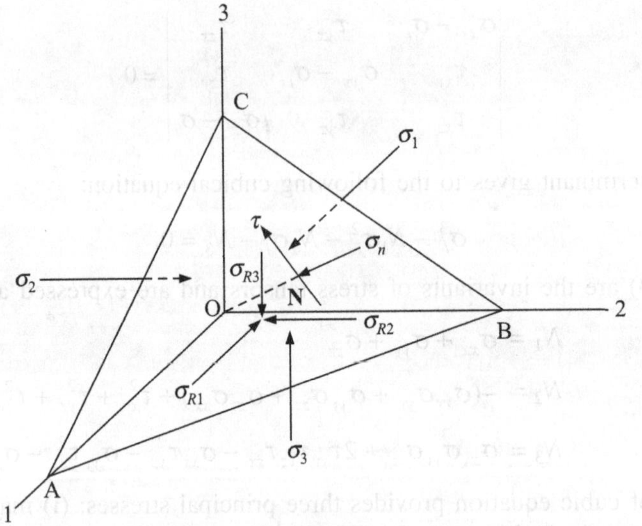

FIGURE 4.13 Normal and shear stress in 3D.

The magnitude of shear stress can be obtained from the following:

$$\tau^2 = \sigma_R^2 - \sigma_n^2 = (l\sigma_1)^2 + (m\sigma_2)^2 + (n\sigma_3)^2 - (l^2\sigma_1 + m^2\sigma_2 + n^2\sigma_3)^2$$

$$\tau^2 = l^2m^2(\sigma_1 - \sigma_2)^2 + m^2n^2(\sigma_2 - \sigma_3)^2 + n^2l^2(\sigma_3 - \sigma_1)^2 \tag{4.48}$$

EXAMPLE 4.4 Find the magnitude and direction of principal stresses for Example 4.3.

Solution: Using Eq. 4.36, we find σ_1 = 15.5 MPa and σ_3 = 2.5 MPa. Now, for the direction of σ_1 with *x*-axis, we find

$$m = \frac{12.5}{2.5} l = 5l \quad \text{and} \quad (1 + 5^2)l^2 = 1,\ l = 0.196$$

Hence, the angle between the normal of a plane where the major principal stress acts (major principal axis) and *x*-axis is 78.69° or 258.69° measured in counter-clockwise direction.

EXAMPLE 4.5 A body shown in the figure below is subjected to biaxial loading:

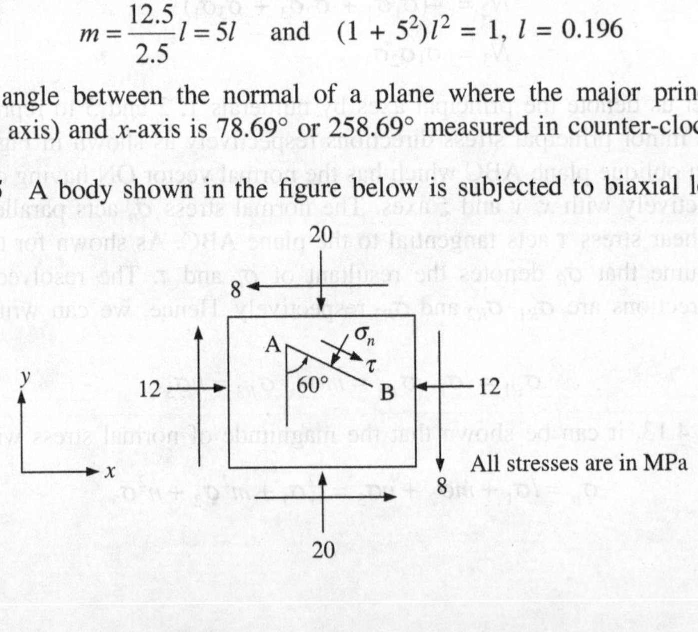

Determine the normal and shear stresses on the plane denoted by AB. Determine the magnitude of principal stresses and direction cosines of the major principal stress.

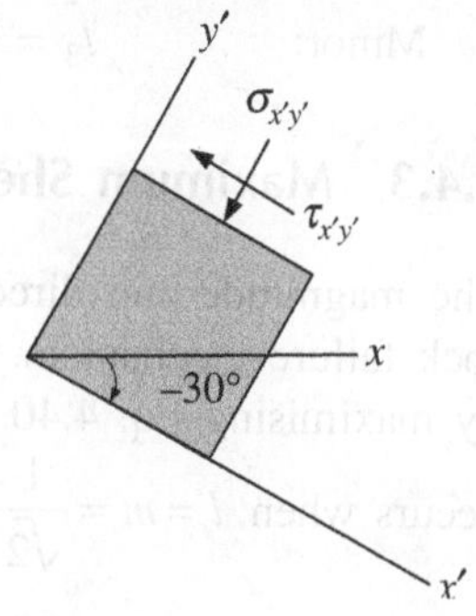

Solution: Here, σ_{xx} = 12 MPa, σ_{yy} = 20 MPa, τ_{xy} = 8 MPa and θ = –30°. Using Eq., 4.30, $\sigma_n = \sigma_{y'y'}$ = 24.93 MPa and $\tau = \sigma_{x'y'}$ = –0.54 MPa.

Using Eq. 4.36, principal stresses are obtained: σ_1 = 24.93 MPa and σ_3 = 7.05 MPa. Considering l and m as the direction cosines of the normal of the plane where major principal stress acts, m/l = 1.62 and $(1 + 1.62^2)l^2 = 1$, $l = \pm 0.526$. Hence, the angle between the x-axis and the normal is 58.28° or 238.28° measured counter-clockwise direction.

EXAMPLE 4.6 The six stress components are given below. Estimate the values of principal stresses and their directions.

σ_{xx} = 20.5 MPa $\quad$ σ_{yy} = 34.8 MPa $\quad$ σ_{zz} = 8.1 MPa
τ_{xy} = –2.4 MPa $\quad$ τ_{yz} = –8.0 MPa $\quad$ τ_{zx} = 5.2 MPa

Solution: The stress invariants are:

$$N_1 = \sigma_{xx} + \sigma_{yy} + \sigma_{zz} = 63.4$$
$$N_2 = -(\sigma_{xx}\sigma_{yy} + \sigma_{yy}\sigma_{zz} + \sigma_{zz}\sigma_{xx}) + \tau_{xy}^2 + \tau_{yz}^2 + \tau_{zx}^2 = -1064.53$$
$$N_3 = \sigma_{xx}\sigma_{yy}\sigma_{zz} + 2\tau_{xy}\tau_{yz}\tau_{zx} - \sigma_{xx}\tau_{yz}^2 - \sigma_{yy}\tau_{zx}^2 - \sigma_{zz}\tau_{xy}^2 = 3678.57$$

Hence, the cubic equation of principal stress becomes

$$\sigma_i^3 - 63.4\sigma_i^2 + 1064.53\sigma_i - 3678.57 = 0$$

Solution of this equation can be obtained using iterative (Newton-Raphson) method and the results are given below:

Major: σ_1 = 37.83 MPa
Intermediate: σ_2 = 20.92 MPa
Minor: σ_3 = 4.65 MPa

Let us assume that the direction cosines of the major principal stress σ_1 with respect to x, y and z axes are l, m and n. Then the system of equations as given in Eq. 4.42 becomes

$$17.332l + 2.4m - 5.2n = 0$$
$$2.4l + 3.032m + 8n = 0$$
$$-5.2l + 8m + 29.732n = 0$$

From these equations, following relationships can be obtained:

$$m = -4.32l$$
$$n = 1.34l$$

Using the equality condition $l^2 + m^2 + n^2 = 1$, direction cosines are calculated as:

$$l_1 = 0.22\ (77°),\quad m_1 = -0.93\ (158°),\quad n_1 = 0.29\ (73°)$$

Similarly, direction cosines for intermediate and minor principal stresses are:

Intermediate: $l_2 = 0.94$, $m_2 = 0.28$, $n_2 = 0.20$
Minor: $l_3 = 0.27$, $m_3 = -0.23$, $n_3 = -0.93$

4.4.3 Maximum Shear Stress

The magnitude and direction of the maximum shear stress are important for the analysis of rock failure mechanism. In 2D, the magnitude of the maximum shear stress can be obtained by maximising Eq. 4.40 with respect to the angle θ. For this case, the maximum shear stress occurs when $l = m = \frac{1}{\sqrt{2}}$ and hence,

$$\tau_{max} = \frac{\sigma_1 - \sigma_3}{2} \tag{4.49}$$

For 3D cases, the maximum shear stress at *x-y* plane is found by letting $n = 0$ and then maximising Eq. 4.48 as:

$$(\tau_{max})_{xy} = \frac{\sigma_1 - \sigma_2}{2} \tag{4.50}$$

Similarly, for *y-z* and *z-x* planes, considering $l = 0$ and $m = 0$ respectively, the maximum shear stresses can be computed from

$$(\tau_{max})_{yz} = \frac{\sigma_2 - \sigma_3}{2} \text{ and } (\tau_{max})_{zx} = \frac{\sigma_3 - \sigma_1}{2} \tag{4.51}$$

Note that the direction of the maximum shear stress is 45° from any principal stress direction in a given plane.

4.5 MOHR'S CIRCLE OF STRESS IN TWO DIMENSIONS

Mohr's circle of stress was first presented by Otto Mohr in 1882. This method applies graphical technique to represent two-dimensional stress fields in order to evaluate the orientation and magnitude of normal stress, shear stress and principal stresses.

The concept of Mohr's circle of stress is imbedded in the set of equations given in Eq. 4.30. Manipulating Eq. 4.30a and taking square in both sides and adding it with the square of Eq. 4.30c, the following equation is obtained.

$$\left[\sigma_{x'x'} - \left(\frac{\sigma_{xx} + \sigma_{yy}}{2}\right)\right]^2 + \tau_{x'y'}^2 = \left[\left(\frac{\sigma_{xx} - \sigma_{yy}}{2}\right)\cos 2\theta + \tau_{xy}\sin 2\theta\right]^2 + \left[-\left(\frac{\sigma_{xx} - \sigma_{yy}}{2}\right)\sin 2\theta + \tau_{xy}\cos 2\theta\right]^2 \tag{4.52a}$$

After simplification,

$$\left[\sigma_{x'x'} - \left(\frac{\sigma_{xx} + \sigma_{yy}}{2}\right)\right]^2 + \tau_{x'y'}^2 = \left(\frac{\sigma_{xx} - \sigma_{yy}}{2}\right)^2 + \tau_{xy}^2 \tag{4.52b}$$

Assuming centre $C = \left(\frac{\sigma_{xx} + \sigma_{yy}}{2}\right)$ and radius $R = \sqrt{\left(\frac{\sigma_{xx} - \sigma_{yy}}{2}\right)^2 + \tau_{xy}^2}$, Eq. 4.52(b) can be rewritten as $(\sigma_{x'x'} - C)^2 + \tau_{x'y'}^2 = R^2$. Hence, it is proved that Eq. 4.29 is a set of equations which produces a circle in normal and shear stress plane. Similarly, it can be shown that $(\sigma_{y'y'} - C)^2 + \tau_{x'y'}^2 = R^2$. This circle of stress is termed as Mohr's circle of stress.

4.5.1 Steps for Constructing Mohr's Circle

Step 1: Normal stress is plotted on the horizontal axis (σ_n) and the vertical axis (τ) refers to the shear stress. The scale of the vertical axis is the same as that of normal stress axis.

Step 2: Shear stress is plotted with positive sign if it tends to rotate the rock element counter-clockwise direction as shown in Figure 4.9. In this figure, the shear stress acting on the horizontal plane will be designated by positive sign. On the contrary, if shear stress acting on a plane rotates the element clockwise direction, it will be marked with a negative sign while constructing the Mohr's circle. In the same figure, shear stress acting on the vertical plane will be designated with negative sign.

Step 3: Plot point A(σ_{xx}, $-\tau_{xy}$) and A′(σ_{yy}, τ_{xy}) point on a graph paper with proper scale and sign convention as mentioned above. In this case, point A refers the stresses acting on the vertical plane and point A′ signifies the same of the horizontal plane (Figure 4.14).

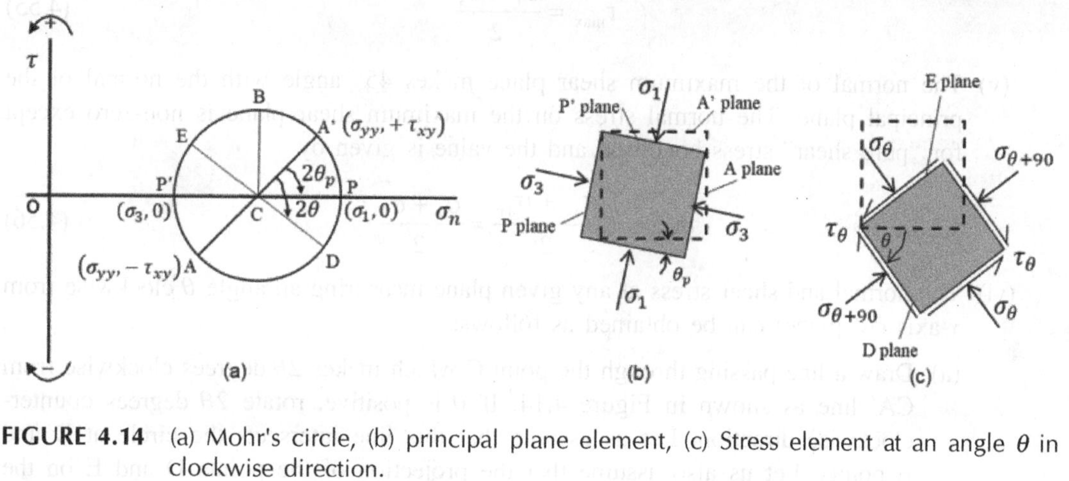

FIGURE 4.14 (a) Mohr's circle, (b) principal plane element, (c) Stress element at an angle θ in clockwise direction.

Step 4: Join the AA′ line and the intersection point between that line and the horizontal axis is the centre of the Mohr's circle. It can also be shown that the coordinate of this centre point will be

$$OC = \frac{\sigma_{xx} + \sigma_{yy}}{2} \tag{4.53}$$

Step 5: Draw the circle with the centre at point C and diameter as AA′. This completes the construction of the Mohr's circle with any state of stress condition in two dimensions.

Step 6: If the principal stresses are used to draw the Mohr's circle then P and P′ points will lie on the horizontal axis. Repeat steps 4 and 5 to draw the Mohr's circle.

4.5.2 Inference from the Mohr's Circle of Stress

(i) Each point of the circumference of Mohr's circle represents a state of stress condition with normal and shear stress.

(ii) The intersection points between the horizontal axis and zero shear stress plane are marked by points, P and P′. The magnitude of stresses represented by OP′ and OP equals to the major and minor principal stresses, respectively considering compressive as positive normal stress.

(iii) The direction of the major principal stress with respect to horizontal plane (A′plane) can also be estimated by measuring the angle A′CP′ and divided it by 2 as shown in the figure.

$$\theta_1 = \frac{\angle A'CP'}{2} \tag{4.54}$$

(iv) The magnitude of the maximum shear stress, τ_{max} is the radius of this circle and can be given by CB. It can also be shown that the magnitude of τ_{max}

$$\tau_{max} = \frac{\sigma_1 - \sigma_3}{2} \tag{4.55}$$

(v) The normal of the maximum shear plane makes 45° angle with the normal of the principal plane. The normal stress on the maximum shear plane is non-zero except for "pure shear" stress condition and the value is given by

$$\sigma_{\tau max} = \frac{\sigma_{xx} + \sigma_{yy}}{2} = \frac{\sigma_1 + \sigma_3}{2} \tag{4.56}$$

(vi) The normal and shear stress at any given plane measuring an angle θ clockwise from x-axis (A′ plane) can be obtained as follows:

(a) Draw a line passing through the point C which makes 2θ degrees clockwise from CA′ line as shown in Figure 4.14. If θ is positive, rotate 2θ degrees counter-clockwise direction. Let us assume that this line intersects the circle at D and E points. Let us also assume that the projection of the points D and E on the horizontal axis are D′ and E′ respectively.

(b) The magnitude of OD′ and OE′ represent the values of normal stress of σ_θ and $\sigma_{\theta+90}$ respectively. The magnitude of shear stress, τ_θ is equal to the line DD′ or EE′.

4.5.3 Mohr's Circle of Special Stress Conditions

(i) **Uniaxial compression:** In uniaxial compression test, core sample is loaded axially and the failure stress σ_1 is recorded. No external lateral stress σ_3 and shear stress act on the sample. Figure 4.15(a) shows the Mohr's circle of stress for uniaxial compression loading condition. The maximum shear stress developed in the sample will be $\tau_{max} = \sigma_1/2$ and acts 45° from the horizontal plane of the specimen.

(ii) **Uniaxial tension:** In direct uniaxial tension test, the rock sample is pulled with an axial load. The tensile failure stress σ_1 is recorded. The Mohr's circle is drawn with $\sigma_3 = 0$ and $\sigma_1 = F$ where F is the failure stress as shown in Figure 4.15(b).

(iii) **Triaxial compression:** In this test, the rock specimen is loaded both in axial and lateral directions. The failure stress σ_1 is recorded for a given lateral stress σ_3, generally applied by placing the sample in an oil chamber. The Mohr's circle is drawn with centre as $(\sigma_1 + \sigma_3)/2$ and the radius as $(\sigma_1 - \sigma_3)/2$ [Figure 4.15(c)].

(iv) **Pure shear:** Pure shear condition occurs when the rock sample is twisted with equal moment from the two ends. A small rectangular element on the sample experiences compressive normal stress in one plane and equal tensile normal stress in the orthogonal direction of the compressive plane as shown in Figure 4.15(d). Hence, $\sigma_3 = -\sigma_1$. This condition can also prevail when normal stress vanishes in the maximum shear plane as shown in Figure 4.15(e). Figure 4.15(d) and 4.15(e) shows the Mohr's circle of stress of pure shear condition when center of the circle lies at the origin. Here, $\tau_{max} = \sigma_1$.

(v) **Hydrostatic stress condition:** Hydrostatic stress field means that the equal compression loading acts on the rock sample from all directions. Hence, $\sigma_1 = \sigma_3$ and thus, Mohr's circle vanishes to a point as shown in Figure 4.15(e).

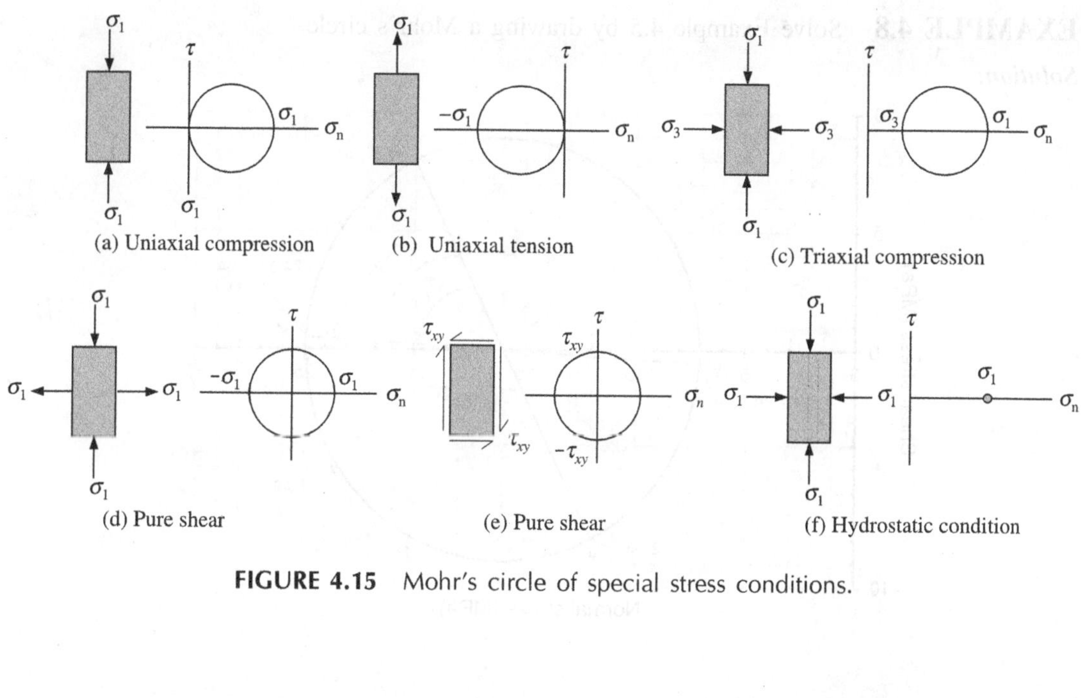

FIGURE 4.15 Mohr's circle of special stress conditions.

EXAMPLE 4.7 Determine the principal stresses and their directions using Mohr's circle of Example 4.3.

Solution: Solution with Mohr's circle is shown below. The magnitude and direction of principal stresses are also shown. The dotted line indicates the state of stress after rotating of –30°.

The major principal stress acts normally on a plane which is oriented 11.31° from the *x*-axis or major principal axis measured in clockwise direction.

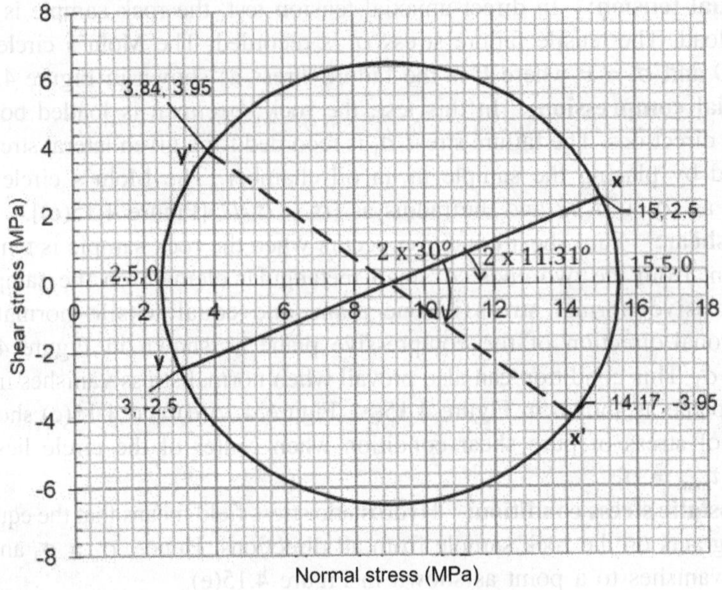

EXAMPLE 4.8 Solve Example 4.5 by drawing a Mohr's circle.

Solution:

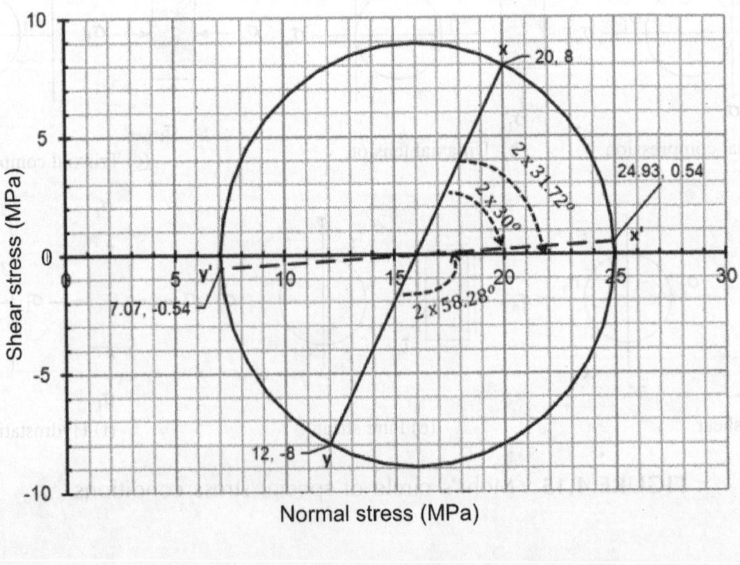

The major principal stress acts normally on a plane which is oriented 31.72° from the x-axis measured clockwise direction. Hence, the normal of that plane makes 90 – 31.72 = 58.28° with the y-axis in counter-clockwise direction.

4.6 ANALYSIS OF STRAIN

The procedure of strain analysis is similar to that of stress analysis. In Section 4.3.3, transformation procedure of strain tensors is explained. In this section, the concept of the principal strains and graphical representation of strains using Mohr's circle are illustrated with examples.

4.6.1 Principal Strains in Two Dimensions

The principal strains are normal strains that act on the planes where shear strains are absent. In this respect, the computational procedure of principal strains is the same as that of principal stresses. The procedures described in Section 4.4.1 are directly applicable for the estimation of the principal strains. Thus, principal strains can be computed by solving the following equation:

$$\varepsilon_i^2 - M_1\varepsilon_i - M_2 = 0 \tag{4.57}$$

where M_1 and M_2 are invariants of strains and are computed based on

$$M_1 = \varepsilon_{xx} + \varepsilon_{yy} \tag{4.58a}$$

$$M_2 = -\varepsilon_{xx}\varepsilon_{yy} + \frac{\gamma_{xy}^2}{4} \tag{4.58b}$$

Solution of Eq. 4.57 results in two principal strains as:

$$\varepsilon_{1,3} = \frac{\varepsilon_{xx} + \varepsilon_{yy}}{2} \pm \frac{1}{2}\sqrt{(\varepsilon_{xx} - \varepsilon_{yy})^2 + \gamma_{xy}^2} \tag{4.59}$$

Comparison between Eqs. 4.36 and 4.59 shows that for computation of principal strains normal stresses are replaced by normal strains but shear stress is replaced by 1/2 of shear strain. It can also be shown easily that the direction of principal stresses and respective principal strains are the same.

EXAMPLE 4.9 Determine the principal strains and the direction of the major principal strain using the data given in Example 4.2.

Solution: The principal strains are estimated using Eq. 4.59 as:

$$\varepsilon_{1,3} = \left(\frac{4+4}{2} \pm \frac{1}{2}\sqrt{(4-4)^2 + (-8)^2}\right) \times 10^{-5} = (4 \pm 4) \times 10^{-5}$$

Hence, $\varepsilon_1 = 8 \times 10^{-5}$, $\varepsilon_3 = 0$

We find that $(4 - 8)l - 4m = 0$ or $l = -m$

Therefore, $l = \pm\frac{1}{\sqrt{2}}$ or the direction of major principal strain is 45° counter-clockwise from the x-axis.

4.6.2 Principal Strains in Three Dimensions

The procedures outlined in Section 4.4.2 are directly applicable for the estimation of principal strains in 3D. As before for equations mentioned in that section, normal stresses can be replaced by normal strains but all shear stresses must be replaced by 1/2 of shear strains. For example, the cubical equation for estimating principal strains will be

$$\varepsilon_i^3 - M_1\varepsilon_i^2 - M_2\varepsilon_i - M_3 = 0 \tag{4.60}$$

where invariants of strains are:

$$M_1 = \varepsilon_{xx} + \varepsilon_{yy} + \varepsilon_{zz} \tag{4.61a}$$

$$M_2 = -(\varepsilon_{xx}\varepsilon_{yy} + \varepsilon_{yy}\varepsilon_{zz} + \varepsilon_{zz}\varepsilon_{xx}) + \frac{\gamma_{xy}^2}{4} + \frac{\gamma_{yz}^2}{4} + \frac{\gamma_{zx}^2}{4} \tag{4.61b}$$

$$M_3 = \varepsilon_{xx}\varepsilon_{yy}\varepsilon_{zz} + \frac{\gamma_{xy}\gamma_{yz}\gamma_{zx}}{4} - \frac{\varepsilon_{xx}\gamma_{yz}^2}{4} - \frac{\varepsilon_{yy}\gamma_{zx}^2}{4} - \frac{\varepsilon_{zz}\gamma_{xy}^2}{4} \tag{4.61c}$$

4.6.3 Maximum Shear Strain

The magnitude of one-half of the maximum shear strain in x–y, y–z and z–x planes can be obtained in terms of principal strains as:

$$\left(\frac{\gamma_{max}}{2}\right)_{xy} = \frac{\varepsilon_1 - \varepsilon_2}{2} \tag{4.62a}$$

$$\left(\frac{\gamma_{max}}{2}\right)_{yz} = \frac{\varepsilon_2 - \varepsilon_3}{2} \tag{4.62b}$$

$$\left(\frac{\gamma_{max}}{2}\right)_{zx} = \frac{\varepsilon_3 - \varepsilon_1}{2} \tag{4.62c}$$

Again note that the direction of the maximum shear strain will be 45° from the principal strain axes.

4.6.4 Mohr's Circle of Strain in Two Dimensions

Mohr's circle can also be constructed for graphical representation of two-dimensional strain tensors ε_{xx}, ε_{yy} and ε_{zz}. The construction procedure of Mohr's circle for strains is similar to that of stresses with following exceptions:

(i) The horizontal axis is represented by the normal strain and the vertical axis is denoted by half of the shear strain. The scale of the vertical axis will be the same as the horizontal or normal strain axis.

(ii) The location of shear strain in the plot, i.e., above or below the normal strain axis depends on the sign of the shear strain. As mentioned earlier, if the right angle increases, the sign of shear strain is negative, then half of the shear strain is plotted below the horizontal axis. On the contrary, if the right angle decreases, the sign of the shear strain is positive. Then the same is plotted above the horizontal strain axis (Figure 4.16).

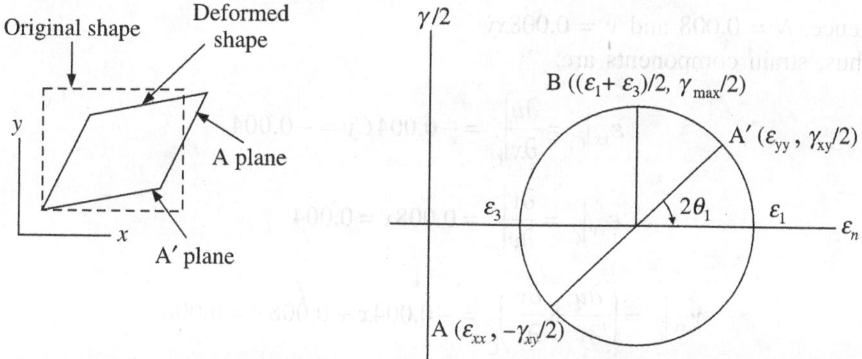

FIGURE 4.16 Mohr's circle of strain.

Consider the schematic drawing shown in Figure 4.16. Assume that the normal strains ε_{xx} and ε_{yy} are both compressive strains or positive quantities and act in A and A′ planes respectively. The shear strain γ_{xy} is negative in A plane but positive in A′ plane. Thus, Mohr's circle of strains can be plotted with points A and A′ as shown in the figure. Note that one-half of the magnitude of shear strains has to be considered while plotting the Mohr's circle.

EXAMPLE 4.10 A rectangle shown in figure below is deformed into the shape indicated by dashed lines. The displacements are found as $u = Mxy$ and $v = Nxy$, where M and N are constant.

(a) Determine the state of strain at point C when coordinates of point C* for the deformed body are (0.502, 0.996) m.

(b) Determine the normal strain at C in the direction of line CA.

(c) Determine the shearing strain at C for the undeformed line CA and a line perpendicular to line CA.

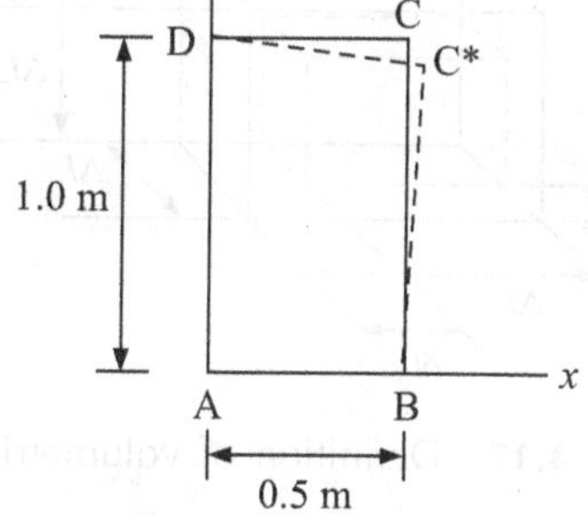

Solution:

(a) The magnitude of M and N are obtained from the known displacement at point C. We have

$$M \times 0.5 = -0.002$$

Hence, $M = -0.004$ and $u = -0.004xy$

$$N \times 0.5 = 0.004$$

Hence, $N = 0.008$ and $v = 0.008xy$

Thus, strain components are:

$$\varepsilon_{xx}\big|_C = \frac{\partial u}{\partial x}\bigg|_C = -0.004Cy = -0.004$$

$$\varepsilon_{yy}\big|_C = \frac{\partial v}{\partial y}\bigg|_C = 0.008x = 0.004$$

$$\gamma_{xy}\big|_C = \left(\frac{\partial u}{\partial y} + \frac{\partial v}{\partial x}\right)_C = -0.004x + 0.008y = 0.006$$

(b) Let x' denotes the line from C to A. Hence, the direction cosines of the line CA will be $c = -0.447$, $s = -0.894$

Using Eq. 4.25, the magnitude of normal strain in x' direction is computed as:

$$\varepsilon_{x'x'} = c^2\varepsilon_{xx} + s^2\varepsilon_{yy} + cs\gamma_{xy} = 0.0048$$

(c) The shearing strain $\gamma_{x'y'}$ will be:

$$\gamma_{x'y'} = -2cs\varepsilon_{xx} + 2cs\varepsilon_{yy} + (c^2 - s^2)\gamma_{xy} = 0.0028$$

4.6.5 Volumetric Strain

Volumetric strain refers to the change of volume to the original volume of the body due to orthogonal straining. For example, let us consider a parallelogram with sides, Δl_x, Δl_y, and Δl_z as shown in Figure 4.17. The original volume of this parallelogram is given as:

$$\Delta V = \Delta l_x \Delta l_y \Delta l_z \tag{4.63}$$

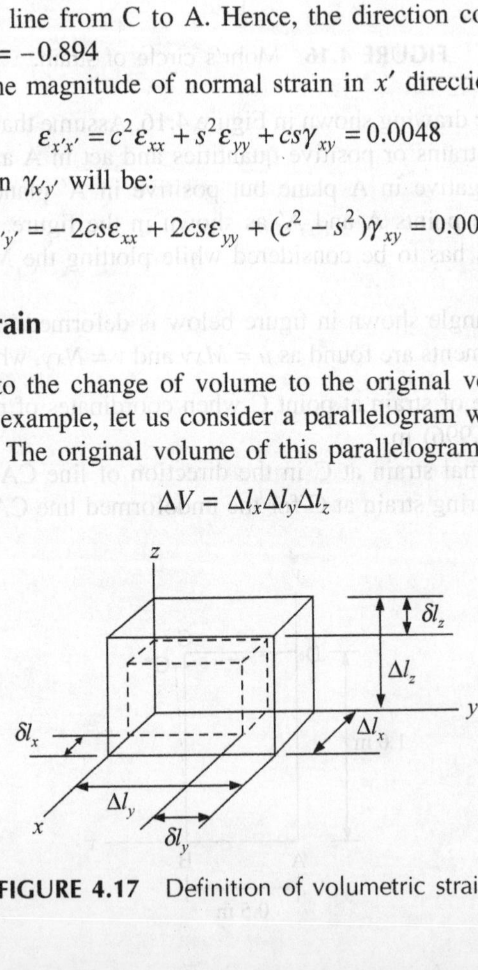

FIGURE 4.17 Definition of volumetric strain.

Let us also assume that due to normal or orthogonal strains in three coordinate directions, the volume changes to $\Delta V'$. Then the volumetric strain will be

$$\varepsilon_v = \frac{\Delta V - \Delta V'}{\Delta V} = 1 - \frac{\Delta V'}{\Delta V} \tag{4.64}$$

Now, $\Delta V' = (\Delta l_x - \delta l_x)(\Delta l_y - \delta l_y)(\Delta l_z - \delta l_z)$ and $\varepsilon_{xx} = \dfrac{\delta l_x}{\Delta l_x}$ and so on. Thus, Eq. 4.64 can be rewritten as:

$$\varepsilon_v = 1 - \frac{(\Delta l_x - \delta l_x)(\Delta l_y - \delta l_y)(\Delta l_z - \delta l_z)}{\Delta l_x \Delta l_y \Delta l_z} = 1 - (1 - \varepsilon_{xx})(1 - \varepsilon_{yy})(1 - \varepsilon_{zz}) \tag{4.65}$$

By multiplying all strain components and neglecting the product of strains, we find that

$$\varepsilon_v = \varepsilon_{xx} + \varepsilon_{yy} + \varepsilon_{zz} \tag{4.66}$$

Thus, the volumetric strain of a body is equal to the sum of the normal or orthogonal strains.

4.7 STRESS EQUILIBRIUM EQUATIONS

4.7.1 Cartesian Coordinate System

Equilibrium of an element either in 2D or in 3D requires that at least the resultant forces acting in the element must be zero in two or three axes directions respectively. Let us consider a rectangular element in two dimensions comprising of edges dx and dy. The stresses acting on the surfaces are shown in Figure 4.18. For example, the force $\sigma_{xx}dy$ acts on the left side of

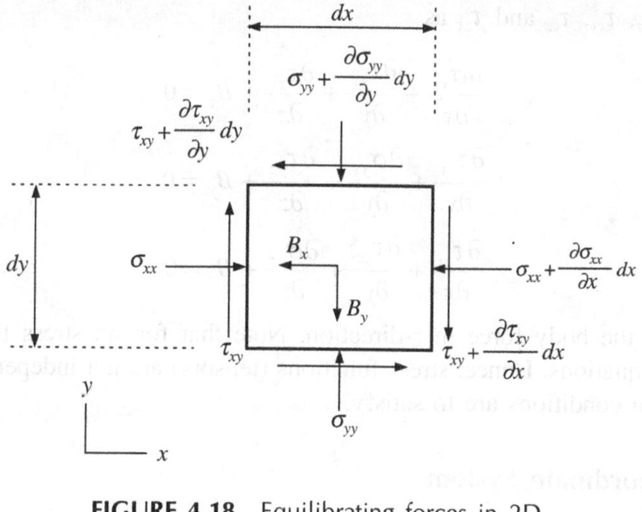

FIGURE 4.18 Equilibrating forces in 2D.

the rectangle while force $\left(\sigma_{xx} + \frac{\partial \sigma_{xx}}{\partial x} dx\right) dy$ acts on the right side. This is required to maintain the continuity of the stresses. In other words, the continuity of the stresses satisfies the increment of stresses from one surface to the opposite surface. The body forces are $B_x\, dxdy$ and $B_y\, dxdy$ in x and y directions respectively. The equilibrium of the body requires that the resultant force acting in x and y directions are zero. Hence, it can be written that

$$\sum F_x = \sum F_y = 0 \tag{4.67}$$

Thus, summation of all forces in x and y directions are:

$$\sum F_x = \sigma_{xx} dy - \left(\sigma_{xx} + \frac{\partial \sigma_{xx}}{\partial x} dx\right) dy + \tau_{xy} dx - \left(\tau_{xy} + \frac{\partial \tau_{xy}}{\partial y} dy\right) dx - B_x dxdy = 0 \tag{4.68a}$$

$$\sum F_y = \sigma_{yy} dx - \left(\sigma_{yy} + \frac{\partial \sigma_{yy}}{\partial y} dy\right) dx + \tau_{xy} dy - \left(\tau_{xy} + \frac{\partial \tau_{xy}}{\partial x} dx\right) dy - B_y dxdy = 0 \tag{4.68b}$$

Dividing each equation with *dxdy*, we can obtain two equilibrium conditions:

$$\frac{\partial \sigma_{xx}}{\partial x} + \frac{\partial \tau_{xy}}{\partial y} + B_x = 0 \tag{4.69a}$$

$$\frac{\partial \tau_{xy}}{\partial x} + \frac{\partial \sigma_{yy}}{\partial y} + B_y = 0 \tag{4.69b}$$

The equilibrium conditions expressed in Eqs. (4.69a and 4.69b) mean that the stress functions σ_{xx}, σ_{yy} and τ_{xy} cannot be selected independently. Rather they are generally defined based on the geometry and boundary conditions of the problem.

The stress equilibrium conditions are also obtained for volume element in 3D using stress tensors σ_{xx}, σ_{yy}, σ_{zz}, τ_{xy}, τ_{yz} and τ_{zx} as

$$\frac{\partial \sigma_{xx}}{\partial x} + \frac{\partial \tau_{xy}}{\partial y} + \frac{\partial \tau_{zx}}{\partial z} + B_x = 0 \tag{4.70a}$$

$$\frac{\partial \tau_{xy}}{\partial x} + \frac{\partial \sigma_{yy}}{\partial y} + \frac{\partial \tau_{yz}}{\partial z} + B_y = 0 \tag{4.70b}$$

$$\frac{\partial \tau_{zx}}{\partial x} + \frac{\partial \tau_{yz}}{\partial y} + \frac{\partial \sigma_{zz}}{\partial z} + B_z = 0 \tag{4.70c}$$

where $B_z dxdydz$ is the body force in z-direction. Note that for six stress functions, there are three equilibrium equations. Hence, stress functions (tensors) are not independent to each other if force equilibrium conditions are to satisfy.

4.7.2 Polar Coordinate System

In polar coordinates, the stress equilibrium equations can be derived considering a $drd\theta$ element in radial and tangential direction as shown in Figure 4.7. Considering a radial component of body force, B_r, the stress equilibrium equations are derived (Obert and Duval, 1967) and are reproduced in the following:

$$\frac{\partial \sigma_{rr}}{\partial r} + \frac{\sigma_{rr} - \sigma_{\theta\theta}}{r} + \frac{1}{r}\frac{\partial \tau_{r\theta}}{\partial \theta} + B_r = 0 \tag{4.71a}$$

$$\frac{1}{r}\frac{\partial \sigma_{\theta\theta}}{\partial \theta} + \frac{\partial \tau_{r\theta}}{\partial r} + \frac{2\tau_{r\theta}}{r} = 0 \tag{4.71b}$$

EXAMPLE 4.11 A steel rope of length L is hung from the ceiling. If the density of the rope is ρ, find the longitudinal stress in the rope. Neglect shear stresses.

Solution: Let us assume the vertical direction is denoted by y-axis from the ceiling. We assume that only σ_{yy} acts on the rope and the shear stresses τ_{yx} and τ_{zx} are zero. Hence applying stress equilibrium equation, we find that

$$\frac{\partial \sigma_{yy}}{\partial y} + \rho g = 0$$

Hence by integrating, we fined

$$\therefore \qquad \sigma_{yy} = -\rho g y + c$$

Now, the stress at $y = 0$ will be zero and hence $c = 0$. Therefore, the expression of longitudinal stress will be

$$\sigma_{yy} = -\rho g y$$

4.8 STRAIN COMPATIBILITY CONDITIONS

4.8.1 Cartesian Coordinate System

As mentioned earlier, displacement fields u and v are functions of coordinates x and y and strain tensors are the gradient of displacement fields. At every point inside a two-dimensional body, there exist two orthogonal displacements u and v and three in-plane strains ε_{xx}, ε_{yy} and γ_{xy}, which are derived from these displacement fields. Now, the dilemma is that three strain tensors are derived from two displacement fields. Thus, strain tensors cannot be independent to each other. There must exits a relationship between the strain tensors so that continuity of strains remains at every point of the body, which means physically that the body must deform in a compatible manner, i.e., without developing cracks or overlaps. For any admissible displacement fields, u and v, it can be easily derived that

$$\frac{\partial^2 \varepsilon_{xx}}{\partial y^2} + \frac{\partial^2 \varepsilon_{yy}}{\partial x^2} = \frac{\partial^2}{\partial x \partial y}\left(\frac{\partial u}{\partial y} + \frac{\partial v}{\partial x}\right) = \frac{\partial^2 \gamma_{xy}}{\partial x \partial y} \tag{4.72}$$

Hence, Eq. 4.72 must be satisfied at every point in the body and is generally termed as *strain compatibility condition* in two dimensions.

In a three-dimensional body, u, v and w represent the displacement fields in x, y and z directions respectively. In 3D problems, there must be three strain compatibility conditions to maintain the continuity of strains at every point of the body. Equation 4.72 can be extended for other dimensions as

$$\frac{\partial^2 \varepsilon_{xx}}{\partial y^2} + \frac{\partial^2 \varepsilon_{yy}}{\partial x^2} = \frac{\partial^2 \gamma_{xy}}{\partial x \partial y} \tag{4.73a}$$

$$\frac{\partial^2 \varepsilon_{yy}}{\partial z^2} + \frac{\partial^2 \varepsilon_{zz}}{\partial y^2} = \frac{\partial^2 \gamma_{yz}}{\partial y \partial z} \tag{4.73b}$$

$$\frac{\partial^2 \varepsilon_{zz}}{\partial x^2} + \frac{\partial^2 \varepsilon_{xx}}{\partial z^2} = \frac{\partial^2 \gamma_{zx}}{\partial z \partial x} \tag{4.73c}$$

EXAMPLE 4.12 Verify strain compatibility condition for the data given in Example 4.2.

Solution: The strains are: $\varepsilon_{xx} = y(2x - y) \times 10^{-5}$, $\varepsilon_{yy} = x(2y - x) \times 10^{-5}$, and $\gamma_{xy} = (x^2 + y^2 - 4xy) \times 10^{-5}$. From Eq. 4.72, it can be easily shown that both left and right hand sides of the equation equals to –4.

4.8.2 Polar Coordinate System

From strain displacement relationship given Eq. 4.17, strain compatibility equation in polar coordinate system can be expressed as:

$$\frac{\partial^2 r\gamma_{r\theta}}{\partial \theta \partial r} = \frac{\partial^2 \varepsilon_{rr}}{\partial \theta^2} + r\frac{\partial^2 r\varepsilon_{\theta\theta}}{\partial r^2} - r\frac{\partial \varepsilon_{rr}}{\partial r} \tag{4.74}$$

It may be noted that Eq. 4.74 must be satisfied if the strains are to be compatible with the displacement such that displacements are continuous functions of the space coordinates, r and θ (Obert and Duvall, 1967).

The compatibility equation can be written in terms of the stresses rather than the strain by recalling the constitutive equation (Eq. 5.5 in Chapter 5). Substituting the stress relationship in Eq. 4.72 results into a stress compatibility (Eq. 5.20) for plane stress as:

$$\frac{\partial^2}{\partial y^2}(\sigma_{xx} - \nu\sigma_{yy}) + \frac{\partial^2}{\partial x^2}(\sigma_{yy} - \nu\sigma_{xx}) = 2(1+\nu)\frac{\partial^2 \gamma_{xy}}{\partial x \partial y} \tag{4.75}$$

Stresses are satisfied by Eq. 4.75 guarantee the compatibility of strain. But, the stress must also satisfy the equilibrium equations, which in two dimensions can be written as:

$$\frac{\partial \sigma_{xx}}{\partial x} + \frac{\partial \tau_{xy}}{\partial y} = 0 \tag{4.76a}$$

$$\frac{\partial \tau_{xy}}{\partial x} + \frac{\partial \sigma_{yy}}{\partial y} = 0 \tag{4.76b}$$

Airy defined a stress function from which the stress tensor can be obtained by differentiation, which is discussed in next section.

4.9 AIRY STRESS FUNCTION

The stress tensors can be expressed by means of differentiation of a stress function, commonly known as *Airy stress function*. In the following, Airy stress function will be elaborated considering body force components B_x and B_y. In general, body forces are derivable from a potential function V such that

$$B_x = -\frac{\partial V}{\partial x} \tag{4.77a}$$

$$B_y = -\frac{\partial V}{\partial y} \tag{4.77b}$$

Inserting Eq. 4.77 into Eq. 4.68, the stress equilibrium equations become

$$\frac{\partial(\sigma_{xx} - V)}{\partial x} + \frac{\partial \tau_{xy}}{\partial y} = 0 \tag{4.78a}$$

$$\frac{\partial \tau_{xy}}{\partial x} + \frac{\partial(\sigma_{yy} - V)}{\partial y} = 0 \tag{4.78b}$$

It is observed that Eq. 4.78 will be identically satisfied by choosing a representation of

$$\sigma_{xx} = \frac{\partial^2 \Phi}{\partial y^2} + V \tag{4.79a}$$

$$\sigma_{yy} = \frac{\partial^2 \Phi}{\partial x^2} + V \tag{4.79b}$$

$$\tau_{xy} = -\frac{\partial^2 \Phi}{\partial x \partial y} \tag{4.79c}$$

where, $\Phi = \Phi(x, y)$ is an arbitrary function form known Airy stress function. Similarly, the stress tensors in terms of Airy stress function can be expressed in polar coordinate system as:

$$\sigma_{rr} = \frac{1}{r}\frac{\partial \Phi}{\partial r} + \frac{1}{r^2}\frac{\partial^2 \Phi}{\partial \theta^2} + V \tag{4.80a}$$

$$\sigma_{\theta\theta} = \frac{\partial^2 \Phi}{\partial r^2} + V \tag{4.80b}$$

$$\tau_{xy} = -\frac{\partial}{\partial r}\left(\frac{1}{r}\frac{\partial \Phi}{\partial \theta}\right) \tag{4.80c}$$

EXAMPLE 4.13 Assuming Airy stress function, $\Phi = a_0 + a_1 \ln r + a_2 r^2 + a_3 r^2 \ln r$, determine the stress tensor in polar coordinate system.

Solution: Applying Eq. 4.80, resulting stress tensors can be written as:

$$\sigma_{rr} = 2a_3 \ln r + \frac{a^2}{r^2} + a_3 + 2a_2$$

$$\sigma_{rr} = 2a_3 \ln r - \frac{a_1}{r^2} + 3a_3 + 2a_2$$

$$\tau_{r\theta} = 0$$

EXERCISES

1. Show that $\tau_{\max} = \frac{1}{2}(\sigma_1 - \sigma_3)$ and that the plane at which the $\tau_{\max}$ acts makes a 45° with the direction of the largest and the smallest principal stresses.
2. Use matrix multiplication to transform the following stress and strain states to axes rotated by θ = 30° from the original *x-y* axis.

   (a) $\sigma = \begin{bmatrix} 2.0 \\ -5.0 \\ 3.0 \end{bmatrix}$ (b) $\varepsilon = \begin{bmatrix} 0.002 \\ -0.005 \\ 0.003 \end{bmatrix}$

3. The component of a stress tensor at a certain point in a 2D body is represented by

   $$\begin{bmatrix} 5 & 1 \\ 1 & 2 \end{bmatrix} \text{MPa}$$

   If the plane $2x + y = 5$ passes through the point in question, what is the stress tensor acting on the outside (the side away from the origin) of the plane? What are the normal and tangential components of the stress vector on this plane?
4. The state of stress at a point (*a*, *b*, *c*) is given by the following tensors:

   $$\begin{bmatrix} 2 & -1 & 2 \\ -1 & 6 & 1.5 \\ 2 & 1.5 & 8 \end{bmatrix} \text{MPa}$$

   (a) Determine the magnitude and direction of the principal stresses at the point (*a*, *b*, *c*).
   (b) Based on the principal stresses obtain in part (a), find the stress vector and the magnitude of the normal and shear stress acting on the plane $(x - a) + (y - b) + (z - c) = 0$.
   (*Hint:* assume $a = 1, b = 2, c = 3$ and another point (2, 3, 1) such that line connecting these points denotes one shear direction).

5. In a body the stresses in the *x*-*y* plane are as shown in figure below on each stress element.
   (a) Determine the normal and shear stresses on plane A.
   (b) Determine the principal stresses at the point.
   (c) Determine the maximum shear stress at the point.
   (d) Draw the principal element.

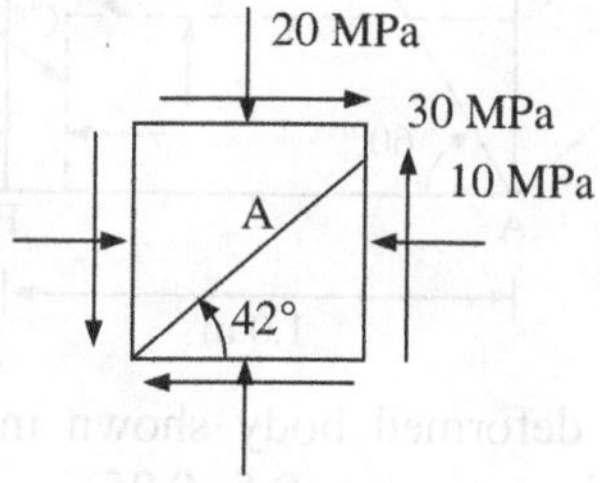

6. The state of stress at a point is given by the following tensors:

$$\begin{bmatrix} 8 & -1.6 & 2.6 \\ -1.6 & 2.6 & 2.8 \\ 2.6 & 2.8 & 3.6 \end{bmatrix} \text{MPa}$$

   Determine the normal and shearing stress acting on the plane perpendicular to x'-axis whose direction cosines are $l_1 = m_1 = n_1 = \frac{1}{3}$ and also assume $l_2 = m_2$.

7. The stress tensor given in Exercise 6, determine the magnitude and direction of the principal stresses and the magnitude of the maximum shearing stress.
8. Construct Mohr's circle solutions for the transformations of Exercise 2.
9. Solve Exercise 3 using Mohr's circle of stress.
10. Draw the Mohr's circles and determine the magnitudes of the principal stresses for the following states of stress. Denote the principal stress state on a suitably rotated stress square.
    (a) $\sigma_{xx} = 25$ MPa, $\sigma_{yy} = -10$ MPa, $\tau_{xy} = 20$ MPa
    (b) $\sigma_{xx} = -10$ MPa, $\sigma_{yy} = -20$ MPa, $\tau_{xy} = -20$ MPa
11. Using data given in Exercise 6, compute the ratio of magnitude of octahedral shear stress to the maximum shear stress.
12. The stress at a point in a body has the following components:

$$\begin{bmatrix} 2 & 0 & -2 \\ 0 & 2 & 0 \\ -2 & 0 & 2 \end{bmatrix} \text{MPa}$$

    Find the value of stress invariants N_1, N_2 and N_3 and also determine the magnitude of principal stresses.

13. The rectangular plate given in figure below is in the state of plane strain, i.e., $\varepsilon_{zz} = \gamma_{zx} = \gamma_{yz} = 0$. The plate is deformed as shown by the dotted lines. Determine the strain tensors at point C. Also determine the strain tensors at point C in (x', y') axes.

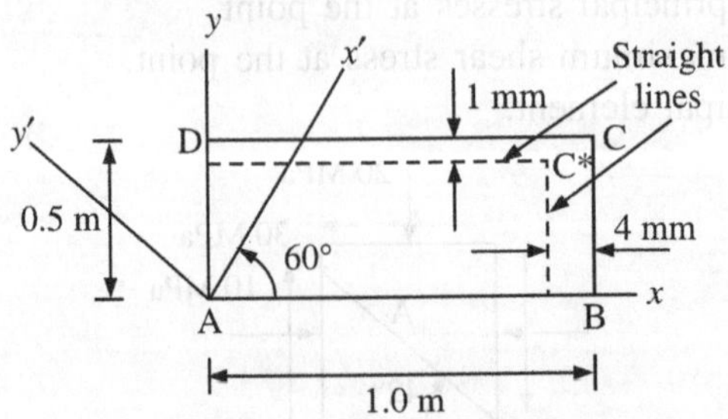

14. Solve Exercise 13 for the deformed body shown in figure below. Also estimate the magnitude of principal strains at point (0.5, 0.25).

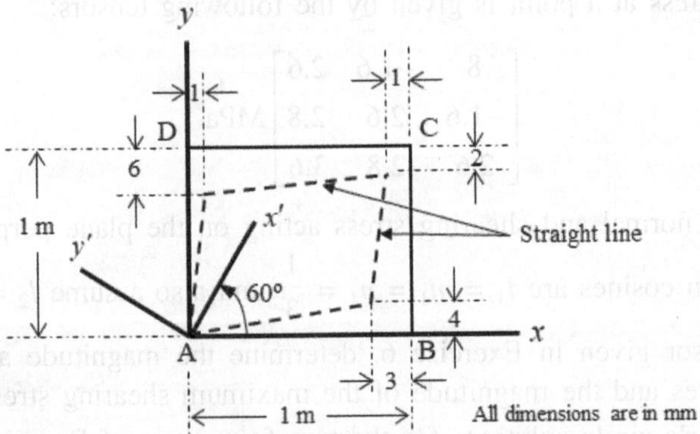

15. For the given stress function $\Phi = a_1x^4 + a_2x^3y + a_3x^2y^2 + a_4xy^3 + a_5y^4$ in x–y coordinate system, stress tensors are obtained as

$$\sigma_{xx} = \frac{\partial^2\Phi}{\partial y^2},\ \sigma_{yy} = \frac{\partial^2\Phi}{\partial x^2},\ \text{and}\ \tau_{xy} = -\frac{\partial^2\Phi}{\partial x\partial y}$$

(a) Show that the stress equilibrium equations are satisfied (ignore body forces).

(b) If only a_4 is non-zero quantity, draw the σ_{xx} and τ_{xy} on the side AB of the element shown in figure given below.

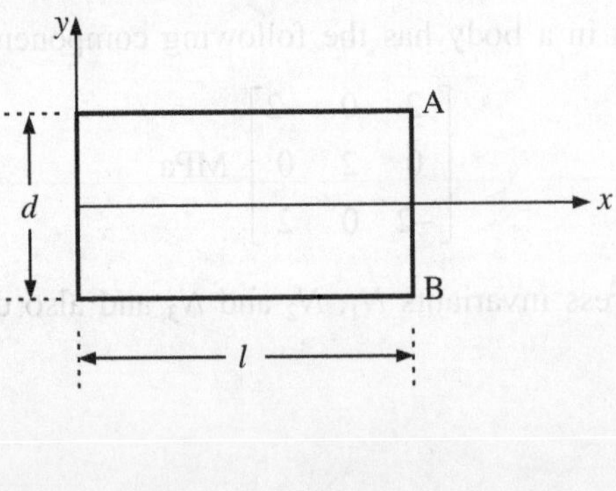

16. Given the strain at a point

$$\varepsilon = \begin{bmatrix} ax^2 + b & exy & fyz \\ exy & a + by^2 & czx \\ fyz & czx & dzx \end{bmatrix} \times 10^{-2}$$

Find the condition under which given strain state is compatible.

17. The stress state at a point is

$$\sigma = \begin{bmatrix} x^2 + y & xy^2 & yz^2 \\ xy^2 & x + y^2 & x^2 z \\ yz^2 & x^2 z & z^2 + x \end{bmatrix}$$

Find the body force under which the given stress field maintain the equilibrium.

18. Given a displacement at a point

$$\boldsymbol{u} = \{(2x^2 + y^2)\boldsymbol{e}_x + (3y^2 + yz)\boldsymbol{e}_y + (3x + z^2)\boldsymbol{e}_z\} \times 10^{-5} \text{ m}$$

where $\mathbf{e}_i$ is the unit vector along the direction i. What are the principal strains at the point (1, 0, 2)?

Chapter 5

Stress-Strain Relationships

5.1 INTRODUCTION

Rocks are mostly elastic-plastic material. The term *elasticity* refers to a particular engineering property of rock having mostly linear variations between strains and stresses. An elastic rock must also exhibit that upon removal of applied loads, the deformed body returns precisely to its original geometry. In general, elasticity is of two types: linear and non-linear. In linear elasticity, the magnitude of strain linearly varies with amount of applied stress. The concept of linear stress-strain relationship is enumerated for linear isotropic and transverse isotropic behaviour of rocks. The stress-strain behaviour of a three-dimensional body can be effectively approximated in two dimensions by imposing proper boundary conditions. These procedures of in-plane stress analyses are discussed for approximations viz. *plane stress*, *plane strain* and *axisymmetric* conditions. The concept of Airy stress function given in Chapter 4 is extended further for determination of stress, strain and displacement functions. Anisotropic rock properties, though linear, are not same in every direction. A brief description of anisotropy in rocks is presented in this chapter. In non-linear elasticity, strain is assumed to be polynomial function of stress. However, this concept is complex and hardly being used for the analysis of stress-strain behaviour of rocks.

Like most materials rock also exhibits yielding characteristics and achieves ultimate strength before yielding/failure occurs. Figure 5.1 shows schematic of stress-strain relationships under unaxial compressive loading of rock sample. Linear portion of the curve represents the elastic response of the rock until its yield strength (σ_Y) is reached. If the stress is increased beyond that point, permanent or irreversible deformation occurs in the sample. In some rocks, failure occurs suddenly after yielding having little or no permanent (plastic) deformation. After that residual stress is attained in the sample. This type of stress-strain behaviour is termed as *elastic-brittle plastic* behaviour of rock. Examples of such rock type are granite, gneiss, hard sandstone, slate, etc. Ductile rocks can sustain permanent deformation beyond yielding. Stress increases with further straining of the rock sample. This property signifies *strain-hardening* characteristics of rock. Weak rocks such as shale, mudstone, coal etc. exhibit such behaviour upon large confining stress. Ideally, *elastic-perfectly plastic* behaviour is defined when stress does not increase with strain increment in the plastic regime. In other words, stress remains the same but plastic strain continues to increase. However, in the plastic domain most rocks show *strain-softening* characteristics where stress decreases upon further strain increment. Shale, limestone, siltstones are some of the example of this type of stress-strain behaviour.

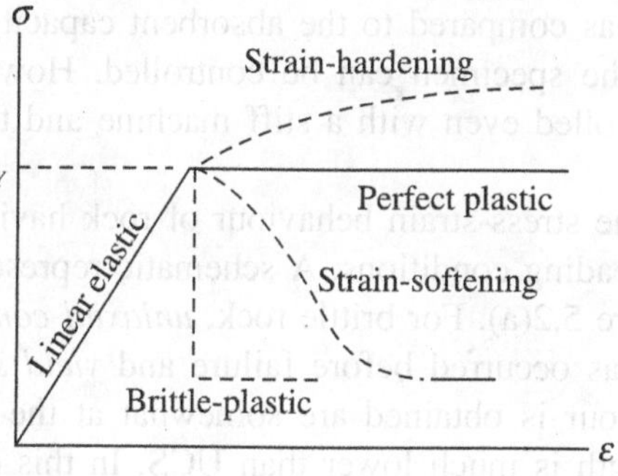

FIGURE 5.1 Stress-strain relationship of rock.

5.2 STRENGTH AND ELASTICITY OF ROCK

International Society of Rock Mechanics (ISRM) in 2007 has suggested the standard procedure for testing rock under uniaxial loading conditions (ISRM, 2007). In nutshell, the following salient points are to be maintained while cylindrical rock sample is to be tested using a compression or universal testing machine. For details, kindly see the above reference or Brady and Brown, 2007.

(i) Test specimen should be right cylinder with length to diameter ratio (*L*/*D*) ranges between 2.5:1 and 3.0:1. It is required that uniform stress develops along the length of the specimen throughout a test. Hence, a sufficient length (minimum *L*/*D* ratio is 2.5:1) is required to eliminate the effect of steel platens at both ends. Moreover, sample should fail along a shear plane which may develop across the specimen.

(ii) The end of the specimen should be flat within 0.02 mm between any two points on the surface. This is required to minimise "bedding-down" effect which is obtained at the beginning of the experiment and also shown in the initial portion of the stress-strain graph. In addition, loading platens may not come in full contact with the uneven end surfaces of rock specimen causing stress concentration and premature failure of the specimen.

(iii) A constant loading rate of 0.5 to 1.0 MPa s^{-1} or a constant strain rate of 10^{-5} to 10^{-4} s^{-1}, is to be maintained throughout the experiment. In other words, the peak strength should attain within 5–10 min. If the rate is increased beyond these limits, it may not be pertained to static loading condition rather it will be considered as pseudo-dynamic or dynamic loading conditions.

(iv) Axial load and axial and radial or circumferential strains or deformations need to be recorded throughout the test.

In addition to the above points, type of testing machine, *soft* or *stiff* with respect to rock specimen, also affects the post failure behaviour of the specimen. The stress-strain behaviour up to the peak strength does not affect by type of machine. During post peak deformation, machine unloads linearly while specimen sheds load in a non-linear fashion. The released energy of a soft machine in the post peak region (due to unloading) is higher as compared to the absorbent capacity of the specimen. As a result, excess energy is transferred to the specimen from the machine and hence, catastrophic failure occurs. On the other hand, a stiff machine releases less stored energy as compared to the absorbent capacity of the specimen. Therefore, post failure behaviour of the specimen can be controlled. However, post peak behaviour of brittle rock cannot be controlled even with a stiff machine and thus a *servo-controlled testing machine* may be required.

Figure 5.2(b) depicts the stress-strain behaviour of rock having brittle and strain softening behaviour under uniaxial loading conditions. A schematic representation of deformation of the specimen is shown in Figure 5.2(a). For brittle rock, *uniaxial compressive strength* (UCS), i.e., the maximum stress that has occurred before failure and *yield strength,* i.e., maximum stress up to which elastic behaviour is obtained are somewhat at the same point. On the contrary, for softer rock, yield strength is much lower than UCS. In this case, a slight strain hardening behaviour observed between the yield point to the UCS point.

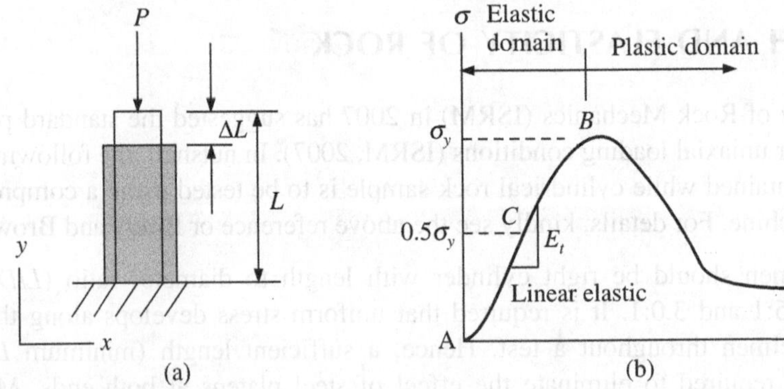

FIGURE 5.2 Stress-strain relationship in uniaxial compression loading.

5.3 STRESS-STRAIN RELATIONSHIP OF ELASTIC MATERIAL

Figure 5.2(b) clearly shows that stress-strain behaviour is linear during elastic deformation of the specimen. The figure also shows that load is applied in the axial *y*-direction of the rock specimen. In elastic domain, the linear portion of the curve, stress is directly proportional to the strain and can be expressed mathematically as:

$$\sigma_{yy} \propto \varepsilon_{yy} \tag{5.1}$$

$$\sigma_{yy} = E\varepsilon_{yy} \text{ or } \varepsilon_{yy} = \frac{\sigma_{yy}}{E} \tag{5.2}$$

where

σ_{yy} = axial normal stress, N/m^2

ε_{yy} = axial normal strain, dimensionless

E = proportionality constant or modulus of elasticity, N/m^2

The above stress-strain relationship is known as Hooke's law (1660) for one-dimensional loading condition. Modulus of elasticity can be computed in three different ways. *Average modulus of elasticity* is the slope of the stress-strain curve measuring from point A to B as shown in Figure 5.2(b). *Secant modulus* is defined by the slope of the curve from point A to C; where the point C denotes the 50% of the compressive strength. *Tangent modulus* is measured as the slope of the stress-strain curve at point C. In general, the slope of the linear portion of the curve is also denoted as the tangent modulus.

Due to axial compressive loading (*y*-direction), length of the sample decreases whereas diameter increases causing lateral or circumferential strain. This phenomenon is termed as *Poisson's effect*. Mathematically, this relationship can be written as:

$$\varepsilon_{xx} = -\nu\varepsilon_{yy} = -\nu\frac{\sigma_{yy}}{E} \tag{5.3}$$

where the parameter ν is called *Poisson's ratio* and defined as the ratio of *lateral strain to axial or longitudinal strain.* The negative sign signifies that lateral strain has the opposite direction that of axial strain. In the above example, axial strain is compressive or positive while lateral strain is negative or tensile.

EXAMPLE 5.1 A copper cable hangs down in a vertical mine shaft. If the maximum permissible stress is 8 MPa, determine the maximum length of the cable for copper density of 9000 kg/m^3. Also determine the maximum deflection occurred due to self-weight. Assume E of copper is 100 GPa.

Solution: Assuming length of the copper cable is l, the maximum stress is:

$$\sigma_{\max} = \rho g l = 8 \times 10^6,\ \text{hence,}\ l = 90.61 \text{ m}$$

Using Eq. 5.2, maximum strains can be estimated as:

$$\varepsilon_{\max} = \frac{\sigma_{\max}}{E} = \frac{8}{100 \times 10^3} = 8 \times 10^{-5}$$

The maximum deflection at bottom end of the cable will be

$$\delta = \varepsilon_{\max} l = 7.23 \text{ mm}$$

EXAMPLE 5.2 A concrete pillar of 0.5 m diameter is reinforced with steel rods. The total cross-sectional area of the rods is 5×10^{-3} m^2. The modulus of elasticity of steel and concrete is 200 GPa and 25 GPa respectively. Considering composite behaviour, determine the fraction of load carried by steel rods if an axial compressive stress of 2 MPa is applied on the pillar.

Solution: Assuming composite behaviour of the reinforced concrete pillar, it can be written that

Deflection of steel rods = Deflection of concrete

Again assuming one-dimensional stress field and deformation (ignoring Poisson's effect), for l length of pillar,

$$\frac{\sigma_s l}{E_s} = \frac{\sigma_c l}{E_c} \quad \text{or} \quad \sigma_s = \sigma_c \frac{E_s}{E_c}$$

The axial force applied to the concrete pillar is shared by both concrete and steel rods.

Hence, $$P = P_s + P_c = \sigma_s A_s + \sigma_c A_c$$

From the above, it can be derived that

$$\sigma_s = \frac{PE_s}{(A_s E_s + A_c E_c)} \quad \text{and} \quad \sigma_c = \frac{PE_c}{(A_s E_s + A_c E_c)}$$

Now, the total cross-sectional area of the pillar is

$$A = A_s + A_c = \frac{\pi}{4}(0.5)^2 = 0.1963 \text{ m}^2$$

Hence, $$A_c = 0.1963 - 0.005 = 0.1913 \text{ m}^2$$

$$P = 2 \times 0.1963 = 0.3927 \text{ MN}.$$

Then axial stress in steel rods and concrete can be estimated as:

$$\sigma_s = \frac{0.3927 \times 200}{(0.005 \times 200 + 0.1913 \times 25)} = 13.58 \text{ MPa}$$

$$\sigma_c = \frac{0.3927 \times 25}{(0.005 \times 200 + 0.1913 \times 25)} = 1.69 \text{ MPa}$$

Load on the steel rods will be $P_s = 13.58 \times 0.005 = 0.0679$ MN. Therefore, steel rods will carry 0.173 times of the total load.

The term *isotropy* refers to uniform material properties in all directions. Thus, linear isotropic elasticity signifies uniform linear stress-strain behaviour of material in all directions and upon unloading the deformed body returns to its original shape.

Applying the similar principle in three-dimensional stress conditions where all three normal stress tensors σ_{xx}, σ_{yy}, and σ_{zz} act on x, y and z directions respectively, the infinitesimal strains in three directions can be estimated as follows:

Strains developed due to σ_{xx}: $$\varepsilon_{xx} = \frac{\sigma_{xx}}{E}, \quad \varepsilon_{yy} = -\nu\frac{\sigma_{xx}}{E}, \quad \varepsilon_{zz} = -\nu\frac{\sigma_{xx}}{E} \tag{5.4a}$$

Strains developed due to σ_{yy}: $$\varepsilon_{xx} = -\nu\frac{\sigma_{yy}}{E}, \quad \varepsilon_{yy} = \frac{\sigma_{yy}}{E}, \quad \varepsilon_{zz} = \nu\frac{\sigma_{yy}}{E} \tag{5.4b}$$

Strains developed due to σ_{zz}: $$\varepsilon_{xx} = -\nu\frac{\sigma_{zz}}{E}, \quad \varepsilon_{yy} = -\nu\frac{\sigma_{zz}}{E}, \quad \varepsilon_{zz} = \frac{\sigma_{zz}}{E} \tag{5.4c}$$

Note that in the above equations E and ν are kept the same in all directions due to isotropic conditions. Using superposition principle of small strain, expression for three normal strains can be written as:

$$\varepsilon_{xx} = \frac{1}{E}\left[\sigma_{xx} - \nu(\sigma_{yy} + \sigma_{zz})\right] \tag{5.5a}$$

$$\varepsilon_{yy} = \frac{1}{E}\left[\sigma_{yy} - \nu(\sigma_{xx} + \sigma_{zz})\right] \tag{5.5b}$$

$$\varepsilon_{zz} = \frac{1}{E}\left[\sigma_{zz} - \nu(\sigma_{xx} + \sigma_{yy})\right] \tag{5.5c}$$

In the similar fashion, shear strain γ also depends linearly on shear stress τ as:

$$\gamma = \frac{\tau}{G} \tag{5.6}$$

where the parameter G is proportionality constant and referred as *shear modulus* or *modulus of rigidity*. The shear modulus G can be estimated from elastic properties of material considering pure shear loading condition as shown in Figure 5.3. From Mohr's circle of stress and strains, we can write

$$\sigma_1 = -\sigma_3 = \tau \tag{5.7a}$$

$$\varepsilon_1 = \frac{\gamma}{2} = \frac{\tau}{2G} \tag{5.7b}$$

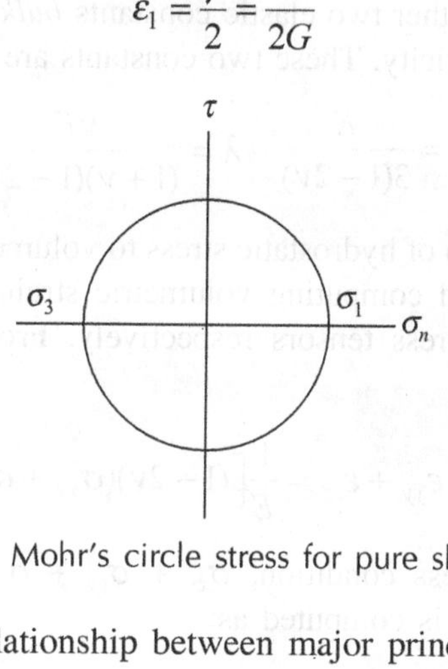

FIGURE 5.3 Mohr's circle stress for pure shear loading.

Again, based on Eq. 5.5, relationship between major principal stress and minor principal strain can be expressed as:

$$\varepsilon_1 = \frac{1}{E}(\sigma_1 - \nu\sigma_3) \tag{5.8}$$

Inserting Eqs. (5.7a and 5.7b) into Eq. 5.8 and rearranging, we find

$$G = \frac{E}{2(1+\nu)} \tag{5.9}$$

For isotropic material, shear strains in three planes *x-y*, *y-z* and *z-x* can be computed from the shear stresses as:

$$\gamma_{xy} = \frac{\tau_{xy}}{G},\ \gamma_{yz} = \frac{\tau_{yz}}{G},\ \gamma_{zx} = \frac{\tau_{zx}}{G} \tag{5.10}$$

Now, combining Eqs. 5.5 and 5.9, the general Hooke's law relating strain and stress tensors can be written in the matrix form as:

$$\begin{Bmatrix} \varepsilon_{xx} \\ \varepsilon_{yy} \\ \varepsilon_{zz} \\ \gamma_{xy} \\ \gamma_{yz} \\ \gamma_{zx} \end{Bmatrix} = \frac{1}{E}\begin{bmatrix} 1 & -\nu & -\nu & 0 & 0 & 0 \\ -\nu & 1 & -\nu & 0 & 0 & 0 \\ -\nu & -\nu & 1 & 0 & 0 & 0 \\ 0 & 0 & 0 & 2(1+\nu) & 0 & 0 \\ 0 & 0 & 0 & 0 & 2(1+\nu) & 0 \\ 0 & 0 & 0 & 0 & 0 & 2(1+\nu) \end{bmatrix}\begin{Bmatrix} \sigma_{xx} \\ \sigma_{yy} \\ \sigma_{zz} \\ \tau_{xy} \\ \tau_{yz} \\ \tau_{zx} \end{Bmatrix} \tag{5.11}$$

or

$$\boldsymbol{\varepsilon} = \mathbf{C}\boldsymbol{\sigma}$$

Here, the matrix **C** is called the *compliance matrix* of the material and contains only two material constants as *E* and *ν*. The compliance matrix is symmetric and contains 9 non-zero position in the upper-half of the diagonal. The interaction between normal strain tensors and shear stress tensors or between shear strain tensors and normal stress tensors are neglected.

Apart from, *E*, *ν* and *G*, another two elastic constants *bulk modulus*, *K* and *Lame constant*, *λ* are also used in theory of elasticity. These two constants are related with *E* and *ν* as follows:

$$K = \frac{E}{3(1-2\nu)},\ \lambda = \frac{\nu E}{(1+\nu)(1-2\nu)} \tag{5.12}$$

The bulk modulus is the ratio of hydrostatic stress to volumetric strain. The above expression of bulk modulus can be derived computing volumetric strain and hydrostatic stress in terms of normal strain and normal stress tensors respectively. From Eq. 5.5, volumetric strain is obtained as:

$$\varepsilon_v = \varepsilon_{xx} + \varepsilon_{yy} + \varepsilon_{zz} = \frac{1}{E}\left[(1-2\nu)(\sigma_{xx} + \sigma_{yy} + \sigma_{zz})\right] \tag{5.13}$$

Considering hydrostatic stress condition, $\sigma_{xx} + \sigma_{yy} + \sigma_{zz} = 3p$, where *p* is the average normal stress, bulk modulus, *K* is computed as:

$$K = \frac{p}{\varepsilon_v} = \frac{E}{3(1-2\nu)} \tag{5.14}$$

The relationships between *E*, *K*, *G* and *λ* are given in Table 5.1. Stresses in terms of strains, are obtained by inverting the matrix **C**, and then the relationship between stress and strain tensors becomes

$$\begin{Bmatrix} \sigma_{xx} \\ \sigma_{yy} \\ \sigma_{zz} \\ \tau_{xy} \\ \tau_{yz} \\ \tau_{zx} \end{Bmatrix} = \frac{E}{(1+\nu)(1-2\nu)} \begin{bmatrix} 1-\nu & \nu & \nu & 0 & 0 & 0 \\ \nu & 1-\nu & \nu & 0 & 0 & 0 \\ \nu & \nu & 1-\nu & 0 & 0 & 0 \\ 0 & 0 & 0 & \frac{(1-2\nu)}{2} & 0 & 0 \\ 0 & 0 & 0 & 0 & \frac{(1-2\nu)}{2} & 0 \\ 0 & 0 & 0 & 0 & 0 & \frac{(1-2\nu)}{2} \end{bmatrix} \begin{Bmatrix} \varepsilon_{xx} \\ \varepsilon_{yy} \\ \varepsilon_{zz} \\ \gamma_{xy} \\ \gamma_{yz} \\ \gamma_{zx} \end{Bmatrix} \quad (5.15)$$

or

$$\boldsymbol{\sigma} = \mathbf{D}\boldsymbol{\varepsilon}$$

The matrix **D** is called the *elasticity* or *constitutive matrix* of the material. This constitutive matrix is also symmetrical and has 9 non-zero terms.

The relationship between stress and strain tensors can also be represented using λ and G as

$$\begin{Bmatrix} \sigma_{xx} \\ \sigma_{yy} \\ \sigma_{zz} \\ \tau_{xy} \\ \tau_{yz} \\ \tau_{zx} \end{Bmatrix} = \lambda \varepsilon_v \begin{Bmatrix} 1 \\ 1 \\ 1 \\ 0 \\ 0 \\ 0 \end{Bmatrix} + 2G \begin{Bmatrix} \varepsilon_{xx} \\ \varepsilon_{yy} \\ \varepsilon_{zz} \\ \gamma_{xy}/2 \\ \gamma_{yz}/2 \\ \gamma_{zx}/2 \end{Bmatrix} \quad (5.16)$$

TABLE 5.1 Conversion of Elastic Parameters

Known quantity	*Derived quantity*
E, ν	$G = \frac{E}{2(1+\nu)}, \quad K = \frac{E}{3(1-2\nu)}, \quad \lambda = \frac{\nu E}{(1+\nu)(1-2\nu)}$
G, K	$\nu = \frac{3K-2G}{2G+6K}, \quad E = \frac{9GK}{G+3K}, \quad \lambda = \frac{3K-2G}{2}$

Based on the above discussions, the complete stress-strain relationships can be expressed as:

$$\begin{aligned} \boldsymbol{\sigma} &= \mathbf{D}(\boldsymbol{\varepsilon} - \boldsymbol{\varepsilon}_0) + \boldsymbol{\sigma}_0 \\ \boldsymbol{\varepsilon} &= \mathbf{C}(\boldsymbol{\sigma} - \boldsymbol{\sigma}_0) + \boldsymbol{\varepsilon}_0 \end{aligned} \quad (5.17)$$

where $\boldsymbol{\sigma}_0$ and $\boldsymbol{\varepsilon}_0$ are vectors which contain initial stress and strain tensors respectively. Unless specified, for most of the cases, they are considered to be zero or null vector.

Similarly, in polar coordinate system, stress-strain relationships are expressed as:

$$\varepsilon_{rr} = \frac{1}{E}(\sigma_{rr} - \nu\sigma_{\theta\theta}) \quad (5.18a)$$

$$\varepsilon_{\theta\theta} = \frac{1}{E}(\sigma_{\theta\theta} - \nu\sigma_{rr}) \tag{5.18b}$$

$$\gamma_{r\theta} = \frac{2(1+\nu)}{E}\tau_{r\theta} \tag{5.18c}$$

5.3.1 Relationship between Principal Stresses and Principal Strains

Since the material constants do not change with the direction, the relationship between principal stresses and principal strains is straightforward. Recalling shear stresses do not exist in principal plane, the compliance matrix and constitutive matrix can be obtained by truncating Eqs. 5.11 and 5.15 respectively as:

$$\begin{Bmatrix} \varepsilon_1 \\ \varepsilon_2 \\ \varepsilon_3 \end{Bmatrix} = \frac{1}{E}\begin{bmatrix} 1 & -\nu & -\nu \\ -\nu & 1 & -\nu \\ -\nu & -\nu & 1 \end{bmatrix}\begin{Bmatrix} \sigma_1 \\ \sigma_2 \\ \sigma_3 \end{Bmatrix} \tag{5.19}$$

$$\begin{Bmatrix} \sigma_1 \\ \sigma_2 \\ \sigma_3 \end{Bmatrix} = \frac{E}{(1+\nu)(1-2\nu)}\begin{bmatrix} 1-\nu & \nu & \nu \\ \nu & 1-\nu & -\nu \\ \nu & \nu & 1-\nu \end{bmatrix}\begin{Bmatrix} \varepsilon_1 \\ \varepsilon_2 \\ \varepsilon_3 \end{Bmatrix} \tag{5.20}$$

EXAMPLE 5.3 Refer to Example 4.6 and calculate the principal strains. Assume isotropic rock with E = 20 GPa and ν = 0.25.

Solution: Since in the principal stress planes, no shear stress exists, Eq. 5.19 can be written as:

$$\begin{Bmatrix} \varepsilon_1 \\ \varepsilon_2 \\ \varepsilon_3 \end{Bmatrix} = \frac{1}{E}\begin{bmatrix} 1 & -\nu & -\nu \\ -\nu & 1 & -\nu \\ -\nu & -\nu & 1 \end{bmatrix}\begin{Bmatrix} \sigma_1 \\ \sigma_2 \\ \sigma_3 \end{Bmatrix}$$

or

$$\begin{Bmatrix} \varepsilon_1 \\ \varepsilon_2 \\ \varepsilon_3 \end{Bmatrix} = \frac{1}{20,000}\begin{bmatrix} 1 & -0.25 & -0.25 \\ -0.25 & 1 & -0.25 \\ -0.25 & -0.25 & 1 \end{bmatrix}\begin{Bmatrix} 37.83 \\ 20.64 \\ 4.64 \end{Bmatrix}$$

Hence, $\varepsilon_1 = 1.575\times10^{-3}, \varepsilon_2 = 5.011\times10^{-4}, \varepsilon_3 = -4.988\times10^{-4}$

Here the major and intermediate principal strains are compressive but minor principal strain is tensile.

5.3.2 Stress-Strain Relationship in Two Dimensions

Stress analysis in two dimensions is an approximation of three-dimensional cases. It is performed by analysing stresses in a suitable section of a 3D problem. In mining, civil or geotechnical engineering problems, three types of approximations, *plane stress, plane strain* and *axisymmetric* conditions are assumed to represent a 3D problem in a plane. In the following, they are described in details.

Plane Stress Conditions

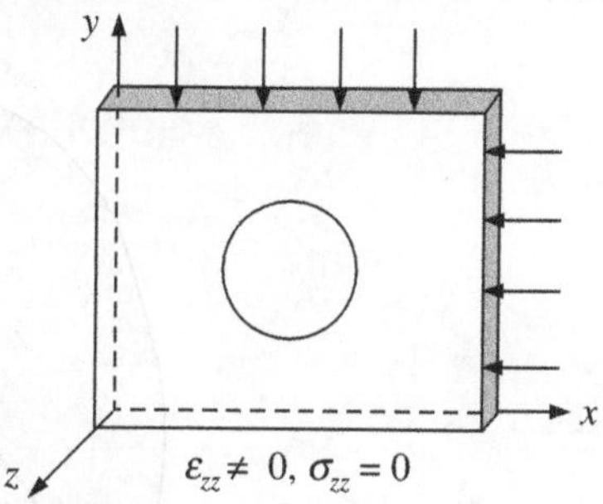

FIGURE 5.4 Plane stress condition.

In this case, loads are only applied to *x-y* plane and *z*-direction is not restrained so that $\sigma_{zz} = \tau_{yz} = \tau_{zx} = 0$. Examples of this type of analysis are thin pipe, hole in a thin plate and others under biaxial stress field. An example of biaxial stress field applied to a thin plate having circular hole is shown in Figure 5.4. For the analysis of stresses around underground excavations, plane stress analysis is not generally considered since the length of excavation is much greater as compared to the excavation width or height. Normal strain in out-of-plane or *z*-direction does exist and is computed using stresses σ_{xx} and σ_{yy}. The compliance matrix can be obtained by simply omitting appropriate locations of Eq. 5.11 as:

$$\begin{Bmatrix} \varepsilon_{xx} \\ \varepsilon_{yy} \\ \gamma_{xy} \\ \varepsilon_{zz} \end{Bmatrix} = \frac{1}{E} \begin{bmatrix} 1 & -\nu & 0 & -\nu \\ -\nu & 1 & 0 & -\nu \\ 0 & 0 & 2(1+\nu) & 0 \\ -\nu & -\nu & 0 & 1 \end{bmatrix} \begin{Bmatrix} \sigma_{xx} \\ \sigma_{yy} \\ \tau_{xy} \\ 0 \end{Bmatrix} \tag{5.21}$$

Note that in the above equation, out-of-plane tensor is placed at the last. The constitutive matrix for plane stress condition is then obtained as:

$$\begin{Bmatrix} \sigma_{xx} \\ \sigma_{yy} \\ \tau_{xy} \\ 0 \end{Bmatrix} = \frac{E}{(1+\nu)(1-2\nu)} \begin{bmatrix} 1-\nu & \nu & 0 & \nu \\ \nu & 1-\nu & 0 & \nu \\ 0 & 0 & \frac{(1-2\nu)}{2} & 0 \\ \nu & \nu & 0 & 1-\nu \end{bmatrix} \begin{Bmatrix} \varepsilon_{xx} \\ \varepsilon_{yy} \\ \gamma_{xy} \\ \varepsilon_{zz} \end{Bmatrix} \tag{5.22}$$

From, Eq. 5.21, it can be easily shown that the normal strain in the out-of-plane direction is

$$\varepsilon_{zz} = -\frac{\nu}{1-\nu}(\varepsilon_{xx} + \varepsilon_{yy}) = -\frac{\nu}{E}(\sigma_{xx} + \sigma_{yy}) \tag{5.23}$$

Plane Strain Conditions

Around a long tunnel or mine gallery, deformation along *z*-direction is assumed to be zero except at the end boundaries (Figure 5.5). Displacement fields, *u* and *v* in *x-y* plane are dependent on *x* and *y* coordinates only. Since it is assumed that no deformation exists in the out-of-plane direction, i.e., along *z*-axis, $\varepsilon_{zz} = \gamma_{yz} = \gamma_{zx} = 0$. Then the constitutive matrix is obtained by truncating Eq. 5.15 as:

$$\begin{Bmatrix} \sigma_{xx} \\ \sigma_{yy} \\ \tau_{xy} \\ \sigma_{zz} \end{Bmatrix} = \frac{E}{(1+\nu)(1-2\nu)} \begin{bmatrix} 1-\nu & \nu & 0 & \nu \\ \nu & 1-\nu & 0 & \nu \\ 0 & 0 & \frac{(1-2\nu)}{2} & 0 \\ \nu & \nu & 0 & 1-\nu \end{bmatrix} \begin{Bmatrix} \varepsilon_{xx} \\ \varepsilon_{yy} \\ \gamma_{xy} \\ 0 \end{Bmatrix} \tag{5.24}$$

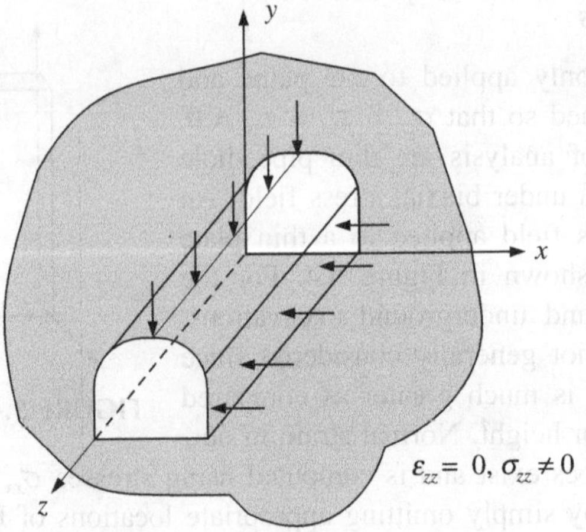

FIGURE 5.5 Plane strain condition.

As before, note that out-of-plane component is placed at the last row and column. The compliance matrix is then obtained as follows:

$$\begin{Bmatrix} \varepsilon_{xx} \\ \varepsilon_{yy} \\ \gamma_{xy} \\ 0 \end{Bmatrix} = \frac{1}{E}\begin{bmatrix} 1 & -\nu & 0 & -\nu \\ -\nu & 1 & 0 & -\nu \\ 0 & 0 & 2(1+\nu) & 0 \\ -\nu & -\nu & 0 & 1 \end{bmatrix}\begin{Bmatrix} \sigma_{xx} \\ \sigma_{yy} \\ \tau_{xy} \\ \sigma_{zz} \end{Bmatrix} \tag{5.25}$$

From Eq. 5.24, it can be easily shown that the out-of-plane stress which develops in z-direction is

$$\sigma_{zz} = \nu(\sigma_{xx} + \sigma_{yy}) = \lambda(\varepsilon_{xx} + \varepsilon_{yy}) \tag{5.26}$$

EXAMPLE 5.4 Show that ratio of uniform horizontal stress to vertical stress at a point below the ground equals to $\dfrac{\nu}{1-\nu}$.

Solution: Assume that the horizontal stress in x and z directions at a point below the ground is σ_h and the vertical stress in y-direction at the same point equals to σ_v. Considering plane strain condition, we can write from Eq. 5.5(c) as:

$$\varepsilon_{zz} = \frac{1}{E}\left[\sigma_{zz} - \nu(\sigma_{xx} + \sigma_{yy})\right] = 0$$

Hence, $\sigma_h - \nu(\sigma_h + \sigma_v) = 0$ and $\dfrac{\sigma_h}{\sigma_v} = \dfrac{\nu}{1-\nu}$

EXAMPLE 5.5 A parallelepiped rock sample is loaded as shown in the figure. Determine the 2D strain tensors and out-of-plane normal stress considering (i) w = 1 cm, and (ii) 100 cm. Assume E and ν of rock are 10 GPa and 0.25 respectively.

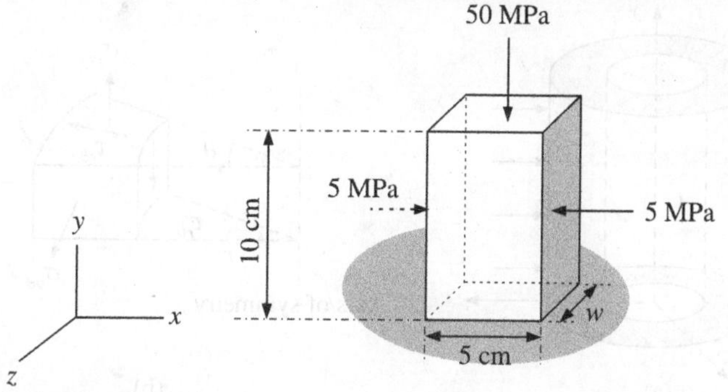

Solution:

(i) Considering z as out-of-plane axis, for w =1 cm, the problem can be approximated to plane stress condition. Using Eq. 5.21, strain tensors can be obtained as:

$$\begin{Bmatrix} \varepsilon_{xx} \\ \varepsilon_{yy} \\ \gamma_{xy} \\ \varepsilon_{zz} \end{Bmatrix} = \frac{1}{10000}\begin{bmatrix} 1 & -0.25 & 0 & -0.25 \\ -0.25 & 1 & 0 & -0.25 \\ 0 & 0 & 2(1+0.25) & 0 \\ -0.25 & -0.25 & 0 & 1 \end{bmatrix}\begin{Bmatrix} 5 \\ 50 \\ 0 \\ 0 \end{Bmatrix} = \begin{Bmatrix} -7.5 \\ 48.75 \\ 0.0 \\ -13.75 \end{Bmatrix} \times 10^{-4}$$

Since plane stress condition is assumed, hence the out-of-plane stress is zero.

(ii) For w =10 cm, the problem can be approximated to plane strain condition. Using Eq. 5.25, strain tensors can be obtained as:

$$\begin{Bmatrix} \varepsilon_{xx} \\ \varepsilon_{yy} \\ \gamma_{xy} \\ 0 \end{Bmatrix} = \frac{1}{10000}\begin{bmatrix} 1 & -0.25 & 0 & -0.25 \\ -0.25 & 1 & 0 & -0.25 \\ 0 & 0 & 2(1+0.25) & 0 \\ -0.25 & -0.25 & 0 & 1 \end{bmatrix}\begin{Bmatrix} 5 \\ 50 \\ 0 \\ 13.75 \end{Bmatrix} = \begin{Bmatrix} -10.9375 \\ 44.8125 \\ 0.0 \\ 0.0 \end{Bmatrix} \times 10^{-4}$$

The out-of-plane stress is found by Eq. 5.26 as 13.75 MPa.

Axisymmetric Conditions

Axisymmetric analysis is performed for structures that are rotationally symmetric about an axis. Examples of axisymmetric structure in geotechnical engineering are circular tunnels and shafts, tunnel lining, timber supports and others. If the loading on the structure is symmetric as shown in Figure 5.6(a), a two-dimensional analysis can be performed considering a unit radian of the structure. Here, y-axis is considered to be axis of symmetry. Stress notations is given in polar coordinate system as:

$$\sigma^T = \{\sigma_{rr} \quad \sigma_{\theta\theta} \quad \tau_{r\theta} \quad \sigma_{yy}\}$$

where, σ_{rr} = radial stress, $\sigma_{\theta\theta}$ = tangential or hoop stress, $\tau_{r\theta}$ = shear stress and σ_{yy} = longitudinal stress.

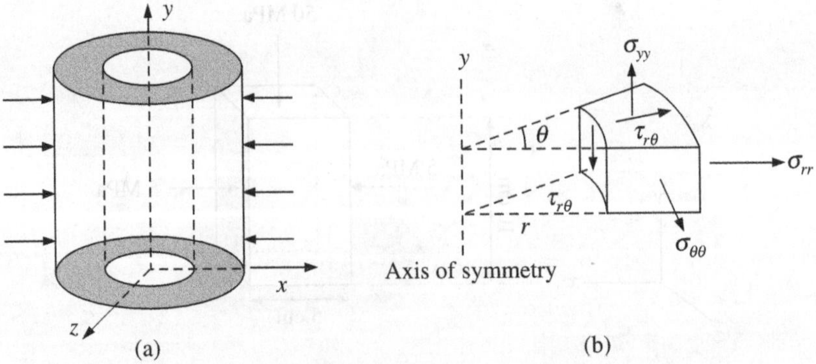

FIGURE 5.6 Axisymmetric condition.

Symmetrical loads are applied in radial or longitudinal directions. If the loading on the structure is non-symmetrical then the choice of analysis lies between fully three-dimensional, in which substructuring or cyclic symmetry are used and a Fourier decomposition of the loads with a Fourier axisymmetric solution (Bathe, 1982). In this text, we will discuss axisymmetric analysis with symmetric loading condition only.

Consider the stress conditions of a small element as shown in Figure 5.6(b). Let us assume the radial and longitudinal displacement fields are u_r and v respectively. Then strain tensors can be obtained as:

$$\varepsilon_{rr} = \frac{\partial u_r}{\partial r}, \quad \varepsilon_{\theta\theta} = \frac{2\pi(r+u_r) - 2\pi r}{2\pi r} = \frac{u_r}{r}, \quad \gamma_{r\theta} = \frac{\partial u_r}{\partial y} + \frac{\partial v}{\partial r}, \quad \varepsilon_{yy} = \frac{\partial v}{\partial y} \tag{5.27}$$

It may be noted that for axisymmetric condition, θ dependency is ignored in strain tensors expressed in polar coordinate system Eq. 4.16. For linear elastic material, the constitutive relations between stress and strains are given as:

$$\begin{Bmatrix} \sigma_{rr} \\ \sigma_{\theta\theta} \\ \tau_{r\theta} \\ \sigma_{yy} \end{Bmatrix} = \frac{E}{(1+\nu)(1-2\nu)} \begin{bmatrix} 1-\nu & \nu & 0 & \nu \\ \nu & 1-\nu & 0 & \nu \\ 0 & 0 & \frac{1-2\nu}{2} & 0 \\ \nu & \nu & 0 & 1-\nu \end{bmatrix} \begin{Bmatrix} \varepsilon_{rr} \\ \varepsilon_{\theta\theta} \\ \gamma_{r\theta} \\ \varepsilon_{yy} \end{Bmatrix} \tag{5.28}$$

From Eq. 5.28, compliance matrix is obtained as:

$$\begin{Bmatrix} \varepsilon_{rr} \\ \varepsilon_{\theta\theta} \\ \gamma_{r\theta} \\ \varepsilon_{yy} \end{Bmatrix} = \frac{1}{E} \begin{bmatrix} 1 & -\nu & 0 & -\nu \\ -\nu & 1 & 0 & -\nu \\ 0 & 0 & 2(1+\nu) & 0 \\ -\nu & -\nu & 0 & 1 \end{bmatrix} \begin{Bmatrix} \sigma_{rr} \\ \sigma_{\theta\theta} \\ \tau_{r\theta} \\ \sigma_{yy} \end{Bmatrix} \tag{5.29}$$

EXAMPLE 5.6 A 45° rectangular rosette is used to estimate *in situ* stresses in a rock. The normal strains obtained as shown in the figure below. If the modulus of elasticity and Poisson's ratio of that rock is 5 GPa and 0.25 respectively, determine the major, and minor principal stresses. Assume plane strain condition.

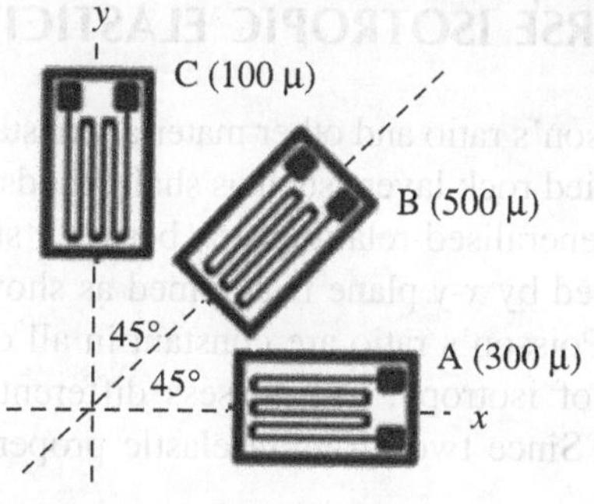

Solution: Given data are:

$$\varepsilon_{45} = 500\mu = \varepsilon_{x'x'},\ \ \varepsilon_{yy} = 100\mu,\ \ \varepsilon_{xx} = 300\mu,\ \ E = 5\text{ GPa},\ \ \nu = 0.25$$

We know,
$$\varepsilon_{x'x'} = \frac{\varepsilon_{xx} + \varepsilon_{yy}}{2} + \frac{\varepsilon_{xx} - \varepsilon_{yy}}{2}\cos 2\theta + \frac{\gamma_{xy}}{2}\sin 2\theta$$

Letting the values in the above equation, we find

$$500 = \frac{300 + 100}{2} + \frac{300 - 100}{2}\cos 90^\circ + \frac{\gamma_{xy}}{2}\sin 90^\circ$$

$$\therefore \qquad \gamma_{xy} = 600\mu$$

Considering plane strain condition and using Eq. 5.24, we have

$$\begin{Bmatrix} \sigma_{xx} \\ \sigma_{yy} \\ \tau_{xy} \\ \sigma_{zz} \end{Bmatrix} = \frac{E}{(1+\nu)(1-2\nu)} \begin{bmatrix} 1-\nu & \nu & 0 & \nu \\ \nu & 1-\nu & 0 & \nu \\ 0 & 0 & \dfrac{(1-2\nu)}{2} & 0 \\ \nu & \nu & 0 & 1-\nu \end{bmatrix} \begin{Bmatrix} \varepsilon_{xx} \\ \varepsilon_{yy} \\ \gamma_{xy} \\ 0 \end{Bmatrix}$$

$$\begin{Bmatrix} \sigma_{xx} \\ \sigma_{yy} \\ \tau_{xy} \\ \sigma_{zz} \end{Bmatrix} = 8000 \begin{bmatrix} 0.75 & 0.25 & 0 & 0.25 \\ 0.25 & 0.75 & 0 & 0.25 \\ 0 & 0 & 0.25 & 0 \\ 0.25 & 0.25 & 0 & 0.75 \end{bmatrix} \begin{Bmatrix} 300 \\ 100 \\ 600 \\ 0 \end{Bmatrix} = \begin{Bmatrix} 2.0 \\ 1.2 \\ 1.2 \\ 0.8 \end{Bmatrix} \text{MPa}$$

Calculation of the maximum and minimum principal stress

$$\sigma_1 = \frac{\sigma_{xx} + \sigma_{yy}}{2} + \frac{1}{2}\sqrt{(\sigma_{xx} - \sigma_{yy})^2 + 4\tau_{xy}^2}$$

$$= \frac{2.0 + 1.2}{2} + \frac{1}{2}\sqrt{(2.0 - 1.2)^2 + 4(1.2)^2} = 2.86\text{ MPa}$$

Similarly, $\sigma_3 = 0.34$ MPa.

Assuming direction cosines of the normal to the plane where the major principal stress acts as l and m, $m = 0.72l$, and $l = 0.8115$. Hence, the normal makes and angle of 35.71° with x-axis measured in counter-clockwise or anti-clockwise direction.

5.4 LINEAR TRANSVERSE ISOTROPIC ELASTICITY

The modulus of elasticity, Poisson's ratio and other material constants can be different in different directions, especially for stratified rock layers such as shale, mudstone, siltstones, etc. Transverse isotropy is a special case of generalised relationships between stress and strain tensors. In this case, a plane of isotropy denoted by *x*-*y* plane is assumed as shown in Figure 5.7. In this plane, the modulus of elasticity and Poisson's ratio are constant in all directions. Along the outer axis (*z*-axis), normal to the plane of isotropy, can possess different elastic parameters than those exist in the plane of isotropy. Since two different elastic properties can exist in perpendicular

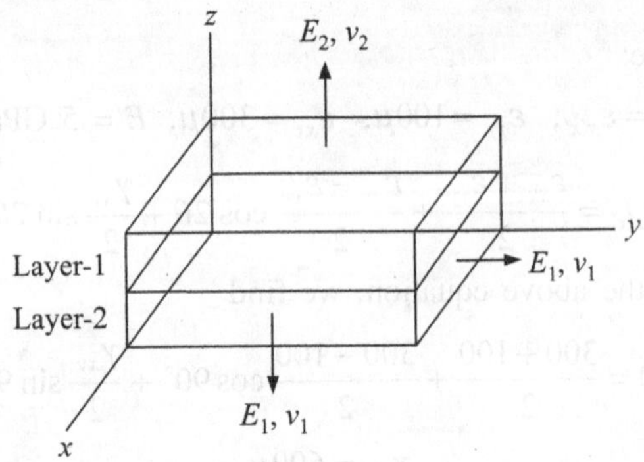

FIGURE 5.7 Transverse isotropic condition for stratified rock layers.

directions, this property of material is termed as *transverse isotropic material*. Assuming E_1, ν_1 and E_2, ν_2 are the elastic modulus and Poisson's ratio in the plane of isotropy and normal to that plane respectively, the compliance matrix can be expressed as:

$$\begin{Bmatrix} \varepsilon_{xx} \\ \varepsilon_{yy} \\ \varepsilon_{zz} \\ \gamma_{xy} \\ \gamma_{yz} \\ \gamma_{zx} \end{Bmatrix} = \frac{1}{E_1} \begin{bmatrix} 1 & -\nu_1 & -\nu_2 & 0 & 0 & 0 \\ -\nu_1 & 1 & -\nu_2 & 0 & 0 & 0 \\ -\nu_2 & -\nu_2 & E_1/E_2 & 0 & 0 & 0 \\ 0 & 0 & 0 & 2(1+\nu_1) & 0 & 0 \\ 0 & 0 & 0 & 0 & E_1/G_2 & 0 \\ 0 & 0 & 0 & 0 & 0 & E_1/G_2 \end{bmatrix} \begin{Bmatrix} \sigma_{xx} \\ \sigma_{yy} \\ \sigma_{zz} \\ \tau_{xy} \\ \tau_{yz} \\ \tau_{zx} \end{Bmatrix} \tag{5.30}$$

where G_2 is defined as the shear modulus along perpendicular direction. However, Batugin and Nirenburg (1972) found that G_2 is dependent on the elastic parameters E_1, E_2 and ν_2 for different rock types such as limestone, sandstone, granite, slatestone, phyllite and granodiorite and the relationship is given as (Kwasniewski, 1984)

$$G_2 = \frac{E_1 E_2}{E_1 + (1 + 2\nu_2) E_2} \tag{5.31}$$

It can be noted that the compliance matrix is symmetric just as in the case of linear isotropic elasticity. The constitutive matrix is reproduced from Brady and Brown (2007) and given as follows:

$$\begin{Bmatrix} \sigma_{xx} \\ \sigma_{yy} \\ \sigma_{zz} \\ \tau_{xy} \\ \tau_{yz} \\ \tau_{zx} \end{Bmatrix} = \frac{E_2}{(1+\nu_1)A} \begin{bmatrix} n(1-n\nu_2^2) & n(\nu_1+n\nu_2^2) & n\nu_2(1+\nu_1) & 0 & 0 & 0 \\ n(\nu_1+n\nu_2^2) & n(1-n\nu_2^2) & n\nu_2(1+\nu_1) & 0 & 0 & 0 \\ n\nu_2(1+\nu_1) & n\nu_2(1+\nu_1) & (1-\nu_1^2) & 0 & 0 & 0 \\ 0 & 0 & 0 & \frac{nA}{2} & 0 & 0 \\ 0 & 0 & 0 & 0 & m(1+\nu_1)\times A & 0 \\ 0 & 0 & 0 & 0 & 0 & m(1+\nu_1)\times A \end{bmatrix} \begin{Bmatrix} \varepsilon_{xx} \\ \varepsilon_{yy} \\ \varepsilon_{zz} \\ \gamma_{xy} \\ \gamma_{yz} \\ \gamma_{zx} \end{Bmatrix} \tag{5.32}$$

where $A = (1-\nu_1-2n\nu_2^2)$, $n = E_1/E_2$ and $m = E_2/G_2$. It can be noted that E_1, E_2, ν_1 and ν_2 are supposed to be independent to each other. However, the value of n and m can be restricted due to constraints of positive definiteness of the constitutive matrix (Brady and Brown, 2007).

5.5 ANISOTROPY

The degree of anisotropy of transversely isotropic material is generally denoted by the ratio of elastic moduli as:

$$k_E = \frac{E_2}{E_1} \tag{5.33}$$

However, stresses and displacements occurred in anisotropic rock mass are function of coefficients of elastic moduli and Poisson's ratio determined in various directions. Hence, the knowledge of only k_E may not be sufficient to define degree of anisotropy. Considering two-dimensional problems, Lekhnitskii (1968) suggested the parameters of anisotropy for transversely isotropic material as (Kwasniewski, 1984)

$$n = \sqrt{2k+m} \tag{5.34}$$

where $k = \sqrt{\dfrac{E_1/E_2 - \nu_2^2}{1-\nu_1^2}}$ and $m = \dfrac{E_1/G_1 - 2\nu_2(1+\nu_1)}{1-\nu_1^2}$.

For problems related to state of stresses of rock mass around cross-headings, diagonal headings, roadway openings, an acute angle in relation to the strike of rock strata occurs. Due to this reason, an additional parameter l is defined by the following equation:

$$l = \sqrt{\frac{E_1}{2(1+\nu_1)G_2}} \tag{5.35}$$

The degree of anisotropy of transversely isotropic material is then estimated by the deviation of values of the above mentioned parameters namely k, n and l from their isotropic values of

$k = 1$, $n = 2$ and $l = 1$. The classification of transverse isotropy is given based on the value of parameter n and is shown in Table 5.2 (Kwasniewski, 1984).

TABLE 5.2 Classification of Anisotropy

Parameter, n	*Class of anisotropy*
<2.1	Quasi-isotropy
Between 2.1 and 2.5	Poor anisotropy
Between 2.5 and 3.0	Moderate anisotropy
>3.0	Strong anisotropy

5.6 NON-LINEAR ELASTICITY

The major difference between linear and non-linear elasticity is that for later case, the relationship between stresses and strains are not linear in the elastic regime. Non-linear elasticity can be best represented by the schematic diagram as shown in Figure 5.8. The loading curve is non-linear depicting stress is a polynomial function of strain. The unloading curve may not coincide with the loading curve and thus exhibits hysteresis effect which corresponds to the imperfect elastic behaviour. In general, non-linear elastic behaviour is not a common property for rocks but certain hard rock may show this type of behaviour (Jeremic, 1985).

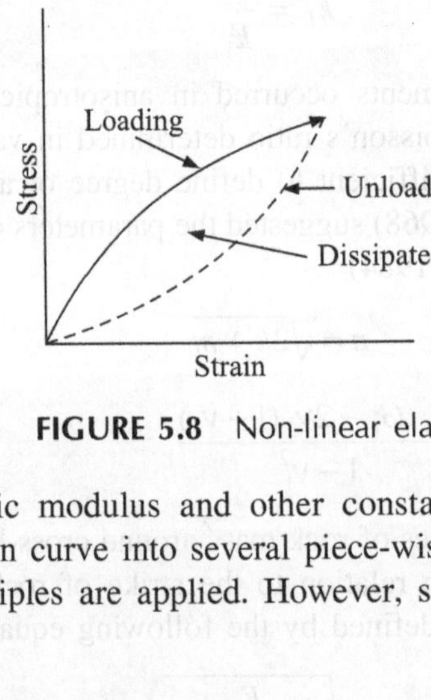

FIGURE 5.8 Non-linear elasticity.

Determination of elastic modulus and other constants are performed by dividing the hysteresis of the stress-strain curve into several piece-wise linear divisions and then for each division linear elastic principles are applied. However, superposition strain theory cannot be applied in such cases.

5.7 PLASTICITY

Plasticity results from permanent deformation or strain in material causing micro- to macro-cracks. Let us revisit the case of uniaxial compression loading on a rock specimen as described in Figure 5.1. At the initial stages of compressive loading test, most rocks show linear elastic constitutive behaviour at least up to 50% of yield load. As the loading increases, material departs from elastic property once the applied stress exceeds the yield strength of the material. Micro- or macro-cracks develop in the rock specimen causing irrecoverable strains. Gramberg commented that under uniaxial compressive loading, micro-cracks initiate in rock samples at about 50% of yield load and secondary fracture phenomena develops after 75% of yield load (Gramberg, 1989). In another study, it is found that micro-cracks initiates at 50% of peak load if sample is tested in triaxial condition. Crack damage stress occurs at about 80% of the peak stress.

In the plastic domain, stress, σ and strain, ε are no longer related by matrix of constants such as constitutive matrix, **D**. The relationship is based on rather strain-dependent constitutive matrix. The computational difficulty in plastic domain is that equilibrium equations must be formulated using material properties that depend on strains, but strains are not known is advance (Cook et al., 1989). For this reason, incremental theory of stress and strain is adopted. The strain increment is divided into two parts: elastic and plastic strain increments as given below:

$$d\varepsilon^t = d\varepsilon^e + d\varepsilon^p \tag{5.36}$$

Here the superscripts *t*, *e* and *p* represent total, elastic and plastic cases respectively. Elastic strain component is responsible for incrementing the stress and thus the relationship between stress and strain increment can be expressed as

$$d\sigma = \mathbf{D}d\varepsilon^e = \mathbf{D}(d\varepsilon^t - d\varepsilon^p) \tag{5.37}$$

In elastic domain, plastic strain increment is zero and thus total strain increment is equal to the elastic strain increment only. In the plastic domain, both elastic and plastic strain increments exist except for ideal plastic case where elastic component of strain increment is zero.

5.7.1 Perfect Plasticity

Figure 5.9 shows the diagram of idealised plastic behaviour of the material for one-dimensional loading condition. Once the stress exceeds the yield strength, no stress increment occurs in the material while plastic strain continues to develop. For ideal plasticity problem, elastic strain increment is zero and hence no stress increment occurs in the plastic domain. If the applied load is removed at point A, the original geometry cannot be recovered due to permanent strain developed on the specimen.

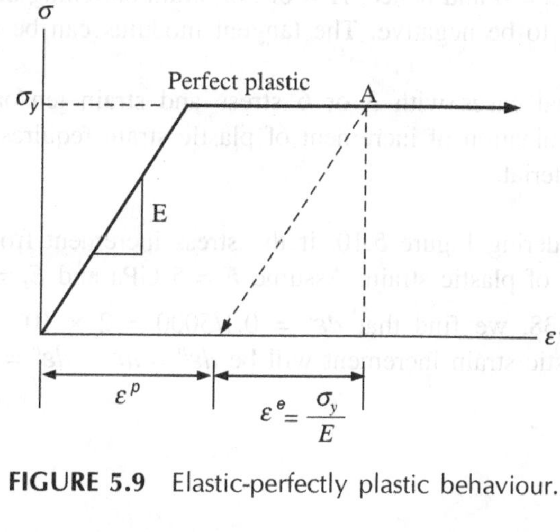

FIGURE 5.9 Elastic-perfectly plastic behaviour.

5.7.2 Elastoplasticity

Material behaviour in the plastic regime is shown in Figure 5.1. In this section, relationship between stress and strain is illustrated for strain hardening behaviour due to one-dimensional loading condition as shown in Figure 5.10. For simplicity, linear stress-strain relationship is considered in the plastic region. Let us assume that in the plastic region, stress is increased from point A to point B giving stress increment, $d\sigma$. Then $d\sigma$ can be expressed in different forms as follows:

$$d\sigma = Ed\varepsilon^e = E_t d\varepsilon^t = Hd\varepsilon^p \tag{5.38}$$

Here, E_t and H are called ***tangent modulus*** and ***strain-hardening parameter*** respectively. By manipulating Eq. 5.37, it can be written as:

$$H = \frac{E_t}{1 - \frac{E_t}{E}} \quad \text{or} \quad E_t = E\left(1 - \frac{E}{E + H}\right) \tag{5.39}$$

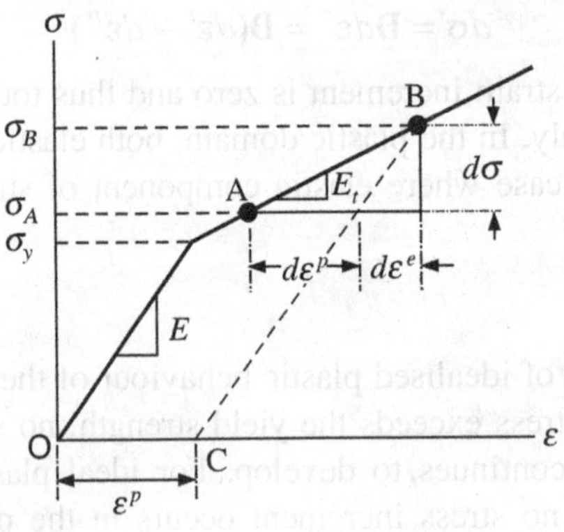

FIGURE 5.10 Elastic-plastic stress-strain behaviour in one dimension.

For ideal plasticity, $E_t = 0$ and hence, $H = 0$. For strain-softening behaviour, the hardening parameter H is assumed to be negative. The tangent modulus can be obtained from uniaxial test of rock samples.

For multi-dimensional cases with 3 or 6 stress and strain tensors, similar approach is applied. However, determination of increment of plastic strain requires special assumption on the behaviour of the material.

EXAMPLE 5.7 Considering Figure 5.10, if the stress increment from A to B is 0.1 MPa, determine the increment of plastic strain. Assume E = 5 GPa and E_t = 1.8 GPa.

Solution: From Eq. 5.38, we find that $d\varepsilon^e = 0.1/5000 = 2 \times 10^{-5}$ and $d\varepsilon^t = 0.1/1800 = 5.55 \times 10^{-5}$. Hence, plastic strain increment will be $d\varepsilon^p = d\varepsilon^t - d\varepsilon^e = 3.55 \times 10^{-5}$.

EXERCISES

1. A rock sample is tested in uniaxial compression by connecting strain gages in axial and lateral directions. The results of the test are: σ_c = 40 MPa, E = 5.7 GPa and ν = 0.28. Determine the volumetric strain at the time of failure of the sample.
2. Considering the results of Exercise 13 of Chapter 4, determine the stress tensors if the modulus of elasticity and Poisson's ratio of the plate are 5 GPa and 0.25 respectively.
3. Considering the results of Exercise 14 of Chapter 4, determine the stress tensors if the modulus of elasticity and Poisson's ratio of the plate are 5 GPa and 0.25 respectively.
4. Show that for linear isotropic material the parameters k = 1, n = 2 and l = 1 as given in Eqs. 5.33 and 5.34.
5. A copper core (density ρ_c, cross-sectional area A_c) electrical cable is surrounded co-axially by a steel sheath (density ρ_s, cross-sectional area A_s) so that they act as composite bar. The cable hangs vertically downward in a mineshaft. Prove that the maximum axial stress in copper core for l m of cable length is

$$\sigma_c = \frac{E_c(\rho_c A_c + \rho_s A_s)gl}{(A_c E_c + A_s E_s)}$$

6. If $G_1 = 2G_2$, $\nu_1 = 0.5$ and $\nu_2 = 0.2$, classify the degree of anisotropy.
7. Measurement of principal stresses is conducted along a vertical plane with 60° rosette. The normal strains are obtained as given in Figure below. Determine the magnitude of the principal stresses and the direction of major principal stress with the x axis. Assume that the modulus of elasticity and Poisson's ratio of rock are 5 GPa and 0.25 respectively. Consider plane strain condition.

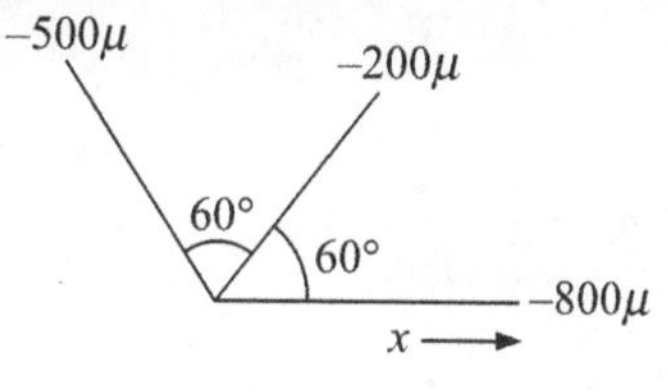

8. In Figure 5.10, assume that σ_A = 41.9 MPa, σ_B = 45.2 MPa and σ_Y = 42 MPa. Determine the elastic and plastic strains when load is increased from A to B. Assume E = 40 GPa and E_t = 5.5 GPa.
9. The line PQR signifies the axial stress-strain behaviour of a material as given in the Figure. Determine the following quantities:
   (a) Yield strength of the material.
   (b) Elastic and plastic strain increments when stress is increased from point A to point B by 2MPa.
   (c) Strain Hardening Parameter.

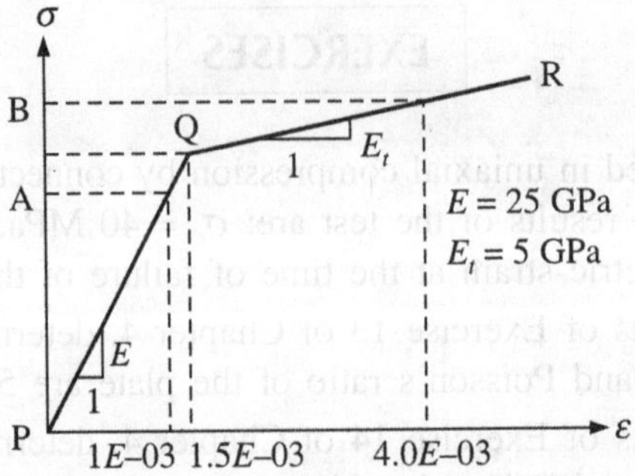

10. A rock sample of 60 mm diameter is subjected to axial load as shown in the figure. Ignoring the Poisson effect, determine the elongation or contraction of the rock sample. Assume modulus of elasticity is 5.0 GPa.

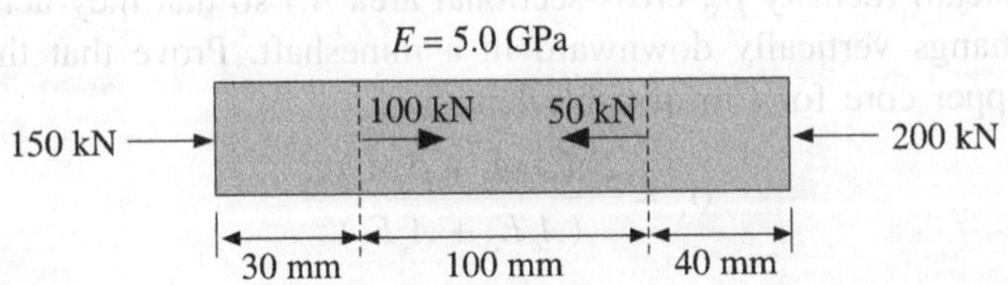

11. Uniaxial compressive strength of a rock sample is tested to be 58 MPa at 0.82% axial and 0.28% lateral strains. Determine the Poisson's ratio, modulus of rigidity and volumetric strain.

12. If the inequality $E > G > K$ is true, what is the range of Poisson's ratio?

Chapter 6

Measurement of *In situ* Stresses and Deformation

6.1 INTRODUCTION

Determination of stress fields is fundamental to rock mechanics as it provides means to analyse the mechanical behaviour of rock, serves as boundary condition in rock-engineering problems, helps to understand the groundwater flow, and at large scales, sheds light on the mechanisms causing tectonic plates those move and faults those rupture (Amini et al., 2009). The state of stress in terms of three principal stresses in subsurface varies as a function of depth. The values of three principal stresses and their associated directions do not generally follow any pattern as the depth increases. These stresses are believed to be influenced by the topography, tectonic forces, constitutive behaviour of rock, and local geological history. Complex relationship between these factors and the *in situ* stress (geostatic, tectonic, frozen, structural, thermal) usually prohibits reliable estimation of rock stress. In addition, stress cannot be measured directly without disturbing the rock at locations of measurement, and therefore, rock stress determination techniques rely on the measurement of responses (e.g., displacement, strain, deformation) those are induced by a disturbance of the rock mass. Therefore, the measured response of rock includes both *in situ* and *induced* stresses. Hence, it is important to disturb the location of measurement to its lowest. Majority of the stress measuring technique advocates the measurement of *in situ* stresses in drill holes of minimum diameter that extends into an undisturbed region inside the rock mass. This chapter discusses the measurement of stresses and deformation in *in situ* and *induced* conditions.

6.2 DEFINITION OF *IN SITU* AND INDUCED STRESSES

6.2.1 *In situ* Stress

The original, undisturbed state of stress in the rock mass is called *in situ stress*. This is simply the natural state of stress (ISRM, 1987) and are cumulative product of events from the geological history of the rock, e.g., gravitational, tectonic, residual, and terrestrial stresses (Amadei and Stephansson, 1997).

6.2.2 Geostatic Stress

The Swiss geologist Heim (1878) first proposed that vertical stress component at a depth below ground is related to the weight of the overlaying rock mass (Jaeger, 1979). He also proposed that horizontal stress may be the same as the vertical stress. Later Terzaghi (1952) derived that horizontal stress is also related to Poisson's ratio and is discussed below.

Below ground surface the geostatic stress is developed due to gravitational loading and lateral constraints to a rock. It has been observed that in general, the *in situ* vertical stress increases linearly with the depth. However, the ratio of the horizontal stress to the vertical stress decreases as the depth increases. In general, for an average unit weight of overburden rock, γ, vertical and horizontal stresses at a depth of H are denoted as (derived in Example 5.4).

$$\sigma_v = \gamma H, \quad \text{and} \quad \sigma_h = \frac{\nu}{1-\nu}\sigma_v \tag{6.1}$$

However, it is also found that the horizontal to vertical stress ratio can be as high as 5.0 near the surface and it becomes close to unity after around a depth of 1500 m (Herget, 1988; Hoek, 2007) as shown in Figure 6.1.

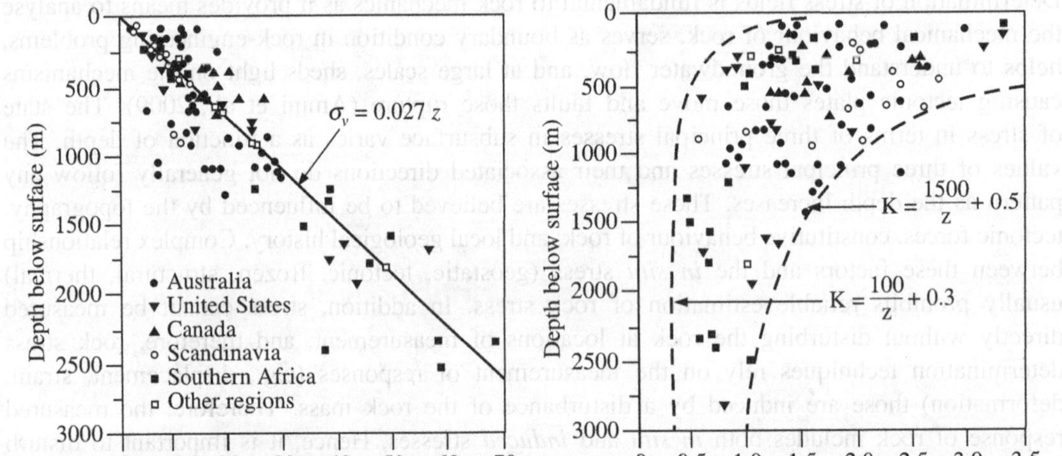

FIGURE 6.1 Profile of (a) *In situ* vertical, and (b) Horizontal stress with depth (Brown and Hoek, 1978).

6.2.3 Tectonic Stress

Plate tectonics deals with dynamic movement of Earth's crust. Plates moves relative to each other and interact at their boundaries causing convergence, divergence and slips (Palmstrom and Stille, 2010). Tectonic stresses can be active to cause earthquake, folding and faulting which are developed due to sudden or gradual release of high energy at the boundary of the lithospheric plates of the earth's crust. However, the passive tectonic stress consists of orogenic forces which may impose higher vertical and horizontal stresses.

6.2.4 Residual Stress

Residual stresses are 'self-equilibrating stresses' that remain in a structure if external forces and moments are removed (Voight, 1966). In rock mechanics, they are also called internal or 'locked-in' stresses. The presence of residual stresses and strains leads to the existence of internal residual strain energy which can be critical for the stability of underground openings and surface excavations in rock. Residual stresses are believed to be responsible (in part) for phenomena such as rockbursts, rock surface spalling and sheet jointing (Varnes, 1970) and time-dependent movements of excavations such as those encountered in the Niagara Falls area (Lo et al., 1975).

6.2.5 Frozen Stress

Frozen stresses are usually found in coal basin and occur due to erosion or glaciation process. In a deep basin, a hydrostatic stress condition may exit once the coal measure strata are formed. Due to erosion, some part of the overlaying rock strata are removed causing the reduction in vertical stress. In this process, the horizontal stress which may be as high as the original stress is locked in the coal formation because the possibility of lateral expansion is far less than that of vertical erosion (Jeremic, 1985). Thus, frozen stress is also termed as locked-in-stress which is locked in the coal formation after the erosion process takes place. Due to this phenomenon, the vertical stress reduces giving higher ratio of horizontal stress to the vertical stress at a shallow depth.

Let us assume that millions of years ago, the vertical and horizontal stresses at a point below H_b (Figure 6.2) before erosion are:

$$\sigma_{vb} = \gamma H_b, \quad \text{and} \quad \sigma_{hb} = K_b \sigma_{vb} \tag{6.2}$$

where, γ is the average unit weight of overburden rock and K_b denotes the stress ratio before erosion. Now assume that due to erosion such as glacier retreat, the depth of cover of the same point becomes H_a. Hence, the vertical stress changes to

$$\sigma_{va} = \gamma H_a \tag{6.3}$$

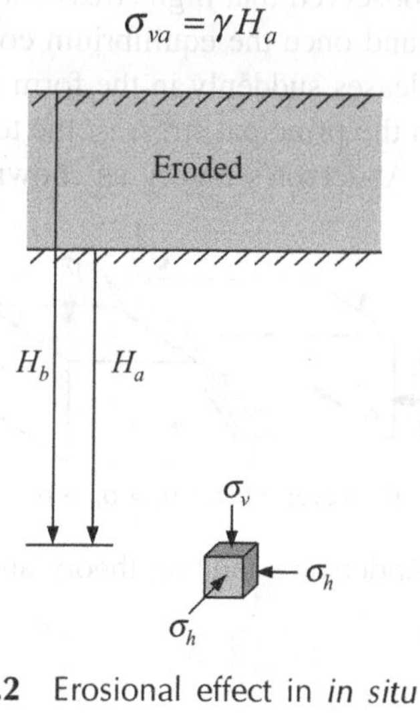

FIGURE 6.2 Erosional effect in *in situ* stresses.

Since a portion of horizontal stress can be "locked-in", it is assumed that change of horizontal stress results from the change in vertical stress. Hence, after erosion, horizontal stress remains as:

$$\sigma_{ha} = \sigma_{hb} - \Delta\sigma_h = \sigma_{hb} - \frac{\nu}{1-\nu}\Delta\sigma_v = K_b\sigma_{vb} - \frac{\nu}{1-\nu}\gamma(H_b - H_a) \tag{6.4}$$

where ν = Poisson's ratio of the overburden rock. Hence, the stress ratio after erosion will be

$$K_a = \frac{\sigma_{ha}}{\sigma_{va}} = \frac{K_b\gamma H_b - \left(\frac{\nu}{1-\nu}\right)\gamma(H_b - H_a)}{\gamma H_a} = K_b\left(\frac{H_b}{H_a}\right) - \left(\frac{\nu}{1-\nu}\right)\left(\frac{H_b}{H_a} - 1\right) \tag{6.5}$$

EXAMPLE 6.1 The magnitudes of vertical stress and horizontal stress at a depth of 2000 m were the same prior to the erosion of 1200 m of overburden. Considering Poisson's ratio and unit weight of overburden rock as 0.25 and 25 kN/m^3 respectively, estimate horizontal stress at the present location.

Solution: It is given that K_b = 1, now using Eq. 6.5, we find

$$K_a = \frac{\sigma_{ha}}{\sigma_{va}} = 1\left(\frac{2000}{800}\right) - \left(\frac{0.25}{0.75}\right)\left(\frac{2000}{800} - 1\right) = 2.0$$

Hence, present horizontal stress is $\sigma_{ha} = 2.0 \times 25.0 \times 800 = 40.0$ MPa

6.2.6 Structural Stress

There could be stresses due to structural formation of rock strata such as stresses due to faults, folds and due to overthrusted strata. Plate tectonics deals with dynamic movement of Earth's crust. Plates moves relative to each other and interact at their boundaries causing convergence, divergence and slips (Palmstrom and Stille, 2010). The stresses developed due to these structural defects are localised in nature but may cause tremendous damage once an excavation is driven near to such discontinuities. It is observed that high stress concentration occurred in the vicinity of such structural discontinuities and once the equilibrium condition is disturbed due to mining or tunnelling activities, energy releases suddenly in the form of major deformations. In general, based on the relationship between the principal stresses, the tectonic environments are classified into three fault regime based on Anderson's theory as shown in Figure 6.3.

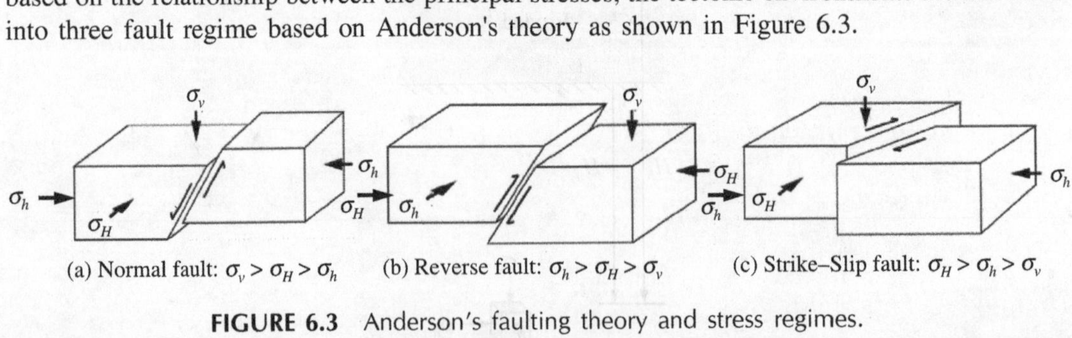

(a) Normal fault: $\sigma_v > \sigma_H > \sigma_h$ (b) Reverse fault: $\sigma_h > \sigma_H > \sigma_v$ (c) Strike–Slip fault: $\sigma_H > \sigma_h > \sigma_v$

FIGURE 6.3 Anderson's faulting theory and stress regimes.

6.2.7 Thermal Stress

They are induced by diurnal and seasonal variations of temperatures, moon pull, and the Coriolis force. A rock containing minerals with different coefficients of thermal expansion undergoes a uniform change in temperature resulting in non-uniform strains.

6.2.8 Induced Stress

The stress around an opening resulting from various man-made activities is termed "induced stress" (ISRM, 1987). They are superimposed on *in situ* stresses which exist prior to excavation. In Chapter 7, analytical solution for induced stresses due to circular and elliptical opening in an underground opening is given.

6.3 TECHNIQUES OF *IN SITU* STRESS MEASUREMENT

Over the years, several methods have been developed for estimating *in situ* stresses below ground. ISRM has recommended six methods for *in situ* stress measurement based on the measurement of induced displacements or strains in the walls of the drill hole when stresses are relieved or on the principle of fracturing the rock strata. These methods are:

(i) Stress compensation method: Flat-jack method
(ii) Stress relieving method or overcoring technique
  (a) USBM[1] borehole deformation gauge (ISRM, 1987)
  (b) CSIR[2] or CSIRO[3] cell with 9 or 12 strain gauge (ISRM, 1987)
  (c) Borre probe cell (ISRM, 2007)
(iii) Hydraulic method
  (a) Conventional or hydrofrac or minifrac (ISRM, 1987)
  (b) Hydraulic test on pre-existing fractures (ISRM, 1987)

6.3.1 Flat-jack Method

In this method, rock stresses are measured in a direction perpendicular to the plane of the cut-slot near the exposed rock surface. All six stress tensors are determined in three dimensions at a zone inside the rock mass by conducting a minimum of six tests in three arbitrary planes.

This method involves the observation of movement of two or more pair of nails grouted on each side of a proposed slot as shown in Figure 6.4(a). Let us assume that the initial distance between the two vertically aligned pins is d. A slot is then cut at the mid distance between the two pins. The pins deform up to $2\Delta d$ due to stress relief. Then a flat-jack is inserted into the slot and by applying pump pressure, the internal surface of the slot is pushed back to regain the original distance d between the pins as shown in Figure 6.4(b).

1. United States Bureau of Mines. Now, NIOSH: National Institute of Occupational Safety and Health.
2. Commonwealth Scientific and Industrial Research.
3. Commonwealth Scientific and Industrial Research Organization (CSIRO) of Australia.

Flat-jack consists of two flat flexible sheets made up of steel or other suitable material welded together around the edge to form a flat envelope of at least 0.1 m^2 area and incorporated with a hydraulic inlet tube for connection with a hydraulic pump. The shape of the flat-jack depends on the method chosen to cut the slot. A hydraulic pump, operated either manually or electrically, with an attachment to a load maintainer for a period of at least 5 min. The pressure should be measured on gauge having an accuracy of at least 5% of the estimated stress. The typical diameter and length of a pin are 12 mm and 150 mm respectively, actual dimensions will depend on rock quality. When the near surface rock appears damaged by the excavation works, it is preferable to measure the displacements at a sufficient depth to avoid the damaged rock.

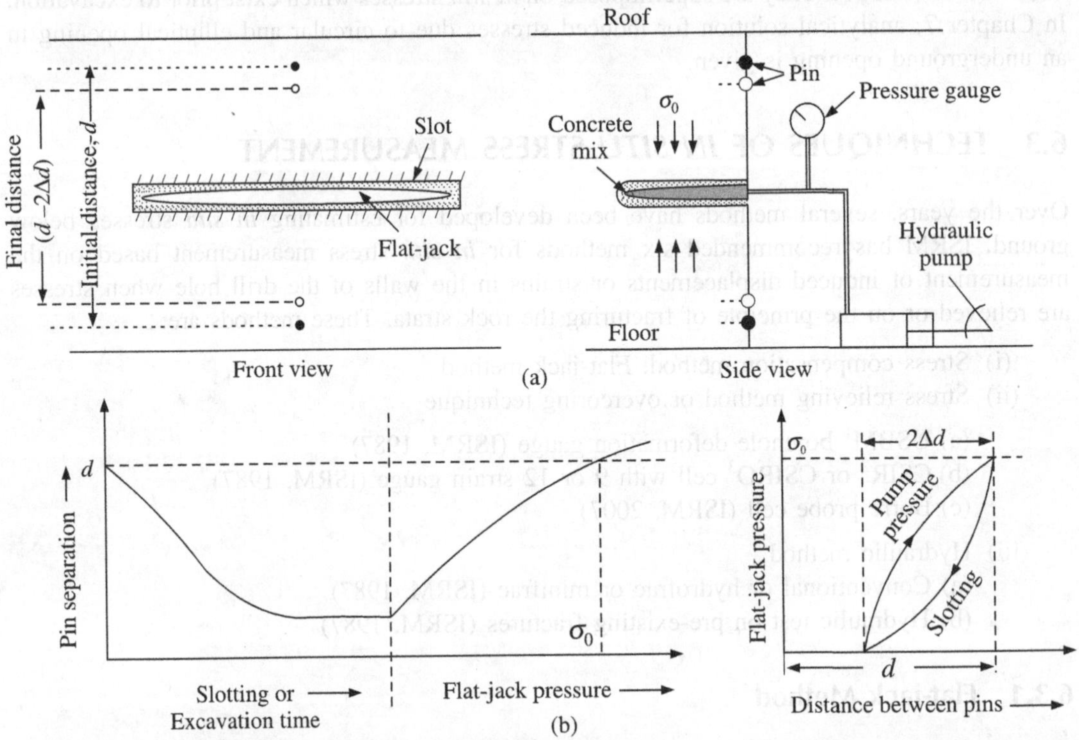

FIGURE 6.4 Stress measurement by flat-jack; view is normal to the free surface of the pre-existing excavation.

The method can also use extensometer for measurement of displacement but no special tools and transducers are required. In this method, knowledge of rock properties is not required to estimate *in situ* stresses. The inherent shortcoming of this method lies with *in situ* stress determination near the excavation surface and hence, the measured stress includes both *in situ* and induced stresses. Hysteresis characteristics of rock mass during loading and unloading of slot results into difference in pressure in flat-jack. Hence, the flat-jack pressure may not represent the actual *in situ* stresses.

EXAMPLE 6.2 Three flat-jack tests have been made close to each other in the wall of a long straight tunnel, the axis of which dips at 7°. The measurement position is approximately 300 m below the ground surface and it is assumed that the flat-jacks operate in the same stress field. The slots for the flat-jacks are cut normal to the wall of the tunnel, and are oriented relative to the tunnel axis as shown. The cancellation pressure for each of the flat-jacks A, B and C is 7.56 MPa, 6.72 MPa and 7.50 MPa, respectively. Compute the principal stresses and their directions, and ascertain whether they accord with worldwide trend if the average unit weight of overlying rock is 22.5 kN/m^3.

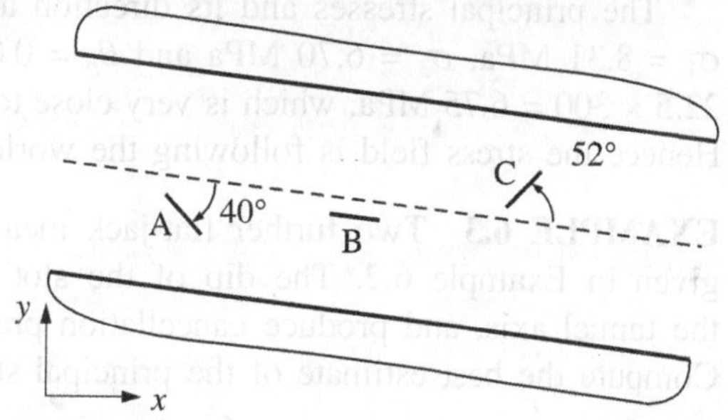

Solution: We know that angle measured in anti-clockwise direction is +ve

Inclination of slot with respect to horizontal (+90°) = Inclination of tunnel axis w.r.t. *x*-direction + Inclination of slot w.r.t. tunnel axis + 90°

$\theta_A = -7° - 40° + 90° = 43°$; $\theta_B = -7° + 90° = 83°$; $\theta_C = -7° + 52° + 90° = 135°$.

Using stress transformation Eq. 4.28, induced normal and shear stresses acting in a slot axis oriented θ degrees from the horizontal axis can be written as:

$$\sigma_\theta = \sigma_{xx}\cos^2\theta + \sigma_{yy}\sin^2\theta + \tau_{xy}\sin 2\theta$$

$$\tau_\theta = \left(\frac{\sigma_{xx} - \sigma_{yy}}{2}\right)\sin 2\theta - \tau_{xy}\cos 2\theta$$

where, σ_{xx}, σ_{yy} and τ_{xy} are *in situ* stress tensors in a plane whose normal is perpendicular to the slot axis. Hence, in three slot planes A, B and C, we can write:

$$\sigma_{\theta A} = \sigma_{xx}\cos^2 43° + \sigma_{yy}\sin^2 43° + 2\tau_{xy}\sin 43°\cos 43°$$
$$\sigma_{\theta B} = \sigma_{xx}\cos^2 83° + \sigma_{yy}\sin^2 83° + 2\tau_{xy}\sin 83°\cos 83°$$
$$\sigma_{\theta C} = \sigma_{xx}\cos^2 135° + \sigma_{yy}\sin^2 135° + 2\tau_{xy}\sin 135°\cos 135°$$

In matrix form, the above equations can be written as:

$$\begin{bmatrix} 7.56 \\ 6.72 \\ 7.50 \end{bmatrix} = \begin{bmatrix} \cos^2 43° & \sin^2 43° & 2\sin 43°\cos 43° \\ \cos^2 83° & \sin^2 83° & 2\sin 83°\cos 83° \\ \cos^2 135 & \sin^2 135° & 2\sin 133°\cos 135° \end{bmatrix} \begin{bmatrix} \sigma_x \\ \sigma_y \\ \tau_{xy} \end{bmatrix}$$

$$\boldsymbol{\sigma}_{\text{jack}} = \mathbf{R}\boldsymbol{\sigma}_{in\ situ}$$

$$\boldsymbol{\sigma}_{in\ situ} = \mathbf{R}^{-1}\boldsymbol{\sigma}_{\text{jack}}$$

$$\begin{bmatrix} \sigma_{xx} \\ \sigma_{yy} \\ \tau_{xy} \end{bmatrix} = \begin{bmatrix} 1.093 & -0.952 & 0.860 \\ -0.134 & 1.021 & 0.113 \\ 0.479 & 0.034 & -0.514 \end{bmatrix} \begin{bmatrix} 7.56 \\ 6.72 \\ 7.50 \end{bmatrix} = \begin{bmatrix} 8.308 \\ 6.696 \\ 0.002 \end{bmatrix} \text{MPa}$$

The principal stresses and its direction are calculated using Eq. (4.35) and Eq. (4.32) as σ_1 = 8.31 MPa, σ_3 = 6.70 MPa and θ_p = 0.07°. The vertical stress is $\sigma_v = \gamma H$. Hence, σ_v = 22.5 × 300 = 6.75 MPa, which is very close to one of the principal stress, i.e., σ_{yy} = 6.696 MPa. Hencce the stress field is following the world-wide trend.

EXAMPLE 6.3 Two further flat-jack measurements are made in the wall of the tunnel as given in Example 6.2. The dip of the slot axis are 20° and 90° anti-clockwise relative to the tunnel axis, and produce cancellation pressures of 7.38 MPa and 7.86 MPa, respectively. Compute the best estimate of the principal stresses.

Solution:

$$\sigma_{\text{jack}} = \begin{bmatrix} \sigma_A \\ \sigma_B \\ \sigma_C \\ \sigma_D \\ \sigma_E \end{bmatrix}, R = \begin{bmatrix} \cos^2\theta_A & \sin^2\theta_A & 2\sin\theta_A\cos\theta_A \\ \cos^2\theta_B & \sin^2\theta_B & 2\sin\theta_B\cos\theta_B \\ \cos^2\theta_C & \sin^2\theta_C & 2\sin\theta_C\cos\theta_C \\ \cos^2\theta_D & \sin^2\theta_D & 2\sin\theta_D\cos\theta_D \\ \cos^2\theta_E & \sin^2\theta_E & 2\sin\theta_E\cos\theta_E \end{bmatrix}, \text{ and } \sigma_{in\ situ} = \begin{bmatrix} \sigma_{xx} \\ \sigma_{yy} \\ \tau_{xy} \end{bmatrix}$$

This shows that there are five equations for three unknowns and in such situations, the solution is found as:

$$\boldsymbol{\sigma}_{\text{jack}} = \mathbf{R}\boldsymbol{\sigma}_{in\ situ}$$

$$\mathbf{R}^T\boldsymbol{\sigma}_{\text{jack}} = \mathbf{R}^T\mathbf{R}\boldsymbol{\sigma}_{in\ situ}$$

Multiplying by $\mathbf{R}^T$ on both side

$$\boldsymbol{\sigma}_{in\ situ} = (\mathbf{R}^T\mathbf{R})^{-1}(\mathbf{R}^T\boldsymbol{\sigma}_{\text{jack}})$$

$$\boldsymbol{\sigma}_{in\ situ} = (\mathbf{R}^T\mathbf{R})^{-1}\mathbf{R}^T\boldsymbol{\sigma}_{\text{jack}} = \begin{bmatrix} 7.91 \\ 7.03 \\ -0.08 \end{bmatrix} \text{MPa}$$

The principal stresses and its direction are σ_1 = 7.92 MPa, σ_3 = 7.02 MPa and θ_p = −4.83°.

6.3.2 United States Bureau of Mines (USBM) Borehole Deformation Gauge

USBM overcoring method is a suggested method of *in situ* stress measurement (ISRM, 1987) as well as American Society for Testing and Materials (ASTM D4632-16). Readers are encouraged to refer either of these two literatures for detail test procedures. Here, this method is illustrated in brief.

The Principle of Overcoring Method

As mentioned in the above literatures, the overcore test is performed to measure the diametrical deformation in three directions of a small-diameter borehole cored coaxially to a larger diameter hole. It is known that the measured borehole deformation can be related to the change in stress in a plane perpendicular to the borehole axis if the elastic modulus of rock is known. This change in stress is assumed to be numerically equal, although opposite in sense to the stresses existing in the parent rock mass (ASTM D4632-16). In order to estimate a complete three-dimensional state of stress in the rock mass, deformation measurements are normally done in three perpendicular boreholes.

The Method

In USBM overcoring method, two concentric boreholes of diameter d_o = 150 mm and d_i = 38 mm are drilled coaxially within ±14 mm. The depth of hole depends on site of stress measurement, but the smaller diameter hole is more than 2 m deep, drilled in a continuous rock mass at the end of large hole. The maximum depth of hole is usually less than 30 m although measurement up to 70 m deep is made in vertical downhole direction.

The borehole deformation gauge with sensitivity of at least 1 mm in 10^5 mm is inserted in a pilot hole of sufficient length which has a diameter of 38 mm or EX size. The gauge is about 30 cm long and has six projected buttons in the front which are pressed against the borehole wall. Three strain gauges are bonded to steel cantilever supporting the buttons as shown in Figure 6.7(a). These buttons are positioned 60° apart from each other and strains are measured along the perpendicular plane whose normal is the axis of the hole [Figure 6.7(b)].

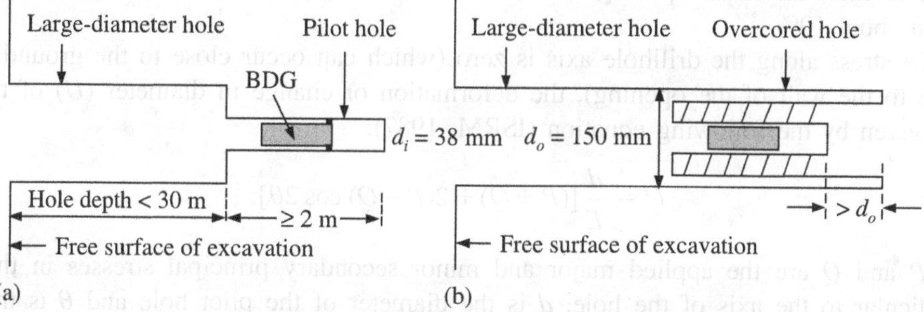

FIGURE 6.5 Stress measurement by USBM gauge.

After the placement of the gauge in the hole, strain data are collected using a strain measurement unit located in the roadway. After stabilisation of deformation, overcoring is done with 150 mm diameter drill rod as shown in Figure 6.5(b). This process creates an annular rock region that is essentially free of stress. The overcoring should be long enough to obtain gauge reading which do not change with further coring. The overcore rock should not split or fracture into pieces of length less than twice its diameter. Any fracture zones encountered by the pilot hole should be avoided during placing of the instrument.

As mentioned before, a single borehole allows measurement of stress field in a 2D plane perpendicular to borehole axis. To define stress tensor in three dimensions at a zone, three perpendicular boreholes will be sufficient as shown in Figure 6.6.

Calculations

The deformations measured by the borehole deformation gauge (BDG) are used to compute planar principal stresses in a plane perpendicular to the axis of the borehole. Since the BDG does not measure rock deformation along the axis of the borehole, an exact calculation of the

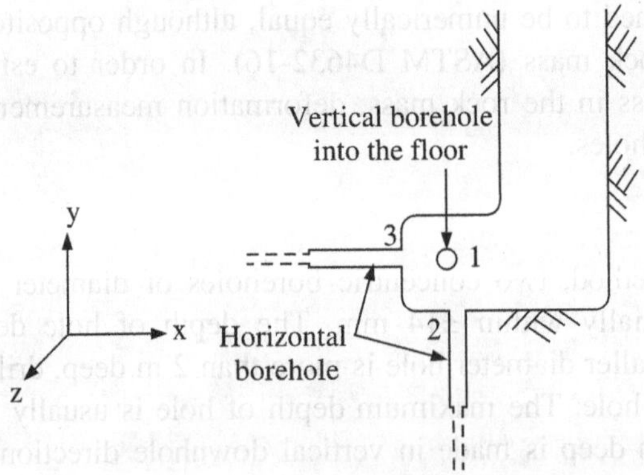

FIGURE 6.6 Plan of the configuration of boreholes 1, 2, and 3 along with opening.

secondary principal stresses is not possible unless some additional data are obtained regarding either axial deformation or axial stress. If the simplifying assumption that zero stress exists along the axis of the borehole is made, then the secondary stresses can be calculated using a plane stress solution. The assumed stress conditions might occur close to the ground surface or close to the wall of an opening. This solution, although not exact, results in an error of less than about 10%.

If the stress along the drillhole axis is zero (which can occur close to the ground surface or close to the wall of the opening), the deformation or change in diameter (U) of the pilot hole is given by the following equation (ISRM, 1987):

$$U = \frac{d}{E}[(P+Q) + 2(P-Q)\cos 2\theta] \tag{6.6}$$

where P and Q are the applied major and minor secondary principal stresses in the plane perpendicular to the axis of the hole, d is the diameter of the pilot hole and θ is the angle between the direction of P and the direction of U, the measured deformation.

In the case of the USBM-type gauge which has three sensor axes, U_1, U_2, and U_3, spaced 60° apart in a anti-clockwise direction (looking into the drillhole), the major and minor principal stresses and their orientations as shown in Figure 6.7(b) are determined by the following expressions (ISRM, 1987):

$$P, Q = \frac{E}{6d}\left[(U_1 + U_2 + U_3) \pm \frac{1}{\sqrt{2}}\left[(U_1 - U_2)^2 + (U_2 - U_3)^2 + (U_3 - U_1)^2\right]^{1/2}\right] \tag{6.7}$$

$$\theta_P = \frac{1}{2}\tan^{-1}\left(\frac{\sqrt{3}(U_2 - U_3)}{2U_1 - U_2 - U_3}\right) \tag{6.8}$$

where the angle θ_P is measured from P and Q in anti-clockwise direction. Depending on relative magnitudes of deformation, the range of this angle is determined as follows:

(i) $U_2 > U_3$ and $U_2 + U_3 < 2U_1$ then θ_P is between 0° and 45°.

(ii) $U_2 > U_3$ and $U_2 + U_3 > 2U_1$ then θ_P is between 45° and 90°.
(iii) $U_2 < U_3$ and $U_2 + U_3 > 2U_1$ then θ_P is between 90° and 135°.
(iv) $U_2 < U_3$ and $U_2 + U_3 < 2U_1$ then θ_P is between 135° and 180°.

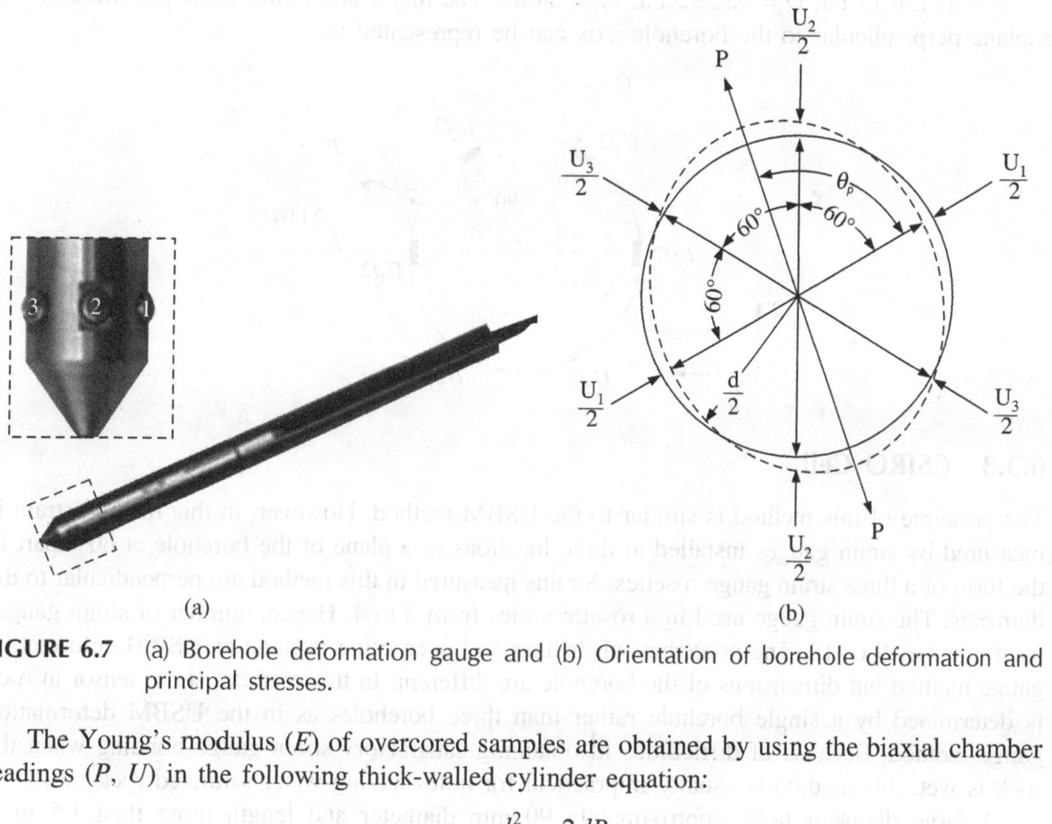

(a) (b)

FIGURE 6.7 (a) Borehole deformation gauge and (b) Orientation of borehole deformation and principal stresses.

The Young's modulus (E) of overcored samples are obtained by using the biaxial chamber readings (P, U) in the following thick-walled cylinder equation:

$$E = \frac{d^2}{d^2 - b^2} \frac{2dP}{U} \tag{6.9}$$

where, d is diameter of overcore, b is diameter of pilot hole, P is applied radial pressure, U is the measured change in pilot hole diameter.

In case where the rock is anisotropic, the measured value of U along different orientation will vary depends on the change in elastic parameters.

EXAMPLE 6.4 Deformations are measured in a borehole drilled perpendicular to a tunnel wall by USBM gauge. The measuring pilot hole diameter is 38 mm and 2.0 m deep, in a test gallery. The overcoring process results into the movement of pair 1 outward by 0.077 mm; pair 2 by 0.055 mm outward; and pair 3 by 0.025 mm inward. It is given that the first pair of buttons is horizontal. If modulus of elasticity of surrounding rock is 15 GPa and Poisson's ratio is 0.25, determine the stress components in the plane perpendicular to the borehole axis. The major and minor normal stresses in the plane and their directions. (Assume that initial stress parallel to the borehole axis is insignificant or zero.)

Solution: It is given that

$U_1 = 0.077$ mm; $U_2 = 0.055$ mm; $U_3 = -0.025$ mm; $d = 38$ mm; $E = 15$ GPa; $\nu = 0.25$.

Using Eqs. 6.8 and 6.9, the major and minor stresses and their direction are:

$P = 13156.13$ Pa; $Q = 922.82$ Pa; $\theta_P = 24.09$. The major and minor principal stress across a plane perpendicular to the borehole axis can be represented as:

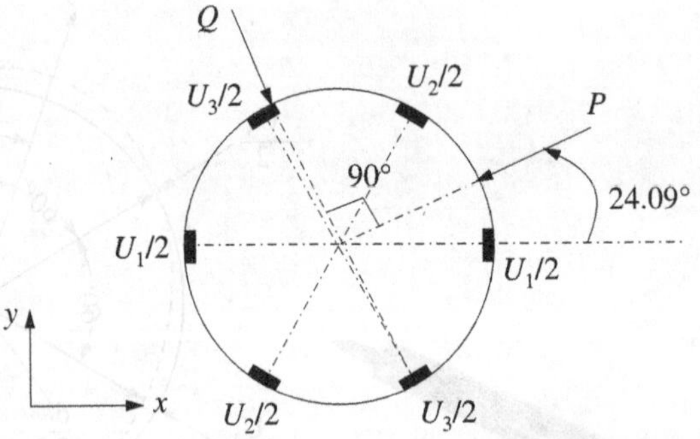

6.3.3 CSIRO Cell

The principle of this method is similar to the USBM method. However, in this method strain is measured by strain gauges installed at three locations in a plane of the borehole at 60° apart in the form of a three strain gauge rosettes. Strains measured in this method are perpendicular to the diameter. The strain gauge used in a rosette varies from 3 to 4. Hence, number of strain gauges used is either 9 or 12. The procedure of drilling and overcoring is same as USBM deformation gauge method but dimensions of the borehole are different. In this method, stress tensor in rock is determined by a single borehole rather than three boreholes as in the USBM deformation gauge method. Because of difficulties in obtaining satisfactory strain gauge bonding when the rock is wet, this method is usually impractical for holes having water saturated rock mass.

A large diameter hole, approximately 90 mm diameter and length more than 1.5 m is drilled into the rock mass to the point ranging between 30 and 50 m from working faces, where the stress in the rock is to be measured (Figure 6.8). In order to avoid water problems, the boreholes are drilled slightly upwards and dip toward the opening. A pilot hole (38 mm diameter and 450 mm long) is drilled at the end of the large diameter borehole. If the core obtained from pilot hole is in one piece and is free from planes of weakness, measurement can be carried out. If broken core are found, then 90 mm borehole needs to be advanced by 0.5 m and then pilot holes are drilled. It will be repeated until a position is found where solid unbroken core is present. A thorough examination of extent and severity of fracturing should be done, before rejecting a potential test zone.

A "strain cell" designed to glue a number of strain gauges onto the sidewall of the pilot hole. To avoid the difficulties caused by the non-homogeneous, anisotropic nature of rocks, the electric resistance strain gauges should be glued as close together as possible so that only a small volume of rock is involved. The strain gauges should be mutually orientated in such a way that six independent strain measurements may be obtained, (e.g., using three standard strain gauge rosettes, each consisting of three or four gauges in an arrangement as shown in

Figure 6.9). This arrangement of gauges yields an additional three to six strain measurements which may be used for cross checking or for error estimation. Each individual strain gauge should have a gauge length of at least 10 mm. If anisotropic rock conditions exist, strains cells containing 12 strain gauges provide a more suitable analysis system.

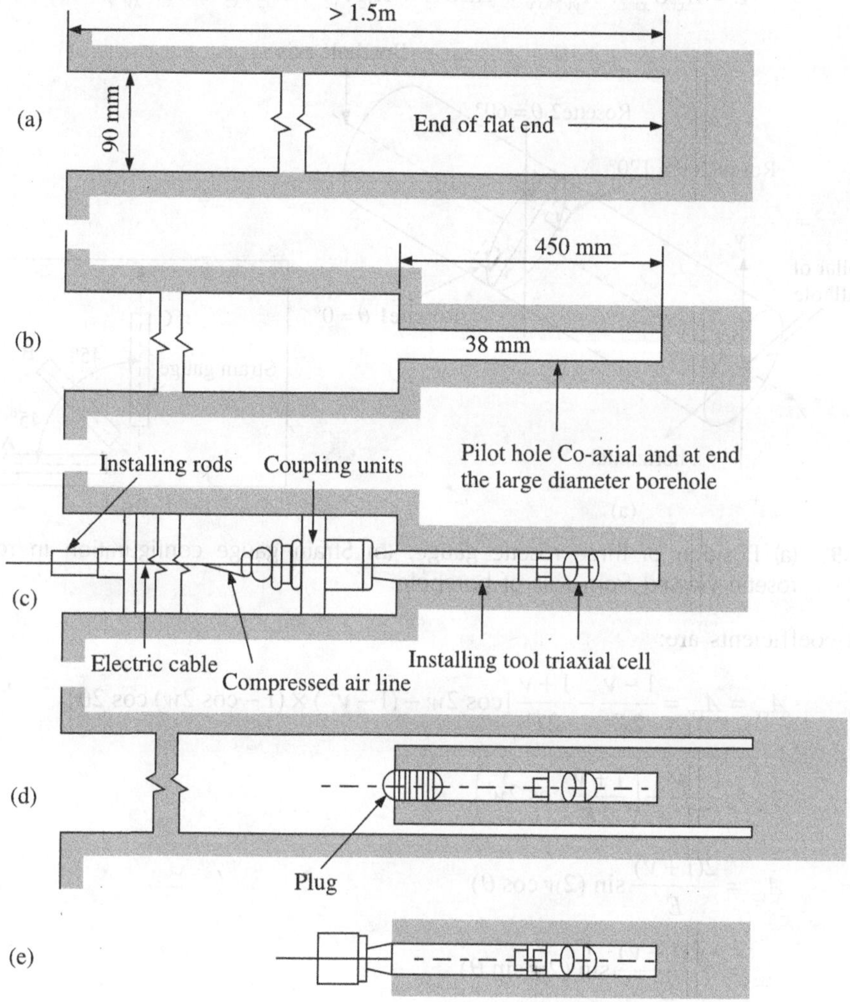

FIGURE 6.8 Description of the use of triaxial strain cell (Leeman, 1971).

Initial strain reading is taken, the installing tool is removed from the hole and the pilot hole is sealed by means of the hole plug. When taking readings, a balance on each gauge should be made twice and the average of the two readings should be used in the calculations. The strain cell is then overcored and the core containing the cell is removed from the bore. This relieves the stresses in the rock to which the strain gauge rosettes are attached. Further rounds of readings are taken, as discussed above to determine a stable level of post-overcore strain for each strain gauge.

Calculations

Strain relief values are computed by subtracting average readings before and after overcoring. Mathematically, normal strain can be expressed as (Amadei, 1983)

$$\varepsilon = A_{xx}\sigma_{xx} + A_{yy}\sigma_{yy} + A_{zz}\sigma_{zz} + A_{yz}\tau_{yz} + A_{xz}\tau_{xz} + A_{xy}\tau_{xy} \quad (6.10)$$

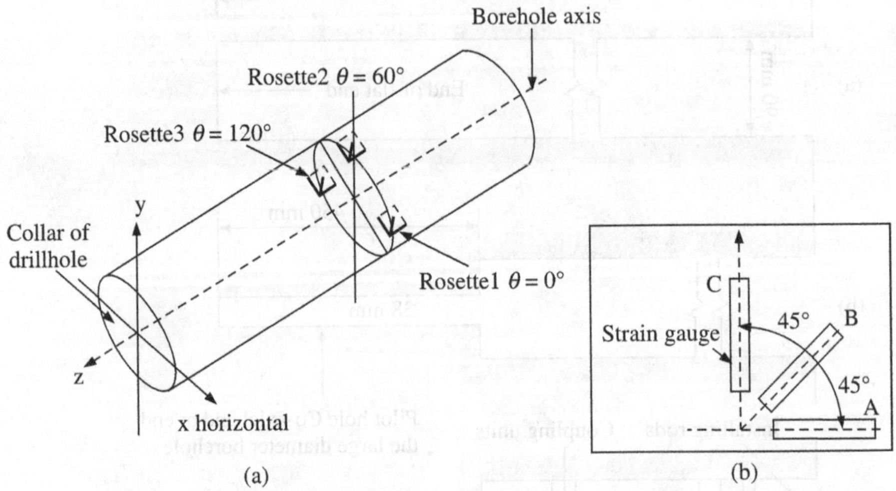

FIGURE 6.9 (a) Position of three rosette gauge, (b) Strain gauge configuration in rectangular rosette viewed from axis of borehole.

The A-coefficients are:

$$A_{xx} = A_{yy} = \frac{1-\nu}{2E} - \frac{1+\nu}{2E}[\cos 2w - (1-\nu^2)\times(1-\cos 2w)\cos 2\theta]$$

$$A_{zz} = \frac{1-\nu}{2E} + \left(\frac{1+\nu}{2E}\cos 2w\right)$$

$$A_{yz} = \frac{2(1+\nu)}{E}\sin(2w\cos\theta)$$

$$A_{xz} = \frac{-2(1+\nu)}{E}\sin(2w\sin\theta)$$

$$A_{xy} = \frac{-2(1-\nu^2)}{E}(1-\cos 2w)\sin 2\theta$$

where w and θ are defined in Figure 6.9, E is the Young's modulus and ν is the Poisson's ratio of the rock and obtained by laboratory testing of specimen taken from the cores or from a biaxial or triaxial test of the core containing the strain gauges. Tensor components σ_{xx}, σ_{yy}, σ_{zz}, τ_{yz}, τ_{xz}, and τ_{xy}, which best fit the measured data are obtained by means of a least square fitting procedure. Principal stresses, σ_1, σ_2, and σ_3 and their direction cosines are then to be calculated from the tensor components of the best fit by means of elasticity equations. The main advantage of this method is that only one borehole is required to determine the complete stress field. The main disadvantages lies with (1) the cleaning of the pilot hole walls, (2) in the standard CSIR cell no monitoring of the strain gages during overcoring is possible, (3) the cell is not recoverable unless one uses the recovery method of Cai and Blackwood (1991), (4) it is a quarter bridge instrument, (5) the time required for curing of the glue can vary from 1 to 10 or even 20 h depending on the hole conditions and the type of glue being used, and (6) long overcores are necessary, which may be difficult in weak or bedded rocks and rocks under high stresses.

6.3.4 Borre Probe

The Borre probe cell includes a built-in datalogger, which runs on batteries and permits a continuous logging of the strain gauges before, during and after the overcoring process, which enhances the evaluation process. The Borre probe carries nine electrical resistance strain gauges mounted in three rectangular rosettes. Thus, the nine strain gauges of the Borre probe from an array representing seven spatially different directions. All strain gauges are mounted at a depth of 160 mm in the pilot hole. The strain gauges are connected to a data-logger inside the probe. The procedure for a single stress measurement using Borre probe is briefly summarised in Figure 6.10.

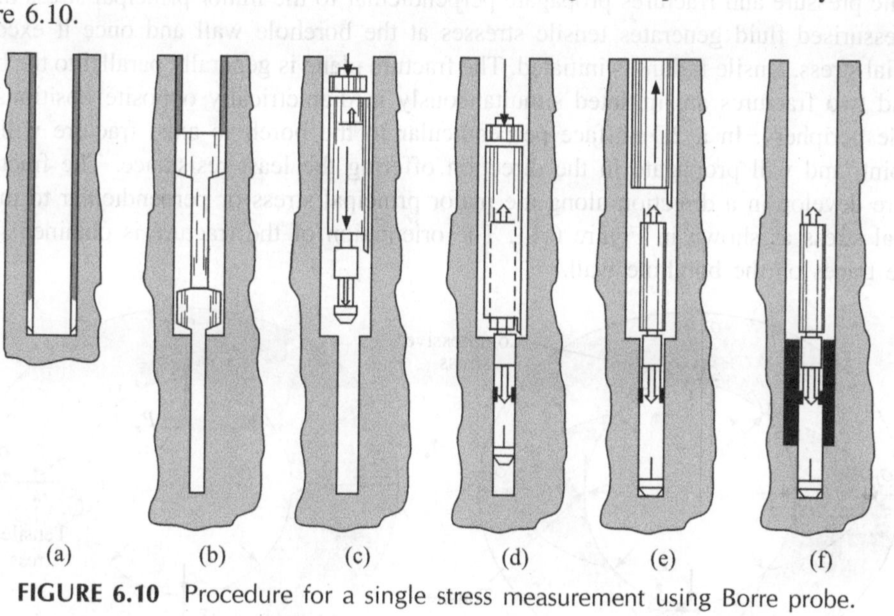

FIGURE 6.10 Procedure for a single stress measurement using Borre probe.

Measurement procedures using Borre probe:

(a) Advance 76 mm diameter main borehole upto the depth of strain measurement and grind the hole bottom using the planing tool.
(b) Drill 36 mm diameter pilot hole and recover core for appraisal. Flush the borehole to remove drill cuttings.
(c) Prepare the probe for measurement and apply glue to strain gauges. Insert the probe in installation tool into hole.
(d) Tip of probe with strain gauges enters the pilot hole. Probe releases from installation tool through a latch, which also fixes the compass, thus recording the installed probe orientation. Gauges bonded to pilot hole wall under pressure from the nose cone.

(e) Pull out installation tool and retrieve to surface. The probe is bonded in place.
(f) Allow glue to harden overnight. Overcore the probe and record strain data using the built-in data logger. Break the core after completed overcoring and recover in core barrel to surface (ISRM, 2007).

First stage of this test is to drill large diameter borehole to check the rock mass homogeneity. Homogeneity is established by continuous borehole log. Once it is established, concentric pilot hole of 36 mm diameter is drilled and core is recovered for determination of strength parameter. The stages of the method are explained above with the help of Figure 6.10. Finally, ISRM recommends to flush 10–15 min before the commencement of overcoring and continue for at least 5–10 min after completed overcoring.

6.3.5 Hydraulic Fracturing Method

Principle of the Method

The major concept behind this method is that rock surrounding a borehole can be fractured with hydraulic pressure and fractures propagate perpendicular to the minor principal stress direction. The pressurised fluid generates tensile stresses at the borehole wall and once it exceeds the tangential stress, tensile failure is initiated. The fracture plane is generally parallel to the borehole axis and two fractures are initiated simultaneously in diametrically opposite positions on the borehole periphery. In a 2D surface perpendicular to the borehole axis, fracture will initiate at a point, and will propagate in the direction offering the least resistance. The fracture will therefore develop in a direction along the major principal stress or perpendicular to the minor principal stress as shown in Figure 6.11. The orientation of the fracture is obtained from the fracture traces on the borehole wall.

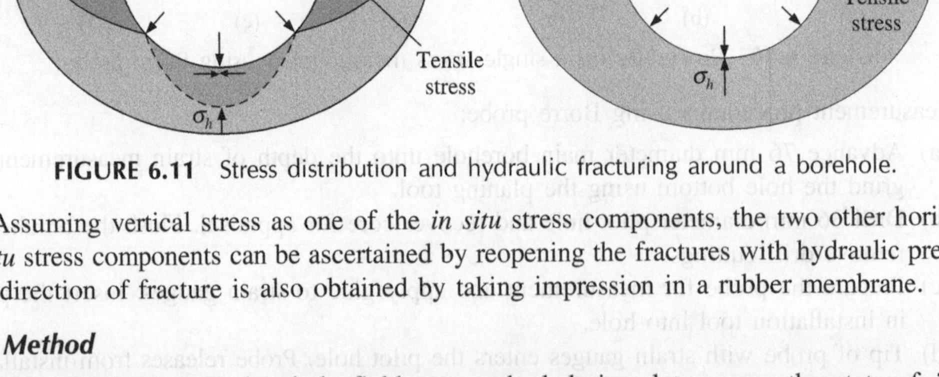

FIGURE 6.11 Stress distribution and hydraulic fracturing around a borehole.

Assuming vertical stress as one of the *in situ* stress components, the two other horizontal *in situ* stress components can be ascertained by reopening the fractures with hydraulic pressure. The direction of fracture is also obtained by taking impression in a rubber membrane.

The Method

Hydraulic fracturing is a borehole field test method designed to assess the state of *in situ* stress in the earth's crust. Hydraulic test estimates the state of *in situ* stress (both magnitude and directions) in the plane perpendicular to the axis of the borehole. The method uses the analytical relationship to find the stresses around opening. There are two methods that use same hydraulic based apparatus to fracture rock around the borehole to determine the stress field:

(i) Hydraulic fracturing (HF)
(ii) Hydraulic test on pre-existing fractures (HTPF)

The test provides the magnitude and directions of the maximum and minimum stresses in the plane perpendicular to the borehole. Hence, to find the complete stress tensor six tests are required. This test is recommended for vertical borehole with a deviation of ±15°.

For both HF and HTPF methods, a section of a borehole is sealed off (at least 6 times the diameter of the borehole) by two inflatable rubber packers (called Straddle packers). These packers are connected mechanically as well as hydraulically to form one unit as shown in Figure 6.12. These packers are pressured at 2 MPa higher than the internal pressure to ensure isolation. Now, pressure between the straddle packers is raised with hydraulic fluid (typically water) under constant and predetermined flow rate. Note that the flow rate may depend on rock permeability and the aim is to raise the internal pressure steadily so that the peak pressure at which rock at the borehole wall fractures, termed as *breakdown pressure*, P_B, or pre-existing opening pressure.

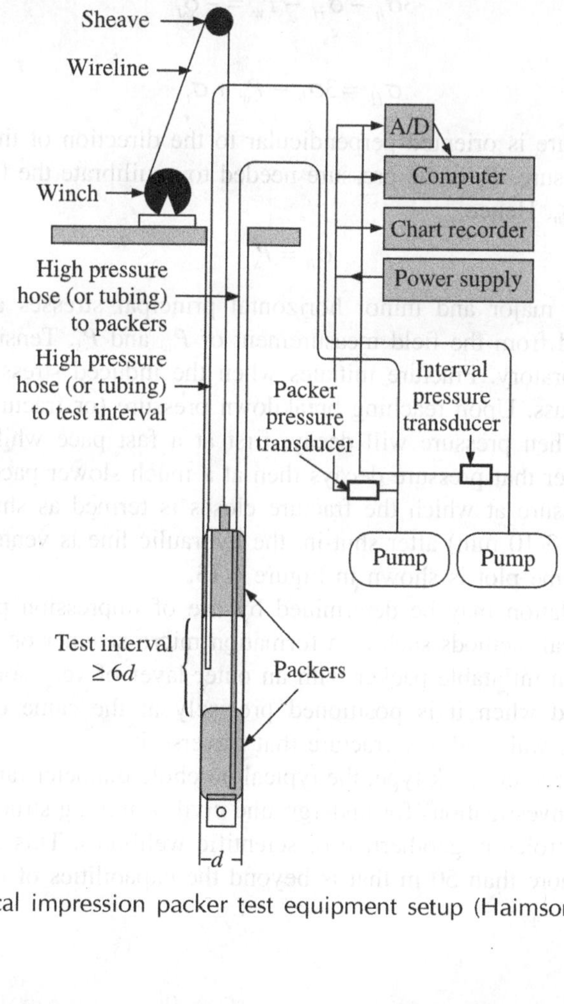

FIGURE 6.12 Typical impression packer test equipment setup (Haimson and Cornet, 2003).

From Kirsch Eq. 7.1 (Chapter 7), the tangential stresses around a circular opening without internal pressure in a stress field of (σ_H, σ_h) at $\theta = 0°$, $\theta = 90°$ are given by

$$\sigma_0 = 3\sigma_h - \sigma_H \tag{6.11}$$

$$\sigma_{90} = 3\sigma_H - \sigma_h \tag{6.12}$$

With the application of internal pressure (P_w) in the hydraulic chamber for fracturing the induced stress at σ_θ at $\theta = 90°$ will be more than $\theta = 0°$ (Figure 6.11). The stresses at the time of breakdown can be given as:

$$\sigma_0 = 3(\sigma_h - P_w) - (\sigma_H - P_w) + P_w \tag{6.13}$$

$$\sigma_0 = 3\sigma_h - \sigma_H - P_w \tag{6.14}$$

$$\sigma_{90} = 3\sigma_H - \sigma_h - P_w \tag{6.15}$$

Since $\sigma_{90} > \sigma_0$, tensile crack may generate at $\theta = 0°$ or in other words, along the direction of major principal stress. The tensile fracture begins once the induced stress equals the tensile strength (σ_t) of the rock mass. At the time of shut down of pressure,

$$3\sigma_h - \sigma_H - P_w = -\sigma_t \tag{6.16}$$

Rearranging

$$\sigma_H = 3\sigma_h - P_w + \sigma_t \tag{6.17}$$

Note that the fracture is oriented perpendicular to the direction of the minimum principal stress. The shut-in pressure, P_s is the pressure needed to equilibrate the fracture-normal stress, which in this case is σ_h. Hence,

$$\sigma_h = P_s \tag{6.18}$$

The magnitude of major and minor horizontal principal stresses are determined using Eqs. 6.17 and 6.18 and from the field measurement of P_w and P_s. Tensile strength of rock is determined in the laboratory. Fracture initiates when the induced stress equals to the tensile strength of the rock mass. Upon reaching breakdown pressure (or fracture opening), pumping of water is stopped. Then pressure will decay, first at a fast pace while the fracture is still open and growing. After that pressure decays then at a much slower pace, after the fracture is closed again. The pressure at which the fracture closes is termed as shut-in pressure (P_s). A few minutes (typically 3-10 min) after shut-in, the hydraulic line is vented. A typical pressure and flow rate versus time plot is shown in Figure 6.13.

The fracture orientation may be determined by use of impression packer and a compass or by use of geophysical methods such as a formation micro-scanner or a borehole televiewer. Impression packer is an inflatable packer with an outer layer of very soft semi-curved rubber. This packer is inflated when it is positioned precisely at the same depth, resulting in an impression of borehole wall and any fracture that traverse it.

Depending on project and rock type, the typical borehole diameter range from 76 to 96 mm (N- to H-type) in site investigations for underground civil or mining structures, and may extend to 180 mm in deep petroleum, geothermal or scientific wellbores. This method is more suited to measure at depth more than 50 m that is beyond the capabilities of other techniques.

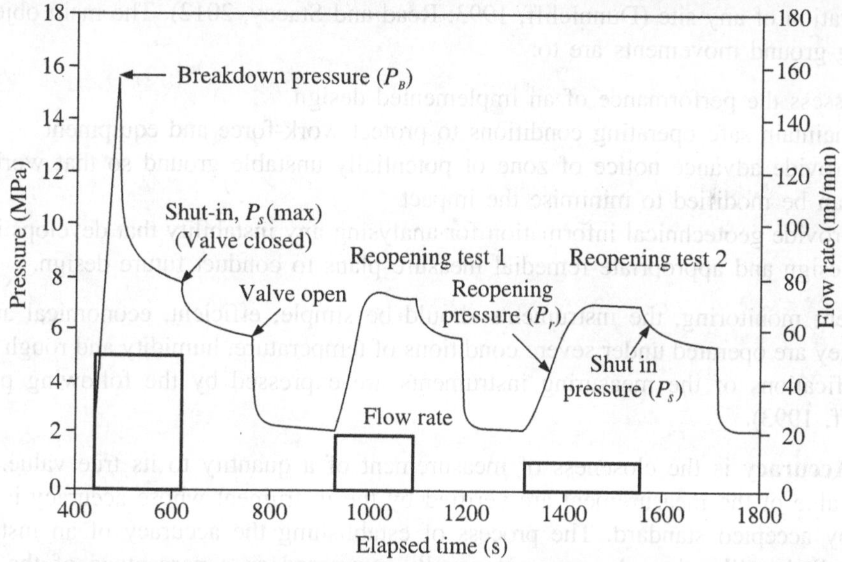

FIGURE 6.13 Pressure-time record for hydraulic pressure measurement.

Assumption and Limitations

There is no theoretical limit to the depth at which this measurement can be done. However, it is necessary to verify that only one single fracture has been opened. For complete stress tensor determination, minimum six tests are required for HF method without determination of material properties. However, HTPF method provides an evaluation of the complete stress tensor (6 components), independent of borehole orientation and material property.

EXAMPLE 6.5 *In situ* stress in a granitic rock mass has been measured by the hydraulic fracturing technique at a depth of 500 m from the surface and the breakdown pressure is 14 MPa and shut-in pressure is 8 MPa. If tensile strength of the rock at 500 m depth is 10 MPa and the average density of overlying rock is 2500 kg/m^3 then estimate the principal stress at test location.

Solution: The principal stresses at test locations are the vertical stress, major horizontal stress and minor horizontal stress. Vertical stress are determined by using Eq. (6.1) whereas Eqs. (6.17) and (6.18) are used to find the major and minor horizontal stress as:

For 500 m: $\sigma_V = 2500 \times 9.81 \times 500 = 12.26$ MPa; $\sigma_h = P_s = 8.0$ MPa; $\sigma_H = 20$ MPa

6.4 TECHNIQUES FOR INDUCED STRESS AND DEFORMATION MEASUREMENT

Monitoring is an invaluable tool for assessing design performance and risk of failure. Monitoring of ground movement not only aids in identification of hazard and risk but also improves safety and health of workers. The monitoring instruments can be installed prior, during and post excavation of any site (Dunnicliff, 1993; Read and Stacey, 2012). The main objectives of monitoring ground movements are to:

(i) assess the performance of an implemented design
(ii) maintain safe operating conditions to protect work-force and equipment
(iii) provide advance notice of zone of potentially unstable ground so that working plan can be modified to minimise the impact
(iv) provide geotechnical information for analysing any instability that develops in current design and appropriate remedial measure plans to conduct future design.

For efficient monitoring, the instruments should be simple, efficient, economical and sturdy because they are operated under severe conditions of temperature, humidity and rough handling. The specifications of the measuring instruments are expressed by the following parameters (Dunnicliff, 1993).

(i) **Accuracy** is the closeness of measurement of a quantity to its true value. The true value of the measurement are verified by the instrument whose accuracy is indicated by accepted standard. The process of establishing the accuracy of an instrument is called *calibration.* Accuracy is usually expressed as a percentage of the full-scale reading. For an example, a 10 MPa pressure gauge having an accuracy of 1% would be accurate to within ±100 kPa over the entire range of the gauge.
(ii) **Precision** is a measure of the ability of the instrument to reproduce a certain reading. It is synonymous with reproducibility and repeatability. Precision is expressed as a ± number. The number of significant figures after decimal point quoted for a measurement reflects the precision of measurement; thus, ±1.00 indicates a higher precision than ±1.0. The difference between accuracy and precision is illustrated in Figure 6.14.

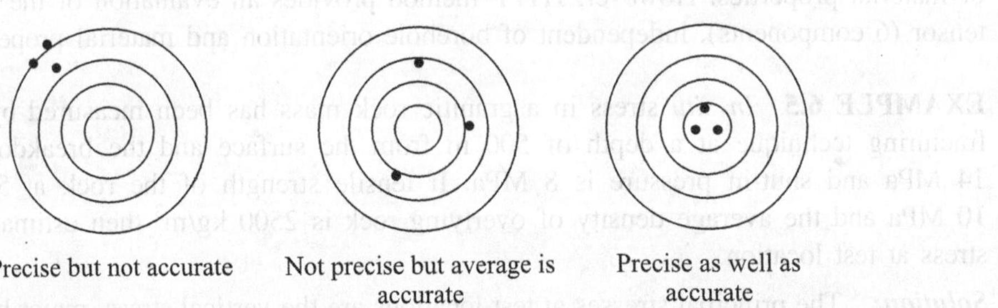

FIGURE 6.14 Concept of accuracy and precision (Dunnicliff, 1993).

(iii) **Resolution** or Least Count (LC) is the smallest unit of measurement detectable by the instrument.
(iv) **Sensitivity** of an instrument is variously defined as the ratio of the movement on the readout unit to the change in the measured variable causing the output, the input to output ratio to the smallest unit of the measurement detectable by the instrument. In other words, it refers to the amount of output response that an instrument or transducer produces when an input quantity is applied to it. Hence, high sensitivity does not imply high accuracy or precision. Sensitivity then depends on the readability of the readout unit, and the least count (the smallest difference between two indications that can be detected on the readout scale) (Brady and Brown, 2007).

(v) An instrument is said to be linear when it's indicated measured values are directly proportional to the quantity being measured.

(vi) The difference between the increasing and decreasing measurement is called *hysteresis*. Instruments with large hysteresis are not suitable for measurement.

(vii) **Noise** is used to cover random measurement variations caused by external factors, creating lack of precision and accuracy.

(viii) **Error** is the difference between an observed or calculated value and the true value. The error may be gross, systematic, conformance, environmental, observational and sampling.

The monitoring instruments are classified based on (i) electrical signal, (ii) optical signal.

The output of a electrical signal based device is either change in voltage or resistance. The devices measuring the change in resistance are strain gauges, load cell, etc. and the devices measuring the change in voltage are LVDT, potentiometer, etc. The electrical signal based instruments consists of three basic components:

(i) **Transducer or sensor or detector:** It is a device to measure variables such as displacement, load, stress and converts into a corresponding output signal. Transducers are categorised as mechanical, hydraulic, pneumatic and electrical transducer based on their principle of working. Sensors are classified as contact or non-contact type depending upon whether the sensors are in contact or not with the object while measurement. Contact type sensors are strain gauge, load cells, LVDT, etc. whereas the non-contact type technique such as optical image based digital image correlation (DIC) uses camera. Along with optical image based displacement, strain measurement is gaining popularity with the use of DIC method.

(ii) **Transmitting or communication system:** It is a device to transmit the information to a readout unit. Examples: rods, electrical cables or radio-telemetry.

(iii) **Data acquisition system (DAS):** It ranges from simple portable readout and/or recording unit to complex automatic systems to give a digital or graphical display of the measured quantity. The readout unit may consist of a Analog to Digital Converter (ADC), display unit only, or may include facilities for recording data on paper, magnetic tape, disk, pen drive, data card, or wireless/bluetooth transmission. The recording may be continuous, in response to a command signal or at preset intervals. The readout may be located close to each instrument station or remote, transmitting the readings to a central control house by a pneumatic, hydraulic or electrical system.

The output of an optical signal based device are either DIC or photoelasticity based. The basic component of these type of instruments are based on the technique used. For example digital image corelation (DIC) uses (i) speckle pattern, (ii) camera, (iii) mathematical algorithms.

6.5 DISPLACEMENT AND STRESS MONITORING DEVICES

In underground and surface excavations, convergence in the roof or side walls and pressures in the natural pillars are monitored to assess the stability of the ground. Apart from that artificial supports are also monitored to obtain load bearing capacity of these supports. Depending on the applications, monitoring devices are divided into three categories:

(i) Displacement monitoring device
   (a) Optical surveying
   (b) Dial indicator
   (c) Strain gauge (or gage)
   (d) Linear variable differential transformer (LVDT)
   (e) Potentiometer
   (f) Vibrating wire transducer

(ii) Displacement measuring instruments for field applications
   (a) Tape extensometer
   (b) Borehole extensometers
   (c) Telltale devices

(iii) Load or stress monitoring device
   (a) Flat jacks
   (b) Piezo/Strain gauge/Vibrating wire pressure gauge type load cell
   (c) Earth pressure cell

In subsequent sections, principles of the most commonly used instruments are discussed.

6.5.1 Displacement Monitoring Device

Convergence or relative displacement of two points either on the boundary of an excavation or inside the rock mass are measured with telescopic rod, invar bar or tape under constant tension, placed between two or multiple points. An optical surveying, dial gauge, strain gauge, or an electrical device such as an LVDT, potentiometer or vibrating wire gauge are used to obtain the measurement of relative displacement. Relative displacement inside rock mass is measured by driving a borehole and installing a borehole deformation gauge in single or multiple points.

Optical Surveying

This method employs high quality surveying techniques to measure displacements or deformation of ground. This method is one of the most popular techniques for measurement of surface subsidence. Disadvantage of the method is that it tends to interfere with the day to day mining activities if interval of the measurement is frequent. Also, small deformation may not be possible to measure using this technique. Survey instruments such as levels, theodolites, total stations, GPS/DGPS receivers, photogrammetric cameras or a combination of these instruments are generally used to collect deformation data in predefined intervals. Displacement of the surveyed points can be determined by comparing the coordinates from two or more survey periods.

Dial Indicator

A mechanical device used to convert linear movement of a spring-loaded plunger to large and more visible movement of a pointer that rotates above a dial. Dial gauge consists of a rack and pinion and gear train. Accuracies of dial gauge are generally ±0.025 mm or ±0.0025 mm for a typical range of 50 mm. Dial indicators are fragile and susceptible to the effect of dirt entering the gear train. These days, sealed and water proofing dial gauges are also available and widely used in the laboratories. A manual and digital dial gauge is shown in Figure 6.15.

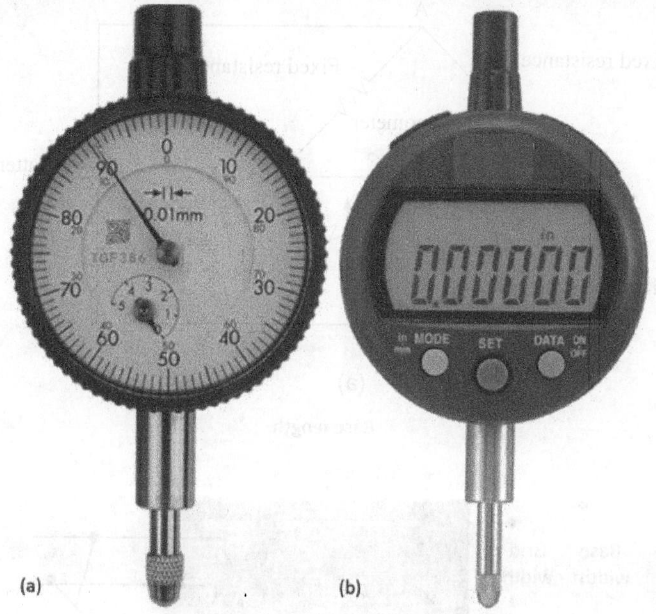

FIGURE 6.15 A dial gauge (a) manual type, (b) digital type.

Strain Gauge

An electrical resistance strain gauge is a conductor with the property that resistance changes in direct proportion to change in length. This resistance change, usually measured using a Wheatstone bridge, is related to the strain by the quantity known as strain sensitivity or *gauge factor* (*GF*) (Figure 6.16). The relationship between resistance change ΔR and length change ΔL is given by gauge factor, where

$$GF = \frac{\Delta R}{R} \times \frac{L}{\Delta L} = \left(\frac{\Delta R}{R}\right) \Big/ \varepsilon \tag{6.19}$$

The Wheatstone bridge circuit has four "arms" formed by resistances, R_1, R_2, R_3 and R_4 where R_4 represents the strain gauge. A voltage is applied between A and B and the resistance R_4 is altered until no current flows between C and D. At this point, the needle of the galvanometer is not deflected, i.e., $\Delta V = 0$ and the bridge is balanced. Under these conditions, since, R_1, R_2, and R_3 are known, R_4 can be calculated using Eq. (6.20).

$$\frac{R_1}{R_2} = \frac{R_3}{R_4} \tag{6.20}$$

This indicates the initial condition. Now, if length of the strain gauge changes due to applied load or pressure, resistance R_4 changes to $R_4 \pm \Delta R$ either in negative (compression) or positive (tension). As a result, voltage between points C and D becomes non-zero, $\Delta V \neq 0$. This change in voltage is directly related to the change in strain or resistance R_4 using the following equation.

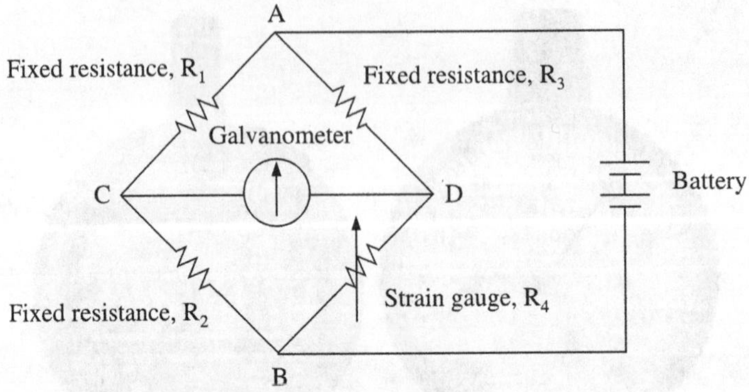

(a)

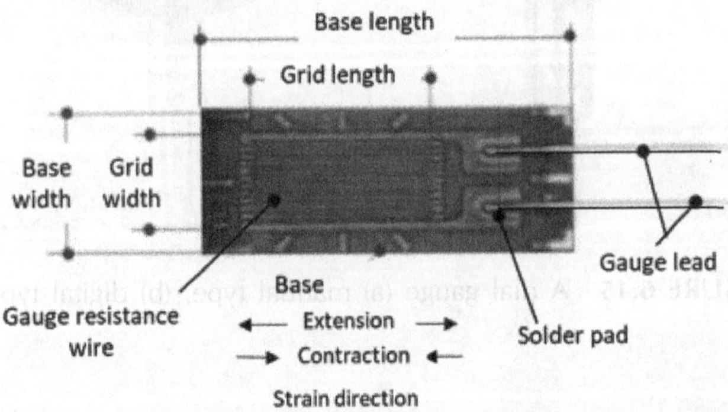

(b)

FIGURE 6.16 (a) Wheatstone bridge circuit, (b) Bonded wire type strain gauge.

$$\Delta V = \left(\frac{R_4}{R_3 + R_4} - \frac{R_2}{R_1 + R_2} \right) V \tag{6.21}$$

Assuming all resistances are having the same values in the initial stage, i.e. R, we can write

or, $$\Delta V = \left(\frac{(R + \Delta R_4)}{R + (R + \Delta R_4)} - \frac{1}{2} \right) V = \left(\frac{\Delta R_4}{4R + 2\Delta R_4} \right) V$$

Assuming $R >> 2\Delta R_4$ and applying Eq. 6.19, we can write

$$\varepsilon = \frac{4}{GF} \left(\frac{\Delta V}{V} \right) \tag{6.22}$$

The strain gauge readout box converts this change in voltage into strain and displays it. Five basic types of electrical resistance strain gauge are namely bonded wire, unbounded wire, bonded foil, semiconductor and weldable. The gauge factor for bonded foil and bonded wire gauges is usually close to 2. Semiconductor strain gauges have much larger gauge factor, between 50 and 200.

EXAMPLE 6.6 The resistance of a balanced Wheatstone bridge is 120 Ω with gauge factor of 2.2. An input voltage of 6 VDC is applied to the bridge. One of the strain gauge (R_4) attached by two pairs of 10 m long lead wire having resistance 0.025 Ω/m through the Wheatstone bridge. Find:

(a) the voltage output up to 3 decimal places in mV, and
(b) the strain gauge, i.e., R_4 with the lead wire is bonded in the middle of the cylindrical rock sample (as shown in Figure) in lateral direction to measure the Poisson's effect while uniaxial loading of the sample till failure. If Poisson's ratio of the rock is estimated to be 0.25 and axial strain is 8.8 micron, find the output voltage.

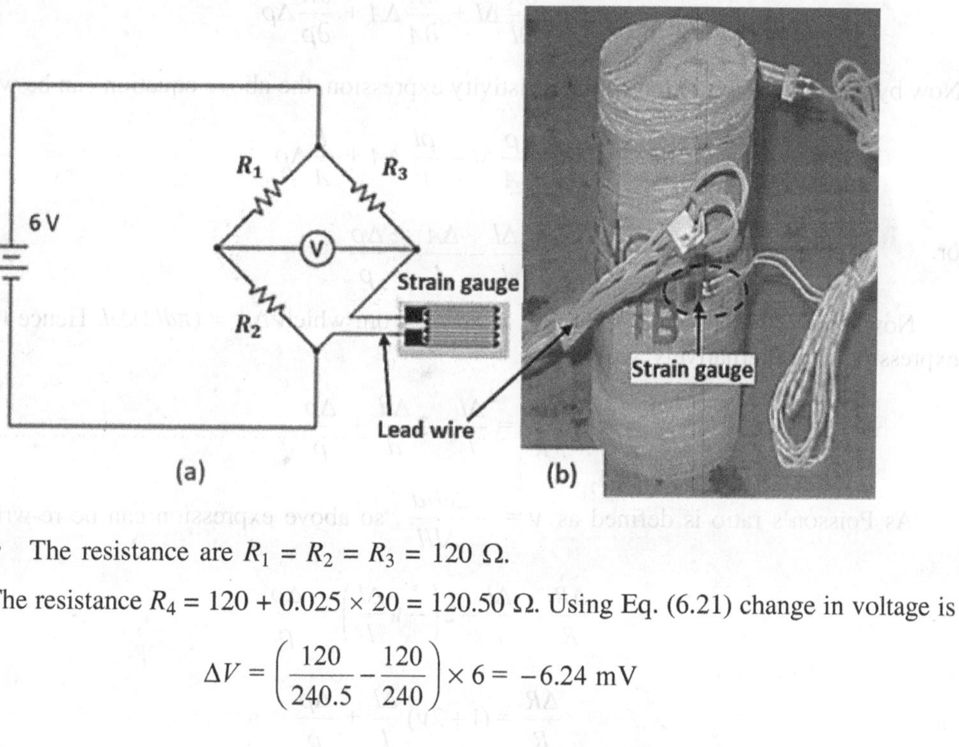

Solution: The resistance are $R_1 = R_2 = R_3 = 120\ \Omega$.

(a) The resistance $R_4 = 120 + 0.025 \times 20 = 120.50\ \Omega$. Using Eq. (6.21) change in voltage is

$$\Delta V = \left(\frac{120}{240.5} - \frac{120}{240}\right) \times 6 = -6.24 \text{ mV}$$

(b) $\varepsilon_l = -8.8\mu$, $\nu = 0.25$. The axial strain is $\varepsilon_a = -\frac{\varepsilon_l}{\nu} = \frac{-8.8\mu}{0.25} = 35.2\,\mu$

Change in resistance in axial and transverse direction is

$$\Delta R_a = GR_4\varepsilon_a = 2.2 \times 120 \times 35.2\mu = 0.009293\ \Omega$$

$$\Delta V_a = \left(\frac{R_4}{R_3 + R_4} - \frac{R_2}{R_1 + R_2}\right)V = \left(\frac{120 + 0.009293}{240.009293} - \frac{1}{2}\right) \times 6 = 0.116 \text{ mV}$$

$$\Delta R_l = GR_0\varepsilon_l = -2.2 \times 120 \times 8.8\mu = -0.002323\ \Omega$$

$$\Delta V_l = \left(\frac{R_4}{R_3 + R_4} - \frac{R_2}{R_1 + R_2}\right)V = \left(\frac{120 - 0.002323}{240 - 0.002323} - \frac{1}{2}\right) \times 6 = -0.029 \text{ mV}$$

EXAMPLE 6.7 Prove that the gauge factor can be written as $1 + 2\nu + \frac{1}{\varepsilon_l}\frac{\Delta\rho}{\rho}$. Here, ρ is the resistivity of the wire. Show that for metal the overall gauge factor is around 2.0.

Solution: Let us consider a long wire of length l having circular cross section with diameter d. When the wire is subjected to a tensile force, a strain will be generated and as a results, the dimension of the wire changes. If length is increasing then diameter is decreasing or vice-versa. If the resistivity of the wire is ρ, then the resistance of the wire is given by $R = \rho l/A$. From this expression, the change in resistance due to strain:

$$\Delta R = \frac{\partial R}{\partial l}\Delta l + \frac{\partial R}{\partial A}\Delta A + \frac{\partial R}{\partial \rho}\Delta\rho$$

Now by partial differentiation of the resistivity expression, the above equation can be written as:

$$\Delta R = \frac{\rho}{A}\Delta l - \frac{\rho l}{A^2}\Delta A + \frac{l}{A}\Delta\rho$$

or,

$$\frac{\Delta R}{R} = \frac{\Delta l}{l} - \frac{\Delta A}{A} + \frac{\Delta\rho}{\rho}$$

Now, for a circular cross section, $A = \pi d^2/4$, from which $\Delta A = (\pi d/2)\Delta d$. Hence the above expression can alternatively written as:

$$\frac{\Delta R}{R} = \frac{\Delta l}{l} - 2\frac{\Delta d}{d} + \frac{\Delta\rho}{\rho}$$

As Poisson's ratio is defined as $\nu = -\frac{\Delta d/d}{\Delta l/l}$, so above expression can be re-written as:

$$\frac{\Delta R}{R} = \frac{\Delta l}{l} - 2\left(-\nu\frac{\Delta l}{l}\right) + \frac{\Delta\rho}{\rho}$$

or,

$$\frac{\Delta R}{R} = (1 + 2\nu)\frac{\Delta l}{l} + \frac{\Delta\rho}{\rho}$$

or,

$$GF = \frac{\Delta R/R}{\Delta l/l} = 1 + 2\nu + \frac{1}{\varepsilon_l}\frac{\Delta\rho}{\rho}$$

For most metal $\nu \approx 0.3$ and the term $\frac{1}{\varepsilon_l}\frac{\Delta\rho}{\rho}$ representing strain-induced changes in resistivity is small. So, that the overall GF is around 2.0.

Linear Variable Differential Transformer (LVDT)

A LVDT consists of a movable magnetic core passing through one primary and two secondary coils as shown in Figure 6.17. An AC voltage, called the excitation voltage, is applied to the primary coil, thereby inducing an AC voltage in each secondary coil, with a magnitude that

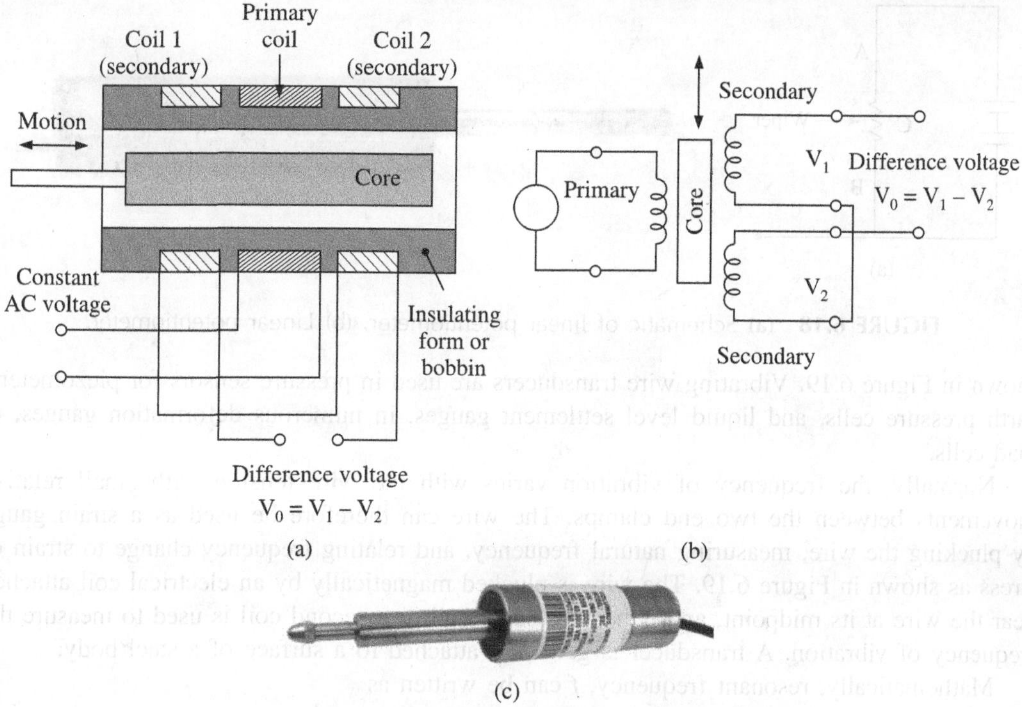

FIGURE 6.17 (a) and (b) Schematic of LVDT, (c) LVDT.

depends on the proximity between the magnetic core and each secondary coil. The secondary voltages are connected in series but opposite in direction so that the net output of the LVDT is the difference between these two voltages. When the core is at its mid-position (equidistant from the secondary coils), the net output voltage is zero. As the core moves off the centre, the output voltage increases linearly in magnitude with a polarity depending on the direction of core displacement. Since the core of an LVDT does not contact the coils, friction is avoided. There is no hysteresis and LVDTs are particularly suitable for measuring dynamic motions and very small displacements.

Potentiometer

Potentiometer is a device with a movable slider, usually called wiper that makes electrical contact along a fixed resistance strip. Unlike LVDT, potentiometer device is contact type. As shown in Figure 6.18, a regulated DC voltage is applied to the two ends of the resistance strip and the voltage between B and C is measured as the output signal. The voltage varies between the voltage at A and the voltage at B as the wiper moves from A to B. When the device is used for measurement of linear deformation and the relationship between wiper position and output signal is proportional, the device is called a *linear potentiometer*.

Vibrating Wire Transducer

Vibrating wire transducer consists of a length of a steel wire and it is mounted between two mounting blocks. The wire is tensioned so that it is free to vibrate at its natural frequency as

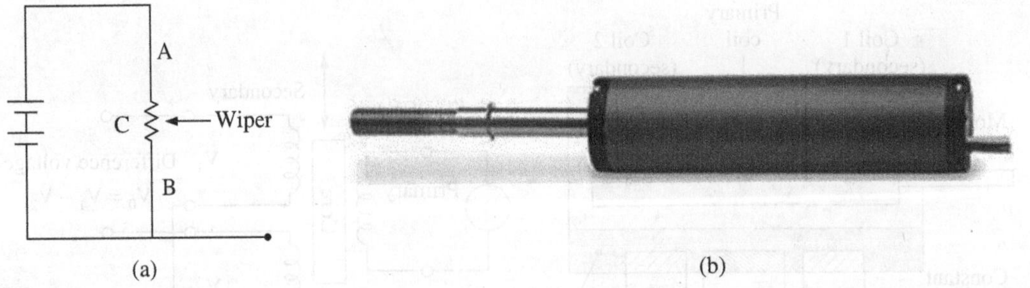

FIGURE 6.18 (a) Schematic of linear potentiometer, (b) Linear potentiometer.

shown in Figure 6.19. Vibrating wire transducers are used in pressure sensors for piezometers, earth pressure cells, and liquid level settlement gauges, in numerous deformation gauges, in load cells.

Normally, the frequency of vibration varies with the wire tension with small relative movements between the two end clamps. The wire can therefore be used as a strain gauge by plucking the wire, measuring natural frequency, and relating frequency change to strain or stress as shown in Figure 6.19. The wire is plucked magnetically by an electrical coil attached near the wire at its midpoint, and either this same coil or a second coil is used to measure the frequency of vibration. A transducer is generally attached to a surface of a steel body.

Mathematically, resonant frequency, f can be written as

$$f = \frac{1}{2L}\sqrt{\frac{\sigma g}{\rho}} \tag{6.23}$$

where, L = length of the wire, σ = tensile stress (N/m^2), ρ = unit weight of the wire material (N/m^3) and g = acceleration due to gravity (m/s^2)

$$\sigma = \frac{4L^2\rho}{g} f^2 = Kf^2 \tag{6.24}$$

The constant K is known as Gauge Factor. It means that if frequency increases, tension changes as square of it. During installation, if the initial frequency is f_o, then change in stress after installation would be:

$$\Delta\sigma = 1000\,Kz = K'z \tag{6.25}$$

A digit is then written as, $z = \frac{(f^2 \ f_o^2)}{1000}$ and K' is adjusted Gauge Factor. Once the stress is known, displacement, d can be estimated since modulus of elasticity of the wire, E is known. It is given as

$$d = \frac{\Delta\sigma L}{E} \tag{6.26}$$

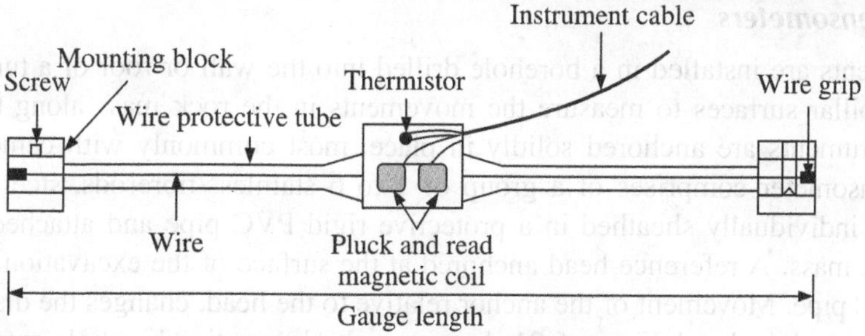

(a)

(b)

FIGURE 6.19 (a) Schematic of surface-mounted vibrating wire strain gauge, (b) Vibrating wire strain gauge.

6.5.2 Displacement Measuring Instruments for Field Applications

Displacement measurements between mine roof and floor, squeezing of tunnel walls and movement of slope face are measured *in situ* using several instruments which are explained below. It may be noted that there are many other sophisticated instruments available such as 3D laser scanner, slope stability radar (SSR) and differential global positioning system (DGPS) used for deformation measurement of rock surfaces. Readers are encouraged to refer appropriate literature for such instruments.

Tape Extensometer

Convergence relative to two opposite surfaces of excavation is measured by means of a tape or rod extensometer by installing targets attached to the walls and roof/floor. The instrument consists of adjustable length, with a dial gauge, Vernier scale, or micrometer attached at one end. The measuring stations are established by inserting an anchor in a hole into the roof or floor, vertically aligned to each other. Figure 6.20 shows a tape extensometer with digital readout unit.

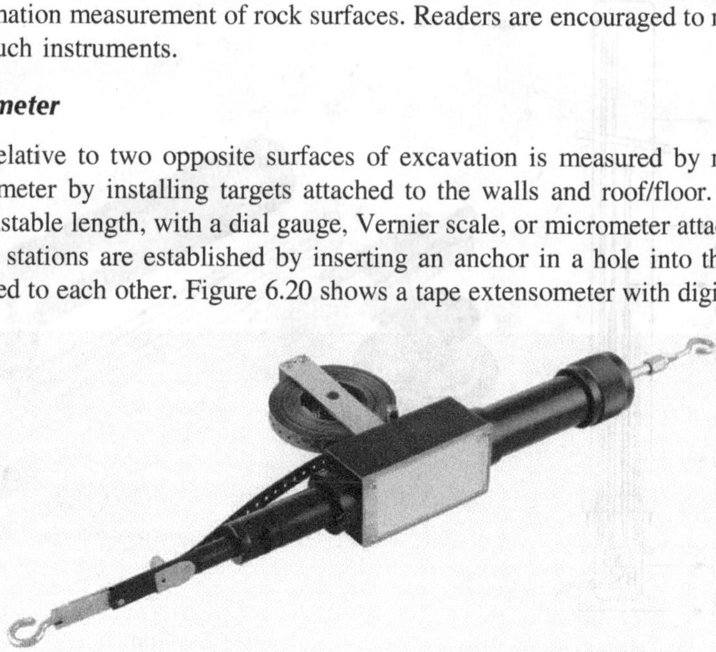

FIGURE 6.20 A digital tape extensometer.

Borehole Extensometers

These instruments are installed in a borehole drilled into the wall or roof of a tunnel/mines, on rock face, or pillar surfaces to measure the movements in the rock mass along the axis of the hole. The instruments are anchored solidly in place, most commonly with cement grout. The borehole extensometer comprises of a group of 1 to 6 stainless/fibrerods, steel wire ropes or steel tape reel individually sheathed in a protective rigid PVC pipe and attached to an anchor inside the rock mass. A reference head anchored at the surface of the excavation terminates the protective PVC pipe. Movement of the anchor relative to the head, changes the distance between the anchor point and the head. Figure 6.21 shows a multiple-position borehole extensometer with different components. The change in distance is measured either by mechanical devises (i.e., dial gauge) or electrical devices (i.e., LVDT). The data can be read manually or automatically in real-time. There are two types of extensometers:

(i) **Single-position borehole extensometers (SPBX):** It usually consists of an end-anchored rod equipped with a measuring facility at the hole collar or head. This instrument measures the relative axis displacement between anchor point and head along the length of a borehole. Assume that point A is the anchor point and LVDT is attached with the head at point B (Figure 6.21). If there is any relative movement between these points due to fracture or otherwise, LVDT records this movement with time.

(ii) **Multiple-position borehole extensometers (MPBX):** It records differential movements of anchor points installed at various heights from the excavation surface. In other words, these extensometers can measure the strata deformation at various points along a borehole providing differential settlement data below the anchor points.

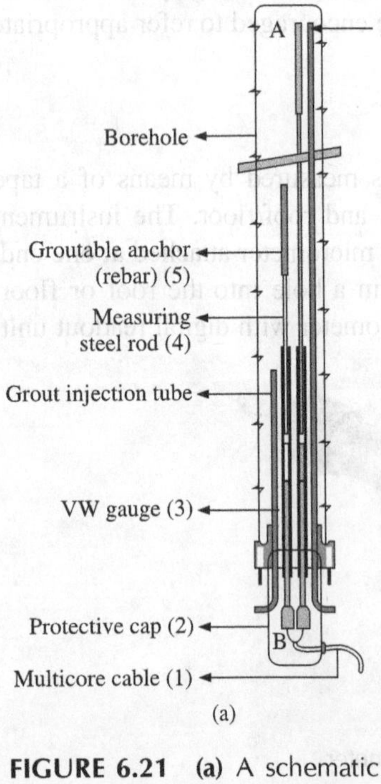

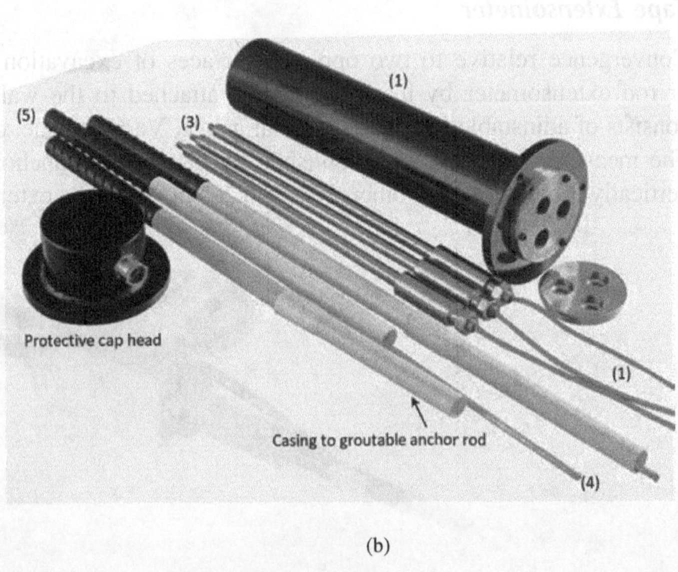

FIGURE 6.21 (a) A schematic diagram of multiple-position borehole extensometer, (b) MPBX.

EXAMPLE 6.8 A multiple-position borehole extensometer is installed to identify the location of a single discontinuity plane by measuring deformations at three locations as shown in the Figure. The absolute readings of deformations measured on two different time span are given in the Table. Based on the measured data locate the position of discontinuities.

Reading	*Anchor*-1 (mm)	*Anchor*-2 (mm)	*Anchor*-3 (mm)
Initial	1.37	7.51	22.94
Final	3.02	14.35	35.42

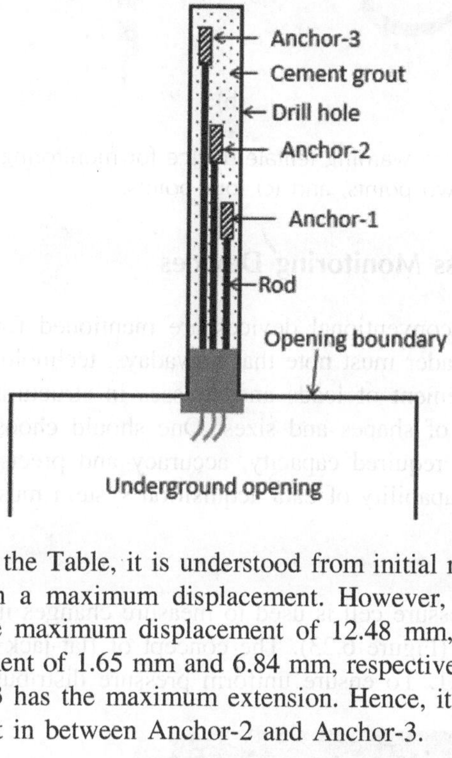

Solution: From Figure and the Table, it is understood from initial reading that the Anchor-3 is deep inside the roof with a maximum displacement. However, the final reading shows that the Anchor-3 shows the maximum displacement of 12.48 mm, where as Anchor-1 and Anchor-2 recorded displacement of 1.65 mm and 6.84 mm, respectively. These data imply that rod connected with Anchor-3 has the maximum extension. Hence, it can be assumed that the discontinuity plane may exist in between Anchor-2 and Anchor-3.

Telltale Devices

A telltale is an unstressed or un-tensioned wire, rod or pipe installed inside rock mass for measuring deformation of rock. The far end of telltale is anchored mostly by a spring in the rock mass. The other end of the telltale is connected to a vibrating wire gauge transducer or any visual display unit attached near the excavation surface (Figure 6.22). The display unit is secured at the excavation by a reference tube. As the rock deforms, the top end of the telltale moves to a new location. If the telltale is connected to mine roof, movement will be downward direction. The movement of anchor point along the borehole is assumed to be equal to the deformation of rock. In mining applications, movement of telltale is marked by green, yellow and red zones using millimeter scale. Nowadays, multi-point telltale is also available with automatic data acquisition system. The top anchor of a dual point telltale is placed at a height twice the bolt length and the bottom one is anchored in between the bolt length. The accuracy of such system is in the range of 0.5 mm.

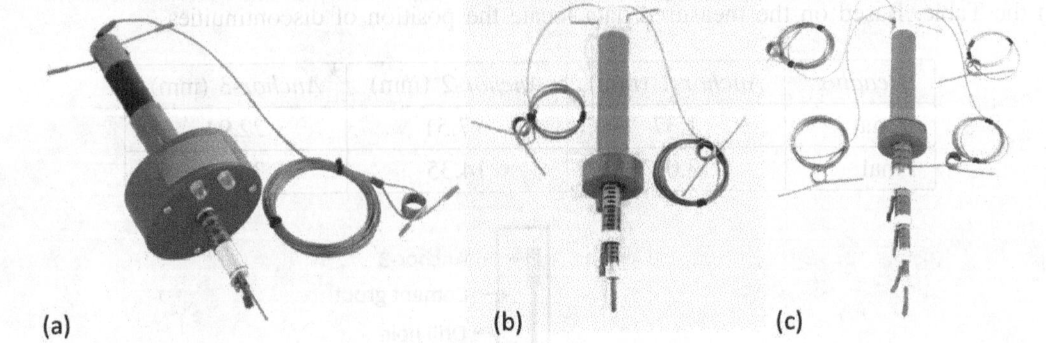

FIGURE 6.22 A typical auto warning telltale device for monitoring roof deformation at (a) single point, (b) two points, and (c) four points.

6.5.3 Load and Stress Monitoring Devices

In the following, a few conventional devices are mentioned for measurement of loads and stresses in rock mass. Reader must note that nowadays, technologically advanced instruments are available for measurement of loads and stresses in structural members. The sensors are also available in variety of shapes and sizes. One should choose a sensor according to the application based on the required capacity, accuracy and precision. Accordingly, shape and size of the sensors and capability of data acquisition system must be selected.

Flat-jack

Flat-jack or hydraulic pressure cell is used to measure changes in total normal stress between rock and support system (Figure 6.23). The concept of flat-jack and one of its application is explained in Section 6.3.1. To ensure uniform pressure distribution during installation, each

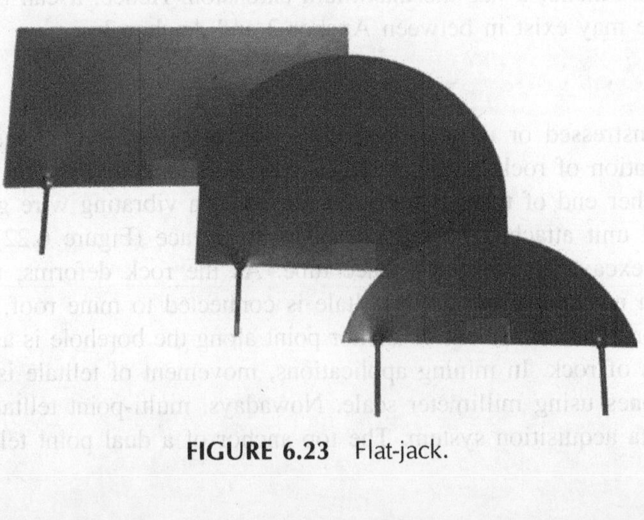

FIGURE 6.23 Flat-jack.

pressure cell is sandwiched between two pieces of flat steel plates before installation. Normal stress transferred from the surrounding filling, rock, shotcrete are measured by balancing the fluid pressure applied to the reverse side of the diaphragm (ISRM, 1987).

Piezo Load Cell

Piezoelectric load sensor is used to measure load applied to any material where the sensor is placed. Figure 6.24(a) illustrates the piezoelectric effect with the help of a compression disk. A compression disk looks like a capacitor with a piezoceramic material sandwiched between two electrodes. A force applied perpendicular to the disk causes voltage difference at the electrodes. The voltage difference (V) at the electrodes is represented as:

$$V = \frac{CFd}{A} \tag{6.27}$$

where, C = piezo-constant, F = applied force, d = thickness and A = electrode area.

By measuring the voltage, force F can be known for a given piezo sensor. Nylon shielding is done to filter out the required signal from noise. This type of load sensors are used for many different applications in the field of geotechnical, civil, mining, mechanical and aerospace engineering and also medical technology. A typical button type piezo load cell is shown in Figure 6.24(b).

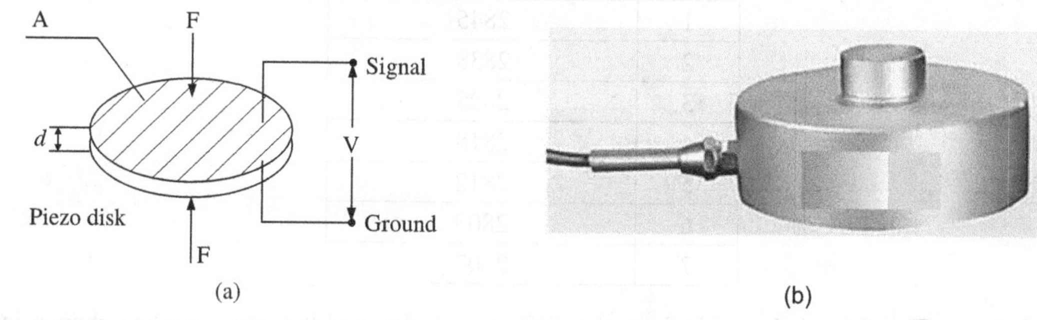

FIGURE 6.24 (a) Working principle of piezoelectric sensor and (b) Button type piezo sensor.

Earth Pressure Cell

Earth pressure cell is made from two stainless steel thin plates welded together around the periphery so as to keep a narrow space between them. This space is filled with de-aired hydraulic oil. It is assumed that the external pressure exerted by the rock mass above it is balanced by the internal fluid pressure. The developed internal pressure is measured using a pressure transducer where the oil pressure is converted to an electrical signal mostly using a vibrating wire gauge system. Figure 6.25 shows a typical earth pressure cell used for geotechnical applications such as measurement of pressure inside a dam, mine pillars and others. These pressure cells are firmly grouted inside a mine pillar or soil structure for proper contact with the surrounding rock or soil strata.

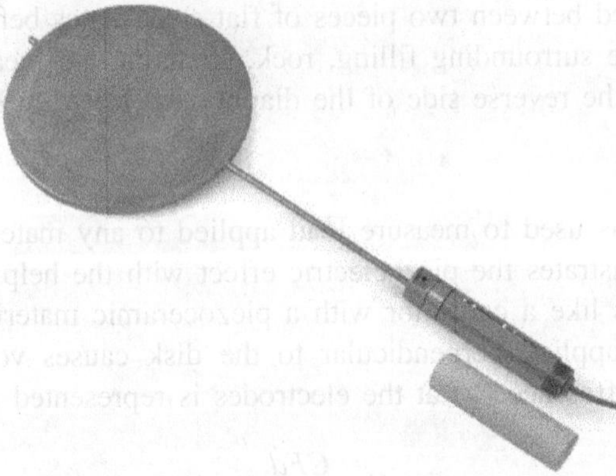

FIGURE 6.25 A circular earth pressure cell with steel plate and VW gauge sensor.

EXAMPLE 6.9 A vibrating wire transducer based stress measurement cell is installed inside a rib pillar. The frequencies measured from the day of installation for 7 days are given below. Assume the Gauge factor is 1.28×10^4 Pa/s^2. Plot the trend of stress change in the pillar.

Day	*Frequency* (s^{-1})
0	2854
1	2845
2	2838
3	2825
4	2818
5	2812
6	2809
7	2807

Solution: The Table below calculates the digit in each day and the corresponding change in stress based on Eq. 6.23.

Day	*Frequency* (s^{-1})	*Digit*	*Change of stress* (MPa)
0	2854	0	0.00
1	2845	–51.291	–0.66
2	2838	–91.072	–1.17
3	2825	–164.691	–2.11
4	2818	–204.192	–2.61
5	2812	–237.972	–3.05
6	2809	–254.835	–3.26
7	2807	–266.067	–3.41

The plot below shows the trend of stress change in the pillar. Negative values represent compressional stress.

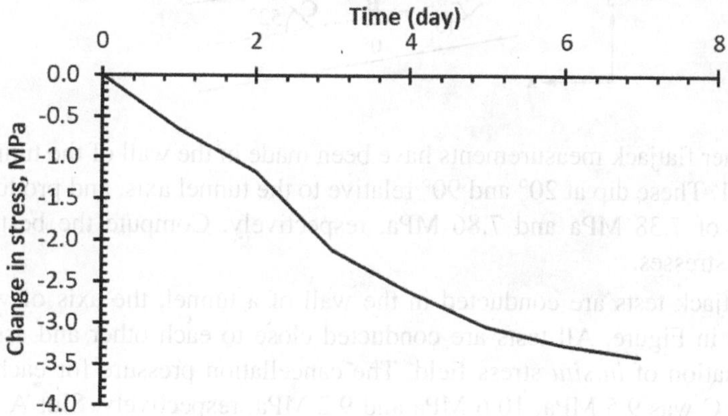

6.6 DATA ACQUISITION SYSTEM (DAS)

It is impractical to take manual readings from multiple sensors which are installed in mines. Sensors may be of different types and are installed at different locations in drives/galleries and at different levels/horizons. An ideal data acquisition system should acquire data from sensors, check data integrity while transmission and store them in a data storage facility. A feedback type data acquisition system can also be designed to send signal to the machine/support where sensor is installed for appropriate response to the acquired data. This type of DAS required real-time data analysis, setting up appropriate trigger levels based on the acquired data and a control system. Generally, a DAS is comprised of six stages: (i) signal (data) acquisition, (ii) signal conditioning, (iii) analog to digital conversion, (iv) signal transmission, (v) signal processing, and (vi) data storage. All data acquisition systems consist of two distinct components: (i) hardware incorporating microprocessor, sensors, amplifiers, counters and cables, and (ii) software providing protocols and instructions and visual display.

EXERCISES

1. Three flatjack tests have been made close to each other in the wall of a long, straight tunnel, the axis of which dips at 10° as shown in Figure. It is assumed that the flatjacks are in the same stress field. The slots for the flatjacks were cut normal to the wall of the tunnel, and were oriented relative to the tunnel axis as shown. The cancellation pressure for each of the flatjacks A, B and C was 7.56 MPa, 6.72 MPa and 7.50 MPa, respectively. Compute the principal stresses and their directions.

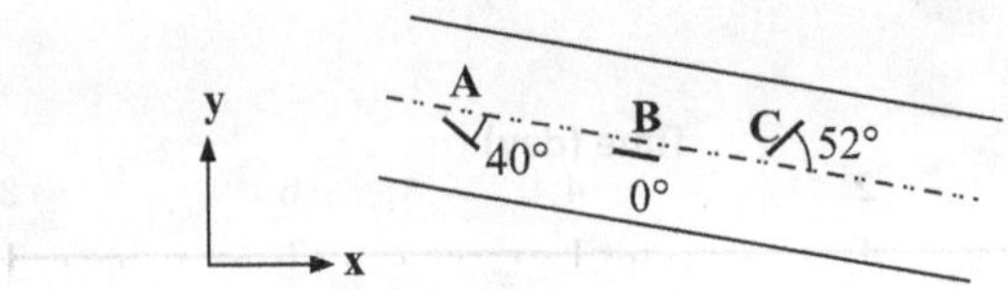

2. Two further flatjack measurements have been made in the wall of the tunnel considered in Exercise 1. These dip at 20° and 90° relative to the tunnel axis, and produced cancellation pressures of 7.38 MPa and 7.86 MPa, respectively. Compute the best estimate of the principal stresses.
3. Three flatjack tests are conducted in the wall of a tunnel, the axis of which rise at 10° as shown in Figure. All tests are conducted close to each other and assumed that there is no variation of *in situ* stress field. The cancellation pressure for each of the flatjacks A, B and C was 9.5 MPa, 10.6 MPa and 9.2 MPa, respectively. Slot A is perpendicular, Slot B is along and Slot 3 is at 45° to the tunnel axis. Compute the normal and shear stresses as shown in the Cartesian co-ordinate system.

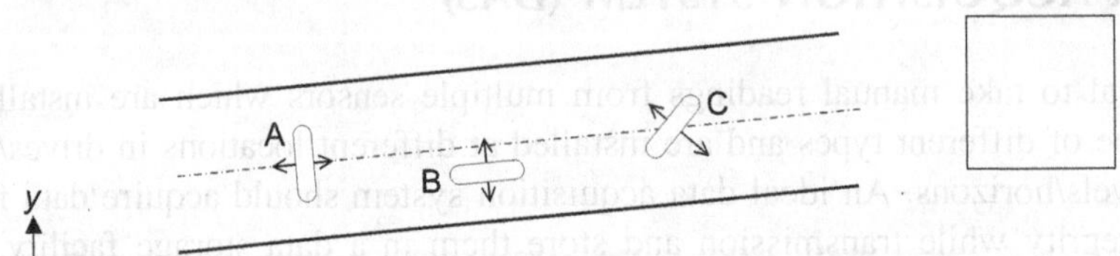

4. A horizontal tunnel rectangular in shape is excavated at a depth of 500 m below the surface. The average density of overlying rock mass is 2250 kg/m^3.
   (a) Three flat jack tests are carried out on the surface of the heading as shown in Figure. The cancellation pressure of the flatjacks A, B and C are 11.5 MPa, 22.0 MPa and 16.75 MPa, respectively. Plane of slot A is horizontal, slot B is vertical and slot C is inclined at 45° with *x*-axis. Determine the vertical and horizontal stresses. Find the principal stresses in X-Y plane.
   (b) One more flat jack test is carried on the side wall of the tunnel with cancellation pressure of 24.75 MPa. Then find the complete stress tensor.

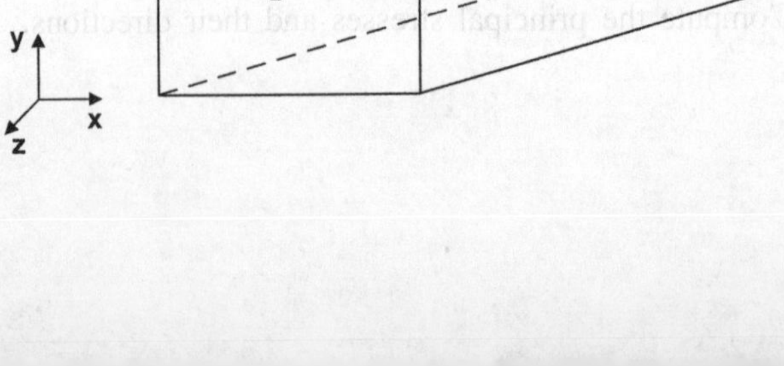

5. A horizontal tunnel rectangular in shape is excavated at a depth of 500 m below the surface. The average density of overlying rock mass is 2250 kg/m^3.

(a) Three flat jack tests are carried out on the surface of the heading as shown in Figure. The cancellation pressure of the flatjacks A, B and C are 13.7 MPa, 12.3 MPa and 6.8 MPa, respectively. Determine the vertical and horizontal stresses. Find the principal stresses in x-y plane.

(b) Three more tests are required to be conducted to find the complete stress tensor at this location. Two tests are conducted in *y–z* plane with cancellation pressure of slot D and E as 11.9 MPa and 8.0 MPa. Slot D is along y-axis whereas slot D is inclined at an angle of 45° with *z*-axis. Third flat jack test was conducted in the tunnel roof and equally inclined with *x* and *z*-axis. The cancellation pressure for slot F is 5.6 MPa. Then find the 3D stress tensor.

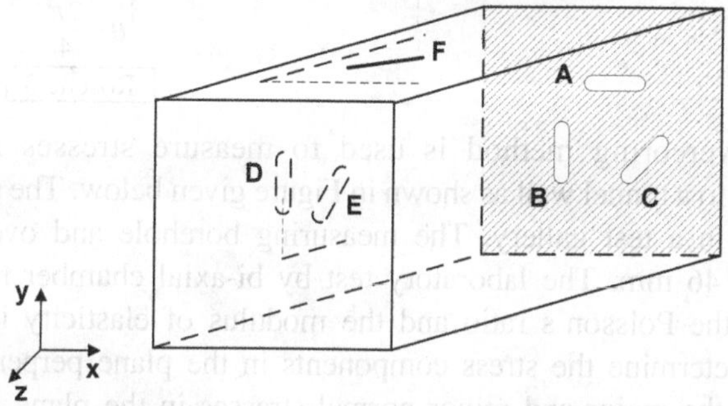

6. CSIR/CSIRO cell is used to estimate the *in situ* stress in a rock mass. The strain measured by CSIR strain gauge installed at three locations in a plane of the borehole at 60° apart in the form of a three strain gauge rosettes as shown in Figure. Gauge A is always oriented to measure in the axial direction of borehole. Gauge C is always oriented to measure in the positive θ direction. The measurement of strain by three rectangular rosettes are measured as follows:

	Gauge	*Strain* $\times 10^{-6}$, m/m
	A	204
Rosette 1	B	55
	C	276
	A	102
Rosette 2	B	49
	C	52
	A	35
Rosette 3	B	131
	C	–18

The laboratory tests indicate that the rock is isotropic with the elastic modulus 800 MPa and Poisson's ratio 0.25. Assume plane stress condition. Determine the magnitude and direction of major, minor principal stresses for individual rosette.

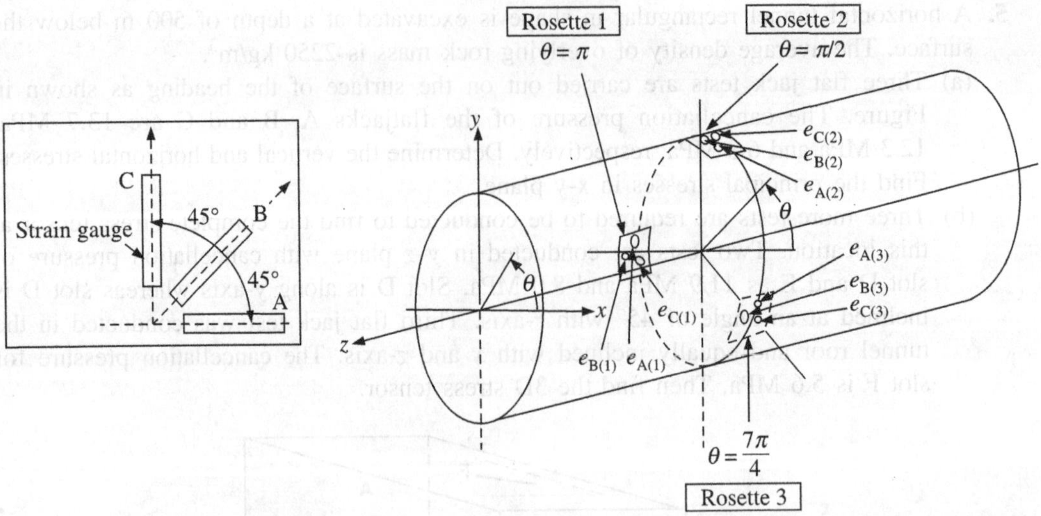

7. The USBM overcoring method is used to measure stresses in a borehole drilled perpendicularly to a tunnel wall as shown in Figure given below. The site of the measurement is 15 m deep, in a test gallery. The measuring borehole and overcoring diameters are 31.75 mm and 46 mm. The laboratory test by bi-axial chamber reveals that the rock is isotropic with the Poisson's ratio and the modulus of elasticity is 0.20 and 13.79 GPa respectively. Determine the stress components in the plane perpendicular to each of the boreholes, and the major and minor normal stresses in the plane and their directions.

Borehole No.	*Deformation of button pair* (mm) *and their orientation with horizontal*			*Plane of measurement*
	U_1	U_2	U_3	
1.	0.0762	0.0508	0.0254	*x-y*
2.	0.0869	0.0464	0.0328	*x-z*
3.	0.0694	0.0674	0.0394	*y-z*

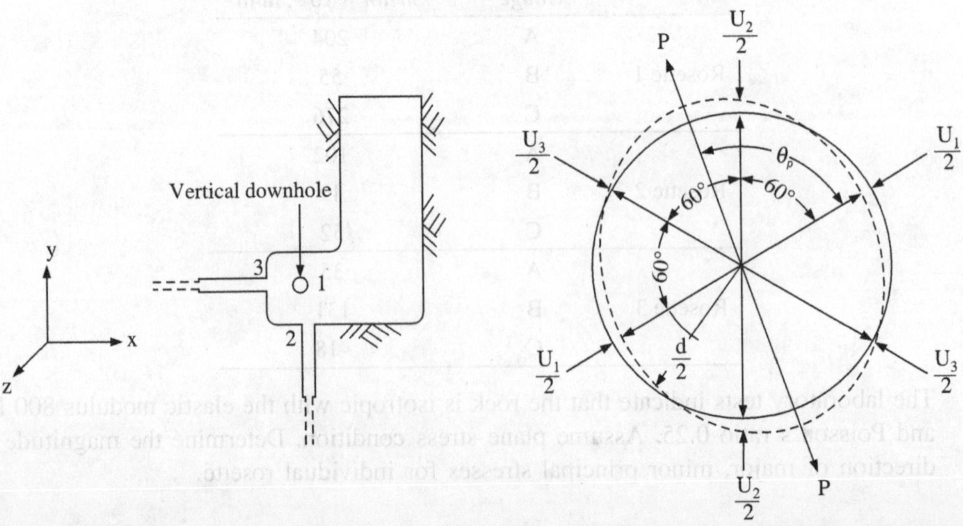

8. In a typical field investigation by overcoring test of dolomite using the USBM deformation gauge shows $U_1 = 11.7\ \mu m$, $U_2 = 11.9\ \mu m$, and $U_3 = 13.1\ \mu m$. The pilot hole and overcore diameter of the borehole is 38 mm and 150 mm respectively. The laboratory test reveals the deformation modulus, E and Poisson's ratio of the rock mass as 71 GPa and 0.3 respectively. Find the secondary major and minor principal stresses and their direction.

9. Deformations are measured in a borehole drilled perpendicular to a tunnel wall by USBM gage. The measuring borehole diameter is 38 mm and 1.524 m deep, and overcore diameter of 150 mm. The overcoring process results into the movement of Pair 1 outward by 0.0146 mm; Pair 2 by 0.0508 mm; and Pair 3 by 0.0254 mm. It is given that the first pair of buttons is horizontal. If modulus of elasticity is 20 GPa and Poisson's ratio is 0.22, determine the stress components in the plane perpendicular to the borehole, and the major and minor normal stresses in the plane and their directions. (Assume that initial stress parallel to the borehole is significant).

10. The stress in a granitic rock mass has been measured by the hydraulic fracturing technique. Two tests were conducted in a vertical borehole: one test at a depth of 500 m, and the other test at a depth of 1000 m. The results were as follows:

Depth (m)	*Breakdown pressure, P_B* (MPa)	*Shut-in pressure, P_s* (MPa)
500	14.0	8.0
1000	24.5	16.0

Given that the tensile strength, σ_t, of the rock is 10 MPa, estimate and list the values of σ_1, σ_2 and σ_3 at the two depths. Average unit weight of the rock mass is 26 kN/m^3. State all of the assumptions.

11. In a gold mine, stresses are measured at a depth of 3352 m despite complications due to thermal effects, borehole closure, and core disking. The vertical stress was measured to be 91.0 MPa. The north-south and east-west horizontal components are 75.5 MPa and 71.6 MPa, respectively. The major principal stress was calculated to be 142 MPa with a dip of 47°. The tensile strength of the rock is 2 MPa. Estimate the shut-in-pressure and breakdown pressure during hydraulic fracturing.

12. A 350 Ω strain gauge of gauge factor 2.2 is strained by 0.2×10^{-3}. Find the change in resistance.

13. A metallic strain gauge resistance of 120 Ω and a gauge factor of 2.0 is pasted longitudinally on a cylindrical sandstone sample of diameter 54 mm. After uniaxial compressive strength (UCS) test on the sandstone sample, the UCS of the sample is found to be 130 MPa and average young's modulus is 10 GPa. Determine the change is resistance of the gauge that is caused because of the loading the sandstone sample upto failure.

14. A strain gauge is bonded to a steel cylindrical post of 4 m in length and 100 mm in diameter installed in the underground gallery to measure its axial load. Modulus of elasticity of the steel beam is 207 GN/m^2. The unstrained resistance of the strain gauge

350 Ω and gauge factor of 2.2. When the load is applied by the overlying strata, the gauge resistance changes by 0.015 Ω. Calculate the change in length of the steel beam and the stress applied.

15. A multiple-position vibrating wire borehole extensometer is installed in an underground opening. Anchor points are a depth of 5 m, 10 m and 15 m from the collar of a drill hole into the roof as shown in Figure. The measured data over a period of 450 days are shown in the table. Plot the deformation versus time of each sensors and comment on the roof characteristics.

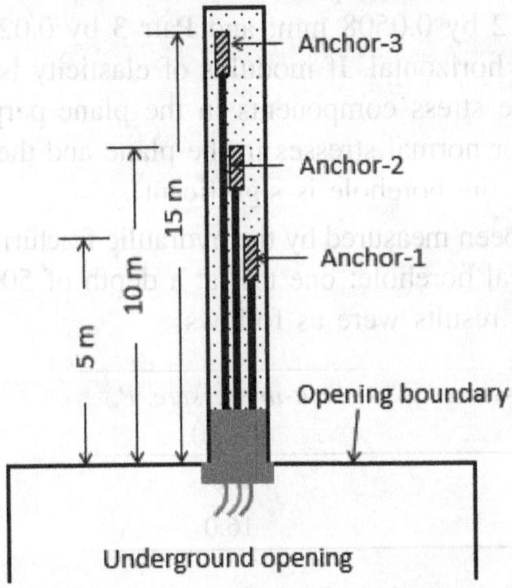

Time, day	Displacement in mm		
	Anchor-1	Anchor-2	Anchor-3
0	0	0	0
60	0.41	2.18	5.31
121	0.96	4.52	16.17
182	1.37	7.51	22.95
304	21.02	70.35	53.42
364	22.46	73.68	55.21
395	22.51	73.90	57.32
450	22.84	76.84	58.66

16. A vibrating wire pressure cell is installed in a metal mine pillar to measure the developed stress over a period of 500 days. Just after the installation of the pressure cell, the frequency of the vibrating wire, i.e., the initial reading is 2873.8 Hz followed by eight subsequent readings as shown in the Table. If the gauge factor (G) of the instrument provided by the manufacturer is –0.9618 MPa/digit. Find the maximum pressure developed and infer based on the pressure vs time plot.

Sl. No.	*Day*	*Frequency* (Hz)
R_0	0	2873.8
R_1	50	2878.8
R_2	100	2886.8
R_3	250	2919.7
R_4	300	2927.7
R_5	350	2939.5
R_6	400	2941.6
R_7	450	2944.6
R_8	500	2944.6

Chapter 7

Stresses Around Underground Opening

7.1 INTRODUCTION

Stresses and displacements around an underground opening are required to be known for design of tunnels, underground spaces for civil use and mine openings for mineral extraction or other practical problems in the field of geotechnical, petroleum and mining engineering. Analytical solutions for estimating stresses and displacement have been derived by considering material properties, constitutive relationship and far-field boundary conditions. The analytical solutions are useful to understand the effect of individual parameter on the geometry of underground opening. In the year 1898, Kirsch had first derived an analytical solution of stresses and displacements around circular opening in plane strain condition. This solution has further been improved by incorporating internal support pressure around the tunnel boundary and joint near opening.

It has been noticed that different shapes and sizes of the opening such as circular, D or horse-shoe, elliptical, rectangular, etc. do not produce similar stress pattern around the openings. This chapter first elaborates the concepts and application of Kirsch solution for a circular opening of radius (r) and far-field (*in situ*) stresses as p_0 and kp_0, where k denotes a ratio between horizontal to vertical stresses called *stress ratio*. Then the effect of different shapes and sizes are discussed mainly focusing the changes in stress region around the tunnel boundary. In underground mining, multiple parallel openings are frequently developed to extract coal or minerals. It has been found that induced stresses around opening would influence the stress region of other if this opening is located within the 'zone of influence'. The concept of zone of influence and estimation of its radius are also discussed in this chapter.

Finally, effect of joints in the vicinity of a circular tunnel/shaft are also estimated with practical examples.

7.2 CIRCULAR OPENING

Circular or semi-circular openings are preferred due to its symmetry, stability and possibility for faster excavation by sophisticated tunnel boring machines. Circular shape is always preferred in squeezing/loose grounds (Ramamurthy, 2014). An analytical relationship for stress and displacement distribution around a long circular tunnel was developed in an ideal rock mass by a German engineer Kirsch (1898). The relationship is valid for an ideal rock mass, defined by continuous, homogeneous, isotropic and linear-elastic (CHILE) material. Interested reader can refer to **Appendix B** for derivation of these relationships.

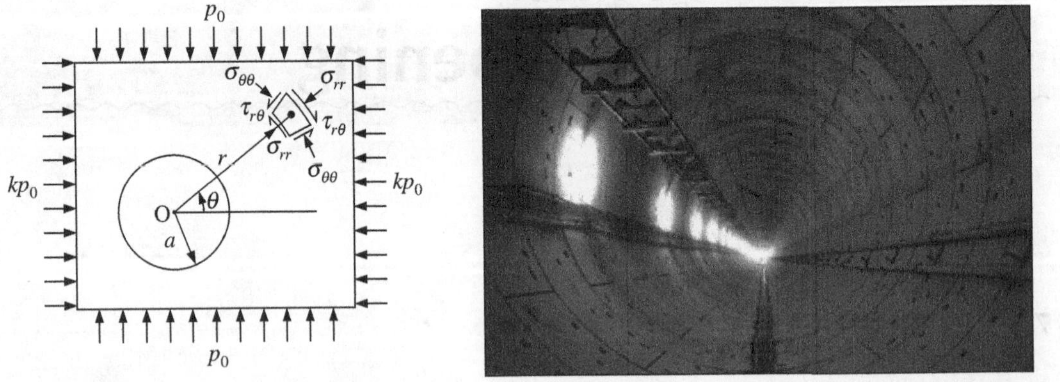

FIGURE 7.1 (a) Real tunnel and (b) Stress distribution around a circular opening.

Mathematically, a point (r, θ) defined by a polar coordinate system near a circular opening having radius (a) driven in biaxial stress fields (p_0) and (kp_0) has radial (σ_{rr}), tangential ($\sigma_{\theta\theta}$) and shear ($\tau_{r\theta}$) stresses as given by

$$\sigma_{rr} = \frac{p_0}{2}\left\{(1+k)\left(1-\frac{a^2}{r^2}\right)-(1-k)\left(1-\frac{4a^2}{r^2}+\frac{3a^4}{r^4}\right)\cos 2\theta\right\} \tag{7.1a}$$

$$\sigma_{\theta\theta} = \frac{p_0}{2}\left\{(1+k)\left(1+\frac{a^2}{r^2}\right)+(1-k)\left(1+\frac{3a^4}{r^4}\right)\cos 2\theta\right\} \tag{7.1b}$$

$$\tau_{r\theta} = \frac{p_0}{2}\left\{(1-k)\left(1+\frac{2a^2}{r^2}-\frac{3a^4}{r^4}\right)\sin 2\theta\right\} \tag{7.1c}$$

It is noticed that the stress fields are dependent on the radius of the tunnel, far-field stresses and orientation with respect to horizontal axis. Elastic stress fields do not depend on the material properties such as modulus of elasticity (E) and Poisson's ratio (ν). It may be noticed that a vertical section of a long tunnel Figure 7.1(b) is depicted in Figure 7.1(a). The tunnel axis is then directed towards out-of-plane or in-plane and the far-field stress along that direction is omitted in Eq. 7.1 considering plane strain conditions.

The radial and tangential displacements are obtained for plane strain condition as:

$$u_r = \frac{p_0 a^2}{4Gr}\left\{(1+k)-(1-k)\left[4(1-\nu)-\frac{a^2}{r^2}\right]\cos 2\theta\right\} \tag{7.2a}$$

$$u_\theta = \frac{p_0 a^2}{4Gr}\left\{(1-k)\left[2(1-2\nu)+\frac{a^2}{r^2}\right]\sin 2\theta\right\} \tag{7.2b}$$

where, $k = \dfrac{\sigma_H}{\sigma_V}$; shear modulus $(G) = \dfrac{E}{2(1+\nu)}$, ν = Poisson's ratio

EXAMPLE 7.1 A circular horizontal N-S tunnel of 2 m radius will be constructed at a depth of 560 m below surface as shown in figure given below. The vertical stress gradient is found to be 0.025 MPa/m. The horizontal stresses in north and east directions are σ_N = 13 MPa and σ_E = 14 MPa. Estimate radial and tangential stresses for $\theta = 0°$ and $\theta = 90°$ from the east axis.

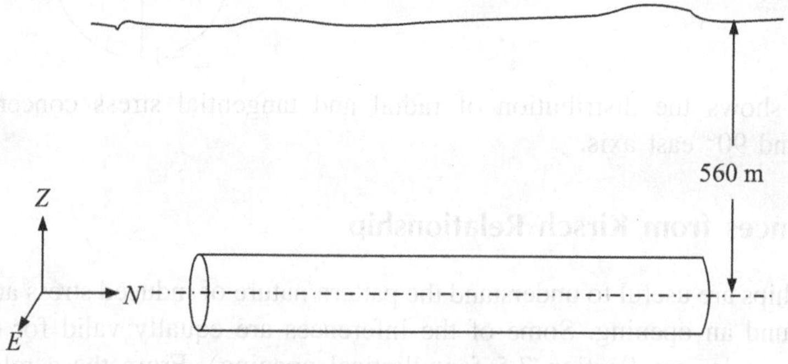

Solution: In this problem, the *in situ* stress fields are defined. Vertical stress (σ_V) is developed due to weight of overlying rock mass. The stress field (σ_N) is parallel to the axis of the tunnel and hence it has no effect on induced stresses around opening. Here, ratio $k = (\sigma_E/\sigma_V)$ and $p_0 = \sigma_V$ = 14 MPa. Using the above values, σ_{rr} and $\sigma_{\theta\theta}$ are estimated and given in the following table.

	$\theta = 0°$		$\theta = 90°$	
r/a	σ_{rr}/p_0	$\sigma_{\theta\theta}/p_0$	σ_{rr}/p_0	$\sigma_{\theta\theta}/p_0$
1.0	0.000	2.000	0.000	2.000
1.5	0.556	1.444	0.556	1.444
2.0	0.750	1.250	0.750	1.250
2.5	0.840	1.160	0.840	1.160
3.0	0.889	1.111	0.889	1.111
4.0	0.938	1.063	0.938	1.063
5.0	0.960	1.040	0.960	1.040

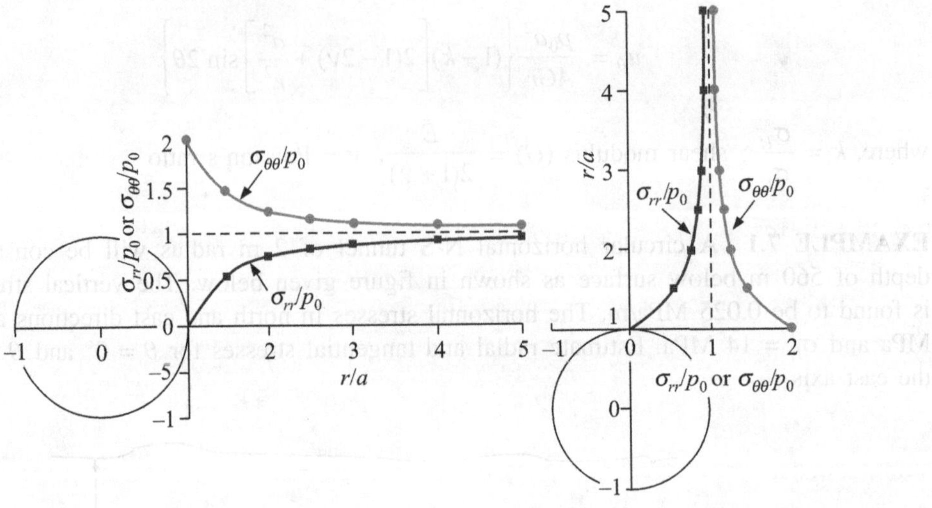

The figure shows the distribution of radial and tangential stress concentration around opening at 0° and 90° east axis.

7.2.1 Inferences from Kirsch Relationship

Kirsch relationships are useful to understand the pattern/nature of induced stress and displacement distribution around an opening. Some of the inferences are equally valid for other shapes of opening (for example, see Section 7.5 for elliptical opening). From these relationships, it is easy to understand the effect of various parameters like material properties, far-field stress, size and shape of the opening on stress and displacement distribution. The inferences drawn from Kirsch equation are as follows:

(i) Elastic stress distributions are independent of elastic constants, i.e., Young's modulus and Poisson's ratio.

(ii) Stresses at a point around an opening is a function of the ratio of radial to radius distance (r/a), a dimensionless quantity. It means that the stresses at the boundary of excavation are independent of radius or size of the opening. Further, it can be interpreted as, if the ratio (r/a) is the same for two different excavations (i.e., for radius a = 1 m and a = 10 m), the induced stresses around the opening will be same for both the cases.

(iii) For $r = 5a$, the ratio of induced to applied stresses are 96% of the *in situ* stress value as shown in Example 7.1. It means that the induced stresses are close to *in situ* stress for distance greater than or equal to five time the radius of opening.

(iv) For $r = a$, i.e., at the periphery of the opening, Kirsch equation for radial (σ_{rr}), and shear ($\tau_{r\theta}$) stresses are zero. These are the stress boundary conditions imposed while deriving the solution (**Appendix B**). The tangential ($\sigma_{\theta\theta}$) stresses at the boundary is expressed as:

(a) $\sigma_{\theta\theta} = p_0\,[1 + k + 2(1 - k)\cos 2\theta]$

(b) for $\theta = 0°$; $\sigma_{\theta\theta} = p_0(3 - k)$ and for $\theta = 90°$; $\sigma_{\theta\theta} = p_0(3k - 1)$

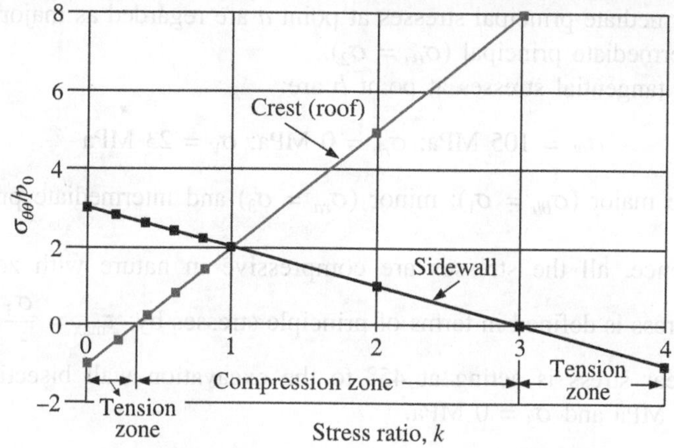

FIGURE 7.2. Variation of the ratio of tangential stress to *in situ* stress w.r.t. stress ratio.

This result can be interpreted as: for any shape of excavation, only tangential stress ($\sigma_{\theta\theta}$) exists at the boundary. The tangential stress at sidewall and crest (roof) are given by above expression. Figure 7.2 shows the variation of tangential stress with respect to the stress ratio (k). For $0 < k \leq 0.333$, crest is in tension. For $k \geq 3$, tensile stress occurs in sidewall. So, the boundary of the opening can experience tensile stress even if far-field stresses are in compression depending the value of k. Figure 7.2 also depicts that for $k = 0$, $\sigma_{\theta\theta} = 3p_0$ in sidewall and for $k = 3$, $\sigma_{\theta\theta} = 8p_0$ in the roof. Similarly, for $k = 1$, the stress concentration ($\sigma_{\theta\theta}$) is twice the *in situ* stress at sidewall, roof and circumference of the opening.

EXAMPLE 7.2 In a vertical circular shaft, the principle stresses are oriented E-W and N-S at 850 m depth. The magnitudes of principle stresses in E-W and N-S directions are 40 MPa and 15 MPa respectively. Determine the maximum compressive stress, the maximum tensile stress and the maximum shear stress at 850 m horizon on shaft boundary. Consider the vertical stress gradient as 0.027 MPa/m.

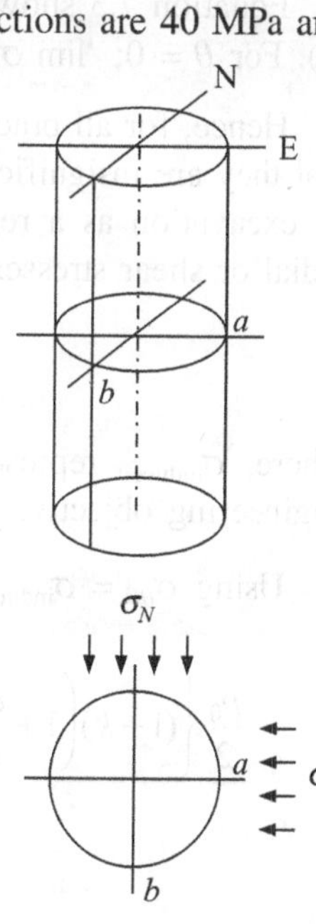

Solution: At 850 m horizon, the vertical stress is

$$\sigma_V = 850 \times 0.027 = 23 \text{ MPa}$$

No stress concentration occurs along the straight shaft wall in vertical direction (σ_V) since axis of the shaft is vertical.

As $\sigma_{rr} = 0$, the stress ratio is estimated as:

$$\sigma_{HE} = k \times \sigma_{hN}$$

$$k = 2.667$$

From Kirsch equation, the radial and tangential stresses at point a are:

$$\sigma_{\theta a} = 5 \text{ MPa}; \ \sigma_{ra} = 0 \text{ MPa}; \ \sigma_V = 23 \text{ MPa}$$

The stresses ($\sigma_{\theta a}$, σ_{ra}, σ_V) at point a are compressive in nature. The directions of the stresses are principal direction. So, maximum, minimum and intermediate principal stresses at point *a* are regarded as major ($\sigma_V = \sigma_1$); minor ($\sigma_{ra} = \sigma_3$) and intermediate principal ($\sigma_{\theta a} = \sigma_2$).

The radial and tangential stresses at point *b* are:

$$\sigma_{\theta b} = 105 \text{ MPa}; \ \sigma_{rb} = 0 \text{ MPa}; \ \sigma_V = 23 \text{ MPa}$$

At point *b*, the major ($\sigma_{\theta b} = \sigma_1$); minor ($\sigma_{ra} = \sigma_3$) and intermediate principal ($\sigma_V = \sigma_2$) stresses.

On circumference, all the stresses are compressive in nature with zero tensile stress. Maximum shear stress is defined in terms of principle stresses by: $\tau_{max} = \dfrac{\sigma_3 - \sigma_1}{2} = 52.5$ MPa.

The maximum shear stress is acting at 45° to the excavation wall bisecting the 90° angle between σ_1 = 105 MPa and σ_3 = 0 MPa.

7.2.2 Zone of Influence

Mathematically, induced stresses defined by Eq. 7.1 become equal to corresponding *in situ* stresses at infinite distance away from the tunnel centre. It can be easily shown by taking limit of Eq. 7.1a with respect to $r \to \infty$ as:

$$\lim_{r\to\infty} \sigma_{rr} = \lim_{r\to\infty} \frac{p_0}{2}\left\{(1+k)\left(1-\frac{a^2}{r^2}\right)-(1-k)\left(1-\frac{4a^2}{r^2}+\frac{3a^4}{r^4}\right)\cos 2\theta\right\}$$

$$= \frac{p_0}{2}\{(1+k)-(1-k)\cos 2\theta\} \tag{7.3}$$

Equation 7.3 shows, radial stress at a point located far away from excavation at an angle (θ). For $\theta = 0$; $\lim_{r\to\infty} \sigma_{rr} = kp_0$. Similarly, tangential stress for $\theta = 0$; $\lim_{r\to\infty} \sigma_{\theta\theta} = p_0$.

Hence, for all practical purposes, we are interested to find the changes in stress fields such that they are insignificant at a finite distance (r). The zone of influence is then defined around an excavation as a region within which at least one component of stress tensor (tangential, radial or shear stresses) is more by $d\%$ of its *in situ* value, expressed mathematically as

$$|\sigma_{\text{induced}} - \sigma_{in\ situ}| \geq \frac{d}{100}\sigma_{in\ situ} \tag{7.4}$$

where, σ_{induced} represents any component of induced stress, and $d\%$ value relevant to the engineering objective.

Using $\sigma_{\theta\theta} = \sigma_{\text{induced}}$ in above equation, and $\sigma_{in\ situ} = \dfrac{p_0}{2}\{(1+k)+(1-k)\cos 2\theta\}$ then

$$\frac{\dfrac{p_0}{2}\left\{(1+k)\left(1+\dfrac{a^2}{r^2}\right)+(1-k)\left(1+\dfrac{3a^4}{r^4}\right)\cos 2\theta\right\}-\dfrac{p_0}{2}\{(1+k)+(1-k)\cos 2\theta\}}{\dfrac{p_0}{2}\{(1+k)+(1-k)\cos 2\theta\}} = 0.01\,d$$

$$(1+k)\left(\frac{a^2}{r^2}-0.01\,d\right)+(1-k)\cos 2\theta\left(\frac{3a^4}{r^4}-0.01\,d\right)=0 \quad (7.5)$$

Iteratively solve the above equation to find the radius of zone of influence (r_{ZOI}). For k = 1 and d = 5%, the above equation becomes

$$r_{ZOI} = a\sqrt{20} \quad (7.6)$$

EXAMPLE 7.3 A circular opening of radius 2 m is excavated in rock mass subjected to *in situ* stress fields (p_0) and (kp_0). Plot the radius of zone of influence with respect to (θ) for k = 0.50, 0.75, 1.00, 1.50, and 2.0. Assume d = 5.

Solution: From Eq. 7.5,

$$(1+k)\left(\frac{a^2}{r^2}-0.05\right)+(1-k)\cos 2\theta\left(\frac{3a^4}{r^4}-0.05\right)=0$$

Solve the above equation iteratively by Newton–Raphson method to find the zone of influence for k = 0.50, 0.75, 1.00, 1.50, and 2.0 as shown in figure given below.

$$f(r) = (1+k)\left(\frac{a^2}{r^2}-0.05\right)+(1-k)\cos 2\theta\left(\frac{3a^4}{r^4}-0.05\right)=0$$

$$f'(r) = -(1+k)\frac{2a^2}{r^3}-(1-k)\frac{12a^4}{r^5}\cos 2\theta=0$$

From Newton–Rapshson method,

$$r_{n+1} = r_n - \frac{f(r_n)}{f'(r_n)}$$

For k = 0.5, θ = 0° and a = 2.0. Start with a guess value of r_0 = 2.

1st iteration

n = 0 then

$$r_1 = r_0 - \frac{f(r_0)}{f'(r_0)}$$

$$f(r_0) = (1+k)\left(\frac{a^2}{r_0^2}-0.05\right)$$

$$f'(r_0) = -\frac{2(1+k)a^2}{r_0^3}$$

$$r_1 = r_0 - \frac{f(r_0)}{f'(r_0)} = 2.95$$

2nd iteration

$$f(r_1) = (1+k)\left(\frac{a^2}{r_1^2} - 0.05\right)$$

$$f'(r_1) = -\frac{2(1+k)a^2}{r_1^3}$$

$$r_2 = r_1 - \frac{f(r_1)}{f'(r_1)} = 4.26$$

3rd iteration

$$f(r_2) = (1+k)\left(\frac{a^2}{r_2^2} - 0.05\right)$$

$$f'(r_2) = -\frac{2(1+k)a^2}{r_2^3}$$

$$r_3 = r_2 - \frac{f(r_2)}{f'(r_2)} = 5.91$$

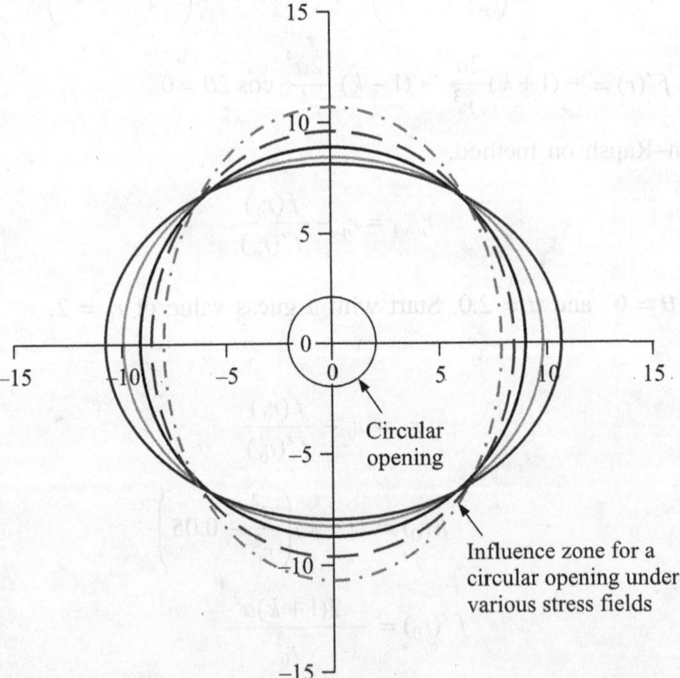

Similarly, the procedure is followed for other values of k. The figure shows the plots of zone of influence around circular opening. For hydrostatic condition (k = 1), the shape of zone of influence is circular. Whereas, for other stress field, it is elliptical in shape with major axis along maximum stress field direction or in other words, minor axis of the ellipse are along minimum stress field. The same observations are true for three-dimensional stress field conditions, where shape will be spherical or ellipsoid.

EXAMPLE 7.4 A circular opening of radius 2 m is excavated in rock mass. Find the stress distribution at the periphery of the opening for k = 0.25, 0.33, 0.50, 0.75, 1.0, 2.0, 3.0 and 4.

Solution: Figure given below shows the variation of tangential stress around the periphery of the opening. Tangential stresses at the periphery for k > 0.33, are compressive in nature; for k = 0.33, are neither in tension nor in compression and for k < 0.33, there exist several points (or line) in tension in crest, whereas side walls are highly compressed. The stress concentrations in the sidewall for k = 0.25, 0.3, 0.50, 0.75 and 1.0 are 2.75, 2.67, 2.5, 2.25 and 2.0 respectively. For crest the stress concentration for k = 0.25, 0.3, 0.50, 0.75 and 1.0 are – 0.25, – 0.01, 0.5, 1.25 and 2.0 respectively.

From the following figure, the tangential stress around the periphery of the opening for 1 ≥ k > 3; is compressive in nature and is same all around the opening. For k > 3, there exist several points (or line) in tension in crest whereas sidewalls are highly compressed. The stress concentrations in the sidewall for k = 1, 2, 3 and 4 are 2, 1, 0 and 1 respectively. For crest, the stress concentration for k = 1, 2, 3 and 4 are 2, 5, 8 and 11 respectively.

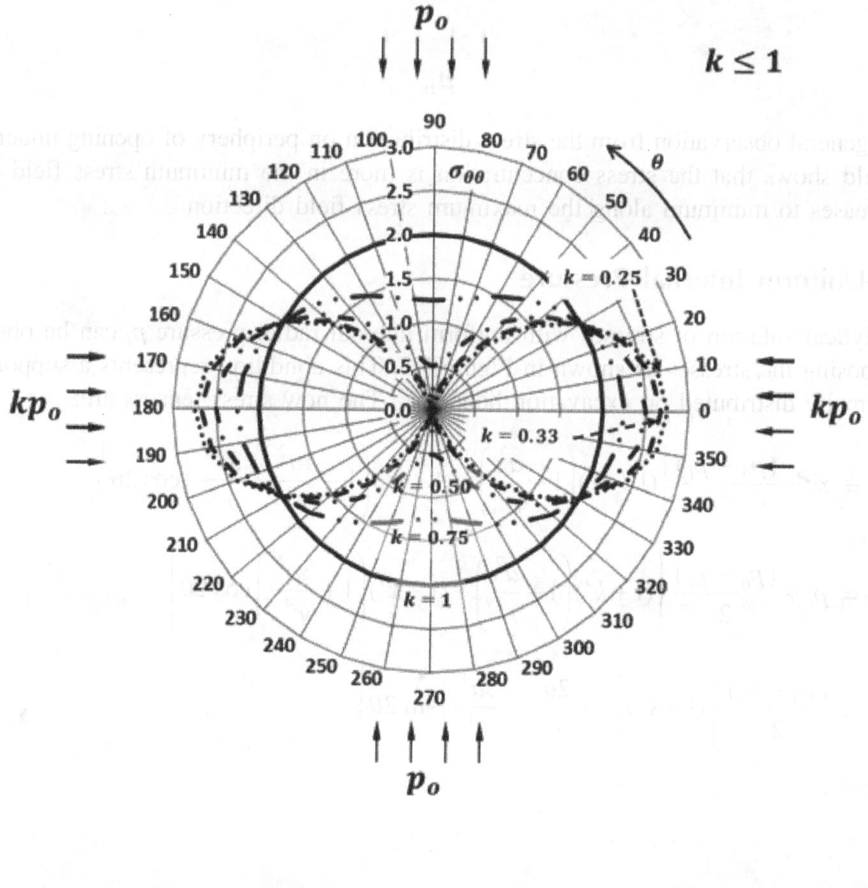

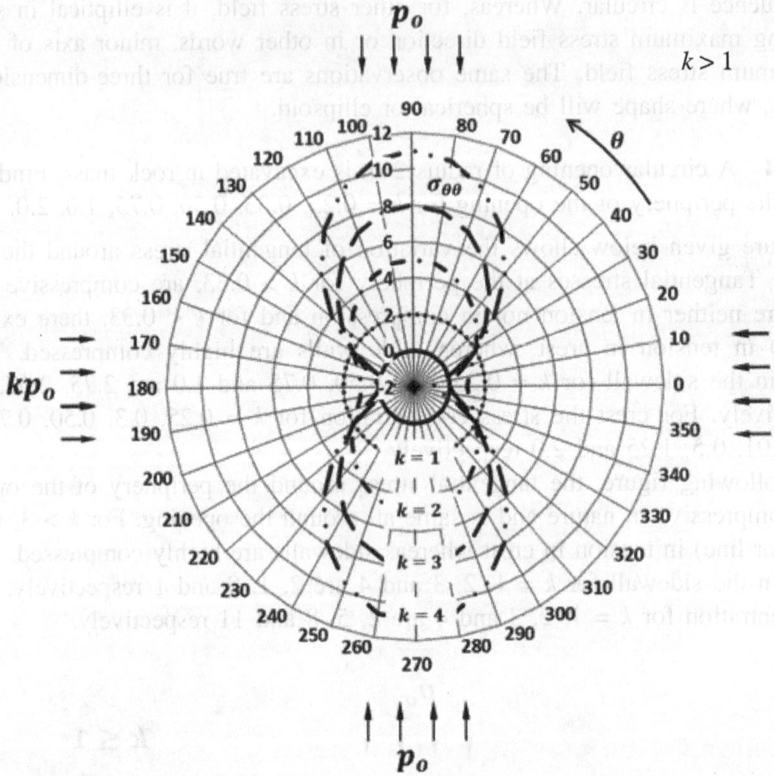

The general observation from the stress distribution on periphery of opening under varying stress field shows that the stress concentration is more in the minimum stress field direction and decreases to minimum along the maximum stress field direction.

7.2.3 Uniform Internal Pressure

The analytical solution of stresses with uniform internal radial pressure p_i can be obtained by superimposing the stresses as shown in Figure 7.3. This condition represents a support system is uniformally distributed on excavation boundary. The new stress tensors are:

$$\sigma_{rr} = p_i + \frac{(p_0 - p_i)}{2}\left\{(1+k')\left(1-\frac{a^2}{r^2}\right) - (1-k')\left(1-\frac{4a^2}{r^2}+\frac{3a^4}{r^4}\right)\cos 2\theta\right\} \tag{7.7a}$$

$$\sigma_{\theta\theta} = p_i + \frac{(p_0 - p_i)}{2}\left\{(1+k')\left(1+\frac{a^2}{r^2}\right) + (1-k')\left(1+\frac{3a^4}{r^4}\right)\cos 2\theta\right\} \tag{7.7b}$$

$$\tau_{r\theta} = \frac{(p_0 - p_i)}{2}\left\{(1-k')\left(1+\frac{2a^2}{r^2}-\frac{3a^4}{r^4}\right)\sin 2\theta\right\} \tag{7.7c}$$

where $k' = \dfrac{kp_0 - p_i}{p_0 - p_i}$. It may be noted the above solution may also be obtained from Airy stress function imposing boundary condition $\sigma_{rr} = p_i$ at $r = a$.

For a special case of $k = 1$, the above equations for $\theta = 0°$ become

$$\sigma_{rr} = p_i + (p_0 - p_i)\left(1 - \frac{a^2}{r^2}\right) \tag{7.8a}$$

$$\sigma_{\theta\theta} = p_i + (p_0 - p_i)\left(1 + \frac{a^2}{r^2}\right) \tag{7.8b}$$

$$\tau_{r\theta} = 0 \tag{7.8c}$$

The solution of radial displacement becomes

$$u_r = \frac{(p_0 - p_i)a^2}{2Gr} \tag{7.9}$$

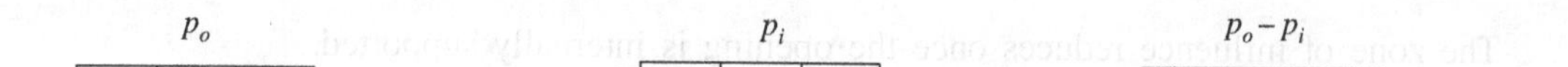

FIGURE 7.3 Stress equivalent with internal support pressure.

EXAMPLE 7.5 A circular tunnel of 2 m radius is excavated horizontally at a depth of 200 m in hydrostatic stress field (k = 1) with vertical stress gradient of 0.025 MPa/m. Find the stress distribution at the periphery of the opening when internal pressure p_i = 1 MPa. Compare the results with the condition of a tunnel without internal pressure, and comment on the change in zone of influence of tunnel with or without support.

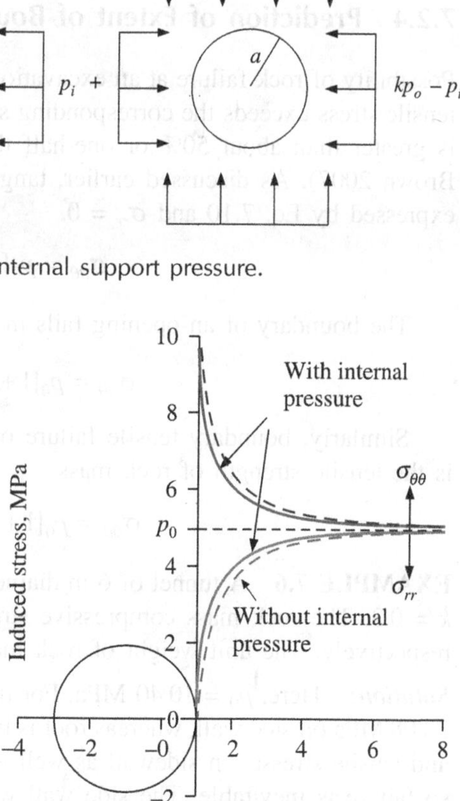

Solution: In case of hydrostatic stress field condition, the stress distribution around the periphery is independent of angular position and the shear stresses are zero. For given stress field of p_0 = 5 MPa, the radial and tangential stress distributions based on Eq. 7.8 are shown in the figure. The firm and dotted line shows the stress distribution with or without internal pressure respectively with numerical values are shown in table given below. It can be seen that tangential stress at the boundary of the tunnel is 1.75 times the far-field stress and gradually decreases with radial distance away from the tunnel boundary. The radial stress at the boundary is equal to the internal pressure of 1 MPa. This figure shows the maximum radial deformation occurs at the tunnel boundary.

r/a	*With Internal Pressure* (MPa)		*Without Internal Pressure* (MPa)	
	σ_{rr}	$\sigma_{\theta\theta}$	σ_{rr}	$\sigma_{\theta\theta}$
1.0	1.00	9.00	0.00	10.00
1.5	3.22	6.78	2.78	7.22
2.0	4.00	6.00	3.75	6.25
3.0	4.56	5.44	4.44	5.56
4.0	4.75	5.25	4.69	5.31
5.0	4.84	5.16	4.80	5.20

The zone of influence reduces once the opening is internally supported.

7.2.4 Prediction of Extent of Boundary Failure

Possibility of rock failure at an excavation boundary can exist if magnitude of compressive and tensile stress exceeds the corresponding strength. When the tangential stress around an opening is greater than about 50% or one-half the UCS (σ_{ci}), cracks may begin to form (Brady and Brown 2007). As discussed earlier, tangential stress ($\sigma_{\theta\theta}$) at an excavation boundary, can be expressed by Eq. 7.10 and $\sigma_{rr} = 0$.

$$\sigma_{\theta\theta} = p_0[1 + k + 2(1 - k)\cos 2\theta] \tag{7.10}$$

The boundary of an opening fails in compression when the following relation holds good.

$$\sigma_{\theta\theta} = p_0[1 + k + 2(1 - k)\cos 2\theta] \geq \sigma_{ci} \tag{7.11a}$$

Similarly, boundary tensile failure occurs over interval satisfying the condition, where σ_t is the tensile strength of rock mass.

$$\sigma_{\theta\theta} = p_0[1 + k + 2(1 - k)\cos 2\theta] \leq -\sigma_t \tag{7.11b}$$

EXAMPLE 7.6 A tunnel of 6 m diameter is driven at a depth of 400 m in a stress field with $k = 0.3$. The rock mass compressive strength and tensile strength are 15 MPa and 0.25 MPa respectively. The unit weight of rock mass is 26 kN/m^3. Find the extent of boundary failure.

Solution: Here, $p_0 = 10.40$ MPa. For the given stress field condition, there is compression of 28.08 MPa on side wall, whereas roof is in tension with –0.312 MPa. As the induced compressive and tensile stresses in sidewall as well as in the roof are less than the corresponding strength, so failure is inevitable. The side wall will experience the spalling, whereas tensile crack will generate in roof. The extent of boundary failure can be computed by using Eq. 7.11.

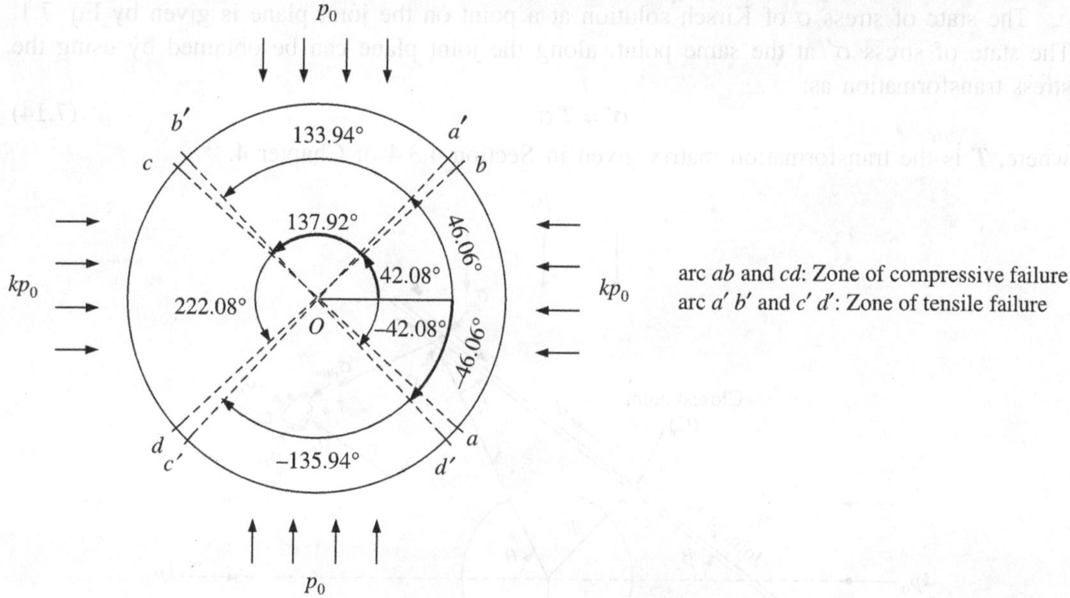

$$p_0[1 + k + 2(1 - k)\cos 2\theta] \geq 15$$

$$\theta \geq n\pi \pm \frac{\cos^{-1} 0.1017}{2} \equiv n\pi \pm 42.08°$$

$$-42.08° \geq \theta \geq 42.08° \quad \text{and} \quad 222.08° \geq \theta \geq 137.92°$$

$$p_0[1 + k + 2(1 - k)\cos 2\theta] \leq -0.312$$

$$\theta \leq \frac{\cos^{-1}(-0.037)}{2} = 46.06° \equiv n\pi \pm 46.06°$$

$$46.06° \leq \theta \leq 133.94° \quad \text{or} \quad -46.06° \leq \theta \leq -135.94°$$

7.2.5 Analytical Solution of a Joint Near a Circular Opening

In this section, derivation of normal and shear stresses along a joint plane is given based on the work done by Das, 2012. Consider a joint plane, the strike runs parallel to the axis of the circular-shaped tunnel, inclined at an angle β to the horizontal axis. The closest point C on the joint plane is located at a distance h from the centre of the circular opening as shown in Figure 7.4. A point located at a distance d measured along the joint plane from its closest point to the circular opening gives a radial distance r from the centre of the opening as:

$$r = \sqrt{h^2 + d^2} \tag{7.12}$$

The angle ξ between the joint plane and the radial line and the angular coordinate θ can be expressed as:

$$\xi = \tan^{-1}\frac{h}{d}; \quad \text{and} \quad \theta = \xi + \beta \tag{7.13}$$

The state of stress σ of Kirsch solution at a point on the joint plane is given by Eq. 7.1. The state of stress σ' at the same point, along the joint plane can be obtained by using the stress transformation as:

$$\sigma' = T\sigma \tag{7.14}$$

where, $\boldsymbol{T}$ is the transformation matrix given in Section 4.3.4 of Chapter 4.

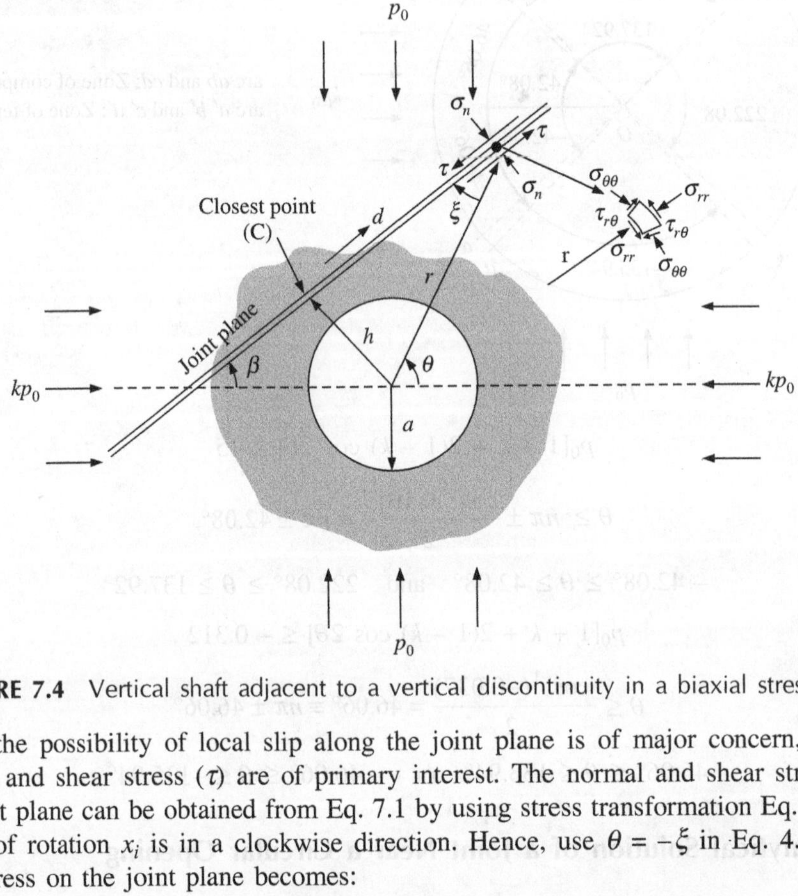

FIGURE 7.4 Vertical shaft adjacent to a vertical discontinuity in a biaxial stress field.

Since the possibility of local slip along the joint plane is of major concern, the normal stress (σ_n) and shear stress (τ) are of primary interest. The normal and shear stresses acting on the joint plane can be obtained from Eq. 7.1 by using stress transformation Eq. 4.29, where the angle of rotation x_i is in a clockwise direction. Hence, use $\theta = -\xi$ in Eq. 4.29, then the induced stress on the joint plane becomes:

$$\sigma_n = \sigma'_{\theta\theta} = \frac{1}{2}(\sigma_{rr} + \sigma_{\theta\theta}) - \frac{1}{2}(\sigma_{rr} - \sigma_{\theta\theta})\cos 2\xi + \tau_{r\theta}\sin 2\xi \tag{7.15a}$$

$$\tau = \sigma'_{r\theta} = \frac{1}{2}(\sigma_{rr} - \sigma_{\theta\theta})\sin 2\xi + \tau_{r\theta}\cos 2\xi \tag{7.15b}$$

Substituting stresses from Eq. 7.1 into Eq. 7.15 and after simplification, normal and shear stresses are given by

$$\sigma_n = \frac{p_0}{2}\left[(1+k)\left(1+\frac{a^2}{r^2}\cos 2\xi\right) + (1-k)\left\{\cos 2\beta + \frac{2a^2}{r^2}\cos 2(\xi+\beta) + \left(\frac{3a^4}{r^4} - \frac{2a^2}{r^2}\right)\cos 2(2\xi+\beta)\right\}\right] \tag{7.16a}$$

$$\tau = -\frac{p_0}{2}\left[(1+k)\frac{a^2}{r^2}\sin 2\xi - (1-k)\left\{\sin 2\beta + \left(\frac{2a^2}{r^2} - \frac{3a^4}{r^4}\right)\sin 2(2\xi+\beta)\right\}\right] \tag{7.16b}$$

The shear strength of the joint plane is described by the linear Mohr-Coulomb criterion (discussed in Chapter 9) and the mobilised friction angle is expressed as:

$$\varphi_m = \tan^{-1}\left(\frac{\tau - c_j}{\sigma_n}\right) \tag{7.17}$$

If the mobilised friction angle (φ_m) given in Eq. 7.17 exceeds the friction angle (φ_j) of the joint plane, joint plane starts to slip and stress redistribution occurs along the joint plane. Figure 7.4 shows the generalised condition in which the discontinuity is inclined with the *in situ* stress. Four orientation of discontinuity with respect to opening can be derived by readers for (i) $\beta = 0$, $h = 0$, (ii) $\beta = 90°$, $h = 0$, (iii) $\beta = 0$, $h > 0$, (iv) $\beta = 90°$, $h > 0$.

EXAMPLE 7.7 A vertical shaft of 3 m radius is to be excavated through a sandstone horizon in which the remote horizontal principal field stresses are 3.8 MPa and 5.7 MPa as shown in figure given below. The figure shows a vertical discontinuity at its closest point A, located a distance of 4.0 m from the shaft centre. Discontinuity is inclined at an angle 7° with maximum principle stress in anti-clockwise direction. The discontinuity has Coulomb shear strength parameter cohesion of 0.02 MPa and angle of internal friction is 24°. Check the stability of discontinuity at point B that may not able to sustain the stress distribution predicted by Kirsch equation.

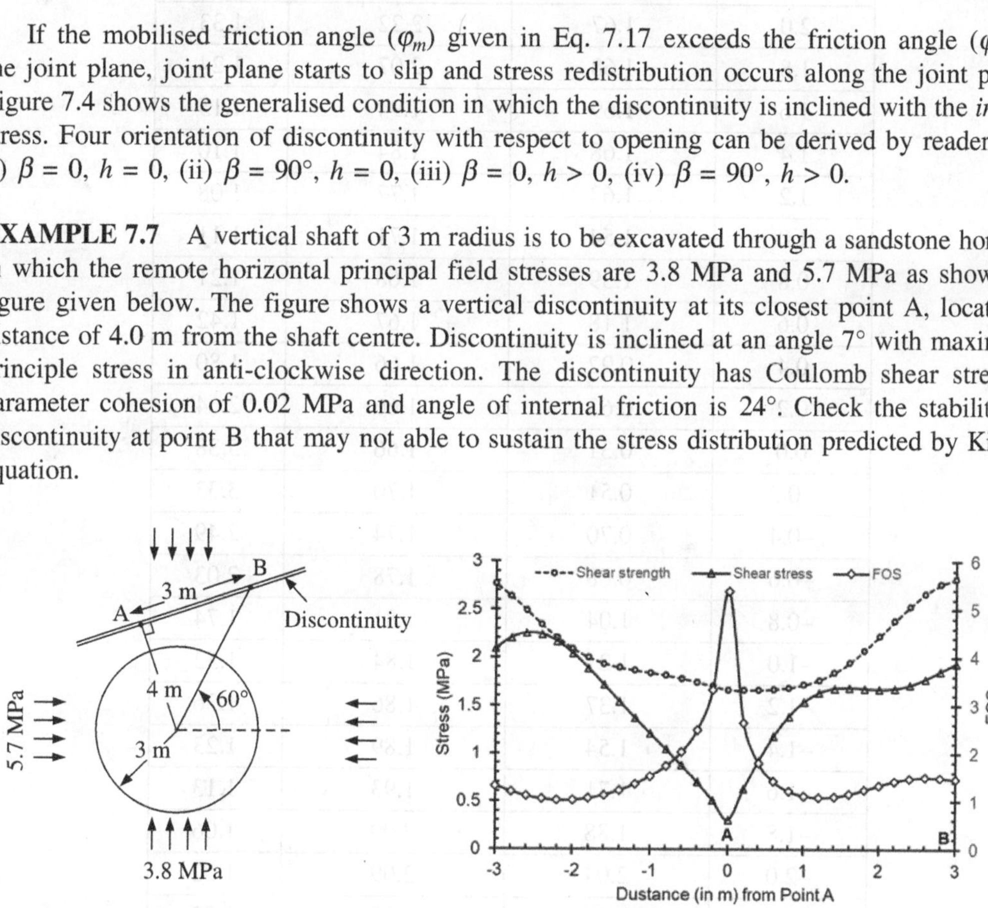

Solution: The shear stress and shear strength distribution along the line of discontinuity is shown in table and graph given below. Graph shows that the shear stress along line of discontinuity exceeds the strength in two regions 0.5 m to 3 m on left of point A and from 1.5 m to 4.5 m on right from point A. The factor of safety (FOS) is 0.62 (lowest) at 1 m left of point A, thus, means that the failure starts from this point onwards. The FOS is defined as follows:

$$\text{FOS} = \frac{\text{Shear strength}}{\text{Shear stress}} = \frac{|c_j + \sigma_n \tan\varphi_j|}{\tau}$$

Distance (*d*)	*Shear stress* (MPa)	*Shear strength* (MPa)	*FOS*
3.0	1.94	2.82	1.46
2.8	1.85	2.77	1.49
2.6	1.78	2.67	1.50
2.4	1.71	2.54	1.48
2.2	1.68	2.38	1.42
2.0	1.67	2.22	1.33
1.8	1.68	2.07	1.24
1.6	1.69	1.94	1.15
1.4	1.68	1.84	1.10
1.2	1.63	1.77	1.08
1.0	1.54	1.71	1.11
0.8	1.39	1.68	1.21
0.6	1.18	1.67	1.42
0.4	0.92	1.66	1.80
0.2	0.63	1.66	2.64
0.0	0.31	1.66	5.38
–0.2	0.51	1.70	3.33
–0.4	0.70	1.74	2.49
–0.6	0.88	1.78	2.03
–0.8	1.04	1.81	1.74
–1.0	1.21	1.84	1.52
–1.2	1.37	1.86	1.36
–1.4	1.54	1.89	1.23
–1.6	1.71	1.93	1.13
–1.8	1.88	2.00	1.06
–2.0	2.04	2.09	1.02
–2.2	2.16	2.20	1.02
–2.4	2.24	2.34	1.05
–2.6	2.25	2.49	1.11
–2.8	2.20	2.64	1.20
–3.0	2.09	2.76	1.32

The table shows the shaded region in which the discontinuity is unstable or in other words, the factor of safety of the discontinuity against shear is less than one.

7.3 MULTIPLE OPENINGS

In case of underground activities, there are many instances when two or more openings are sufficiently close together and the introduction of one opening affects the stress concentration around the other. This condition is of primary importance, because increase in stress concentrations around multiple opening will affect ground stability around the openings. Hence, in such situations, we are interested to find the maximum stress concentration, as well as stress distribution in rock mass formed between two or more proximate openings.

In case of two or more circular opening, solution of stresses and displacements around multiple opening can be found by multiple application of the Kirsch solution using principle of superposition (Obert and Duvall, 1967). In this method, stresses at a point are determined by each excavation and then with the vector addition, the final stress can be determined. The procedure to solve the problem of multiple opening is demonstrated using the following example.

EXAMPLE 7.8 At a depth of 450 m, a 3 m diameter circular tunnel is driven in rock having unit weight of 26 kN/m^3 and uniaxial compressive and tensile strengths of 40.0 MPa and 2.0 MPa, respectively. If the stress field ratio is 0.3. Find:

(a) The induced stress at points (1), (2) and (3) and corresponding factor of safety.

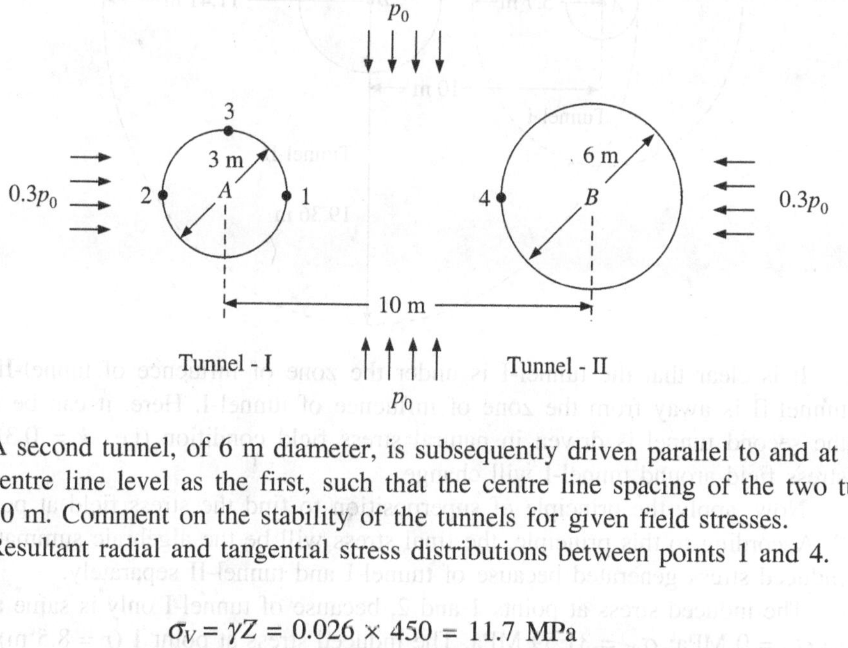

(b) A second tunnel, of 6 m diameter, is subsequently driven parallel to and at the same centre line level as the first, such that the centre line spacing of the two tunnels is 10 m. Comment on the stability of the tunnels for given field stresses.

(c) Resultant radial and tangential stress distributions between points 1 and 4.

Solution:

$$\sigma_V = \gamma Z = 0.026 \times 450 = 11.7 \text{ MPa}$$

$$\sigma_h = 0.3 \times 11.7 = 3.51 \text{ MPa}$$

(a) The radial stresses is zero at three locations. The tangential stress at points (1) and (2) is same because of symmetry. The tangential stress and corresponding FOS is:

$$\sigma_{0^\circ} = \sigma_{180^\circ} = 31.59 \text{ MPa}; \quad \text{FOS} = 40/31.59 = 1.27$$

The tangential stress and FOS at point (3) is:

$$\sigma_{90^\circ} = -1.17 \text{ MPa}; \quad \text{FOS} = -2/-1.17 = 1.71$$

For all locations, FOS is more than one, hence stable.

(b) After the second tunnel has been driven, identify the zone of influence of each tunnel by considering 5% zone of influence. Based on Eq. 7.5, the zone of influence is elliptical with major and minor axis for the first tunnel is 5.7 m and 9.68 m; whereas it is 11.41 m and 19.36 m for second tunnel. The orientation of the major and minor axes is vertical and horizontal respectively.

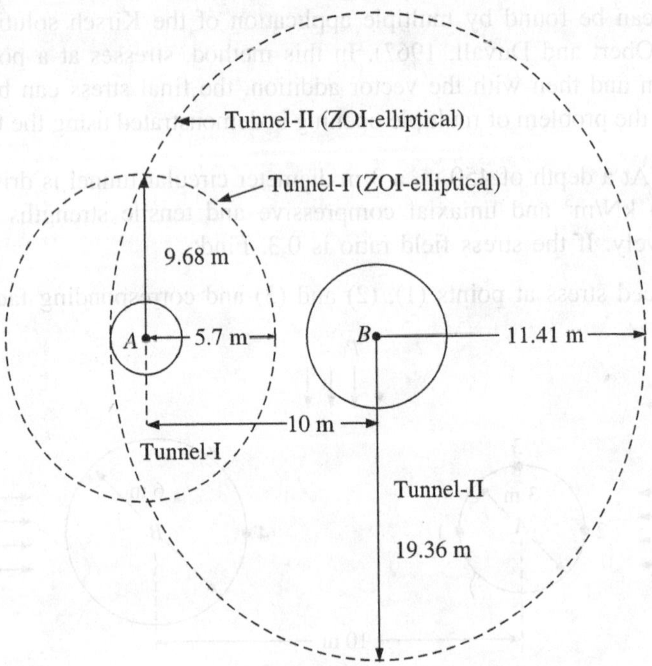

It is clear that the tunnel-I is under the zone of influence of tunnel-II, whereas tunnel-II is away from the zone of influence of tunnel-I. Here, it can be noted that the second tunnel is driven in natural stress field condition (i.e., k = 0.3), whereas stress field around tunnel-I will change.

Now, apply the principle of superposition to find the stress field at points 1 and 2. According to this principle, the final stress will be the algebraic summation of the induced stress generated because of tunnel-I and tunnel-II separately.

The induced stress at points 1 and 2, because of tunnel-I only is same and equals to $\sigma_{r\text{I}} = 0$ MPa; $\sigma_{\theta\text{I}} = 31.59$ MPa. The induced stress at point 1 (r = 8.5 m) and point 2 (r = 11.5 m) due to tunnel-II will be equal to $\sigma_{r\text{I}} = 4.41$ MPa; $\sigma_{\theta\text{I}} = 12.84$ MPa and $\sigma_{r\text{II}} = 4.05$ MPa; $\sigma_{\theta\text{II}} = 12.27$ MPa respectively.

The final stresses at point 1 due to construction of both the tunnels are:

Point 1: $\sigma_r = 0 + 4.41 - 3.51 = 0.9$ MPa; $\sigma_{\theta\text{I}} = 31.59 + 12.84 - 11.7 = 32.73$ MPa

$$\text{FOS} = 40/32.73 = 1.22$$

Here, it is important to mention that under the influence of two openings, the radial stress at point 1 is not zero. This radial stress may contribute to the generation of tensile cracks around the opening.

The final stresses at point 2 due to construction of both the tunnels are:

Point 2: $\sigma_r = 0$ MPa; $\sigma_{\theta II} = 31.59$ MPa

$$\text{FOS} = 40/31.59 = 1.27$$

Point 2 is outside the elliptical ZOI of tunnel-II. Hence, the induced stress at this point is only due to tunnel-I.

It is to be noted that the *in situ* stress is once deducted to find the final induced stress. This is because in calculation, for both the tunnel and the *in situ* stress are added twice, and hence one *in situ* stress is deducted from the final calculation.

Comments from the above analysis:

- Induced stresses at both the location will alter due to construction of second tunnel. Earlier in case of only one tunnel-I, the stresses at both the points are same.
- As the first tunnel is under the influence of second tunnel, the stress field around the first tunnel changes. For the two points under consideration, the FOS decreases substantially.
- Here, design engineer can take a decision which tunnel should be driven first followed by the other. If only two tunnels are under consideration, driving second tunnel followed by first is the best option as far as ground stability is concerned. In this situation, both excavations will be created in stress fields that will not be subsequently disturbed or in other words both the tunnels will be driven in a stabilised stress field condition.

(c) The radial and tangential stress distributions between points 1 and 4 by individual tunnels and their results are shown in figure given below. Principle of superposition is used to find the resultant of the two induced stresses as discusses above.

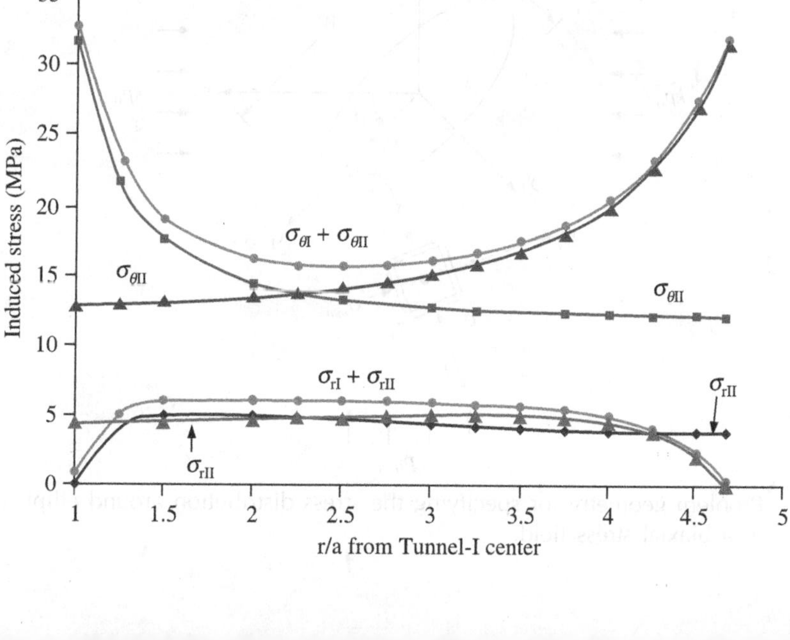

7.4 ELLIPTICAL OPENING

The size and shape of an opening depends primarily on (i) the purpose it is supposed to serve, (ii) the depth of its location below the ground surface, and (iii) the condition of rock mass. In general, D-shape or horse-shoe shape tunnel is adopted in relatively non-squeezing grounds for railways and roadways. Narrow elliptical shape is chosen in case the vertical stress is much larger than the horizontal stress if the opening will be located at greater depth. Wider elliptical shape tunnel is preferred when horizontal stress is much larger than the vertical stress. It may be noted that stress concentration around an elliptical opening is relatively less the major axis aligned in the major principal stress direction.

Solutions for the stress distribution in this case are quoted by Poulos and Davis (1974) and Jaeger et al. (2007). In both cases, the solutions are expressed in terms of elliptical curvilinear co-ordinates and hence their practical use is somewhat cumbersome. Bray (1986) formulae for calculation of the state of stress at points surrounding an elliptical opening is considerably simplified and discussed here. Utility of the solution of elliptical opening is more than circular opening because this provides a first approximation to a wide range of engineering geometrics, especially with high width/height ratio, for example, mine stopes in underground metalliferous mines and caverns in underground power house.

The problem geometry is defined in Figure 7.5, with the global x-axis parallel to the field stress component kp_0, and with an axis of the ellipse defining the local x_1-axis for the opening. The width, W, of the opening is measured in the direction of the x_1-axis, and the height, H, in the direction of the local y_1-axis. The attitude of the ellipse in the biaxial stress field is described by the angle β between the global x-axis and local x_1-axis. The position of any point (A) is defined by its local Cartesian co-ordinates (x_1, y_1).

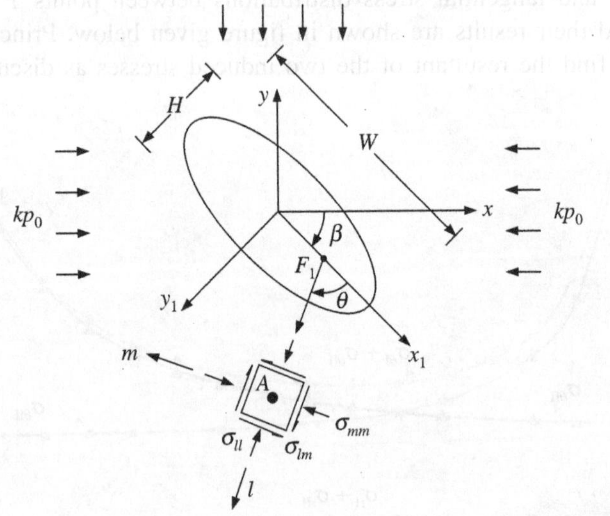

FIGURE 7.5 Problem geometry for specifying the stress distribution around elliptical excavation in a biaxial stress field.

Bray's solution specifies the state of stress at a point in terms of geometrical parameters, and relative to local axes (l, m), centred on the point of interest. The various geometric parameters are defined as follows (Brady and Brown, 2007):

$$\sigma_{ll} = \frac{p_0(e_0 - e)}{J}\left[(1+k)(e^2-1)\frac{c}{2e_0} + (1-k)\left[\left\{\frac{J}{2}(e-e_0) + Ce\right\}\cos 2(\psi+\beta) - C\cos 2\beta\right]\right] \tag{7.18}$$

$$\sigma_{mm} = \frac{p_0}{J}\left[(1+k)(e^2-1) + 2(1-k)e_0[e\cos 2(\psi+\beta) - \cos 2\beta]\right] - \sigma_{ll} \tag{7.19}$$

$$\sigma_{lm} = \frac{p_0(e_0 - e_1)}{J^2}\left[(1+k)\frac{Ce}{e_0}\sin 2\psi + (1-k)\right.$$

$$\left.\left[e(e_0+e)\sin 2\beta + e\sin 2(\psi-\beta)\left\{\frac{J}{2}(e_0-e) + e^2 e_0\right\}\sin 2(\psi+\beta)\right]\right] \tag{7.20}$$

Where the following geometrical parameters are:

$$e_0 = \frac{(W+H)}{(W-H)},\quad b = \frac{4(x_1^2 - y_1^2)}{(W^2-H^2)},\quad d = \frac{8(x_1^2-y_1^2)}{(W^2-H^2)} - 1,\quad u = b + \frac{e_0}{|e_0|}\sqrt{b^2-d},$$

$$e = u + \frac{e_0}{|e_0|}\sqrt{u^2-1},$$

$$\psi = \tan^{-1}\left\{\left(\frac{e+1}{e-1}\right)\frac{y_1}{x_1}\right\},\quad \theta = \tan^{-1}\left\{\left(\frac{e+1}{e-1}\right)^2\frac{y_1}{x_1}\right\},\quad C = 1 - ee_0,\ J = 1 + e^2 - 2e\cos 2\psi$$

where, p_0 is the far-field stress attached with the global reference axes (x, y), k is the ratio of horizontal to vertical stresses, $q = W/H$, β is the inclination of the major axis to the horizontal, and ψ is a parameter for locating position on the boundary and physically it represents the angle the tangent to the boundary of the opening makes to the horizontal. θ is the orientation of the local reference axes (l, m) relative to the ellipse local axes (x_1, y_1).

The boundary stresses around an elliptical opening with axes inclined to the field stress directions are obtained by selecting values of x_1, y_1 which fall on the boundary contour. For this case, $e = e_0$, $\sigma_{ll} = \sigma_{lm} = 0$ and the x-axis is directed normal to the boundary. The boundary stress is given by

$$\sigma_{\theta\theta} = \frac{p_0}{2q}[(1+k)\{(1+q^2) + (1-q^2)\cos 2(\psi-\beta)\} - (1-k)\{(1+q)^2\cos 2\psi$$
$$+ (1-q^2)\cos 2\beta\}] \tag{7.21}$$

When the axes of the ellipse are oriented parallel to the field stress direction ($\beta = 0°$), Eq. 7.21 reduces to:

$$\sigma_{\theta\theta} = \frac{p_0}{2q}[(1+k)\{(1+q^2) + (1-q^2)\cos 2\psi\} - (1-k)\{(1+q)^2\cos 2\psi + (1-q^2)\}] \tag{7.22}$$

Determination of angle ψ

From Figure 7.5, the local and global co-ordinates are related as:

$$x_1 = x \cos \beta + y \sin \beta \tag{7.23a}$$

$$y_1 = -x \sin \beta + y \cos \beta \tag{7.23b}$$

Here, β is positive if the angle is measured counter-clockwise. The equation of ellipse is given as:

$$\frac{x_1^2}{W^2} + \frac{y_1^2}{H^2} = \frac{1}{4} \tag{7.24}$$

Replacing Eq. 7.23 into Eq. 7.24, the equation becomes,

$$Ax^2 + By^2 + Cxy = 1 \tag{7.25}$$

where

$$A = 4\left(\frac{\cos^2\beta}{W^2} + \frac{\sin^2\beta}{H^2}\right), B = 4\left(\frac{\sin^2\beta}{W^2} + \frac{\cos^2\beta}{H^2}\right) \text{ and } C = 8 \cos \beta \sin \beta\left(\frac{1}{W^2} - \frac{1}{H^2}\right)$$

Now taking derivative of Eq. 7.25 with respect to x, we find

$$\tan \psi = \frac{dy}{dx} = -\left(\frac{2Ax + Cy}{2By + Cx}\right) \tag{7.26}$$

EXAMPLE 7.9 An elliptical opening of $W/H = 2.0$ is to be constructed in hydrostatic stress field of 20 MPa. Determine the magnitude and location of the maximum tangential stresses at the boundary if the major axis of the opening is horizontal.

Solution: Use the numerical values (like $q = W/H = 2.0$; $\beta = 0°$; $k = 1$; $p_0 = 20$ MPa) given in Eq. 7.22 for tangential stress around the periphery of the elliptical opening oriented parallel to the stress field. Eq. 7.22 reduces to:

$$\sigma_{\theta\theta} = 10(5 - 3 \cos 2\psi)$$

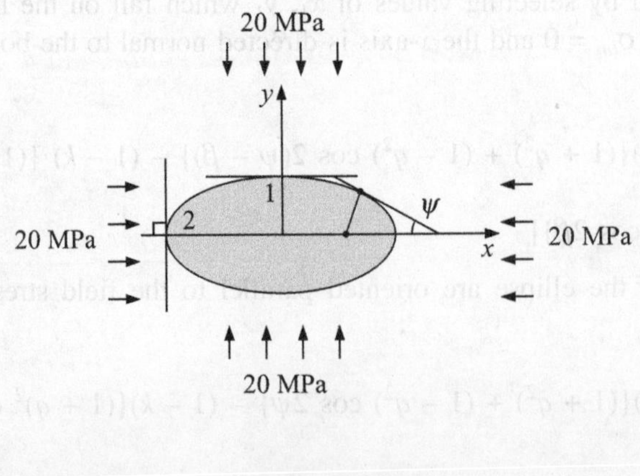

Differentiating $\sigma_{\theta\theta}$ with respect to ψ and equating to zero for maximum and minimum values of ψ. Physical description of angle ψ is shown in figure given below, represents the inclination of tangent at a point on the boundary of the opening with the horizontal.

$$\frac{d\sigma_{\theta\theta}}{d\psi} = \sin 2\psi \equiv \psi = 0° \text{ or } 90°$$

The tangents to the opening are at points 1 and 2. Tangent at point 1 is parallel to x-axis or inclined 0° or 180° with the horizontal; whereas tangent at point 2 is perpendicular to x-axis or inclined 90° with the horizontal.

For $\psi = 0°$; $\sigma_{\theta\theta} = 20$ MPa and $\psi = 90°$; $\sigma_{\theta\theta} = 60$ MPa

The maximum tangential stress induced at the vertex of the elliptical opening.

EXAMPLE 7.10 An elliptical opening of 6 m width is to be driven horizontally in a vertical stress field of 10 MPa. The horizontal stress ratio is 0.5. Draw tangential stress profiles along the upper half boundary (x = 0 to 3 m) if height of the opening is varied as 3 m, 4 m, 5 m and 6 m.

Solution: Here, $q = W/H$ is varied as 2.0, 1.5, 1.2 and 1.0 for the given heights (H). Here estimate the angle ψ of Eq. 7.26 as:

$$\psi = \tan^{-1}\left(\frac{xH^2}{yW^2}\right)$$

The tangential stress profiles along the curve AB for different W/H ratios are given in the Figure below:

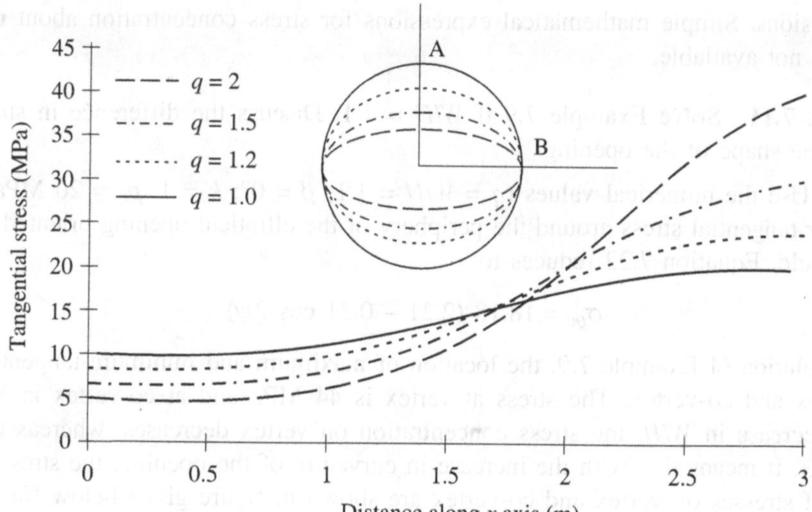

7.4.1 Inference from Elliptical Solution

(i) Elastic stress distributions are independent of elastic constants, i.e., Young's modulus and Poisson's ratio.
(ii) Stresses are function of aspect ratio of the opening.
(iii) In case of hydrostatic pressure, entire boundary is in compression and the maximum stress is always on the vertex of the major axis being independent of its inclination. The minimum stress generates on minor axis and shear stresses are zero everywhere.
(iv) If stress ratio $k < 0$, the critical compressive stress occurs on side and tensile stress at top/bottom of the opening.
(v) For given horizontal stress and aspect ratio of opening, the compressive stress is greatest at $\beta = 0°$ and smallest at $\beta = 90°$.
(vi) Compressive stress increases linearly with increasing aspect ratio for small inclination (β).

7.5 OTHER SHAPES OF OPENING

A perfectly rectangular opening tends to produce very large stress concentration at the right-angled corners, therefore, rounded or filleted corners are assumed. The near infinite stress eventually reduces to finite stress due to rounded or filleted corners specified by radius of curvature. Normal mining conditions probably produce approximately rounded corners in a rectangular opening. The greater the rounding, the smaller is the stress concentration. The maximum corner radius possible is one-half the least dimension of the section and results in an ovaloid. As with ellipse, one may characterise rectangular shafts by an aspect ratio $m = b/a$ where, a is the semi-axes parallel to major principal stress and b is parallel to minor principal stress. A range of m between 1/8 and 8 covers more than the range of practical rectangular shaft dimensions. Simple mathematical expressions for stress concentration about rectangular opening are not available.

EXAMPLE 7.11 Solve Example 7.9 if $W/H = 1.1$. Discuss the difference in stresses with respect to the shape of the opening.

Solution: Use the numerical values ($q = W/H = 1.1$; $\beta = 0°$; $k = 1$; $p_0 = 20$ MPa) given in Eq. 7.22 for tangential stress around the periphery of the elliptical opening oriented parallel to the stress field. Equation 7.22 reduces to:

$$\sigma_{\theta\theta} = 18.18\ (2.21 - 0.21 \cos 2\psi)$$

From solution of Example 7.9, the location of maximum and minimum tangential stresses are at vertex and co-vertex. The stress at vertex is 44 MPa and at co-vertex is 36.36 MPa. With the decrease in W/H, the stress concentration on vertex decreases, whereas it increases on co-vertex. It means that with the increase in curvature of the opening, the stress decreases. Variation of stresses on vertex and co-vertex are shown in figure given below for hydrostatic condition and horizontal opening.

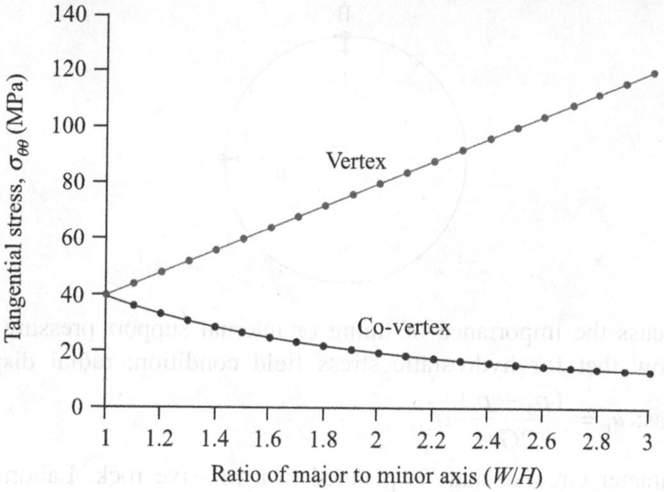

EXERCISES

1. 1.A company is planning to excavate a 5 km long horizontal tunnel of 6 m diameter at a depth of 500 m from ground surface as shown in Figure below. The average unit weight of rock is 26 kN/m³. From *in situ* stress measurement, it is found that the direction of major horizontal stress is coinciding with the tunnel axis with magnitude 30 MPa whereas the minor horizontal stress oriented along the *E*-axis with magnitude 15 MPa. Comment on the stability of the tunnel boundary using analytical solution.

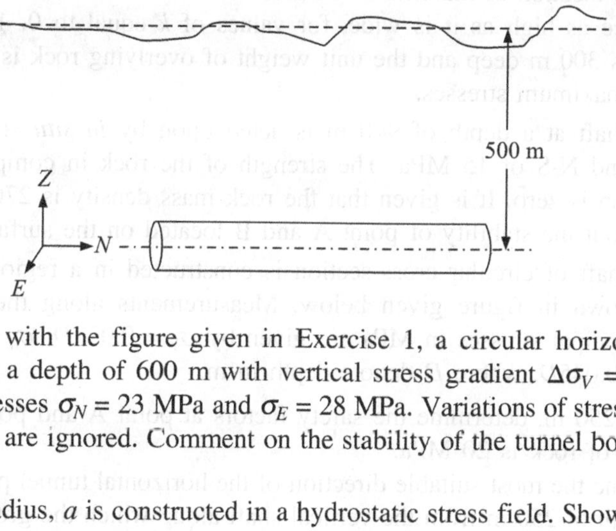

2. In continuity with the figure given in Exercise 1, a circular horizontal N-S tunnel is considered at a depth of 600 m with vertical stress gradient $\Delta\sigma_V = 0.026$ MPa/m and horizontal stresses $\sigma_N = 23$ MPa and $\sigma_E = 28$ MPa. Variations of stress due to dimension of excavation are ignored. Comment on the stability of the tunnel boundary using Kirsh solution.
3. A tunnel of radius, *a* is constructed in a hydrostatic stress field. Show that the tangential stress will be twice the radial stress at $r = \sqrt{3}a$.
4. A circular tunnel of radius *a* is constructed in prevailing far-field stress of σ_v (vertical) and σ_h (horizontal). Tangential stress at point B is twice than that of at point A. Determine the value, σ_h/σ_v.

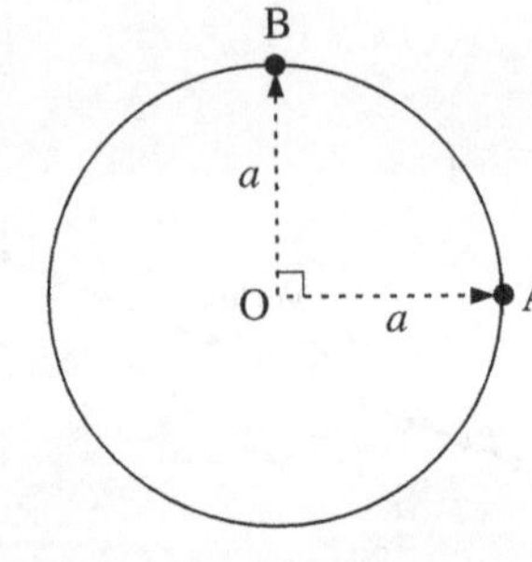

5. Briefly discuss the importance of lining or internal support pressure around a circular tunnel. Show that for hydrostatic stress field condition, radial displacement can be expressed as: $u_r = \frac{(p_o - p_i)}{2Gr} a^2$.

6. A 6 m diameter circular shaft is planned in a massive rock. Laboratory tests on core from exploration drilling show that
σ_c = 152 MPa; σ_t = 8.3 MPa; γ = 23 kN/m^3; E = 34 MPa; G = 13.8 MPa at a depth of 915 m. Site measurements show that no tectonic stresses are present.

   Find:
   (a) The factors of safety with respect to failure in compression and tension.
   (b) If the opening in above question is in an *in situ* stress field such that $\sigma_h = \sigma_v$, then determine safety factors.
   (c) If the opening is in an *in situ* stress field such that $\sigma_h = 0.01\sigma_v$, is there a possibility of failure? Justify your answer.

7. Calculate the location of maximum compressive stress around an elliptical underground opening twice as high as it is wide, for values of k equal to 0, 1/3, 2/3, 1, 2 and 3. The tunnel is 300 m deep and the unit weight of overlying rock is 25 kN/m^3. Find the location of maximum stresses.

8. A circular shaft at a depth of 900 m is acted upon by *in situ* stresses oriented E-W = 30 MPa and N-S of 15 MPa. The strength of the rock in compression is 100 MPa and in tension is zero. It is given that the rock mass density is 2700 kg/m^3. Give your comment about the stability of point A and B located on the surface of the opening.

9. A vertical shaft of circular cross-section is constructed in a region of high horizontal stress as shown in figure given below. Measurements along the shaft indicate that pre-shaft principal stresses in MPa are given by $\sigma_V = 0.027D$, $\sigma_h = 1.5 + 0.035D$ and $\sigma_H = 3.5 + 0.045D$, where D denote depth in m.

   (a) At D = 250 m, determine the safety factors at point A and point B if compressive strength of rock is 20 MPa.
   (b) Determine the most suitable direction of the horizontal tunnel proposed to be driven at a depth of 250 m from the vertical shaft along which the ground control problem will be least.
   (c) If a second vertical shaft of same dimension is planned near the first shaft, what would be the minimum safe distance between these shafts and show in plan their relative location with respect to *in situ* stresses. Comment on the stability of the rock mass in between these two shafts.

(d) Determine the uniform internal pressure (p_i) required to be applied at shaft boundary at D = 250 m depth, such that safety factor at point B becomes unity.

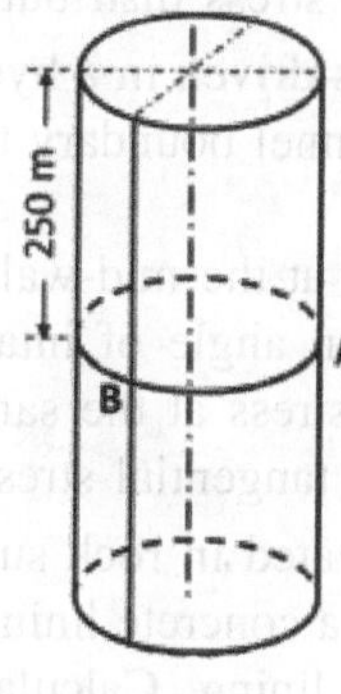

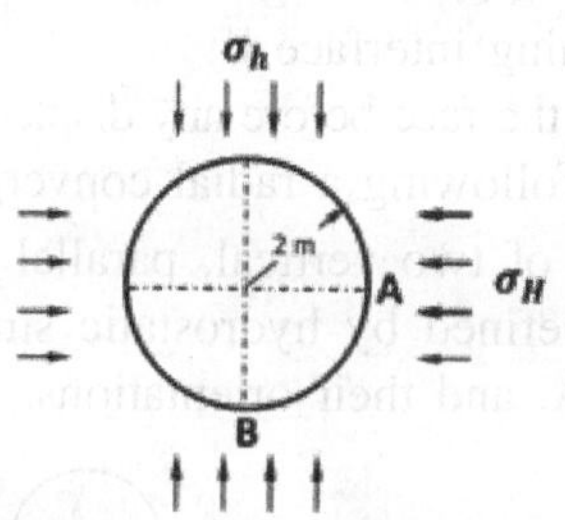

10. A circular tunnel-I of 3 m diameter is driven in a rock at a depth of 250 m from ground surface. The *in situ* stress field is hydrostatic. Rock unit weight, uniaxial compressive and tensile strengths are 25 kN/m^3, 30.0 MPa and 3.0 MPa, respectively. A second tunnel-II, of 3 m diameter, is subsequently driven parallel to and at the same centre line level as the first tunnel is at a distance of 10 m. Check the stability of the circumference of both the tunnels.
11. A 3 m radius vertical shaft is to be excavated through a sandstone in which the *in situ* stresses are 4 MPa and 6 MPa. A vertical discontinuity at its closest point A and inclined at 45 deg in anti-clockwise with major *in situ* horizontal stress direction, located at distance of 1.5 m from the shaft wall and inclined at an angle of 45 deg in anti-clockwise direction from major horizontal stress field. The discontinuity has Coulomb shear strength parameters c = 0.02 MPa, φ = 24° and carries a water pressure of 0.10 MPa in the sandstone horizon. Identify, in terms of the distance d measured from A, the region of the discontinuity that will be unable to sustain the continuum stress distribution predicted by Kirsch equations.
12. A plan view of a vertical shaft of 8 m diameter, with a clay-filled fault located near the shaft boundary is shown in Figure. The far-field stress state at a particular horizon is hydrostatic, of magnitude 8 MPa. The resistance to slip on the fault is purely cohesive, and of magnitude 1.5 MPa. Determine:

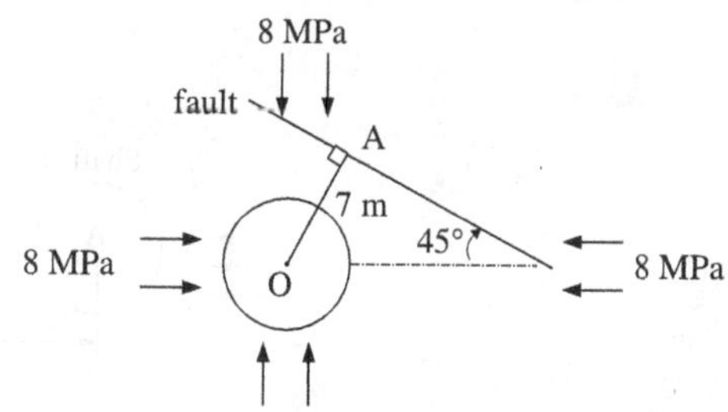

(a) The maximum shear stress generated on the fault and its location after development of the shaft, and

(b) The elastic (shear and normal) stress distribution along the fault.

13. A horizontal tunnel of radius 4 m is driven in a hydrostatic stress field of 15 MPa. Rock bolts are installed all around the tunnel boundary to provide a uniform internal pressure of 1 MPa.

(a) Determine the tangential stress at the mid-wall of the tunnel boundary, i.e., $\theta = 0°$.

(b) Assuming cohesion and friction angle of intact rock and yielded rock are 1 MPa and 30°, determine tangential stress at the same location given in part (a).

(c) Comment on the difference in tangential stresses in part (a) and part (b).

14. A tunnel of radius 1.85 m is excavated in rock subjected to an initial hydrostatic stress field of 20 MPa and provided with a concrete lining of internal radius 1.70 m. Assuming elastic behaviour of the rock and lining. Calculate the radial pressure and the radial displacement at the rock-lining interface if:

(a) The lining is installed at the face before any displacement of the rock has occurred; and

(b) The lining is installed following a radial convergence of 1 mm.

15. Figure shows the locations of two vertical, parallel shafts, each 4 m in diameter. The pre-mining stress field is defined by hydrostatic stress field of 20 MPa. Estimate the principal stresses at point A, and their orientations.

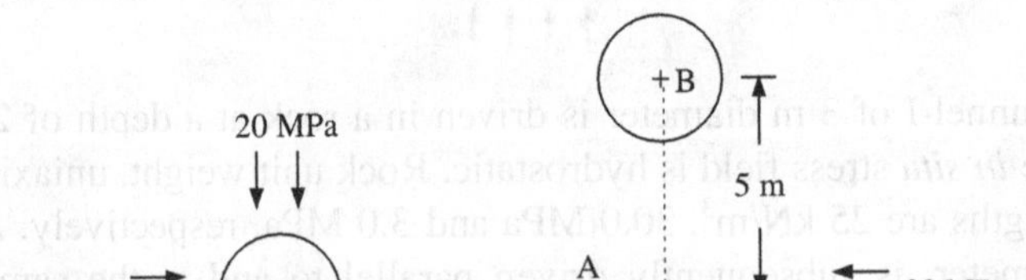

16. Figure shows the locations of two vertical, parallel shafts, each 4 m in diameter. The pre-mining stress field is shown in figure. Estimate approximate values for the boundary stresses around each opening, and calculate the principal stresses at point A and their orientations.

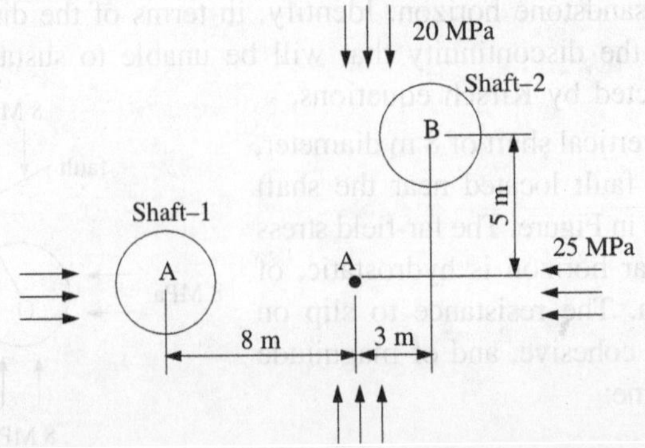

17. A 5 m thick copper ore body dips at 80° as shown in the Figure. The depth of the ore body is 500 m from surface and extend deep below the surface. The average unit weight of rock mass is 29 kN/m^3. If the compressive and tensile strengths of the rock are 145 MPa and 6 MPa respectively, what would be the maximum extraction height of the ore body for which the rock around the excavation fails?

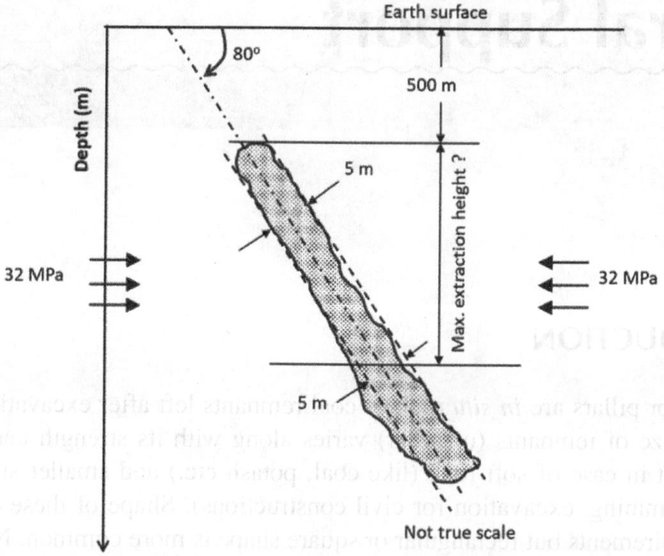

18. A vertical shaft of circular cross section having radius of 3 m is constructed in hilly area. *In situ* stress measurements shows that the vertical stress (in MPa) varies with depth, D (expressed in m) as $\sigma_v = 0.027D$ MPa. Minor and major horizontal stress ratios are $\sigma_h/\sigma_v = 0.5 + 100/D$ and $\sigma_H/\sigma_v = 2.5 + 1000/D$, respectively. Determine the safety factors at points A and B for depth of 400 m. if the compressive and tensile strengths of rock are 40 MPa and 4 MPa respectively.

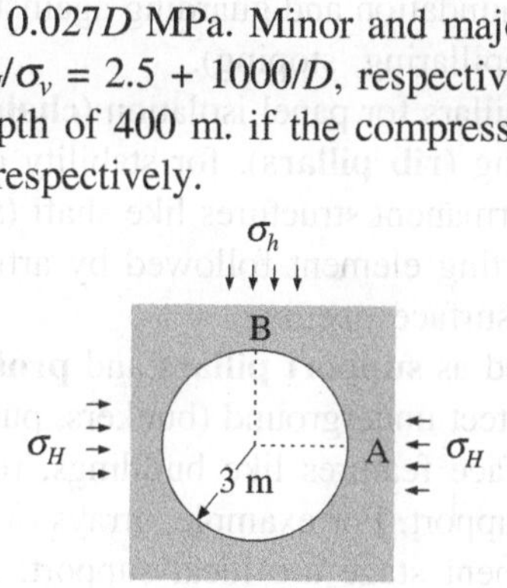

19. In continuation of Exercise 18, a uniform internal pressure, p_i is applied in the tunnel boundary such that radial displacement at Point A is reduced by 25% than that of unsupported tunnel. Assuming modulus of elasticity and Poisson's ratio of rock are 5.0 GPa and 0.25, determine the required p_i in MPa.

20. A circular tunnel of radius a is constructed in prevailing far field stress of σ_v (vertical) and σ_h (horizontal). Tangential stress at point B is thrice times than that of at point A. Determine the value, σ_h/σ_v.

Chapter 8

Natural Support

8.1 INTRODUCTION

Natural supports or pillars are *in situ* rock or coal remnants left after excavation of ore block or coal seam. The size of remnants (or pillar) varies along with its strength and shape. Pillars of larger size are left in case of soft rock (like coal, potash etc.) and smaller size is preferred for hard rock (metal mining, excavation for civil constructions). Shape of these pillars depends on the structural requirements but rectangular or square shape is more common. Natural support has its purpose such as protecting roadways or gallery, access way (shafts, incline or adit), surface features (natural or man made), underground mine operations like panel isolation, current mine working from dangers of water inundation and guarding against roof collapse during excavation activities (roadway extension, depillaring, stoping).

There are different types of pillars for panel isolation (**chain/barrier pillars**), for protection against the goaf during depillaring (**rib pillars**), for stability of upper level stopes (**crown or sill pillars**) and for protecting permanent structures like shaft (**shaft pillars**). Natural pillars are the primary and essential supporting element followed by artificial support to improve safety and stability of underground or surface opening.

Natural supports are classified as **support pillars** and **protection pillars**. Protection pillars are permanent pillars used to protect underground (bunkers, pump stations, service excavations, etc.) and surface structures (surface features like buildings, road, shaft, etc.) whereas support pillars are for local or regional support. For example, arrays of pillars formed in bord and pillar mining method during development stage are local support, designed to provide high safety factor. Local supports are of temporary use and are extracted once they fulfil their purpose. Barrier pillars are typical example of regional support. In hard rock excavation (like metal mining and excavation for civil purpose), natural pillars like crown, rib, sill and post pillars are used for regional support. Shapes of these pillars can be square, rectangular or irregular.

In this chapter, different types of pillars are introduced with their design principles. Different parameters affecting the strength of pillars are discussed along with the major empirical strength relationships. Pillar stresses during development (tributary area method), and depillaring operations (advanced or dynamic pillar design approach) are also discussed with examples. Pillar stabilisation by bolting and backfilling/stowing are introduced in this chapter. Some of the specialised methods based on practical approach like yield pillar technique is also discussed.

8.2 PILLAR DESIGN APPROACH

In bord and pillar coal mining, pillar design is an essential exercise for estimating safety factor of underground workings during development and depillaring operations. Development is defined as the process of creation of required underground structure by removal of its surrounding *in situ* material. *Depillaring* is a process of removal of developed pillars systematically without compromising safety of the remaining underground structure. Similarly, in geotechnical application of major underground power plant, two or three large and wide chambers are constructed parallel to one another and rock pillars are left in between them. These pillars are not removed rather strengthened further using artificial support elements. On the contrary, in coal mining application, arrays of pillars are developed for its partial removal during depillaring operations.

Classical approach to the pillar design is to determine its strength and to compare it with the applied stress. The ratio of estimated strength to the applied stress is expressed as *factor of safety* of the pillar (FOS). The design procedure involves estimating the required FOS by altering dimension of a pillar. For local stability, FOS is generally taken to be more than one. Mathematically, factor of safety during development is defined as:

$$\text{FOS} = \frac{\text{Pillar strength}}{\text{Pillar stress}} = \frac{S_p}{\sigma_p} \qquad (8.1)$$

In order to estimate the safety factor, pillar stress and pillar strength have to be known. There are two fundamental pillar design approaches:

(i) Stiff pillar approach: It is based on ultimate strength of a pillar to fully support the overburden. The design determines the strength of a pillar on the basis of its geometry, size and the compressive strength of the rock or coal material. The strength is compared with the expected load on pillar to find its ultimate strength to determine FOS. The main assumption of this approach is that, once the ultimate strength is overcome the pillar will have zero strength, which is not strictly true in reality especially if the material is weak rock or coal.

(ii) Yield pillar approach: Yield pillar approach is based on progressive yielding of a pillar and the transferring of shaded load onto adjacent stiff or abutment pillar. The design assumes smaller dimension of pillar so as it may yield once ultimate load is applied. The yielding begins at the point of yield strength, and gradually progresses to ultimate failure. This is probably a more realistic representation of the *in situ* condition.

8.2.1 Pillar Stress

The load on pillars comprises of vertical virgin stress in level or near-level pillars and horizontal stress (in inclined pillars). In case of horizontal pillars, the effect of horizontal stresses are negligible. Pillar stresses are calculated by tributary area method, beam theory, numerical methods, photoelastic techniques, or pressure arc theory. Analytically, tributary area method is widely used to estimate the pillar stress. Numerical methods are used to determine pillar stress based on material behaviour, i.e., elastic, elasto-plastic and visco-plastic. Figure 8.1 shows a typical two-dimensional numerical model for analyzing stresses on the coal pillars in a flat seam. Figure 8.2 shows a typical trend of pillar stress due to tributary method and numerical method based on elastic or elasto-plastic material for fixed and horizontal stress boundary conditions. The figure shows the following observation regarding pillar stress by these two methods:

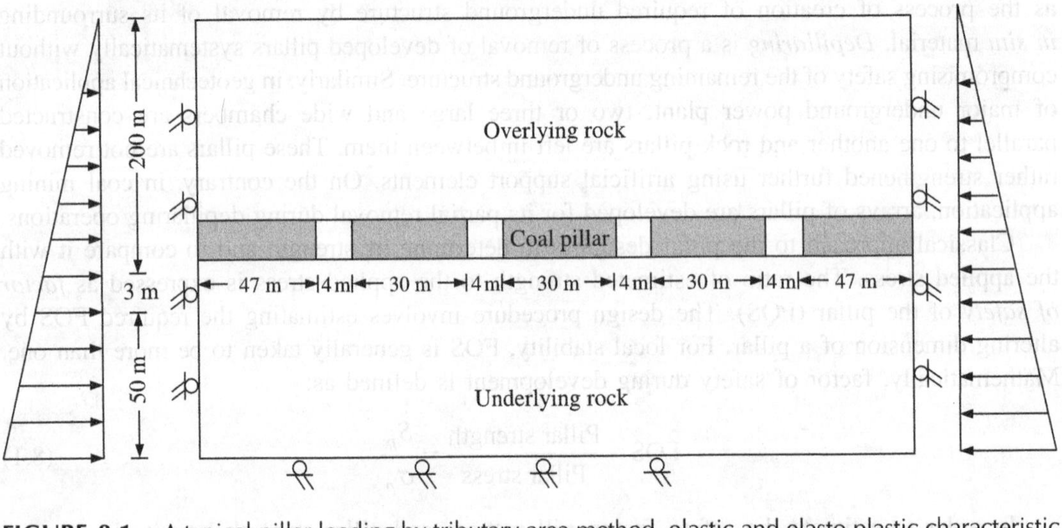

FIGURE 8.1 A typical pillar loading by tributary area method, elastic and elasto-plastic characteristic of rock mass with or without horizontal stresses.

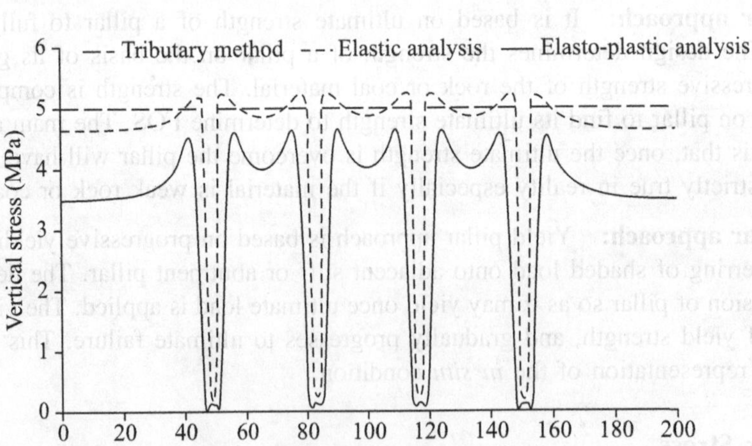

FIGURE 8.2 Vertical stress distribution on pillar by tributary area method, elastic and elasto-plastic analysis.

(i) Tributary area method shows that the pillar stresses are same across the pillar dimension.
(ii) Numerical method based on elastic analysis shows that pillar stresses are maximum at the pillar corner and reduces to a minimum at the middle of the pillar.
(iii) Numerical method based on elasto-plastic analysis shows that the maximum stress is concentrated little inside the pillar corner and reduces to minimum at the middle of the pillar. The shifting of the stress concentration (or maximum stress) from pillar corner (as in elastic analysis) represents the failure or fractured zone around the pillar boundary.

The results show the limitation of the tributary area method for estimating the pillar stress as results based on numerical methods are more realistic. It overestimates the pillar stress and cannot explain the progressive failure of the pillar.

Vertical Pillar

Let us assume a regular array of more or less uniform shape and size of pillars are developed in a flat deposit in a vertical stress field of σ_V as shown in Figure 8.3. The stress on a pillar (σ_p) can be expressed as:

$$\sigma_p = \frac{\gamma H(w_p + w_g)^2}{w_p^2} = \frac{\gamma H}{1 - R} \qquad (8.2)$$

where w_p is pillar width; w_g is gallery width; H is depth of cover; γ is average unit weight of overburden, and R is area extraction ratio and is defined as:

$$R = 1 - \left\{ w_p^2 / (w_p + w_g)^2 \right\}$$

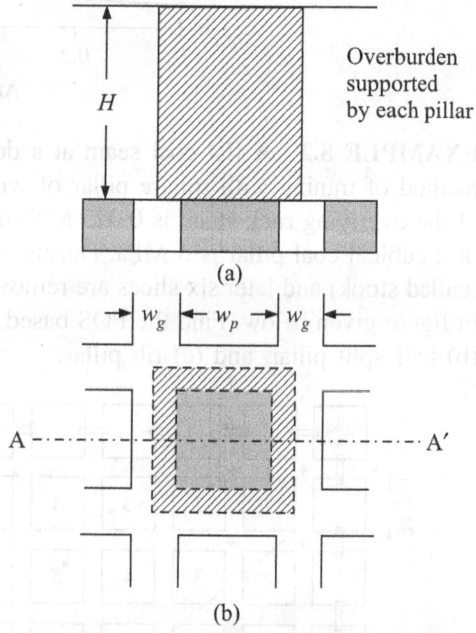

FIGURE 8.3 Tributary area method for pillar stress in flat opening: (a) Vertical section along line (A–A′) and (b) Top view of pillars.

The above method of pillar stress calculation is termed as tributary area theory which calculates the load developed on pillars during development operation only. Equation 8.2 works very well in practice for regular arrays. In general, stress on the pillar corner is higher as compared to the middle of the pillar. However, it is found that tributary area method over estimates the pillar stress followed by elastic and elastic-plastic stress analysis based on numerical method (Figure 8.2).

EXAMPLE 8.1 Find the effect of area extraction ratio on pillar stress.

Solution: From Eq. 8.2, it is clear that pillar stress increases hyperbolically with extraction ratio. In the following figure, it is seen that increase of pillar stress is moderate upto R = 0.5 and pillar stress is twice as much the *in situ* stress.

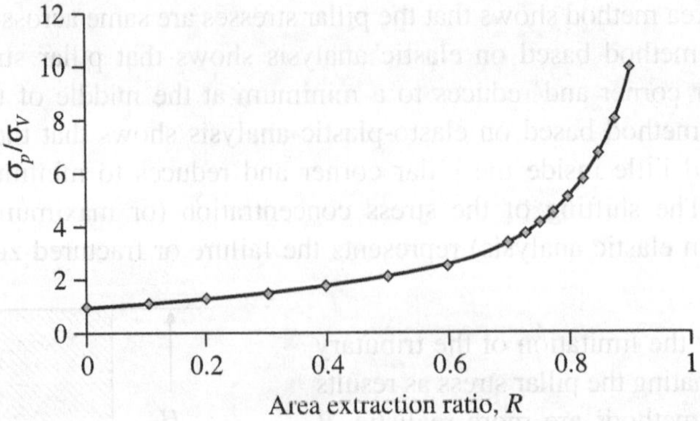

EXAMPLE 8.2 A flat coal seam at a depth of 80 m is being developed by bord and pillar method of mining with square pillar of width 30 m and gallery width 4.2 m. The unit weight of the overlying rock strata is 0.025 MN/m^3. Assume mining height of 4 m and *in situ* strength of a cubical coal pillar is 3 MPa. During depillaring operation, pillars are split into two halves (called stook) and later six slices are removed in sequence by leaving intact rib pillars as shown in figure given below. Find the FOS based on tributary area method for: (a) development pillar, (b) half split pillar, and (c) rib pillar.

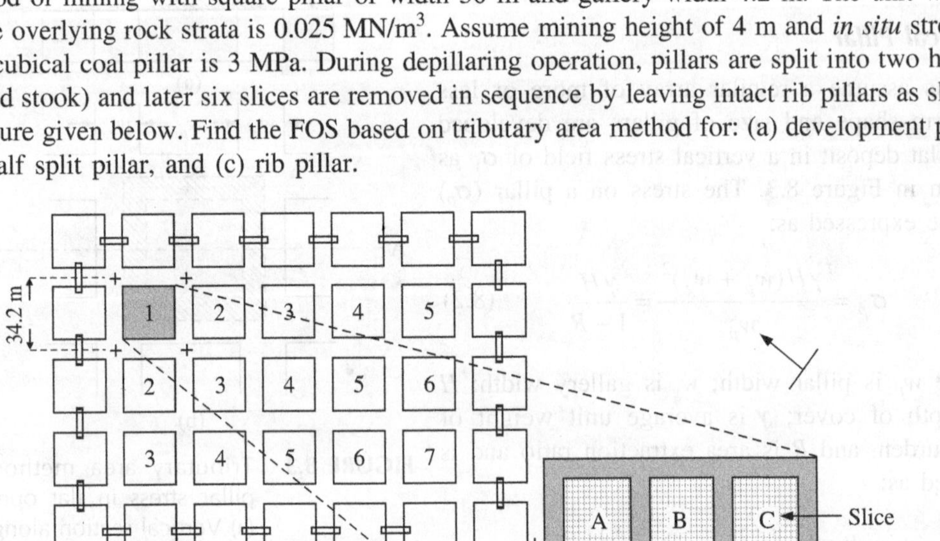

Solution: During depillaring process, the coal is extracted in different stages, first by dividing pillars into stooks; followed by removing them in slices and leaving ribs along the goaf side. The gravitational stress on pillar, stook and rib is γH = 2 MPa.

Strength of individual square pillar is obtained using Bieniawski's pillar strength formula (given in Eq. 8.29) and found to be is S_p = 10.02 MPa. The parameter $(1 - R)$ is 0.77.

For development pillar, pillar stress and FOS are:

$$\sigma_p = \frac{2}{0.77} = 2.60 \text{ MPa}; \quad \text{FOS}_p = \frac{10.02}{2.60} = 3.86$$

After pillar splitting, the pillar becomes rectangular having length 30 m and width 12.9 m. Strength of individual rectangular pillar is found using Bieniawski's rectangular pillar strength formula (given in Eq. 8.34) is S_p = 6.40 MPa. Extraction ratio is 0.34 and pillar stress in stook (σ_s) is 2/0.66 = 3.02 MPa. Hence,

$$\text{FOS}_{\text{stook}} = \frac{6.40}{3.02} = 2.12$$

The safety factor of rib pillars can be calculated either on a strength or load basis. The calculation based on strength is cumbersome because of the irregular shape of rib pillars; hence factor of safety is calculated based on load basis. Based on load, the factor of safety of the rib pillar is defined as the ratio of load applied to the maximum load a rib pillar can bear. A load applied on the rib pillar is $L_{rp} = \sigma_v \times A_{\text{ZOI}} = 1169.64$ MN, where, σ_v = 2 MPa is vertical stress and $A_{\text{ZOI}} = 34.2 \times 17.1 = 584.82$ m^2 is the zone of influence of the pillar and the maximum load the rib pillar can bear is $L_{\max} = \Sigma S_{rp} \times A_{rp} = 514.37$ MN. S_{rp} is the strength of rib pillar and A_{rp} rib pillar cross-section. The strength of rib pillars is found by dividing them into rectangular pillars in two ways: (i) 1 pillar of 30 m × 2.5 m, and 4 pillars of 10.4 m × 2.5 m; (ii) 4 pillars of 12.9 m × 2.5 m, and 3 pillars of 6.67 m × 2.5 m.

Based on the first division, the maximum load on the rib pillar is

$$L_{\max} = 2.90 \times 30 \times 2.5 + 4 \times 2.85 \times 10.4 \times 2.5 = 514.37 \text{ MN}$$

Based on the second division, the maximum load on the rib pillar is

$$L_{\max} = 4 \times 2.87 \times 12.9 \times 2.5 + 3 \times 2.81 \times 6.67 \times 2.5 = 510.15 \text{ MN}$$

The FOS of rib pillar is

$$\text{FOS}_{\text{rib}} = \frac{514.37}{1169.64} = 0.44$$

After comparing the results, it is found that the pillar FOS decreases from full size pillar to rib size pillar due to progressive coal extraction from the developed pillar.

Inclined Pillar

Let us assume again that regular array of more or less uniform shape and size of pillars are developed in an inclined opening in an *in situ* stress field (σ_H, σ_V) as shown in Figure 8.4(a). Pillars in inclined opening are kept perpendicular to the roof and floor. If dip direction is more, pillars are developed in apparent dip direction. Figure 8.4(a) shows a pillar inclined at an angle (θ) either in true dip or apparent dip direction. The pillar and gallery width are w_p and w_g respectively. Inclination in counter-clockwise is consdered positive.

The stress components normal and shear to the inclined pillar (σ_n, τ_s) are found by stress transformation (using Eqs. 4.29(b) and 4.29(c) in Section 4.3.4, and τ_{xy} = 0 as σ_H and σ_V are in the principal stresses) as:

$$\sigma_n = \sigma_H \sin^2\theta + \sigma_H \cos^2\theta = \frac{\sigma_H + \sigma_V}{2} - \frac{\sigma_H - \sigma_V}{2} \cos 2\theta \quad (8.3a)$$

$$\tau_s = -\sigma_H \sin\theta \cos\theta + \sigma_V \sin\theta \cos\theta = -\frac{\sigma_H - \sigma_V}{2} \sin 2\theta \quad (8.3b)$$

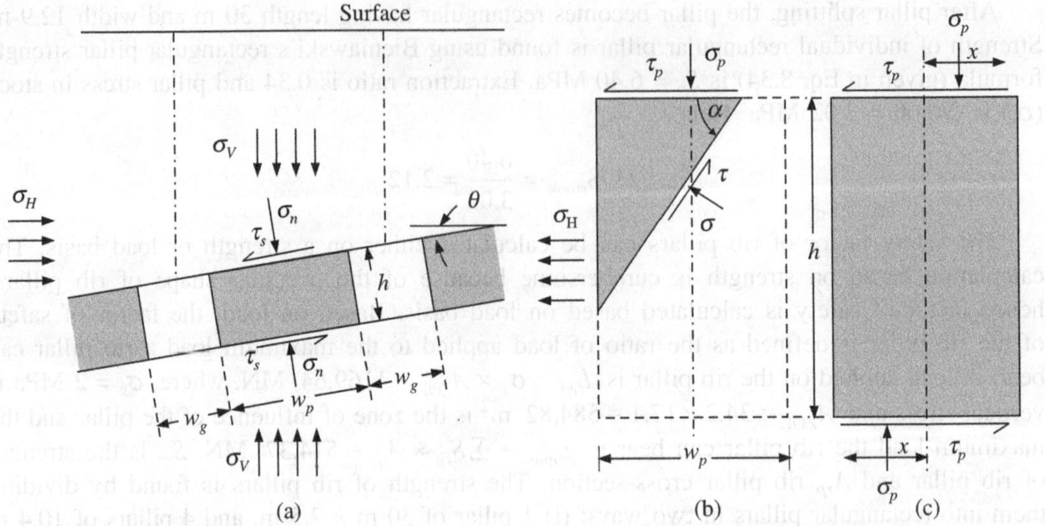

FIGURE 8.4 Tributary area method for pillar stress in inclined opening.

Finally, the induced stress on pillar along normal and shear stress direction is related with the area extraction ratio as:

$$\sigma_p = \frac{\sigma_n}{1 - R} \tag{8.4a}$$

$$\tau_p = \frac{\tau_s}{1 - R} \tag{8.4b}$$

The above relation reduces to Eq. 8.2 for flat seam by using $\theta = 0$ in Eq. 8.3. Force equilibrium along and perpendicular to an inclined plane having inclination α is shown by a free body diagram in Figure 8.4(b). The normal (σ) and shear (τ) stresses on any plane intersecting the pillar is found by force equilibrium equation as:

$$\sigma = \frac{\sigma_p}{2}(1 + \cos 2\alpha) - \frac{\tau_p}{2}\sin 2\alpha \tag{8.5a}$$

$$\tau = \frac{\sigma_p}{2}\sin 2\alpha + \frac{\tau_p}{2}(1 + \cos 2\alpha) \tag{8.5b}$$

where, α is the angle between the direction of σ_p and the normal to the inclined surface being considered.

Moment equilibrium of the forces acting on pillars as shown in Figure 8.4(c) gives following condition for stability against overturn and fails by toppling. Normal stress can be axisymmetric due to many reasons. Some of them are unequal excavation on the rise and dip side and unequal overburden stresses.

$$\frac{2x}{h} = \frac{\tau_p}{\sigma_p} \tag{8.6}$$

Here, $x < \frac{w_p}{2}$ or $2x \leq w_p$. Condition for not toppling is then

$$\frac{w_p}{h} \geq \frac{\tau_p}{\sigma_p} \tag{8.7}$$

Thus, when an offset of normal stress is required for moment equilibrium, the pillar width to height must satisfy the above inequality.

The induced pillar stress given in Eq. 8.5 can be rearranged to express as a Mohr's circle.

$$\left(\sigma - \frac{\sigma_p}{2}\right) = \frac{\sigma_p}{2}\cos 2\alpha - \frac{\tau_p}{2}\sin 2\alpha \tag{8.8a}$$

$$\left(\tau - \frac{\tau_p}{2}\right) = \frac{\sigma_p}{2}\sin 2\alpha + \frac{\tau_p}{2}\cos 2\alpha \tag{8.8b}$$

Equation 8.8 can be written in the form of a Mohr's circle as:

$$\left(\sigma - \frac{\sigma_p}{2}\right)^2 + \left(\tau - \frac{\tau_p}{2}\right)^2 = \left(\frac{\sigma_p}{2}\right)^2 + \left(\frac{\tau_p}{2}\right)^2 \tag{8.9}$$

Equation 8.9 is an equation of circle with centre ($\sigma_p/2$, $\tau_p/2$) and radius $r = \sqrt{(\sigma_p/2)^2 + (\tau_p/2)^2}$.

For a particular excavation conditions, pillar stress (σ_p, τ_p) changes with area excavation ratio as shown in Figure 8.5. With the increase in stresses, radius of stress circle increases and the centre of circle moves outward from the origin along a line inclined to the normal axis at an angle β, where,

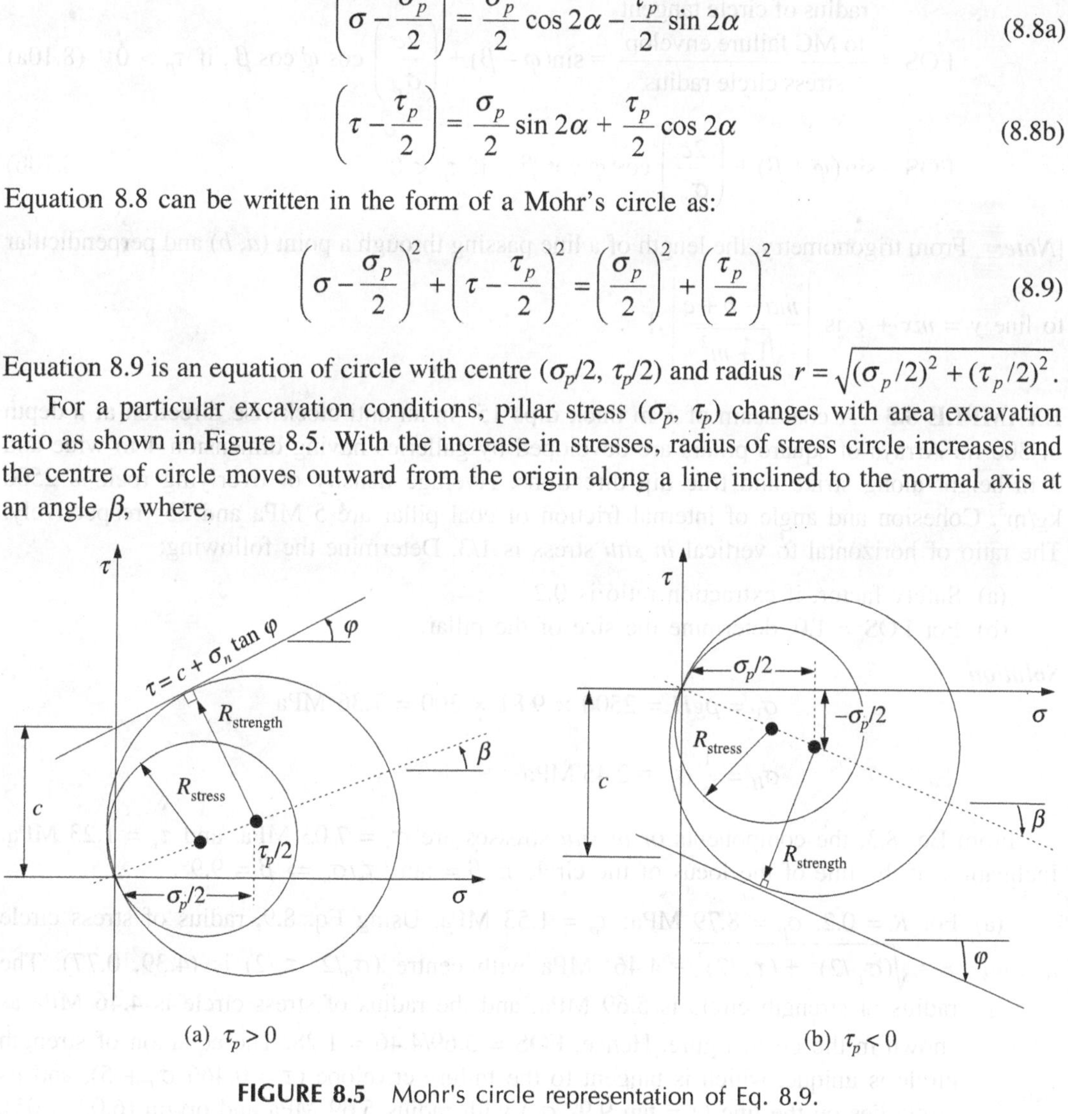

FIGURE 8.5 Mohr's circle representation of Eq. 8.9.

$$\tan\beta = \frac{(\tau_p/2)}{(\sigma_p/2)} = \frac{\tau_p}{\sigma_p}$$

or by using Eq. (8.4), $\tan\beta = \dfrac{\tau_s}{\sigma_n}$

The strength of pillar is defined by the radius of the circle tangent to the Mohr-Coulomb failure envelope (discussed in detail in Chapter 9). The factor of safety of inclined pillar is defined by ratio of strength to the radius of particular stress circle, i.e.,

$$\text{FOS} = \frac{\begin{array}{c}\text{radius of circle tangent}\\ \text{to MC failure envelop}\end{array}}{\text{stress circle radius}} = \sin(\varphi - \beta) + \left(\frac{2c}{\sigma_p}\right)\cos\varphi\cos\beta,\ \text{if } \tau_p > 0 \quad (8.10a)$$

$$\text{FOS} = \sin(\varphi + \beta) + \left(\frac{2c}{\sigma_p}\right)\cos\varphi\cos\beta,\ \text{ if } \tau_p < 0 \quad (8.10b)$$

[*Note:* From trigonometry, the length of a line passing through a point (a, b) and perpendicular to line $y = mx + c$ is $\left|\dfrac{ma - b + c}{\sqrt{1+m^2}}\right|$.]

EXAMPLE 8.3 A coal seam of 3 m thick dips 15° in an anti-clockwise direction at a depth of 300 m. Arrays of square pillars are developed by galleries having dimension 4 m wide and 3 m height along strike and true dip directions. Average density of overlying rock is 2500 kg/m^3. Cohesion and angle of internal friction of coal pillar are 5 MPa and 25° respectively. The ratio of horizontal to vertical *in situ* stress is 1/3. Determine the following:

(a) Safety factor, if extraction ratio is 0.2.
(b) For FOS = 1.0, determine the size of the pillar.

Solution:

$$\sigma_V = \rho gH = 2500 \times 9.81 \times 300 = 7.36 \text{ MPa}$$

$$\sigma_H = \frac{1}{3}\sigma_V = 2.45 \text{ MPa}$$

From Eq. 8.3, the components or *in situ* stresses are $\sigma_n = 7.03$ MPa, and $\tau_s = 1.23$ MPa. Inclination of the line of the locus of the circle is $\beta = \tan^{-1} \tau_s/\sigma_n \Rightarrow \beta = 9.9°$.

(a) For $R = 0.2$, $\sigma_p = 8.79$ MPa; $\tau_p = 1.53$ MPa. Using Eq. 8.9, radius of stress circle $r = \sqrt{(\sigma_p/2)^2 + (\tau_p/2)^2} = 4.46$ MPa with centre $(\sigma_p/2,\ \tau_p/2)$ is (4.39, 0.77). The radius of strength circle is 5.69 MPa, and the radius of stress circle is 4.46 MPa as shown in the given figure. Hence, FOS = 5.69/4.46 = 1.28. The equation of strength circle is unique, which is tangent to the failure envelope ($\tau = 0.466\ \sigma_n + 5$), and its circle lies on the line ($\tau = \tan 9.9°\ \sigma_n$) with radius 5.69 MPa and origin (6.03, 1.05). The FOS can also be found by using Eq. 8.10(a).

(b) For FOS = 1, Using Eq. 8.10(a), normal pillar stress σ_p = 12.07 MPa and shear stress on pillar $\tau_p = \sigma_p \times (\sigma_n/\tau_s) = 12.07 \times (7.03/1.23) = 2.11$ MPa. Using Eq. 8.9, radius of stress circle, $r = \sqrt{(12.07/2)^2 + (2.11/2)^2} = 6.13$ MPa with centre (6.04, 1.05). Hence, FOS = 6.13/4.46 = 1.37.

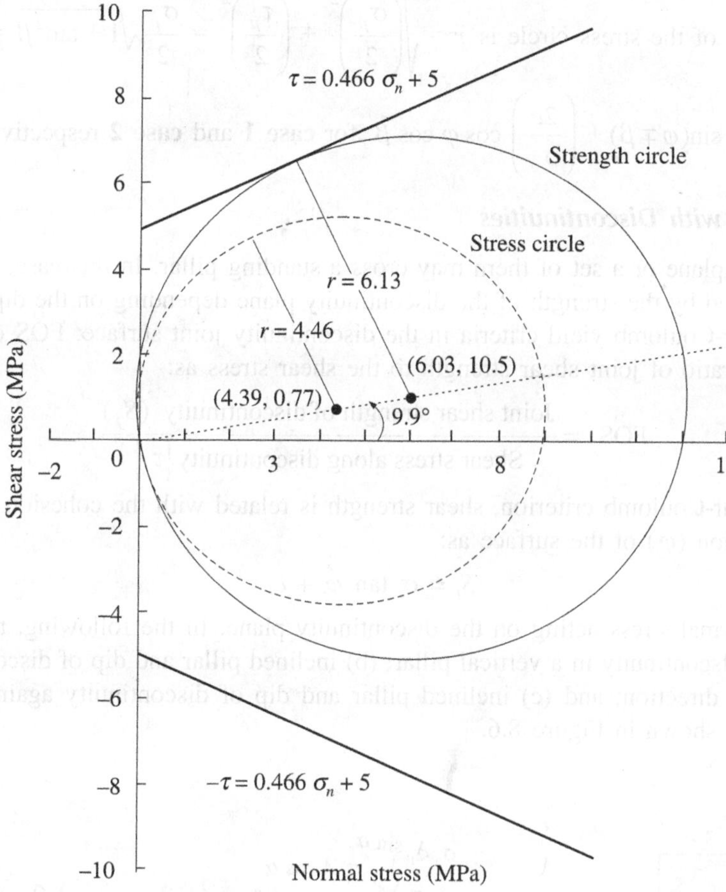

Derivation of Equation of 8.10

***Case* 1:** $\tau_p > 0,\ \sigma_p > 0,\ c \geq 0$

$\tau_p = \sigma_p \tan\beta$ and Mohr-Coulomb criterion becomes as $\sigma \tan\varphi + c - \tau = 0$.

Since, centre of the circle is $(\sigma_p/2,\ \tau_p/2)$, the distance from the centre to the tangent point between Mohr-Coulomb and the strength circle is

$$d = \frac{\sigma_p/2 \tan\varphi - \tau_p/2 + c}{\sqrt{1 + \tan^2\varphi}} = \sigma_p/2 \sin(\varphi - \beta)/\cos\beta + c\cos\varphi$$

***Case* 2:** $\tau_p < 0,\ \sigma_p > 0,\ c \geq 0$

Mohr Coulomb criterion becomes as $-\sigma\tan\varphi - c + \tau = 0$ and $\tau_p = \sigma_p \tan(-\beta) = -\sigma_p \tan\beta$.

The distance from the centre to the tangent point between Mohr-Coulomb and the strength circle is

$$d = \frac{\sigma_p/2 \tan\varphi + \tau_p/2 + c}{\sqrt{1+\tan^2\varphi}} = \sigma_p/2 \sin(\varphi+\beta)/\cos\beta + c\cos\varphi$$

Now, radius of the stress circle is $r = \sqrt{\left(\frac{\sigma_p}{2}\right)^2 + \left(\frac{\tau_p}{2}\right)^2} = \frac{\sigma_p}{2}\sqrt{1+\tan^2\beta} = \frac{\sigma_p}{2\cos\beta}$

$$\text{FOS} = \frac{d}{r} = \sin(\varphi \mp \beta) + \left(\frac{2c}{\sigma_p}\right)\cos\varphi\cos\beta \text{ for } \textbf{case 1} \text{ and } \textbf{case 2} \text{ respectively.}$$

Pillar Strength with Discontinuities

A discontinuity plane or a set of them may cross a standing pillar. In this case, pillar strength will be influenced by the strength of the discontinuity plane depending on the dip of the plane. Assuming Mohr-Coulomb yield criteria in the discontinuity joint surface, FOS of the pillar is defined by the ratio of joint shear strength to the shear stress as:

$$\text{FOS}_j = \frac{\text{Joint shear strength of discontinuity } (S_j)}{\text{Shear stress along discontinuity } |\tau_j|} \tag{8.11}$$

As per Mohr-Coulomb criterion, shear strength is related with the cohesion (c_j) and angle of internal friction (φ_j) of the surface as:

$$S_j = \sigma_j \tan\varphi_j + c_j \tag{8.12}$$

where, σ_j is normal stress acting on the discontinuity plane. In the following, three cases are discussed: (a) discontinuity in a vertical pillar; (b) inclined pillar and dip of discontinuity along the general dip direction; and (c) inclined pillar and dip of discontinuity against the general dip direction as shown in Figure 8.6.

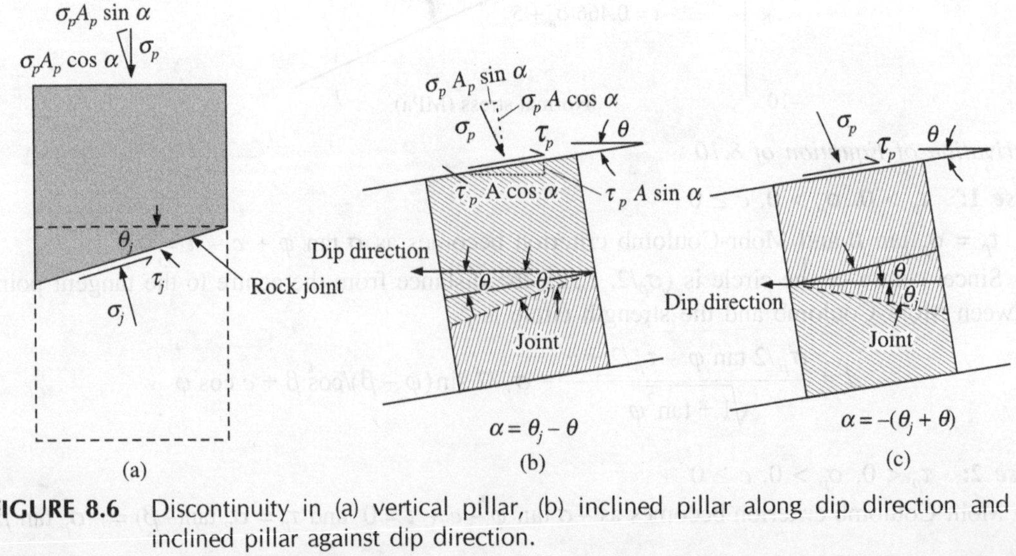

FIGURE 8.6 Discontinuity in (a) vertical pillar, (b) inclined pillar along dip direction and (c) inclined pillar against dip direction.

Let us assume that a discontinuity plane having an inclination α exists in a vertical pillar as shown in Figure 8.6(a). For a inclined pillar, discontinuity plane makes an angle θ_j with the dip direction. Assuming σ_p applies on the top surface having area A, the following normal and shear stresses on the discontinuity occur:

(a) Vertical pillar and dip angle of discontinuity (θ_j)

$$\sigma_j = \frac{\sigma_p}{2}(1 + \cos 2\theta_j) \tag{8.13a}$$

$$\tau_j = \frac{\sigma_p}{2}\sin 2\theta_j \tag{8.13b}$$

(b) Inclined pillar (inclination: θ) and dip angle of discontinuity (θ_j) and $\alpha = \theta_j - \theta$

$$\sigma_j = \frac{\sigma_p}{2}(1 + \cos 2\alpha) - \frac{\tau_p}{2}(\sin 2\alpha) \tag{8.14a}$$

$$\tau_j = \frac{\sigma_p}{2}(\sin 2\alpha) + \frac{\tau_p}{2}(1 + \cos 2\alpha) \tag{8.14b}$$

(c) Inclined pillar (inclination: θ) and dip of discontinuity ($-\theta_j$) and $\alpha = -(\theta_j - \theta)$

$$\sigma_j = \frac{\sigma_p}{2}(1 + \cos 2\alpha) + \frac{\tau_p}{2}(\sin 2\alpha) \tag{8.15a}$$

$$\tau_j = -\frac{\sigma_p}{2}(\sin 2\alpha) + \frac{\tau_p}{2}(1 + \cos 2\alpha) \tag{8.15b}$$

EXAMPLE 8.4 Average vertical stress on a vertical coal pillar is found to be 29 MPa. A discontinuity plane cuts the pillar at an angle φ_i with the horizontal axis. Assuming Mohr-Coulomb yield criteria for discontinuity surface having friction angle 35° and cohesion 4 MPa. What will be the range of φ_j so that pillar is unstable?

Solution: For an unsafe pillar, FOS < 1. Using Eqs. 8.12 and 8.13, we have

$$\frac{\sigma_p}{2}(1 + \cos 2\theta_j)\tan\varphi_j + c_j < \frac{1}{2}\sigma_p \sin 2\theta_j$$

$$\sin(2\theta_j - 35) > 0.8$$

$$44.04° < \theta_j < 80.96°$$

EXAMPLE 8.5 A coal seam having an inclination of 15° clockwise from the horizontal axis is located at a depth of 300 m. The ratio of horizontal stress to vertical stress is 0.33 and unit weight of overburden rock is 25 kN/m^3. The coal seam is developed by bord and pillar method with an area extraction ratio of 0.25. Assume that a discontinuity plane intersects a pillar along the dip of the seam and another discontinuity plane cuts another pillar against the dip of the seam with an inclination of 35° for both the cases. The cohesion and angle of internal friction of discontinuity plane are 0.75 MPa, 20° respectively. Find the stability of the pillar for both the cases.

Solution: From the given data, the vertical and horizontal stresses are 7.5 MPa and 2.48 MPa respectively. Using Eqs. 8.3 and 8.4, the normal and shear stresses on pillar are $\sigma_p = 9.55$ MPa; $\tau_p = -1.68$ MPa respectively.

For discontinuity plane along the dip of the seam (Eq. 8.14), normal and shear stresses along discontinuity plane are $\sigma_j = 6.34$ MPa and $\tau_j = 4.59$ MPa respectively. The shear strength of the discontinuity plane is determined using Mohr-Coulomb failure criteria (Eq. 8.12) as 3.06 MPa. Finally Eq. 8.11 provides FOS = 0.67.

For discontinuity plane against the dip of the seam (Eq. 8.15), $\sigma_j = 3.38$ MPa and $\tau_j = 4.64$ MPa. Shear strength is 1.98 MPa then Eq. 8.11 provides FOS = 0.43.

8.2.2 Pressure Arch Load

Arching is a process of stress redistribution due to relaxation of strata immediately above an excavation and increased compression occurring over pillars and abutments (Herget, 1988). Tributary area method assumes stresses in the pillar but ignores deformations of roof, floor and pillars which are continuously changing the stresses on the pillars. Due separation of roofs, stress relaxation occurs and a pressure arch is formed above the excavated zone [Figure 8.7(a)]. The stress trajectories bypass the relaxation zone and is passed through the abutments or neighbouring pillars. This condition is termed as *pressure arch* as if a compressed zone is balanced between two abutments causing an arching effect above the excavation. Figure 8.7(b) further demonstrates this theory. The top figure of Figure 8.7(b) shows two small pressure arches above the excavations. Now if the middle pillar is mined out, further relaxation occurs in the roof due to further convergence and separation. As a result, a new arch is formed which is balanced by the two side pillars. In this way, height and width of the arch grows with excavation of pillars and finally it may show up on the surface in the form of subsidence.

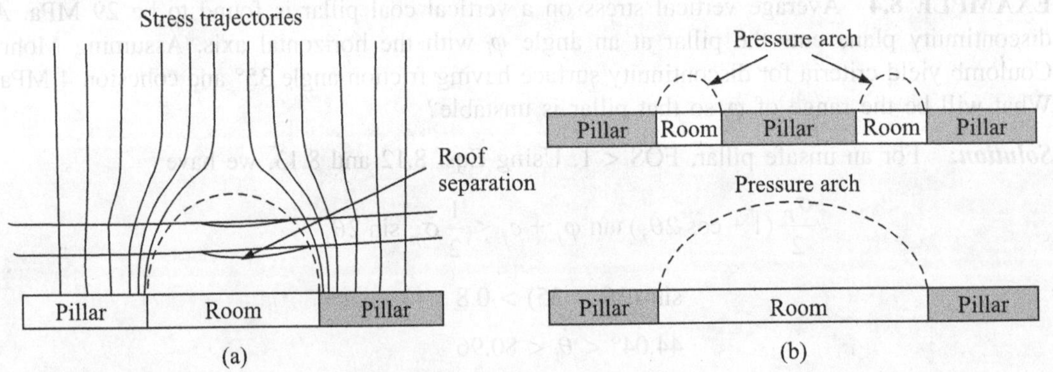

FIGURE 8.7 Concept of pressure arch theory.

The width or span of pressure arch depends on the depth of cover, H. Empirically, it is given as:

$$w_a = 0.15H + 60 \text{ (all units are in feet)}$$

Poulsen (2010) estimated load transfer distance (LTD) in terms of depth of cover as:

$$\text{LTD} = -1 \times 10^{-4} H^2 + 0.2701H \tag{8.16}$$

Pillars within LTD will be able to interact and shed load as shown schematically in Figure 8.8. Typically a stiffer pillar, generally of greater width to height ratio such as barrier pillar, will carry a greater proportion of the load than smaller production pillars within its zone of influence. From Figure 8.8, it can be seen that a pillar of effective width (w_e), given in Eq. 8.17, a lateral zone of influence (ZI) is measured from the pillar centroid.

$$\text{ZI} = 2\text{LTD} + \frac{w_e}{2} \tag{8.17}$$

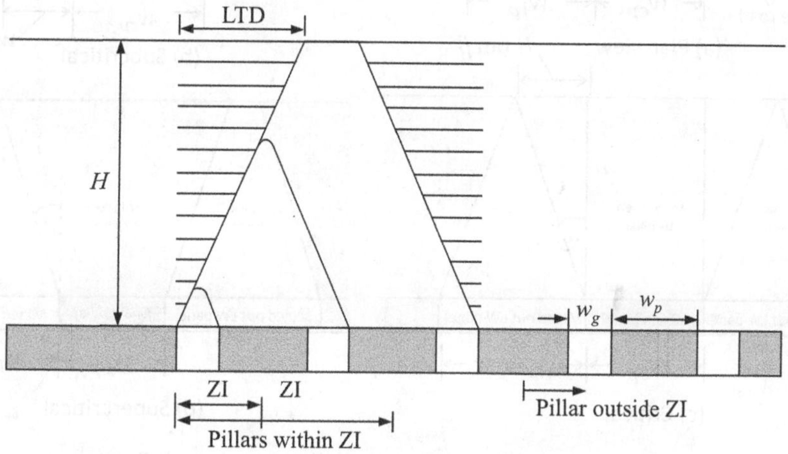

FIGURE 8.8 Load transfer between pillars within the pressure arch.

Hence, an average vertical force in a pillar's zone of influence is defined by Eq. 8.18.

$$F_{zz} = \rho g H \left(2\text{LTD} + \frac{w_e}{2}\right)^2 \tag{8.18}$$

8.2.3 Stress Distribution on Chain or Barrier Pillar

Chain and Barrier pillars are typically an *in situ* coal or rock block left in between two subsequent excavations for proper ventilation and ground control. As for example, chain pillars are left between two adjacent longwall (LW) panels to protect the gate roads from excessive deformation, and to protect the surface from undue subsidence. Figure 8.9(a) shows plan view of typical mined out longwall panels or goaf and chain pillar in between and Figures 8.9(b), (c), and (d) depict the vertical section of the same for subcritical, critical and supercritical conditions, respectively (for more details, check Section 12.3.3 of Chapter 12 on Subsidence). Figure 8.9(b), (c) and (d) shows the overburden rock responsible for loading the pillars in subcritical, critical and supercritical condition respectively. After excavation of longwall panel width, the total mined out width of longwall panel includes the longwall panel width plus the width of main gate and tail gate. Finally, as the width of the longwall panel (200 m to 250 m) is almost 50 times more than the width of the main and tail gate (4 to 5 m). Hence by ignoring the loading based on tributary area method (as discussed in Section 8.2.1) total load is calculated from the pillar corner. These pillars can be square, rectangular or diamond in shape configured in single or double rows.

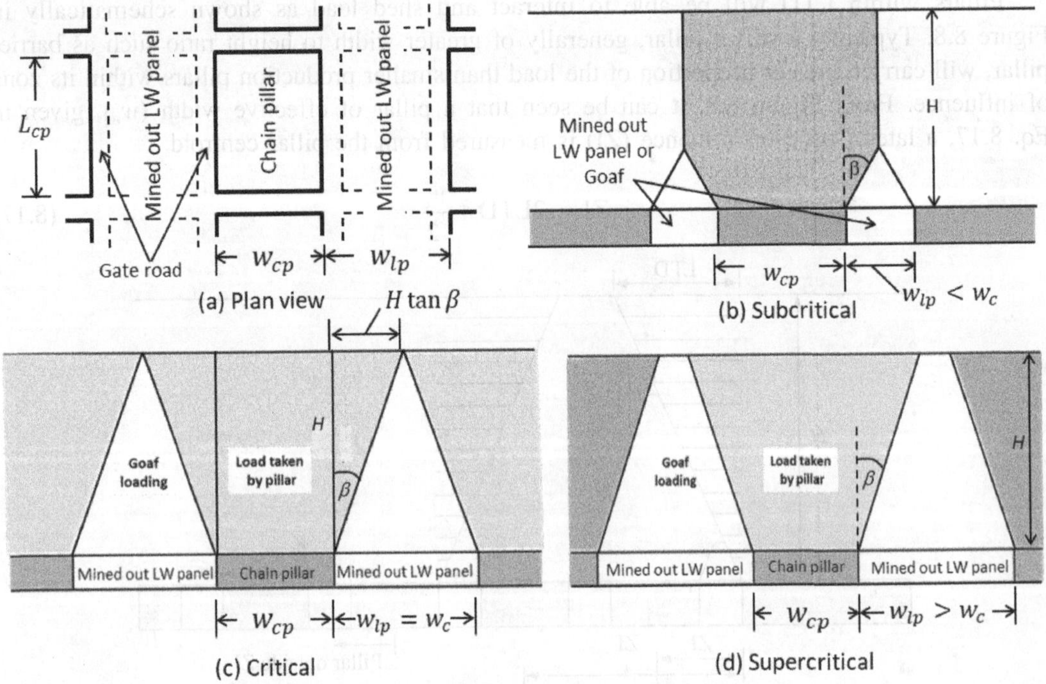

FIGURE 8.9 Loading on a chain pillar in a longwall mine.

For critical and supercritical conditions, the vertical stress for unit length of chain or barrier pillar can be estimated as:

$$\sigma_{cp} = \gamma H \left(1 + \frac{H}{w_{cp}} \tan\beta\right) \tag{8.19}$$

For subcritical condition, the stress on pillar is

$$\sigma_{cp} = \gamma H \left[1 + \frac{w_{lp}}{w_{cp}} - \frac{\cot\beta}{4Hw_{cp}} w_{lp}^2\right] \tag{8.20}$$

where H is depth, β is abutment angle L_{cp} is length of a chain pillar, w_{lp} is width of mined out longwall panel or goaf with gallery width, and w_{cp} is width of chain pillar.

The above Eqs. (8.19) and (8.20) can be used to design chain and barrier pillars.

EXAMPLE 8.6 A single line of square chain pillars separates two longwall panels as shown in a vertical section (see figure given below). A 3.0 m thick coal seam is to be mined at a depth of 300 m below surface and the longwall face is 200 m wide. The average abutment angle is 31°, width of the roadways is 6 m, and average unit weight of overlying rock is 25 kN/m^3. Find the minimum width of the chain pillar while considering that both side longwall panels are mined. It is given that the strength of a chain pillar is 42.5 MPa.

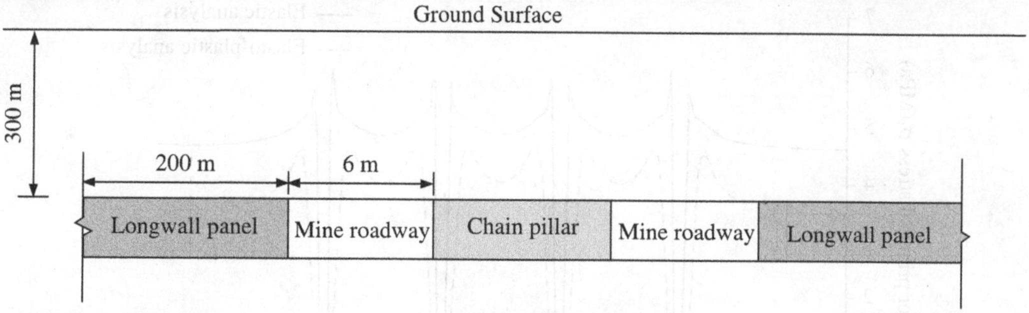

Solution: After extraction of panels on both sides, the goaf, i.e., the width of the longwall panel plus the width of gate road is 206, which is less than $2H \tan 31° = 360.52$ m. Hence the chain pillar loading will be under subcritical condition as shown in the Figure below. Figure shows the geometrical shape of the overburden loading the chain pillar.

Now, Eq. 8.20 can be used to find the stress on chain pillar. In the question, it is mentioned to find the minimum width of the chain pillar which is possible while considering the factor of safety one. Hence the stress on chain pillar can be equated with the strength with the given data as:

$$\gamma H\left[1 + \frac{w_{lp}}{w_{cp}} - \frac{w_{lp}^2 \cot \beta}{4Hw_{cp}}\right] = 42.5 \times 10^6$$

$$w_{cp} = 31.51 \text{ m}$$

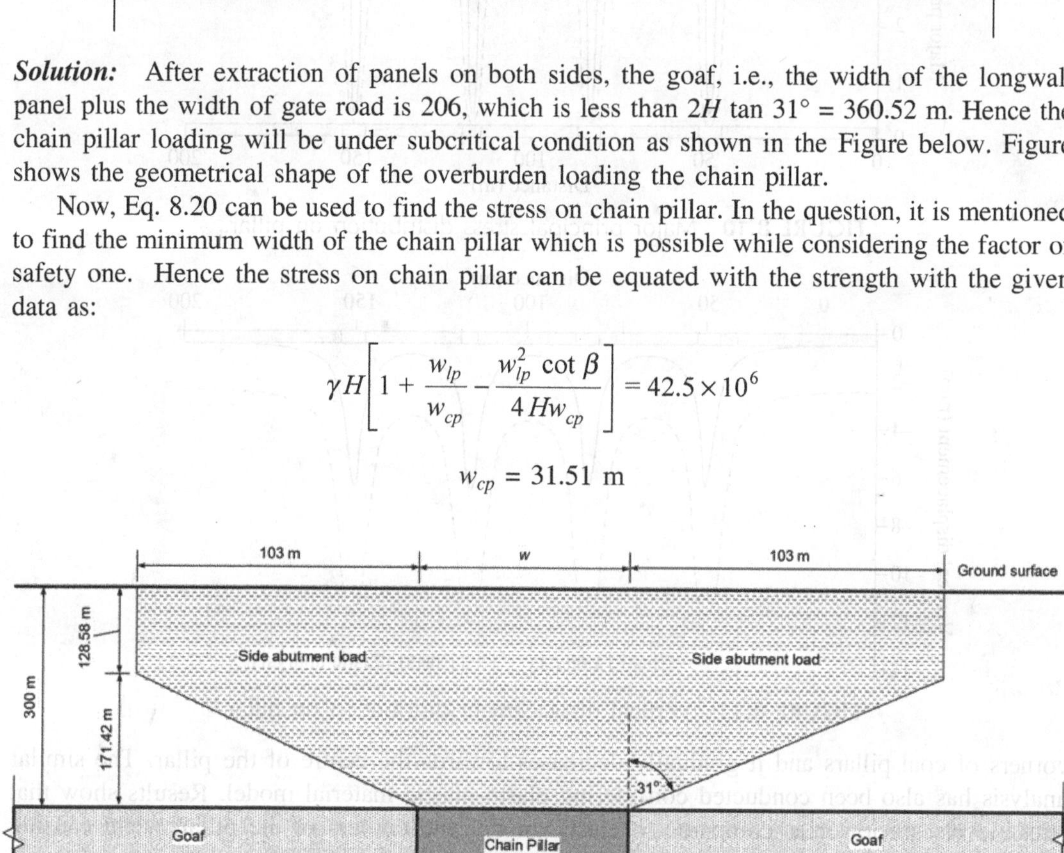

8.2.4 Stress and Displacement Distribution on Pillars

Stress and displacement distribution on pillar depend on the pillar stress and elastic-plastic properties of rock or coal. Figures 8.2, 8.10, and 8.11 show the distribution of vertical stress, major principal stress and vertical displacement respectively on a series of coal pillars as shown in Figure 8.1. This stress and displacement distribution shown in Figures 8.2, 8.10 and 8.11 are based on numerical analysis conducted using numerical modelling technique and is briefly mentioned in Chapter 14. It can be seen that for elastic model, the peak stress occurs at the

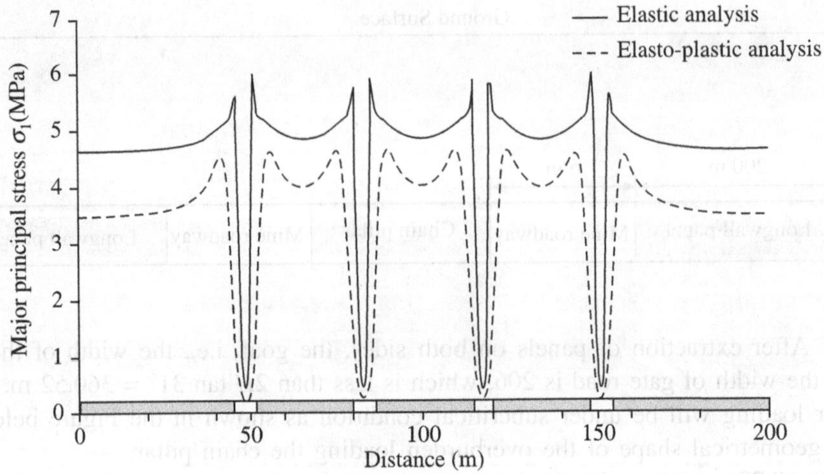

FIGURE 8.10 Major principal stress distribution on pillar.

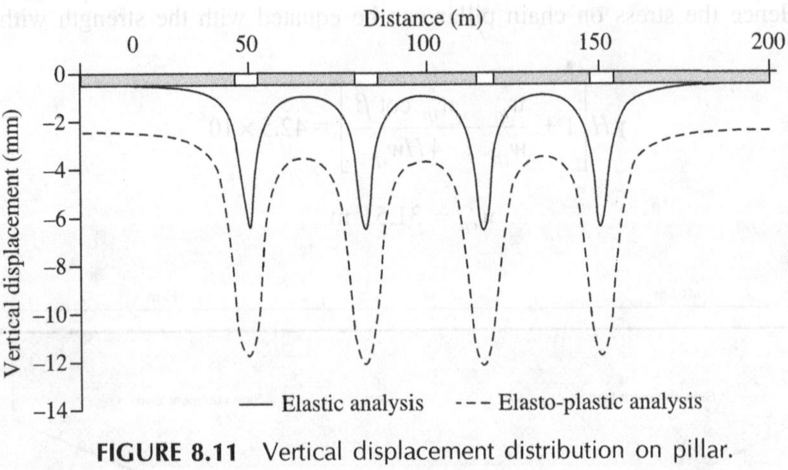

FIGURE 8.11 Vertical displacement distribution on pillar.

corners of coal pillars and it gradually decreases towards the centre of the pillar. The similar analysis has also been conducted considering elasto-plastic material model. Results show that stress levels are lower as compared to elastic model and corners of the pillars yield causing peak stress to shift inside the pillar.

Vertical displacement profile shows the maximum value occurs at the centre of the mid gallery and the minimum value is obtained at the centre of the pillar. As the loading condition is beyond elastic limit displacement increases further. The results show that the tributary area theory has limitations and merely estimates the average stress on pillar. On the contrary, using numerical methods, pattern of stress and displacement distribution can be obtained for various combination of loading and boundary conditions.

8.2.5 Pillar Strength

Vertical overburden load on pillar represents uniaxial compressive loading, hence the pillar strength is generally established by uniaxial compressive strength test of rock/coal samples in laboratory. It is not possible to estimate *in situ* strength of a coal pillar since destructive tests cannot be conducted in the field on actual size of pillar. Various researchers have established the factors those influences the pillar strength such as uniaxial and triaxial coal strength; width to height ratio of pillar; pillar size and shape; pre-excavation horizontal stresses; end conditions or condition at the roof-pillar and/or floor-pillar contact and also presence of dirt bands in the seam; water and weathering underground; and method of drivage either mechanical or manual. Laboratory tests reveal that strength of cubical or prismatic rock/coal samples is affected by its size and shape. However, it is possible to extrapolate laboratory data for *in situ* field conditions. In the following, effects of shape and size on pillar strength are discussed in details.

Size Effect on Strength

The effect of specimen size on strength is clearly seen in Figure 8.12, where strength reduces drastically due to larger volume and geological discontinuities. According to Hustrulid (1976), size effect is negligible when length of cubic size coal sample is 0.9 m or more. Mathematically, the variation of strength with length l of cubic sample is given by power law as:

$$S_l = kl^{-a} \tag{8.21}$$

where k is a constant defined as UCS of a cubic sample and a is a constant varies between 0.2 and 0.5. The above equation can be rewritten by taking natural log as:

$$\ln S_l = \ln k - a \ln l \tag{8.22}$$

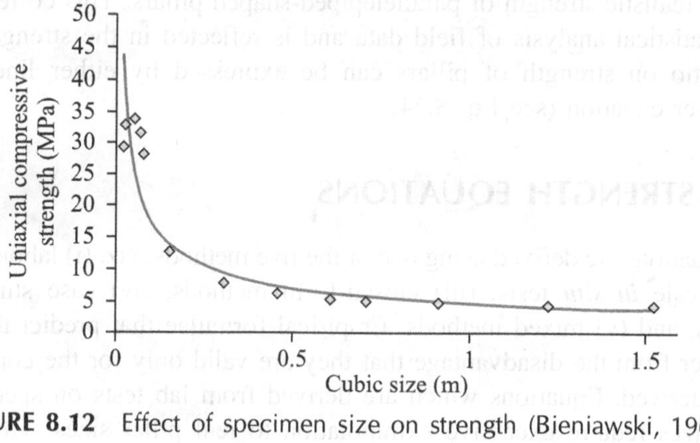

FIGURE 8.12 Effect of specimen size on strength (Bieniawski, 1968).

This implies that $\ln S_l$ and $\ln l$ has linear relationship and hence, by obtaining strength of different cubical size samples, the parameter k and a can be ascertained. Once these parameters are obtained, strength (S_l) of 0.9 m cubical size sample is determined using Eq. 8.22.

Shape Effect on Strength

Laboratory investigation on variations of shape effect, width to height ratio, w/h = 1.0 ~ 13.5 are shown by different stress strain curves in Figure 8.13. After comparing the post failure behaviour of all these curves, three marked characteristics are identified. In the post failure region, one set of characteristics curves shows reduction in stress with increase in strain; second, insignificant change in stress with increase in strain; and third, increases in stress with increase in strain. Based on these behaviours, pillars are characterised respectively as slender or slim, intermediate and squat (or flat) pillars.

The above result conceptualises that pillars having *w*/*h* ratios between 15 and 20 are virtually indestructible. Actually, the pillars at these *w*/*h* ratios do fail but still retain about the same resistance after failure. These pillars may be termed as *indestructible pillars or mechanically shows strain hardening characteristics*. However, pillars having *w*/*h* = 4 to 8 are considered to be intermediate pillars shown in elastic-brittle-plastic characteristics and mostly developed in coal mines. In metal mines, pillars are slim having *w*/*h* between 1 and 3.

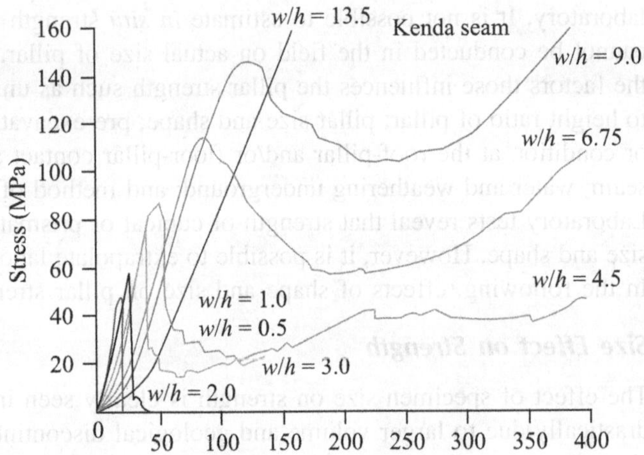

FIGURE 8.13 Complete stress strain curve for Indian coal specimen (Das, 1986).

In the laboratory, cubical size coal samples are prepared and tested in uniaxial compression. However, in the field, pillars are not cubical in shape. Hence, a shape correction needs to be applied to obtain realistic strength of parallelepiped-shaped pillars. This correction is generally obtained from statistical analysis of field data and is reflected in the strength equations. The effect of *w*/*h* ratio on strength of pillars can be expressed by either linear equation (see Eq. 8.23) or power equation (see Eq. 8.24).

8.3 PILLAR STRENGTH EQUATIONS

Pillar strength equations are derived using one of the five methods, viz. (i) laboratory compression tests, (ii) large-scale *in situ* tests, (iii) closed-form methods, (iv) case studies of collapsed and stable pillars, and (v) mixed methods. Empirical formulae that predict the strength of the whole pillar suffer from the disadvantage that they are valid only for the conditions for which they have been derived. Equations which are derived from lab tests on specimens of various *w*/*h* ratios and sizes require excessive extrapolation to real pillar sizes. This drawback may, however, be overlooked provided these equations fit the case studied.

The strength equations developed so far are expressed either in linear form or power law form.

$$S_p = S_l\left(a + b\frac{w}{h}\right) \tag{8.23}$$

$$S_p = kh^{\alpha}w^{\beta} \tag{8.24}$$

where, a and b are dimensionless and $a + b = 1$, α and β are dimensionless parameters. An examination of the formula published by various authors shows that the value α varies from -1.0 to -0.66, whereas β is found to be in between 0.46 and 0.6. Several pillar strength formulas are developed like Bauschinger (1876); Holland-Gaddy (1964), Obert-Duvall (1967). In following section few of the important pillar strength formula are discussed.

8.3.1 Strength of Square Pillar

Salamon-Munro Formula: The original analytical mine pillar strength formula of Salamon-Munro (1967) is shown in Eq. 8.25.

$$S_p = 7.2\frac{w^{0.46}}{h^{0.66}} \text{ MPa} \tag{8.25}$$

The above relation was developed for room and pillar mining of horizontal coal seams, typically for South African coal mines. In 1996, a similar database of Australian pillars was compiled and Galvin, Hebblewhite, and Salamon developed the following coal pillar strength formula:

$$S_p = 8.60\frac{w^{0.51}}{h^{0.84}} \text{ MPa} \tag{8.26}$$

It was found that the formula derived from the Australian database resembles closely to the results obtained by Salamon and Munro. The similarity in the pillar strength formulae derived from the two national databases in 1999, produced the following formula:

$$S_p = 6.88\frac{w^{0.50}}{h^{0.70}} \text{ MPa} \tag{8.27}$$

Comparison of above three coal pillar strength formulae, for width-to-height ratio from 2 to 4, it is found that the coal pillar strength calculated by these three formulae resembled each other closely and the differences is less than 8.3%. Galvin, et al. (1999) has suggested that the application of Eqs. 8.25, 8.26, and 8.27 are suitable only within the range $w/h > 2$ to $w/h < 5$.

The dimensionally correct form of each is

$$S_p = \frac{k(w/w_0)^{\alpha}}{(h/h_0)^{\beta}} \tag{8.28}$$

where k is the strength of a cube of width w_0 and height h_0, and hence the parameter k in the above equation is the strength of a representative sample of unit volume and not necessary the mass strength of coal. Salamon describes k as multiplier, thus, it is not a property of the material (i.e., of the coal), but the property of the involved system.

Bieniawski's Formula: This formula is based on large scale testing of *in situ* coal samples in South Africa and in USA and it is expressed as

$$S_p = S_l\left(0.64 + 0.36\frac{w}{h}\right) \tag{8.29}$$

where S_l is the strength of a *in situ* cubical coal pillar.

CIMFR Formula: This formula is given based on the experience of Indian coal mines as:

$$S_p = 0.27\sigma_c h^{-0.36} + \frac{H}{160}\left(\frac{w}{h} - 1\right) \tag{8.30}$$

where, H is depth of seam.

Holland and Gaddy: The pillar strength equation given by Holland and Gaddy (1964) is as follows:

$$S_p = k\frac{\sqrt{w}}{h} \tag{8.31}$$

where, $k = \sigma_c/\sqrt{d}$ is Gaddy constant. It is the strength of a 2.5 cm cube coal sample. σ_c is the uniaxial compressive strength of a cubic sample of size d inch cube.

8.3.2 Strength of Rectangular or Irregular Pillar

The above strength equations are developed for square pillar only. For a rectangular pillar having width w and length l, the equivalent width w_e is determined by area equivalence with a square pillar as:

$$w_e = \sqrt{wl} \tag{8.32}$$

In a detailed study of the failure process of coal pillars, Wagner (1974) showed that the failure commences at the circumference of the pillar and migrates inwards. An interesting observation made during these studies was that, at the time of the overall structural failure of the pillar, the central position of the pillar does not yet reached its full load-bearing potential. On the basis of these observations, it is suggested that the ratio of the area, A, to the circumference, C, of a coal pillar has a strong influence on the pillar strength. Hence, the concept of hydraulic radius in introduced to define the effective width w_e as:

$$w_e = \frac{4A}{C} \tag{8.33a}$$

where, w_e can be used as equivalent width of square pillar. For rectangular pillar,

$$w_e = \frac{2wl}{w+l} \tag{8.33b}$$

Mark-Bieniaswski (Mark and Chase, 1997) equation for rectangular pillar indicates that the strength of rectangular pillar depends on the width-to-height and width-to-length ratio. Mark-Bieniaswski equation for rectangular pillar is as follows:

$$S_p = S_l\left(0.64 + 0.54\frac{w}{h} - 0.18\frac{w^2}{lh}\right) \quad \text{where, } l > w \tag{8.34}$$

EXAMPLE 8.7 The depth of a bord and pillar mine is 150 m and average unit weight of overburden is 25 kN/m^3. The gallery width and height of mining are 4.6 m and 2.4 m respectively. Upon testing of cubical coal samples, the following data are obtained.

Size of the sample, mm	10	50	90	120
Strength, MPa	18	10	7	6

Find the width of a square pillar if the safety of the pillar is 1.5.

Solution: From the given laboratory test data, the plot shows the relation between log S_l and log l. Which can be simplified as: $S_l = 10^{7.7}\, l^{-0.44}$. Comparing with the derived equation with the size effect Eq. (8.21), $k = 10^{7.7}$ and $a = 0.44$.

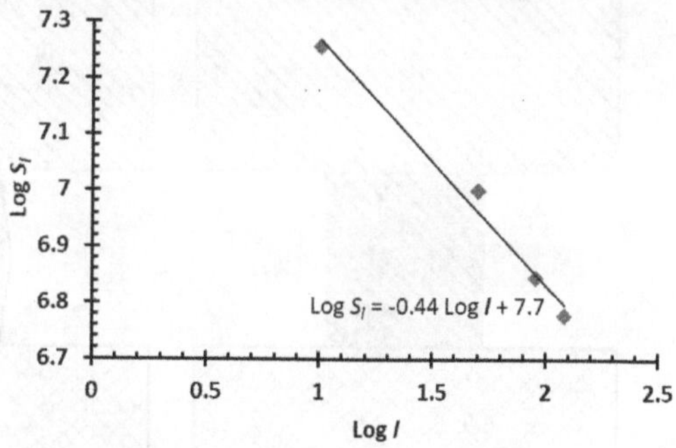

For 900 mm cubic block, the strength parameter $S_l = 10^{7.7}\ 900^{-0.44} = 2.51$ MPa. The stress on the pillar is found by Eq. 8.2 as $\gamma H \frac{(w_p + w_g)^2}{w_p^2}$ and the pillar strength by using Using Bieniawski's strength equation: $S_p = 2.51\left(0.64 + 0.36\frac{w_p}{2.4}\right)$. Now for FOS = 1.5, we can use Eq. 8.1 as:

$$1.5 = \frac{2.51\left(0.64 + 0.36\frac{w}{2.4}\right)w_p^2}{3.75(w_p + 4.6)^2}$$

or

$$0.38w^3 - 4.03w^2 - 51.75w - 476.10 = 0$$

By iterative solution, the width of the pillar is $w = 20.44 \approx 21$ m, Other two solutions are complex number, which is unacceptable.

8.4 MODES OF PILLAR FAILURE

Theoretically, a pillar passes through different phases of loading conditions. Mortazavi et al., (2009) categorised these transition phases as follows:

(i) Virgin conditions
(ii) Stresses rises at pillar sides with core stress relatively unchanged
(iii) Sides fractured and failed and core stress rises
(iv) Side fracture zones grow larger while core approaching yield conditions
(v) Convergence of fractured zones
(vi) Ground distressing around failed pillar

In relatively soft and weak ground, the full overburden weight is transferred to the pillar system. However, in stiffer and stronger rocks, a greater portion of the overburden load is transferred to the unmined coal barriers, and the pillar stress is reduced. The amount of pillar compression is determined by the relative stiffness of the pillars and the surrounding strata. Stiffer pillars carry more stress, while stiffer surrounding strata will deflect less and impose smaller load on the pillars. The concept of the pillar stiffness and strata stiffness is well established in the field of pillar design and has been used to evaluate the potential for violent pillar collapse. Failure of pillar includes roof damage or floor punching and not only the failure of the pillar as a structure. Figure 8.14 shows some of the principal modes of failure of the pillars.

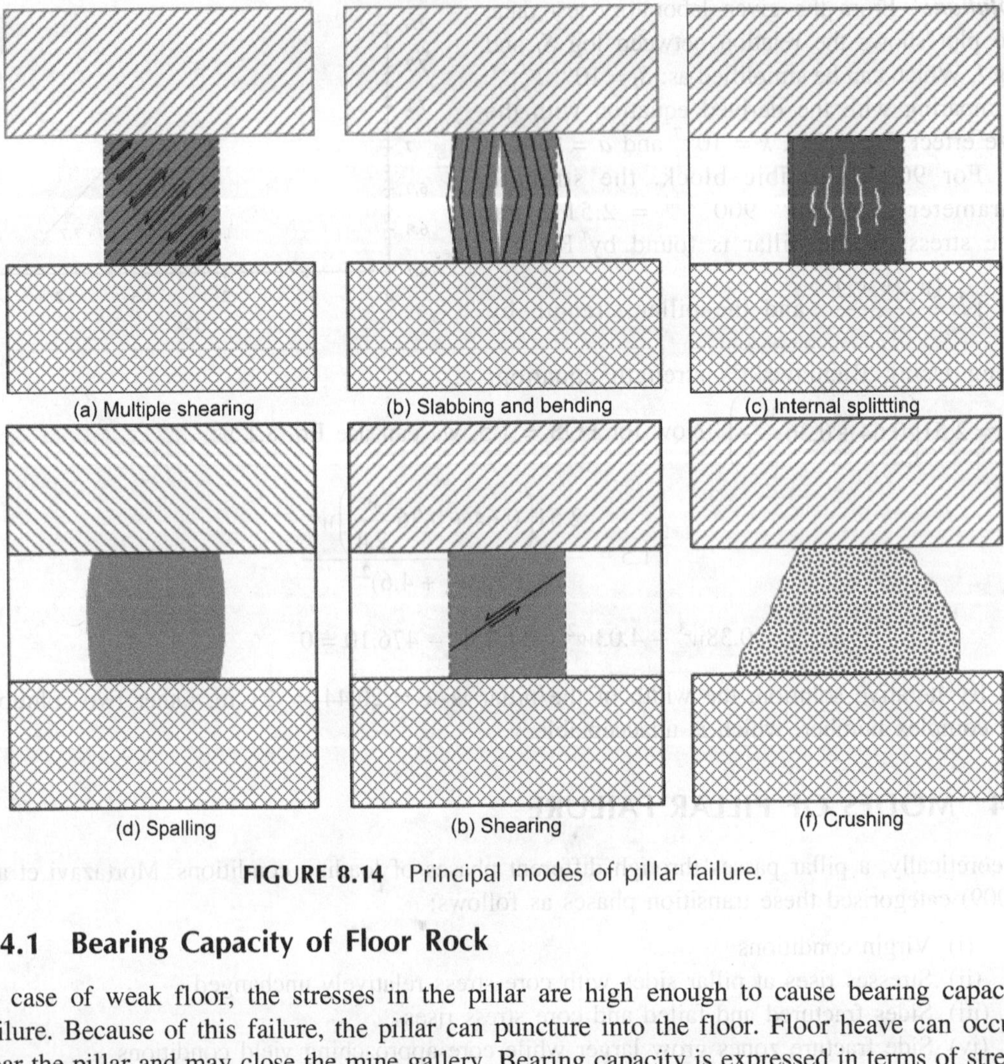

FIGURE 8.14 Principal modes of pillar failure.

8.4.1 Bearing Capacity of Floor Rock

In case of weak floor, the stresses in the pillar are high enough to cause bearing capacity failure. Because of this failure, the pillar can puncture into the floor. Floor heave can occurs near the pillar and may close the mine gallery. Bearing capacity is expressed in terms of stress. For square pillar uniform strip loading on a half-space, bearing capacity is given by classical plastic analysis (Terzaghi, 1943) as:

$$q_b = 0.5\gamma w N_\gamma + cN_c \tag{8.35}$$

where γ is the unit weight of floor, w is width of pillar, c is the cohesion, and N_c and N_γ are bearing capacity factors. The bearing capacity factors are defined as:

$$N_c = (N_q - 1)\cot\varphi$$

$$N_\gamma = 1.5\,(N_q - 1)\tan\varphi$$

where φ is the angle of friction of floor, and N_q is given by

$$N_q = e^{\pi \tan\varphi}\tan^2\left[\left(\frac{\pi}{4}\right)+\left(\frac{\varphi}{2}\right)\right]$$

Equation 8.36 describes the bearing capacity developed under a long chain/barrier pillar. For pillars of width w and length l, the expression for bearing capacity is

$$q_b = 0.5\gamma w N_\gamma S_\gamma + c \cot \varphi \, N_q S_q - c \cot \varphi \tag{8.36}$$

where, $S_\gamma = 1.0 - 0.4\left(\frac{w}{l}\right)$; $S_q = 1.0 + \left(\frac{w}{l}\right)\sin \varphi$

The factor of safety against bearing capacity failure is given by

$$\text{FOS} = \frac{q_b}{\sigma_p} \tag{8.37}$$

EXAMPLE 8.8 Use the data of Example 8.7 to find the factor of safety of the pillar against the bearing capacity of floor. The floor properties like unit weight, cohesion and angle of internal friction are 2.5 kN/m^3, 5 MPa, and 25° respectively.

Solution: From Example 8.7, pillar depth is 150 m, unit weight of overburden is 25 kN/m^3, gallery width is 4.6 m, and pillar width is 43 m.

Using tributary area method, pillar stress is σ_p = 5.57 MPa. Where, vertical stress is σ_V = 3.75 MPa and pillar extraction ratio is 0.33.

Now, for square pillar use Eq. 8.36, then N_q = 10.66, S_γ = 0.6 and S_q = 1.42, N_γ = 6.76, then q_b = 152.98 MPa. By using Eq. 8.37, the FOS against bearing capacity of floor is 27.45.

8.5 YIELD PILLAR TECHNIQUE

A yield pillar is defined as a pillar that yields or fails gradually thereby allows roof deformation in a controlled manner. The subsequent transfer of the overburden load onto the neighbouring mine structure (or abutment pillars) happens gradually (Haramy and Kneisley, 1990). Yield pillar design uses the concept of progressive failure based on pressure arch theory. The pillar size is much less as compared to normal chain pillars. The yield pillars are designed not to take full load instead, they are slightly under-designed. This will cause pillar to yield, thereby transferring/shed load to the barrier/abutment pillars or to larger pillar nearby. The advantage of such design of yield pillar rather than large pillars results in improved ground condition and may increase coal recovery.

8.6 PILLAR STABILISATION

Pillars can be stabilised artificially by providing confinement. This helps slowing down the slabbing or spalling process. There are two ways of pillar stabilisation: (i) stowing around pillar, (ii) pillar bolting.

8.6.1 Stowed Pillar

There is no real substitute for filling the voids. In case, sand is not available, alternative materials like fly ash or mill tailings can be used. Nowadays, cemented paste backfill materials are being used for filling mine voids and strengthening remnant pillars.

Backfill Effects on Pillar Safety Factor

Let us assume that a bord/room and pillar mine needs to be filled as shown in Figures 8.15(a) and 8.15(b). Two cases are considered: (i) filling is done in one side of the pillar Figure 8.15(a), and (ii) filling done in opposite two sides of the pillar Figure 8.15(b). It is obvious that fill material provides lateral confinement to the pillar apart from the support to the roof if 100% filling is done. In this section, effect of lateral confinement is discussed with an example.

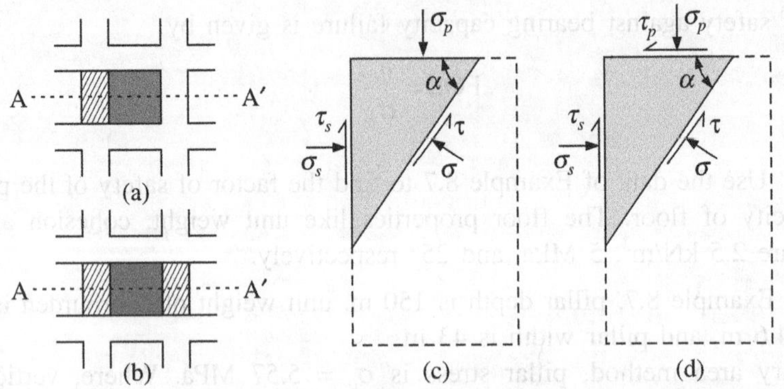

FIGURE 8.15 (a) Pillar stowed in one side, (b) Pillar stowed on both sides, (c) Force body diagram for pillar stowed on both side, and (d) Force body diagram for pillar stowed on one side.

Figure 8.15(c) shows stress equilibrium condition at a vertical section of the pillar. The normal stress σ_s and shear stress τ_s are provided by the fill material.

It is assumed that due to stowing or backfilling horizontal stress is re-established either fully or partially depending on the strength and stiffness of the backfill materials. Assuming $p\%$ of *in situ* horizontal stress is recovered due to backfilling, it may be written that

$$\sigma_s = \left(\frac{p}{100}\right)\sigma_H \cos^2\theta \tag{8.38}$$

$$\tau_s = \left(\frac{p}{100}\right)\sigma_H \sin\theta\cos\theta \tag{8.39}$$

Here, angle θ denotes the inclination of the seam. Once the backfill materials hardens and make firm contact with the pillar materials, shear stress τ_s has the limiting value

$$\tau_s = \mu\sigma_s$$

where, μ is the frictional coefficient of the contact surface. As before, taking moment about any point at the bottom surface as shown in Figure 8.15(d), we find condition for not toppling as:

$$\sigma_p w_p - \tau_p h + \tau_s \frac{h}{2} + \sigma_s \frac{h}{2} > 0 \tag{8.40}$$

Considering σ_p acts on the edge of the pillar, the above equation gives,

$$\frac{w_p}{h} > \tau_p \frac{\frac{\sigma_s}{2}}{\frac{\tau_s}{2}} \tag{8.41}$$

If stowing is to be conducted in both sides of the pillar, it is expected that σ_p will act through the middle of the pillar ($x = 0$) since moment due to σ_s and τ_s will be nullified (Figure 8.15c). In this case, $\tau_p = 0$ for any inclination of the pillar.

In Figure 8.15(d), the opposite side of the pillar is not filled, hence no lateral confinement pressure acts from that side. Hence, τ_p will be developed at the top surface for maintaining equilibrium condition in the pillar. Assuming an arbitrary plane having inclination α, force equilibrium equations give rise to

(i) Filling in one side

$$\sigma - \frac{\sigma_p + \sigma_s}{2} = \frac{\sigma_p - \sigma_s}{2}\cos 2\alpha - \frac{\tau_s}{2}\sin 2\alpha \tag{8.42}$$

$$\tau + \frac{\tau_s}{2} = \frac{\sigma_p - \sigma_s}{2}\sin 2\alpha + \frac{\tau_s}{2}\cos 2\alpha \tag{8.43}$$

(ii) Filling in both sides

$$\sigma - \frac{\sigma_p + \sigma_s}{2} = \frac{\sigma_p - \sigma_s}{2}\cos 2\alpha - \frac{\tau_p + \tau_s}{2}\sin 2\alpha \tag{8.44}$$

$$\tau - \frac{\tau_p - \tau_s}{2} = \frac{\sigma_p - \sigma_s}{2}\sin 2\alpha + \frac{\tau_p + \tau_s}{2}\cos 2\alpha \tag{8.45}$$

The pair of equations given above can be represented by circle in $\tau - \sigma$ plane as:

(i) *Filling in one side*

$$\left(\sigma - \frac{\sigma_p + \sigma_s}{2}\right)^2 + \left(\tau + \frac{\tau_s}{2}\right)^2 = \left(\frac{\sigma_p - \sigma_s}{2}\right)^2 + \left(\frac{\tau_s}{2}\right)^2 \tag{8.46}$$

(ii) *Filling in both sides*

$$\left(\sigma - \frac{\sigma_p + \sigma_s}{2}\right)^2 + \left(\tau - \frac{\tau_p - \tau_s}{2}\right)^2 = \left(\frac{\sigma_p - \sigma_s}{2}\right)^2 + \left(\frac{\tau_p + \tau_s}{2}\right)^2 \tag{8.47}$$

EXAMPLE 8.9 A coal seam 3 m thick dip anti-clockwise at 15° is developed by arrays of square pillars with gallery of width 6 m and height 3 m. The depth of the seam is 300 m. The centre to centre distance between pillars is 34 m. Average density of overlying rock is 2500 kg/m^3. Cohesion and angle of internal friction of coal are 2.5 MPa and 25° respectively. The ratio of horizontal to vertical *in situ* stress is 1/3 and after filling, 60% of horizontal stress is developed. Assume frictional coefficient between the fill material and *in situ* pillar rock is 0.7. Estimate FOS of the (a) developed pillars, (b) stowed in one side, and (c) stowed in both sides.

Solution: The vertical stress is calculated using Eq. 6.1 as $\sigma_v = \rho g H = 2500 \times 9.81 \times 300 = 7.36$ MPa, and the horizontal stress $\sigma_h = (1/3)\,\sigma_v = 2.45$ MPa. The normal and shear stress on pillar is calculated using Eq. 8.3 as, $\sigma_n = 7.03$ MPa, and $\tau_s = -1.23$ MPa by using $\theta = 15°$. As the center-to-center distance between the pillars is, 34 m, and the gallery width is 4 m. The corner to corner pillar width is calculated as 30 m and by using Eq. 8.2, the extraction ratio, $R = 0.32$. The stress on pillar is calculated using Eq. 8.4, then $\sigma_p = 10.36$ MPa and $\tau_p = 1.81$ MPa.

(a) The FOS of the developed pillar is determined using Eq 8.10(a) or graphically by plotting two Mohr's circles, one representing pillar strength and another pillar stress. The center and radius of the stress circle in MPa unit are $(\sigma_p/2, \tau_p/2) = (5.18, 0.90)$ and 5.26 respectively. The perpendicular distance of MC line from center of the stress circle is 5.90 MPa. Hence, FOS of developed pillar is 5.90/5.26 = 1.12.

(b) For one side stowing, and assuming 60% recovery of horizontal stresses, the normal stress, $\sigma_s = \left(\dfrac{p}{100}\right)\sigma_H \cos^2\theta = 0.6 \times 2.45 \times \cos^2 15 = 1.37$ MPa, $\tau_s = \mu\sigma_s = 0.7 \times 1.37 = 0.61$. Using Eq. 8.47, the radius of stress circle is 4.54 MPa with center at (5.87, 1.21). The perpendicular distance of MC line from center of the stress circle is 5.91 MPa. Hence, FOS of developed pillar is 5.91/4.54 = 1.30.

(c) For one side stowing, and assuming 60% recovery of horizontal stresses, the normal stress, $\tau_s = \mu\sigma_s = 0.7 \times 1.37 = 1.66$. Using Eq. 8.47, the radius of stress circle is 4.51 MPa with center at (5.87, 0.31). The perpendicular distance of MC line from center of the stress circle is 6.73 MPa. Hence, FOS of developed pillar is 6.73/4.51 = 1.49.

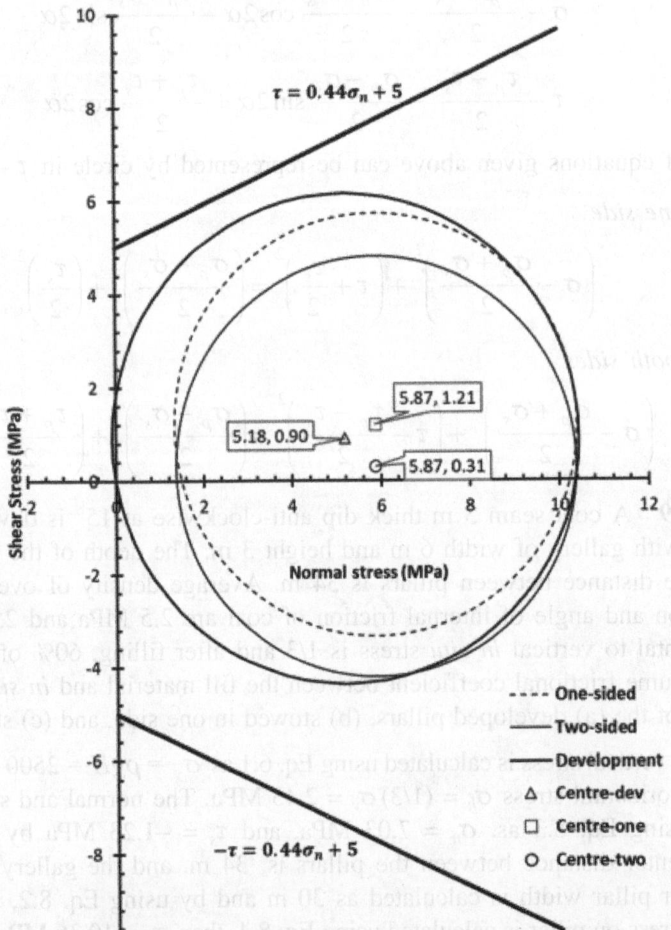

Filled Pillar Strength

The strength of pillar increases due to lateral confinement as empirically given by Wang et al. (2011) for friction angle of 42°.

$$S_p = k\frac{w^{\alpha}}{h^{\beta}} + 0.842e^{2.361(h_{bf}/h)} \tag{8.48}$$

In case of non-cohesive backfill material having friction angle 42°, the strength of pillar changes to

$$S_p = k\frac{w^{\alpha}}{h^{\beta}} + 0.33e^{2.256(h_{bf}/h)} \tag{8.49}$$

where h_{bh} is height of backfill in gallery, h is height of gallery, UCS of pillar material of 0.9 m cube side, k, α and β are constants.

The strength of pillar increases with increasing roadway backfill, while the percentage increase of pillar strength is greater for taller pillars than squat pillars. In general, cohesive backfill is more effective than non-cohesive backfill material.

8.6.2 Bolted Pillars

Rock bolting produces an improvement in the strength properties of the rock mass and also reduces rock movement. Mine pillars are reinforced by rock bolts and wire rope when the pillars are susceptible to failure or contain joints and fractures or when they have partly failed without much slabbing. Lateral bolting of pillars is an alternative to filling (Figure 8.16). It improves the load bearing capacity of the outer zones of fractured pillars. Rock bolts and wire rope bind the fractured surface of the pillar to the competent inner core. As the pillar dilates, the reinforced portion of the pillar provides confining pressure on the core resulting in increased strength. Wire rope provides a transverse confining pressure, simulating triaxial conditions, resulting in an increase in the vertical compressive strength of the pillar. Grouted bolt with wire mesh may be used to prevent partial spalling of pillar. Another good alternative is grouting of used wire ropes like "stitching", in combination with short timber for tightening. The advantage with stitching is that the rope can be taken round pillar corners, which are most susceptible to spalling.

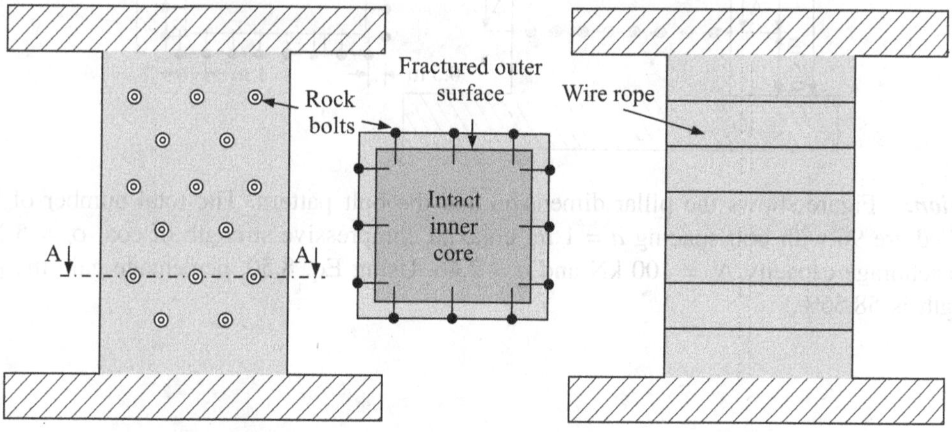

FIGURE 8.16 Pillar reinforcing with bolts and wire rope.

Sheorey (1988) has given the following relation based on the concept of yield zone formation proposed by Wilson to find the probable gain in strength of bolted pillars, which are intact before bolting.

$$\Delta S = \frac{20A_s(q-1)}{\sigma_c a^2} \times 100\% \tag{8.50}$$

where, the unbolted pillar strength is defined by any acceptable pillar strength formula. A_s is anchorage strength of each bolt, a is bolt spacing, σ_c is uniaxial compressive strength, and $q = \left(\sqrt{\mu^2+1}+\mu\right)^2$, μ is coefficient of internal friction (tan φ).

Above equation indicates that the increase in load carrying capacity is dependent on the angle of internal friction and bolt anchorage strength. The gain in strength decreases with the bolt spacing and increase in compressive strength of pillar material. Further Eq. 8.50 is independent of size and shape of the pillar, hence, can be used for any size of pillar of shape square or rectangular.

EXAMPLE 8.10 A pillar is bolted using rock bolts with square spacing of 1 m having anchorage capacity 100 kN. If the uniaxial compressive strength of 0.9 m cubic sample of the pillar material is 5 MPa and angle of internal friction is 25°, then find the percentage gain in pillar strength due to bolting.

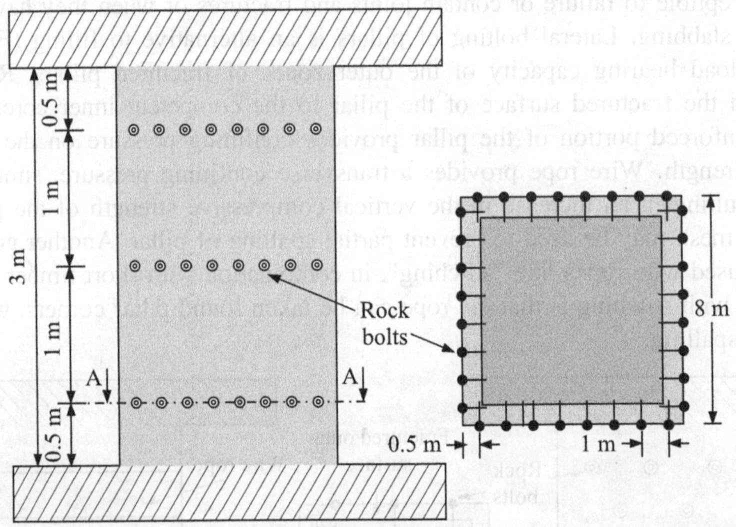

Solution: Figure shows the pillar dimension and the bolt pattern. The total number of bolts installed are 96 with bolt spacing a = 1 m, uniaxial compressive strength of coal σ_c = 5 MPa, bolt anchorage capacity A_s = 100 kN and q = 2.46. Using Eq. 8.50, percentage gain the pillar strength is 58.56%.

8.7 ADVANCED OR DYNAMIC PILLAR DESIGN APPROACH

In previous sections, pillar design approaches are applicable for developed pillars which are static in nature. It means that dimensions of the pillars do not change with time. In addition, effect of excavations in neighbouring panels does not alter the loading pattern on the pillar and hence FOS remains unaltered. However, during depillaring, or longwall operations, load on the pillars are continuously changing owing to goaf formation. In depillaring operation, dimension of pillars are continually changing by maintaining a line of extraction called *goaf line*. During this time, stress concentration on group of production pillars increases due to load shedding from front and side abutment. Hence, during depillaring operation safety factor of a group of pillars are of more interest as opposed to that of a single pillar. Figure 8.17 shows a group of pillar being loaded by front abutment and development loads. The area in which group of pillars are overloaded is called *abutment zone* and the area in which 90% of the abutment load are concentrated called *active mining zone* (AMZ) as shown in Figure 8.18 (Mark, 1997) Figure 8.18 shows distribution or load shedding in abutment influence zone. Similar plots can be drawn for side abutment conditions as well. The distribution of abutment stresses in AMZ is a function of the extent of the abutment influence zone (L) and extent of AMZ (L_{AMZ}). A mathematical function called R-factor for front abutment and side abutment is given by Eqs. 8.51 and 8.52. The parameter R_f is the fraction of the total front abutment load that is applied to the AMZ, whereas, R_s determines the distribution of side abutment load in AMZ which is affected by the width of the barrier pillar. w_{bar} and w_g are width of chain or barrier pillar and width of gallery respectively.

$$R_f = 1 - \left(\frac{L - L_{\text{AMZ}}}{L}\right)^3 \tag{8.51}$$

$$R_s = 1 - \left(\frac{L - w_{\text{bar}} - 0.5w_g}{L}\right)^3 \tag{8.52}$$

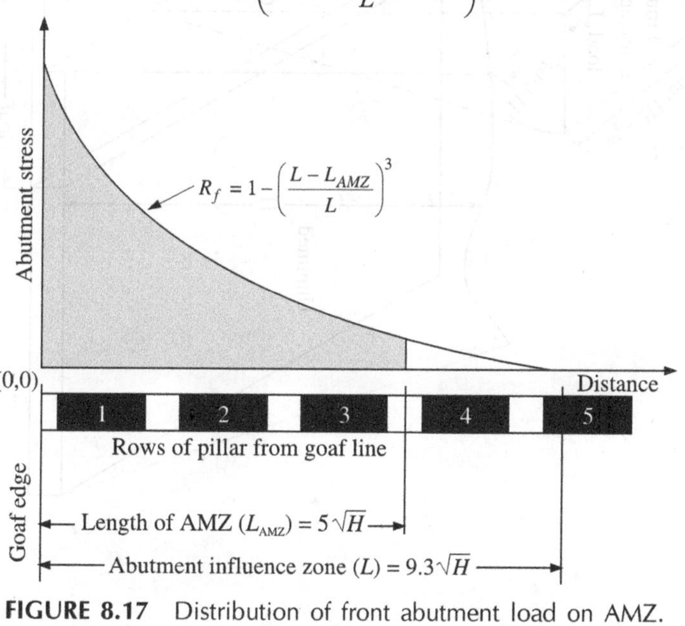

FIGURE 8.17 Distribution of front abutment load on AMZ.

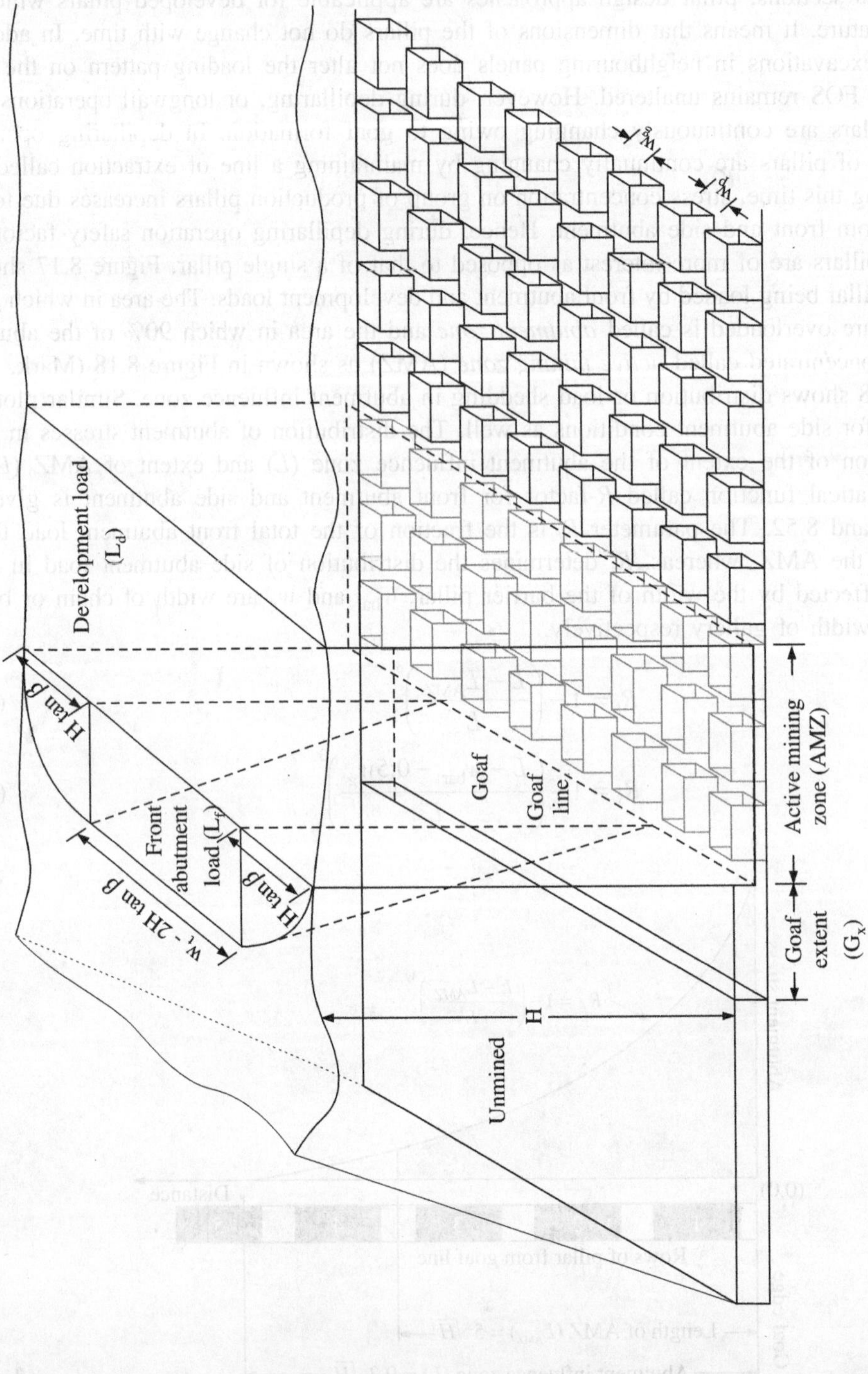

FIGURE 8.18 Conceptualisation of active mining zone and pillar loading.

8.7.1 Distribution of Front Abutment Load

Peng and Chiang (1984) determined the extent of the abutment influence zone (L), defined as the distance from the goaf edge to the line of zero stress concentration, is 9.3 times the square root of the depth of cover (H), in ft, is given by Eq. 8.53.

$$L = 9.3\sqrt{H} \tag{8.53}$$

As per definition, the area of AMZ which carries about 90% of the abutment load can be found by using Eq. 8.54. The extent of AMZ is $5\sqrt{H}$, H measured in ft, as shown in Figure 8.18. Hence, the area of AMZ is:

$$A_{\text{AMZ}} = L_{\text{AMZ}} \times w_t = 5n\sqrt{H}(w_p + w_g) \tag{8.54}$$

where, w_t is width of AMZ or panel and equal to $n(w_p + w_g)$ and n is number of pillars in a row.

The safety factor in AMZ is defined as the ratio of the load bearing capacity (LBC) to the total load (TL) applied

$$\text{FOS} = \frac{\text{LBC}}{\text{TL}} \tag{8.55}$$

LBC is the maximum load carried by the systems of pillar and is defined as:

$$\text{LBC} = mS_pA_p \tag{8.56}$$

where, m is the number of pillars in AMZ, S_p is the strength of individual pillar with pillar cross-section area, $A_p = w_pl_p$ for a rectangular pillar and $A_p = w_p^2$ for square pillar.

8.7.2 Pillar Load

The total load applied on the pillars in AMZ includes the development load, front and side abutment loads. Four different loading conditions are shown in Figure 8.19.

(a) Development load
(b) Development and front abutment load
(c) Development, front abutment and one side load abutment load
(d) Development, front abutment and two side abutment loads

The following sub-sections discuss the load developed by development load, front abutment load and side abutment load individually. The total load for the above four cases are determined by algebraic additions of individual loads assuming that the load addition follows principle of superposition.

Development Load

Development load (L_d) occurs due to the weight of the overburden directly above the pillars before any depillaring operation takes place. Figure 8.20 shows the development load on a single (pillar 1) and group of pillars (4, 5, and 6). It is calculated based on tributary area method as

$$L_d = H\gamma A_{\text{AMZ}} \tag{8.57}$$

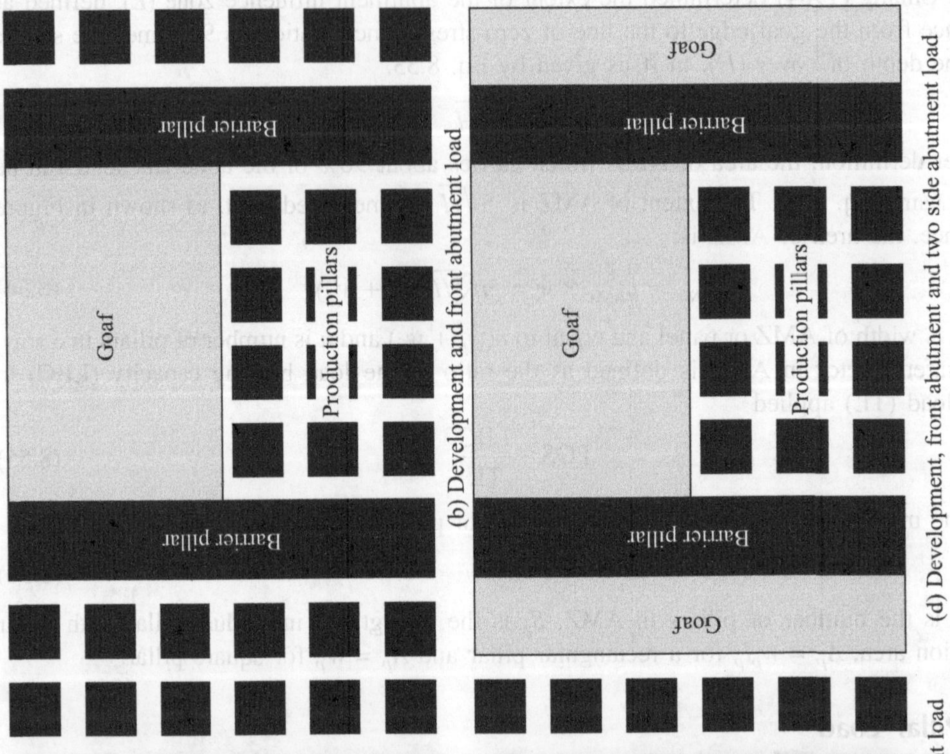

FIGURE 8.19 Different loading conditions: (a) Development load only, (b) Development and front abutment load, (c) Development, front abutment load and one side abutment load, and (d) Development, front abutment load and two side abutment load.

Using Eq. 8.50, Eq. 8.53 becomes

$$L_d = 5n\gamma(w_p + w_g)H^{3/2} \tag{8.58}$$

where, H is depth and γ is the average unit weight of overburden.

Front Abutment Load

The magnitude of the front abutment load is the weight of overburden confined within the boundaries of abutment line (line pq), extent of goaf (line qr) and the depth of cover (line rp) as shown in Figure 8.20. Figure 8.21 shows the load transferred to the pillars in AMZ depending on width of AMZ and goaf extent due to removal of pillars. Figure 8.21(a) shows the situation when the goaf is fully developed and width of AMZ is sufficient for transferring of the maximum load. This condition resembles supercritical or critical condition. Figure 8.21(b) refers to a situation of narrow goaf extent indicating subcritical condition. Similarly, Figure 8.21(c) shows narrow AMZ width and may represents subcritical condition.

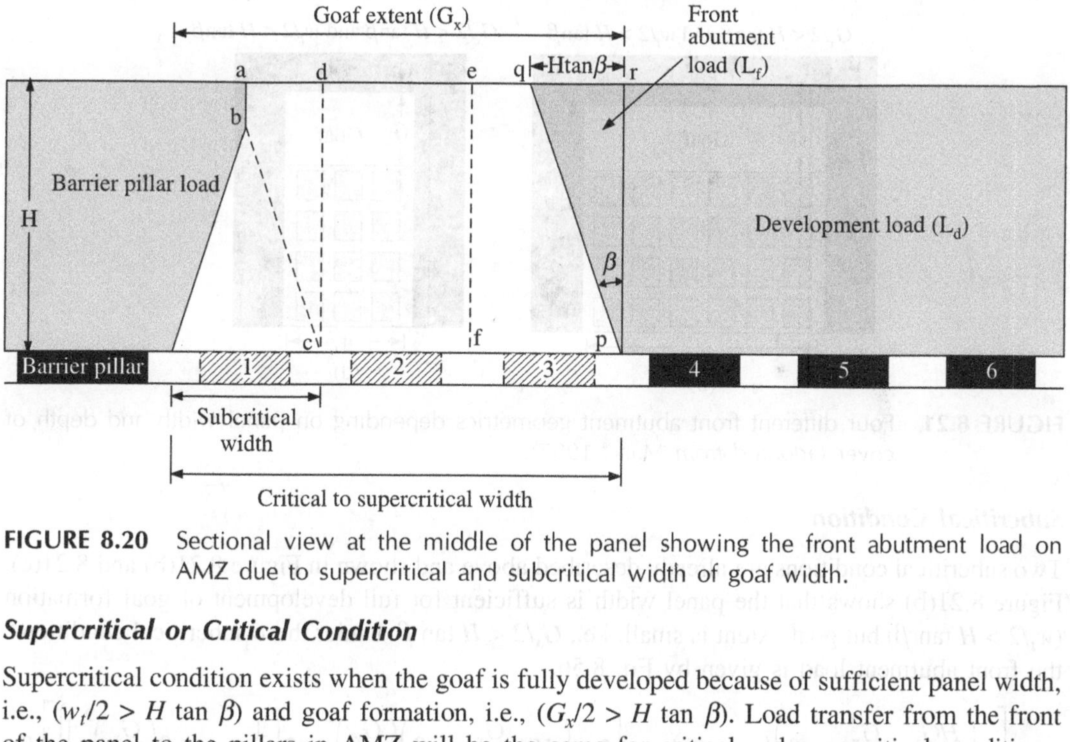

FIGURE 8.20 Sectional view at the middle of the panel showing the front abutment load on AMZ due to supercritical and subcritical width of goaf width.

Supercritical or Critical Condition

Supercritical condition exists when the goaf is fully developed because of sufficient panel width, i.e., ($w_t/2 > H \tan \beta$) and goaf formation, i.e., ($G_x/2 > H \tan \beta$). Load transfer from the front of the panel to the pillars in AMZ will be the same for critical and supercritical conditions. Referring to Figure 8.21(a) the total front abutment load is the weight of one triangular prism and two quater cone at the corner of the panel as shown in Figure 8.17. Hence, the total front abutment load that acts on the pillars in AMZ is

$$L_f = \left\{\frac{\gamma H^2 \tan \beta}{2}(w_t - 2H \tan \beta) + \left[2\left(\frac{1}{3} \times H \times (H \tan \beta)^2 \times \gamma\right)\right]\right\} \times R_f$$

or

$$L_f = 0.5\gamma H^2 \tan \beta \left(w_t - \frac{2}{3} H \tan \beta\right) R_f \tag{8.59}$$

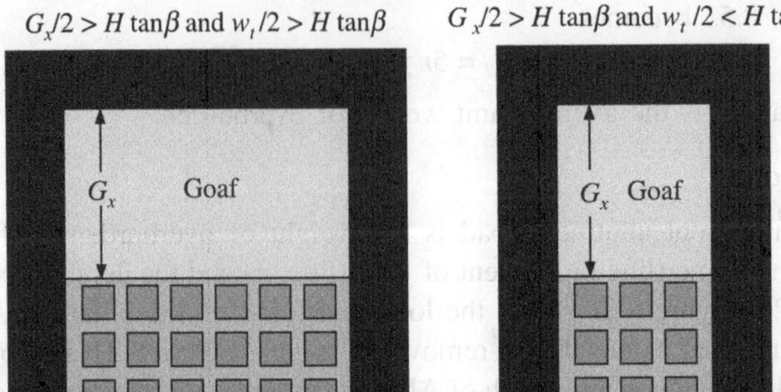

FIGURE 8.21 Four different front abutment geometrics depending on panel width and depth of cover (*adopted from* Mark, 1997).

Subcritical Condition

Two subcritical conditions are already described above and shown in Figure 8.21(b) and 8.21(c). Figure 8.21(b) shows that the panel width is sufficient for full development of goaf formation ($w_t/2 > H \tan \beta$) but goaf extent is small, i.e., $G_x/2 < H \tan \beta$. Under these subcritical conditions, the front abutment load is given by Eq. 8.56.

$$L_f = \left[\gamma\left(\frac{HG_x}{2} - \frac{G_x^2}{8}\cot\beta\right)(w_t - G_x) + \gamma\left\{2\times\left(H - \frac{G_x}{2}\cot\beta\right)\left(\frac{G_x}{2}\right)^2 + 2\left(\frac{1}{4}G_x\cot\beta\left(\frac{G_x}{2}\right)^2\right)\right\}\right]R_f$$

$$L_f = \frac{\gamma}{2}\left(Hw_tG_x - \frac{w_tG_x^2}{4}\cot\beta - \frac{1}{12}G_x^3\cot\beta\right)R_f \tag{8.60}$$

In other case, although $G_x/2 \geq H \tan \beta$, but $w_t/2 < H \tan \beta$ (Figure 8.21c). Total front abutment load is given as:

$$L_f = \gamma\left\{\left(H - \frac{w_t}{2}\cot\beta\right)\left(\frac{w_t}{2}\right)^2 + 2\left(\frac{1}{3}\left(\frac{w_t}{2}\right)^3 \cot\beta\right)\right\} R_f$$

$$L_f = \frac{\gamma w_t^2}{4}\left\{H - \frac{w_t}{6}\cot\beta\right\} R_f \tag{8.61}$$

Side Abutment Load

Side abutment load is mainly carried by the barrier pillar. The rest is transferred to the pillars in AMZ as given below:

Supercritical condition:

$$L_s = (0.5\gamma H^2 \tan\beta)\, L_{\text{AMZ}} R_s \tag{8.62}$$

Subcritical condition:

$$L_s = \gamma G_x\left[\frac{H}{2} - \frac{G_x}{8}\cot\beta\right] L_{\text{AMZ}}\, L R_s \tag{8.63}$$

Therefore, total load for the four cases as discussed above are:

(a) Development load only

$$TL = L_d \tag{8.64a}$$

(b) Development and front abutment load

$$TL = L_d + L_f \tag{8.64b}$$

(c) Development, front abutment and one side load abutment load

$$TL = L_d + L_f + L_s \tag{8.64c}$$

(d) Development, front abutment and two side abutment load

$$TL = L_d + L_f + 2L_s \tag{8.64d}$$

EXAMPLE 8.11 A bord and pillar panel is developed with 35 square pillars (5 pillars in dip and 7 pillars in strike direction). Thickness of coal seam is 3 m and located at a depth of 200 m (or 656 ft). The pillars are extracted along dip line extraction as shown in figure given below. The width of the pillar and gallery are 22.5 m and 3 m respectively. The average unit weight of overlying rock strata is 25.45 kN/m^3. Strength of 0.9 m cubic coal sample is 6 MPa and abutment angle is 21°.

Assume that there is no excavation nearby. Then find the following:

(a) FOS of developed pillar.

(b) After complete removal of row 1, find the FOS of AMZ.

(c) After complete removal of row 1, find the FOS of pillar zone in rows 2, 3, and 4.

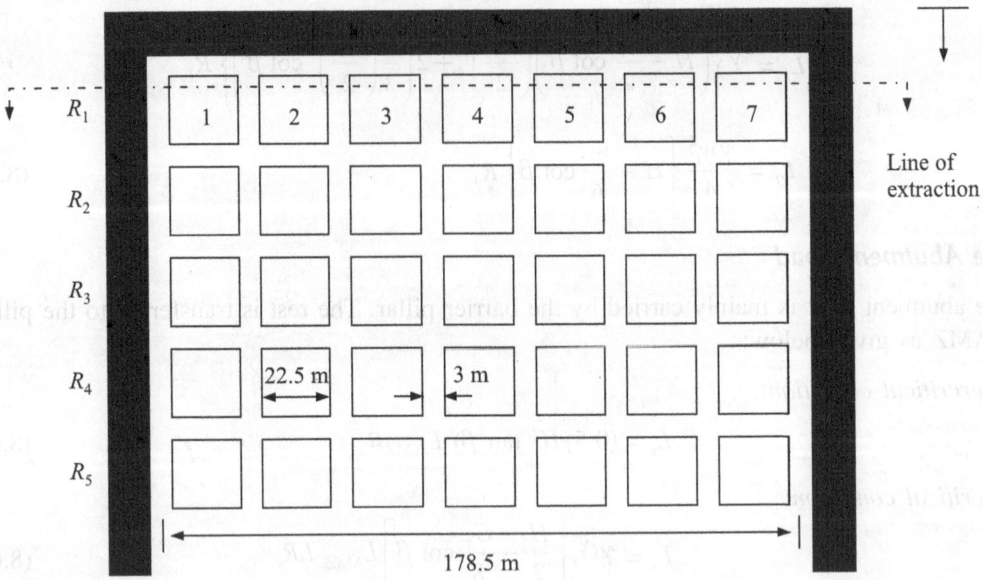

Solution:

(a) Strength of individual pillar is found using Bieniawski's pillar strength formula (Eq. 8.27) as S_p = 20.04 MPa. Individual pillar stress is calculated using Eq. 8.2 as σ_p = 6.54 MPa. Hence, the FOS of the developed pillar is found using Eq. 8.1 as

$$\text{FOS} = \frac{20.04}{6.54} = 3.07$$

(b) The calculated abutment influence zone (L) is 73 m (238.2 ft) it means that the load shading is upto 4th row of pillars. The breadth of abutment zone (L_{AMZ}) is 39 m (or 128 ft) it means that upto 3rd row of pillars are bearing about 90% of the abutment load. Now considered the complete 2nd and 3rd row of pillars in abutment zone, the modified breadth of AMZ is 51 m. Hence, the modified L_{AMZ} = 51 m, width of AMZ is 178.5 m and area of AMZ using Eq. 8.51 as $A_{AMZ} = L_{AMZ} \times w_t = 9103.5$ m^2. Development load is found by using Eq. 8.52 as L_d = 46.34 GN. Here, G_x = 25.5 m, w_t = 178.5 m, $H \tan \beta$ = 76.77 m, $G_x/2 < H \tan \beta$, and $w_t/2 > H \tan \beta$. Hence,it satisfies subcritical condition. Equation 8.56 gives the value of front abutment as L_f = 4.07 GN. As no side abutment is present, the total load is sum of the development load and front abutment load as TL = 50.41 GN. Load bearing capacity of AMZ is found using Eq. 8.52 as LBC = 14 × 22.5^2 × 20.04 = 142.03 GN and 142.03

$$\text{FOS} = \frac{\text{LBC}}{\text{TL}} = \frac{142.03}{50.41} = 2.82$$

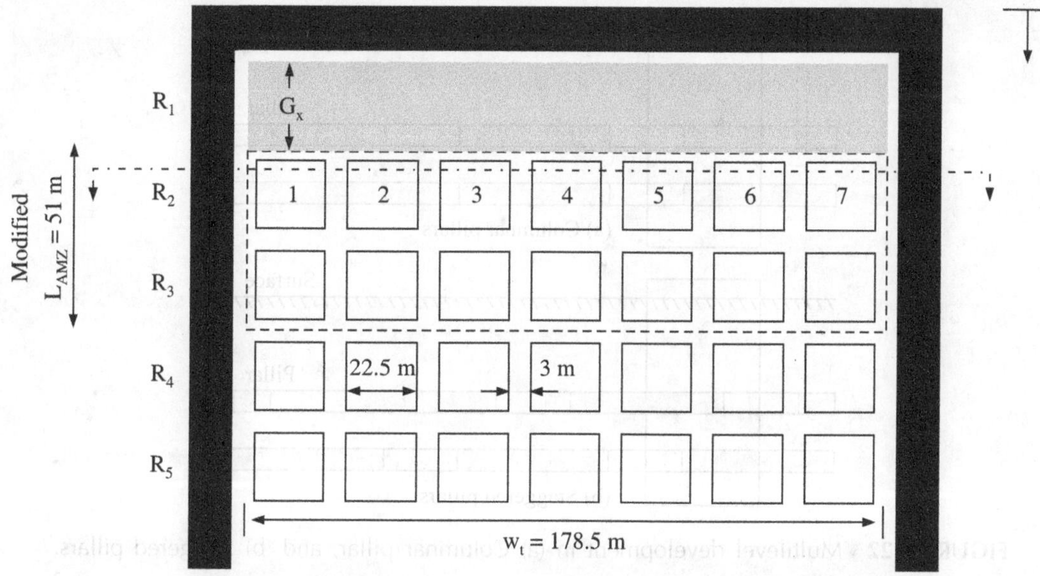

(c) The development load and LBC for each row will be same as L_d = 25.45 × 200 × 25.5 × 178.5 = 23.17 GN, and LBC = 7 × 22.5^2 × 20.04 = 71.02 GN respectively.

R_f	R_f *for Individual rows of pillars*	L_f (GN)	TL (GN)	FOS
2nd row	0.72	3.03	26.20	2.71
3rd row	0.25	1.04	24.21	2.93

Comparing the results in section (a) and (b), it is found that the pillar FOS is more during development stage than the depillaring stage. Section (c) shows that farther the distance of pillars from line of extraction the FOS increases.

8.8 PILLARS IN MULTILEVEL OPERATIONS

In multi-level working either in hard rock mining or in multi-seam working in coal mining, it is important to follow a basic design rule. There are two methods of pillar alignments: (a) columnar, and (b) staggered as shown in Figure 8.22. In columned cases, the pillars are aligned along line of gravitation as shown in the figure. Hence, overburden and excavation stresses are transferred to the lower level through the pillars without experiencing high stress concentrations. In staggered case, the upper and lower pillars are offset with each other. It is preferable to follow columnar pillars rather than staggered for better ground control.

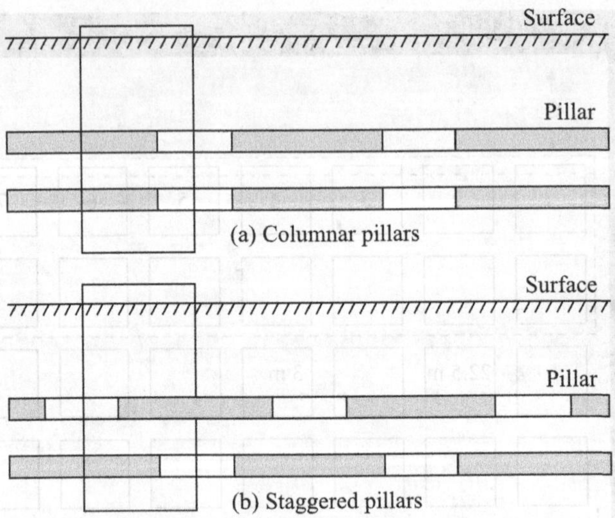

FIGURE 8.22 Multilevel development in (a) Columnar pillar, and (b) Staggered pillars.

EXERCISES

1. A board and pillar operation is planned for a depth of 300 m in a coal seam. The pillar strength is estimated to be 12.40 MPa. If the gallery width is planned to be 6 m and vertical stress gradient to be 0.025 MPa/m. Then find the (a) maximum possible area extraction ratio, and (b) pillar width.
2. A coal pillar is located at a depth of 80 m from surface with average unit weight of overburden 22 kN/m^3. It is given that the height of the pillar is 3 m and width of pillar is 18 m. The uniaxial compressive strength of the coal samples estimated from laboratory tests are 30 MPa and 18 MPa for 2.5 cm and 12 cm cubic samples, respectively. For safe pillar, find the maximum possible gallery width.
3. Laboratory tests to find the uniaxial compressive strength are conducted on a prismatic rock sample by varying their width and height ratio (w/h). The results are shown in table below. Develop the strength relationship.

w/h	UCS (MPa)
0.5	35
1	45
2	60
3	75
4	85
7	115
9	150

4. In continuation of Exercise 3, additional test are conducted on cubical coal samples are to determine their uniaxial compressive strength (UCS). The test results are shown below. Determine the valid pillar strength formula.

Side of a cubic coal sample (cm)	UCS (MPa)
1.0	30
2.0	20
3.0	15
5.0	10
10.0	6.0
15.0	5.5
20.0	5.2
25.0	5.1
30.0	5.0
100.0	4.8

5. A horizontal stratiform at a depth of 150 m below ground surface is planned for extraction 3 m thick coal seam using 6.0 m gallery and pillars 7.0 m square in plan. The unit weight of the overburden rock is 22.5 kNm^{-3}. Analysis of pillar failure indicates that pillar strength is defined by $S_p = 10.44\, w_p^{0.5} h^{-0.7}$,where S_p in MPa, h is pillar height in m and w_p is pillar width in m. Determine the factor of safety of the pillar against compressive failure. If the factor of safety is less than 1.6, propose a mining layout that will achieve an extraction ratio for the factor of safety of 1.6.
6. Three proposals are provided to increase the extraction ratio in a Bord and pillar method of mining from 50% to 75%. In this situation, the ratio of gallery width to pillar width will increase from 0.5 to 1.0 by:

   Proposal 1: Increasing gallery width to 9 m from 6 m while decreasing pillar width. Spacing between pillar to pillar are held constant.

   Proposal 2: Keeping pillar width constant while increasing gallery width to 12 m.

   Proposal 3: Keeping gallery width constant while decreasing pillar width to 6 m.

   Which proposal is preferable from the ground control point of view and why?
7. A Board and pillar mining is planned at a depth of 100 m in a seam dipping 10°. An extraction percentage of 50% is anticipated. The ratio of *in situ* horizontal to vertical stress is estimated to be 1.5. A joint set is present that dips 60° in the same direction as the dip of the seam. Rock mass cohesion and angle of internal friction are 15 MPa and 35°, respectively. Whereas Joint cohesion and angle of internal friction are 0.16 MPa and 25°, respectively. If the average unit weight of the overlying rock mass is 22 kN/m^3. Determine:

   (a) if there is a threat to pillar stability,

   (b) the range of joint dips that would allow its slip within the excavation plan.
8. A coal seam 3.5 m thick dips 15° south at a depth of 250 m below the surface. Level galleries of 6 m wide are developed along the east. Pillars are three times as longer

than its width and separated by a 5 m wide gallery along dip direction. The longer side of the pillar is oriented along the dip direction. The average compressive and tensile strength of coal is 20 MPa and tensile strength is 2 MPa respectively. Determine the maximum extraction ratio possible and the width of the pillar. Assume that the *in situ* horizontal stress is 3/4th the vertical stress and the average density of overlying rock strata is 2400 kg/m^3.

9. A room and pillar excavation plan is executed in a flat lying, thick, strong, limestone deposit with square pillars. Vertical stress on the roof level is 3.75 MPa, room width is 8.5 m wide, Rock mass compressive strength is 18 MPa and the ratio of compressive to tensile strength is 8. If factor of safety 1.1 is required, determine pillar width.

10. In continuation of Exercise 8, a joint with dip angle 55°, cohesion is 0.75 MPa, and friction angle 20° intersects the pillar along its true dip. Assume the length and width of the pillar as 86 m and 29 m respectively. Determine: (a) the pillar safety factor without a joint when the extraction ratio is 0.4, and (b) the safety factor of the pillar with joint.

11. In continuation of Exercise 10(b), find the pillar factor of safety of the pillar if the pillar is reinforced by the bolt pattern shown in Example 8.10 with the anchorage capacity of the bolt is 100 kN and the spacing of 1 m. Also, determine the gain in joint shear strength if the friction angle of the joint remains the same after bolting.

12. Long-chain pillars of height 3.5 m situated in a flat coal seam at a depth of 400 m from the surface. Calculate the safe width of the chain pillar when caving is allowed. The pillars are desired to be of square shape connected by 6 m wide cross-cut. The unit weight of the overburden is 22 kN/m^3, the abutment angle is 20°, and the chain pillar strength is 25 MPa.

13. Calculate the safety factor of a long barrier pillar situated at a depth of 250 m and maintaining a uniform width of 28 m with a height of 3.5 m. Assume that caved zone or goaf is fully developed around the pillar, abutment angle is 21° and average unit weight of overburden rock is 27 kN/m^3. It is given that the strength of the barrier pillar is 35 MPa.

14. A board and pillar panel is developed at a depth of 150 m from ground surface with a pillar dimension of 20 m × 20 m × 3 m with 5 pillar in each row. The width of the galleries are 4.5 m. It is given that the average unit weight of overburden, *in situ* strength of coal, abutment angle and goaf extent are 24 kN/m^3, 6 MPa, 21°, 100 m respectively. Estimate the safety factor of developed pillars and factor of safety of AMZ after complete extraction of all five pillars in row one.

15. A transverse stopes of 20 m width, 60 m length and 45 m height are extracted. Rib pillars of 15 m width is left in both sides of a stope. Assume cohesion and friction angle of pillar rock is 5 MPa and 25° respectively. If the *in situ* vertical and horizontal stresses are 12 MPa and 14 MPa respectively, what's the safety factor of the pillar. If stope is filled with a paste of maximum compressive strength of 1.2 MPa and friction coefficient at the contact surface is 0.75, determine the safety factor (a) before filling, (b) filling on one side, and (c) filling on both sides of the stope.

Chapter 9

Rock and Rock Mass Failure Criteria

9.1 INTRODUCTION

Load-bearing capacity or limit of stresses is dependent on the material properties of rock or rock mass. Here, the term 'rock' signifies the intact rock free from any discontinuity. The limit of stress of a rock specimen can be tested in the laboratory in uniaxial, biaxial, triaxial, shear and tensile mode. For uniaxial compression or tension loading cases, rocks generally fail if applied stress exceeds its corresponding strengths. If failure occurs along the principal stress plane, it is called the *extension of fractures* since no shear stress acts on that plane. The fracture plane(s) oriented other than principal stress plane is developed due to shear stresses and is called *shear fracture(s)*. It is generally occurs in rocks during triaxial loading.

9.2 TRIAXIAL STRENGTH OF ROCK

Triaxial tests of rock samples (mostly cylindrical in shape) are conducted to determine the effect of confinement on the strength of rocks. Triaxial tests are conducted by applying a constant confining pressure in lateral (diametrical) direction and failure load is applied in the axial direction as shown in Figure 9.1. In this case, the stress at failure termed as *triaxial strength* (σ_1) is compared with confining stresses as, $\sigma_1 > \sigma_3 = \sigma_2$, where, later two stresses act as the confinement. Rock sample and loading platens are covered with a rubber membrane such that oil does not come in contact with the rock sample during the test. Lateral pressure is applied using hydraulic pressure control system based on the procedure outlined in ISRM standard (ISRM 2007). Stress at failure is recorded for a given confining pressure (σ_3). Generally, tests are repeated for different σ_3 for the same rock type and σ_1 is recorded for each case. Then pair data of σ_1 and σ_3 are plotted to determine the failure envelope. Figure 9.2(a) shows stress-strain relationships for four different σ_3 (0, 3 MPa, 6 MPa and 9 MPa) of artificial rock samples are prepared with sand and cement mortar. Each test is conducted with a strain rate of 10^{-5}/s. From the figure, it is noticed that stress at failure increases with increasing confinement.

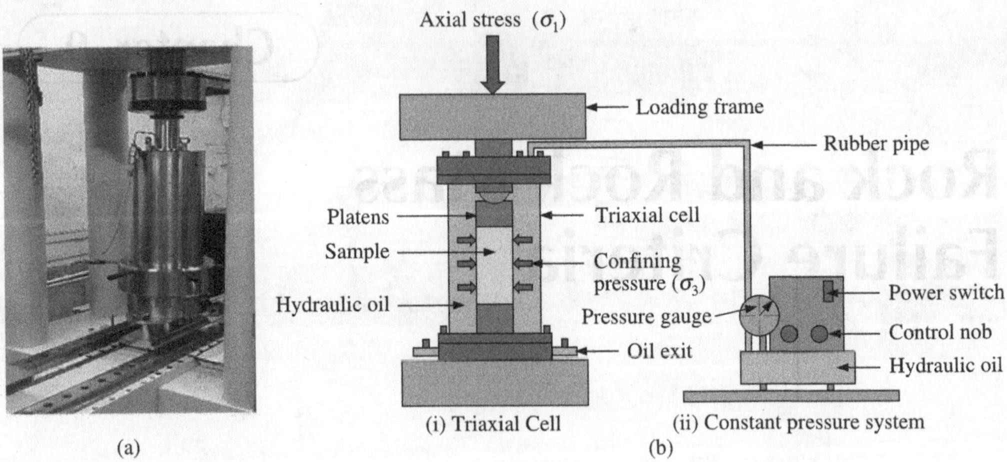

FIGURE 9.1 (a) Triaxial cell under loading condition, and (b) Schematic of experimental setup of triaxial tests.

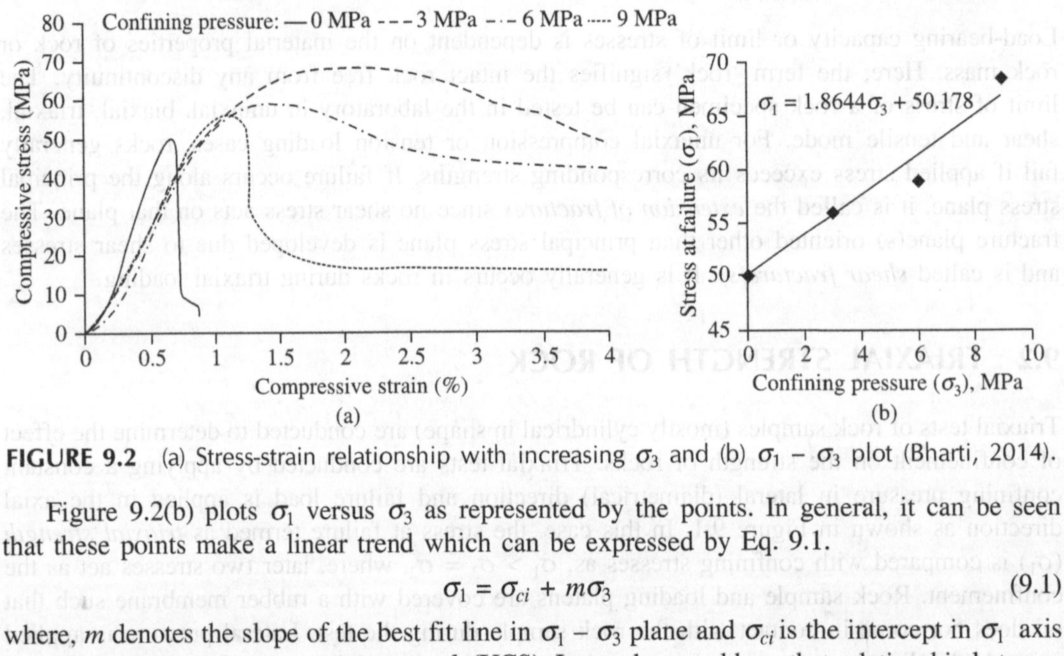

FIGURE 9.2 (a) Stress-strain relationship with increasing σ_3 and (b) $\sigma_1 - \sigma_3$ plot (Bharti, 2014).

Figure 9.2(b) plots σ_1 versus σ_3 as represented by the points. In general, it can be seen that these points make a linear trend which can be expressed by Eq. 9.1.

$$\sigma_1 = \sigma_{ci} + m\sigma_3 \tag{9.1}$$

where, m denotes the slope of the best fit line in $\sigma_1 - \sigma_3$ plane and σ_{ci} is the intercept in σ_1 axis is known as *uniaxial compressive strength* (UCS). It may be noted here that relationship between σ_1 and σ_3 may not be linear for all values of σ_3. These cases will be discussed in Section 9.6.

Now, considering n number of tests, the parameters σ_{ci} and m can be estimated by solving Eqs. 9.2 and 9.3 simultaneously obtained by applying least square method as:

$$\sum_{k=1}^{n} \sigma_{1k} - m\sum_{k=1}^{n} \sigma_{3k} - n\sigma_{ci} = 0 \tag{9.2}$$

$$\sum_{k=1}^{n} (\sigma_{1k}\sigma_{3k}) - m\sum_{k=1}^{n} \sigma_{3k}^2 - \sigma_{ci}\sum_{k=1}^{n} \sigma_{3k} = 0 \tag{9.3}$$

The parameters, m and σ_{ci} are related with *cohesion* and *angle of internal friction* of the material if the linear failure envelope shown in Figure 9.2(b) is assumed to be obtained based on Mohr–Coulomb failure criterion. The details of this derivation is given in Section 9.6. Therefore, angle of internal friction (φ) and cohesion (c) of rock can be determined using Eqs. 9.4 and 9.5.

$$\varphi = \sin^{-1}\left(\frac{m-1}{m+1}\right) \tag{9.4}$$

$$c = \frac{\sigma_{ci}(1-\sin\varphi)}{2\cos\varphi} \text{ or } \frac{\sigma_{ci}}{2\tan\left(\frac{\pi}{4}+\frac{\varphi}{2}\right)} \tag{9.5}$$

Similar to Figure 9.2(a), Figure 9.3 is plotted for triaxial test confinement upto 28 MPa and deviatoric stress vs longitudinal and transverse strain for Makrana marble of Rajasthan, India.

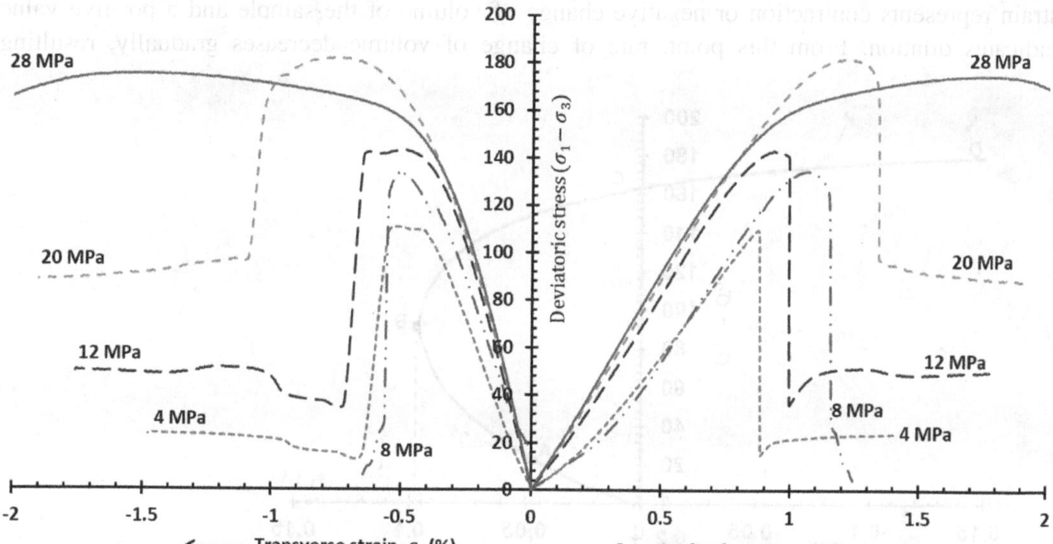

FIGURE 9.3 Triaxial test at high confinement (Sarkar and Kumar, 2024).

9.2.1 Dilatancy of Rock in Triaxial Condition

Dilation is defined as the increment of the volume of rock due to plastic deformation. Vermeer and de Borst (1984) deduced a formula to estimate the dilation angle in plastic regime, as:

$$\sin\psi = \frac{\varepsilon_v^p}{-2\varepsilon_1^p + \varepsilon_v^p} \tag{9.6}$$

where, ε_v^p and $\varepsilon_1{}^p$ are volumetric and axial plastic strain increments, respectively, and

$$\varepsilon_1^p = \varepsilon_1 - \varepsilon_1^e$$

$$\varepsilon_1^e = \frac{\sigma_1 - 2\nu\sigma_3}{E}$$

$$\varepsilon_v^p = \varepsilon_v - \varepsilon_v^e$$

$$\varepsilon_v^e = \frac{(1-2\nu)(\sigma_1 + 2\sigma_3)}{E}$$

where, ε_1 is total incremental axial strain, ε_1^e is axial incremental elastic strain, ε_v is total incremental volumetric strain and ε_v^e is incremental elastic volumetric strain; E and ν are Young's modulus and Poisson's ratio, respectively.

Zhao and Cai (2010) defined failure process of brittle rock can be divided into the following stages: (1) crack closure; (2) linear elastic deformation; (3) crack initiation; (4) stable crack growth; (5) crack coalescence and damage; (6) unstable crack growth; (7) failure; (8) post peak behaviour. The onset of dilation (i.e. crack initiation threshold) is marked as the point where volumetric strain curves depart from linearity (point A) and generally starts at stress levels of approximately 0.3–0.5 times UCS as shown in Figure 9.4(a). Here, a positive volumetric strain represents contraction or negative change of volume of the sample and a positive value indicates dilation. From this point, rate of change of volume decreases gradually, resulting

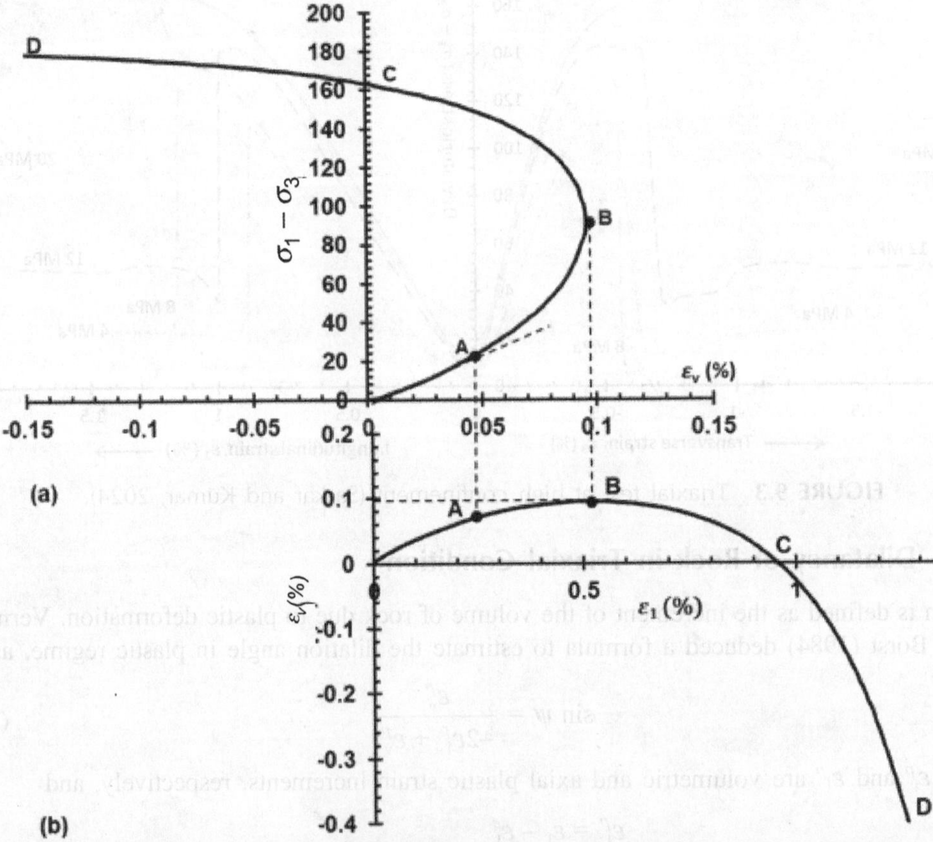

FIGURE 9.4 Triaxial test at 20 MPa conferment for Marble plot. (a) Deviatoric stress versus volumetric strain. (b) Volumetric strain verses axial strain. (Sarkar and Kumar 2024).

from the propagating existing cracks and new cracks initiate in a stable fashion. It means that contribution of samples gradually decreases, however the rate of change is still positive. Hence, the volume of the rock sample reaches to a minimum value at point B. At this point, crack density is sufficient for cracks to coalesce to form tensile spalls or shear band, and volumetric strain reversal occurs and unstable crack growth begins. Onset of crack damage starts at stress levels of approximately 0.7–0.8 times the peak strength. With further increase of the differential stress, rock dilation increases substantially following macrocracks or shear bands formation. As the load further increases, the dilation will not increase at the same rate as before but gradually decrease, and finally reach a constant value even if axial deformation still increases (point D). The volumetric strain at point C is null, which reveals that the compaction of initially existing volume of pores, flaws or microcracks in the rock sample is equal to the volume of newly generated cracks in the course of deformation. From then on, the volume of the rock sample shows a dilation trend.

9.3 POLYAXIAL STRENGTH OF ROCK

Polyaxial also termed as "true" triaxial strength of rock sample is determined by varying confinement stress in two perpendicular directions as $\sigma_1 > \sigma_2 > \sigma_3$. The vertical cross-section of the sample is taken to be rectangular to apply two different confining stresses as shown in Figure 9.5. Tests are conducted mostly to ascertain the influence of intermediate stress σ_2, on the failure stress σ_1.

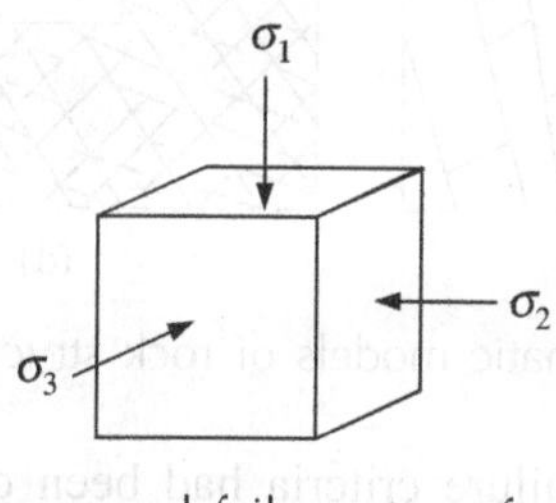

FIGURE 9.5 Confinement pressure and failure stress for polyaxial test ($\sigma_1 > \sigma_2 > \sigma_3$).

9.4 FAILURE MODES OF INTACT ROCK AND ROCK MASS

In Chapter 8, possible failure modes of rock mass are depicted in Figure 8.14. Rock fails gradually or violently depending on the brittle or ductile nature, type of testing machine (soft or stiff) and the strain energy absorbed in the sample before failure. The failure mode of intact rock specimen also depends on the level of confinement stresses, i.e., whether sample is tested in uniaxial, triaxial or ployaxial conditions.

On the other hand, the failure mechanism of rock mass is rather complex and many times, it is almost impossible to identify the root cause of a failure. Based on the displacements induced by excavation, four conceptual models are proposed for possible failure of rock mass as given in Figure 9.6 (Brady, 1987). The displacement field in Figure 9.6(a) can be considered continuous and thus, elastic-plastic analysis can be applied for such conditions. Figure 9.6(b) shows large discrete joints or plane of weakness but within the two such joints, the rock can be considered continuous. Hence, for this type of model, near-field stress conditions can also be regarded as elastic-plastic. If rock mass is frequently jointed around an excavation, displacements near the opening will be dependent on the condition of joints, and rigid body movements (translation and rotation) [Figure 9.6(c)]. Application of elastic-plastic stress analysis in such conditions may not be appropriate. Heavily jointed rock mass can be regarded as pseudo-continuous as shown in Figure 9.6(d). This condition is termed as *ubiquitous joint model* and elastic-plastic analysis of equivalent material is appropriate to define near-field stress conditions.

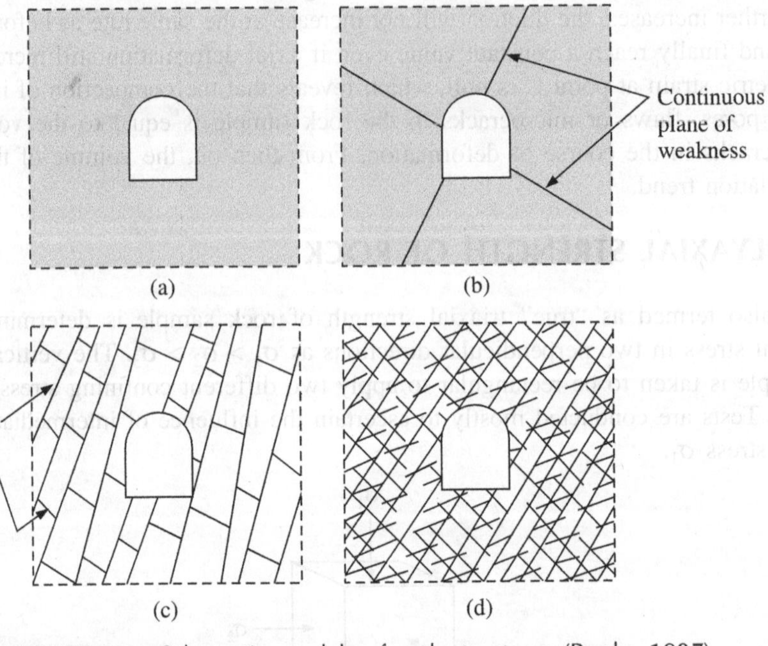

FIGURE 9.6 Schematic models of rock structures (Brady, 1987).

Various rock and rock mass failure criteria had been developed over the years and they are given in Table 9.1 and Table 9.2. Most of these criteria are developed based on the peak triaxial compressive strength or peak shear strength on the failure surface. Thus, most of the strength criteria are expressed in terms of σ_1 and σ_3 and considered to be independent of intermediate principal stress, σ_2. In general, failure criteria are expressed using stress invariants so that transformation of axes does not alter the criteria. Since principal stresses are one such stress invariants, most of these criteria are expressed using principal stresses. However, computation of magnitude and direction of principal stresses in three-dimension is cumbersome and can be avoided if the yield or failure criteria are expressed using invariants of deviatoric stress. The relationship between principal stresses and invariants of deviatoric stress is presented in the following section. In this chapter, the most common shear failure criteria such as Mohr–Coulomb, Drucker–Prager, Hoek–Brown, alternative criteria based on Hoek–Brown and tensile failure criteria are discussed in detail.

TABLE 9.1 Failure Criteria for Intact Rock (Edelbro, 2004)

Failure equation	*Development/comments*	*Author, criterion first published*
$(\sigma_1 - \sigma_3)^2 = a + b(\sigma_1 + \sigma_3)$	An empirical generalisation of Griffith theory of intact rock	Fairhurst (1964)
$\sigma_1 = \sigma_c + \sigma_3 + F\sigma_3^f$	Empirical test data fitting for intact rock	Hobbs (1964)
$\sigma_1 = \sigma_c + a\sigma_3^b$	Empirical test data fitting for intact rock	Murrel (1965)
$\dfrac{\tau_m - \tau_o}{\sigma_c} = D\left\lvert\dfrac{\sigma_m}{\sigma_c}\right\rvert^C$	Empirical curve fitting for intact rock	Hoek (1968)
$\sigma_1 = \sigma_c + a\sigma_3$	Triaxial tests on soft rock	Bodonyi (1970)
$\sigma_1 = \sigma_3 + \sigma_c^{1-B}(\sigma_1 + \sigma_3)^B$	Empirical curve fitting for 500 rock specimens	Franklin (1971)
$\sigma_1 = \sigma_3 + (m\sigma_c\sigma_3 + s\sigma_c^2)^{1/2}$	Application of Griffith theory and empirical curve fitting. Both for intact and heavily jointed rock masses	Hoek and Brown (1980)
$\dfrac{\sigma_1}{\sigma_c} = a + b\left(\dfrac{\sigma_c}{\sigma_3}\right)^\alpha$	Empirical curve fitting for 700 rock specimens. Both for intact and heavily jointed rock masses	Bieniawski (1974), modified by Yudhbir et al. (1983)
$\sigma_1 = \sigma_3 + a\sigma_3\left(\dfrac{\sigma_3}{\sigma_c}\right)^b$	Applied to 80 rock samples	Ramamurthy et al. (1995)
$\sigma'_{1n} = \left(\dfrac{M}{B}\sigma'_{3n} + 1\right)^B$	Empirical curve fitting for both soil and rock specimens	Johnston (1985)
$\sigma_1 = \sigma_c\left(1 + \dfrac{\sigma_3}{\sigma_t}\right)^b$	Both for intact and heavily jointed rock masses	Sheorey et al. (1989)
$\sigma_1 = \sigma_3 + A\sigma_c\left(\dfrac{\sigma_3}{\sigma_c} - S\right)^{1/B}$	*A*, *B* and *S* are strength parameters	Yoshida et al. (1990)
$\sigma_1 - \dfrac{\beta}{1+b}(b\sigma_2 + \sigma_3) = \sigma_t$ if $\sigma_2 \le \dfrac{\sigma_1 + \beta\sigma_3}{1+\beta}$ $\dfrac{\beta}{1+b}(b\sigma_1 + \sigma_2) = -\beta\sigma_3 = \sigma_t$ if $\sigma_2 \ge \dfrac{\sigma_1 + \beta\sigma_3}{1+\beta}$ $\beta = \dfrac{\sigma_t}{\sigma_c}$	It reflects the effect of intermediate principal stress on the strength of intact rock specimens. It is called unified strength criterion. The parameter *b* ranges from 0 to 1 depending on the effect of intermediate stress.	Yu et al. (1991), modified by Yu et al. (2002) ***Note:*** This criterion was not reported in the reference mentioned in the title of the table.

where, $\tau_m = (\sigma_1 - \sigma_3)/2$, $\sigma_m = (\sigma_1 + \sigma_3)/2$, σ_1 = major principal stress, σ_3 = minor principal stress, σ'_{1n} = major normalised effective principal stress, σ_t = uniaxial tensile strength, σ'_{3n} = minor normalised effective principal stress, σ_c = uniaxial compressive strength and *a*, *b*, *B*, *f*, *F*, *C*, *D*, *M* and α are constants.

TABLE 9.2 Rock Mass Failure Criteria (Edelbro, 2004)

Failure equation	*Development/comments*	*Author, criterion first published*
$\sigma_1' = \sigma_3' + \sigma_{ci}\left(m_b \frac{\sigma_3}{\sigma_{ci}} + s\right)^a$	2002 version	Hoek–Brown (1980)
$\sigma_1 = A\sigma_{ci} + B\sigma_{ci}\left(\frac{\sigma_3}{\sigma_{ci}}\right)^\alpha$	A is dimensionless parameter and B is a rock material constant, α is suggested = 0.65	Yudhbir and Pronzl (1983)
$\sigma_1 = \sigma_{cm}\left(1 + \frac{\sigma_3}{\sigma_{tm}}\right)^{b_m}$	Use RMR_{76} value	Shcorey et al. (1989)
$\sigma_1' = \sigma_3' + \sigma_3 B_j\left(\frac{\sigma_{cj}}{\sigma_3}\right)^{a_j}$	2001 version	Ramamurthy (1995)

where, σ_1' = major effective principal stress, σ_{tm} = tensile strength of rock mass, σ_3' = minor effective principal stress, σ_{cm} = compressive strength of rock mass, σ_{ci} = uniaxial compressive strength of intact rock and a_j, b_m, m_b, A and B_j are constants.

9.5 INVARIANTS OF DEVIATORIC STRESS

The significance of deviatoric stress in formulation of yield functions will be discussed in the later sections. Here, the definition of deviatoric stress and its computational procedure will be outlined. Let us represent a state of stress denoted by P(σ_1, σ_2, σ_3) in three perpendicular axes with the principal stresses taken as rectangular coordinates as shown in Figure 9.7. In the figure, three axes represent compressive principal stresses. Assume that the state of stress at a point (P) can be represented by a vector originating from the origin O and denoted by line OP. Now, imagine a line OA inclined to the three axes with equal direction cosines as

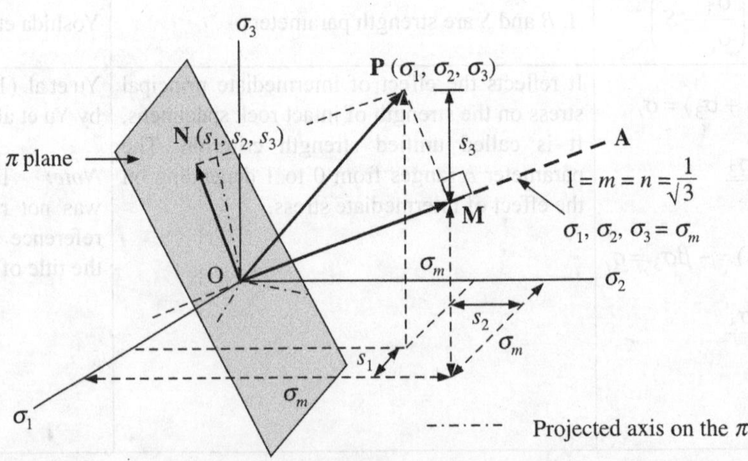

FIGURE 9.7 Definition of deviatoric stress.

$1/\sqrt{3}$. This line indicates the mean stress path since each stress axis is inclined equally with this line. The stress vector **OP** can be resolved into vector **OM** on the line OA and another vector **ON** perpendicular to the line **OA**. This signifies that the **OP** vector has two components viz. (i) mean stress or hydrostatic stress vector **OM**, and (ii) deviatoric stress vector **ON** or deviation from the mean stress. The magnitude of the vector **OM** can be estimated considering projected length of OP along OA as:

$$|\mathbf{OM}| = \mathbf{OP}\cdot\hat{\mathbf{u}}_{\mathrm{OA}} = (\sigma_1\hat{\mathbf{i}} + \sigma_2\hat{\mathbf{j}} + \sigma_3\hat{\mathbf{k}})\cdot\frac{1}{\sqrt{3}}(\hat{\mathbf{i}} + \hat{\mathbf{j}} + \hat{\mathbf{k}}) \tag{9.6a}$$

$$|\mathbf{OM}| = \frac{1}{\sqrt{3}}(\sigma_1 + \sigma_2 + \sigma_3) = \frac{1}{\sqrt{3}}(3\sigma_m) = \sqrt{3}\sigma_m \tag{9.6b}$$

where mean stress, $\sigma_m = \frac{1}{3}(\sigma_1 + \sigma_2 + \sigma_3)$ and $\hat{\mathbf{u}}_{\mathrm{OA}}$ is the unit vector of **OA** vector. The point N can be represented by coordinates (s_1, s_2, s_3) where

$$\left.\begin{aligned} s_1 &= \sigma_1 - \sigma_m \\ s_2 &= \sigma_2 - \sigma_m \\ s_3 &= \sigma_3 - \sigma_m \end{aligned}\right\} \tag{9.7}$$

and symbol s denotes deviatoric stress tensor. From the above relationship, invariants of principal deviatoric stresses can also be obtained using Eq. 4.43, as:

$$\left.\begin{aligned} J_1 &= s_1 + s_2 + s_3 = 0 \\ J_2 &= \frac{1}{2}(s_1^2 + s_2^2 + s_3^2) \\ J_3 &= \frac{1}{3}(s_1^3 + s_2^3 + s_3^3) \end{aligned}\right\} \tag{9.8}$$

The deviatoric stress tensor for any stress tensor is given by the following relationship as:

$$S_{ij} = \sigma_{ij} - \sigma_m\delta_{ij} \tag{9.9}$$

δ_{ij} is the Kronecker delta which takes a value of unity when $i = j$ and zero when $i \neq j$. Here, σ_m is the mean hydrostatic stress and expressed as $\sigma_m = \frac{1}{3}(\sigma_{xx} + \sigma_{yy} + \sigma_{zz})$. In Cartesian coordinate system, deviatoric stress tensor and principal deviatoric stress can be expressed as:

$$\left.\begin{aligned} s_{xx} &= \sigma_{xx} - \sigma_m \\ s_{yy} &= \sigma_{yy} - \sigma_m \\ s_{zz} &= \sigma_{zz} - \sigma_m \\ s_{xy} &= \tau_{xy} \\ s_{yz} &= \tau_{yz} \\ s_{zx} &= \tau_{zx} \end{aligned}\right\} \tag{9.10}$$

Then the invariants of deviatoric stress can be obtained by replacing stress tensor by deviatoric stress tensors in Eq. 9.8 as:

$$\left.\begin{aligned} J_1 &= s_{xx} + s_{yy} + s_{zz} = 0 \\ J_2 &= \frac{1}{2}(s_{xx}^2 + s_{yy}^2 + s_{zz}^2) + s_{xy}^2 + s_{yz}^2 + s_{zx}^2 \\ J_3 &= \frac{1}{3}(s_{xx}^3 + s_{yy}^3 + s_{zz}^3) - s_{xx}s_{yz}^2 - s_{yy}s_{zx}^2 - s_{zz}s_{xy}^2 + 2s_{xy}s_{yz}s_{zx} \end{aligned}\right\} \tag{9.11}$$

The magnitude of the vector **ON** can be estimated as

$$r = |\mathbf{ON}| = \sqrt{s_1^2 + s_2^2 + s_3^2} = \sqrt{2J_2} \tag{9.12}$$

For any given state of stress, deviatoric stress vector, **ON** lies on the plane passing through origin and perpendicular to the line **OA**. This plane is called *deviatoric plane* or *π plane* and its equation is given as $\sigma_1 + \sigma_2 + \sigma_3 = 0$ in the principal stress space.

The relationship between principal stresses and invariants of deviatoric stresses can be found as (Deb, 2012)

$$\begin{Bmatrix} \sigma_1 \\ \sigma_2 \\ \sigma_3 \end{Bmatrix} = \frac{2\sqrt{J_2}}{\sqrt{3}} \begin{Bmatrix} \cos(\pi/6+\theta) \\ \sin\theta \\ -\cos(\pi/6-\theta) \end{Bmatrix} + \sigma_m \begin{Bmatrix} 1 \\ 1 \\ 1 \end{Bmatrix} \tag{9.13}$$

where, angle θ is called *Lode angle* and is considered as another invariant of deviatoric stress relating second and third invariants of deviatoric stress namely J_2 and J_3 respectively as (Chakrabarty, 1987)

$$\sin 3\theta = -\frac{3\sqrt{3}J_3}{2J_2^{3/2}} \tag{9.14}$$

Lode angle ranges between –30° (uniaxial tensile, $\sigma_1 = \sigma_2 = 0$, $\sigma_3 < 0$) to +30° (uniaxial compression, $\sigma_2 = \sigma_3 = 0$, $\sigma_1 > 0$) and 0 represents pure shear condition with $\sigma_2 = \frac{1}{2}(\sigma_1 + \sigma_3)$. Lode angle represents the angle between the pure shear line and the deviatoric stress vector **ON**.

EXAMPLE 9.1 Plot the variation of Lode angle for the following condition. Assume $\sigma_3 = 0$ for all cases.

Sl. no.	σ_1 (MPa)	σ_2 (MPa)
1	100	0
2	75	25
3	50	50
4	25	75
5	0	100

Solution:

Sl. no.	σ_1 (MPa)	σ_2 (MPa)	σ_3 (MPa)	σ_m (MPa)	J_2 (MPa)^2	J_3 (MPa)^3	θ (deg)
1	100	0	0	33.33	3333.33	74074.07	–30.00
2	75	25	0	33.33	1458.33	11574.07	–10.89
3	50	50	0	33.33	833.33	–9259.26	–30.00
4	25	75	0	33.33	1458.33	11574.07	–10.89
5	0	100	0	33.33	3333.33	74074.07	–30.00

9.6 MOHR–COULOMB (MC) YIELD CRITERION

Mohr–Coulomb shear failure criterion was first proposed by Coulomb in 1773. It was suggested that shear stress developed in rocks is related to cohesion, angle of internal friction of the material and is also dependent on the applied normal stress. In the limiting case, when magnitude of shear stress is equal to the shear strength, the criterion reduces to

$$|\tau| = c + \sigma_n \tan \varphi \tag{9.15}$$

where

τ = shear stress, N/m^2, absolute value will be considered
c = cohesion, N/m^2
σ_n = normal stress, N/m^2 and positive value is considered (compressive)
φ = angle of internal friction, degree

Equation 9.15 suggests that shear stress at failure is linearly related with the normal stress. In 1900, Mohr proposed that shear failure takes place across a plane where shear stress is a function of normal stress. He added that the shear stress at failure could be linearly related with normal stress or could be any polynomial function.

9.6.1 MC Criterion in Principal Stress Space—Two Dimensions

Consider that a rock sample is failed at stress σ_1 for a confining stress of σ_3. Using the concept outlined in Chapter 4, Mohr's circle of stress can be plotted in τ–σ_n plane as shown in Figure 9.8. It is postulated that at failure, Coulomb criterion makes a tangent with the Mohr's circle at the failure stress point (σ_n, τ). On the Mohr's circle, Coulomb's failure criterion is drawn to show the shear and normal stresses at failure. The definition of cohesion and angle of internal friction is also given in the figure. From the figure, it can be found that normal and shear stresses at failure are related to the principal stresses as given below:

$$\sigma_n = \left(\frac{\sigma_1 + \sigma_3}{2}\right) - r \cos 2\alpha \tag{9.16a}$$

$$\tau = r \sin 2\alpha \tag{9.16b}$$

where r is the radius of the Mohr's circle and absolute value is considered given in the following equation:

$$r = \frac{\sigma_1 - \sigma_3}{2} \tag{9.17}$$

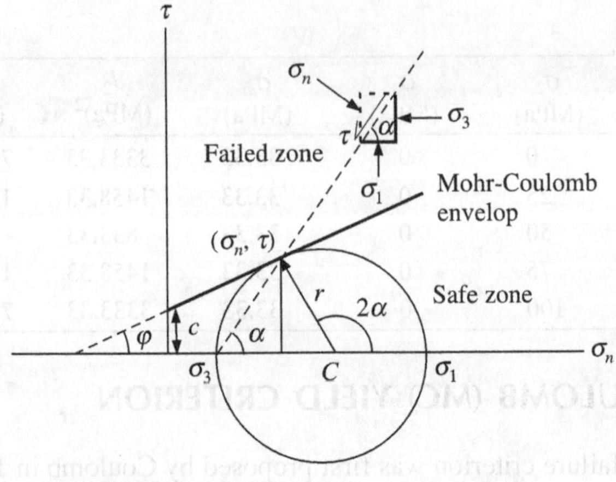

FIGURE 9.8 Mohr–Coulomb yield criterion.

At failure, the angle 2α equals to 90° + φ and thus, Eq. 9.16 becomes

$$\sigma_n = \left(\frac{\sigma_1 + \sigma_3}{2}\right) + \left(\frac{\sigma_3 - \sigma_1}{2}\right) \sin\varphi \tag{9.18a}$$

$$\tau = \left(\frac{\sigma_1 - \sigma_3}{2}\right) \cos\varphi \tag{9.18b}$$

Substituting Eq. 9.18 into Eq. 9.15, Mohr–Coulomb failure criterion in $\sigma_1 - \sigma_3$ plane is obtained as given in Eq. 9.19 and shown in Figure 9.10.

$$\Phi = \sigma_1 - \sigma_3 N_\varphi - 2c\sqrt{N_\varphi} = 0 \tag{9.19}$$

where N_φ is the slope (as mentioned in Eq. 9.19) and is called *triaxial factor*. It is computed as $N_\varphi = \dfrac{1+\sin\varphi}{1-\sin\varphi} = \tan^2\left(\dfrac{\pi}{4} + \dfrac{\varphi}{2}\right)$

If the value of Φ is positive or zero, rock yields/fails under the specified stress condition. For a negative value of Φ, rock is safe or elastic. From Eq. 9.19, the theoretical uniaxial compressive strength of rock can be obtained by considering $\sigma_3 = 0$ as:

$$\sigma_{ci} = \sigma_1 = 2c\sqrt{N_\varphi} \tag{9.20}$$

The theoretical maximum tensile strength is estimated by letting $\sigma_1 = 0$ in Eq. 9.19 and the magnitude is given as:

$$\sigma_t^{\max} = -\frac{2c}{\sqrt{N_\varphi}} \tag{9.21}$$

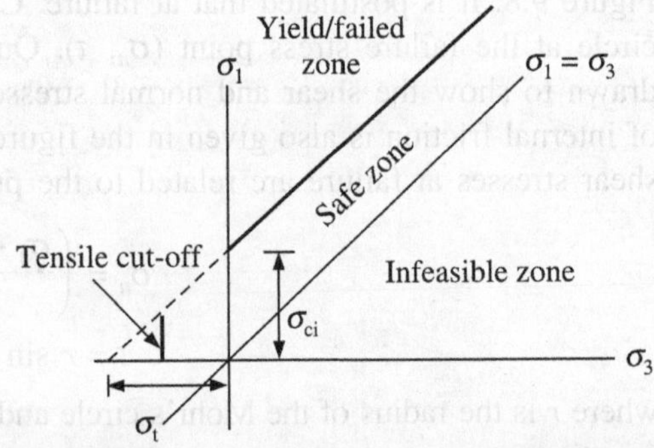

FIGURE 9.9 Mohr-Coulomb failure envelop in $\sigma_1 - \sigma_3$ plane.

However, the experimental or measured tensile strength σ_t of rock is generally less than the theoretical tensile strength. In this case, σ_t will be considered as *tensile cut-off* and thus, the failure envelope will be truncated as shown in Figure 9.9.

EXAMPLE 9.2 A series of triaxial tests of shale reveals the following results. Determine the uniaxial compressive strength and tensile strength of rock.

Test no.	*Confining stress* (MPa)	*Failure stress* (MPa)
1	2	45
2	4	56
3	6	63
4	8	75

Solution: Uniaxial compressive strength and slope N_φ can be estimated based using Eq. 9.20. They are σ_{ci} = 35.5 MPa and N_φ = 48.5. Hence, φ = 41.15° and c = 4.40 MPa. Using Eq. 9.21, the maximum tensile strength of shale can be found as 3.99 MPa.

9.6.2 MC Criterion in Deviatoric Plane

In this section, Mohr–Coulomb shear yielding envelope is reformulated using invariants of deviatoric stress. In Eq. 9.19, σ_1 and σ_3 can be replaced based on Eq. 9.13 as given below:

$$\Phi = \frac{2\sqrt{J_2}}{\sqrt{3}}\cos\left(\frac{\pi}{6}+\theta\right)+\sigma_m+\left[\frac{2\sqrt{J_2}}{\sqrt{3}}\cos\left(\frac{\pi}{6}-\theta\right)-\sigma_m\right]N_\varphi-2c\sqrt{N_\varphi}=0 \qquad (9.22)$$

Using trigonometric relationship and then by rearranging the above equation, we get

$$\Phi = \sqrt{J_2}\left[(1+N_\varphi)\cos\theta+(N_\varphi-1)\frac{\sin\theta}{\sqrt{3}}\right]-\sigma_m(N_\varphi-1)-2c\sqrt{N_\varphi}=0 \qquad (9.23)$$

The above equation represents Mohr–Coulomb yield/failure criterion in deviatoric stress space. This criterion is dependent on all three invariants of deviatoric stress. In this case also, if the value of Φ in Eq. 9.23 is positive or zero, rock yields/fails under the specified stress condition. Else, rock is safe. Figure 9.10 shows a typical plot of this yield/failure surface in principal stress space having $\sigma_1 \geq \sigma_2 \geq \sigma_3$. It can be seen that Mohr–Coulomb criterion gives a conical yield surface whose normal section at any point is an irregular hexagon. The conical shape (other than cylindrical shape) of yield criterion is the manifestation of the influence of hydrostatic stress on yielding. This is clearly seen by the second term of Eq. 9.23. On the hydrostatic axis, that is for $\sigma_1 = \sigma_2 = \sigma_3 = \sigma_m$, second invariant of deviatoric stress, $J_2 = 0$. For this case, Eq. 9.22 gives $\sigma_m = c \cot \varphi$, which signifies that the apex of the conical yield surface lies at a point where $\sigma_1 = \sigma_2 = \sigma_3 = c \cot \varphi$. Then the distance of the apex from the origin will be $\sqrt{3}c \cot \varphi$.

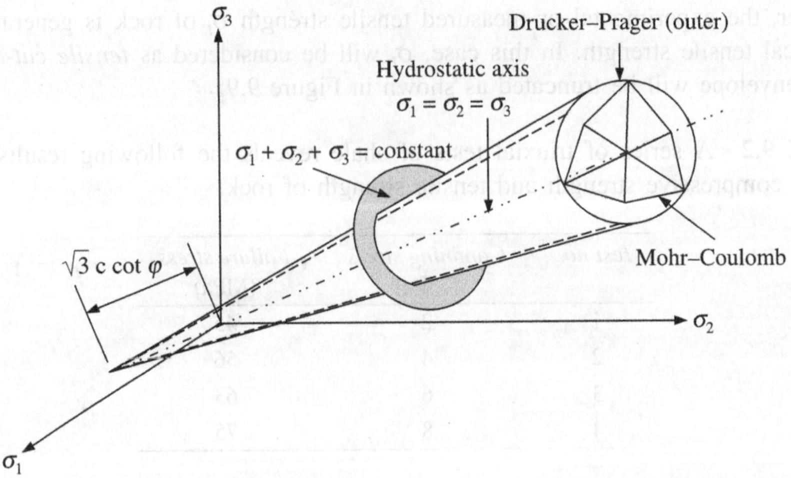

FIGURE 9.10 Mohr–Coulomb yield surface in principal stress space.

Figure 9.11 shows the two-dimensional plot of Mohr–Coulomb yield surface in deviatoric plane. For the similar stress level, the shape and size of hexagonal yield surface depends on cohesion and angle of internal friction of rock.

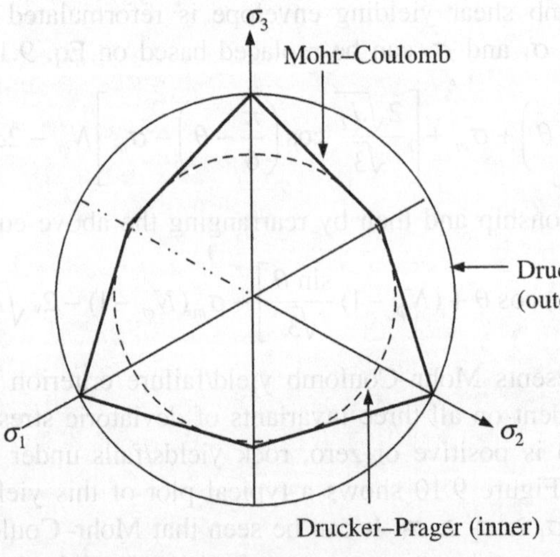

FIGURE 9.11 Mohr–Coulomb yield surface in deviatoric plane.

EXAMPLE 9.3 Stresses at a point is given by $\sigma^T = \{3\ \ 6\ \ 14\ \ -1\ \ 0\ \ 0\}$ MPa, decide whether the rock fails based on Mohr–Coulomb criterion. Assume cohesion and friction angle of rock are 2 MPa and 25° respectively.

Solution: From the data following parameters can be estimated as:

$$N_\varphi = \frac{1 + \sin 25}{1 - \sin 25} = 1.73,\ \ \sigma_m = 7.667 \text{ MPa},\ \ \sqrt{J_2} = 5.774 \text{ MPa},\ \ J_3 = 42.926,\ \ \theta = -0.206 \text{ rad}$$

Hence, using Eq. 9.22 we find that

$$\Phi = \sqrt{J_2}\left[(1+N_\varphi)\cos\theta + (N_\varphi - 1)\frac{\sin\theta}{\sqrt{3}}\right] - \sigma_m(N_\varphi - 1) - 2c\sqrt{N_\varphi} = 6.68$$

Since the value is positive, rock yields/fails under this stress condition.

9.7 DRUCKER–PRAGER CRITERION

Drucker–Prager yield criterion (1952) is an approximation of Mohr–Coulomb criterion and is derived using invariants of deviatoric stress. Drucker–Prager yield surfaces resembles circle in deviatoric plane and are obtained by coinciding outer and inner apices of Mohr–Coulomb hexagon as shown in Figure 9.11. Drucker–Prager yield surfaces are circular conical pyramid and the distance of apex from the origin is also $\sqrt{3}c\cot\varphi$. The loci of the circles, coinciding outer and inner apices of Mohr–Coulomb hexagon, can be obtained by replacing θ with –30° or 30° respectively in Eq. 9.23 and are given below:

$$\Phi = \sqrt{\frac{J_2}{3}}(2+N_\varphi) - \sigma_m(N_\varphi - 1) - 2c\sqrt{N_\varphi} = 0 \text{ coinciding with outer apices} \quad (9.24)$$

$$\Phi = \sqrt{\frac{J_2}{3}}(1+2N_\varphi) - \sigma_m(N_\varphi - 1) - 2c\sqrt{N_\varphi} = 0 \text{ coinciding with inner apices} \quad (9.25)$$

The above equations can be written in compact form as:

$$\Phi = \alpha\sqrt{J_2} - \beta\sigma_m - 2c\sqrt{N_\varphi} = 0 \quad (9.26)$$

where α and β are given in Table 9.3.

TABLE 9.3 Parameters of Drucker–Prager Criterion

Drucker–Prager criterion	α	β
Outer	$\frac{(2+N_\varphi)}{\sqrt{3}}$	$N_\varphi - 1$
Inner	$\frac{(1+2N_\varphi)}{\sqrt{3}}$	$N_\varphi - 1$

EXAMPLE 9.4 Recalculate Example 9.3 based on Drucker–Prager yield criterion. In this case, consider the cohesion of rock is 6 MPa with similar friction angle.

Solution: Taking the calculated stress dependent parameters from Example 9.1, we can estimate the functional value as given in Eqs. 9.24 and 9.25 as:

$$\sqrt{\frac{J_2}{3}}(2+N_\varphi) - \sigma_m(N_\varphi - 1) - 2c\sqrt{N_\varphi} = -1.09$$

and $$\sqrt{\frac{J_2}{3}}(1+2N_\varphi)-\sigma_m(N_\varphi-1)-2c\sqrt{N_\varphi} = 1.34$$

Hence, rock fails if we consider the yield surface which coincide inner apices and it is safe for the yield surface coinciding outer apices.

9.8 HOEK–BROWN (HB) CRITERION

The Hoek–Brown yield criterion was first introduced in early eighties. Hoek–Brown (1980) found that the peak triaxial compressive strength of isotropic rock is related to confining stress based on rock type designated by m_i (Brady and Brown, 1985). This relationship is given below:

$$\sigma_1=\sigma_3+\sqrt{m_i\sigma_3\sigma_{ci}+\sigma_{ci}^2} \tag{9.27}$$

where, σ_1 and σ_3 are the peak triaxial compressive strength and confining stress respectively and σ_{ci} is the uniaxial compressive strength of intact rock. The parameter m_i depends on rock type. This criterion was then extended for rock mass and another parameter namely s was introduced in Eq. 9.27 along with reduced value of m_i designated by m. The modified failure criterion for rock mass is given as:

$$\sigma_1=\sigma_3+\sqrt{m\sigma_3\sigma_{ci}+s\sigma_{ci}^2} \tag{9.28}$$

The parameter m and s are constants depending on rock type and the extent to which it had broken before being subjected to failure stresses. It was suggested that m and s are related to Bieniawski's rock mass rating (RMR) and equations for estimating m and s are proposed (Hoek and Brown, 1980). Since then this criterion was widely used in geotechnical and rock engineering problems. The strengths and limitations of this criterion were again proposed in an update in 1988 (Hoek and Brown, 1988). It was then proposed that the correct use of this criterion should be decided based on the nature of rock mass which does not contain any structural discontinuities and the volume of rock under consideration might contain four or more closely spaced and almost uniform discontinuity sets. In 1997, a generalised Hoek–Brown failure criterion was published for accommodating hard and soft rock masses (Hoek and Brown, 1997) and the use of geological strength index (GSI) was proposed instead of Bieniawski's rock mass rating (RMR). The estimation of RMR for very weak rock masses is cumbersome and some cases it is too difficult to evaluate. The details about the GSI is given in Section 3.3.4 of Chapter 3. The generalised criterion is again revised in the year 2002 and is the latest update of Hoek–Brown failure criterion (Hoek et al., 2002).

The generalised Hoek–Brown rock mass failure criterion for drained condition is proposed as follows (Hoek et al., 2002):

$$\Phi=\sigma_1-\sigma_3-\sigma_{ci}\left(m_b\frac{\sigma_3}{\sigma_{ci}}+s\right)^a=0 \tag{9.29}$$

where $\sigma_{1,3}$ are principal stresses and should be arranged as $\sigma_1\ge\sigma_2\ge\sigma_3$ (considering compressive stress as positive); σ_{ci} is uniaxial compressive strength of intact rock; m_b is reduced material constant based on intact property of rock m_i, and s and a are rock mass properties based on GSI as given below:

$$m_b = m_i \exp\left(\frac{\text{GSI}-100}{28-14D}\right) \tag{9.30}$$

$$s = \exp\left(\frac{\text{GSI}-100}{9-3D}\right) \tag{9.31}$$

$$a = \frac{1}{2} + \frac{1}{6}\left[\exp(-\text{GSI}/15) - \exp(-20/3)\right] \tag{9.32}$$

The range of m_i ranges between 5 and 40. Table 9.4 provides some typical values of m_i for different rock types.

TABLE 9.4 Typical m_i Values of Various Rock Types

Rock type	m_i	*Rock type*	m_i	*Rock type*	m_i
Sedimentary		**Igneous**		**Metamorphic**	
Anhydrite	12 ± 2	Agglomerate	19 ± 3	Amphybolites	26 ± 6
Breccias	20 ± 2	Andesite	25 ± 5	Gneiss	28 ± 5
Chalk	7 ± 2	Basalt	25 ± 5	Hornfels	19 ± 4
Claystones	4 ± 2	Breccia	19 ± 5	Marbles	9 ± 3
Conglomerates	21 ± 3	Dacite	25 ± 3	Metasandstone	19 ± 3
Crystalline limestones	12 ± 3	Diabase	15 ± 5	Migmatite	29 ± 3
Dolomite	9 ± 3	Diorite	25 ± 5	Phyllites	7 ± 3
Greywackes	18 ± 3	Dolerite	16 ± 5	Quartzite	20 ± 3
Gypsums	12 ± 2	Gabbro	27 ± 3	Schists	10 ± 3
Marls	7 ± 2	Granite	32 ± 3	Slates	7 ± 4
Micritic limestones	8 ± 3	Granodiorite	29 ± 3		
Sandstones	17 ± 4	Norite	20 ± 5		
Shales	6 ± 2	Obsidian	19 ± 3		
Siltstones	7 ± 2	Peridotite	25 ± 5		
Sparitic limestones	10 ± 5	Porphyries	20 ± 5		
		Rhyolite	25 ± 5		
		Tuff	13 ± 5		

In Eqs. 9.30 and 9.31, D is a factor ranges from 0 to 1 and depends on the degree of disturbances due to blasting and stress relaxation (Hoek and Brown, 1997). For hard rock and tunnelling without blasting, the value of D is assumed as 0. For tunnelling in poor rock with blasting $D = 0.8$ and for large open pit slopes with huge blasting D can be considered as 1.0.

9.8.1 HB Criterion in Principal Plane: Two Dimensions

Equation 9.29 is plotted in $\sigma_3 - \sigma_1$ plane assuming three different rock mass strength properties as shown in Figure 9.13. In the figure, three rock mass types are assumed resembling

(i) *Strong:* GSI = 100, m_i = 25 and σ_{ci} = 100 MPa
(ii) *Moderately strong:* GSI = 50, m_i = 10 and σ_{ci} = 30 MPa
(iii) *Weak:* GSI = 25, m_i = 7 and σ_{ci} = 10 MPa.

In general, the relationship between σ_1 and σ_3 is non-linear, however, for stronger rock mass, it is almost linear for lower confining stress, σ_3. From Figure 9.12, it is clear that effect of uniaxial compressive strength as well as GSI have significant role on rock mass strength.

The compressive strength of rock mass is evaluated by letting $\sigma_3 = 0$ in Eq. 9.29 as given below:

$$\sigma_{cm} = \sigma_{ci}(s)^a \tag{9.33}$$

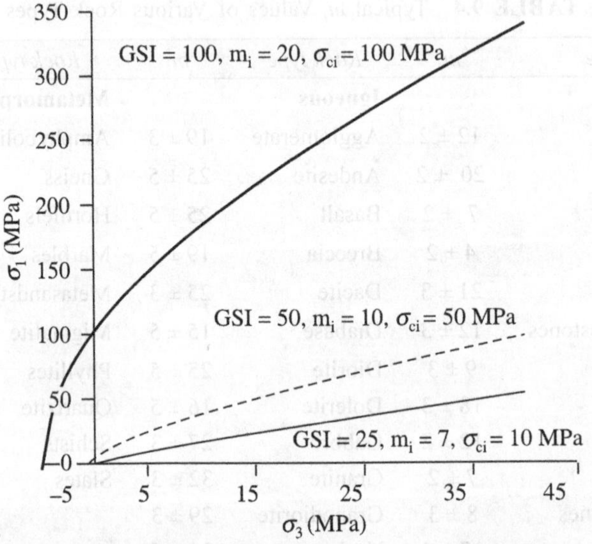

FIGURE 9.12 Hoek–Brown shear yield envelope.

Similarly, tensile strength of rock mass σ_{tm} is estimated by letting $\sigma_1 = 0$ in Eq. 9.29 and then solving σ_3 iteratively as given in the following equation:

$$(\sigma_{ci})^{1/a}(\sigma_3)^{1/a} - \left(\frac{m_b}{\sigma_{ci}}\right)\sigma_3 - s = 0 \tag{9.34}$$

Alternatively, for biaxial stress field having $\sigma_1 = \sigma_3$, the tensile strength can also be evaluated from

$$\sigma_{tm} = \frac{-s\sigma_{ci}}{m_b} \tag{9.35}$$

EXAMPLE 9.5 Determine the compressive strength and tensile strength of limestone rock mass for GSI = 56, σ_{ci} = 90 MPa, m_i = 12 and D = 0.

Solution: Rock mass strength is estimated from Eq. 9.33 (compressive) and Eq. 9.35 (tensile). The Hoek and Brown parameters are:

$$m_b = 12 \times \exp\left[\frac{56-100}{28}\right] = 2.49$$

$$s = \exp\left[\frac{(56-100)}{9}\right] = 0.00753$$

$$a = 0.5 + \frac{1}{6}\left[\exp\left(\frac{-56}{15}\right) - \exp\left(\frac{-20}{3}\right)\right] = 0.503$$

Hence, rock mass strength (compressive) will be $\sigma_{cm} = 90 \times 0.00753^{0.503} = 7.69$ MPa and the magnitude of tensile strength is $\sigma_{tm} = \dfrac{0.00753 \times 90}{2.49} = 0.272$ MPa.

9.8.2 HB Criterion in Deviatoric Plane

As shown for Mohr–Coulomb criterion, Hoek–Brown yield surface (Eq. 9.29) can also be derived using invariants of deviatoric stress as given in the following equation:

$$\Phi = \alpha\left(\sqrt{J_2}\right)^{1/a} + \beta\sqrt{J_2} - m_b\sigma_m - s\sigma_{ci} = 0 \qquad (9.36)$$

where, the expression for parameters α and β are given in Table 9.5. These two parameters are dependent on the type of rock mass and Lode angle. It is clear from Eq. 9.36 that the generalised Hoek–Brown yield function is dependent on σ_m, J_2 and θ. The yield function is plotted in three mutually perpendicular principal stress axes ($\sigma_1 \geq \sigma_2 \geq \sigma_3$) as shown in Figure 9.13. In principal stress space, the yield locus is conical whose normal section at any point is an irregular hexagon just as Mohr–Coulomb yield surface. Again the conical shape of

TABLE 9.5 Parameters of Hoek–Brown Yield Functions

Yield criterion	α	β
Hoek–Brown	$(2\cos\theta)^{1/a}(\sigma_{ci})^{(a-1)/a}$	$m_b\left(\cos\theta + \dfrac{\sin\theta}{\sqrt{3}}\right)$
Alternative–1	$\left(\sqrt{3}\right)^{1/a}(\sigma_{ci})^{(a-1)/a}$	$\dfrac{m_b}{\sqrt{3}}$
Alternative–2	$\left(\sqrt{3}\right)^{1/a}(\sigma_{ci})^{(a-1)/a}$	$\dfrac{2m_b}{\sqrt{3}}$

this yield surface is the evidence that hydrostatic stress does influence the yielding as shown in the third term of Eq. 9.36. In addition, the apex of the hexagonal pyramid can be obtained when $J_2 = 0$ with a mean hydrostatic stress, $\sigma_m = \dfrac{s\sigma_{ci}}{m_b}$ and hence, the distance of the apex from the origin will be $\dfrac{\sqrt{3}s\sigma_{ci}}{m_b}$.

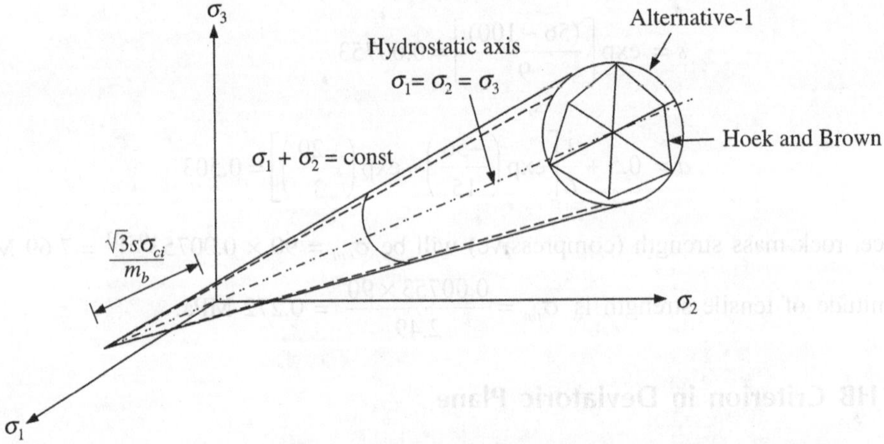

FIGURE 9.13 Generalised Hoek–Brown yield surface in principal stress space.

Figure 9.14 shows the two-dimensional, deviatoric or π-plane ($\sigma_1 + \sigma_2 + \sigma_3 = 0$), representing the generalised Hoek–Brown yield surface. The yield surface is convex and has six sharp corners or points. In all other points, the yield surface is regular. The shape and size of the yield surface varies with rock mass parameters GSI, m_i and σ_{ci}.

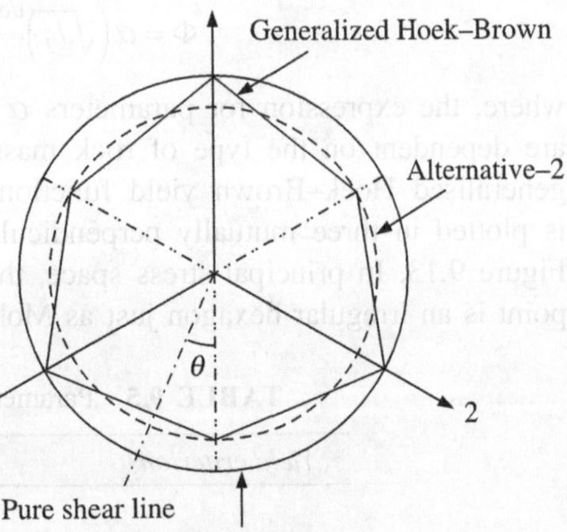

FIGURE 9.14 Yield surfaces in deviatoric plane.

Figure 9.15 shows two yield surfaces for moderately weak rock mass with GSI = 50, $m_i = 10$, $\sigma_{ci} = 30$ MPa and strong rock mass with GSI = 100, $m_i = 25$, $\sigma_{ci} = 100$ MPa for the same state of stress. It can be seen that a rock having GSI = 100 (intact rock) and $\sigma_{ci} = 100$ MPa, yield surface becomes flatter if $\theta = 30°$ and σ_3 is low. This confirms the fact that for stronger rock mass, the relationship between σ_1 and σ_3 in Eq. 9.29 is almost linear for lower σ_3.

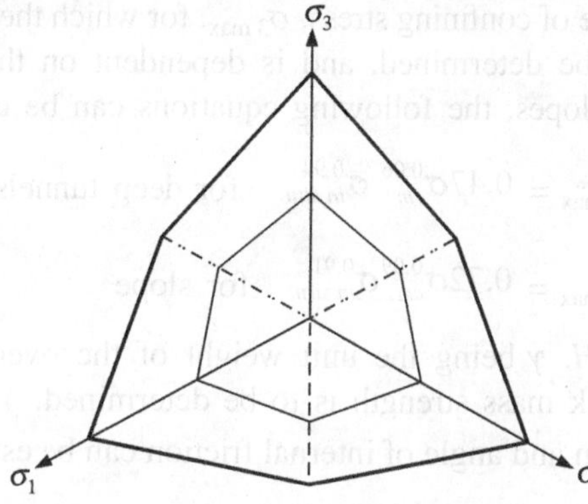

FIGURE 9.15 Yield surfaces for different parameters of rock mass.

EXAMPLE 9.6 Decide whether limestone rock mass yields based on Hoek–Brown criterion for the data given in Example 9.4. Assume GSI of rock mass is 56, $m_i = 12$ and $\sigma_{ci} = 40$ MPa.

Solution: The Hoek and Brown parameters are: $m_b = 2.49$, $s = 0.00753$, $a = 0.503$. From Example 9.4, $\sigma_m = 7.667$ MPa, $\sqrt{J_2} = 5.774$ MPa, $J_3 = 42.926$, $\theta = -0.206$ rad. From Table 9.5, it is found that $\alpha = 0.1$, $\beta = m_b\left(\cos\theta + \dfrac{\sin\theta}{\sqrt{3}}\right) = 2.145$. Hence, Hoek–Brown criterion gives $\Phi = \alpha\left(\sqrt{J_2}\right)^{1/a} + \beta\sqrt{J_2} - m_b\sigma_m - s\sigma_{ci} = -3.77$. Rock does not yield based on Hoek–Brown criterion.

9.8.3 Rock Mass Strength and Equivalent MC Parameters

Rock mass strength is of utmost interest to geotechnical and mining engineers. In the laboratory, uniaxial or triaxial strength of intact rock can be tested. An empirical estimation of rock mass strength is given in Eq. 9.33 based on the Hoek–Brown yield criterion. However, this is a conservative estimate. Hoek et al., 2002 presented a practical approach to estimate rock mass strength, equivalent cohesion and angle of internal friction of Mohr–Coulomb yield criterion of rock mass based on Hoek–Brown yield criterion. Authors have derived these equations by fitting an average linear relationship of the curve presented in Eq. 9.36. While fitting the equivalent Mohr–Coulomb line, authors have balanced the areas above and below the line inscribed by the Hoek–Brown curve. Readers may see the reference for details. In the following steps, the relevant equations are reproduced.

Step 1: Estimated rock mass strength is

$$\sigma_{cm} = \sigma_{ci}\frac{(m_b + 4s - a(m_b - 8s))(m_b/4 + s)^{a-1}}{2(1+a)(2+a)} \tag{9.37}$$

Step 2: The maximum value of confining stress, $\sigma_{3\,max}$, for which the equivalent Mohr–Coulomb parameters are to be determined, and is dependent on the type of application. For deep tunnels and slopes, the following equations can be used

$$\sigma_{3max} = 0.47\sigma_{cm}^{0.06}\,\sigma_{in\,situ}^{0.94} \quad \text{for deep tunnels} \tag{9.38}$$

$$\sigma_{3max} = 0.72\sigma_{cm}^{0.09}\,\sigma_{in\,situ}^{0.91} \quad \text{for slope} \tag{9.39}$$

where $\sigma_{in\,situ} = \gamma H$, γ being the unit weight of the overburden rock and H is the depth at which rock mass strength is to be determined.

Step 3: Equivalent cohesion and angle of internal friction can be estimated using the following equations

$$\varphi_m = \sin^{-1}\left[\frac{6am_b(s+m_b\sigma_{3n})^{a-1}}{2(1+a)(2+a)+6am_b(s+m_b\sigma_{3n})^{a-1}}\right] \tag{9.40}$$

$$c_m = \left[\frac{\sigma_{ci}\left[(1+2a)s+(1-a)m_b\sigma_{3n}\right](s+m_b\sigma_{3n})^{a-1}}{(1+a)(2+a)\sqrt{1+(6am_b(s+m_b\sigma_{3n})^{a-1})/((1+a)(2+a))}}\right] \tag{9.41}$$

where $\sigma_{3n} = \dfrac{\sigma_{3max}}{\sigma_{ci}}$.

EXAMPLE 9.7 Determine rock mass strength, equivalent cohesion and angle of internal friction of limestone at a depth 200 m. Use the data given in Example 9.6.

Solution: Using the equations given in Eq. 9.37 to Eq. 9.41, the following data are obtained $\sigma_{3\,max}$ = 2.70 MPa, σ_{cm} = 8.663 MPa, c_m = 1.036 MPa and φ_m = 44.69°. The figure shows the plot of HB and equivalent MC envelope of the data. For equivalent MC plot, the following equation is used:

$$\sigma_1 = \sigma_3\left(\frac{1+\sin\varphi_m}{1-\sin\varphi_m}\right) + \frac{2c_m\cos\varphi_m}{1-\sin\varphi_m}$$

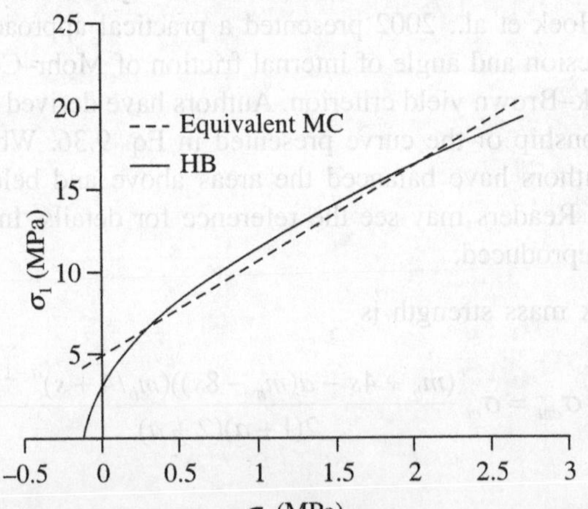

9.8.4 Instantaneous Cohesion and Angle of Internal Friction of Rock Mass

Cohesion and angle of internal friction of rock mass can be estimated from the generalised Hoek and Brown criterion. Estimation of Mohr–Coulomb instantaneous c_i and φ_i of rock masses begins with the fundamental assumption that for a given confining stress σ_3, the tangent of Eq. 9.29 will represent the equivalent Mohr–Coulomb surface as given below:

$$\sigma_1 - \sigma_3 N_\varphi - \sigma_c^M = 0 \tag{9.42}$$

The tangent of both Eqs. 9.42 and 9.36 can be equated as:

$$\frac{d\sigma_1}{d\sigma_3} = N_\varphi = 1 + am_b\left[\left(\frac{m_b\sigma_3}{\sigma_{ci}}\right) + s\right]^{a-1} \tag{9.43}$$

Substituting σ_1 expressed in Eq. 9.42 into Eq. 9.29, σ_c^M and instantaneous cohesion intercept, c_i can be estimated as given in Eqs. 9.44 and 9.45:

$$\sigma_c^M = -\sigma_3(N_\varphi - 1) + \sigma_{ci}\left[\left(\frac{m_b\sigma_3}{\sigma_{ci}}\right) + s\right]^{a} \tag{9.44}$$

$$c_i = \frac{\sigma_c^M}{2\sqrt{N_\varphi}} \tag{9.45}$$

From Eq. 9.43, the instantaneous friction angle, φ_i becomes

$$\varphi_i = 2\tan^{-1}\sqrt{N_\varphi} - \frac{\pi}{2} \tag{9.46}$$

EXAMPLE 9.8 Determine the instantaneous cohesion and angle of internal friction of limestone for GSI = 56, σ_{ci} = 90 MPa and m_i = 12 for a confining stress of 10 MPa.

Solution: Based on the data given above the following parameters are estimated using Eqs. 9.43 to 9.46: N_φ = 3.343 MPa, φ_m = 32.65°, σ_c^M = 24.35 MPa and c_m = 6.657 MPa.

9.9 ALTERNATIVE CRITERIA BASED ON HOEK–BROWN YIELD SURFACE

Hoek–Brown yield surface can be approximated using a circular conical pyramid where circle can coincides with outer or inner apices of Hoek–Brown hexagon (Figure 9.14) just as Drucker–Prager criterion is for Mohr–Coulomb yield surface. From now onwards, the yield surface coinciding outer and inner apices will be termed as Alternative–1 and Alternative–2 yield surfaces respectively. These alternative yield surfaces are developed based on Hoek–Brown strength parameters (m_b, s, a and σ_{ci}) and can also be used as independent yield functions (Deb and Choi, 2005a). The approximated yield surfaces for alternatives can be written using Eqs. 9.47 and 9.48.

$$\Phi = \left(\sqrt{3}\right)^{1/a}(\sigma_{ci})^{(a-1)/a}\left(\sqrt{J_2}\right)^{1/a} + \frac{m_b}{\sqrt{3}}\sqrt{J_2} - m_b\sigma_m - s\sigma_{ci} = 0 \text{ when } \theta \approx -\frac{\pi}{6} \text{ (outer)} \tag{9.47}$$

$$\Phi = \left(\sqrt{3}\right)^{1/a} (\sigma_{ci})^{(a-1)/a} \left(\sqrt{J_2}\right)^{1/a} + \frac{2m_b}{\sqrt{3}}\sqrt{J_2} - m_b\sigma_m - s\sigma_{ci} = 0 \text{ when } \theta \approx \frac{\pi}{6} \text{ (inner)} \quad (9.48)$$

It can be seen that alternative yield criteria are functions of J_2 and σ_m only. Hence these yield functions can be plotted in $\sigma_m - \sqrt{J_2}$ plane. Figure 9.16 and Figure 9.17 show the graphical representation of these functions plotted using three different rock mass strength parameters.

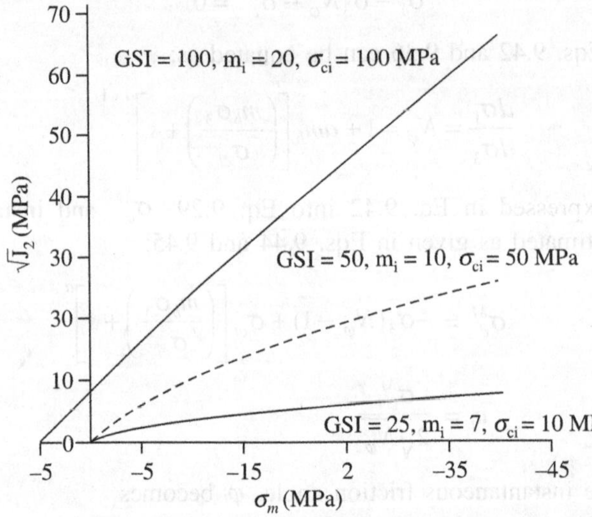

FIGURE 9.16 Alternative–1 yield function in $\sigma_m - \sqrt{J_2}$ plane.

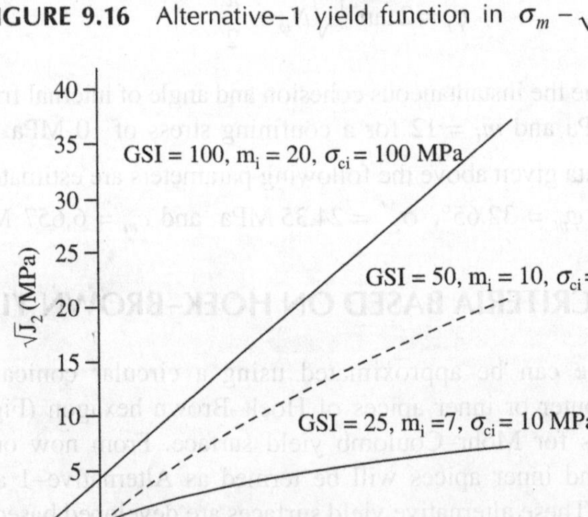

FIGURE 9.17 Alternative–2 yield function in $\sigma_m - \sqrt{J_2}$ plane.

This relationship is comparable with Eq. 9.36 if plotted in $\sigma_1 - \sigma_3$ plane (Figure 9.12). For Alternative–2 yield criterion, the deviatoric stress is smaller as compared to Alternative–1 yield surface for the same mean hydrostatic stress, since the safe zone covered by the former is smaller than the later. The difference between deviatoric stresses for Alternative–1 and Alternative–2 is significant (almost twice as much as Alternative–2) especially in case of stronger rock mass.

9.10 TENSILE YIELD CRITERION BASED ON GRIFFITH CRACK EXTENSION THEORY

Rocks are, in general, weak in tension. Tensile stress can extend crack growth or cause complete failure. Griffith crack extension theory (1924) postulates that cracks in brittle material extend from the end of microscopic flaws in the material. The extension of cracks drives due to tensile stress exceeding the tensile strength of the material. These microscopic defects in intact rock could be random small cracks, fissures or grain boundaries (Edelbro, 2004). Assuming a plane state of stress within a plate, Griffith's crack extension theory postulates that elliptical cracks will propagate from the point of maximum tensile stress concentration (Brady and Brown, 1985). The following criterion was given for crack extension in plane compression:

$$\Phi^t = (\sigma_1 - \sigma_3)^2 - 8\sigma_t(\sigma_1 + \sigma_3) = 0 \text{ when } \sigma_1 + 3\sigma_3 < 0 \quad (9.49)$$

$$\Phi^t = -\sigma_3 + \sigma_t = 0 \quad \text{when} \quad \sigma_1 + 3\sigma_3 > 0 \quad (9.50)$$

where σ_t is the tensile strength of rock. In rock engineering problems, Eq. 9.50 is extensively applied with numerical modelling for both tension and no-tension ($\sigma_t = 0$) material. Hence, rocks are considered to be in yielded or failed state if minor principal stress exceeds tensile strength of the rock. Expanding σ_3 using invariants of deviatoric stress, generalise tensile yield criterion becomes

$$\Phi^t = \sqrt{J_2}\left(\cos\theta + \frac{\sin\theta}{\sqrt{3}}\right) + \sigma_m - \sigma_t = 0 \quad (9.51)$$

Griffith's crack theory has limited applications since it describes failure of brittle materials only. However, Eq. 9.51 is applied to determine onset of yielding due to tensile stress. Using Mohr's circle of stress, a strength envelope is derived as given below:

$$\tau^2 = 4\sigma_t(\sigma_n + \sigma_t) \quad (9.52)$$

where τ is the shear stress along the plane of failure and σ_n is the stress acting normal to the plane of failure. According to the theory, the uniaxial compressive strength is eight times the uniaxial tensile strength.

9.11 TUNNEL EXAMPLE: ELASTIC-PLASTIC ANALYSIS

9.11.1 Analytical Solution: Mohr–Coulomb Media

It was shown in Chapter 7 that high stress concentration occurs at the boundary of the circular tunnel, especially on the side walls. Rocks surrounding the boundary of the tunnel may yield/

fail due to high tangential (failure) stress concentration and negligible radial (confining) stress. As a result, a plastic zone may develop around excavation boundary as shown in Figure 9.18. In the plastic zone, stress concentration is relaxed. For geotechnical and mining engineers, it is imperative to estimate the extent of the plastic zone around an excavation boundary so that judicious decision can be made for designing the support requirement. In this section, the

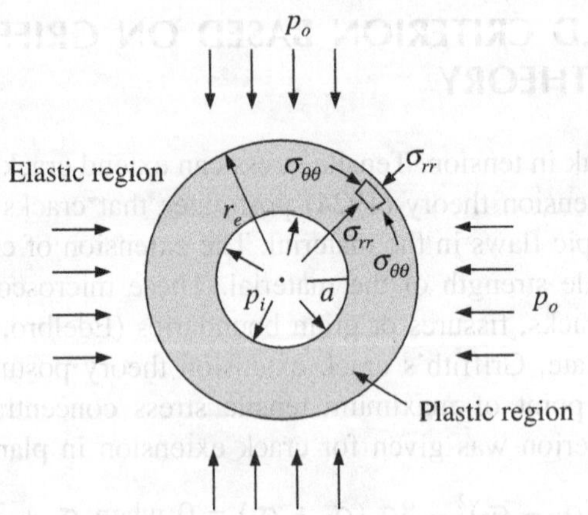

FIGURE 9.18 Elastic-plastic analysis of a circular tunnel.

extent of plastic zone (radius of elastic-plastic boundary, r_e) and stress conditions are evaluated considering Mohr–Coulomb yield criterion. For this purpose, uniform hydrostatic far field stress (p_o) condition is assumed. A uniform internal support pressure of p_i is also applied at the boundary of the tunnel as internal artificial support.

Stresses in the Elastic and Plastic Regions

For hydrostatic condition, since $k = 1$, $\tau_{r\theta} = 0$. Hence, the stress equilibrium condition in polar coordinates ignoring body force component in radial direction given in Eq. 4.70.

$$\frac{d\sigma_{rr}}{dr} = \frac{\sigma_{\theta\theta} - \sigma_{rr}}{r} \tag{9.53}$$

In the absence of shear stress, normal stresses acting in an element in the plastic zone are principal stresses. Assuming Mohr–Coulomb yield criterion, the following relationship can be obtained:

$$\sigma_{\theta\theta} = \sigma_{cf} + N_f \sigma_{rr} \tag{9.54}$$

where, $N_f = \dfrac{1+\sin\varphi_f}{1-\sin\varphi_f}$, φ_f = angle of internal friction of fractured rock, $\sigma_{cf} = \dfrac{2c_f \cos\varphi_f}{1-\sin\varphi_f}$ and

c_f = cohesion of fractured rock.

Replacing Eq. 9.54 in Eq. 9.53 and then rearranging, it is found that

$$\frac{1}{(N_f - 1)} \int_{\sigma_{cf} + p_i(N_f - 1)}^{\sigma_{cf} + \sigma_{re}(N_f - 1)} \left(\frac{d\chi}{\chi}\right) = \int_a^{r_e} \frac{dr}{r} \tag{9.55}$$

where $\chi = \sigma_{cf} + (N_f - 1)\,\sigma_{rr}$ and σ_{re} = radial stress at the elastic-plastic boundary. The solution of the above integral equation finds the radial stress at elastic-plastic boundary (σ_{re}) known as the critical stress, which is given as:

$$\sigma_{re} = (c_f \cot \varphi_f + p_i)\left(\frac{r_e}{a}\right)^{N_f - 1} - c_f \cot \varphi_f \tag{9.56}$$

Considering superposition concept, tangential and radial stresses in the elastic zone can be estimated as given below:

$$\left.\begin{aligned} \sigma_{\theta\theta} &= \frac{(p_o - \sigma_{re})}{2}\left[2\left(1 + \frac{r_e^2}{r^2}\right)\right] + \sigma_{re} = p_o\left(1 + \frac{r_e^2}{r^2}\right) - \sigma_{re}\left(\frac{r_e^2}{r^2}\right) \\ \sigma_{rr} &= \frac{(p_o - \sigma_{re})}{2}\left[2\left(1 - \frac{r_e^2}{r^2}\right)\right] + \sigma_{re} = p_o\left(1 - \frac{r_e^2}{r^2}\right) + \sigma_{re}\left(\frac{r_e^2}{r^2}\right) \end{aligned}\right\} \text{for } r > r_e \tag{9.57}$$

From Eq. 9.57, we find that tangential stress at the elastic-plastic boundary, $r = r_e$ is:

$$\sigma_{\theta e} = 2p_o - \sigma_{re} \tag{9.58}$$

At the elastic-plastic boundary, stress condition is such that a rock element is in equilibrium and no more yielding is imminent. Hence, it can be written that Mohr–Coulomb criterion is just satisfied at $r = r_e$ as:

$$\sigma_{\theta e} = \sigma_{cr} + N_r \sigma_{re} \tag{9.59}$$

where, $N_r = \dfrac{1 + \sin \varphi_r}{1 - \sin \varphi_r}$, σ_{cr} = uniaxial compressive strength of rock and φ_r = angle of internal friction of rock.

Substituting Eq. 9.59 into Eq. 9.58, it becomes

$$\sigma_{re} = \frac{2p_o - \sigma_{cr}}{1 + N_r} \tag{9.60}$$

The critical pressure expressed by Eq. 9.60 has special significant in rock failure process. If the internal pressure p_i is more than this critical pressure, no failure occurs around the boundary of the tunnel. Replacing Eq. 9.60 into Eq. 9.56, elastic-plastic radius is obtained as given in the following equation.

$$r_e = a\left[\frac{2p_o - \sigma_{cr} + (c_f \cot \varphi_f)(1 + N_r)}{(p_i + c_f \cot \varphi_f)(1 + N_r)}\right]^{1/(N_f - 1)} \tag{9.61}$$

Radial and tangential stresses in the plastic zone are estimated using the equations given below:

$$\left.\begin{aligned} \sigma_{rr} &= (c_f \cot \varphi_f + p_i)\left(\frac{r}{a}\right)^{N_f - 1} - c_f \cot \varphi_f \\ \sigma_{\theta\theta} &= \sigma_{cf} + N_f \sigma_{rr} \end{aligned}\right\} \text{for } r \le r_e \tag{9.62}$$

Strains and Displacements in the Elastic and Plastic Regions

In the elastic region, the radial displacement is obtained as given in Eq. 7.2(a). In this case, $k = 1$ and hence, the radial displacement in the elastic region becomes

$$u_r = \frac{(p_o - \sigma_{re})r_e^2}{2Gr} \quad \text{for } r > r_e \tag{9.63}$$

In the plastic region, $r_0 \le r \le r_e$, the radial (ε_r) and circumferential ($\varepsilon_{\theta\theta}$) strains are decomposed into elastic and plastic parts as given below:

$$\varepsilon_{rr} = \varepsilon_{rr}^e + \varepsilon_{rr}^p \tag{9.64a}$$

$$\varepsilon_{\theta\theta} = \varepsilon_{\theta\theta}^e + \varepsilon_{\theta\theta}^p \tag{9.64b}$$

Appendix B discusses briefly the deformation following yield. Considering radially upward displacements as positive, radial and circumferential strains at any point in the plastic zone can be expressed based on the radial displacement u (Eq. 4.16) as:

$$\varepsilon_{rr} = \frac{du}{dr} \tag{9.65a}$$

$$\varepsilon_{\theta\theta} = \frac{u}{r} \tag{9.65b}$$

Assuming non-associative flow rule the plastic potential function, Q as function of stresses and dilation angle, ψ can be expressed as:

$$Q = \sigma_{\theta\theta} - \sigma_{rr} N_\psi = \text{constant} \tag{9.66}$$

where, $N_\psi = \dfrac{1+\sin\psi}{1-\sin\psi}$. Then, plastic strains are obtained as:

$$\varepsilon_{rr}^p = \frac{\partial Q}{\partial \sigma_{rr}} d\eta = -N_\psi d\eta \tag{9.67a}$$

$$\varepsilon_{\theta\theta}^p = \frac{\partial Q}{\partial \sigma_{\theta\theta}} d\eta = d\eta \tag{9.67b}$$

From Eqs. 9.65–9.67, it can be written that

$$\varepsilon_{rr}^p + N_\psi \varepsilon_{\theta\theta}^p = \frac{du}{dr} + N_\psi \frac{u}{r} - (\varepsilon_{rr}^e + N_\psi \varepsilon_{\theta\theta}^e) = 0 \tag{9.68}$$

The third term of Eq. 9.68 is function of r only and can be derived considering analogy of thick cylinder subjected to radial external and internal pressure. From this concept, the third term is given in the following (Sharan, 2003)

$$(\varepsilon_{rr}^e + N_\psi \varepsilon_{\theta\theta}^e) = \frac{1}{2G}\left[C(1-2v)(1+N_\psi) - \frac{D}{r^2}(N_\psi - 1)\right] \tag{9.69}$$

where, $C = \dfrac{(\sigma_{re} - p_o)r_e^2 - (p_i - p_o)a^2}{r_e^2 - a^2}$, and $D = \dfrac{(p_i - \sigma_{re})a^2 r_e^2}{r_e^2 - a^2}$

Equation 9.68 is solved after applying the boundary conditions which state that (i) at the elastic-plastic boundary, the radial displacement is equal to the elastic displacement obtained from Eq. 9.63 for $r = r_e$, (ii) $u = 0$ at $r = \infty$. Hence, the solution of the differential Eq. 9.68 is obtained as:

$$u = r^{-N_\psi} \int_{r_e}^{r} \left\{ r^{N_\psi} \frac{1}{2G} \left[C(1-2\nu)(1+N_\psi) - \frac{D}{r^2}(N_\psi - 1) \right] \right\} dr + \frac{(p_o - \sigma_{re}) r_e}{2G} \left(\frac{r_e}{r} \right)^{N_\psi}$$

$$u = \frac{r^{-N_\psi}}{2G} \left[C(1-2\nu)(r_e^{N_\psi+1} - r^{N_\psi+1}) - D(r_e^{N_\psi-1} - r^{N_\psi-1}) \right] + \frac{(p_o - \sigma_{re}) r_e}{2G} \left(\frac{r_e}{r} \right)^{N_\psi} \quad (9.70)$$

EXAMPLE 9.9 Plot tangential and radial stresses with respect to radial distance r from a circular tunnel of radius 2.5 m for hydrostatic far field stress of 10 MPa and internal support pressure of 2 MPa. Consider Mohr–Coulomb rock media and assume cohesion and friction angle of intact rock are 2 MPa and 30° respectively and those of fractured rock are 1 MPa and 20°.

Solution: Using Eqs. 9.60 and 9.61, we find σ_{re} = 3.27 MPa and r_e = 3.14 m. Then using Eqs. 9.57 and 9.62, we can plot the tangential and radial stresses as given below.

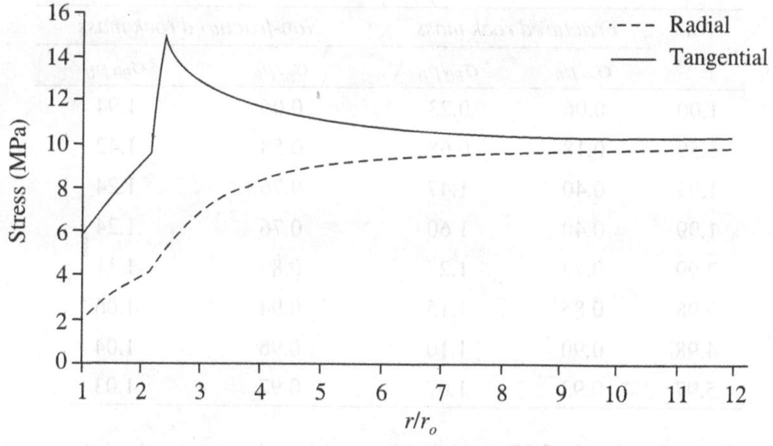

EXAMPLE 9.10 A horizontal tunnel of radius 3 m is driven in a hydrostatic stress field of 15 MPa. Rock bolts are installed all around the tunnel boundary to provide a uniform internal pressure of 0.5 MPa. Find the radial and tangential stress distribution at mid-wall of the tunnel boundary, i.e., θ = 0°, assuming rock mass behaves like (a) elastic material, and (b) elastic-perfectly plastic material. The cohesion and friction angle of intact rock are 4 MPa and 35°, while for fractured rock, cohesion and angle of internal friction are 1 MPa and 15° respectively. Estimate the elastic-plastic radius and compare the results of part (a) and (b).

Solution: Using Eqs. 7.20, 7.21 and 7.26, the following figure and table shows the stress distribution around opening for both elastic and elastic-perfectly plastic material.

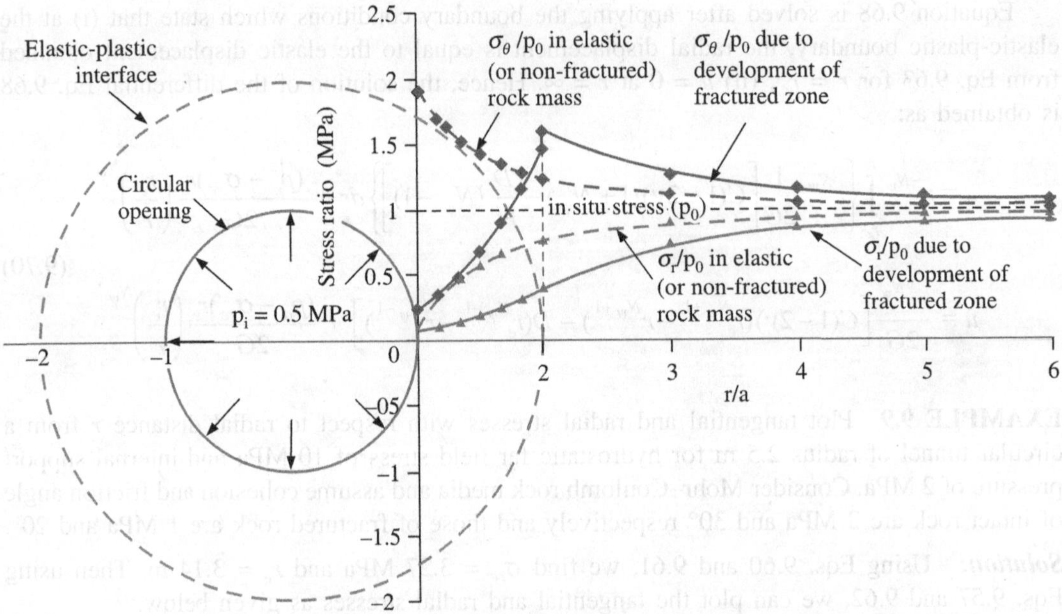

r/a	Fractured rock mass		Non-fractured rock mass	
	σ_{rr}/p_0	$\sigma_{\theta\theta}/p_0$	σ_{rr}/p_0	$\sigma_{\theta\theta}/p_0$
1.00	0.06	0.23	0.06	1.94
1.50	0.18	0.68	0.58	1.42
1.99	0.40	1.47	0.76	1.24
1.99	0.40	1.60	0.76	1.24
2.99	0.73	1.27	0.89	1.11
3.98	0.85	1.15	0.94	1.06
4.98	0.90	1.10	0.96	1.04
5.97	0.93	1.07	0.97	1.03

The radius of plastic zone is 5.97 m. After comparing the results the inference is as follows:

(a) The tangential stresses gradually increases from the boundary of the tunnel up to the elasto-plastic boundary.

(b) For elastic analysis, maximum tangential stress occurs at the boundary of the opening, whereas, for plastic analysis, the maximum stress has shifted inside the intact or rock mass. The peak tangential stress in plastic analysis is less than that of elastic case.

(c) The stored strain energy in case of elastic analysis is more than plastic analysis. For plastic analysis, the energy is wasted in creating fractures around the opening.

9.11.2 Analytical Solution: Hoek–Brown Media

The procedure for obtaining elastic-plastic analytical solutions of a circular tunnel made in Hoek–Brown rock media is similar to that of Mohr–Coulomb media as mentioned above. In this case, the yield criterion given in Eq. 9.29 is applied to estimate the onset of yielding of rock mass. Equation 9.54 then can be written as:

$$\sigma_{\theta\theta} = \sigma_{rr} + \sigma_{ci}\left(\frac{m_f \sigma_{rr}}{\sigma_{ci}} + s_f\right)^{a_f} \tag{9.71}$$

where m_f, s_f and a_f are Hoek–Brown parameters of fractured rock. Replacing Eq. 9.54 in Eq. 9.53 and then rearranging, it is found that

$$\frac{1}{m_b}\int_{m_f p_i/\sigma_{ci}+s_f}^{m_f \sigma_{re}/\sigma_{ci}+s_f}\left(\frac{d\chi}{\chi^{a_f}}\right) = \int_a^{r_e}\frac{dr}{r} \tag{9.72}$$

where $\chi = m_f \sigma_{rr}/\sigma_{ci} + s_f$. The solution of the above integral equation finds the radial stress at elastic-plastic boundary (σ_{re}), known as critical pressure, which is given as:

$$\sigma_{re} = \frac{\sigma_{ci}}{m_f}\left(\left[(1-a_f)m_f \ln\left(\frac{r_e}{a}\right) + N_f^{1-a_f}\right]^{1/(1-a_f)} - s_f\right) \tag{9.73}$$

where $N_f = m_f \dfrac{p_i}{\sigma_{ci}} + s_f$

In the above equation, σ_{re} cannot be obtained since r_e is not known. Hence, the radial stress at elastic-plastic boundary can be obtained by equating Eq. 9.60 with Eq. 9.29. The following equation is solved by iterative method to estimate the value of σ_{re}.

$$\sigma_{re} + \frac{1}{2}\sigma_{ci}\left(\frac{m_b \sigma_{re}}{\sigma_{ci}} + s_r\right)^{a_r} - p_0 = 0 \tag{9.74}$$

If the value of $a_r = 0.5$, the value of σ_{re} is found to be

$$\sigma_{re} = p_0 - M\sigma_{ci} \tag{9.75}$$

where, $M = \dfrac{1}{2}\left(\sqrt{\left(\dfrac{m_b}{4}\right)^2 + \dfrac{m_b p_0}{\sigma_{ci}} + s_r} - \dfrac{m_b}{8}\right)$

Comparing Eq. 9.71 and with the σ_{re} value obtained from Eq. 9.74, we find r_e as:

$$r_e = a\exp\left\{\frac{\left[\dfrac{m_f}{\sigma_{ci}}\left(\sigma_{re} + \dfrac{s_f \sigma_{ci}}{m_f}\right)\right]^{1-a_f} - N_f^{1-a_f}}{(1-a_f)m_f}\right\} \tag{9.76}$$

In the following, the expression for the radial stress in the plastic region is given using Hoek–Brown parameters of fractured rock mass and uniaxial compressive strength of rock.

$$\sigma_{rr} = \frac{\sigma_{ci}}{m_f}\left(\left[(1-a_f)m_f \ln\left(\frac{r}{a}\right) + N_f^{1-a_f}\right]^{1/(1-a_f)} - s_f\right) \quad \text{for } r < r_e \tag{9.77}$$

The tangential or circumferential stress in the plastic region is calculated from Eq. 9.36 for residual rock mass parameters m_f, s_f and a_f. The radial displacement in the plastic region is estimated using Eq. 9.70. In the elastic region, radial displacements and stresses are calculated based on Eqs. 9.63 and 9.57 respectively.

9.11.3 Effect of Gravity

In the above discussions, effect of gravity has not been considered while analysing the radial stresses in the fractured and intact rock mass. It is quite possible that fractured and broken rocks located in the roof can starts to displace due to the effect of gravity. As a result, internal pressure, p_i assumed for the above calculations may not be sufficient to resist the movement of rocks. Again considering a circular tunnel, weight of a broken rock acting downward can be estimated as:

$$w_f = c\gamma_r(r_e - a)s^2 \tag{9.78}$$

where c is a constant quantity ranging between 0 and 1 and depends on the frictional resistance applied by its neighbour blocks and s denotes the width of spacing in both longitudinal and circumferential directions. Hence, the total internal pressure required to support a tunnel roof would be

$$p_T = p_i + w_f \tag{9.79}$$

It may be noted that with increasing p_i the elasto-plastic radius r_e reduces and hence p_T decreases. The value of p_T reaches to a minimum before it increases linearly again with increasing p_i. The design internal support pressure p_i is the value at which p_T is the minimum.

EXAMPLE 9.11 Using the data given in Example 9.8, plot p_T versus p_i, considering gravity effect of c = 1.0 for a chormite ore body having unit weight 36 kN/m^3. Assume side of the square spacing is 2 m.

Solution: For different values of p_i ranging from 0 to 1 MPa, estimate the elasto-plastic radius r_e using Eq. 9.61. Then apply Eqs. 9.80 and 9.81 to determine w_f and p_T respectively. The plot of p_T versus p_i is given below.

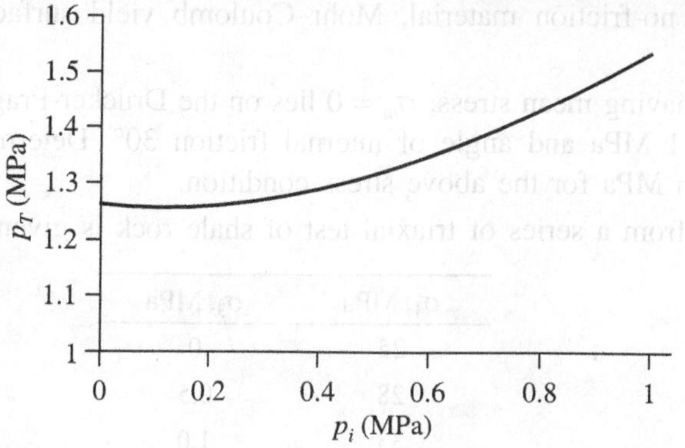

EXERCISES

1. Prove Eq. 9.8 given in this chapter.
2. If point (20, 10, –5) denote the principal stresses (σ_1, σ_2, σ_3) respectively measured in MPa, what will be the corresponding values of principal stresses in the π-plane. Draw the deviatoric stress vector in the π-plane for the above principal stress point. Calculate the lode angle for the above deviatoric stress vector.
3. A series of triaxial test of sandstone rock reveals the following relationship

   $\sigma_1 - 3\sigma_3 - 75 = 0$ (All stress units are in MPa)

   (a) Determine the cohesion and angle of internal friction of sandstone.
   (b) If the state of stress at a point is expressed as σ_1 = 80 MPa and σ_3 = 1.5 MPa, determine whether the point is safe or not.
4. Stress tensors in MPa at a point are given by

$$\begin{bmatrix} 3 & -1 & 0 \\ -1 & 16 & 0 \\ 0 & 0 & 4 \end{bmatrix}$$

   If the cohesion and angle of internal friction of rock are 3 MPa and 35° respectively. Then using Mohr-Coulomb criterion determine whether the point is safe or not.
5. Solve Exercise 4 using Drucker–Prager criterion for both inner and outer circles.
6. Draw the deviatoric stress vector on the π-plane and compute the Lode angle using the date given in Exercise 4.
7. A series of triaxial test of sandstone rock reveals the following relationship between major and minor principal stress

   $\phi = \sigma_1 - \sigma_3 - 50(0.0958\sigma_3 + 0.0117)^{0.511} = 0$ (All stress units are in MPa)

   If the state of stress at a point is expressed as σ_3 is 5.0 MPa and σ_3 is 12.0 MPa, determine whether the point is safe or not.

8. Show that for no-friction material, Mohr–Coulomb yield surface becomes a regular hexagon.

9. A stress point having mean stress, $\sigma_m = 0$ lies on the Drucker-Prager outer yield surface with cohesion 1 MPa and angle of internal friction 30°. Determine the radius of the yield surface in MPa for the above stress condition.

10. Data obtained from a series of triaxial test of shale rock is given below:

σ_1, MPa	σ_3, MPa
25	0
28	0.5
33	1.0
37	1.5
40	2.0
49	3.0
57	4.0

(a) Establish Mohr–Coulomb criterion based on the above data.
(b) Establish Drucker–Prager envelope based on the above data.
(c) Determine cohesion and angle of internal friction of the shale.

11. Show that the ratio of radius of outer to inner circle of Drucker–Prager yield surface is $\dfrac{1+2N_\varphi}{2+N_\varphi}$.

12. The GSI of a limestone strata is estimated to be 59. Measurement of pillar stresses in the limestone bed shows σ_1 = 8.0 MPa, σ_2 = 5.0 MPa and σ_3 = 1.0 MPa. Considering m_i and uniaxial compressive strength of limestone are 20 and 40 MPa respectively, determine whether the pillar is safe.

13. Solve Exercise 9 considering alternative criteria.

14. Solve Exercise 4 considering Hoek and Brown criterion. Assume m_i and *GSI* of shale are 12 and 42 respectively.

15. In continuation of Example 9.9 and considering Hoek and Brown rock media with m_i and GSI value 12 and 42 respectively. Estimate the elasto-plastic radius considering 60% reduction in peak properties.

Chapter 10

Joints in Rock Mass

10.1 INTRODUCTION

Strength and deformability of isotropic and homogeneous material such as intact rock has been enumerated in previous chapters. Presence of bedding or cleavage planes, joints and other natural discontinuities makes rock mass anisotropic. In Section 5.4, the simplest form of anisotropy (linear transverse isotropy) and its constitutive model has been described. This concept is often applied for designing geotechnical structures. The concept and mechanical behaviour of anisotropic rocks have been illustrated in several texts giving detail computational methodologies with field examples (Obert and Duvall, 1967; Lama and Vutukuri, 1978; Jaeger and Cook, 1979; Brady and Brown, 1985; Goodman, 1989; Indraratna and Haque, 2000). In this chapter, strength and deformability aspects of jointed rock mass are elaborated giving emphasis on shear behaviour of joints.

Mechanical properties and behaviour of jointed rock mass are studied by many researchers using natural and artificial or synthetic samples. Some of the researchers who have significantly contributed in understanding of mechanical behaviour of jointed rock mass are Patton (1966), Goodman (1970), Barton and Choubey (1977), Bandis et al. (1981), and others. A detailed review on rock joints and their shear strength was presented by Barton (1976). Gerarrd (1985) and Sun et al. (1985) presented a detailed study on the formulation of compliance matrices of jointed rock mass which have contributed to the understanding of deformability of jointed rock mass. Pande et al. (1990) commented that joints in rock masses can vary widely in their physical state and mechanical behaviour since joints can be fresh or weathered, asperities matching or mismatching, filled or unfilled with gauge material. Recently, Barton has published a comprehensive review of shear strength criterion of rock, rock joints and rock fills stating the advantage and limitation of Mohr–Coulomb criterion for jointed rock mass (Barton, 2013). Thus, mechanical properties such as strength (fracture and frictional) and deformability of jointed rock mass depends on the roughness of joint plane, interlocking properties of asperities, applied normal stress and residual properties of the joint surface.

10.2 FRACTURE STRENGTH OF JOINTED ROCK MASS

Strength of a jointed rock mass can be defined as fracture strength or frictional (shear) strength. The fracture strength is measured by either uniaxial or triaxial tests of jointed rock sample. The shear strength is estimated by direct shear tests or triaxial tests of rock sample under constant normal load (CNL) or constant normal stiffness (CNS). In the following sections, discussions will focus on the fracture strength and shear strength properties of jointed rock mass.

10.2.1 Fracture Strength with Single Joint

Jaeger (1960) presented an instructive example of fracture strength of rock element having a single joint where the joint plane is inclined at an angle β with the x-axis as shown in Figure 10.1.

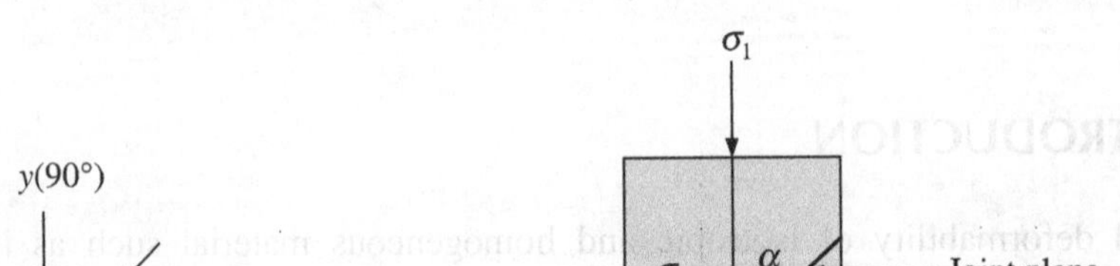

FIGURE 10.1 A single joint plane in rock sample.

The shear strength of the joint is defined by Mohr–Coulomb criterion as

$$|\tau| = c_j + \sigma_n \tan\varphi_j \tag{10.1}$$

where c_j and φ_j are the cohesion and angle of internal friction respectively of the joint plane. Considering compressive stress as positive and then applying the transformation of stresses as defined in Section 4.3 of Chapter 4, it can be written as

$$\sigma_n = \left(\frac{\sigma_1 + \sigma_3}{2}\right) + \left(\frac{\sigma_1 - \sigma_3}{2}\right)\cos 2\beta \tag{10.2}$$

$$|\tau| = \left(\frac{\sigma_1 - \sigma_3}{2}\right)\sin 2\beta \quad \text{for } 0° \le \beta \le 90° \tag{10.3}$$

$$|\tau| = -\left(\frac{\sigma_1 - \sigma_3}{2}\right)\sin 2\beta \quad \text{for } -90° \le \beta \le 0° \tag{10.4}$$

Inserting Eqs. 10.2 and 10.3 into Eq. 10.1 and after rearranging it becomes

$$\sigma_1 = \sigma_3 + \frac{2(\sigma_3 \tan\varphi_j + c_j)}{(1 - \tan\varphi_j \cot\beta)\sin 2\beta} \quad \text{for } 0° \le \beta \le 90° \tag{10.5}$$

Similarly, inserting Eqs. 10.2 and 10.5 into Eq. 10.1 and after rearranging it is found that

$$\sigma_1 = \sigma_3 - \frac{2(\sigma_3 \tan\varphi_j + c_j)}{(1 + \tan\varphi_j \cot\beta)\sin 2\beta} \quad \text{for } -90° \le \beta \le 0° \tag{10.6}$$

Thus, if $\beta = 90°$ or $\beta = \varphi_j$, the stress required to produce the slip along the joint plane is theoretically infinity. Let us define these two limits as $\beta_{max} = 90°$ and $\beta_{min} = \varphi_j$. Then slip would occur if $\beta_{min} < \beta < \beta_{max}$. Figure 10.2 shows a typical plot of difference in principal stresses $\sigma_1 - \sigma_3$ vs. angle of inclination β taking $\varphi_j = 30°$, $\sigma_3 = 2$ MPa and $c_j = 2$ MPa. In this case, it is assumed that for $\beta = 0°$ triaxial strength of rock sample is 49.2 MPa. It can be seen that the peak strength of the sample is affected by the joint plane if angle of inclination falls within φ_j to 90°. If the angle of inclination is less than φ_j, the strength of rock sample is determined by the property of intact rock. Similar analysis can be performed considering the inclination of the joint plane as $-\beta$ degree with the σ_3 axis. In this case, we find that $\beta_{min} = 90°$ and $\beta_{max} = -\varphi_j$. Then the slip would occur if $\beta_{min} < -\beta < \beta_{max}$. Hence, the rock sample will fail due to its intact strength for angle of inclination as $-\varphi_j < \beta < \varphi_j$. For all other cases, fracture strength of the sample is affected by the existence of the joint plane. Note that fracture strength is truncated at $\beta = \pm 33.83°$ in Figure 10.2 since intact rock strength is 49.2 MPa.

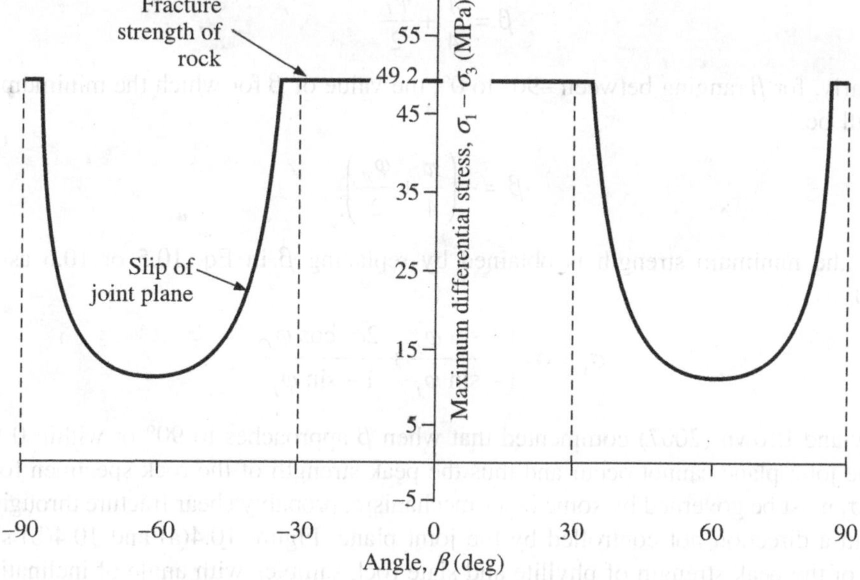

FIGURE 10.2 Variation of peak fracture (slip) strength with inclination of the joint.

Figure 10.3 depicts Mohr–Coulomb envelope of intact rock and jointed rock sample having the minimum strength. In this case, the minimum strength is obtained at $\beta = 60°$. For intact rock (at $\beta = 0°$), friction angle of 45° is assumed. It may be noted that all other intermediate Mohr–Coulomb envelope for $30° \leq \beta \leq 60°$ and $60° \leq \beta \leq 90°$, lie in between the Mohr–Coulomb envelopes shown in the figure.

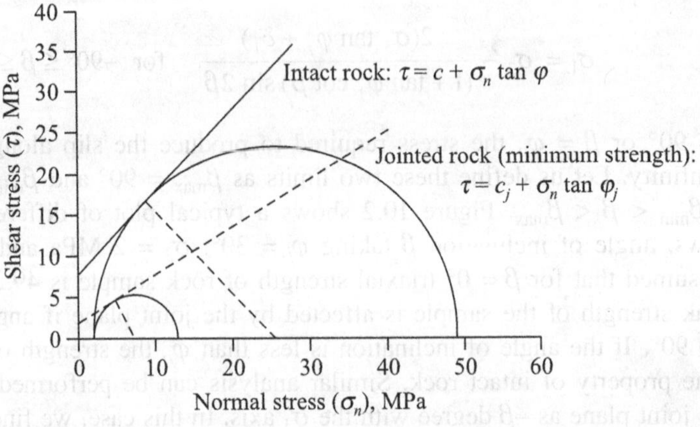

FIGURE 10.3 Mohr's circles of stresses and MC envelope for intact and jointed rock.

By differentiating σ_1 (Eq. 10.5) with respect β and then equating it to zero, we find the value of β, for which the minimum strength occurs, can be obtained as

$$\tan 2\beta = -\cos \varphi_j \tag{10.7}$$

or

$$\beta = \frac{\pi}{4} + \frac{\varphi_j}{2} \tag{10.8}$$

Similarly, for β ranging between $-90°$ to $0°$, the value of β for which the minimum strength occurs will be

$$\beta = -\left(\frac{\pi}{4} + \frac{\varphi_j}{2}\right) \tag{10.9}$$

Thus, the minimum strength is obtained by replacing β in Eq. 10.5 or 10.6 as given in Eq. 10.10

$$\sigma_1 = \sigma_3 \frac{1 + \sin \varphi_j}{1 - \sin \varphi_j} + \frac{2c_j \cos \varphi_j}{1 - \sin \varphi_j} \tag{10.10}$$

Brady and Brown (2007) commented that when β approaches to 90° or within 0 to φ_j, the slip on the joint plane cannot occur and thus the peak strength of the rock specimen for a given value of σ_3 must be governed by some other mechanism, probably shear fracture through the rock material in a direction not controlled by the joint plane. Figure 10.4(a) and 10.4(b) shows the variation of the peak strength of phyllite and slate rock samples with angle of inclination of the minor principal stress to the plane of joint. Comparing Figure 10.2, Figure 10.4(a) and 10.4(b),

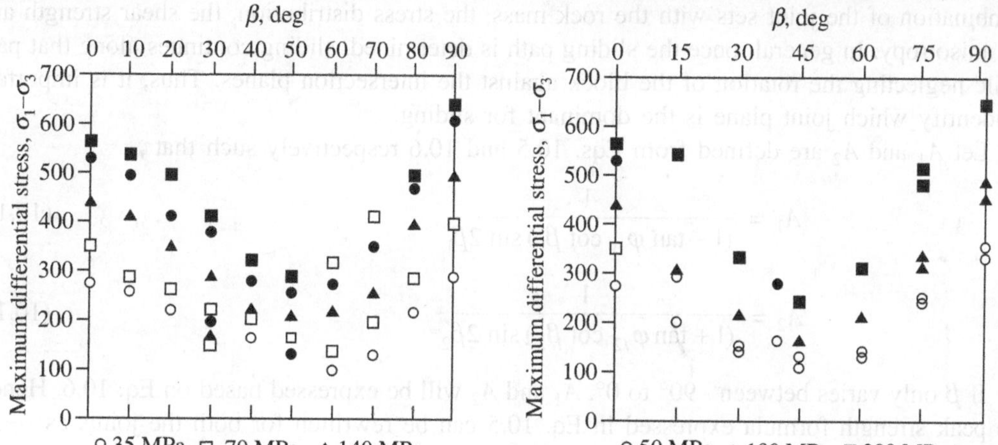

(a) Moretown phyllite (b) Slate

FIGURE 10.4 Variation of peak principal stress with angle of inclination of single joint for the confining mention in the legend (*Data source:* Brady and Brown, 2007).

it can be found that the shape of the peak strength curve with angle β, does not match with actual test data. In particular, the peak strength value for β between 0 to φ_j, is not a plateau as given in Figure 10.2. Based on similar concept, Brady and Brown (2007) commented that theoretical fracture strength of jointed rock sample is the oversimplified representation of peak strength variation in anisotropic rocks. It is proposed that cohesion intercept, c_j and friction angle φ_j are not constant and they may vary with β.

Jennings (1970) extended the concept of Jager (1960) and introduced the concept of rock bridge to calculate the strength of a single joint as follows:

$$\tau = k\tau_j + (1-k)\tau_i \qquad (10.11)$$

where, τ_j and τ_i represent shear strength of the joint segment and intact rock bridge respectively, k is joint continuity factor defined as $k = \dfrac{L_j}{(L_j + L_r)}$, where, L_j, L_r are the lengths of the joint segment and intact rock bridge, respectively.

10.2.2 Fracture Strength with Two Joints

Figure 10.5 represents a rock specimen with two joints with angle of inclinations β_1 and β_2 degrees with the horizontal axis. Zhu et al. (1993) suggested that the most significant factor for two joints case will be the plane of joint where sliding begins. They also suggested that the sliding would depend on the geometry of the

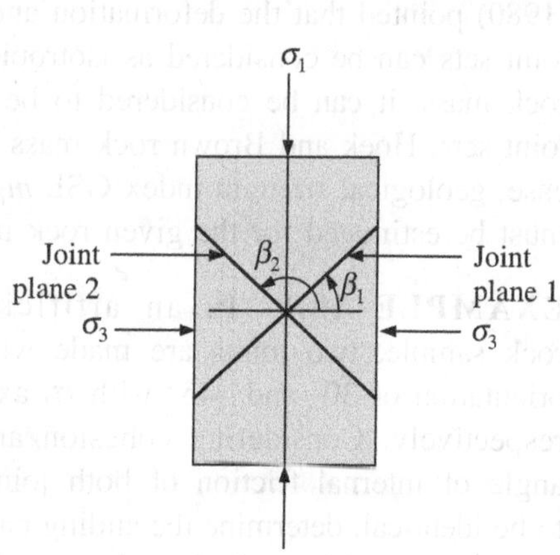

FIGURE 10.5 A rock sample having two joints under triaxial loading.

combination of the joint sets with the rock mass, the stress distribution, the shear strength and the anisotropy. In general, once the sliding path is determined, sliding continues along that path while neglecting the rotation of the block against the intersection planes. Thus, it is important to identify which joint plane is the dominant for sliding.

Let A_1 and A_2 are defined from Eqs. 10.5 and 10.6 respectively such that

$$A_1 = \frac{1}{(1 - \tan \varphi_{j1} \cot \beta_1) \sin 2\beta_1} \tag{10.12}$$

$$A_2 = \frac{1}{(1 + \tan \varphi_{j2} \cot \beta_2) \sin 2\beta_2} \tag{10.13}$$

If β only varies between –90° to 0°, A_1 and A_2 will be expressed based on Eq. 10.6. Hence, the peak strength formula expressed in Eq. 10.5 can be rewritten for both the joints as

For joint 1: $$\sigma_1 = \sigma_3 - 2(\sigma_3 \tan \varphi_{j1} + c_{j1}) A_1 \tag{10.14}$$

For joint 2: $$\sigma_1 = \sigma_3 - 2(\sigma_3 \tan \varphi_{j2} + c_{j2}) A_2 \tag{10.15}$$

If the magnitude of the right hand side expression of Eq. 10.15 is more than that of Eq. 10.14, then sliding will occur along the joint 2 and vice versa. If cohesion and friction angles are the same for both the joints, then same analogy is applied just by comparing the values of A_1 and A_2. For this special case, sliding occurs along the joint plane 1 if $A_1 < A_2$ and thus, the peak strength for the specified σ_3 will be expressed by Eq. 10.14. Similarly, sliding occurs along joint plane 2 if $A_1 > A_2$ and the peak strength of the rock specimen will be computed using Eq. 10.15.

10.2.3 Fracture Strength with Multiple Joints

For three or more joints sets, the concept outlined in Section 10.1 is applicable. Hoek and Brown (1980) pointed that the deformation and strength properties of rock mass having four or more joint sets can be considered as isotropic. Brady (1987) also suggested that for heavily jointed rock mass, it can be considered to be pseudo-continuous [Figure 9.4(d)]. Thus, for multiple joint sets, Hoek and Brown rock mass strength (σ_1) as given in Eq. 9.29 is applicable. In this case, geological strength index GSI, m_i value and uniaxial compressive strength of intact rock must be estimated for the given rock mass condition.

EXAMPLE 10.1 In an artificial rock sample, two joints are made with orientation of 30° and –45° with σ_3 axis respectively. Considering cohesion and angle of internal friction of both joints to be identical, determine the sliding path once axial load is applied on the sample.

Solution: Since, cohesion and friction angle are same for both the joints, we can only compare the magnitude of A_1 and A_2 as given below:

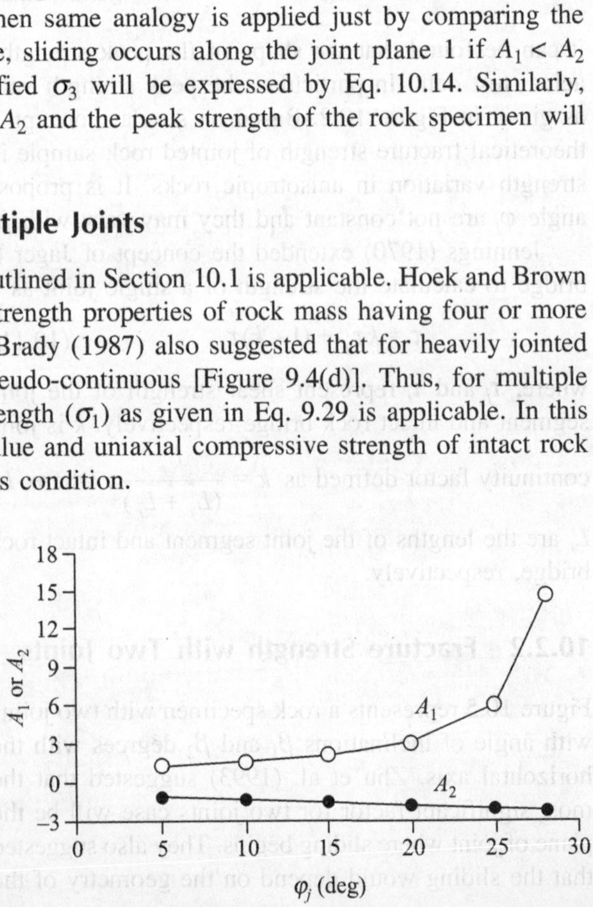

$$A_1 = \frac{1}{(1 - \tan\varphi_{j1}\cot 30°)\sin 60°}$$

$$A_2 = \frac{1}{(1 + \tan\varphi_{j1}\cot(-45°))\sin(-90°)}$$

The Figure given on previous page shows the plot of A_1 and A_2 with respect to friction angle up to 28°. It is clear that magnitude of $A_2 < A_1$ and hence, joint 2 will slide first.

10.3 SHEAR STRENGTH OF ROCK JOINT

Shear strength of a rock joint can be measured using direct shear test apparatus or using triaxial compression test. In the direct shear test, the joint surface is oriented parallel to the direction of applied shear load and the two halves of the sample are fixed inside a shear box, using Cycap, concrete, plaster or epoxy (Goodman, 1989). In general, lower block is fixed for any movement while shear load is applied to the upper block with constant normal load as shown in Figure 10.6(a). In this test, dilation is allowed since normal displacement (δ_n) of upper block is not restricted. This condition resembles free movement of a rock block in slopes. Figure 10.6(b) depicts a 500 kN capacity double acting shear load and 2000 kN normal load direct shear test apparatus for constant normal load (CNL) and constant normal stiffness (CNS) condition. Figure 10.7(a) depicts the stress conditions of a CNL test where normal stress or load remains constant throughout of the test and the normal stiffness may vary. However, when a block of roof rock moves downward in an underground excavation, it shears with neighbouring block of rocks. The neighbouring blocks prohibit dilation and as a result, normal stress may increase with the increase in normal displacement. This phenomenon may resemble the condition of constant normal stiffness (CNS) test where normal movement of the upper block is restricted as shown in Figure 10.7(b). In this test, constant normal stiffness is maintained and thus normal load varies with shear movement. However, mechanics of shear movement of rock block in the field may lie in between these two extreme conditions.

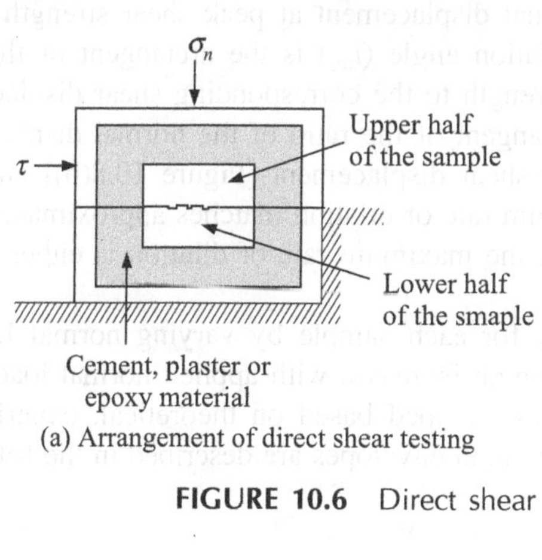

(a) Arrangement of direct shear testing

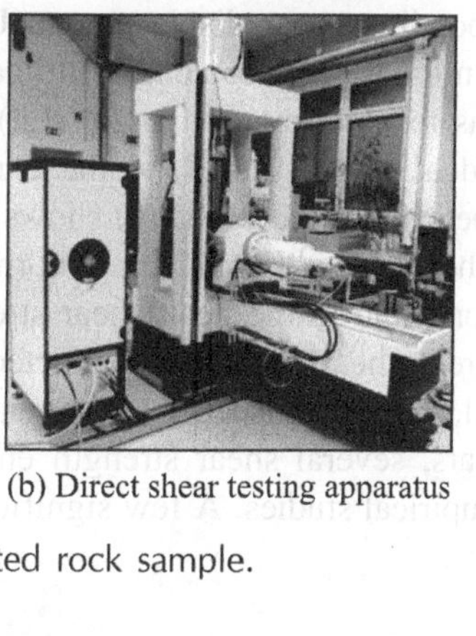
(b) Direct shear testing apparatus

FIGURE 10.6 Direct shear test of jointed rock sample.

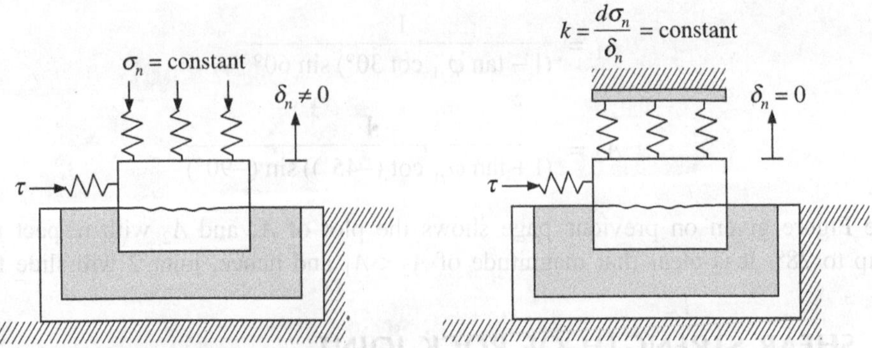

(a) Constant normal load (CNL)
(b) Constant normal stiffness (CNS)

FIGURE 10.7 Conditions for shear testing of rock joints.

A typical shear stress-shear displacement curve is shown in Figure 10.8(a) clearly marking the peak and residual strengths and corresponding displacements. Shear displacement across the joint is designated by δ_s and is the measure of relative displacement of the joint halves measured along the direction of the shear load. Average shear stress is estimated as the ratio of shear load to initial contact area between the upper and lower blocks. From direct shear tests, the normal and shear displacements during shearing are measured after each load increment. After the *peak shear strength* (τ_{peak}) is attained, asperities sheared off and hence, shear stress starts to decrease. After considerable shear movement, the minimum value of shear stress is achieved and it is termed as *residual shear strength* (τ_{res}). From the shear stress-shear displacement curve, shear stiffness, k_s is estimated as shown in Figure 10.8(a). Normal stiffness, k_n of joint surface is measured from the normal stress-normal displacement curve as shown in Figure 10.16. Note that this curve is non-linear and hence, k_n is a function of normal displacement or normal stress.

If the joint plane is rough or undulating, the joint will tend to contract or dilate during shearing. The term dilatancy, is the difference between normal displacements of upper and lower blocks as a result of shearing or shear displacement. In general, the rate of dilatancy is the highest at peak shear strength and measured as peak dilation angle (i_{peak}). Peak dilation angle is the arctangent of the ratio of the normal displacement at peak shear strength to the corresponding shear displacement. Ultimate dilation angle (i_{ult}) is the arctangent of the ratio of the normal displacement at ultimate shear strength to the corresponding shear displacement whereas residual dilation angle (i_{res}) is the arctangent of the ratio of the normal displacement at residual shear strength to the corresponding shear displacement [Figure 10.8(b)] and both must be same. Figure 10.8(c) shows the maximum rate of dilation matches approximately with peak shear stress. For different normal stresses, the maximum rate of dilation is either less or more or equal to the peak shear stresses.

Direct shear tests are conducted separately for each sample by varying normal load. In general, it is established that the peak shear strength increases with applied normal load. Over the years, several shear strength envelopes are developed based on theoretical, experimental and empirical studies. A few significant shear strength envelopes are described in the following sections.

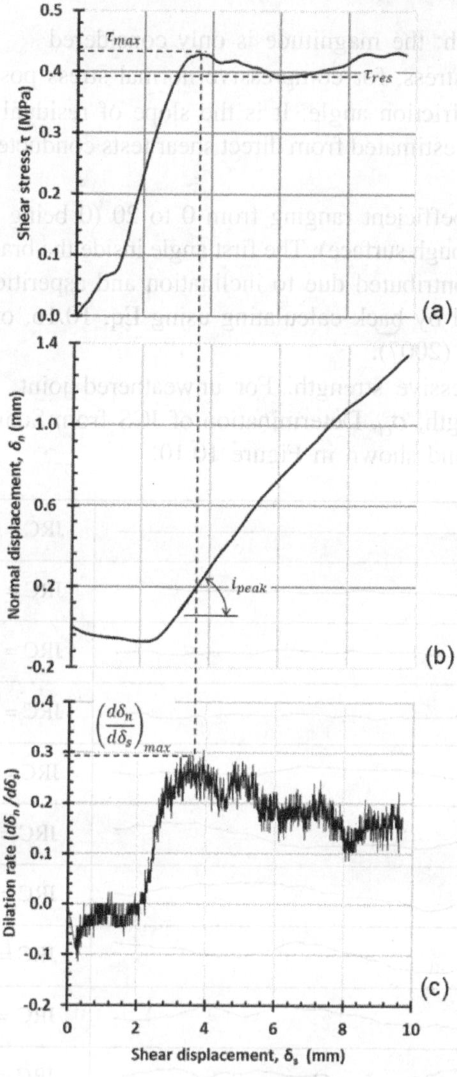

FIGURE 10.8 Relationship between (a) Shear stress-shear displacement, (b) Normal displacement-shear displacement and (c) dilation rate versus shear displacement.

10.3.1 Barton's Empirical Shear Strength Envelope

Barton (1976) proposed an empirical peak shear strength criterion of rock joints based on direct shear test results of natural and artificial specimen. The envelope of peak shear strength is dependent of normal stress, roughness of joint, joint wall strength and residual friction angle as given below:

$$\tau = \sigma_n \tan\left(\text{JRC}\log_{10}\left(\frac{\text{JCS}}{\sigma_n}\right) + \varphi_b\right) \tag{10.16}$$

where

τ = Peak shear strength; the magnitude is only considered

σ_n = Effective normal stress, for compressive normal stress positive value is used

φ_b = Basic or residual friction angle. It is the slope of residual shear strength and normal stress curve and is estimated from direct shear tests conducted on perfectly flat (smooth) joint surface.

JRC = Joint roughness coefficient ranging from 0 to 20 (0 being flat smooth surface and 20 being undulating rough surface). The first angle inside the bracket, i.e., JRC(Log_{10}JCS/σ_n) of Eq. 10.16 is contributed due to inclination and asperities of joint. In general, JRC can be determined by back calculating using Eq. 10.16, or by following a procedure outlined in ISRM (2007).

JCS = Joint wall compressive strength. For unweathered joint, it can be equal to unaxial compressive strength, σ_{ci}. Determination of JCS from Schmidt Hammer tests is given in ISRM (2007) and shown in Figure 10.10.

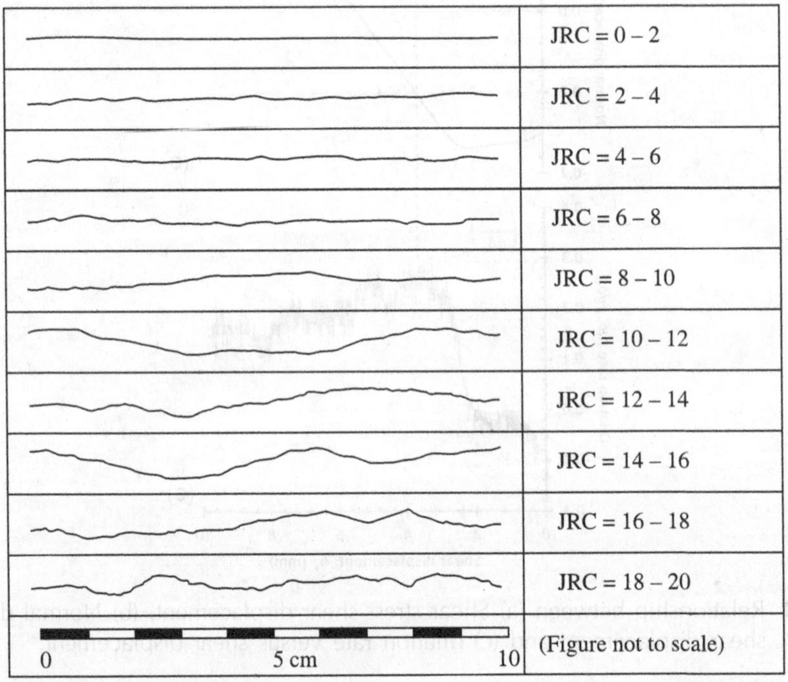

FIGURE 10.9 Roughness profile with corresponding JRC values (Hoek et al., 1998).

Equation 10.16 is used for low level of normal stress ($|\sigma_n| << \sigma_{ci}$) which means that the ratio of JCS/σ_n may vary from about 10^2 to 10^6 and is appropriate for geotechnical problems. However, Barton also proposed modified criterion for high normal stress and will be discussed later. The major features of Eq. 10.16 are summarised as follows:

(i) Barton (1976) commented that for very low effective normal stress, shear strength envelope (Fig 10.8(a)) for rough joints resembles vertical tangent at or close to the shear stress axis and a cohesion intercept is unlikely unless asperities are inclined at

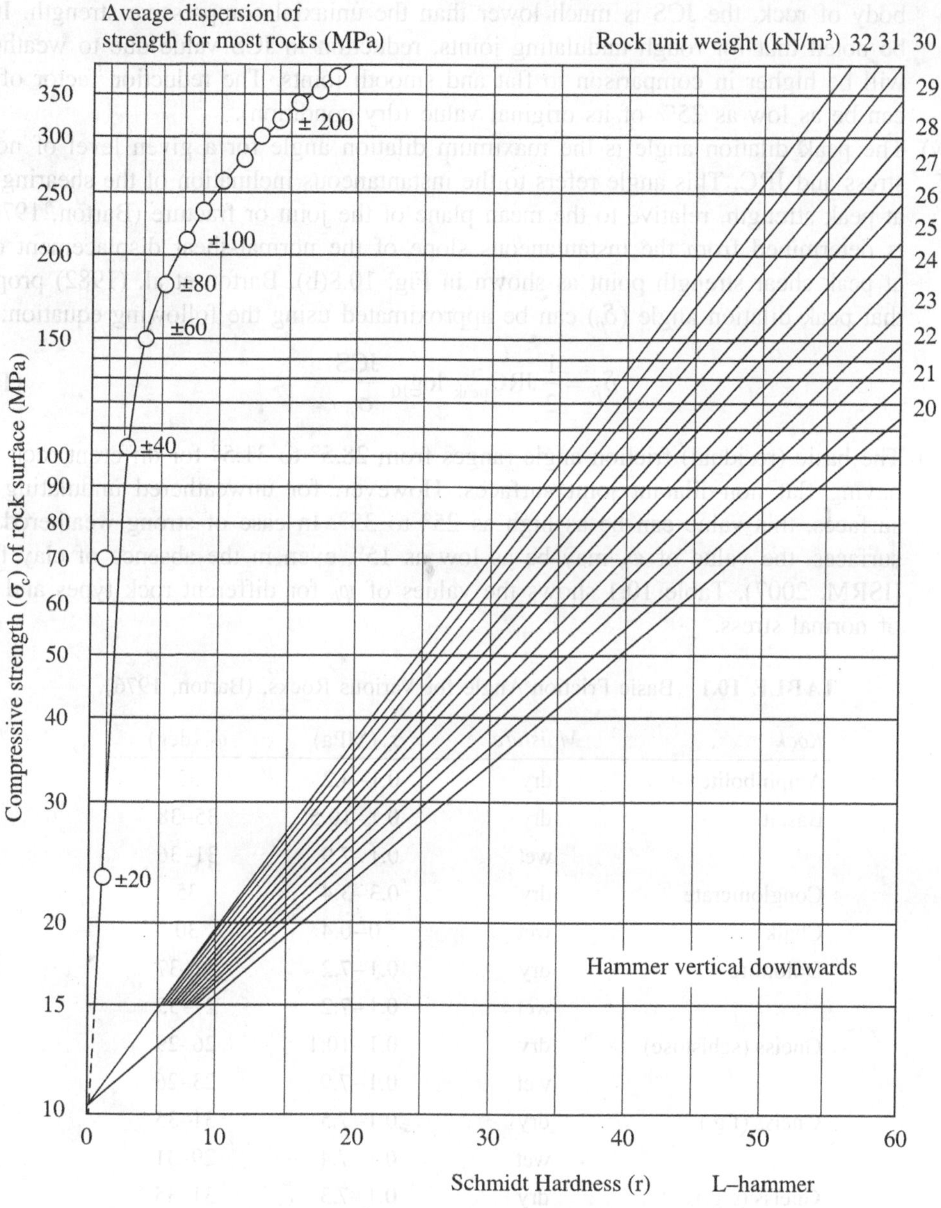

FIGURE 10.10 Relationship between compressive strength of rock surface and Schmidt hardness (*Source:* ISRM, 2007).

more than 60° to the mean joint plane. He has also proposed that cohesion intercept for peak shear strength is inherently dangerous even if the extrapolation is made from the mean effective normal stress level appropriate to the particular engineering problem.

(ii) JRC is the measures of joint roughness as mentioned before.

(iii) For unweathered and dry joint surface, JCS is replaced by the uniaxial compressive strength of rock. Since weathering is more prominent along the joint surface than the body of rock, the JCS is much lower than the uniaxial compressive strength. It can be noted that for rough undulating joints, reduction in JCS value due to weathering will be higher in comparison to flat and smooth joints. The reduction factor of JCS can be as low as 25% of its original value (dry condition).

(iv) The peak dilation angle is the maximum dilation angle for a given level of normal stress and JRC. This angle refers to the instantaneous inclination of the shearing path at peak strength, relative to the mean plane of the joint or fracture (Barton, 1976). It is determined from the instantaneous slope of the normal-shear displacement curve at peak shear strength point as shown in Fig. 10.8(b). Barton et al. (1982) proposed that peak dilation angle (δ_p) can be approximated using the following equation:

$$\delta_p = \frac{1}{2}\,\text{JRC}_{\text{peak}} \log_{10} \frac{\text{JCS}}{\sigma_n} \tag{10.17}$$

(v) The basic (residual) friction angle ranges from 28.5° to 31.5° for different rock types having flat non-dilatant joint surfaces. However, for unweathered undulating rock surfaces, this value can be as high as 25° to 35°. In case of strong weathered joint surfaces, the value of φ_b may be as low as 15°, even in the absence of clay filling (ISRM, 2007). Table 10.1 shows the values of φ_b for different rock types and level of normal stress.

TABLE 10.1 Basic Friction Angle for Various Rocks, (Barton, 1976)

Rock	*Moisture*	σ_n (MPa)	φ_b (deg)
Amphibolite	dry	0.1–4.2	32
Basalt	dry	0.1–10.5	35–38
	wet	0.1–7.9	31–36
Conglomerate	dry	0.3–3.4	35
Chalk	wet	0–0.4	30
Dolomite	dry	0.1–7.2	31–37
	wet	0.1–7.2	27–35
Gneiss (schistose)	dry	0.1–10.1	26–29
	wet	0.1–7.9	23–26
Gneiss (f.g.)	dry	0.1–7.5	31–35
	wet	0.1–7.4	29–31
Gneiss (c.g.)	dry	0.1–7.3	31–35
	wet	0.1–7.5	31–33
Limestone	dry	0–0.5	33–39
	wet	0–0.5	33–36
	dry	0.1–7.1	37–40
	wet	0.1–7.1	35–38
	dry	0.1–10.3	37–39
	wet	0.1–10.3	35

Rock	Moisture	σ_n (MPa)	φ_b (deg)
Porphyry	dry	0–1.0	31
	dry	4.1–13.3	31
Sandstone	dry	0–0.5	26–35
	wet	0–0.5	25–33
	wet	0–0.3	29
	dry	0.3–3.0	31–33
	dry	0.1–7.0	32–34
	wet	0.1–7.3	31–34
Shale	wet	0–0.3	27
Siltstone	wet	0–0.3	31
	dry	0.1–7.5	31–33
	wet	0.1–7.2	27–31
Slate	dry	0–1.1	25–30

Figure 10.11 reproduces peak shear stress-normal stress plots for the four sets of model joint tested by Bandis et al. (1981). This figure shows the peak shear strength envelope using Eq. 10.16 for $\varphi_b = 32°$ and UCS of model material as 2 MPa. The JRC values are back calculated as the best fit curve for JCS = 2 MPa. It can be seen that for higher JRC value, peak shear strength is higher for a given normal stress. Barton (1976) summarised that the values of $\tan^{-1}(\tau/\sigma_n)$ for some common rock type and is reproduced in Table 10.2.

For high normal stress level, i.e., $\sigma_n \geq \sigma_{ci}$, JCS in Eq. 10.16 is replaced by the confined compressive strength $(\sigma_1 - \sigma_3)$ and the Barton's peak shear strength envelope becomes

$$\tau = \sigma_n \tan\left[\text{JRC} \log_{10}\left(\frac{\sigma_1 - \sigma_3}{\sigma_n}\right) + \varphi_b\right] \tag{10.18}$$

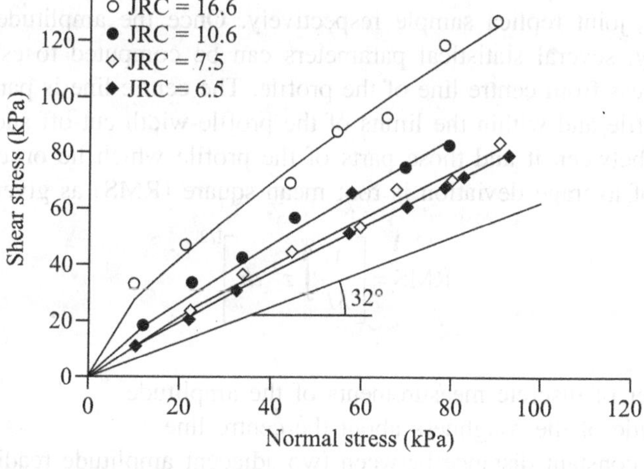

FIGURE 10.11 Peak shear strength envelopes for four sets of model joints. (*Data source:* Bandis et al., 1981).

TABLE 10.2 Maximum Values of Peak (τ/σ_n) Measured on Joints during Tests at Low Normal Stress (Barton, 1976)

Description of discontinuity	(τ/σ_n) (MPa/MPa)	arctan (τ/σ_n) deg
Limestone: Slightly rough bedding surface	0.68/0.16	77
	0.66/0.21	72
	1.68/0.60	71
Limestone: Rough bedding surface	0.68/0.31	66
	2.07/0.68	72
Shale: Closely jointed seam in limestone	0.06/0.02	71
	0.06/0.02	70
Quartzite, gneiss and amphibilite		
discontinuities beneath natural slopes	—	80
discontinuities beneath excavated slopes	—	75
Granite: rough, undulating, artificial extension fractures	0.45/0.15	72
	0.92/0.35	69

10.3.2 Roughness Parameters for Determination of JRC

As mentioned before, JRC can be estimated by experimental methods, Figure 10.9 and procedures outlined in ISRM (2007). Recently, several attempts have been made to estimate JRC from digital elevation model (DEM) of joint surfaces (Indraratna and Haque, 2000; Yu et al., 1991). In this method, a 3D profiler, either laser or imagery based, is used to capture three-dimensional surface of joint plane. A laser-based 3D profiler and associated software have been developed in Mining Engineering Department of IIT Kharagpur (Indian patent no. 372999, Dated 28–07–2021) to generate 3D digital surface of any rock joint of nominal area upto 15 cm × 15 cm. Figure 10.12(a) and (b) depicts the 3D laser-based non-contact type profiler and DEM of a natural or synthetic joint replica sample respectively. Once the amplitude of the roughness is captured digitally, several statistical parameters can be computed to estimate the average deviation of roughness from centre line of the profile. The centre line is parallel to the general direction of the profile and within the limits of the profile-width cut-off such that the sums of the area contained between it and those parts of the profile which lie on either side of it are equal. A measure of average deviation is root mean square (RMS) as given below:

$$\text{RMS} = \left[\frac{1}{M} \int_0^M z^2 \, dx \right]^{1/2} \tag{10.19}$$

where

M is the number of discrete measurements of the amplitude
z is the amplitude of the roughness about the centre line
dx is the small constant distance between two adjacent amplitude readings in x-direction.

It may be noted that x and y coordinates denote the joint plane and z signifies amplitude of roughness as shown in Fig. 10.12(a). Besides RMS, a number of additional parameters have been suggested for describing profile roughness; examples are K (absolute roughness), CLA (center line average), R_p (profile index), R_a (average roughness) Z_2, SF (structure function), RL (real profile length), RL_1, SD_H (standard deviation of height), SD_i (standard deviation of roughness angle i) and others. However, out of these parameters, Z_2, SF and SD_i are found to be suitable for application with joint roughness. Z_2 is unit less, scale dependent and more appropriate for primary asperity of the joint surface. These parameters are defined below.

$$Z_2 = \left(\frac{1}{L}\int_0^L \left(\frac{dz}{dx}\right)^2 dx\right)^{1/2} = \left(\frac{1}{L}\sum_{i=1}^{M-1}\left[\frac{z_{i+1}-z_i}{x_{i+1}-x_i}\right]^2 \Delta x_i\right)^{1/2} \tag{10.20}$$

$$SD_i = \tan^{-1}\left[\frac{1}{L}\int_0^L \left(\frac{dz}{dx} - \tan i_{ave}\right)^2 dx\right]^{1/2} = \tan^{-1}\left[\frac{1}{L}\sum_{i=1}^{M-1}\left(\frac{(z_{i+1}-z_i)}{(x_{i+1}-x_i)} - \tan i_{ave}\right)^2 \Delta x_i\right]^{1/2} \tag{10.21}$$

$$SF = \frac{1}{M}\sum_{i=1}^{M-1}(z_{i+1}-z_i)^2 \Delta x_i \tag{10.22}$$

where

$$i_{\text{ave}} = \frac{1}{L}\int_0^L \tan^{-1}\left(\frac{dz}{dx}\right) dx = \frac{1}{L}\left(\sum_{i=1}^{M-1} \tan^{-1}\left[\frac{z_{i+1}-z_i}{x_{i+1}-x_i}\right]\right)$$

L = Distance over which average is taken and M is number of sampling points.

Once these parameters are estimated along all profiles along x-direction by increasing the interval in y-direction, they can be averaged to get the roughness parameters of the entire sample. Based on the extensive experimental studies, Yu et al. (1991) provided correlation between roughness parameters and JRC based on sampling interval (Δx_i) of 0.25 mm, 0.50 mm and 1.0 mm as given in Table 10.3 (Hong et al., 2008).

TABLE 10.3 Relationship between Roughness Parameters and JRC (Hong et al., 2008)

Δx_i = 0.25 mm	Δx_i = 0.50 mm	Δx_i = 1.0 mm
JRC = 60.32 (Z_2) – 4.51	JRC = 61.79 (Z_2) – 3.47	JRC = 64.22 (Z_2) – 2.31
JRC = 1.12 (SD_i) – 5.06	JRC = 1.14 (SD_i) – 3.88	JRC = 1.17 (SD_i) – 2.57
JRC = 239.27($\sqrt{SF}$) – 4.51	JRC = 121.13($\sqrt{SF}$) – 3.28	JRC = 63.69($\sqrt{SF}$) – 2.31

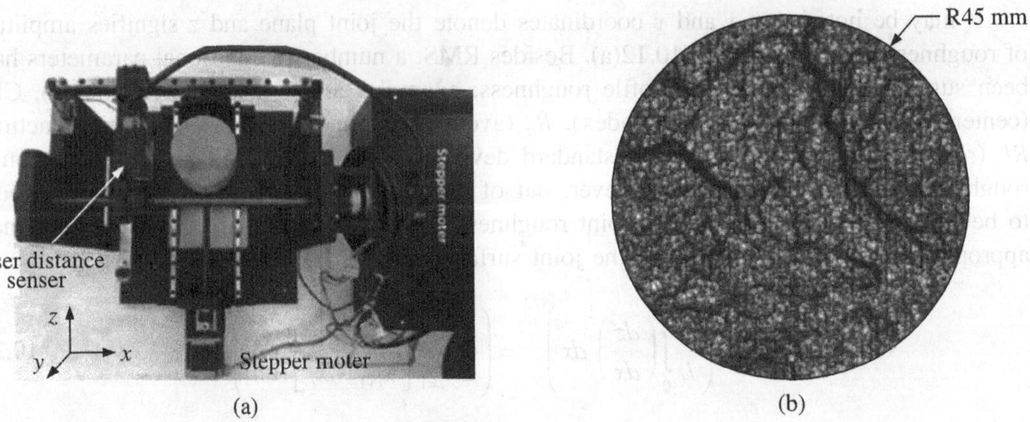

FIGURE 10.12 (a) 3D laser-based surface profiler, and (b) Surface roughness of a natural/synthetic joint replica (Kumar, 2016).

EXAMPLE 10.2 Surface roughness of a joint profile is measured using non-contact type profilometer as shown in the following figure. Determine JRC using Table 10.3.

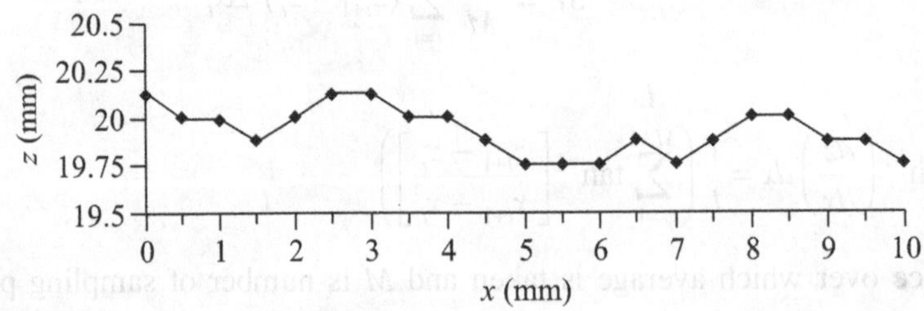

Solution: Using Eqs. 10.20 to 10.22, the following table is prepared.

x (mm)	z (mm)	dz/dx	$(dz/dx)^2 \times 0.5$	$\tan^{-1}(dz/dx)$	$((dz/dx) - \tan i_{ave})^2 \times 0.5$	$(z_{i+1} - z_i)^2 \times 0.5$
0	20.125					
0.5	20	–0.25	0.03125	0.12249	0.084294	0.007813
1	20	0	0	0.00000	0.012896	0.000000
1.5	19.875	–0.25	0.03125	0.12249	0.084294	0.007813
2	20	0.25	0.03125	0.12249	0.003997	0.007813
2.5	20.125	0.25	0.03125	0.12249	0.003997	0.007813
3	20.125	0	0	0.00000	0.012896	0.000000
3.5	20	–0.25	0.03125	0.12249	0.084294	0.007813
4	20	0	0	0.00000	0.012896	0.000000
4.5	19.875	–0.25	0.03125	0.12249	0.084294	0.007813
5	19.75	–0.25	0.03125	0.12249	0.084294	0.007813
5.5	19.75	0	0	0.00000	0.012896	0.000000

(*Contd.*)

x (mm)	z (mm)	dz/dx	$(dz/dx)^2 \times 0.5$	$\tan^{-1}(dz/dx)$	$((dz/dx) - \tan i_{ave})^2 \times 0.5$	$(z_{i+1} - z_i)^2 \times 0.5$
6	19.75	0	0	0.00000	0.012896	0.000000
6.5	19.875	0.25	0.03125	0.12249	0.003997	0.007813
7	19.75	–0.25	0.03125	0.12249	0.084294	0.007813
7.5	19.875	0.25	0.03125	0.12249	0.003997	0.007813
8	20	0.25	0.03125	0.12249	0.003997	0.007813
8.5	20	0	0	0.00000	0.012896	0.000000
9	19.875	–0.25	0.03125	0.12249	0.084294	0.007813
9.5	19.875	0	0	0.00000	0.012896	0.000000
10	19.75	–0.25	0.03125	0.12249	0.084294	0.007813
			$Z_2 = 0.5244$	$i_{ave} = 0.1592$	$SD_i = 15.648$	$SF = 0.00507$
			JRC = 8.98		JRC = 13.958	JRC = 5.345

10.3.3 Scale Effect on JRC and JCS

Effect of sample size on JRC and JCS were investigated by Barton et al. (1982), Bandis et al. (1981) and Verma et al. (2015). It was found that peak shear strength and JRC reduces with larger profile of the same joint. The JCS also reduces with increasing size of the sample. On the basis of extensive laboratory testing of natural and artificial joints and literature reviews, Barton and Bandis (1982) proposed two relationships for scale effect of JRC and JCS as given below:

$$\text{JRC}_n = \text{JRC}_0 \left(\frac{L_n}{L_0} \right)^{-0.02\text{JRC}_0} \tag{10.23a}$$

$$\text{JCS}_n = \text{JCS}_0 \left(\frac{L_n}{L_0} \right)^{-0.03\text{JRC}_0} \tag{10.23b}$$

where JRC_0 refers to the value of JRC for a laboratory sample of 100 mm length (L_0). JRC_n, JCS_n refer to the *in situ* block sample of length L_n.

EXAMPLE 10.3 Estimate the value of JRC for the data obtained from direct shear test of 10 cm square samples as given below:

Normal stress (kPa)	*Exp. peak shear strength* (kPa), τ_e
0	0
10	25
20	43
30	64
40	74
50	88

Assume compressive strength and residual angle of the model material are 4 MPa and 30° respectively.

Solution: By trial and error method, the error of estimation of peak shear stress is estimated as shown in the table below. JRC values are changed gradually to find the minimum error. In this case, it is found that the error is minimum for JRC = 15.10.

Normal stress (kPa)	*Exp. peak shear strength* (kPa), τ_e	*Peak shear stress* (kPa), τ_b	*Error* $(\tau_e - \tau_b)^2$
0	0	0.0	0
10	25	29.2	17.880
20	43	45.7	7.199
30	64	60.4	13.207
40	74	74.0	0.000
50	88	86.8	1.338
		Sum error =	**39.625**

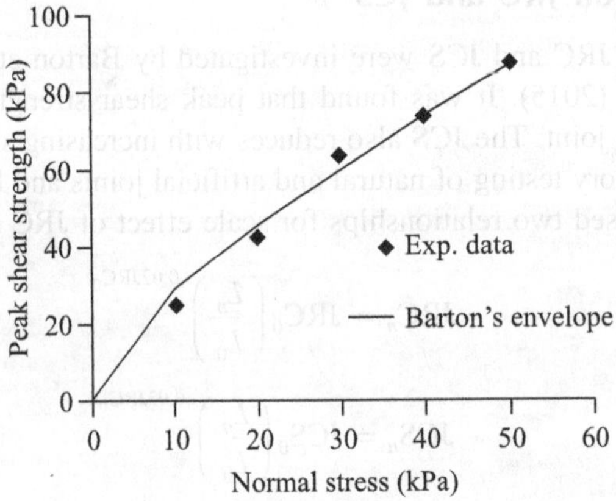

10.3.4 Ladanyi and Archambault's Shear-Strength Envelope

Ladanyi and Archambault (1969) proposed the shear strength envelope as given below (Goodman et al., 1977; Lama and Vutukuri, 1978)

$$\tau = \frac{\sigma_n(1-a_s)(\tan\delta + \tan\varphi_j) + a_s S_R}{1-(1-a_s)\tan\delta\tan\varphi_j} \tag{10.24}$$

where tan δ is the dilation rate at failure and varies with normal stress as given in Eq. 10.33. The shear area ratio of contact, a_s is shown in Figure 10.13 and is expressed empirically as:

$$a_s = 1 - \left(1 - \frac{\sigma_n}{\sigma_T}\right)^{3/2} \quad \text{if } \sigma_n < \sigma_T \tag{10.25}$$

where σ_T can be replaced by UCS of rock. The shear strength S_R of rock varies with σ_n. It can be expressed using Mohr–Coulomb parameters as:

$$S_R = c_o + \sigma_n \tan \varphi_o \tag{10.26}$$

where c_o and φ_o are cohesion and friction angle of rock substance.

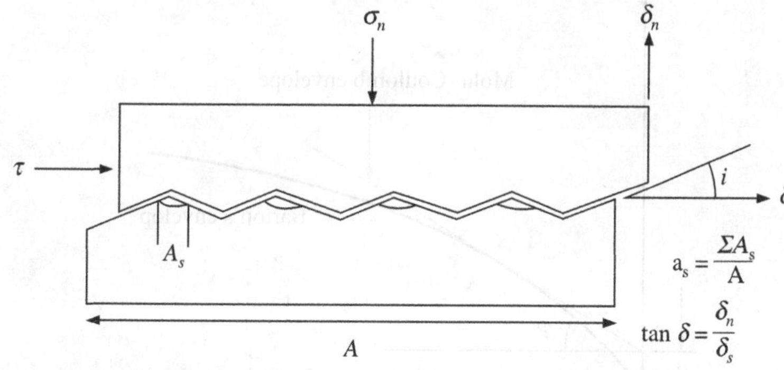

FIGURE 10.13 Definition of dilation rate and shear area ratio.

10.3.5 Mohr–Coulomb Shear Strength Envelope

Mohr–Coulomb shear strength criterion, as described in Chapter 9 for intact or isotropic rocks, is directly applicable for rock joints. The failure criterion for rock joints is expressed in Eq. 10.1. The cohesion intercept and friction angle of joint surface are expressed as c_j and φ_j respectively. A few differences are noticeable between Mohr–Coulomb and Barton's peak shear strength envelopes as given below:

(i) Cohesion intercept is present
(ii) Base or residual friction angle is absent
(iii) Measures of joint roughness is imbedded in the friction angle
(iv) The friction angle in Barton's envelope is dependent on the magnitude of normal stress and thus the envelope becomes non-linear

For elastic-plastic analysis of joint surface, Mohr–Coulomb criterion is being popularly applied due to its simplicity as compared to Barton's envelope.

10.3.6 Instantaneous Cohesion and Friction Angle from Barton's Envelope

Hoek (2007) has derived relationships for estimating instantaneous cohesion intercept and friction angle from Barton's peak shear strength envelope. Instantaneous cohesion intercept and friction angle are obtained by drawing a tangent to the Barton's shear strength envelope for a specified normal stress as shown in Figure 10.14. Mathematically, from Barton's peak shear strength envelope, instantaneous friction angle is obtained as given in Eq. 10.27.

$$\varphi_j = \tan^{-1}\left(\frac{\partial \tau}{\partial \sigma_n}\right) \tag{10.27}$$

From Barton's envelope and using scale effect for both JRC and JCS (Eq. 10.23), we find

$$\frac{\partial \tau}{\partial \sigma_n} = \tan\left(\mathrm{JRC}_n \log_{10}\frac{\mathrm{JCS}_n}{\sigma_n} + \varphi_b\right) - \frac{\pi \mathrm{JRC}_n}{180 \ln 10}\left[\tan^2\left(\mathrm{JRC}_n \log_{10}\frac{\mathrm{JCS}_n}{\sigma_n} + \varphi_b\right) + 1\right] \quad (10.28)$$

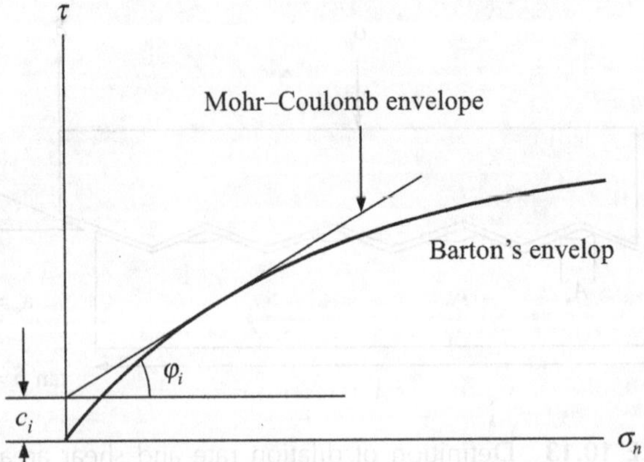

FIGURE 10.14 Determination instantaneous cohesion intercept and friction angle from Barton's envelope.

Once the instantaneous friction angle is found, cohesion intercept can be obtained as:

$$c_i = \tau - \sigma_n \tan \varphi_i \quad (10.29)$$

Hence c_i and φ_i values obtained from Eqs. 10.27 and 10.29, Mohr–Coulomb criterion can be applied as given in Eq. 10.1.

EXAMPLE 10.4 Based on the Barton's envelope obtained in Example 10.3, determine the instantaneous cohesion and friction angle for σ_n = 35 kPa. Assume that, the size of the sample is 10 cm.

Solution: Since the size of the sample is 10 cm, scale effect of JRC and JCS will not be applied. Using Eqs. 10.27 and 10.28, we find the instantaneous friction angle φ_i = 53.67°.

Shear stress for σ_n = 35 kPa is calculated using Eq. 10.16 and found to be τ = 67.28 kPa.

Hence, using Eq. 10.29, instantaneous cohesion will be c_i = 19.68 kPa.

10.4 DEFORMABILITY OF ROCK JOINT

Deformation of joint surface occurs due to both normal and shear stress. The values of deformations are dependent on normal and shear stiffness of the joint surface. In the following sections, the concept and application of joint normal and shear stiffness along with compliance characteristics of joint surface will be discussed.

10.4.1 Joint Normal Deformation

Consider a sufficiently large block of rock which is subjected to normal compression load as shown in Figure 10.15. A typical normal stress versus normal deformation curve is shown by curve A in Figure 10.16(a). It shows that at the beginning of the loading process, for a slight change in normal stress large change in normal deformation may occur. It implies low normal stiffness of rock material at the initial stages of loading. After a while, resistance offered by the block increases rapidly since particles coagulate due to compressional loading. At this time, normal stiffness is high since high normal stress is required to deform the block as interparticle frictional resistance increases due to coagulation. Curve B in Figure 10.16(a) also shows the similar trend between normal stress and normal deformation curve if a horizontal joint exists in the sample. In this case, joint deformation is added with the deformation of rock material. Hence, by subtracting curve A from curve B in Figure 10.16, deformation of joint surface can be obtained as shown by curve C. Figure 10.16(b) shows the normal stress versus normal deformation plot for a natural joint replica as tested in laboratory.

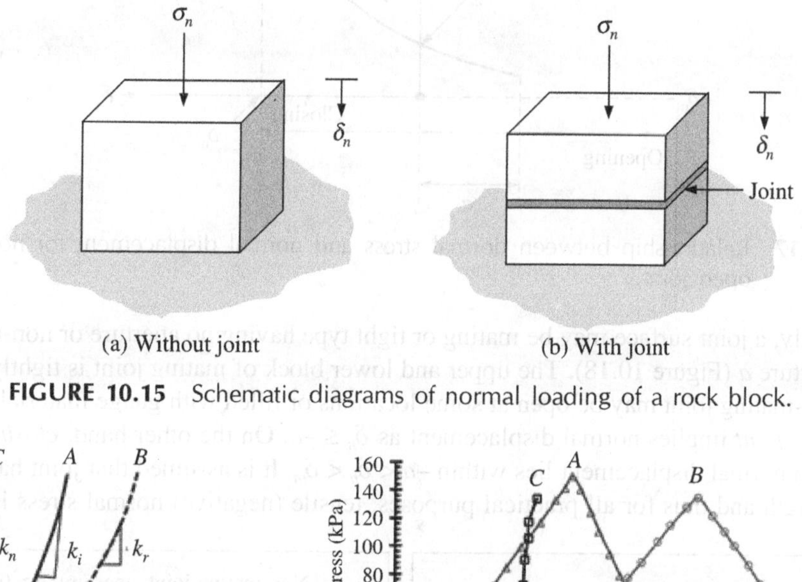

FIGURE 10.15 Schematic diagrams of normal loading of a rock block.

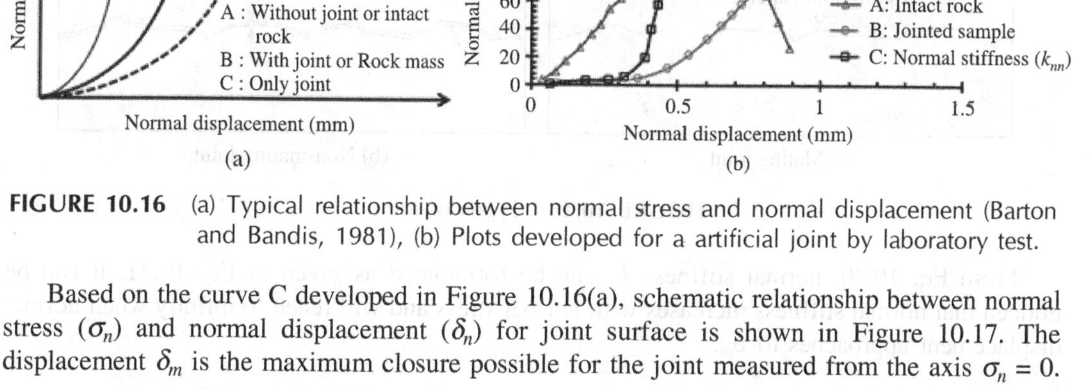

FIGURE 10.16 (a) Typical relationship between normal stress and normal displacement (Barton and Bandis, 1981), (b) Plots developed for a artificial joint by laboratory test.

Based on the curve C developed in Figure 10.16(a), schematic relationship between normal stress (σ_n) and normal displacement (δ_n) for joint surface is shown in Figure 10.17. The displacement δ_m is the maximum closure possible for the joint measured from the axis $\sigma_n = 0$.

The initial normal stress σ_{n0} is measured when $\delta_n = 0$. The relationship shown in the Figure resembles hyperbolic curve as given below (Goodman et al., 1977, Bandis et al., 1983):

$$\sigma_n = \left(\frac{\delta_n}{\delta_m - \delta_n} + 1\right)\sigma_{n0} \tag{10.30}$$

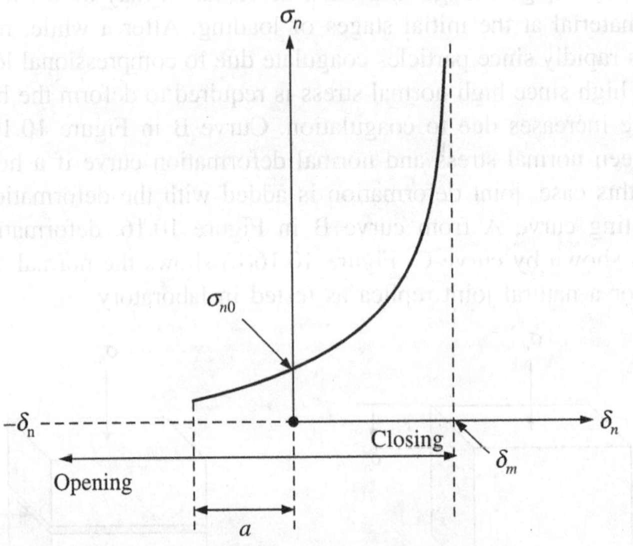

FIGURE 10.17 Relationship between normal stress and normal displacement for non-mating or open joint.

Generally, a joint surface may be mating or tight type having no aperture or non-mating with a finite aperture *a* (Figure 10.18). The upper and lower block of mating joint is tightly connected while a non-mating joint may be open at some locations or filled with gauge material. Therefore, *opening of a joint* implies normal displacement as $\delta_n \leq -a$. On the other hand, *closing of a joint* occurs when normal displacement lies within $-a < \delta_n < \delta_m$. It is assumed that joint has negligible tensile strength and thus for all practical purposes, tensile (negative) normal stress is neglected.

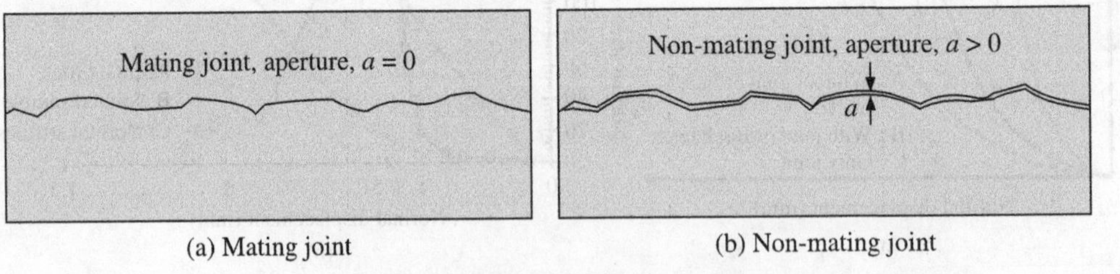

FIGURE 10.18 Types of joint.

From Eq. 10.30, normal stiffness k_n can be formulated as given in Eq. 10.31. It can be noticed that normal stiffness increases with normal stress and will reach to infinity when normal displacement approaches to δ_m.

$$k_n = \frac{\partial \sigma_n}{\partial \delta_n} = \frac{\sigma_n}{\delta_m - \delta_n} \tag{10.31}$$

Hence, the normal stiffness at $\delta_n = 0$ and $\sigma_n = \sigma_{n0}$ is $(k_n)_0 = \dfrac{\sigma_{n0}}{\delta_m}$.

EXAMPLE 10.5 Plot the normal stress vs. normal displacement curve if the magnitude of initial sitting stress and initial normal stress equal to 0.1 MPa. Assume that the maximum closure is 4 mm.

Solution: Using Eq. 10.30, we can plot the normal stress-normal displacement plot as shown below.

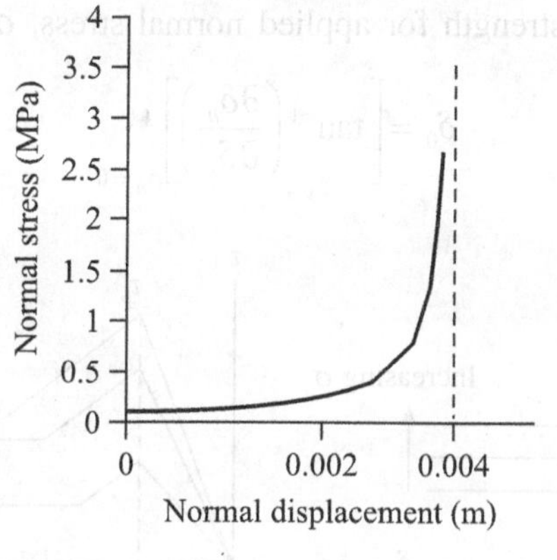

10.4.2 Joint Shear Deformation and Dilatancy

Consider again the shear stress-shear displacement curve shown in Figure 10.8(a) and 10.8(c). It can be seen that shear stress-shear displacement curve has three distinct phases: (i) pre-peak, (ii) peak, and (iii) post-peak. At the onset of shearing, joint contracts first and then dilates having a maximum rate of dilation at the peak shear strength. The peak shear strength, shear stiffness and displacement at peak shear strength are functions of normal stress. It means that peak shear strength increases with applied normal stress but dilation tends to decrease with increasing normal stress. According to Goodman (1977), there are two types of shear stress-shear displacement models: (i) with constant stiffness; where peak and residual displacement changes with normal stress, and (ii) constant displacement; stiffness during pre-peak shearing phase changes with normal stress. Schematic diagrams are plotted to elaborate these two models as shown in Figure 10.19(a) and 10.19(b). For both the models, the following relationship between shear stress and shear displacement is applicable

$$\tau = k_s u \qquad \text{for } \delta < \delta_p \tag{10.32a}$$

$$\tau = \left(\frac{\tau_p - \tau_r}{\delta_p - \delta_r}\right)\delta_s + \left(\frac{\tau_r\delta_p - \tau_p\delta_r}{\delta_p - \delta_r}\right) \qquad \text{for } \delta_p \le \delta \le \delta_r \tag{10.32b}$$

$$\tau = \tau_r \qquad \text{for } \delta > \delta_r \tag{10.32c}$$

In order to quantify the rate of dilation, Ladanyi and Archambault proposed that secant rate of dilatancy which is represented by

$$\frac{\partial \delta_n}{\partial \delta_s} = \tan\delta = \left(1 - \frac{\sigma_n}{\sigma_T}\right)^k \tan\delta_0 \qquad \text{for } \delta \le \delta_r \text{ and } \sigma_n < \sigma_T \tag{10.33}$$

$$\frac{\partial \delta_n}{\partial \delta_s} = 0 \qquad \text{for } \delta > \delta_r \text{ or } \sigma_n \ge \sigma_T \tag{10.34}$$

where σ_T is transitional stress, k an independent material constant and can be obtained from the experimental results (Saeb et al., 1992). However, σ_T is generally assumed as the uniaxial compressive strength of intact rock. The constant k is assumed as 4.0. The angle δ_0 is the dilation angle at peak shear strength for applied normal stress, $\sigma_n = 0$ and is given as:

$$\delta_0 = \left[\tan^{-1}\left(\frac{\partial \delta_n}{\partial \delta_s}\right)\right]_{\sigma_n = 0} \tag{10.35}$$

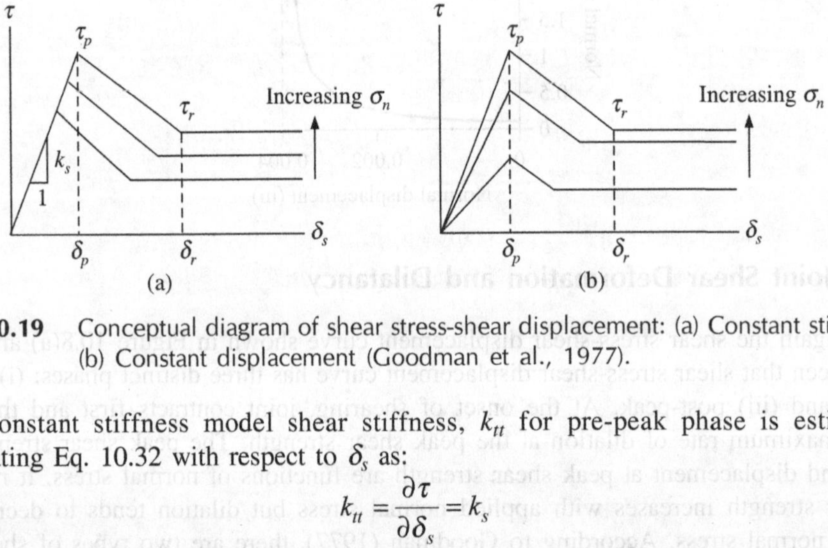

FIGURE 10.19 Conceptual diagram of shear stress-shear displacement: (a) Constant stiffness, and (b) Constant displacement (Goodman et al., 1977).

For constant stiffness model shear stiffness, k_{tt} for pre-peak phase is estimated by differentiating Eq. 10.32 with respect to δ_s as:

$$k_{tt} = \frac{\partial \tau}{\partial \delta_s} = k_s \tag{10.36}$$

Although dilatancy in joint relates normal displacement to shear displacement, for all practical purpose, it is assumed that along the joint surface, shear stress and normal stress occur only due to shear displacement and normal displacement respectively. Hence, the relationship between incremental stresses and displacements becomes

$$\begin{Bmatrix} d\tau \\ d\sigma_n \end{Bmatrix} = \begin{bmatrix} k_{tt} & 0 \\ 0 & k_{nn} \end{bmatrix} \begin{Bmatrix} d\delta_s \\ d\delta_n \end{Bmatrix} \tag{10.37}$$

or

$$\Delta\sigma = k\Delta q \tag{10.38}$$

10.5 CONCEPT OF JOINT COMPLIANCE

Consider a rock mass having n number of horizontal mated joints of negligible thickness as shown in Figure 10.20(a). Joint frequency f is defined as the number of joints per unit length of rock mass.

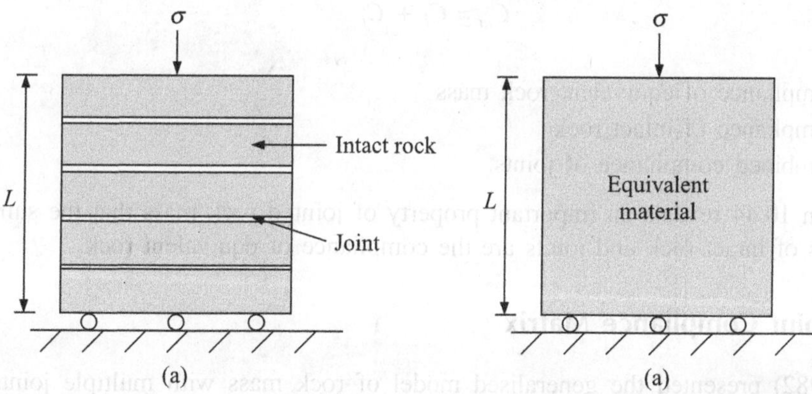

FIGURE 10.20 Concept of equivalent material (a) Rock mass, and (b) Equivalent material.

Hence in this case, $f = n/L$, where L denotes the length of rock mass. The intact rock is the rock matrix between two consecutive joint planes. Let us assume that normal stress σ acts on the rock mass as shown in the figure. It is obvious that the normal stress on the joint plane and on the intact rock is the same. Then the deformation of intact rock, d_i and that of joints, d_j can be obtained as:

$$d_i = \frac{\sigma L}{E_i} \tag{10.39}$$

$$d_j = \frac{\sigma}{k_1} + \frac{\sigma}{k_2} + \ldots + \frac{\sigma}{k_n} = \sigma \sum_{m=1}^{n} \frac{1}{k_m} \tag{10.40}$$

where, E_i is the modulus of elasticity of the intact rock and k_m is the normal stiffness of mth joint. If all normal stiffness are equal as $k_1 = k_2 = \ldots = k_n = k$, then the total displacement including joints and intact rocks is given as follows:

$$d = d_i + d_j = \sigma L\left(\frac{1}{E_i} + \frac{f}{k}\right) \tag{10.41}$$

Now, consider the rock mass can be replaced by an equivalent isotropic and homogeneous rock having equal dimensions and elastic modulus of E_m so that the displacement incurred by the equivalent rock is the same as given in Eq. 10.41 [Figure 10.20(b)]. Hence, displacement occurred in the equivalent material due to normal stress σ can be expressed by

$$d = \frac{\sigma L}{E_m} \tag{10.42}$$

Then the equivalent elastic modulus is obtained by equating Eqs. 10.41 and 10.42 as:

$$\frac{1}{E_m} = \frac{1}{E_i} + \frac{f}{k} \tag{10.43}$$

Equation 10.43 can also be written in the form of compliances as:

$$C_m = C_i + C_j \tag{10.44}$$

where

C_m = compliance of equivalent rock mass
C_i = compliance of intact rock
C_j = combined compliance of joints.

Equation 10.44 reveals an important property of jointed rock mass that the summation of compliances of intact rock and joints are the compliance of equivalent rock.

10.5.1 Joint Compliance Matrix

Gerrard (1982) presented the generalised model of rock mass with multiple joints and also showed examples of special cases of rock mass properties based on compliance matrix of joints. Based on this study and concept given in Pande et al. (1990), joint compliance matrices are determined for both two and three dimensional joints. First, consider a joint plane in two dimensions as shown in Figure 10.21. The joint plane is aligned along the x'-axis and the y'-axis represents the normal direction to that plane. Let us assume that displacement and stress fields along shear and normal directions are δ_s' and δ_n', and $\tau_{x'y'}$ and $\sigma_{y'y'}$ respectively. Thus the relationship between displacements and stresses can be written with joint compliance matrix in the local coordinate system as:

$$\begin{Bmatrix} \delta_n' \\ \delta_s' \end{Bmatrix} = \begin{bmatrix} s_{11} & s_{12} \\ s_{21} & s_{22} \end{bmatrix} \begin{Bmatrix} \sigma_{y'y'} \\ \tau_{x'y'} \end{Bmatrix} \tag{10.45}$$

$$\mathbf{q}' = \mathbf{C}'\boldsymbol{\sigma}' \tag{10.46}$$

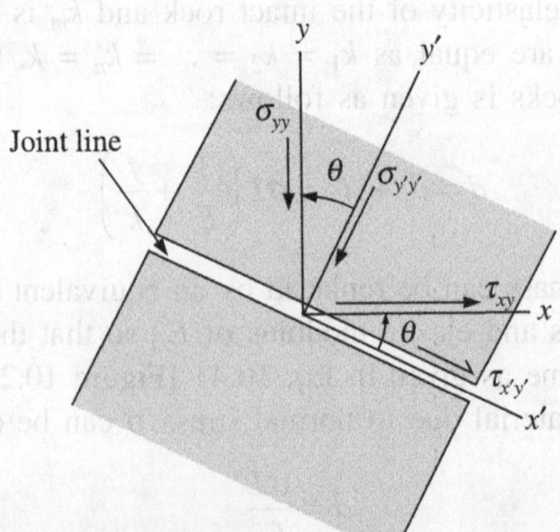

FIGURE 10.21 Transformation of joint compliance matrix.

The compliance matrix, **C**′ can be non-symmetric. However, in practise, the off-diagonal terms (s_{12} and s_{21}) are difficult to estimate and thus, normally they are assumed to be zero. In this case, the unit of compliance components s_{11} and s_{22} is m/Pa and expressed as:

$$s_{11} = \frac{1}{k_n},\ s_{22} = \frac{1}{k_{tt}} \tag{10.47}$$

However, components of the compliance matrix having relationship between strain and stress will be

$$c_{nn} = \frac{1}{dk_n},\ c_{tt} = \frac{1}{dk_{tt}} \tag{10.48}$$

where, $d = 1/f$ = average distance between two consecutive joint planes. The compliance matrix is then generalised by adding another two rows and columns with zeros to represent two-dimensional compliance matrix for plane stress or plane strain conditions as given in Eq. 10.49.

$$\mathbf{C}' = \begin{bmatrix} 0 & 0 & 0 & 0 \\ 0 & c_n & 0 & 0 \\ 0 & 0 & c_s & 0 \\ 0 & 0 & 0 & 0 \end{bmatrix} \tag{10.49}$$

For three-dimensional problems, the joint plane is represented by two perpendicular axes x' and y' and the z'-axis denotes the normal direction to the joint plane. Hence, $\sigma_{z'z'}$ represents the normal stress and $\tau_{x'z'}$ and $\tau_{y'z'}$ signify the shear stresses in x' and y' directions respectively. Now, considering the displacements in the normal direction (z'-axis) and shear directions (x' and y'-axes) as δ'_n, δ'_x and δ'_y respectively, displacement and stress vector is related with compliance matrix as given below:

$$\begin{Bmatrix} \delta'_n \\ \delta'_y \\ \delta'_x \end{Bmatrix} = \begin{bmatrix} s_{11} & 0 & 0 \\ 0 & s_{22} & 0 \\ 0 & 0 & s_{33} \end{bmatrix} \begin{Bmatrix} \sigma_{z'z'} \\ \tau_{y'z'} \\ \tau_{x'z'} \end{Bmatrix} \tag{10.50}$$

where $s_{11} = \dfrac{1}{k_n}$, $s_{22} = \dfrac{1}{k_{sy'}}$, $s_{33} = \dfrac{1}{k_{sx'}}$

The parameters $k_{ty'}$ and $k_{tx'}$ are the shear stiffness along y' and x' directions respectively. For general three-dimensional stress-strain relations in Cartesian coordinate system, the dimension of compliance matrix is 6 by 6 as illustrated as Eq. 5.11 in Chapter 5. In order to keep the parity with that notion, another 3 rows and 3 columns of zeros are added in the compliance matrix as given in Eq. 10.51.

$$C' = \begin{bmatrix} 0 & 0 & 0 & 0 & 0 & 0 \\ 0 & 0 & 0 & 0 & 0 & 0 \\ 0 & 0 & c_n & 0 & 0 & 0 \\ 0 & 0 & 0 & 0 & 0 & 0 \\ 0 & 0 & 0 & 0 & c_{sy} & 0 \\ 0 & 0 & 0 & 0 & 0 & c_{sx} \end{bmatrix} \tag{10.51}$$

Note that in Eq. 10.51, $c_n = \frac{s_{11}}{d} c_{sy} = \frac{s_{22}}{d} c_{sx} = \frac{s_{33}}{d}$.

10.5.2 Equivalent Compliance for Joints Sets

Joint compliance matrix is formed in local (joint) coordinate system. Then this compliance matrix is transformed into global coordinate system. If multiple joint sets are present in rock mass, individual joint compliance matrix of each joint set is transformed into global coordinate system and then they are added together along with compliance matrix of rock substance.

10.5.3 Transformation of Joint Compliance Matrix

Refer to Figure 10.21 where x and y axes define the global coordinate system. Joint compliance matrix, $\mathbf{C}'$ (Eq. 10.49) formed in x'–y' plane is transformed to compliance matrix $\mathbf{C}$ in x–y plane. As mentioned in Section 4.3.4, the strain energy of the body remains same no matter what coordinate system is used for calculation. Let us consider $\delta\boldsymbol{\sigma}$ and $\delta\boldsymbol{\sigma}'$ represent virtual stress tensor in x–y and x'–y' system respectively. Then virtual strain energy estimated along these two systems can be equated as:

$$(\delta\boldsymbol{\sigma})^T\boldsymbol{\varepsilon} = (\delta\boldsymbol{\sigma}')^T\boldsymbol{\varepsilon}' \tag{10.52}$$

Recalling $\boldsymbol{\varepsilon} = \mathbf{C}\boldsymbol{\sigma}$ and $\boldsymbol{\sigma}' = \mathbf{T}\boldsymbol{\sigma}$, where $\mathbf{T}$ is the transformation matrix defined in Eq. 4.28 and is reproduced in Eq. 10.53, Eq. 10.52 is rewritten as given in Eq. 10.54.

$$\mathbf{T} = \begin{bmatrix} c^2 & s^2 & 2cs & 0 \\ s^2 & c^2 & -2cs & 0 \\ -cs & cs & -s^2+c^2 & 0 \\ 0 & 0 & 0 & 1 \end{bmatrix} \tag{10.53}$$

$$(\delta\boldsymbol{\sigma})^T\mathbf{C}\boldsymbol{\sigma} = (\delta\boldsymbol{\sigma})^T\mathbf{T}^T\mathbf{C}'\mathbf{T}\boldsymbol{\sigma} \tag{10.54}$$

Since $\delta\boldsymbol{\sigma} \neq 0$ and assuming non-zero stress tensor $\boldsymbol{\sigma}$, Eq. 10.54 can be written as:

$$\mathbf{C} = \mathbf{T}^{\mathrm{T}}\mathbf{C}'\mathbf{T} \tag{10.55}$$

For 3D cases, the transformation matrix $\mathbf{T}$ is used as given in Appendix A.

10.5.4 Joint Compliances for Multiple Joint Sets

As mentioned before, global compliance matrix is determined first by transforming compliance matrix of each joint set and then adding all together along with compliance matrix of rock substance. For ith joint set, the compliance matrix can be obtained by Eq. 10.56 in global coordinate system as:

$$\mathbf{C}_i = \mathbf{T}_i^T\,\mathbf{C}_i'\,\mathbf{T}_i \tag{10.56}$$

Here $\mathbf{T}_i$ represents the transformation matrix of ith joint set having compliance matrix $\mathbf{C}'$ in local coordinate system. Thus for n number of joint sets, the global compliance matrix will be the summation of all individual compliance matrices as given below:

$$\mathbf{C} = \sum_{i=1}^{n} \mathbf{C}_i \tag{10.57}$$

If the constitutive matrix of intact rock is denoted by $\mathbf{D}_r$, then equivalent compliance of the rock mass will be

$$\mathbf{C}_m = \mathbf{D}_r^{-1} + \mathbf{C} \tag{10.58}$$

Hence, stiffness matrix of equivalent rock mass, $\mathbf{D}_m$ can be obtained by inverting the above matrix as:

$$\mathbf{D}_m = \mathbf{C}_m^{-1} = [\mathbf{D}_r^{-1} + \mathbf{C}]^{-1} \tag{10.59}$$

EXAMPLE 10.6 Determine the equivalent compliance matrix for rock mass having two joint sets oriented 20° (joint 1) and 70° (joint 2) with the horizontal axis. Assume the normal and shear stiffness of both the joints are 10 GPa/m. The joint spacings of joint 1 and joint 2 are 5 m and 2 m respectively. The modulus of elasticity and Poisson's ratio of intact rock are 6 GPa and 0.25 respectively.

Solution: Using Eq. 10.56, we find the compliance for each joint set is:

$$T_1^T C T_1 = \begin{vmatrix} 2.340\text{E-}06 & -4.235\text{E-}22 & -6.428\text{E-}06 & 0 \\ -4.235\text{E-}22 & 1.766\text{E-}05 & -6.428\text{E-}06 & 0 \\ -6.428\text{E-}06 & -6.428\text{E-}06 & 2.000\text{E-}05 & 0 \\ 0 & 0 & 0 & 0 \end{vmatrix}$$

$$T_2^T C T_2 = \begin{vmatrix} 4.415\text{E-}05 & 0 & -1.607\text{E-}05 & 0 \\ 0 & 5.849\text{E-}06 & -1.607\text{E-}05 & 0 \\ -1.607\text{E-}05 & -1.607\text{E-}05 & 5.000\text{E-}05 & 0 \\ 0 & 0 & 0 & 0 \end{vmatrix}$$

For intact rock, the compliance can be obtained Eq. 5.21 as:

$$C_r = \begin{vmatrix} 1.667\text{E-}04 & -4.167\text{E-}05 & 0 & -4.167\text{E-}05 \\ -4.167\text{E-}05 & 1.667\text{E-}04 & 0 & -4.167\text{E-}05 \\ 0 & 0 & 4.167\text{E-}04 & 0 \\ -4.167\text{E-}05 & -4.167\text{E-}05 & 0 & 1.667\text{E-}04 \end{vmatrix}$$

Hence, using Eq. 10.58, we find the equivalent compliance as:

$$C_m = \begin{vmatrix} 2.132\text{E-}04 & -4.167\text{E-}05 & -2.250\text{E-}05 & -4.167\text{E-}05 \\ -4.167\text{E-}05 & 1.902\text{E-}04 & -2.250\text{E-}05 & -4.167\text{E-}05 \\ -2.250\text{E-}05 & -2.250\text{E-}05 & 4.867\text{E-}04 & 0 \\ -4.167\text{E-}05 & -4.167\text{E-}05 & 0 & 1.667\text{E-}04 \end{vmatrix}$$

EXERCISES

1. A series of artificial rock samples having single joint are tested in triaxial compression. The minimum fracture strength is recorded as 6 MPa for a confining stress of 1.5 MPa. If the angle of inclination of the joint plane for the minimum fracture strength is 60°, determine the cohesion and friction angle of the joint plane.

2. A rock sample 100 mm square in cross section is tested in the laboratory to determine its uniaxial compressive strength. It is found that a joint is passing through the sample as shown in the Figure. Joint is not continuous but bridges are formed and which covers about 50% of joint plane. The joint surface is cohesionless with a friction angle of 34° whereas the cohesion and angle of internal friction of the rock bridges are 1 MPa and 40° respectively. Determine the shear strength of the rock bridges if the rock sample failed along the rock bridges at an uniaxial compressive stress of 60 MPa.

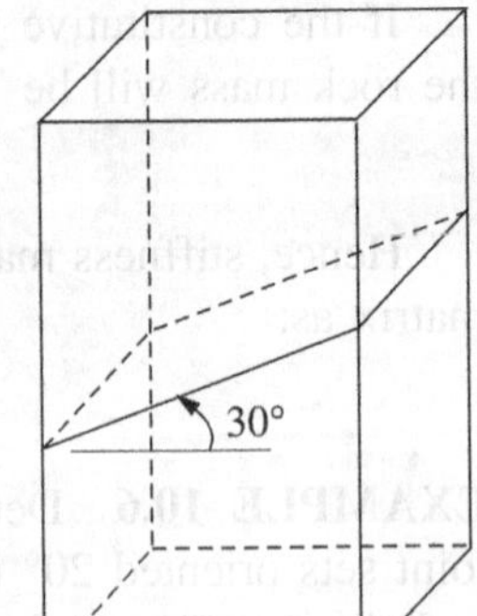

3. Show that for hydrostatic stress condition ($\sigma_1 = \sigma_3 = p$), the internal friction angle of a joint plane is approximated by $\varphi_j = \tan^{-1}\left(\dfrac{c_j}{p}\right)$ where c_j = cohesion of the joint plane.

4. Determine the minimum fracture strength and the direction of slip of a doubly jointed rock sample with Joint plane 1 having c = 1.5 MPa, φ = 35°, β = −60° and Joint plane 2 having c = 2.5 MPa, φ = 20°, β = 30°. Assume uniaxial loading condition.

5. A series of direct shear tests is conducted on similar type of joint surface (roughness, strength, and other properties are same). The peak shear stress, residual shear stress are recorded for different applied normal stress as given in the Table below:

σ_n (MPa)	τ_p (MPa)	τ_r (MPa)
0.50	0.8	0.25
1.00	1.2	0.54
1.50	1.4	0.90
2.00	1.9	1.20
2.50	2.1	1.41
3.00	2.5	1.70
3.50	3.0	2.00

(a) Determine the Mohr-Coulomb peak shear strength envelope of the joint.
(b) Determine the Barton's peak shear strength envelope if uniaxial compressive strength of the joint wall is 30 MPa.
(c) Determine the peak dilation angle for each case.
(d) Plot the peak shear strength with respect to peak dilation angle.

6. Estimate the instantaneous cohesion and friction angle from the Barton's envelop obtained in Exercise 4 for each normal stress. Plot instantaneous cohesion and friction angle with respect to normal stress.

7. (a) Determine the equivalent stiffness matrix in x-y coordinate system of the rock mass shown below:

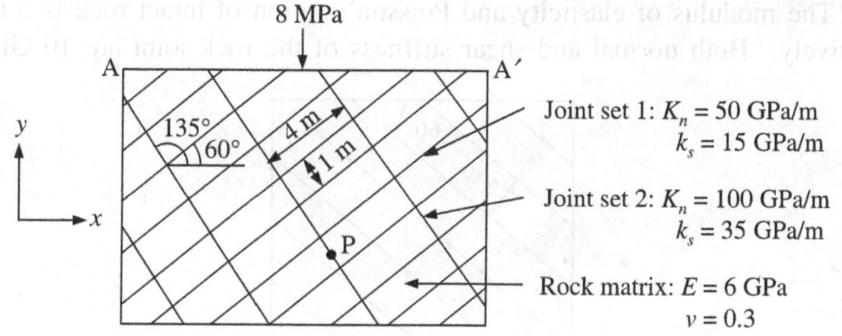

(b) Estimate the strain vector at any point on the AA' line. Assume plane strain condition.

8. In the given Figure, if the normal stress is increased from 0.0 to 0.1 MPa, determine the normal displacement at the joint surface. Compare this displacement with equivalent material model.

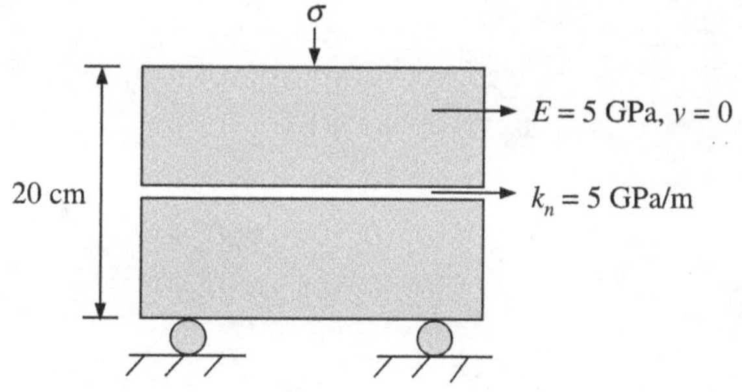

9. If the displacements along x and y directions at point P (see Figure of Exercise 7) are 0.05 mm and 0.15 mm respectively, determine whether the joint slips at point P. Assume JRC = 10.5, JCS = 50 MPa and φ_b = 30°.

10. Assuming $k_n = k$ GPa/m, $E = k$ GPa, and $v = 0$, determine the equivalent elastic stress-strain matrix of the rock mass given below. If $\sigma_{xx} = \sigma_{yy} = k$ MPa and $\tau_{xy} = 0$ at a point inside the equivalent rock mass, determine the strain vector at that point considering plane strain condition.

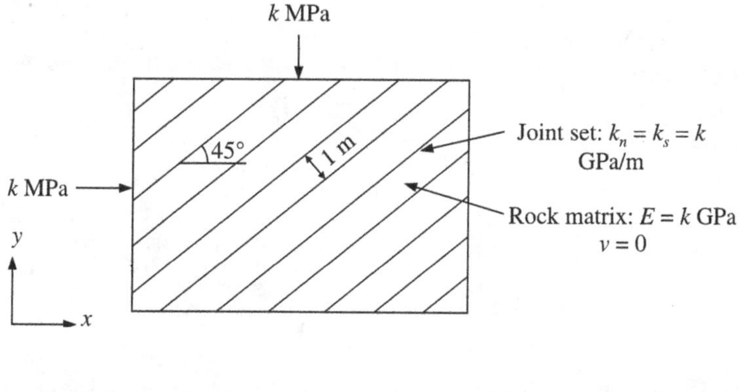

11. A series of rock samples having single joint are tested in triaxial compression. The minimum fracture strength is recorded as 10 MPa for a confining stress of 2.0 MPa. If the angle of inclination of joint plane for the minimum fracture strength is 60°, determine the cohesion and friction angle of the joint plane. .

12. Determine the equivalent compliance matrix of the joined rock mass shown in the figure. The modulus of elasticity and Poisson's ration of intact rock is 5 GPa and 0.25 respectively. Both normal and shear stiffness of the rock joint are 10 GPa/m.

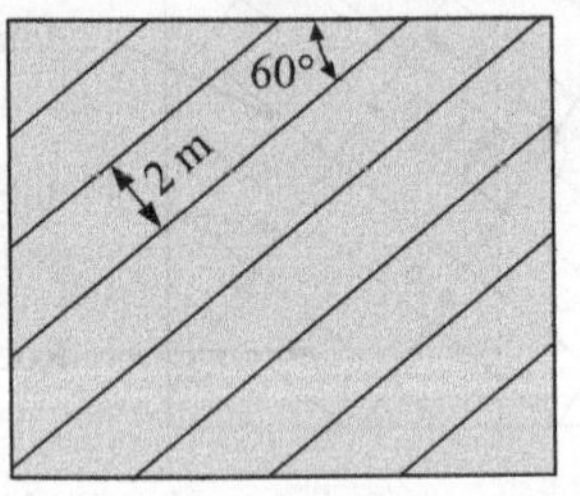

Chapter 11

Artificial Supports

11.1 INTRODUCTION

The artificial supports are made up of materials like wood, steel, glass fibre reinforced plastic (GRP), concrete and brick. The use of different types of materials as support material is based on their strength, durability and operating ease. Wood is the most important material for support until steel has become the primary material. This is because of scarcity, cost, quality, degradation of timber in underground environment. But advantage of wooden supports include its light weight, ease of handling, size manipulation to cut in the form of mine support. Wooden supports also give visual and audible signs before the start of failure. The steel based mine supports have edge over wooden supports because of their superior mechanical strengths (bending, tensional, buckling, shear, compression) and durability in underground environments but high weight, makes size manipulation hazardous and difficult. Nowadays, steel supports (especially bolts) are replaced by glass-fibre reinforced plastic or polymer (GPR) in some of the applications where excavation is done by equipment's like road header, continuous miner or tunnel boring machine. In same application, GPR may have edge over steel based support due to light weight, rust free, low cost and easy size manipulation. Concrete and bricks are used for erecting walls, water dam, machine housing, shelters, etc. In some cases, *in situ*-concrete lining combined with steel arches are used to support mouth of declines, inclines and other openings.

In this chapter, mechanics of different types of reinforcing supports like bolts, shotcrete/guinite, roof/side stitching, roof truss, forepoling, wire nets, W-strap and roof supports like prop, chock (or cogs/chockmate), roof bars, safety support (goal post), steel arches, backfilling/stowing are discussed.

11.2 ROCK MASS STABILISATION APPROACH

There are two approaches to stabilise the rock mass, viz. rock reinforcement and rock support as shown in Figure 11.1. Rock reinforcement is a rock quality improvement technique, where the inherent shear and tensile strength of rock mass are enhanced by inserting bolts, cables and others to control deformation. It is essentially a ground improvement option where excavations are found or predicted to be unstable. Thus, reinforcement action is linked to the interaction of reinforcing elements with rock mass discontinuities and thus modifies the internal strength and deformation characteristics of a rock mass. Examples are roof bolts, cable bolts, grouting, forepoling, roof stitching, etc.

Rock supports on the other hand, are direct support elements introduced from the surface of excavation to maintain deformation within tolerable limits. Note that no external elements are inserted into the rock mass and hence, supports are in contact with rock mass only at the boundaries. Examples are props, arches, cogs, W-strap, wire nets, shotcrete, GRP spray and others. These supports do not improve the intrinsic strength of the rock mass, but do alter the boundary conditions.

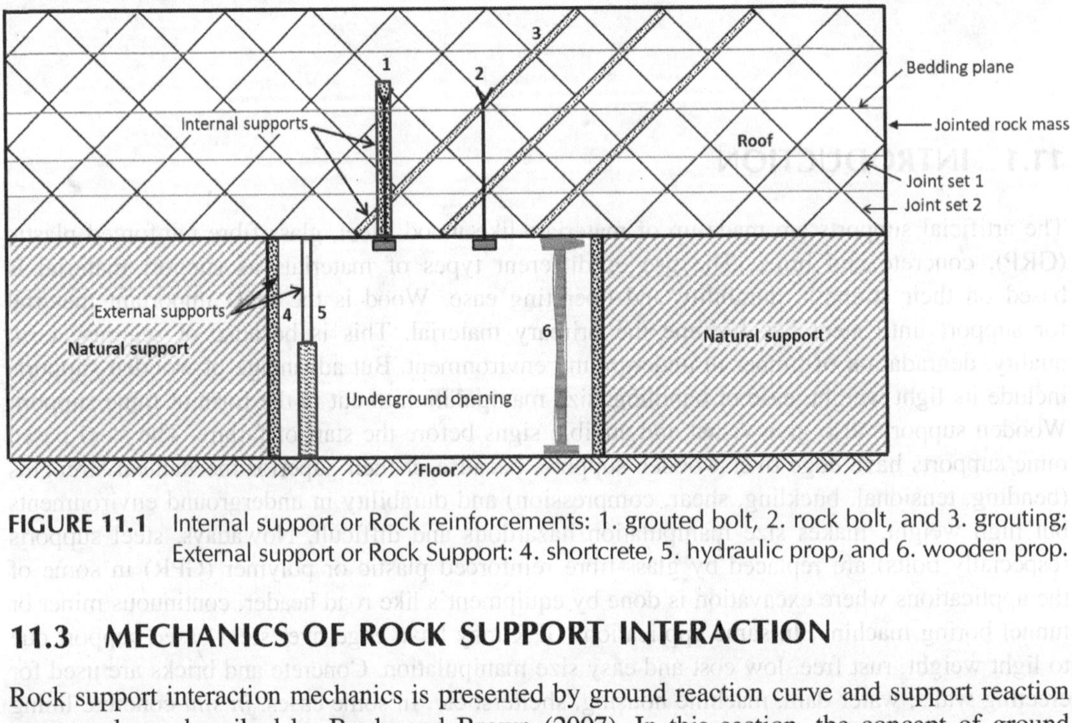

FIGURE 11.1 Internal support or Rock reinforcements: 1. grouted bolt, 2. rock bolt, and 3. grouting; External support or Rock Support: 4. shortcrete, 5. hydraulic prop, and 6. wooden prop.

11.3 MECHANICS OF ROCK SUPPORT INTERACTION

Rock support interaction mechanics is presented by ground reaction curve and support reaction curve and are described by Brady and Brown (2007). In this section, the concept of ground reaction and support reaction curves are reproduced from the above reference. Then a simple algorithm is presented to generate both the curves using simple computer program or a spread sheet program.

Let us assume that a tunnel of radius a is to be driven by drilling and blasting method in a rock mass under hydrostatic stress field. If the tunnel face is away from the observation section, [represented by the A-A′ section in Figure 11.2(a)], the internal pressure, p_i must be the same as the external *in situ* stress, p_0 and hence, radial displacement $u = 0$. This is also represented by a point O (0, p_0) in radial support pressure versus radial displacement plot (Figure 11.3).

Next, the tunnel face has been advanced beyond the observation section and hence, the support pressure p_i previously provided by the rock inside the tunnel, drops down to zero [Figure 11.2(b)]. This is because the rock mass previously supporting the tunnel has been removed. However, the tunnel will not collapse at the observation section since the radial deformation is significantly restricted by the tunnel face located nearby.

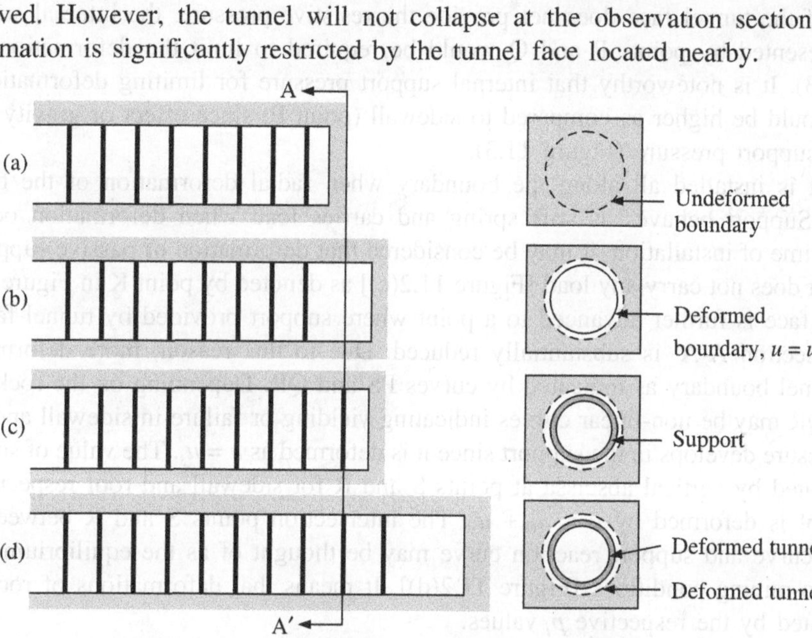

FIGURE 11.2 Ground response curves in (a) stresses at proposed periphery of excavation under equilibrium, (b) unbalanced stresses after excavation, (c) artificial support (such as shotcrete) installed to check rock deformation, and (d) new equilibrium condition occurs between ground and support.

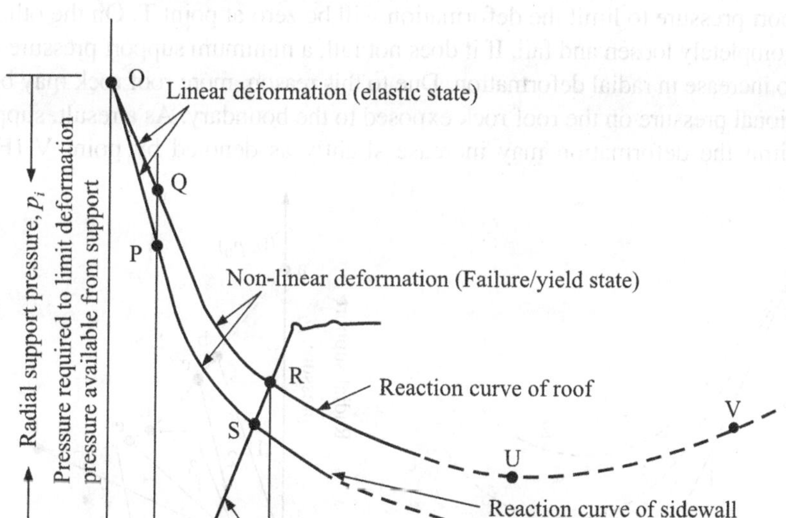

FIGURE 11.3 Radial support versus radial displacement of ground.

In other words, if the tunnel face does not provide the resistive pressure, the internal support pressure, p_i represented by points P and Q would be required to limit the deformation at $u = u_e$ (Figure 11.3). It is noteworthy that internal support pressure for limiting deformation of roof (point Q) would be higher as compared to sidewall (point P) since effect of gravity must be added to the support pressure (Figure 11.3).

Then support is installed all along the boundary when radial deformation of the tunnel boundary is u_0. Support behaves as stiff spring and carries load when deformation occurs. However, at the time of installation, it may be considered that deformation of passive support is zero and hence, it does not carry any load [Figure 11.2(c)] as denoted by point K in Figure 11.3.

Next, tunnel face is further advanced to a point where support provided by tunnel face to the observation section A-A′ is substantially reduced. Due to this reason, more deformation occurs in the tunnel boundary as indicated by curves PS and QR. Depending on the rock type and p_0, PS and QR may be non-linear curves indicating yielding or failure in sidewall and roof respectively. Pressure develops in the support since it is deformed as $u = u_s$. The value of support pressure is indicated by vertical abscissa at points S and R for sidewall and roof respectively. At this time, roof is deformed by $u = u_0 + u_s$. The intersection points S and R between the ground reaction curve and support reaction curve may be thought of as the equilibrium point for the given geo-mining conditions [Figure 11.2(d)]. It means that deformations of roof and sidewall are limited by the respective p_i values.

In a hypothetical situation, if no support is installed at point K, then radial deformation of tunnel would continue to increase for additional excavation of tunnel. The ground reaction curve may take the path ST or RUV for sidewall and roof respectively. Both roof and sidewall would fail or may attain a new state of equilibrium. Sidewall rocks may be completely loosen and hence, required support pressure to limit the deformation will be zero at point T. On the other hand, roof may also be completely loosen and fall. If it does not fall, a minimum support pressure is attained at point U due to increase in radial deformation. Due to this reason, more roof rock may be loosen and provide additional pressure on the roof rock exposed to the boundary. As a result, support pressure required to limit the deformation may increase slightly as denoted by point V (Figure 11.3).

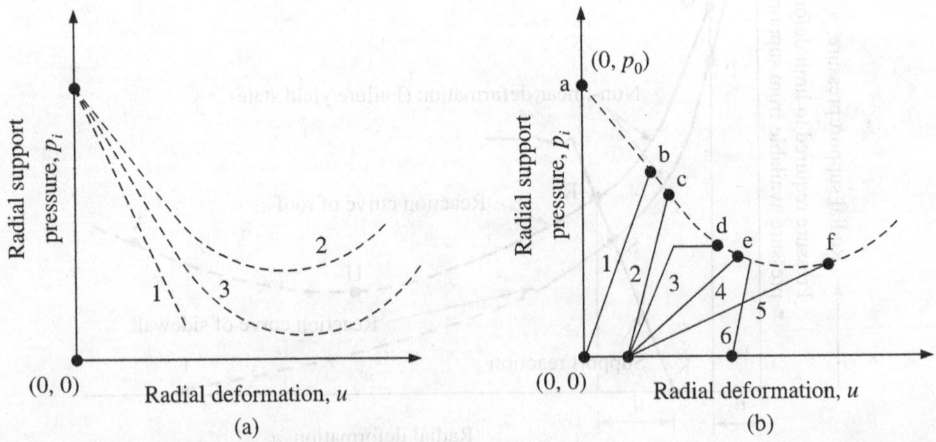

FIGURE 11.4 (a) GRC based on method of excavation, and (b) Interaction of GRC and SRC, influence of support stiffness and timing of their installation on performance.

Figure 11.4(a) shows three ground reaction curves (GRC) for a particular rock mass, depend on method of rock excavation. **Curve 1** is ideal excavation case, in which there is no disturbance of the remaining rock and there is no deviation from CHILE behaviour. But it may deviate slightly from elastic response. **Curve 2** could represent high-quality blasting, where more disturbance is inevitable and the resulting displacements are further increased. **Curve 3** represents the response of poor-quality blasting, and indicates that support is essential if stability of the excavation is to be maintained. Curves 2 and 3 reach the least self-supporting characteristics and later on the load carrying capacity increases due to self-adjustment. Here, self-supporting characteristics of the fractured rock mass is defined as the self strength developed due to inter-block friction and cohesion, develops an equilibrium condition just before the complete failure or collapse of the rock mass.

Figure 11.4(b) shows the pressure developed in the support element as the deformation occurs. Support reaction represented by line 1 signifies that support is installed immediately after the excavation is made. In other words, $u_0 = 0$. Line 2 represents an elastic support installed when $u_0 > 0$ and intersects in the linear part of the ground reaction curve. This is the most realistic response of the support. The support represented by line 3, although installed at the same time as support 2, does not have sufficient strength or load carrying capacity. Hence, it fails before it can provide support to the deformed roof or sidewall. Supports defined by line 4 and line 5 lack in stiffness and hence, intersect the ground reaction curve when roof or sidewall have already failed. Hence, these supports cannot provide adequate support to limit the deformation. Support represented by line 6 is installed too late or after large deformation and hence, will not be effective in providing support to roof or sidewall. Engineering decision for support selection depends not on GRC and SRC but also on support cost. Hence, techno-feasibility of an artificial support depends on SRC and GRC.

11.4 ROCK SUPPORT INTERACTION ANALYSIS

The concept presented above can have variations depending on rock types, *in situ* stress conditions, geometry of the excavation, support conditions, mode of rock failure and others. Hence, it is cumbersome to consider all these variables into a mathematical framework to derive an analytical solution of the above concept. In the following, an attempt is made to derive rock support interaction using simple mathematical equations based on the concept outlined in Brady and Brown, 2007.

The basic assumptions of this analysis are as follows:

(i) Tunnel geometry: circular with undeformed radius, a
(ii) Hydrostatic *in situ* stress, p_0
(iii) Internal support pressure, p_i applied around the tunnel boundary
(iv) Properties of intact rock mass (before failure):
  (a) Mohr–Coulomb (MC) media: c_r, φ_r
  (b) Hoek–Brown (HB) media: $m_r = m_b$, $s_r = s$, a
(v) Properties of fractured rock mass (after failure):
  (a) Mohr–Coulomb media: c_f, φ_f
  (b) Hoek–Brown media: m_f, s_f, a_f

(vi) Elasto-plastic radius is estimated using Eqs. 9.61 and 9.78 for MC and HB media respectively

(vii) For both MC and HB media, non-associative flow rule with dilation is assumed based on Eq. 9.66.

11.4.1 MC Media

In Section 9.11, derivation of elasto-plastic analysis of stresses and deformation are provided in details. Equations 9.60 and 9.61 are reproduced below:

$$\sigma_{re} = \frac{2p_0 - \sigma_{cr}}{1 + N_r} \tag{11.1}$$

$$r_e = a\left[\frac{2p_0 - \sigma_{cr} + (c_f \cot\varphi_f)(1 + N_r)}{(p_i + c_f \cot\varphi_f)(1 + N_r)}\right]^{1/(N_f - 1)} \tag{11.2}$$

From the above two equations, it is clear that if $p_i > \sigma_{re}$, then $r_e < a$, which is not possible in reality. This implies that rock mass behaves elastically for $p_i > \sigma_{re}$. Due to this reason, the radial stress at elasto-plastic boundary is termed as the critical stress. For this condition, the linear relationship between p_i and the radial deformation on boundary u can be written based on Eq. 9.63 as:

$$u = \frac{(p_0 - p_i)a}{2G} \quad \text{for } p_i > \sigma_{re} \tag{11.3}$$

For an internal pressure $p_i < \sigma_{re}$, rock mass fails and hence, plastic deformation occurs. Equation 9.71 describes the deformation u considering non-associative plastic flow having a dilation angle ψ. This equation is reproduced at $r = a$ as given below:

$$u = \frac{a^{-N_\psi}}{2G}[C(1-2\nu)(r_e^{N_\psi+1} - a^{N_\psi+1}) - D(r_e^{N_\psi-1} - a^{N_\psi-1})] + \frac{(p_0 - \sigma_{re})\,r_e}{2G}\left(\frac{r_e}{a}\right)^{N_\psi} \quad \text{for } p_i < \sigma_{re} \tag{11.4}$$

Note that in the above equation, the variables r_e, σ_{re}, C, D are function of p_i.

11.4.2 HB Media

As mentioned above, σ_{re} is estimated iteratively using Newton-Raphson method or other methods. The elasto-plastic radius is then estimated using Eq. 9.77 for different value of p_i. Similar to the above approach, linear elastic response is expected when $p_i > \sigma_{re}$, as expressed by Eq. 9.72. In that case, relationship between p_i and u is given by Eq. 11.3. The expression of r_e is given in Eq. 9.78. Then for $p_i < \sigma_{re}$, relationship between p_i and u is given in Eq. 11.4.

EXAMPLE 11.1 A tunnel of radius 2.0 m is excavated in rock (modulus of elasticity is 5 GPa; Poisson's ratio is 0.25) subjected to an initial hydrostatic stress field of 10 MPa. Assuming c and φ of intact rock are 2 MPa and 30° and those of broken rock are 1 MPa and 20° respectively, plot the ground reaction curve. Assume dilation angle as 20° and gravity effect based on Example 9.11.

Solution: Using the data, σ_{re} = 3.27 MPa, N_r and N_f are 3.0 and 2.04 respectively. Deformation using Eq. 11.3 until p_i > 3.27 MPa. Once p_i < 3.27 MPa, deformation is estimated using Eq. 11.4. However, rock pieces may also deform due to gravity and hence, total internal pressure p_T is estimated based on the data given in Example 9.11. The ground reaction plot is given in the following Figure.

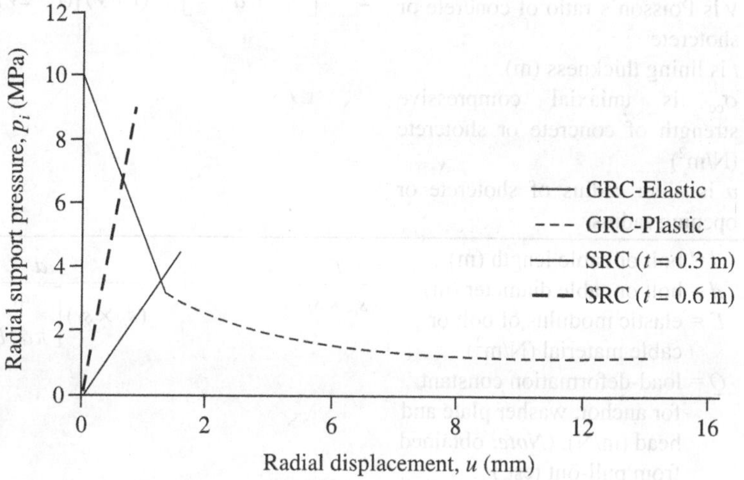

11.4.3 Support Reaction Curve from Single Support

A support is characterised by its stiffness (k expressed in Pa) and the maximum allowable pressure ($p_{\max}$ expressed in Pa). Table 11.1 lists mathematical relations of stiffness and maximum allowable pressure of some typical supports given in reference Hoek et al. (1995). It is noteworthy that supports are generally installed after a small deformation $u = u_0$ occurs in roof or sidewall. This deformation occurs due to delay in installing supports. Delay occurs for the following reasons:

(i) After blasting, re-entry into the working area can only take place after clearance of blast fumes

(ii) Time taken for mucking of blasted material

(iii) Preparation time for installing support

Delay time can be significantly reduced if tunnel boring machine (TBM), road header or other mechanised methods are adopted for the development of tunnels or mine galleries. Nonetheless, due to the delay in installing supports, roof or sidewall may deform. Considering analogy of thick line pipe under radial pressure, for an internal support pressure p_i having stiffness of support k, deformation of roof or sidewall can be expressed as:

$$u = u_0 + u_s = u_0 + \frac{p_i a}{k} \tag{11.5}$$

where u_s is the deformation of the support due to pressure p_i. The schematic relationships between internal pressure and support deformation are shown in Figure 11.4(b). The above equation is valid as long as $p_i < p_{\max}$. After that prediction of deformation of support based on Eq. 11.5 would be erroneous.

TABLE 11.1 Support Capacity and Stiffness of Various Supports

Support type	*Input parameters*	*Support capacity* (p_{max})	*Stiffness* (k)
Concrete and shotcrete lining	E is modulus of elasticity of concrete or shotcrete (N/m^2) v is Poisson's ratio of concrete or shotcrete t is lining thickness (m) σ_{cc} is uniaxial compressive strength of concrete or shotcrete (N/m^2) a is outer radius of shotcrete or opening radius	$\frac{1}{2}\sigma_{cc}\left[1-\frac{(a-t)^2}{a^2}\right]$	$\frac{E[a^2-(a-t)^2]}{(1+v)[(1-2v)a^2+(a-t)^2]}$
Ungrouted mechanically or chemically anchored bolt	l = bolt or cable length (m) d = bolt or cable diameter (m) E = elastic modulus of bolt or cable material (N/m^2) Q = load-deformation constant for anchor, washer plate and head (m/N). (***Note:*** obtained from pull-out test.) T = ultimate failure load from pull-out test (N) S_c = circumferential bolt spacing (m) S_l = longitudinal bolt spacing (m)	$\frac{T}{s_c \times s_l}$	$\frac{a}{(s_c \times s_l)\left[\frac{4l}{\pi d^2 E}+Q\right]}$

11.4.4 Support Reaction Curve from Multiple Supports

In many instances, combination of supports is installed to enhance the overall stiffness of the system. For example, shotcrete with rock bolting or wire mesh is a common form of support in slopes, tunnel and mine galleries. In severe ground conditions, rock bolting and concrete supports are also installed. For this combined support system, stiffness of individual supports are additive as given below.

$$\bar{k} = \sum_{m=1}^{S} k_m \tag{11.6}$$

where k_m is the stiffness of mth support and S denotes number of supports to be installed. It may be noted that all supports should be installed at the same time or after the same ground deformation $u = u_0$.

Considering $p_{\max/m}$ the capacity of mth support, the following conditions are to be checked while estimating ground deformation.

Step 1: Estimate $u_m = ap_{\max/m}/k_m$ for m^{th} support and $\bar{u} = ap_i/\bar{k}$

Step 2: Find $u_{\min} = \min(u_m)$ considering all u_m.

Step 3: If $\bar{u} < u_{\min}$ then $u = u_0 + \frac{p_i a}{\bar{k}}$.

Step 4: Arrange u_m in ascending order from u_1 to u_S. Let u_p denotes deformation of pth support such that $\bar{u} > u_p$ and $u_p < u_m$ (where $p \neq m \propto S$) then $p_{\text{max, combined}} = \dfrac{u_p \bar{k}}{a}$. Hence, this combined support pressure is maintained for all subsequent deformation of roof or wall.

EXAMPLE 11.2 Using the data given in Example 11.1, plot support reaction curve for concrete lining having (a) lining thickness 0.3 m, and (b) lining thickness 0.6 m. Assume modulus of elasticity, Poisson's ratio and compressive strength of concrete 21 GPa, 0.25 and 35 MPa respectively. Also assume that u_0 = 0.2 mm.

Solution: The load deformation curves are plotted based on equation given in Table 11.1. The following table summarises results. The support reaction curves are plotted in Example 11.1.

t = 0.3 m		t = 0.6 m	
u_r (mm)	p_i (MPa)	u_r (mm)	p_i (MPa)
3.5188	3.948	3.0063	7.392
2.7218	3.000	2.8574	7.000
2.3015	2.500	2.6676	6.500
1.8812	2.000	2.4778	6.000
1.4609	1.500	2.2880	5.500
1.0406	1.000	2.0982	5.000
0.6203	0.500	1.9084	4.500
0.2000	0.000	1.7185	4.000
		1.5287	3.500
		1.3389	3.000
		1.1491	2.500
		0.9593	2.000
		0.7695	1.500
		0.5796	1.000
		0.3898	0.500
		0.2000	0.000

EXAMPLE 11.3 Considering data given in Example 11.2, plot combined support reaction curve for a shortcrete support having thickness 0.3 m and ungrouted mechanically anchored rock bolt supports having length 2.4 m, diameter 22 mm, bolt tension 12 t or 12000 N, spacing 0.5 × 0.5 m. Assume elastic modulus of bolt steel as 200 GPa and Q = 0.

Solution: The stiffness and maximum pressure of these two support systems are given in the following Table using Table 11.1 as:

	Stiffness (MPa)	*Maximum pressure* (MPa)
Shotcrete	3813.50	4.86
Bolt	253.42	0.47

The following figure plots the individual support reaction as well as combined one.

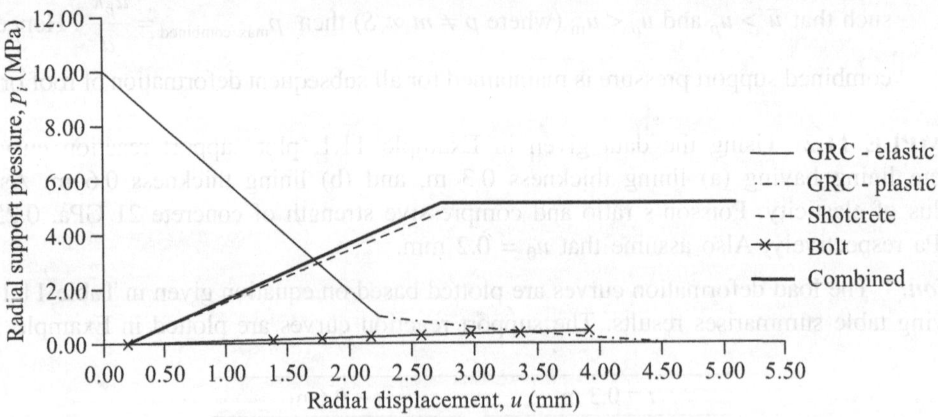

11.5 CLASSIFICATION OF ARTIFICIAL SUPPORTS

Artificial supports are of various forms, sizes and capacities and are used depending on the requirement in a particular workplace. Figure 11.5 shows pictures of various supports used for mining and tunnelling applications. Based on their performance, time of installation and mechanical characteristics, artificial supports are classified in different categories as given below:

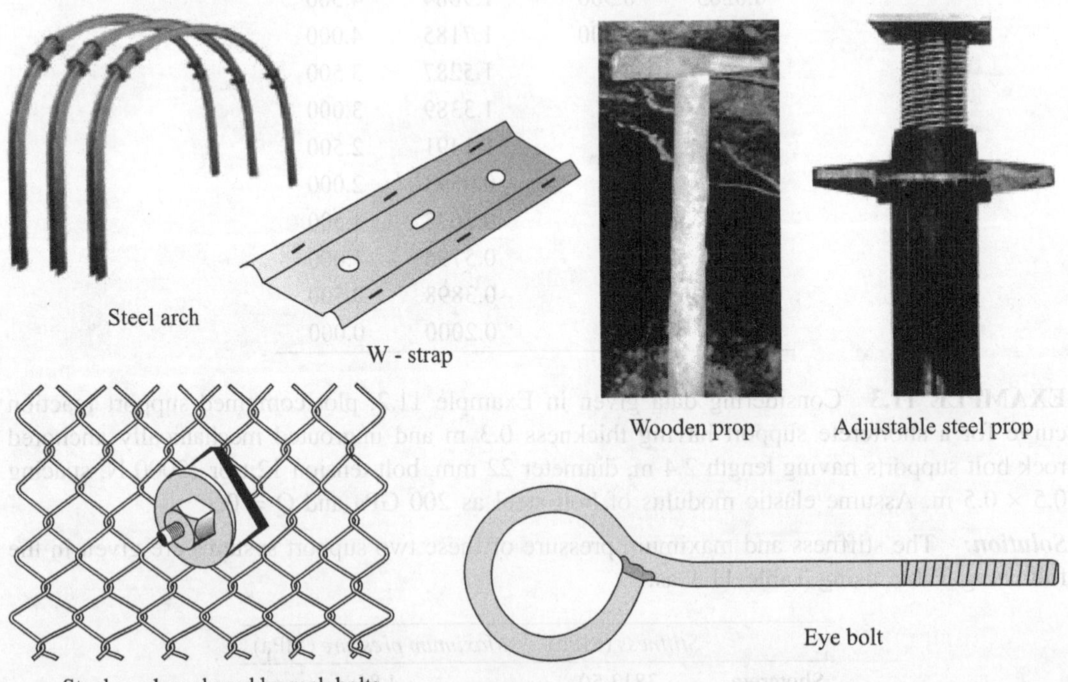

FIGURE 11.5 Different types of support system.

(i) Temporary and Permanent: The temporary and permanent supports are classified based on the installation time elapsed after tunnel/mine gallery is excavated. Temporary supports are of low capacity and installed right after the excavation to prevent deformation of freshly exposed roof. On the other hand, permanent supports are installed to strengthen temporary supports either by adding few more supports or installing different types of supports. Nowadays, excavation-cum-drilling machine can directly install permanent support in spite of poor ground clearance. Sometimes, temporary support is partly or wholly removed to enable the permanent support to be installed. But, this practice negates the advantages that can be obtained by applying principles of rock support interaction mechanics and so should be avoided. Examples of temporary and permanent supports are wooden props and concrete supports respectively.

(ii) Primary and Secondary: Primary supports are the essential ones to ensure safe working conditions during subsequent excavation and to initiate the process of mobilising and conserving rock strength by controlling ground displacement. Whereas, secondary supports are additional support required to supplement primary supports. For example, rock bolts are regarded as the primary support and installed after muck is cleared. If roof deteriorates even after bolting is done, cogs or cribs are installed to prevent rock fall. This support is considered to be secondary support.

(iii) Active and Passive: Based on reaction developed in support, supports are classified as active and passive support. Active support imposes a predetermined load to the rock without any deformation at the time of installation. Examples are tensioned rock bolts or cables, hydraulic prop, and powered support. Passive supports are not installed with an applied loading but as rock mass deforms load develops in the support. For example prop (wooden or steel), steel arches, dowels, timbered sets, untensioned rock bolts (like friction bolt, grouted bolt), reinforcing bars or cables. Active support exerts force on the rock mass before any movement has taken place. Whereas, passive support increases the resisting force provided by shear restraint. Passive support always give lower FOS than active support.

(iv) Rigid and Yieldable: Based on the mechanical behaviour, rigid and yieldable supports are classified. Rigid supports are of high capacity and stiffness and mostly behave in elastic manner. This type of support develops high pressure for very little deformation in the roof. Examples are friction steel props, steel arches, friction bolts and others. On the other hand, yieldable supports behave elasto-plastically and shed load after the maximum capacity is reached. As a result, more deformation occurs. Example of yieldable supports are friction/hydraulic prop, yieldable arches and hydraulic powered support used in longwall mines.

11.6 PRIMARY REINFORCEMENT: ROCK BOLTS

Rock bolts generally consists of a bolt (or shank) with an anchorage at one end and a bearing plate (or face plate), and a tightening nut at the other end. Shank is the threadless part of the bolt in between head and thread as shown in Figure 11.6, whereas in case of threaded part, bolts are called rebar as shown in Figure 11.9. Some bolts tensioned after installation and some are not. Tensioning, approximately 70% of the rock bolt capacity is important to ensure that all components of rock bolts are in contact and that a positive force is applied to the rock. The amount of tension applied may not be so critical and tightening the nut with a conventional wrench or with a pneumatic torque wrench is adequate. The intensity of support provided by a rock bolt system is determined by load-bearing capacity of the individual bolt, bolting pattern, bolt length and diameter and also on grout properties if the bolt is fully grouted. The capacity of the rock bolt is normally determined by the diameter and the grade of the steel. Borehole diameter and annular space between bolt and borehole plays a significant role in load transfer.

A number of different types of rock bolts are in practise in civil and mining industries. Rock bolts are grouped based on their principle of interacting with the surrounding rock mass. In this respect, the following types of rock bolts are considered:

(i) Mechanically anchored rock bolts
(ii) Friction anchored rock bolts
(iii) Grouted rock bolts
(iv) Grouted cable bolts

11.6.1 Mechanically Anchored Rock Bolts

The expansion shell anchored rock bolt, of standard or bail type, is the most common form of mechanically anchored rock bolt (Stillborg, 1986). A wedge is attached to the bolt in the threaded part of the rod. It is pulled into a conical expansion shell when the bolt is rotated and pushed inside the borehole (Figure 11.6). As a result, the shell expands and contact against the wall of the borehole. The void between the bolt and the borehole can be post grouted. In general, mechanical anchor type bolts are design for use in moderately hard to hard rock condition. They are not recommended for very hard or soft rock, since they will prevent the expansion shell from gripping and anchoring.

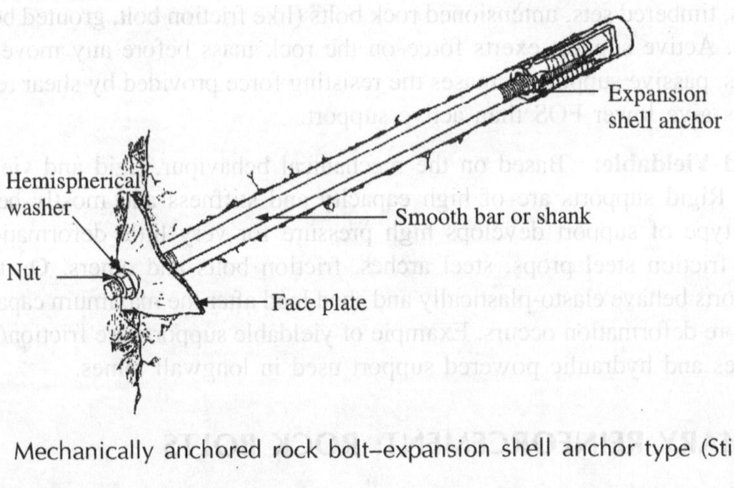

FIGURE 11.6 Mechanically anchored rock bolt–expansion shell anchor type (Stillborg, 1986).

11.6.2 Friction Anchored Rock Bolt

The major difference between mechanically anchored bolts and friction anchored bolts is that in the latter case, frictional resistance to sliding occurs for the entire length of the bolt. Like mechanically anchored bolt, these bolts are also susceptible to corrosion. Although they may be given a pre-tension to ensure that an anchorage is formed, friction bolts are usually not installed with the levels of pre-tension of 5–20 tonnes generally used for other rockbolts. Friction anchored bolts are of two types: (i) Split set, and (ii) Swellex. As the name suggests, the split set bolt expands and anchors borehole rock mass by the primary contact stress [Figure 11.7(a)]. Split set bolts are popular in underground metal mines because they are simple to use and have no moving parts, and can be installed in one step. The maximum load they can support is related to the diameter of the drillhole, diameter of the bolt, and length of the bolt. Generally, the split tube is driven into the borehole for developing load carrying capacity through frictional resistance to sliding. For example, a 38 mm diameter split tube is forced into a 35 mm diameter drillhole. The spring action of the compressed tube applies a radial force against the rock and generates a frictional resistance to sliding of the rock and steel. The drillhole should be proper, it should not be too small or too large.

On the contrary, Swellex satisfies two requirements: (i) a primary contact between the borehole wall and the bolt, and (ii) the secondary contact as bolt fits the irregularities on the borehole wall after expansion. As the Swellex bolt expands by water injection, it accommodates the asperities of borehole wall providing the secondary contact pressure [Figure 11.7(b)]. Figure 11.8 schematically explains the expansion of Swellex bolt. Thus, Swellex bolt is preferred over split set bolts. Both mechanically anchored and friction anchored bolts are considered as active support system since right after installation, bolt starts to provide anchorage against the movement of rock mass.

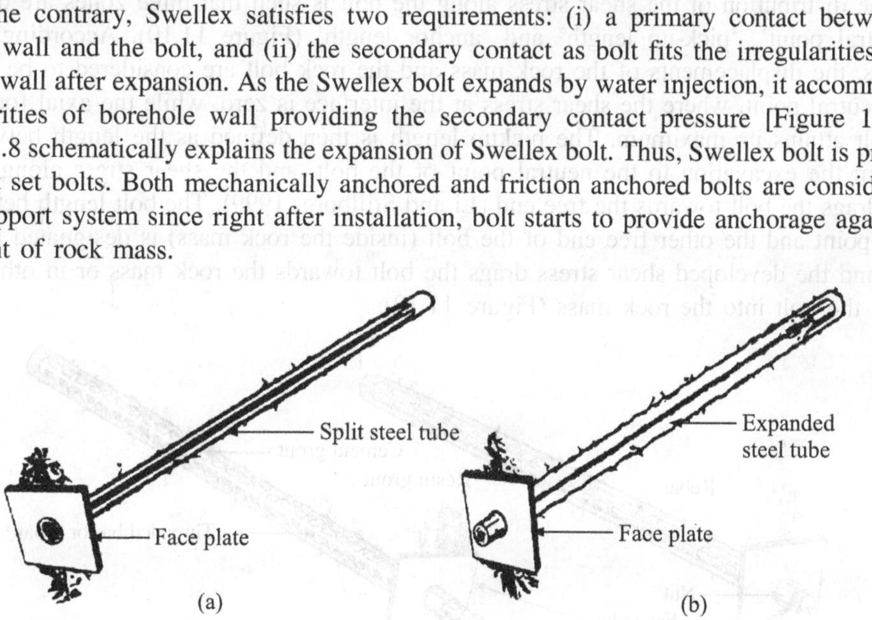

FIGURE 11.7 Friction anchored rock bolt (a) Split set, and (b) Swellex (Stillborg, 1986).

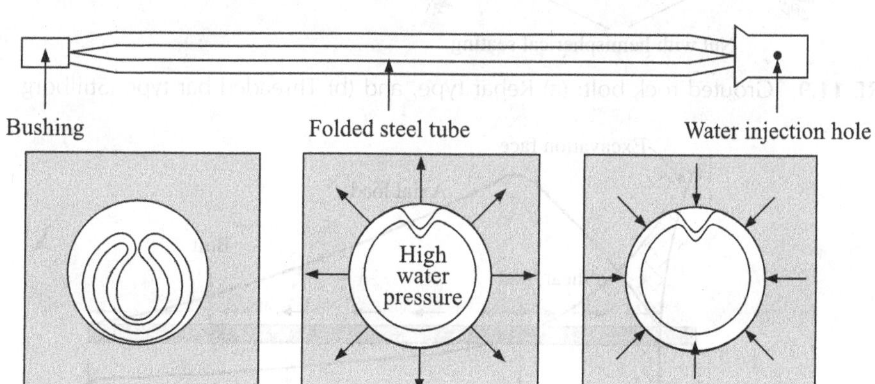

FIGURE 11.8 Sketches illustrating the Swellex bolt and the interaction between the rock and the Swellex bolt (a) placed in the hole, (b) is expanded under high water pressure, and (c) after water pressure is released and the surrounding rock contracts (Stillborg, 1986).

11.6.3 Grouted Rock Bolts

Grouted rock bolts have been widely used worldwide in mining as well as in civil engineering applications. The bolt rod is commonly made of steel rebar or threaded bar and grout material is composed of cement or resin as shown in Figure 11.9. Threaded bar increases the shearing capacity by increase in area of contact by reduction in stress concentration. Whereas, rebar, short form of 'reinforcing bar' is used as a tension device. Shear resistance is developed in both bolt-grout and grout-rock interfaces. It works as a passive support system. The bolt shank acts as a tension bar and at times a face plate is attached to improve the anchoring capacity of the bolt. The distribution of the shear stress along the bolt is such that three zones are developed as 'neutral point', 'pick-up length' and 'anchor length' (Figure 11.10). According to these concepts, the displacements of the rock mass and the rock bolt are considered to be the same at the neutral point, where the shear stress at the interface is zero, while the axial force of the rock bolt attains its maximum. The pickup length is then defined as the length between free end from the excavation to the neutral point of the bolt, and the shear stress along this bolt length drags the bolt towards the free end (Li and Stillborg, 1999). The bolt length between the neutral point and the other free end of the bolt (inside the rock mass) is designated as anchor length and the developed shear stress drags the bolt towards the rock mass or in other words, anchors the bolt into the rock mass (Figure 11.10).

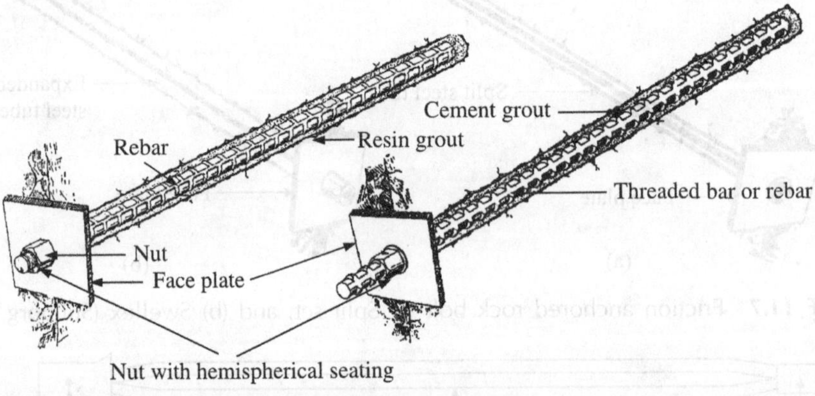

FIGURE 11.9 Grouted rock bolt: (a) Rebar type, and (b) Threaded bar type (Stillborg, 1986).

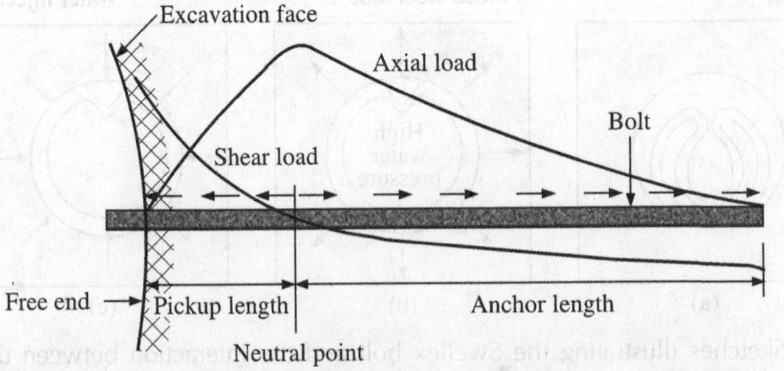

FIGURE 11.10 Schematic of grouted bolt in rock mass.

There may be not one but several neutral points along a rock bolt. After being installed in the rock mass, the rock bolt works compatibly with the surrounding rock mass before the decoupling, and the neutral point is strongly influenced by the deformation of the rock mass. The shear stress at the interface and the axial force of the rock bolt is determined by the deformation of the rock mass. For numerical analysis, most of the research focuses on the mechanical properties of the interface between the rock and bolt.

The primary purpose of grouted mechanically anchored bolts is to prevent corrosion and act as the locking mechanism for the mechanical anchor with the borehole wall. The strength requirement for the grout is not as important as it is in the case of grouted dowels or cables. The grout should be easily pumpable without being too fluid and a typical water/cement ratio of 0.4 to 0.5 is a good starting point for a grout mix for this application. It is most important to ensure that the annular space between the bolt and the drillhole wall is completely filled with grout. Pumping should be continued until there is clear indication that the air has stopped bleeding through the breather tube or that grout is seen to return through this tube.

Generally, cement grout or cement mortar anchor are used with bolts in India. Compared to cement grouted bolts, resin has certain advantages like less effect of vibration which adversely effects the efficacy of anchorage; allow bolt insertion and tensioning in one pass; effective pre-tensioning; suitable for wet conditions; and resin hardens very quickly and develop high strengths.

A typical resin product is made up of two compartment cartridges containing a resin and a catalyst in separate compartment, as shown in Figure 11.11 and is available in various diameters and gel time. Key components of a resin may typically consist of: (i) Polyester resin—28.5%; (ii) Filler (crushed limestone)—66%; and (iii) Accelerator—0.5%. The accelerator assists the reaction between the catalyst and polyster resin so that the mixture cures faster. The cartridges are pushed to the end of the drillhole ahead (angled to form a sharp tapered point) of the bolt rod that is then spun into the resin cartridges by the drill. The plastic sheath of the cartridges break and the resin and catalyst are mixed by this spinning action. Setting of the resin occurs within a few minutes (depending on the specifications of the resin mix) and a very strong anchor is created.

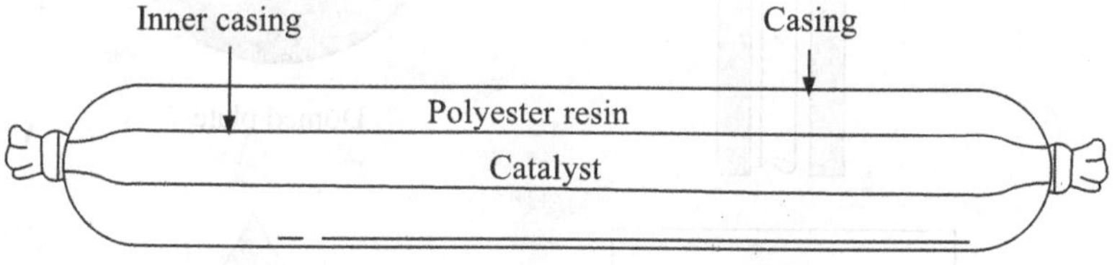

FIGURE 11.11 Resin cartridge.

11.6.4 Grouted Cable Bolts

Grouted cable bolt is used as un-tensioned bolt. Cable bolts are also passive support system and are predominantly used as temporary support. The cable type can be of 15.2 mm diameter, 7-wire steel strand and can have a length of 10–20 m as shown in Figure 11.12. The grout material is generally composed of concrete. The major advantage of cable bolt is that it can suspend long broken rock mass and can be installed in a narrow tunnel or mine gallery. The anchoring mechanics of cable bolts are similar to that of grouted rock bolts.

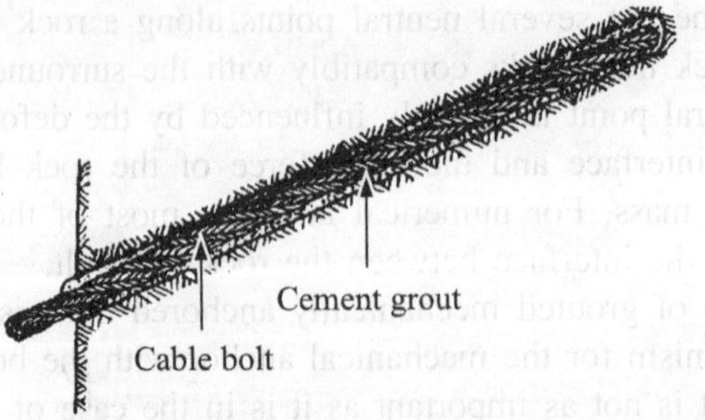

FIGURE 11.12 Grouted cable bolt (Adapted from Stillborg, 1986).

11.6.5 Face Plate

The bearing (face) plate is used to distribute the load from the bolt onto the rock face. The threads on bolt should be as coarse as possible and should be rolled rather than cut. A fine thread is easily damaged and will cause installation problems in a typical underground environment whereas a cut thread weakens the bolt. A face plate plays an important role for distributing load at the bolt head uniformly into the surrounding rock mass. There are several types of face plates used for different applications. The flat plate can be used when the rock surface is smooth and the bolt is installed perpendicular to the surface of the rock mass (Figure 11.13). A hemispherical seating is added to the nut for two reasons: (i) to avoid unfavourable tensile stresses in the bolt, and (ii) to make the nut lie flat against its base.

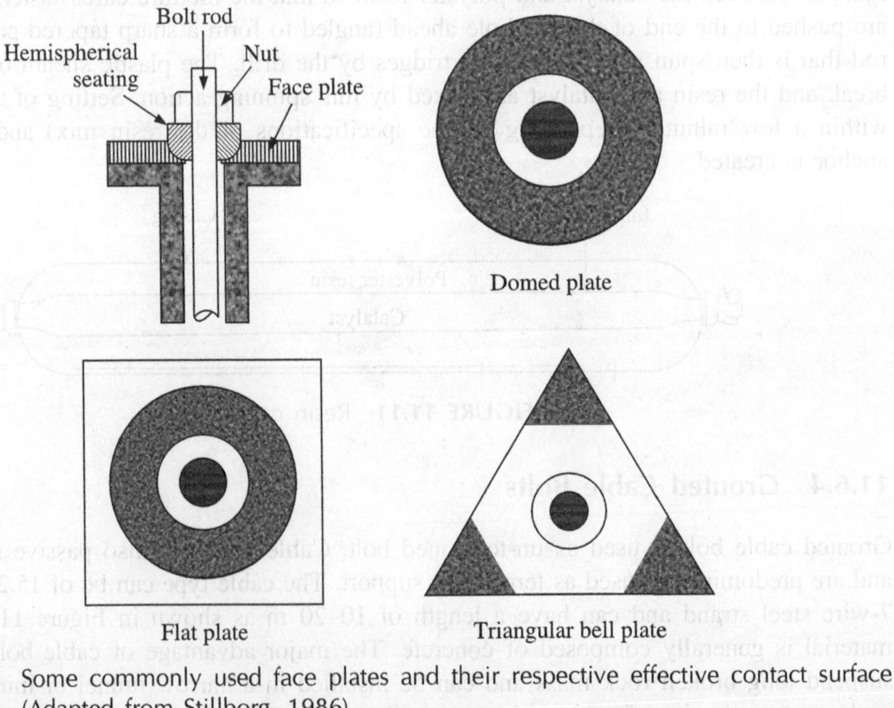

FIGURE 11.13 Some commonly used face plates and their respective effective contact surface (Adapted from Stillborg, 1986).

Under high compressive load, the rock mass may yield or crush at where the face plate is installed. The tension in the rock bolt reduces. It is observed that only 1 to 2 mm setting can reduce the tension in the rock bolt by 20 to 70%. Then bolt must be re-tensioned. This shortcoming in the flat face plate is greatly reduced with the domed or triangular bell plates, which have larger supporting areas. These types of face plate give the rock bolt system greater flexibility.

11.6.6 Design of Rock Bolts—Force Balancing Method

Rock bolt system is designed to minimise the deformation of rock mass to a level that it becomes safe for operation. The first and foremost concept is to restrict the relative movement of rock blocks either by stitching, suspending or improving frictional resistance. In the following, three common examples are described where force balancing method has been applied to estimate the safety factor of the bolt-rock mass system.

Support of Wedges

Let us assume that a wedge or block is formed by two cracks in the roof or side wall of an excavation. The wedge would slide along a fracture at an angle α with the horizontal axis as shown in Figure 11.14. The weight of the block is W. If the shear force along cracks surface exceeds the frictional force, the block slides or caves. In order to resist the movement of the wedge, a series of tension bolts are placed roughly perpendicular to the sliding surface. The resulting normal force on the sliding surface would effectively improve the shear strength of rock mass. At limiting equilibrium, the force, F_n, acting normal to the sliding surface is the sum of the resolved block weight and rock bolt tension components and is given as:

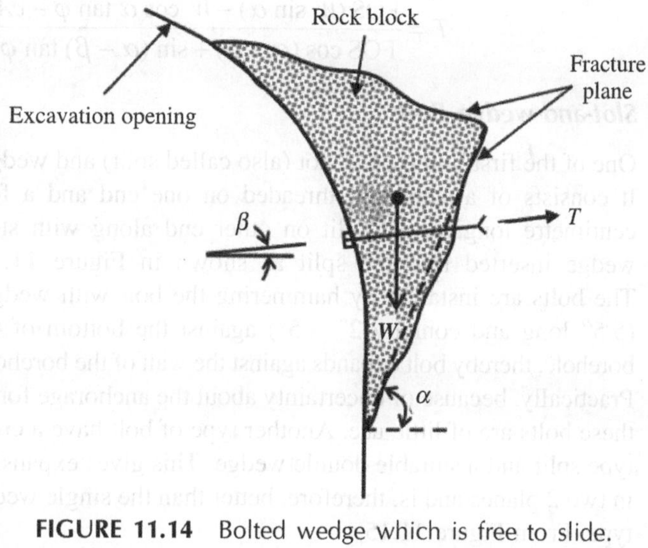

FIGURE 11.14 Bolted wedge which is free to slide.

$$F_n = W\cos\alpha + T\sin(\alpha - \beta) \tag{11.7}$$

where β = angle of the bolt with the horizontal axis and T = combined axial force (tensile) given by the bolt system.

The shear (sliding) force, F_s, acting along the sliding surface is similarly given by

$$F_s = W\sin\alpha - T\cos(\alpha - \beta) \tag{11.8}$$

For a factor of safety (FOS), the limiting equilibrium condition of the wedge-bolt system occurs if

$$\text{FOS}\,[W\sin\alpha - T\cos(\alpha - \beta)] = cA + [W\cos\alpha + T\sin(\alpha - \beta)]\tan\varphi \tag{11.9}$$

where c and φ are the cohesion and friction angle of the interface or sliding surface and A = area of the sliding surface. Equation 11.9 can be rearranged to obtain the bolt tension as:

$$T = \frac{\text{FOS}\,(W \sin \alpha) - W \cos \alpha \tan \varphi - cA}{\text{FOS} \cos (\alpha - \beta) + \sin (\alpha - \beta) \tan \varphi} \tag{11.10}$$

Once the combined bolt force is estimated using Eq. 11.10, bolt diameter can be evaluated for a given spacing and yield strength of bolt steel.

EXAMPLE 11.4 A rock block of weight 40 kN is separated by cracks making an angle $\alpha = 60°$ with horizontal. The cohesion and angle of friction of sliding surface are 6 kPa, and 25° respectively, and the inclination of the bolt is 30° anti-clockwise with horizontal. The area of sliding surface is 1 m^2. Find the applied tightening force if factor of safety of bolt is 2.

Solution: Using Eq. 11.10, the tightening force is:

$$T = \frac{\text{FOS}\,(W \sin \alpha) - W \cos \alpha \tan \varphi - cA}{\text{FOS} \cos (\alpha - \beta) + \sin (\alpha - \beta) \tan \varphi} = \frac{53.95}{2.15} = 33.56 \text{ kN}$$

Slot-and-wedge Bolts

One of the first bolt type is slot (also called split) and wedge. It consists of a steel bolt threaded on one end and a few centimetre longitudinal split on other end along with steel wedge inserted into the split as shown in Figure 11.15. The bolts are installed by hammering the bolt with wedges (5.5″ long and conicity 2° ~ 5°) against the bottom of the borehole, thereby bolt expands against the wall of the borehole. Practically, because of uncertainty about the anchorage force, these bolts are of little use. Another type of bolt have a cross type split and a suitable double wedge. This gives expansion in two 2 planes and is, therefore, better than the single wedge type. From Figure 11.15,

$$T = F_n \sin \alpha + F_s \cos \alpha \tag{11.11}$$

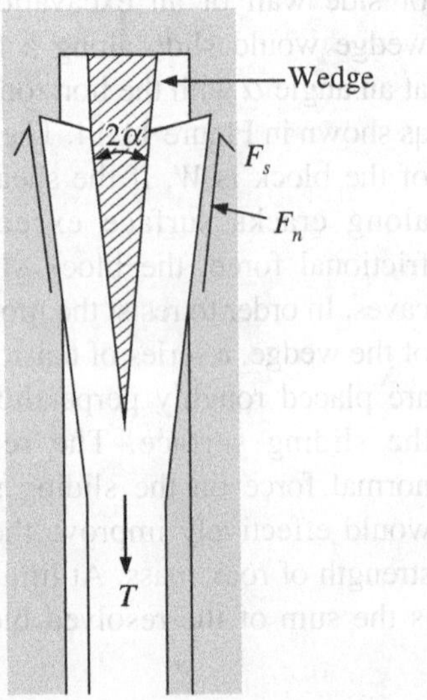

FIGURE 11.15 Slot-and-wedge roof bolts.

where α is half of conicity angle of the wedge measured along the bolt axis. Considering, F_n the normal anchorage force equal to the multiplication of bearing capacity of the rock mass (σ_Y) and area of contact of the wedge (A), and $F_s = F_n \tan \varphi$, the above equation can be rewritten as

$$T = A\sigma_Y (\sin \alpha + \cos \alpha \tan \varphi) \tag{11.12}$$

where φ is the friction angle between wedge and rock mass, and T is anchorage force.

EXAMPLE 11.5 Determine the anchorage force of a slot and wedge bolt installed in rock mass having bearing capacity of 20 MPa. Assume the borehole diameter is 25 mm, bolt contact length is 150 mm, conical angle is 4° and friction angle between bolt and rock mass is 20°.

Solution: The approximate rock-bolt contact are $A = \pi \times 0.025 \times 0.15 = 0.0118$ m^2. Now, using Eq. 11.12, the anchorage force of bolt is 102 kN.

Support Provided by Expansion Shell

As mentioned before, in mechanically anchored bolt, expansion shell makes contact with the rock mass and it provides frictional resistance and tension to the bolt shank. A typical expansion shell bolt is shown in Figure 11.16. It consists of steel rod that holds a conical piece (core) on the threaded end. In general, there are four shells around this central piece (core) those make contact with the rock mass (Biron and Arioglu, 1983). The tightened part is shown in Figure 11.16 along with the free body diagrams of one shell and the core. From free body diagram 11.16(b), it can be written that

$$F_{n1} = F_n \cos \alpha + F_s \sin \alpha \tag{11.13}$$

$$F_{s1} = -F_n \sin \alpha + F_s \cos \alpha \tag{11.14}$$

where α = half of conical angle of the wedge with the bolt axis. Therefore, the anchorage force (T) of the bolt in terms of wedge angle α becomes

$$T = F_{n1} \sin \alpha + F_{s1} \cos \alpha = F_s \tag{11.15}$$

Considering frictional coefficient of expansion shell material as μ_s, from Eq. 11.13 and 11.14, we can get,

$$\tan \varphi = \mu_r = \frac{\sin \alpha + \mu_s \cos \alpha}{\cos \alpha - \mu_s \sin \alpha}$$

Considering, F_n as the bearing capacity of the rock mass σ_Y times the area of contact of the shell (A), and at the limit $F_s = F_n \tan \varphi$, Eq. 11.15 can be rewritten as:

$$T = A\sigma_Y (\cos 2\alpha + \sin 2\alpha \tan \varphi) \tag{11.16}$$

where φ = friction angle between shell and rock mass.

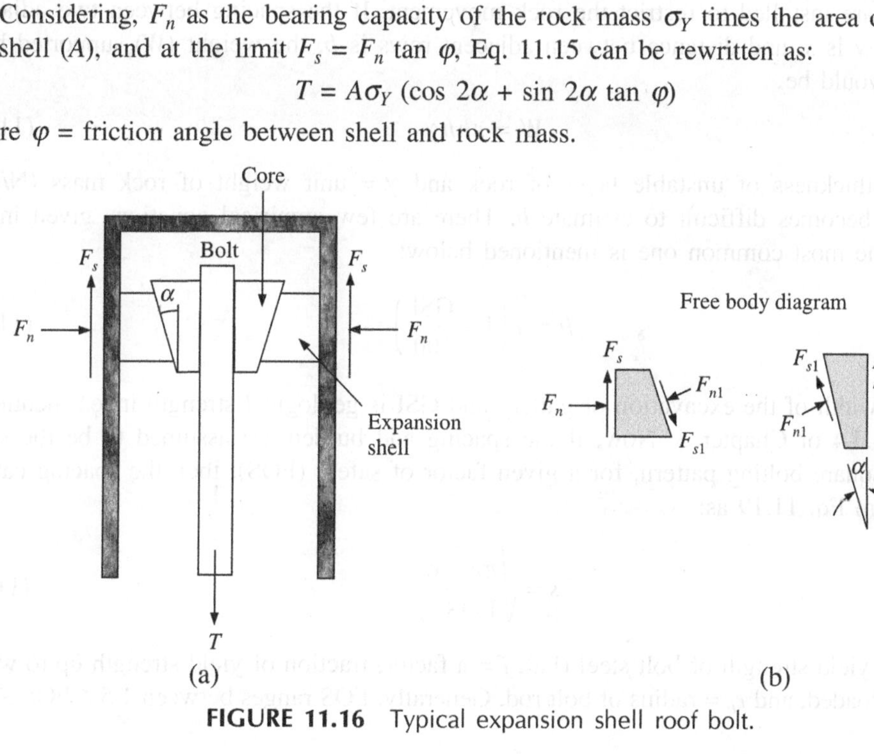

FIGURE 11.16 Typical expansion shell roof bolt.

EXAMPLE 11.6 Determine the anchorage force of a mechanically anchored bolt installed in rock mass having bearing capacity of 20 MPa. Assume the borehole diameter is 25 mm, shell length is 150 mm, conical angle is 4° and friction angle between shell and rock mass is 20°.

Solution: The contact area $A = \pi \times 0.025 \times 0.15 = 0.0118\ \text{m}^2$. Now, using Eq. 11.16, anchorage force of the bolt, $T = A\sigma_Y \tan \varphi$ is 85.76 kN.

EXAMPLE 11.7 An expansion-shell bolt, consists of four pieces of expansion shell each having friction surface area 5 cm^2. The bearing capacity of rock mass is 20 MPa and coefficient of friction between expansion shell and bolt is 0.25. Determine the anchorage force of the expansion shell bolt.

Solution: From Eq. 11.16, the anchorage force is:

$$T = A_n N \sigma_Y \tan \varphi = 5 \times 10^{-4} \times 4 \times 20 \times 10^6 \times 0.25 = 10\ \text{kN}$$

Support of Bedded Rock Mass

Coal mine roofs are normally composed of stratified rock layers. Thickness of such strata may range from few centimetre to a few metre. At the junction of galleries, these strata may segregate from each other and sag due to its weight. Rock bolts are mandatory to stitch these rock layers together with an overlaid competent rock stratum or in other words suspended from the overlain competent rock stratum. In the following calculations, it is assumed that the bolted rock block has been separated from the adjoining strata and hence, spacing of bolt is estimated considering only the weight of the block. Figure 11.17 depicts a coal mine roof strata where bolts are installed to restrict the rock movement. If the spacing between two adjacent bolts in a row is s, and distance between adjacent rows is b, the weight (W) supported by a single bolt would be

$$W = sbh\gamma \tag{11.17}$$

where, h = thickness of unstable layer of rock and γ = unit weight of rock mass (N/m^3). At times, it becomes difficult to estimate h. There are few empirical equations given in the literatures, the most common one is mentioned below:

$$h = w\left(1 - \frac{\text{GSI}}{100}\right) \tag{11.18}$$

where, w = width of the excavation or gallery and GSI is geological strength index mentioned in Section 3.3.4 of Chapter 3. Now, if the spacing and burden are assumed to be the same ($b = s$) for square bolting pattern, for a given factor of safety (FOS), then the spacing can be obtained from Eq. 11.19 as:

$$s = \sqrt{\frac{\pi r_b^2 f \sigma_b}{\text{FOS}\, h\gamma}} \tag{11.19}$$

where, σ_b = yield strength of bolt steel (Pa), f = a factor, fraction of yield strength up to which bolt will be loaded, and r_b = radius of bolt rod. Generally, FOS ranges between $1.5 \le \text{FOS} \le 3.0$.

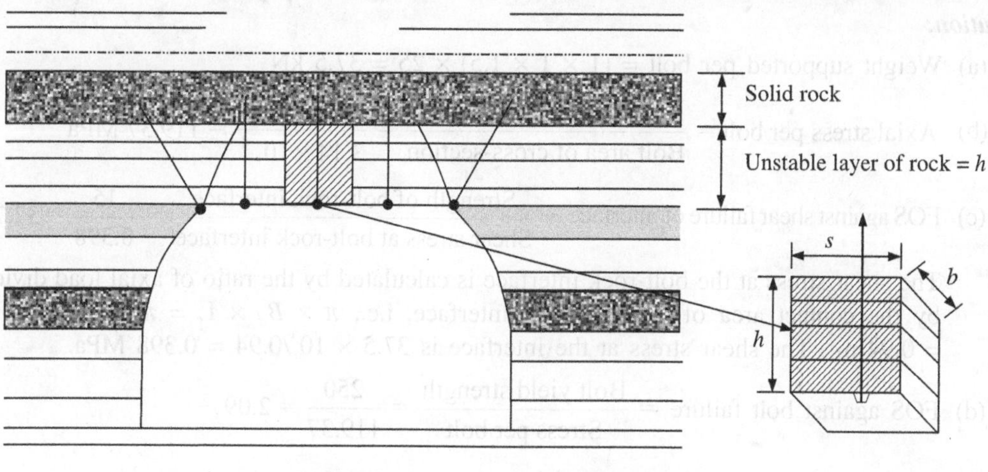

FIGURE 11.17 Design of roof bolt.

EXAMPLE 11.8 A 4 m wide and 20 m long roadway is supported by point anchored bolts in a square pattern with spacing 1 m. If the point of anchorage is 1.5 m above the exposed roof, and average unit weight of roof rock is 25 kN/m^3. Then calculate

(a) load per bolt.

(b) longitudinal stress in bolts if 20 mm diameter (B_d) circular cross-section steel rods are used.

(c) FOS of bolt against shear failure at bolt-rock interface. If breaking strength of interface is 15 MPa.

(d) FOS of bolt against bolt failure if bolt yield strength is 250 MPa in tension.

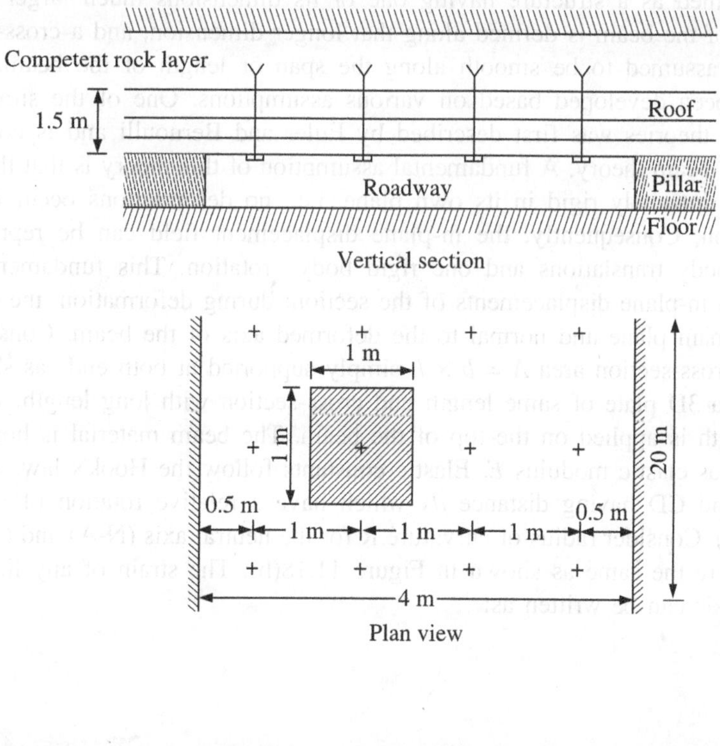

Solution:

(a) Weight supported per bolt = (1 × 1 × 1.5) × 25 = 37.5 kN

(b) Axial stress per bolt $= \dfrac{\text{Weight per bolt}}{\text{Bolt area of cross-section}} = \dfrac{37.5}{3.14 \times 10^{-4}} = 119.37$ MPa

(c) FOS against shear failure of interface $= \dfrac{\text{Strength of bolt-rock interface}}{\text{Shear stress at bolt-rock interface}} = \dfrac{15}{0.398} = 37.70$.

The shear stress at the bolt-rock interface is calculated by the ratio of axial load divided by the contact area of the bolt-rock interface, i.e., $\pi \times B_d \times L = \pi \times 0.020 \times 1.5 = 0.94\ m^2$. The shear stress at the interface is $37.5 \times 10^3/0.94 = 0.398$ MPa.

(d) FOS against bolt failure $= \dfrac{\text{Bolt yield strength}}{\text{Stress per bolt}} = \dfrac{250}{119.37} = 2.09$.

Support to Bedded Strata: Composite Beam

Stratified rock strata are bolted together to reduce overall deformation and tensile stress that develop in individual strata. Bolts bind rock strata together and act by maintaining no slippage or friction along the bedding planes, and thus controlling the dilation of failed roof layers. Roof bolts working in beam building of stratified strata than in suspension must be installed with higher densities. Before analysis of bolted stratified rock strata in terms of displacement and stresses, it is required to have some basic information regarding beam bending due to surface loading.

Beam Theory

A beam is defined as a structure having one of its dimensions much larger than the other two. The axis of the beam is defined along that longer dimension, and a cross-section normal to this axis is assumed to be smooth along the span or length of the beam. Several beam theories have been developed based on various assumptions. One of the simplest and most useful of these theories was first described by Euler and Bernoulli and is commonly called Euler-Bernoulli beam theory. A fundamental assumption of this theory is that the cross-section of the beam is infinitely rigid in its own plane, i.e., no deformations occur in the plane of the cross-section. Consequently, the in-plane displacement field can be represented simply by two rigid body translations and one rigid body rotation. This fundamental assumption deals only with in-plane displacements of the section: during deformation, the cross-section is assumed to remain plane and normal to the deformed axis of the beam. Consider a beam of length L and cross-section area $A = b \times h$ simply supported at both ends as shown in Figure 11.18(a) from a 3D plate of same length and cross-section with long length. A uniform load q per unit length is applied on the top of the beam. The beam material is homogeneous and isotropic and has elastic modulus E. Elastic constants follow the Hook's law. Arbitrary cross-sections AB and CD having distance dx which have a relative rotation of $d\theta$ as shown in Figure 11.18(b). Consider radius of curvature R for the neutral axis (N-A) and (R-y) for the line E_1–E^-, which are the same as shown in Figure 11.18(b). The strain of any line at y distance from neutral axis can be written as:

$$d\theta = \frac{-\varepsilon dx}{y} = \frac{dx}{R}$$

Hence, $$\varepsilon = -\frac{y}{R} \tag{11.20}$$

Strain is negative for compression and possitive for tension.

The below expression is popularly known as bending equation or flexural formula based on Euler-Bernoulli beam theory as:

$$\frac{M}{I} = \frac{\sigma}{y} = \frac{E}{R} \tag{11.21}$$

where, M is bending moment, I is moment of inertia or area moment of inertia or second moment of area about the neutral axis, σ is bending stress, y is distance from the neutral axis, E is the modulus of elasticity, and R is the radius of curvature.

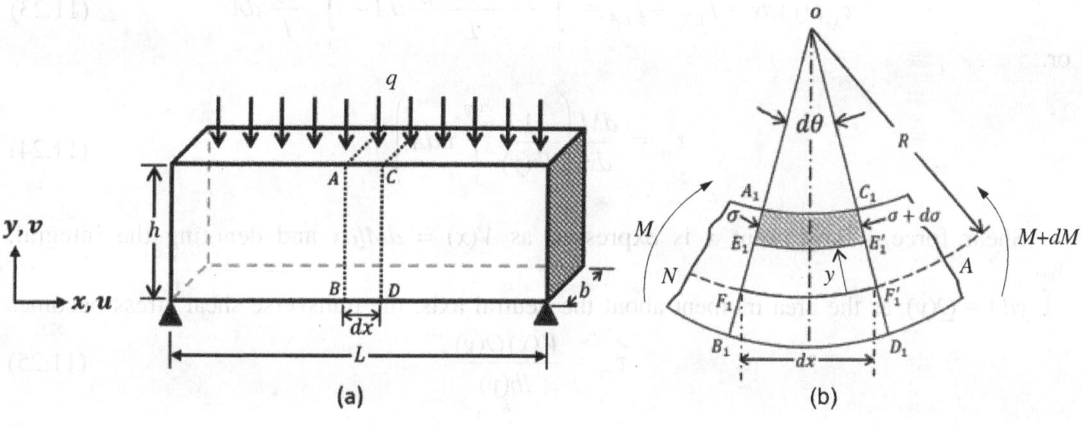

FIGURE 11.18 (a) Beam bending in xy plane with loading and boundary condition; (b) geometry of beam fabric after bending.

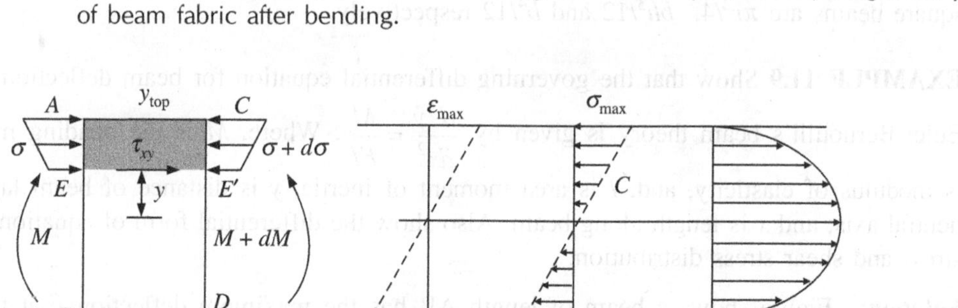

FIGURE 11.19 (a) force and moment acting on beam fabric, (b) Axial strain, axial stress and shear stress distribution.

Now, for Figure 11.19, the sum of the normal forces over the section EA and $E'C$ are:

$$F_{EA} = \int_{y}^{y_{top}} \sigma dA = \int_{y}^{y_{top}} \frac{My}{I} dA \tag{11.22a}$$

$$F_{E'C} = \int_{y}^{y_{top}} (\sigma + d\sigma) dA = \int_{y}^{y_{top}} \frac{(M + dM)y}{I} dA \tag{11.22b}$$

The shear force acting on the lower face (i.e. EE') of the block is $\tau_{xy} b(y) dx$, where τ_{xy} is the transverse shear stress acting on a plane at y distance from the neutral axis, $b(y)$ is beam thickness as a function of y, i.e., distance from the neutral axis. The values of $b(y)$ for different cross-section of the beam varies. For equilibrium of the block under the normal and shear forces

$$\tau_{xy} b(y)\, dx = F_{E'C} - F_{EA} = \int_{y}^{y_{top}} \frac{(M + dM)y}{I} dA - \int_{y}^{y_{top}} \frac{My}{I} dA \tag{11.23}$$

or,

$$\tau_{xy} = \frac{dM}{dx}\left(\frac{1}{Ib(y)} \int_{y}^{y_{top}} y dA\right) \tag{11.24}$$

Shear force at any point x is expressed as $V(x) = dM/dx$ and denoting the integral, $\int_{y}^{y_{top}} y dA = Q(y)$ as the area moment about the neutral axis, the transverse shear stress becomes

$$\tau_{xy} = \frac{V(x)Q(y)}{Ib(y)} \tag{11.25}$$

Note that the area moment of inertia or moment of inertia, I for circular, rectangular, and square beams are $\pi r^2/4$, $bh^3/12$ and $b^4/12$ respectively.

EXAMPLE 11.9 Show that the governing differential equation for beam deflection based on Euler-Bernoulli's beam theory is given by $\frac{d^2 y}{dx^2} = \frac{M}{EI}$. Where, M is the bending moment, E is modulus of elasticity, and, I is area moment of inertia, y is distance of beam layers from neutral axis, and x is length along beam. Also show the differential form of equation for shear stress and shear stress distribution.

Solution: Figure shows a beam of length AB has the maximum deflection δ at the centre. The beam deflects at each point along x-axis having a radius of curvature R. If a small element ds in the deflected beam subtend an angle $d\theta$ at the centre of the bended beam, then $AE = ds = Rd\theta$, $ds = \sqrt{dx^2 + dy^2}$, and $\tan \theta = dy/dx$.

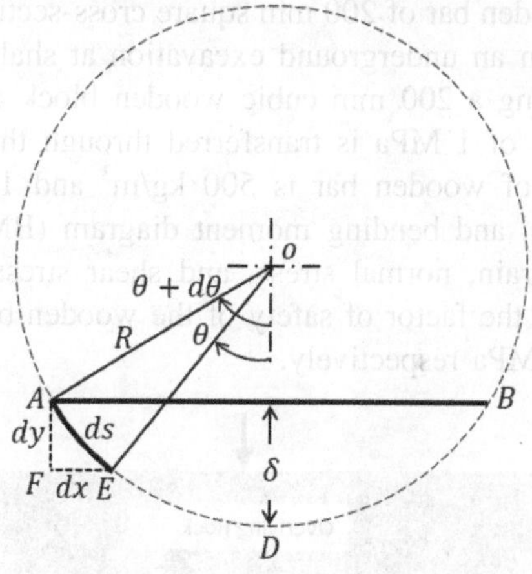

Differentiating tan $\theta = dy/dx$, with respect to x and replacing tan θ gives:

$$d\theta = \frac{d^2y/dx^2}{1 + (dy/dx)^2}dx$$

Using $AE = ds = Rd\theta$, $ds = \sqrt{dx^2 + dy^2}$ in above expression gives,

$$\frac{1}{R} = \frac{d\theta}{dx\sqrt{1 + (dy/dx)^2}}$$

Equating above two expressions it gives,

$$\frac{1}{R} = \frac{d^2y/dx^2}{[1 + (dy/dx)^2]^{3/2}}$$

Using Euler-Bernoulli's equation to replace $1/R = M/EI$ and $dy/dx \approx 0$, we find,

$$\frac{d^2y}{dx^2} = \frac{M}{EI}$$

The relationship between shear force, bending moment and deflection of the beam can be obtained by

$$\frac{dM}{dx} = V = EI\frac{d^3y}{dx^3}$$

where, V represents the shear force and the rate of intensity of loading by shear force or its distribution can be found by further differentiating as:

$$\frac{dV}{dx} = EI\frac{d^4y}{dx^4}$$

EXAMPLE 11.10 A wooden bar of 200 mm square cross-section, 6.5 m in length, is installed in a gallery of 6 m wide in an underground excavation at shallow depth. The wooden bar is tightened with the roof using a 200 mm cubic wooden block as shown in the Figure below. The induced vertical stress of 1 MPa is transferred through the wooden block. The density and modulus of elasticity of wooden bar is 500 kg/m^3 and 10 GPa respectively. Draw the shear force diagram (SFD), and bending moment diagram (BMD) along the beam span and draw deflection, normal strain, normal stress, and shear stress distribution across the beam cross-section. Comment on the factor of safety of the wooden bar if its compressive and shear strength is 44 MPa and 7 MPa respectively.

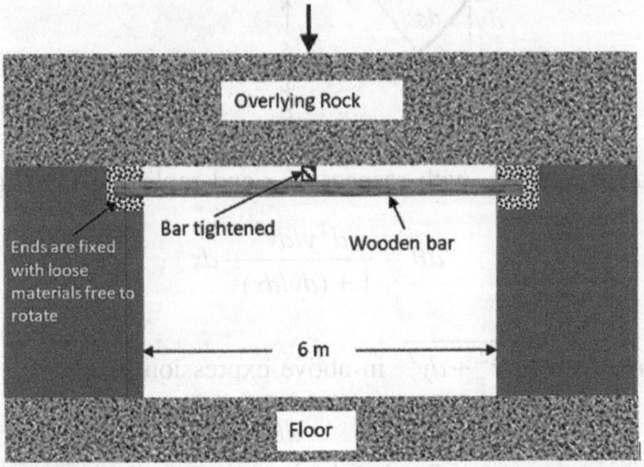

Solution: As the end is not fixed at the end, it is a problem of simply supported beam loaded at the midpoint. Vertical stress of 1 MPa is transferred to the beam through an area of contact $0.2^2 = 0.04$ m^2. Hence, the concentrated load at midpoint is $1 \times 10^6 \times 0.04 = 40000$ N. The FBD shows the support reaction R_A and R_B at both ends. As there is no horizontal forces at the ends so horizontal reactions at both ends are zero.

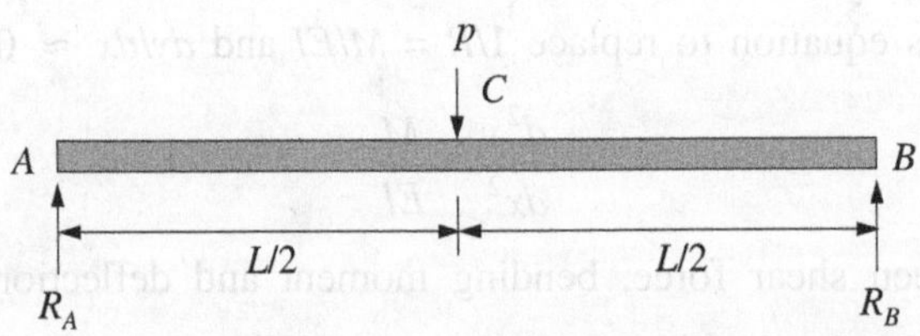

Now, for vertical force equilibrium then $R_A + R_B = p$ and concentrated load is acting at midpoint, hence $R_A = R_B = p/2 = 20000$ N. Let us consider the shear force and bending moment acting on each part of the beam is F_x and M_x respectively. Now consider the shear force acting in between a point A and C which is at a distance of x from point A as shown in Figure below. For shear force calculation, either left or right side of the section can be considered. Let us consider the left part (a) between A and C, $F_x = p/2$; (b) between C and B, $F_x = +\frac{p}{2} - p = -\frac{p}{2} = -20000$ N; (c) at C, shear force F_x changes from $+p/2$ to $-p/2$. Between point A and C, there is no change in forces so shear force remains same and similarly in between point C and B.

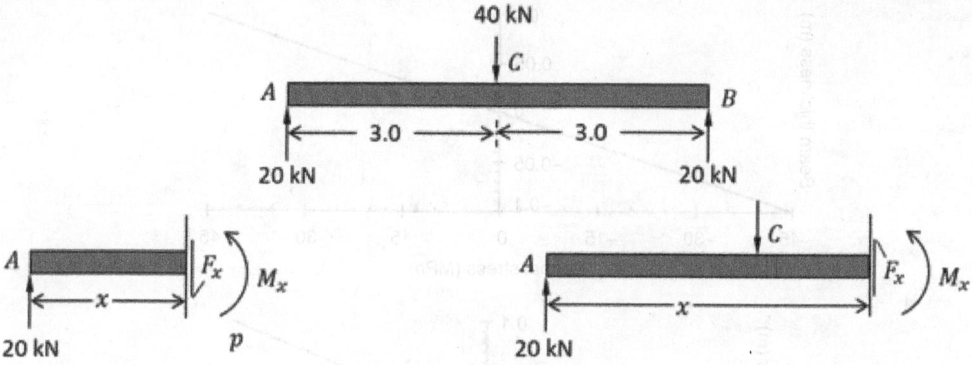

Free Body Diagram (FBD)

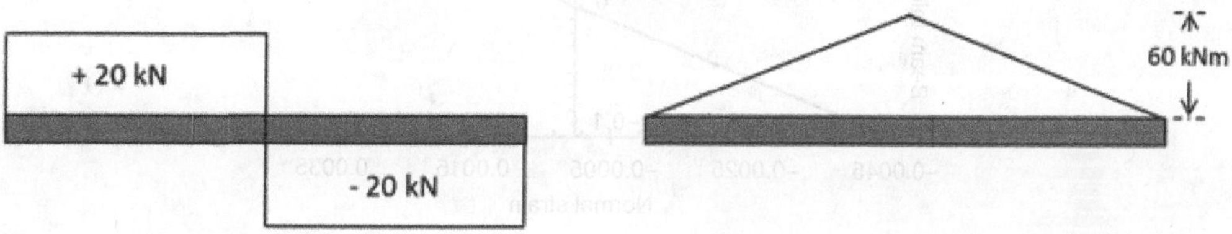

SFD BMD

Considering counter-clockwise moment as posetive, bending moment between A and C, $M_x = R_A x \Rightarrow +\frac{p}{2}x$; at A, $M_A = 0$, and at C, $M_C = +\frac{p}{2}\frac{L}{2} = \frac{pL}{4}$; between C and B, $M_x = R_A x - p\left(x - \frac{L}{2}\right) \Rightarrow \frac{pL}{2} - \frac{px}{2}$; at C, $M_C = \frac{pL}{4} = 60$ kNm , and at B $M_B = 0$.

Considering counter-clockwise moment as positive, bending deflection can be found using Table D.4 **(in Appendix D)** for SS beam with concentrated load at canter then $\delta_x = \frac{Px}{48EI}(3L^2 - 4x^2)$ for $x < l/2$ and $\delta_{max} = \frac{PL^3}{48EI}$ and plotted as:

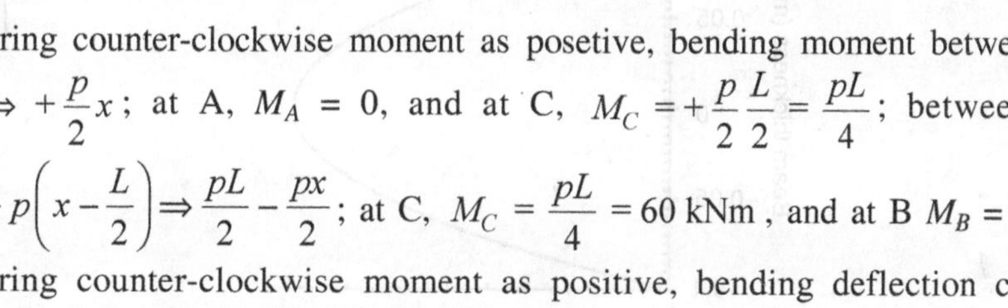

The plots for normal stress is calculated using $\sigma = (M/I)y$, normal strain by $\varepsilon = \sigma/E$, and shear stress using Eq. (11.23) as:

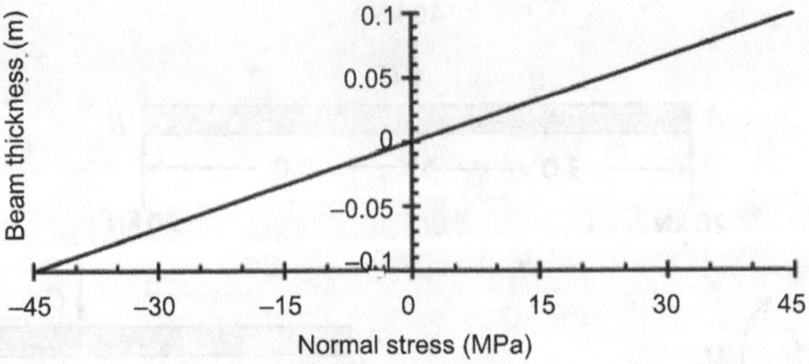

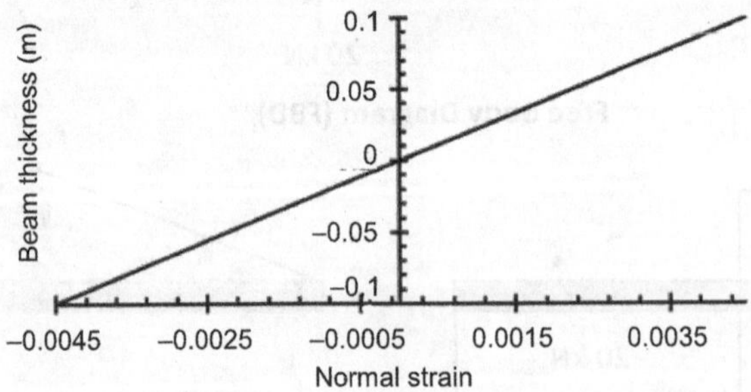

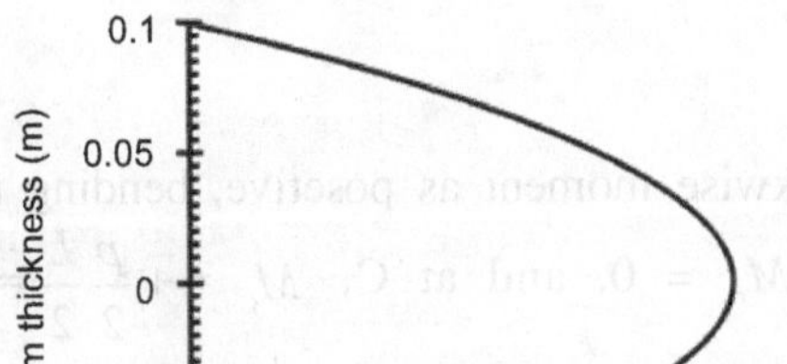

Stratified Composite Beam: Without Rock Bolts

Let us now consider two rock strata of thickness h_1 and h_2 having elastic modulus of E_1 and E_2 respectively are the immediate roof of a coal seam. The width of the opening span in coal seam is L. It is also assumed that a uniform loads q_1 and q_2 in (N/m) act on top of the first and second beams as shown in Figure 11.20. Here, q_1 may be q_2 plus the weight of rock layer 2. It is noticed that both the beams carry a combined load of $q_1 + q_2$. However, in this example, we assume that the external loads are much higher as compared to the weight of the beams and both beams are simply supported at two ends. Assuming uniform thickness

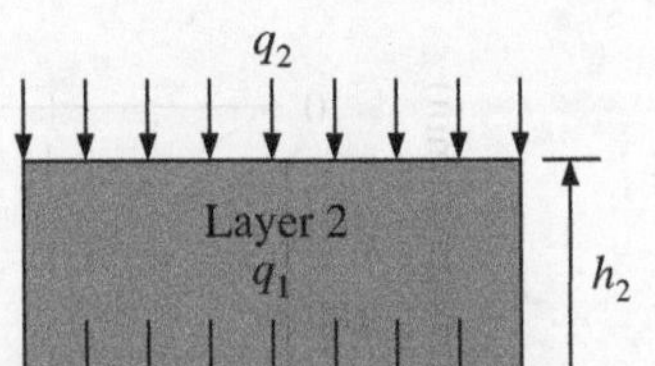

FIGURE 11.20 Two layers of rock strata simply supported.

b for each beam, the neutral axis of each beam will pass through its centre of gravity. The deflection of each beam can be written as follows:

$$v_i = \frac{q_i x}{24E_i I_i}[L^3 - 2Lx^2 + x^3] \text{ for } i = 1, 2 \tag{11.26}$$

where, moment of inertia for rectangular beam $I_i = bh_i^3/12$, b is the width of the beam. The corresponding slope angles of the beams are:

$$\tan \theta_i = \frac{dv_i}{dx} = \frac{q_i}{24E_i I_i}[L^3 - 6Lx + 4x^3] \tag{11.27}$$

It is noticed that slopes at $x = 0$ and $x = L$ are the same and given as:

$$\tan \theta_i \big|_{x=0,L} = \frac{q_i L^3}{24E_i I_i} \tag{11.28}$$

From the above equation, it is clear that for $\tan \theta_1 \leq \tan \theta_2$, both the beams deform together. In other words, the necessary condition for upper beam to be in contact with the lower beam during deformation is:

$$\frac{q_1}{E_1 h_1^3} \leq \frac{q_2}{E_2 h_2^3} \tag{11.29}$$

If the deformation of both the beams is same, the condition is referred as "no interaction case" between the beams meaning one beam is seated on another and act independently. In case, the deformation of the upper beam is higher than that of lower beam, it will be restricted by the lower beam. This condition may resemble to a condition of beam bending in elastic foundation and is beyond the scope of this book. In case, $\tan \theta_1 > \tan \theta_2$, deformation of the lower beam will be higher as compared to the upper beam and hence, the lower beam separates from the upper beam. Hence, a gap between the beams may occur. The expression of the maximum axial and shear stresses and deformation of each beam are given as:

$$\sigma_{i\max} = \frac{M_i y_{i\max}}{I_i} = \frac{3q_i L^2}{4bh_i^2}, \quad \tau_{i\max} = \frac{3q_i L}{4bh_i}, \quad v_{i\max} = \frac{5q_i L^4}{384E_i I_i} \tag{11.30}$$

where, maximum moment $M_i = q_i L^2/8$ and distance to the extreme fibre $y_{i\max} = h_i/2$.

EXAMPLE 11.11 Two rock strata constitute the immediate roof above a coal mine gallery of span 6 m. The thickness and elastic modulus of these two strata from the coal seam are 2 m, 4 m and 5000 MPa and 2000 MPa respectively. Assume that a uniform vertical stress of 2.5 MPa acts on the upper layer and unit weight of both layer is 25 kN/m^3. Show deformation, strain and stress profiles that may generate in each layer.

Solution: We assume $b = 1$ m (this may be the bolt spacing along the axis of the gallery). It is also noted that effect of overlaying strata on these beams are neglected, hence, shear stress on the top fibre of layer 2 is assumed to be zero. From the given data, $q_1 = 2.6$ MN/m (adding weight of 4 m layer with q_2), $q_2 = 2.5$ MN/m, $I_1 = 0.667$ m^4, $I_2 = 5.333$ m^4 and

$L = 6$ m. Using Eq. 11.29, we find that $\tan\theta_1 > \tan\theta_2$ and hence, deformation of layer 1 will be higher as compared to layer 2 and hence, will act independently. Using Eq. 11.30, deformation, shear stress, axial stress and axial strain in both the beams are plotted.

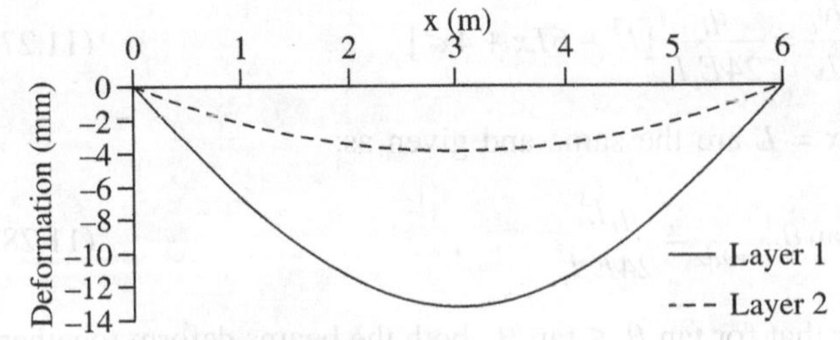

(a) Deformation (layer 1 separates from layer 2)

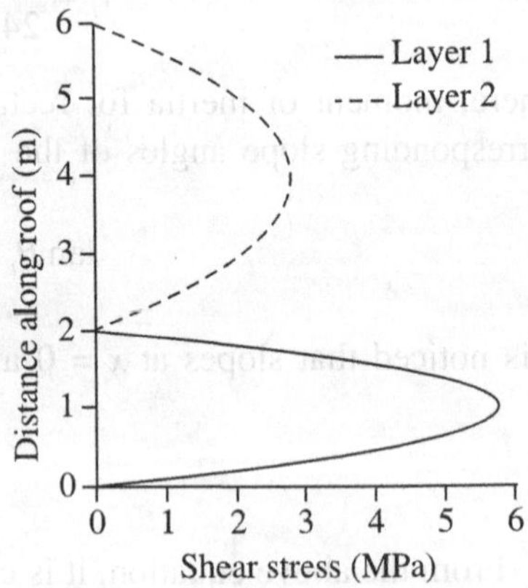

(b) Shear stress

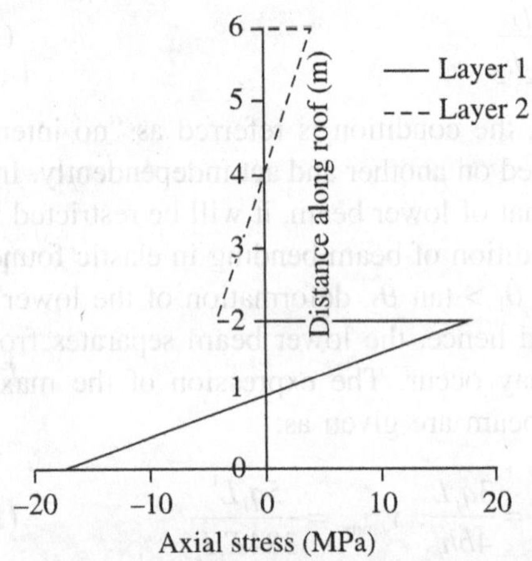

(c) Axial stress (high tensile stress in lower fibre of layer 1)

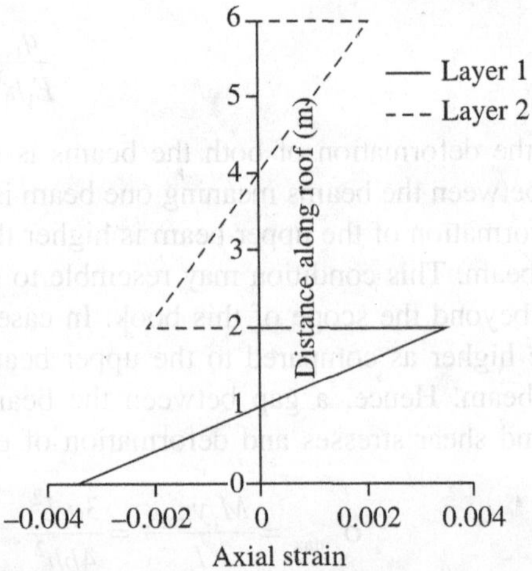

(d) axial strain

Stratified Composite Beam: With Rock Bolts

Installation of rock or cable bolts may bind stratified layers (Figure 11.21) replicating beams into one composite layers/beams within the length of the bolts. In order to achieve this, spacing of bolts must be small. Assuming, uniform load $q = q_1 + q_2$ acts on top of the layer 2, we can obtain the deformation, moment, axial strain, and axial stress and shear stress distributions. It may be noted that axial strain still varies linearly with y, however, axial stress will be discontinuous at the bedding plane since elastic modulus is different for each beam. Distribution of transverse shear stress will be same as a single layer case with exception that a discontinuity of shear jump will occur at the bedding plane.

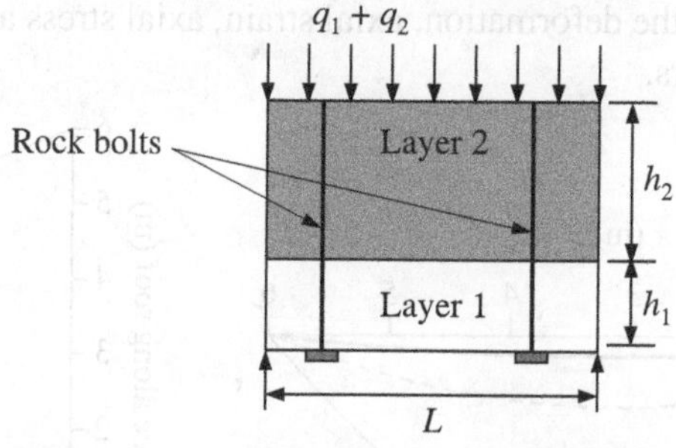

FIGURE 11.21 Composite rock layers made with rock bolts.

The analysis of composite layers is done based on the "transformed area method". In this method, equivalent width of any one of the two layers is determined based on modulus ratio of the layers. For example, if layer 2 having original width b and elastic modulus E_2 is to be converted as layer 1 with elastic modulus E_1, the thickness of the layer remains the same but width is altered. The assumption is that the strain profile and force constant acting on the cross-sections are the same. This results into a new width for layer 2 as:

$$b_2 = \frac{E_2 b}{E_1} \tag{11.31}$$

Now, both the layers are having same elastic modulus but variable widths. Hence, combined moment of inertia and location of neutral axis are required to be estimated before obtaining strain and stress profiles.

EXAMPLE 11.12 Solve Example 11.9 considering bolted strata. Assume all material and geometrical properties remain same.

Solution: We convert layer 2 with the property of Layer 1 and hence, new width of layer 2 would be $b_2 = 2000 \times 1/5000 = 0.4$ m. The figure given below shows the cross-section of layers considering elastic modulus of 5000 MPa. The combined moment of inertia is $I = [(1 \times 6^3)/12] - [2 \times (0.3 \times 4^3/12)] = 14.8$ m^4. The location of neutral axis is at 2.333 m from the bottom of the Layer 1. Considering these geometrical data, first axial strain profile, $\varepsilon(y)$ is estimated using Eq. 11.20. From that axial stress is estimated as $\sigma(y) = E_i \varepsilon(y)$. It is noticed that at the bedding plane, a stress jump will occur since elastic modulus is different for different Layers. For shear stress calculation (Eq. 11.24), width b needs to be considered depending on the distance y.

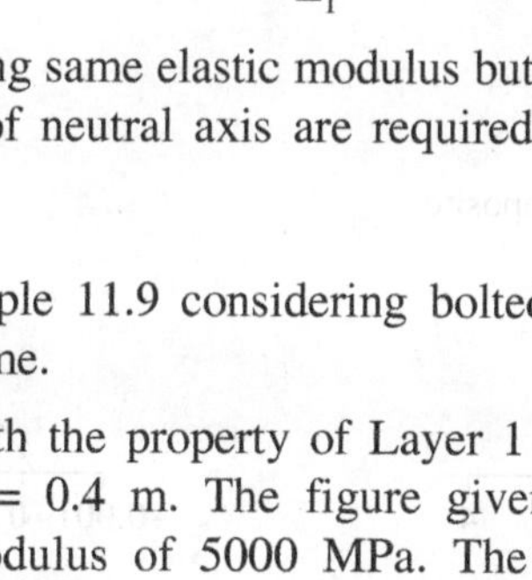

Following figures show the deformation, axial strain, axial stress and shear stress distributions considering composite layers.

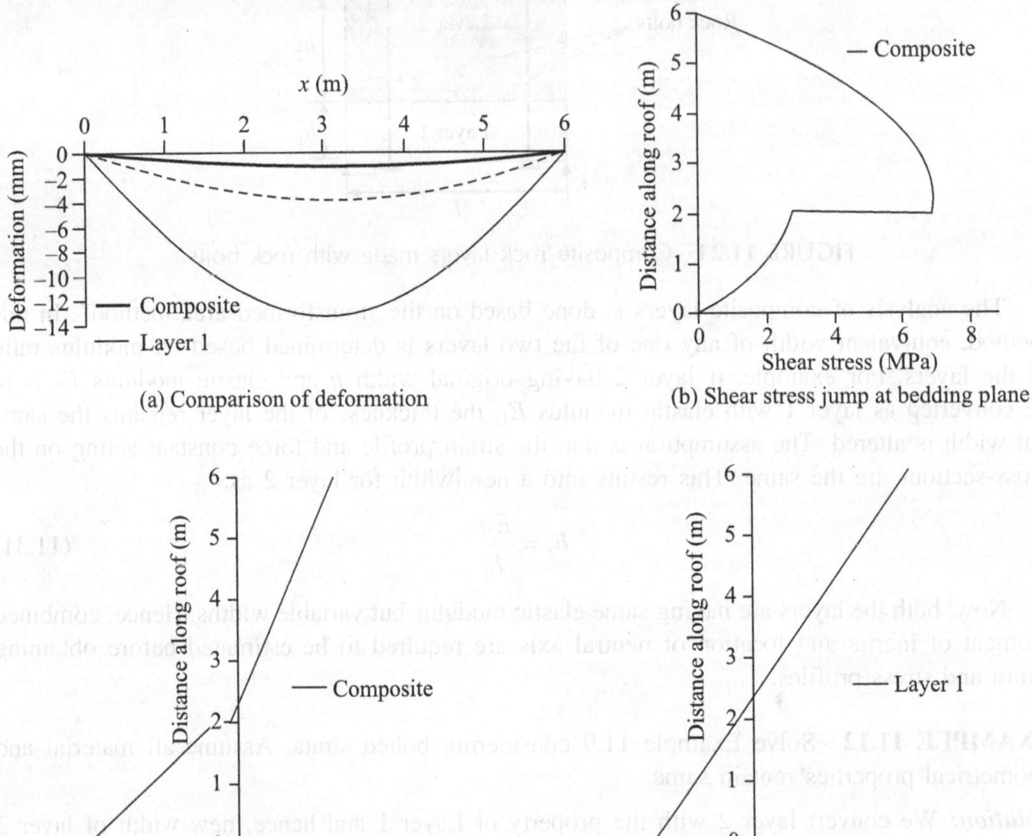

(a) Comparison of deformation

(b) Shear stress jump at bedding plane

(c) Axial stress jump at bedding plane

(d) Axial strain

11.6.7 Fully Grouted Bolts

Rock bolts are grouted either using cement mortar or resins as mentioned above. In the following, mechanics of grouted rock bolts in terms of displacement and load are elaborated starting with equilibrium equation.

A grouted bolt installed in deformable rock mass is subjected to axial loading and it provides resistance to the movement of rock mass through shear stresses developed axially in the bolt-grout interface. In the absence of transverse shear stress (absence of joint), a rock bolt can be represented by one dimensional bar inside rock mass as shown in Figure 11.22(a). Let us assume that *x*-axis represents the axial direction of the bolt. Displacements of bolt rod and that of grout are represented by u_x and u_r respectively. Figure 11.22(b) shows a small (differential) element, dx with stress function in bolt rod, σ_{xx} and shear force per unit length due to bonding, F_s. Hence, the relationship between the shear force and the axial load can be established as:

$$A_b \frac{\partial \sigma_{xx}}{\partial x} dx = -F_s dx \tag{11.32}$$

or

$$\frac{d\sigma_{xx}}{dx} = -\frac{F_s}{A_b} \tag{11.33}$$

where, A_b = area of the cross-section of the bolt = πr_b^2, and r_b = radius of the bolt rod. Now, for an elastic bolt with modulus of E_b,

$$\sigma_{xx} = E_b \frac{du_x}{dx} \tag{11.34}$$

Substituting Eq. 11.34 into Eq. 11.33, it gives

$$\frac{d^2 u_x}{dx^2} = \frac{-F_s}{AE_b} \tag{11.35}$$

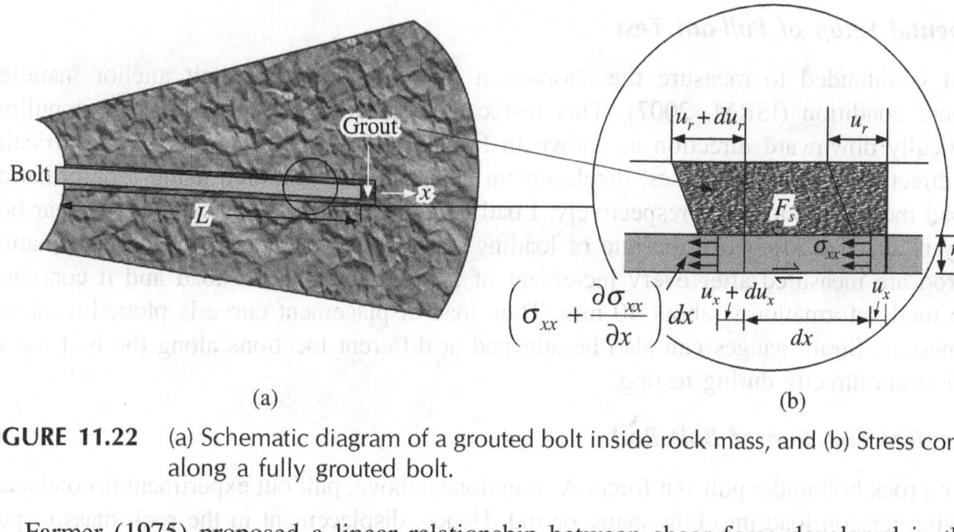

FIGURE 11.22 (a) Schematic diagram of a grouted bolt inside rock mass, and (b) Stress component along a fully grouted bolt.

Farmer (1975) proposed a linear relationship between shear force developed in the bolt-grout interface and relative displacements between bolt and the grout (Tao and Chen, 1984; Hyett et al., 1996). The relationship is expressed as:

$$F_s = 2\pi r_b k\,(u_r - u_x) = k^*(u_r - u_x) \tag{11.36}$$

where k is the shear stiffness (in Pa/m) of the bolt grout interface and can be determined from the stress-displacement response during laboratory/field pull out tests and will be discussed later. In general, k^* depends on the bolt and rock displacements. Displacement function $u_r(x)$ describes the movement of rock mass over the bolt length ($0 < x < L$). In general, u_r will be a monotonically decreasing function with radial distance r measured from a circular tunnel boundary. The form and rate of decrease of u_r will depend on the size and shape of the opening, the strength and structure of the rock mass, stress redistribution occurring around excavation due to the yield of rock mass, and number of bolts installed at the excavation boundary.

Now, combining Eqs. 11.35 and 11.36, the distribution of displacement along the bolt length is described by the second order inhomogeneous linear ordinary differential equation as:

$$\frac{d^2u_x}{dx^2} - \frac{k^*}{AE_b}u_x = -\frac{k^*}{AE_b}u_r \tag{11.37}$$

Equation 11.37 suggests that the solution of u_x can be obtained if u_r is known beforehand. Numerical solution of Eq. 11.37 can be obtained in two ways: (i) inserting the discrete form of the equation into regular finite element procedure and adding additional degree of freedom (nodes) for bolt rod, and (ii) after obtaining the solution of u_r at nodal points, explicit finite difference methods can be applied to solve u_x. Once, u_x is known, strain, function, ε_{xx} and stress function, σ_{xx} can be evaluated.

11.6.8 Pull-out Test of Grouted Rock Bolt

Experimental Setup of Pull-out Test

This test is intended to measure the short-term strength of a rock bolt anchor installed under field condition (ISRM, 2007). This test can be performed in two ways: (i) pulling bolt vertically downward direction as shown in Figure 11.23, and (ii) pulling bolt vertically upward direction. In both the cases, displacement and load are measured using a deformation and a load measurement gauge respectively. Load is applied on the bolt rod only. A gear-box assembly is attached to control the rate of loading. In this test, pull-out load and deformation of bolt rod are measured after every increment of approximately 5 kN load and it continues until the total deformation is about 40 mm. Then load-displacement curve is plotted to obtain the stiffness, k. Strain gauges can also be attached at different locations along the bolt rod to get axial strain directly during testing.

Displacement Function of Bolt Rod

Consider a rock bolt under pull-out force. As mentioned above, pull out experiment is conducted by applying tension load into bolt shank or rod. Hence, displacement in the rock mass to pull out test is assumed to be negligible. In the following analysis, it is also assumed that, the tensile strength (T_y) of the bolt rod is significantly higher compared to the pull-out force, hence, the bolt yielding is not considered or in other word, bolt rod is assumed to behave elastically. Also, it is assumed that no interface slippage will occur. Hence, de-bonding in any part along the bolt rod is neglected.

Now, the pull-out force (T_0) mobilises interface bond shear stress τ, along the length of the rock bolt but the grout/rock is inextensible. In limiting equilibrium, considering a small differential element, similar as describe in previous section, equilibrium equation is obtained by letting $u_r = 0$ in Eq. 11.35 as:

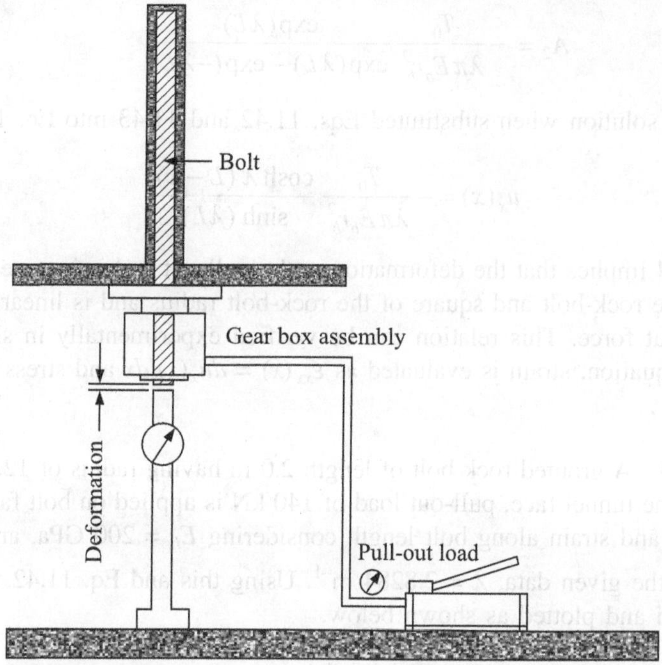

FIGURE 11.23 Field experimental setup of pull-out test.

$$\frac{d^2u_x}{dx^2} - \lambda^2 u_x = 0 \tag{11.38}$$

where $\lambda^2 = \frac{2k}{E_b r_b}$. General solution of Eq. 11.38 is of the form:

$$u_x(x) = A_1 \exp(\lambda x) + A_2 \exp(-\lambda x) \tag{11.39}$$

where A_1 and A_2 are integration constants. Physical meaning of Eq. 11.39 is that displacement (u_x) in bolt rod due to pull-out load varies exponentially along the length x, and also depends upon the value of the factor λ which depends on of the grout and bolt properties. Constants in Eq. 11.39 can be obtained with following boundary conditions:

$$T = T_0 \quad \text{or} \quad \frac{du_x}{dx} = \frac{T_0}{\pi E_b r_b^2} \qquad \text{at } x = 0 \tag{11.40}$$

$$T = 0 \quad \text{or} \quad \frac{du_x}{dx} = 0 \qquad \text{at } x = L \tag{11.41}$$

Using these boundary conditions, constants are obtained as:

$$A_1 = -\frac{T_0}{\lambda \pi E_b r_b^2} \frac{\exp(-\lambda L)}{\exp(\lambda L) - \exp(-\lambda L)} \tag{11.42}$$

$$A_2 = -\frac{T_0}{\lambda \pi E_b r_b^2} \frac{\exp(\lambda L)}{\exp(\lambda L) - \exp(-\lambda L)} \tag{11.43}$$

Therefore, the solution when substituted Eqs. 11.42 and 11.43 into Eq. 11.39 becomes

$$u_x(x) = -\frac{T_0}{\lambda \pi E_b r_b^2} \frac{\cosh \lambda (L - x)}{\sinh (\lambda L)} \tag{11.44}$$

Equation 11.44 implies that the deformation at the pull-out end is inversely proportional to the elasticity of the rock-bolt and square of the rock-bolt radius and is linearly proportional to the applied pull-out force. This relation can be verified experimentally in similar conditions. From the above equation, strain is evaluated as $\varepsilon_{xx}(x) = du_x(x)/dx$ and stress is obtained using Eq. 11.34.

EXAMPLE 11.13 A grouted rock bolt of length 2.0 m having radius of 12.5 mm is installed in rock mass. At the tunnel face, pull-out load of 140 kN is applied on bolt face. Determine the bolt displacement and strain along bolt length considering E_b = 200 GPa, and k = 10 GPa/m.

Solution: From the given data, $\lambda = 2.8284\ \text{m}^{-1}$. Using this and Eq. 11.42, displacement and strain is calculated and plotted as shown below.

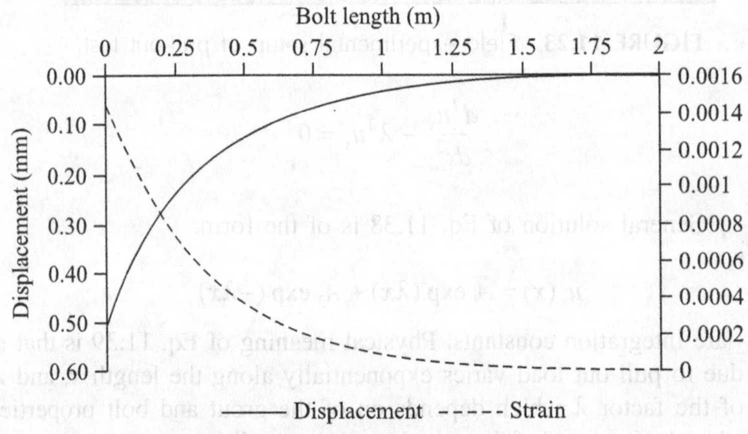

11.6.9 Bolt–Zone of Influence

Roof strata of an excavated gallery or tunnel constructed in a stratified rock deposits deform due to combined effect of their self-weight and overburden pressure. The amount of deformation is the maximum in the centre of the excavation and is minimum at the ends where rock strata are supported by pillars or sidewalls. The amount of maximum deformation varies based on the thickness and stiffness of rock strata, and also on the location of strata from the excavation level. In many cases, bed separation occurs which causes some what destressed zone. The beds within the pressure arch deflect slightly and no longer carry the weight of the superincumbent mass of strata. In an un-reinforced stratified roof, the pressure arch may be located quite deeply in the roof, depending upon the thickness of the beds and other geological conditions. Individual layers without bolting have no interaction with each other, hence, individual pressure arches are formed for every layer. In the case of the bolted beam, all layers reacted together to the applied load and a common pressure arch is formed. The pressure arch formed in bolt-reinforced rock is called an *artificial pressure arch* to distinguish it from the one naturally formed in rock.

The reinforcement effect of an individual bolt is limited in the near field of the bolt and the rock within a certain angle to the bolt ends would be reinforced effectively (Figure 11.24). This angle is called the ''reinforcement angle''. For systematic bolting, the reinforced areas of individual bolts will overlap each other so that an interaction zone is formed in the bolted portion of rock. The thickness of the interaction zone is associated with bolt length and bolt spacing.

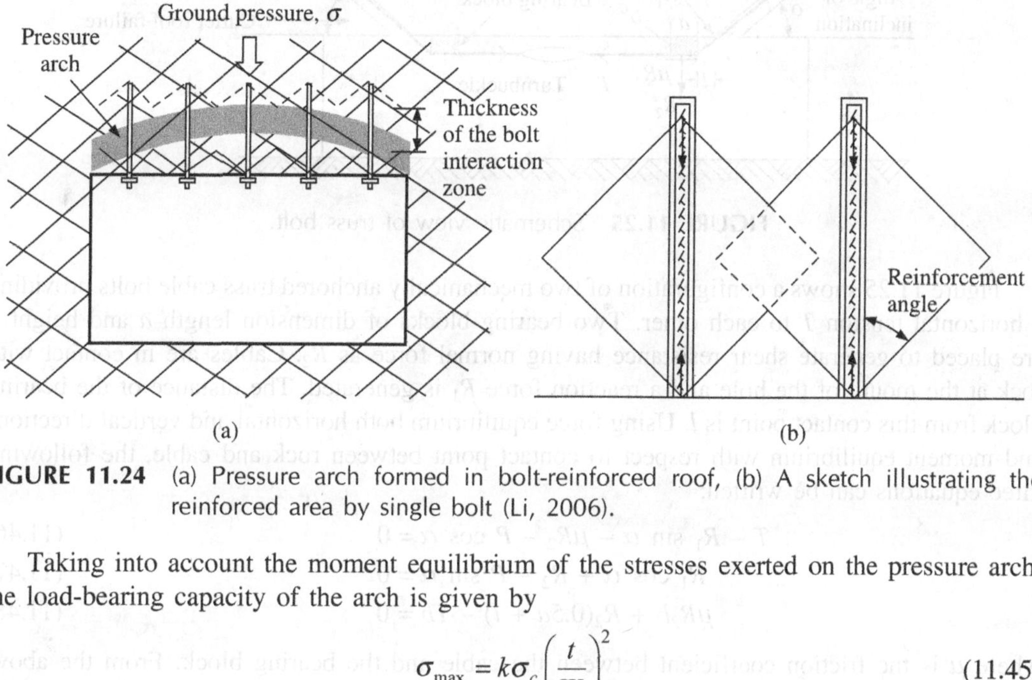

FIGURE 11.24 (a) Pressure arch formed in bolt-reinforced roof, (b) A sketch illustrating the reinforced area by single bolt (Li, 2006).

Taking into account the moment equilibrium of the stresses exerted on the pressure arch, the load-bearing capacity of the arch is given by

$$\sigma_{\max} = k\sigma_c\left(\frac{t}{W}\right)^2 \tag{11.45}$$

where $\sigma_{\max}$ is the maximum lateral ground pressure that the pressure arch could bear, parameter σ_c is the uniaxial compressive strength of the rock, W is the span of the opening, and t is the thickness of the zone of moment interaction. Coefficient k is proportional to the moment arm length in the pressure arch. The theoretical value of k is 1.5.

11.6.10 Truss Bolts

It is used to control the stability of underground excavations in severe ground conditions and cutter roof failure in layered rocks especially in sedimentary deposits (Ghabraie et al., 2013). Truss bolt consists of two inclined members at two top corners and one horizontal member on the roof. A common truss bolt system, known as the Birmingham truss, consists of two long cable bolts which are connected at the middle of the roof. Cables may be anchored mechanically at the end or fully grouted or both. Horizontal tension is applied by means of a turnbuckle at the connection point of the cables at the roof and transferring a compression to the rock. One of the advantages of truss bolt systems is the ability to control the cutter roof failure. Cutter roof is a common type of failure in laminated rock formations in flat roof excavations. In this type of failure, shear cracks propagate from the corners of the roof and as they reach the first bedding plane, a huge block separates from the roof (Su and Peng, 1987).

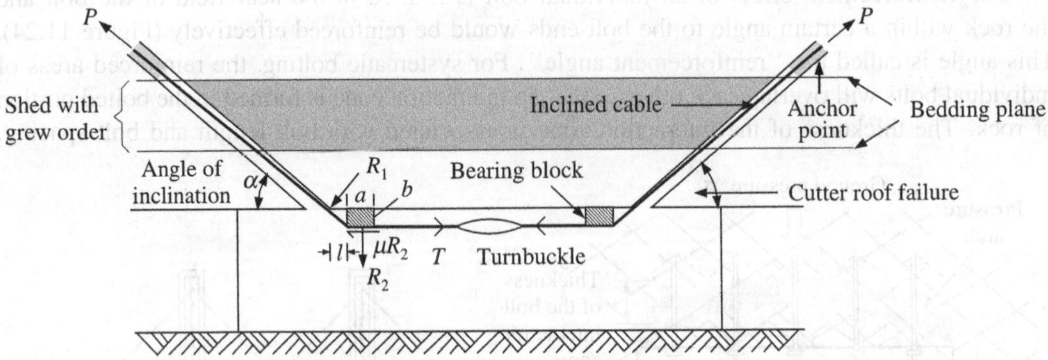

FIGURE 11.25 Schematic view of truss bolt.

Figure 11.25 shows a configuration of two mechanically anchored truss cable bolts providing a horizontal tension T to each other. Two bearing blocks of dimension length a and height b are placed to generate shear resistance having normal force as R_2. Cables are in contact with rock at the mouth of the hole and a reaction force R_1 is generated. The distance of the bearing block from this contact point is l. Using force equilibrium both horizontal and vertical directions and moment equilibrium with respect to contact point between rock and cable, the following three equations can be written:

$$T - R_1 \sin \alpha - \mu R_2 - P \cos \alpha = 0 \tag{11.46}$$

$$R_1 \cos \alpha + R_2 - P \sin \alpha = 0 \tag{11.47}$$

$$\mu R_2 b + R_2(0.5a + l) - Tb = 0 \tag{11.48}$$

where μ is the friction coefficient between the cable and the bearing block. From the above equations, three unknowns, R_1, R_2 and P can be obtained as:

$$R_1 = \frac{T}{(\mu b + 0.5a + l)}[(0.5a + l)\sin \alpha - b\cos \alpha] \tag{11.49}$$

$$R_2 = \frac{Tb}{(\mu b + 0.5a + l)} \tag{11.50}$$

$$P = \frac{T}{(\mu b + 0.5a + l)}[(0.5a + l)\cos \alpha + b\sin \alpha] \tag{11.51}$$

If the bearing blocks are placed very near to the mouth of the hole, then R_1 may be considered to zero since cable may not be in contact with rock. Hence,

$$\tan \alpha = \frac{b}{0.5a}, R_2 = \frac{T}{(\mu + \cot \alpha)}, P = \frac{T}{(\mu \sin \alpha + \cos \alpha)}$$

EXAMPLE 11.14 The fixing force of a cable in the hole of a roof truss tightening force of 10 kN. The holes are driven at an angle of 60°, the block of thickness (8 cm) and length (10 cm) is placed at a distance of 22 cm from hole. The coefficient of friction between block and cable is 0.4. Find anchorage force if the same block is placed at 5 cm from the hole.

Solution: Using Eq. 11.49, we have

$$P = \frac{T}{(\mu b + 0.5a + l)}[(0.5a + l)\cos\alpha + b\sin\alpha] = 9.34 \text{ kN}$$

where, $$\alpha = \tan^{-1}\frac{8}{10+5} = 28°$$

11.6.11 Specialised Bolts

Perfo Rock Bolt

Perfo rock bolts are used for dual purpose to eliminate the loss of grout material in fissured rock and to develop anchorage in fissured rock mass. Perfo rock bolt system consists of thin perforated cylindrical tubes with diameter little less than the borehole diameter along with bolt system as shown in Figure 11.26. These tubes may or may not be split longitudinally into two halves.

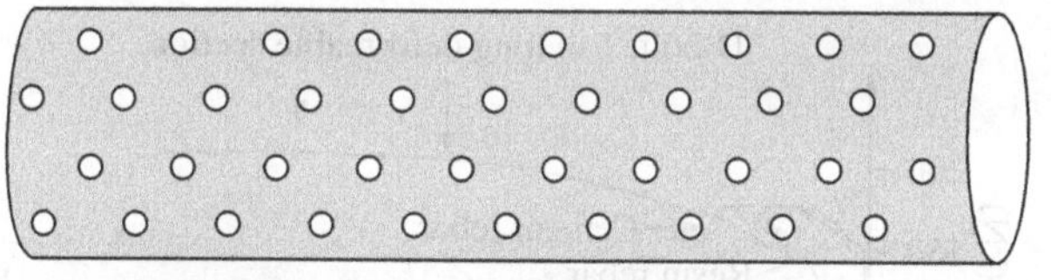

FIGURE 11.26 Perforated rock bolt tube.

Before installation of bolts, the perforated tubes are filled with mortar and inserted into the borehole. A steel bar is then pushed through the perforated tube, forcing the grout material squeeze through perforations, completely filling the annulus between the tube and the borehole wall with grout, thereby anchored in soft or fractured rocks. In case of split tube the grout materials are filled separately and then tied together at the ends with soft iron wire. This procedure will provide an untensioned installation; if a tensioned bolt is required, a threaded bar can be pushed into a short length of perfo tube placed in the back end of the drillhole. The bolt is tensioned after the mortar has hardened.

The perforated system is satisfactory for long-life installations especially in soft rocks that do not provide a normal anchorage for conventional bolts or in harder but jointed or fractured rocks, where anchorage over the length of the hole is desirable. This type of bolt can be equipped with plate and nut for the support of the rock surface.

D-bolt

The dynamic bolt or more commonly termed as D-bolt is made of a smooth steel bar or shank that has number of integrated anchors spaced evenly or unevenly along its length (Figure 11.27).

Anchors are stronger than the shank of the bolt. These bolts can absorb vibration and effective to both burst-prone and squeezing rock conditions. The anchors are firmly fixed within a borehole using either cement grout or resin, while the smooth sections of the bolt between the anchors may freely deform in response to rock deformation. The anchors can have different shapes; two common shapes are paddle and wiggle anchor. Since this bolt is compartmentalised between anchors, failure of one section does not affect the reinforcement performance of other sections. The bolt has large load-bearing and deformation capacities. Static pull tests and dynamic drop tests show that the bolt length elongates by 14–20% at a load level equal to the strength of the bolt material, thereby absorbing a large amount of energy (Li, 2010).

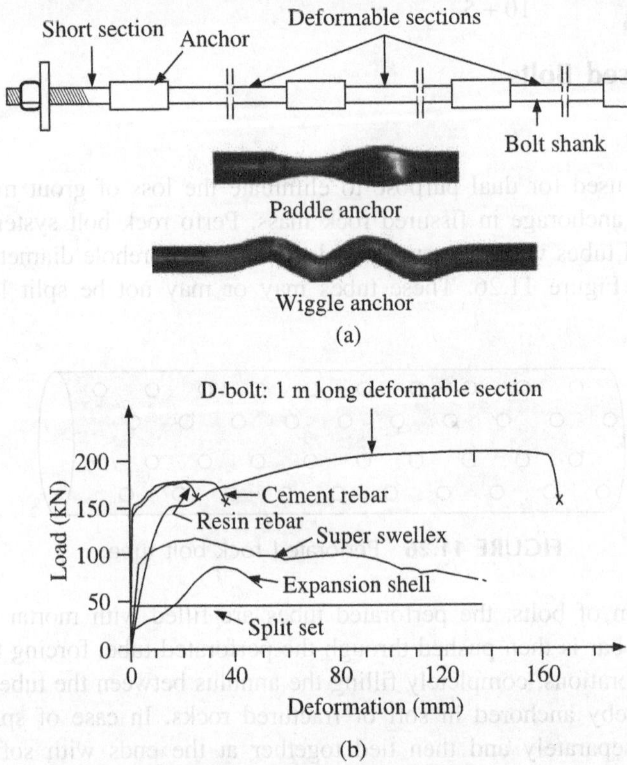

FIGURE 11.27 (a) Layout of D-bolt, and (b) Load–displacement curves of different rock bolts under pull loading (Li, 2010).

11.7 SECONDARY SUPPORT SYSTEMS

11.7.1 W-strap

W-strap are commonly used as secondary support and are required in conjunction with rock bolts. They are made of steel in the shape of straps (Figure 11.28). These straps are pulled into the rock surface by bolts. The main advantage is that they provide large surface confinement to any loose/blocky rock that may form between installed rock bolts. Typically, specifications vary such as length (0.5~8 m), width (155~290 mm) and weight per unit length is about 4.5 kg. Yield strength is estimated based on the minimal cross-section area. Holes of different shapes like circular, elongated, and transverse elongated long with service line slots are made in a multi-hole configuration or in set locations to adjust location of bolts as shown in Figure 11.28, where T, C, L and W are thickness, hole diameter or slot size, width and length respectively.

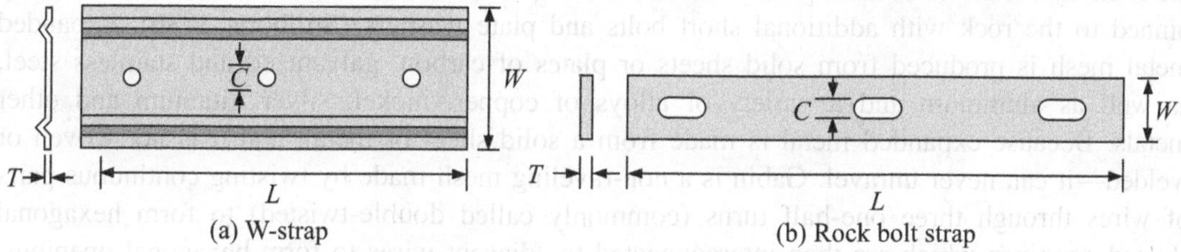

FIGURE 11.28 Straps.

11.7.2 Wire Mesh

Wire mesh is used to support small pieces of loose rock or as reinforcement for shotcrete (Hoek and Brown, 1980) in conjunction with rock bolts. Wire meshes are tied with rock along rock slope or on the roof/sidewall of the underground opening (Figure 11.29). Different types of wire mesh are shown in Figure 11.29, where *T*, *S*, *W*, *L* are wire thickness, opening spacing, width and length respectively.

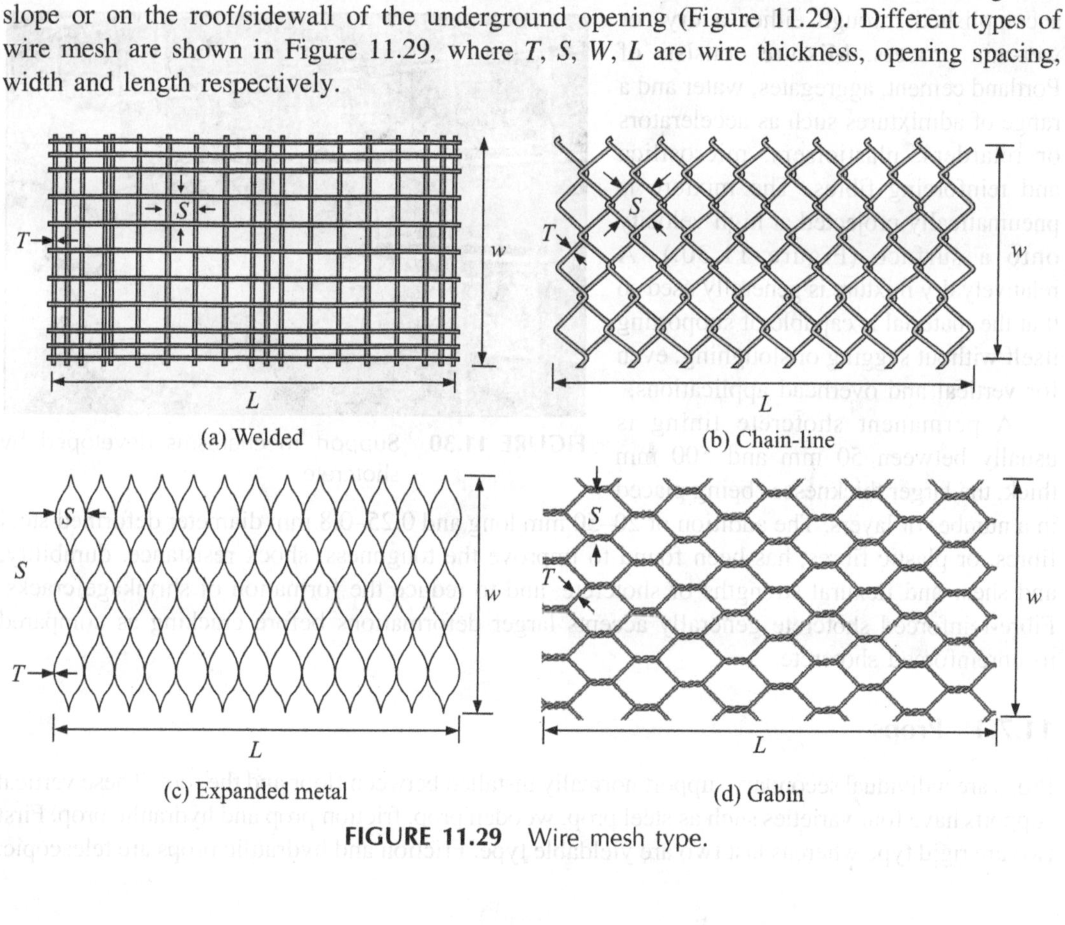

FIGURE 11.29 Wire mesh type.

Chain-link mesh is a woven fabric of wire commonly used for fencing. It is flexible, strong and galvanised for corrosion protection. It is not preferred for reinforcing shotcrete because of the difficulty of getting the shotcrete to penetrate its woven fabric. Weld-mesh consists of steel wires welded in a square or rectangular pattern at each intersection. It is used for reinforcing shotcrete and is much more rigid than the chain-link mesh. Generally, weld-mesh is attached to the rock by means of second plate washer and nut placed on existing rock bolts, alternatively pinned to the rock with additional short bolts and plate washers (Stillborg, 1986). Expanded metal mesh is produced from solid sheets or plates of carbon, galvanized and stainless steel, as well as aluminum and a variety of alloys of copper, nickel, silver, titanium and other metals. Because expanded metal is made from a solid sheet of metal, and it is not woven or welded—it can never unravel. Gabin is a non-raveling mesh made by twisting continuous pairs of wires through three one-half turns (commonly called double-twisted) to form hexagonal shaped openings which are then interconnected to adjacent wires to form hexagonal openings.

11.7.3 Gunite/Shotcrete

Gunite refers to the dry-mix process in which dry mixture of Portland cement and sand thoroughly mixed, conveyed by air through a flexible tube, hydrated at a nozzle prior to application. Shotcrete is an all-inclusive term to describe the spraying of concrete or mortar that may be accomplished through either a dry- or wet-mix process. Mixture consists of Portland cement, aggregates, water and a range of admixtures such as accelerators or retarders, plasticisers, microsilica and reinforcing fibres. The mixture is pneumatically projected at high velocity onto a surface (Figure 11.30.). A relatively dry mixture is generally used so that the material is capable of supporting itself without sagging or sloughing, even for vertical and overhead applications.

FIGURE 11.30 Support mechanisms developed by shotcrete.

A permanent shotcrete lining is usually between 50 mm and 500 mm thick, the larger thicknesses being placed in a number of layers. The addition of 20–50 mm long and 0.25–0.8 mm diameter deformed steel fibres, or plastic fibres, has been found to improve the toughness, shock resistance, durability, and shear and flexural strengths of shotcrete, and to reduce the formation of shrinkage cracks. Fibre-reinforced shotcrete generally accepts larger deformations before cracking as compared to unreinforced shotcrete.

11.7.4 Props

Props are individual secondary support normally installed between floor and the roof. These vertical supports have four varieties such as steel prop, wooden prop, friction prop and hydraulic prop. First two are rigid type whereas last two are yieldable type. Friction and hydraulic props are telescopic, made of two or more pieces gliding one inside the other and set according to the thickness to be supported. The setting of the prop is accomplished either by lock mechanism in friction prop or by hydraulic mechanism in hydraulic support. A hydraulic support system is made by assembling 2 to 6 hydraulic prop, canopy and base in a single unit called walking support or powered support. Rigid supports may cause damage to the roof and floor as it has negligible post failure behaviour.

Wooden Prop

The wooden prop system consists of wedge, lid (or bar) and wooden logs as shown in Fig. 11.5. Lid is used to distribute stresses and the width of the lid and wedge should at least equal to the diameter of the prop. Wedges are generally used to tighten the prop within the roof and floor. In case of in loose ground (like loose coal or sand) flat wooden base is used. A wooden prop of sal wood of 150 mm diameter and 3 m height can bear upto 8 to 10 kN of vertical load. The advantage of wooden props is that they can be cut into proper sizes according to the clearance and they are easy to carry to the site. However, nowadays, it is difficult to get good quality dry wood and hence, usages of wooden support is fast declining.

Steel Prop

Steel prop are rigid and yieldable. Rigid props are of two types: (i) fixed length either solid or hollow cylinder and (ii) adjustable length props. Steel prop are free from natural defects, high modulus of elasticity and compressive strength, not affected in underground by temperature and humidity, and can be reused. There are two variations of yieldable steel props: friction and hydraulic prop which are discussed in subsequent sections.

Friction Prop

As the name suggests, frictional characteristics of two telescopic bodies move with respect to each other is the cause of support resistance. Figure 11.31 shows a single telescopic friction prop and depicts the forces that may act on cylindrical as well as conical piston. Horizontal force (H) is generated by clamping the upper piston as shown in the Figure 11.31.

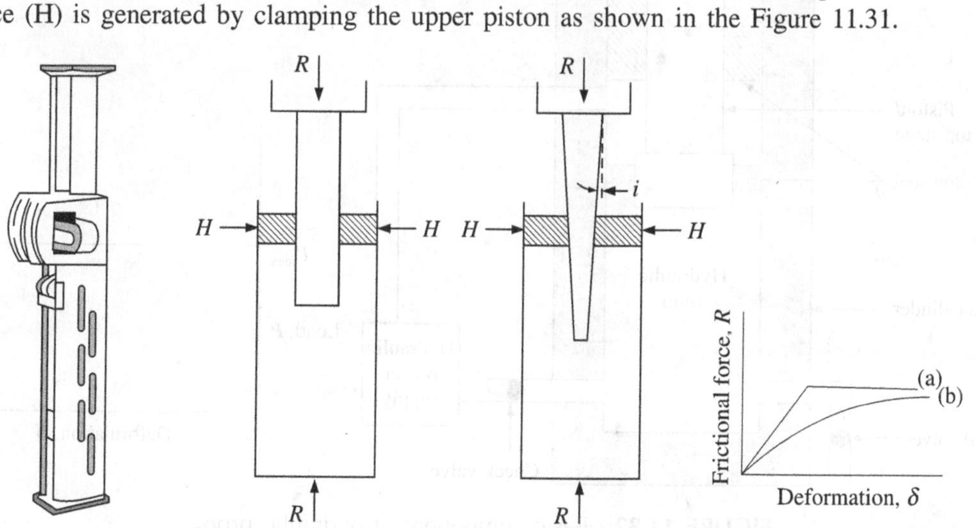

FIGURE 11.31 Friction prop: (a) Cylindrical, and (b) Conical.

The frictional force for both cases can be written as:

$$R = H \tan \varphi \text{ (for cylindrical piston)} \tag{11.52}$$

$$R = H \tan (\varphi + i) \text{ (for conical piston)} \tag{11.53}$$

where, φ and i are frictional angle and angle of conicity respectively. Friction angle may ranges between 15° to 30° while angle of conicity is around 2°.

Hydraulic Prop

Figure 11.32 shows the basic operation of a closed circuit hydraulic prop. Basically there are two ways of building up the oil pressure in the hydraulic prop (a) closed circuit system, (b) open circuit system. A hydraulic power supply pumps fluid into a pressure cylinder(s) on the roof support structure. The fluid acts against piston (inner tube), causing it to displace outward results into very little increase in hydraulic pressure until the support is set against the mine roof and floor. The hydraulic pressure will then increase rapidly until the full pump pressure is reached called the setting pressure. After the pump supply is turned off or isolated from the support, this pressurised hydraulic fluid is trapped inside the cylinder by a pilot-operated check valve. The pressure of the hydraulic fluid inside the cylinder is then intensified in proportion to any increase in roof loading. Yield valve limits the maximum pressure (called yield pressure) P_{max} inside the cylinder to prevent excessive loading that would damage the pressure cylinder. Most state-of-the-art hydraulic cylinders are double acting cylinders, meaning that they are hydraulically powered to both extend and retract. Hydraulic fluid is an oil-in-water emulsion or other fluids approved for use in hydraulic system.

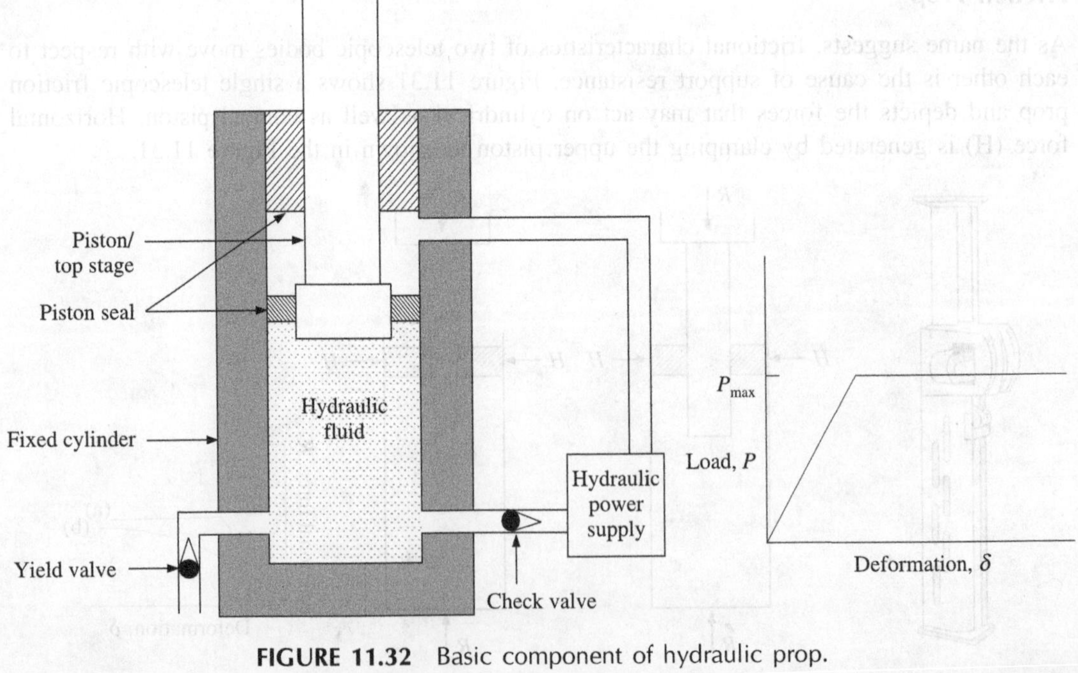

FIGURE 11.32 Basic component of hydraulic prop.

EXAMPLE 11.15 The yield capacity of hydraulic leg of a powered support is 100 kN, the setting pressure is 60% of the yield capacity. The area of cross-section of piston is 0.01 m^2. The inner area of the fixed cylinder is 50% more than the piston. Find the

(a) minimum pump pressure required to install the prop.
(b) thickness of the fixed hydraulic cylinder having FOS 10, tensile strength of cylinder material 400 MPa.

Solution:

(a) The minimum pump pressure (or setting pressure) for installation of hydraulic prop is

$$100 \times \frac{0.6}{0.015} = 4 \text{ MPa.}$$

(b) σ = Working stress = $\frac{400}{10}$ = 40 MPa

The hoop stress is defined as follows:

Hoop stress, $\sigma = p\frac{d_o^2 + d_i^2}{d_o^2 - d_i^2}$, where, d_o is the outer diameter and d_i is the inner diameter, p is hydraulic pressure inside the cylinder. Maximum pump pressure is 0.15 kPa.

$$d_i = \sqrt{\frac{0.015 \times 4}{\pi}} = 0.1382 \text{ m}$$

$$t = \frac{d_o - d_i}{2} = \frac{d_i}{2}\left(\sqrt{\frac{d_i^2\sigma + p}{\sigma - p}} - 1\right) = 7.29 \text{ mm}$$

11.7.5 Steel Arches

Steel arches or steel sets are used where high load are expected. The arch is more efficient than the flat crossbar. The space between the arches may be filled with steel mesh, steel, or timber lagging. It cannot be preloaded against the rock, and their efficiency depends on quality of blocking provided to transmit loads to the steel sets. Improper blocking steel arches are of two type, i.e., rigid arch and yieldable arch.

Rigid arches are designed to support without yielding. Depending on the shape and loading conditions of the arch, the imposed load in the arch is transmitted as compression rather than bending forces. Varieties of configuration are possible but some of the standard one is shown in Figure 11.33.

Yieldable arches consist of arch or ring sets that incorporate a sliding friction assembly to accommodate heavy ground pressure thus prevent distortion of the support. Steel arches, being very ductile, have been used extensively as ground supports in weak rock strata. They are usually fabricated by cold-bending of segments of steel H-section about the flange axis and attaching them by fish plates. Steel arches are passive supports and are of three types viz. rigid, yielding and articulated support types. Yielding arches composed of three sections, the top section slides between two side elements. They can resist vertical and horizontal pressure and can sustain large deformation.

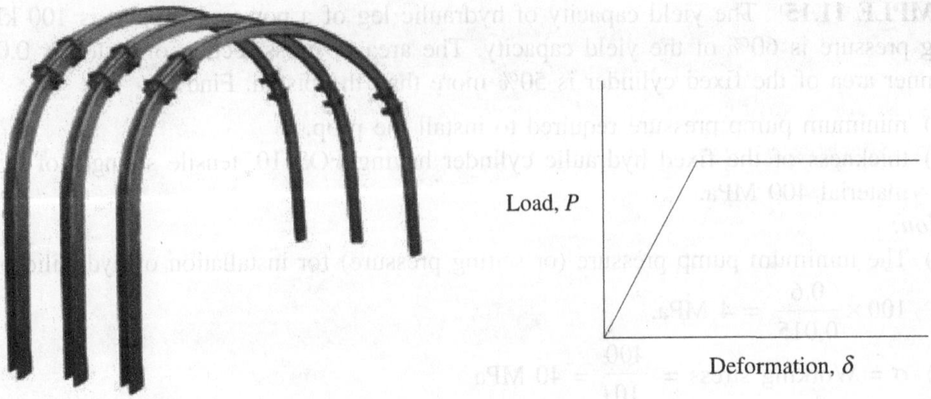

FIGURE 11.33 Yieldable steel arch.

11.7.6 Chock/Cog/Cribs/Chockmate

Chock or cog or crib is generally made of wooden props of square shape, laid in alternate cross-layers perpendicular to each other in square pattern from floor to roof. A typical chock is shown in Figure 11.34. In some instances, inside space of chocks is filled with debris. Load deformation characteristics are found to be elastic-perfectly plastic. The plastic deformation behaviour generally occurs for displacement of 0.4 ~ 0.5 m for well-constructed chocks. The load carrying capability and stiffness of chock reduces as the height increases. The load-carrying capability of wood chocks is a function of the contact area between intersecting chock blocks. Wood chocks are considered to be yieldable supports for vertical loads while remained stable for horizontal loads.

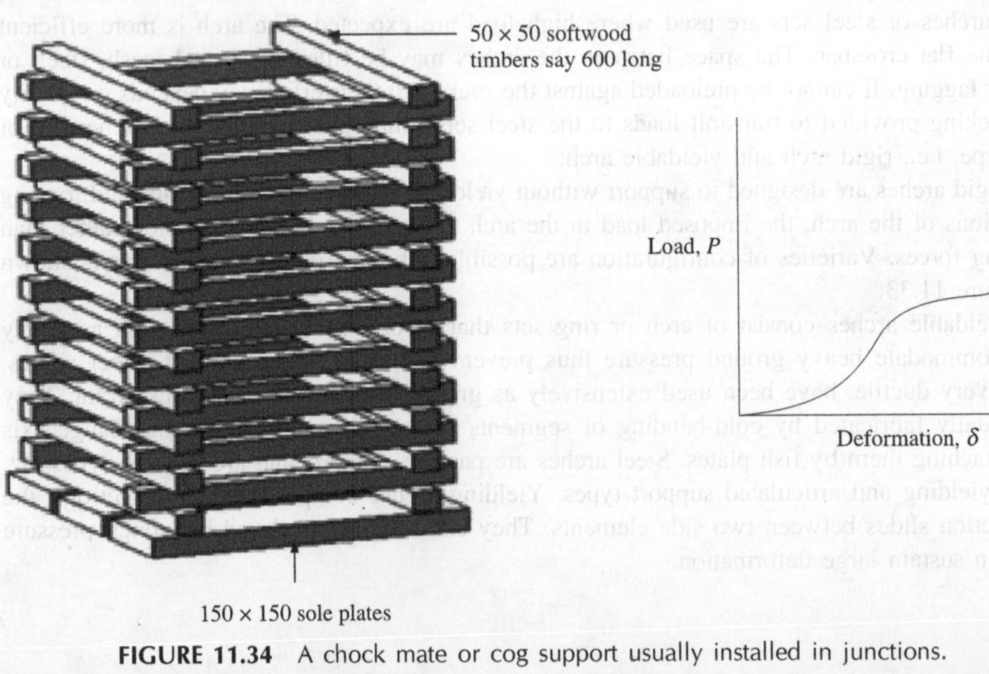

FIGURE 11.34 A chock mate or cog support usually installed in junctions.

Typical timber chock made of seasoned round timber cogs covers the area of 1.2 m × 1.2 m with a variation in height of 3 m and can provide support resistance of 30 t/m^2. Steel chock made of steel cog stool (0.9 × 0.9 × 0.9 m) fabricated from box steel pipes (48.5 × 48.5 mm section, 3.65 mm wall thickness) can also provide support resistance of 30 t/m^2 (DGMS, 1990). A steel cog or a wooden cog provides 30 ~ 40 kN of axial load bearing capacity with an area of 1 m^2 and one cog consists of 60 pieces of rectangular sleeper for a height of 3 m.

11.7.7 Bar or Roof Support

The roof support are made of wood or steel, either installed into the pillar (in case of strong wall rock) or supported on steel prop (in case of loose wall rock) and tightened with wooden wedges. These supports are installed in an area where roof is fractured and broken. Wooden or steel bars are placed in a crisscrossed manner as shown in Figure 11.35.

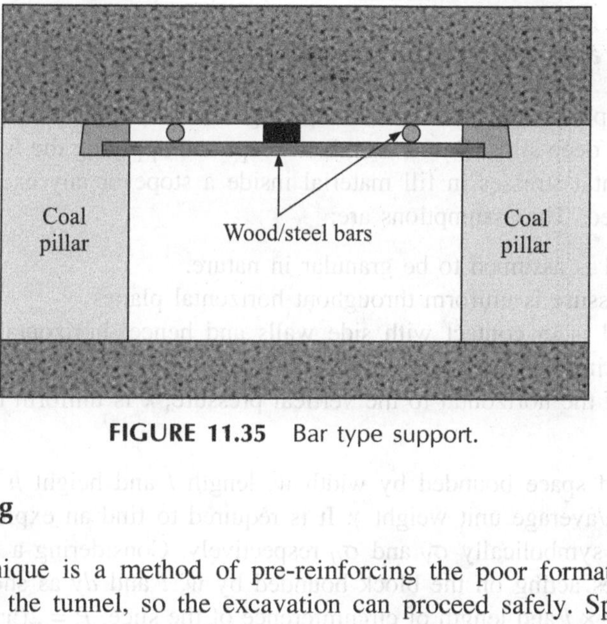

FIGURE 11.35 Bar type support.

11.7.8 Forepoling

The forepoling technique is a method of pre-reinforcing the poor formation that is located ahead of the face of the tunnel, so the excavation can proceed safely. Spiling or forepoling involves installing reinforced rods or bolts or tubes into pre-drilled (or rammed into loose rock) angled-holes (about 5° upward) drilled from the advancing face. In very soft, weathered and poor ground which are normally encountered around decline or incline mouth, the spiles can simply rammed. Apart from bolting, concrete lining and steel lining at every 1 m interval are also installed until competent rock is encountered. The thickness of concrete line may vary from 0.25 to 0.4 m depending upon the rock conditions.

11.8 STOWING AND BACKFILLING

Stowing and backfilling are methods to provide support to pillars and walls in an underground to prevent caving and roof fall and to enhance pillar recovery for improved productivity. The supporting mechanism for both is same, but there is a literal difference. Stowing is defined as filling the void with the material from outside whereas backfilling is defined as filling the void with the *in situ* waste material excavated. The most popular outside materials used are the composition of cement and sand whereas the *in situ* materials are mill tailings, fly ash, waste rock, etc.

There are three basic methods by which waste material can be filled in underground excavations, namely hydraulically, pneumatically or mechanically. Hydraulic filling has certain advantages which include relatively small space requirement (pneumatic filling shares this advantage) and low potential for increasing dust concentrations (compacted with mechanical or pneumatic methods). Additionally, hydraulic fill has a low potential for spontaneous combustion, methane emission can be reduced and the degree of packing is usually denser than with pneumatic systems. There are four types of stowing/backfill materials, i.e., paste backfill, hydraulic sand fill, cemented fill, and dry rock fill. Description of these methods is beyond the scope of this book.

11.8.1 Vertical and Horizontal Stresses in Fill Material

Janssen (1895) proposed equations for computing lateral pressure and vertical pressure of granular material in deep silos. Based on this concept and applying the following assumptions, vertical and horizontal stresses in fill material inside a stope or any excavated area with flat floor can be obtained. The assumptions are:

(i) Fill material is assumed to be granular in nature.
(ii) Vertical pressure is uniform throughout horizontal planes.
(iii) Fill material is in contact with side walls and hence, horizontal pressure is uniform over the perimeter of a cross-section.
(iv) The ratio of the horizontal to the vertical pressure, k is uniform throughout the height of the fill.

Consider a void space bounded by width w, length l and height h to be filled with fill material of uniform/average unit weight γ. It is required to find an expression of vertical and horizontal stresses, symbolically σ_V and σ_H respectively. Considering a vertical height of dy, we find that stresses acting on the block bounded by w, l and dy as shown in Figure 11.36. Area of slice $A = w \times l$ and length of circumference of the slice, $P = 2(w + l)$. Equating forces in vertical direction, we find

$$(\sigma_V + d\sigma_V)A + P\mu\sigma_H dy - \sigma_V A - \gamma A dy = 0 \tag{11.54}$$

$$(d\sigma_V)A + kP\mu\sigma_V dy - \gamma A dy = 0 \tag{11.55}$$

Applying boundary conditions, the above differential equation is solved as:

$$\int_0^{\sigma_V} \frac{d\sigma_V}{\left(\gamma - \dfrac{kP\mu}{A}\sigma_V\right)} = \int_0^h dy \tag{11.56}$$

Assuming $Q = \dfrac{kP\mu}{A} = \dfrac{2k\mu\,(w+l)}{wl}$, the equation of vertical stress is obtained as

$$\sigma_V = \frac{\gamma}{Q}[1 - e^{-Qh}] \tag{11.57}$$

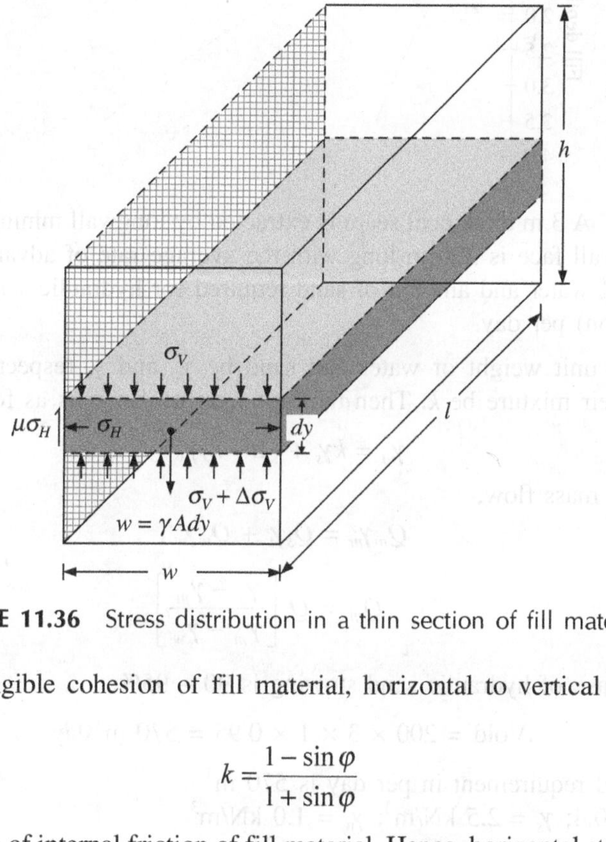

FIGURE 11.36 Stress distribution in a thin section of fill material.

Considering negligible cohesion of fill material, horizontal to vertical stress ratio can be approximated as:

$$k = \frac{1 - \sin\varphi}{1 + \sin\varphi} \tag{11.58}$$

where φ denotes angle of internal friction of fill material. Hence, horizontal stress is estimated as:

$$\sigma_H = k\sigma_V \tag{11.59}$$

EXAMPLE 11.16 A flat coal mine gallery is filled in one side by mechanical sand stowing. The height of the excavation is 3.5 m, gallery width is 4.5 m and width of the pillar is 30 m. The mechanical filling is such that it is just touching the roof. The unit weight of sand is 16 kN/m^3; frictional angle is 30° and coefficient of friction between the coal and sand is 0. Plot the vertical and horizontal stress distributions along fill depth of 3.5 m.

Solution: From the data, we find $k = 1/3$, $A = 135$ m^2, $P = 69$ m and $Q = 0.1363$ m^{-1}. Using these data in Eqs. 11.55 and 11.57, the graphs are plotted as shown the following given Figure.

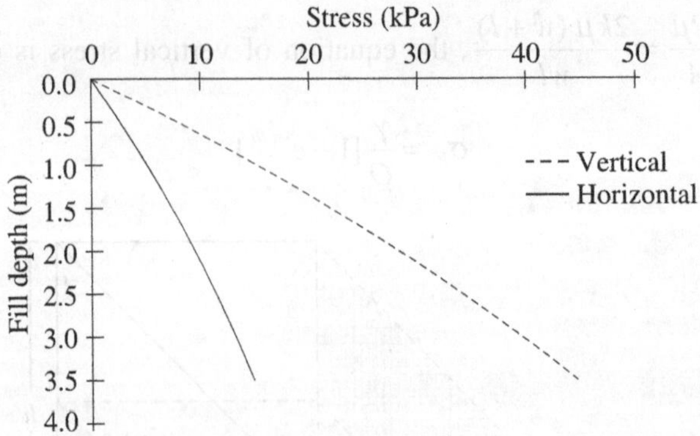

EXAMPLE 11.17 A 3 m thick coal seam is extracted by longwall mining with hydraulic sand stowing. The longwall face is 200 m long with the average rate of advance per day is 1.0 m. Find the quantity of water and amount of sand required for hydraulic sand stowing (assuming 100% coal extraction) per day.

Solution: Let the unit weight of water and sand be γ_w and γ_s respectively and volumetric concentration of their mixture be k. Then the unit weight of mix is as follows

$$\gamma_m = k\gamma_s + (1 - k)\gamma_w$$

From continuity of mass flow,

$$Q_m\gamma_m = Q_s\gamma_s + Q_w\gamma_w$$

$$Q_m = Q_s\left[\frac{\gamma_s - \gamma_m}{\gamma_m - \gamma_w}\right]$$

Void ratio in case of hydraulic sand stowing is 90 – 95%.

$$\text{Void} = 200 \times 3 \times 1 \times 0.95 = 570 \text{ m}^3/\text{day}$$

Volume of sand requirement in per day is 570 m^3

Assuming $k = 0.4$; $\gamma_s = 2.5$ kN/m^3; $\gamma_w = 1.0$ kN/m^3

$$\gamma_m = 1.6 \text{ kN/m}^3;\ Q_w = 855 \text{ m}^3$$

EXERCISES

1. A tunnel of radius 2.00 m excavated in rock (modulus of elasticity 3 GPa; Poisson's ratio 0.25) is subjected to an initial hydrostatic stress field of 15 MPa and provided with a concrete lining of stiffness $k = 2.5$ GPa. Assuming elastic behaviour of the rock and lining, calculate the radial pressure at the rock-lining interface if the lining is installed at the face (a) before any displacement of the rock has occurred; and (b) with a radial deformation of 1.5 mm.

2. Figure given below shows the ground reaction curve (GRC) while excavating rectangular opening into the rock strata. If an active and a passive supports are installed after 2 mm of roof deformation, when stress-deformation characteristics of both the supports are is linear-perfectly-plastic. The stiffness of the support is 1 MPa/mm with yield stress of 6 MPa. The setting pressure of the active support is 1.5 MPa. Draw support reaction curve (SRC) for both supports, and determine roof deformation after their interaction.

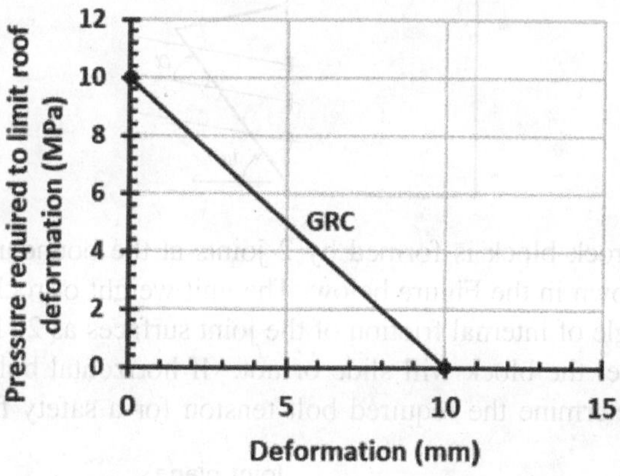

3. In continuation of Exercise 1, if the concrete lining is installed after (a) 3 mm and (b) 5 mm of ground deformation then fined the support pressure and roof deformation in both the conditions.
4. A tunnel of radius 2.0 m is excavated in rock with modulus of elasticity 2 GPa; Poisson's ratio 0.22 is subjected to an initial hydrostatic stress field of 10 MPa. Assuming c_r and φ_r of intact rock are 2 MPa and 25° and for broken rock c_f is 1 MPa and φ_f is 10°. Assume that the dilation angle as 20° and gravity effect based on Example 9.10. Plot the ground reaction curve and determine the radius of plastic zone and radial stress at elastic-plastic interface.
5. A rock block of weight W is bolted to a plane inclined at α to the horizontal axis as shown in the Figure below. Develop an expression for the bolt tension T, in terms of the angle β at which the bolt is installed relative to the plane, the angle of inclination of the plane and the Mohr-Coulomb friction properties c and φ of the block/plane interface.

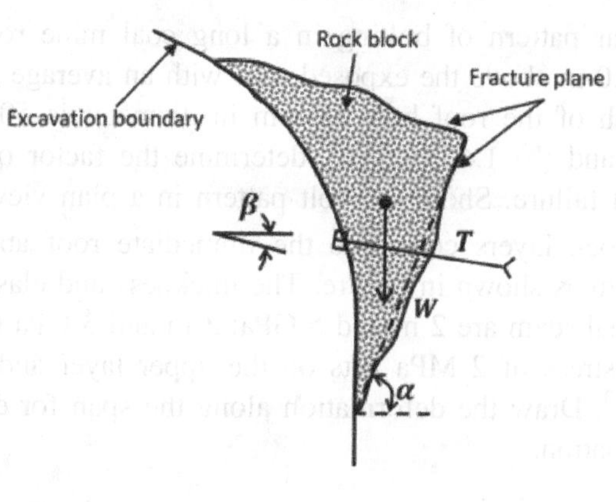

6. A wedge block is formed in the wall of an underground opening and it tends to slide along the lower discontinuity under gravity. The block is stabilized with three bolts which are installed with an angle of α to the discontinuity plane. The sliding of wedge along the discontinuity plane follows Mohr-Coulomb failure criteria. Determine the factor of safety of sliding block if each bolt developed a tension of T MN.

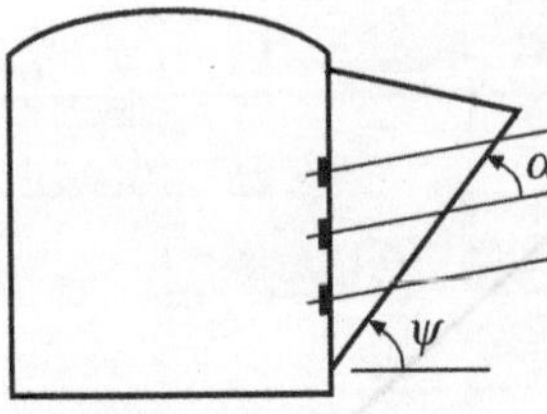

7. A wedge shape rock block is formed by 2 joints at the boundary of a circular tunnel of radius 2 m as shown in the Figure below. The unit weight of rock is 25 kN/m^3. Assuming cohesion and angle of internal friction of the joint surfaces as 20 kPa and 25° respectively, determine whether the block will slide or not. If horizontal bolts are installed as shown in the Figure, determine the required bolt tension for a safety factor of 3.0.

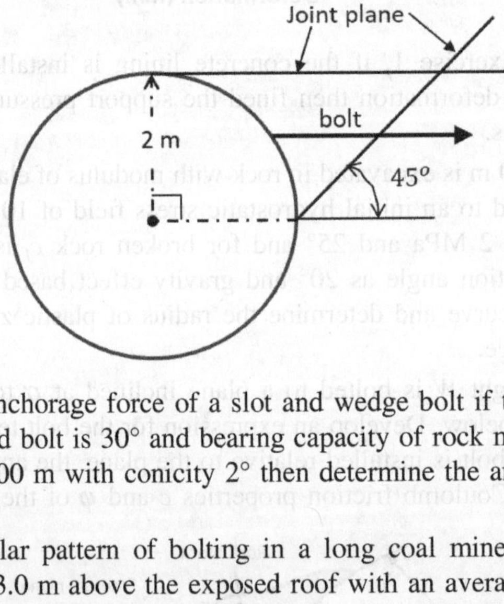

8. Determine the anchorage force of a slot and wedge bolt if the angle of internal friction between rock and bolt is 30° and bearing capacity of rock mass is of 15 MPa. If anchor bolt contact is 200 m with conicity 2° then determine the anchorage force developed in the bolt.

9. Design rectangular pattern of bolting in a long coal mine roadway 6 m wide, if the massive roof is 3.0 m above the exposed roof with an average unit weight of 22 kN/m^3. The yield strength of the roof bolt 22 mm in diameter is 500 MPa. For bolt density of (a) 1 bolt/m^2 and (b) 1.44 bolt/m^2, determine the factor of safety of the rock-bolt interface and bolt failure. Show the bolt pattern in a plan view.

10. Two sandstone rock layers constitute the immediate roof above a coal mine gallery with a span of 5 m as shown in Figure. The thickness and elastic modulus of these two strata from the coal seam are 2 m and 5 GPa; 3 m and 3 GPa respectively. Assume that uniform vertical stress of 2 MPa acts on the upper layer and the unit weight of each layer is 22 kN/m^3. Draw the deformation along the span for each layers and find their maximum deformation.

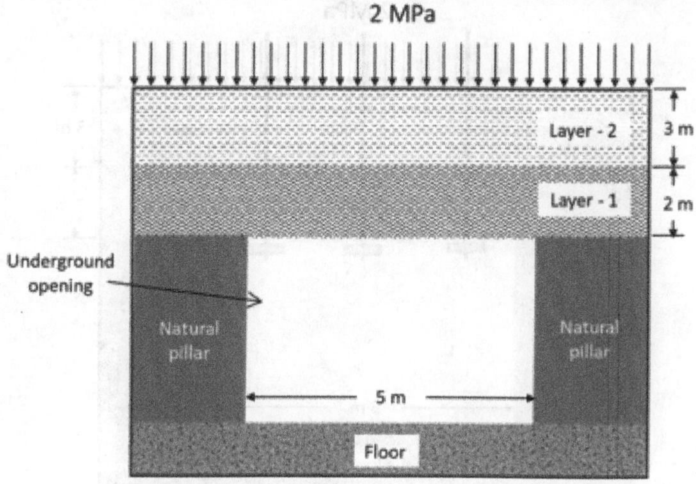

11. A wooden bar and three equally spaced hydraulic props are used in a 6 m wide underground roadway as shown in the Figure. The load on the bar is uniformly distributed and ends are fixed with loose material but free to rotate. The excavation is at a depth of 50 m below the surface in a rock mass with average unit weight of 22 kN/m^3. If the floor of the excavation is rigid, and the density and modulus of elasticity of wooden bar is 500 kg/m^3 and 10 GPa respectively. Determine the maximum deformation in the wooden bar.

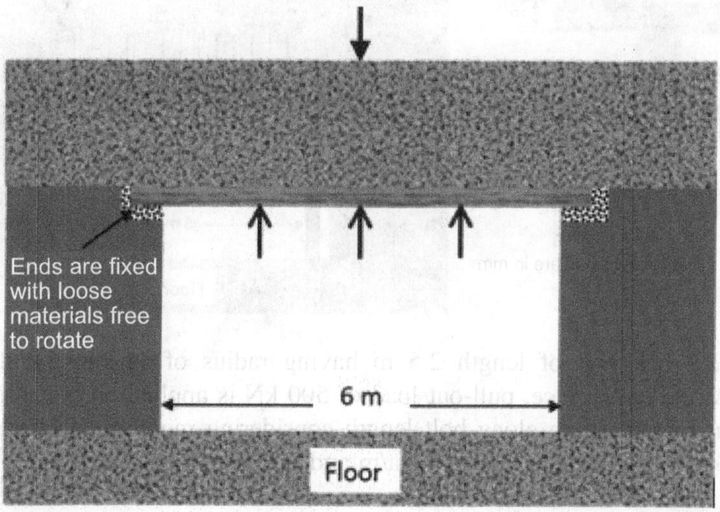

12. In continuation of Exercise 10, if both the layers are bolted together. Assume all material and geometrical properties remain same. Compare the deformation of composite beam with the individual deformation of unbolted layers.

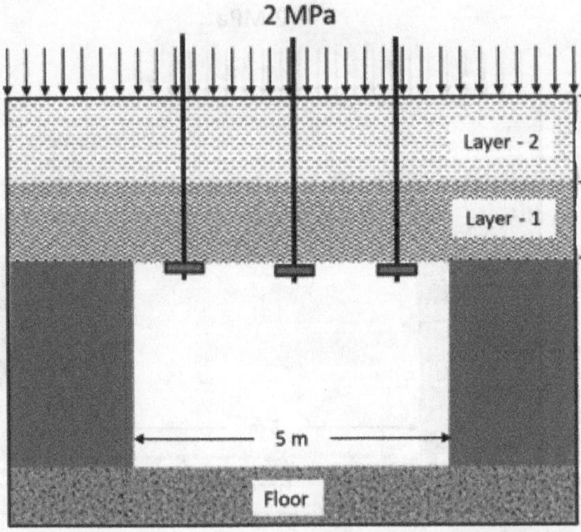

13. An I-section steel bar hinged rigidly at both the ends are used to support the roof of an underground excavation 50 m below the surface as shown in the Figure below. The bar is uniformly loaded by the gravity load. The I-section is made of mild steel A36-ASTM have modulus of elasticity 200 GPa, yield stress is 250 MPa and strain of 0.5%, ultimate strength is 370 MPa and strain of 13.5%. The average unit weight of overlying rock mass is 22 kN/m^3. Draw SFD and BMD.

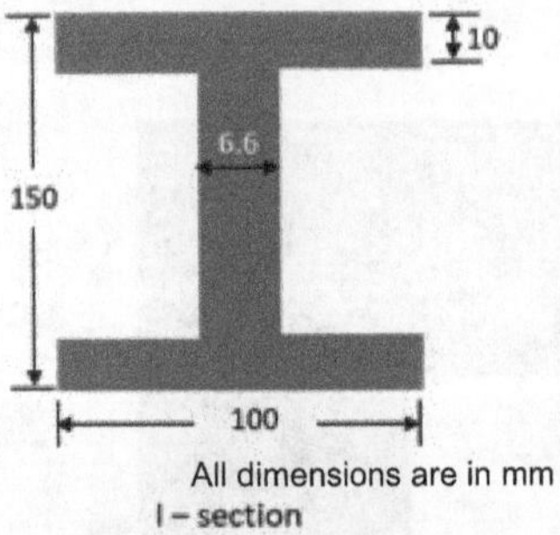

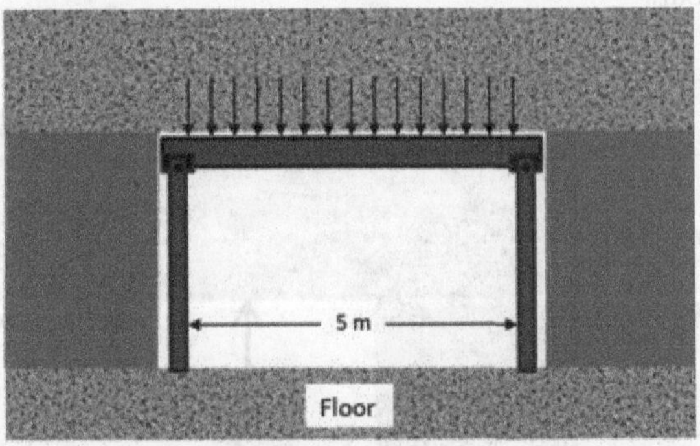

14. A grouted rock bolt of length 2.5 m having radius of 11 mm is installed in rock mass. At the tunnel face, pull-out load of 500 kN is applied on bolt face. Plot the bolt displacement and strain along bolt length considering modulus of elasticity of bolt, E_b is 200 GPa, and stiffness, k is 12 GPa/m and determine the maximum displacement and strain.
15. A stope of a metal mine is filled by cemented paste backfill (CPB) material prepared from mill tailings. The height of the stope is 60 m, length 20 m and width 20 m. The CPB is filled is such that it is just touching the roof. The unit weight, angle of internal friction of CPB material using 8% cement is 15 kN/m^3, 13° and coefficient of friction between the coal and sand is 0. Find the maximum vertical and horizontal stress developed at the bottom of the fill. Draw the stress profile over the depth.

Chapter 12

Subsidence

12.1 INTRODUCTION

The word subsidence usually refers to the downward movement of a ground surface, or "settlement" or "depression" due to caving of underground rock strata. Subsidence is the "result" of activities done in underground. These activities can be manmade or natural. The most common causes of subsidence are: (i) withdrawal of fluids from pores and fractures, and (ii) disturbance of the ground by excavation of minerals. In the first case, pumping of ground water from aquifers and consequent drawdown of the water table is a common cause of surface subsidence or depression. Withdrawal of oil, and to a lesser extent, withdrawal of gas from unconventional reservoir is also a significant source of subsidence in some areas. In these cases, the overburden strata drawdown due to a consolidation process after withdrawal of fluids. Secondly, surface and underground excavation cause subsidence through induced changes in stress, strain, and displacement fields in adjacent soil and rock masses. Sometimes excavation may be a natural process, for example, formation of solution cavities by circulating ground water or flowing water channel in limestone formations. If the induced displacements from underground excavation extend to the surface, then subsidence occurs on the surface. In cases where damage to surface structures or underground utilities occur, there is considerable importance attached to the determination of the cause of damage in the presence of subsidence and the actual subsidence mechanism.

Caving is an often observed phenomenon associated with underground mining and begins with collapse of the back, roof, or hanging wall. Caving may occur unexpectedly or be deliberately induced based on the rock strata and mining methods such as block caving, sub-level caving, longwall mining, bord and pillar mining with pillar extraction. Caving begins with rock fractures and failures just above the excavated zone and extends to the upper rock layers. Finally, broken rock pieces fall to the floor. Failure of the rock mass is greatly assisted by the presence of pre-existing joints, slip planes, shear zones and other discontinuities. Caving process may cease once the caved bulk volume completely occupies the void space and an equilibrium condition occurs. It may be noted that volume of broken rock mass is higher than the *in situ* rock mass. This ratio of volume of broken rock mass to that of volume of *in situ* rock mass is termed as *bulking factor*.

Subsidence profile and its formation depend on type of method of mining, proximity of underground excavation from the surface and condition of overlaying rock strata. In general, following types of subsidence have been observed:

(i) Trough subsidence
(ii) Subsurface caving
(iii) Plug caving
(iv) Chimneying or piping or funnelling or sinkhole

Subsidence over a mine field can be defined as the vertical movement of the ground surface and is the result of re-adjustments of the overburden due to collapse or failure of underground mine workings. Surface subsidence features usually take the form of either sinkholes or troughs. In this chapter, the mechanics of trough subsidence and sinkhole are discussed.

12.2 CONCEPT OF CAVING AND FORMATION OF SUBSIDENCE

Trough type of subsidence starts with the caving of underground rock strata. The *caving zone* is defined as the zone where broken rock materials are fallen and compacted as shown in Figure 12.1. The process of caving has been described in the previous section. Rock layers just above the caving zone settle on top of the broken material having large displacements and strains. They are separated from each other and hence, multiple fractures develop in rock strata. This process continues for a few rock strata above the caving zone and it termed as *fractured zone* (Figure 12.1). Above the fractured zone, rock layers may sag having a continuous deformation with minimal cracks or fractures. They behave like beams bended under own load and load from the overlaying strata. This zone is termed as *continuous deformation zone*. On top of this zone, soil layer may deform causing trough subsidence on the surface.

Caving process for chimney or sinkhole subsidence is quite different and will be discussed in Section 12.5.

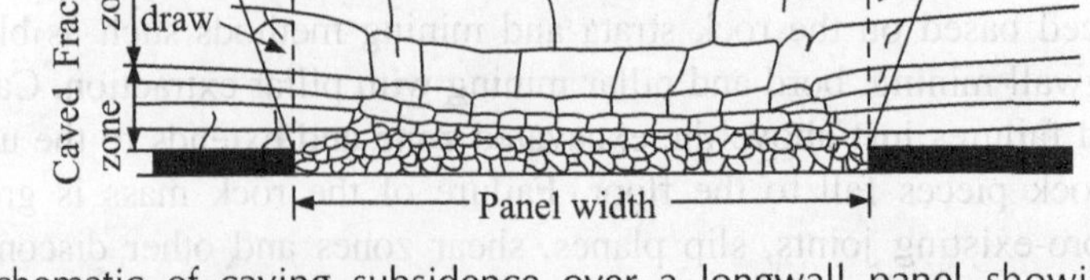

FIGURE 12.1 Schematic of caving subsidence over a longwall panel showing fracture zones (adapted from Peng, 1992).

12.3 TROUGH SUBSIDENCE

The movement of the ground surface results into both horizontal and vertical displacements. The vertical displacement is generally the maximum at the middle of the excavation for a flat deposit. The magnitude of subsidence gradually becomes zero outside the limit of the excavation. Figure 12.2 shows a schematic drawing of influenced zone generated on the surface due to formation of trough over a mined out longwall panel. This influenced zone is termed as *subsidence basin*. The figure also shows the major cross-section and longitudinal section taken along the middle of the subsidence basin. The trough created is confined between the limit lines shown as angle drawn with subsidence occurring outside the excavated area. Traditionally the subsidence profile is drawn with the vertical component of ground movement. Horizontal displacements act towards the area of maximum subsidence and the magnitude are approximately a function of gradient of the subsidence profile. At the centre of the mined out area, where the maximum subsidence occurs, the horizontal displacement is zero.

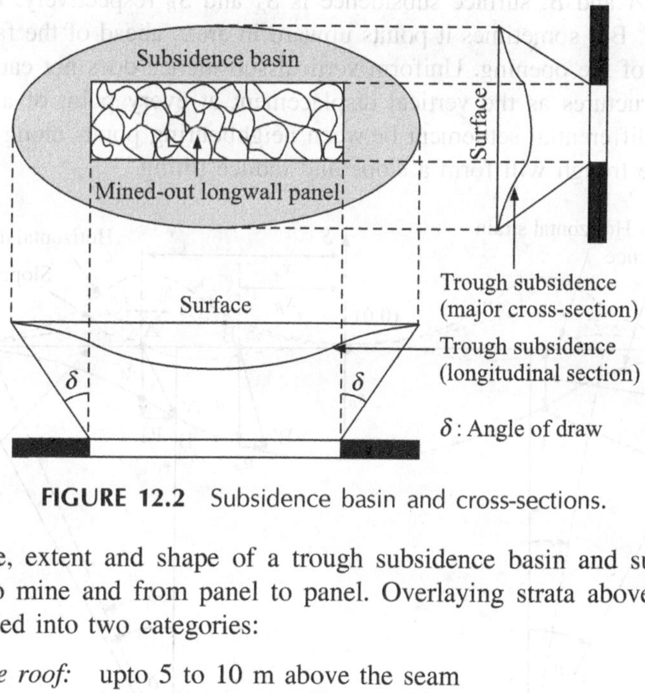

FIGURE 12.2 Subsidence basin and cross-sections.

The magnitude, extent and shape of a trough subsidence basin and surface deformation vary from mine to mine and from panel to panel. Overlaying strata above a longwall panel are normally divided into two categories:

(i) *Immediate roof:* upto 5 to 10 m above the seam
(ii) *Main roof:* upto 20 to 30 m above the immediate roofs.

The factors of immediate and main roofs and others, which affect these properties, are:

(i) Physical and geomechanical properties of immediate and main roofs strata

Immediate roofs	*Main roofs*
• Strong/hard	Strong/hard
• Strong/hard	Weak/soft
• Weak/soft	Strong/hard
• Weak/soft	Weak/soft

(ii) Mined out opening or goaf size
(iii) Mining depth
(iv) Mining height
(v) Topography
(vi) Time

12.3.1 Components of Surface Movement

Figure 12.3 shows typical subsidence profiles for subcritical and supercritical subsidence over a longwall panel. Based on this, the following components are defined. Figure shows that before subsidence the co-ordinate of point A and B are (x_A, 0) and (x_B, 0) respectively whereas after subsidence these points are shifted to A' and B' with co-ordinate (x'_A, s_A) and (x'_B, s_B) respectively.

(i) **Subsidence(s):** On any major cross-section, the vertical component of the surface movement vector (vector AA′ shown in Figure 12.3) is called *surface subsidence.* For point A and B, surface subsidence is S_A and S_B respectively. It generally points downward. But sometimes it points upward in areas ahead of the face line or beyond the edges of the opening. Uniform vertical subsidence does not cause any damage to surface structures as the vertical displacement at every point of a structure. On the contrary, differential settlement between neighbouring points along the surface of the subsidence trough will form a slope and induce tilting.

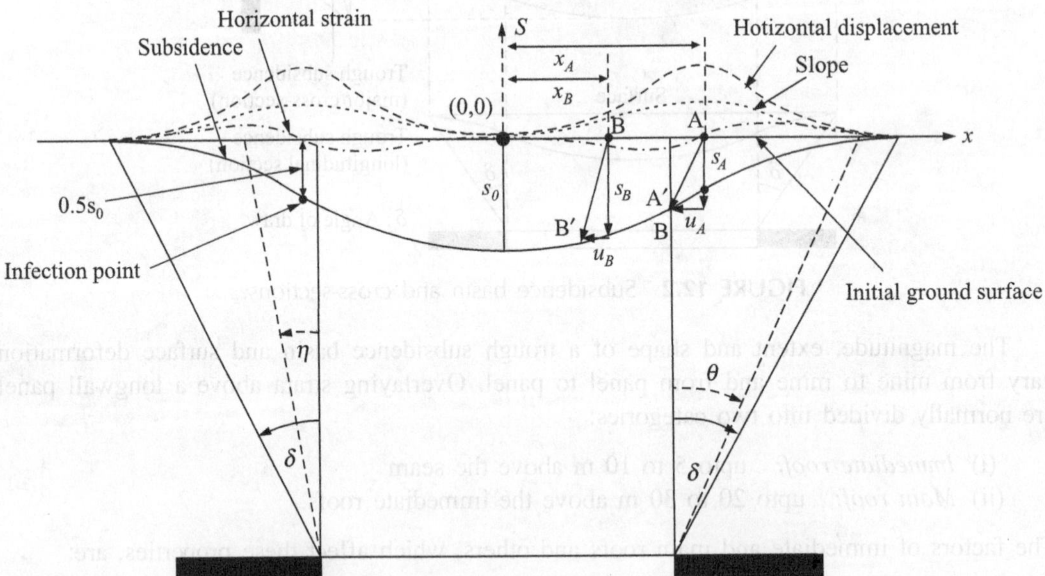

FIGURE 12.3 Terminologies used associated with subcritical subsidence profile.

(ii) **Horizontal displacement (u):** On any major cross-section, the horizontal component of the surface movement vector is called *surface horizontal displacement*. For point A and B, surface horizontal displacement is U_A and U_B respectively. It generally points toward the centre of the subsidence basin. But in steep terrain, it moves along the down dip direction.

(iii) **Slope ($\psi = ds/dx$):** On any major cross-section, the gradient of subsidence curve at any point is denoted as slope at that point. In discrete sense, slope is estimated as the difference in surface subsidence between the two end points of a line segment divided by the horizontal distance between the two points. For example, in Figure 12.3, slope between point A and B will be

$$\psi = \frac{s_B - s_A}{x'_B - x'_A} \tag{12.1}$$

(iv) **Curvature ($\kappa = d^2s/dx^2$):** On any major cross-section, the gradient of slope at any point on the surface is termed as the *curvature* at that point. In discrete sense, curvature is measured as the difference in surface slope between two adjacent line segments divided by the average length of the two line segments. There are two types of curvature: (a) convex or positive curvature, and (b) concave or negative curvature.

(v) **Horizontal strain ($\varepsilon = du/dx$):** On any major cross-section, the gradient of horizontal displacement at any point on the surface is termed as *horizontal strain* at that point. The difference in horizontal displacement between two end points of a line segment on the surface divided by the length of the segment is estimated to be the horizontal strain. Horizontal strain can be positive (compression), negative (tension) or zero (no strain). Due to compressive strain, the ground surface will be depressed and tensile strain creates crack on the ground surface. Horizontal strain between points A and B can be expressed as:

$$\varepsilon = \frac{u_B - u_A}{x_B - x_A} \tag{12.2}$$

It is noteworthy that uniform horizontal displacement may not cause any direct damage to surface structures. However, differential displacement induces compression or tensile strains along the surface of the subsidence trough. Most surface structural damage in a subsidence trough is caused by horizontal strains. When there is differential settlement, the gradient of the ground surface changes and may disrupt and/or alter drainage systems, the flow through cracks, the gravity flow of liquids in pipelines, and the grade of highways or railroad beds. Differential settlement will also induce tilting of tall buildings, towers, and chimneys and cause instability.

12.3.2 Terminologies of Subsidence Profile

(i) Point of inflection: The point that divides the concave and convex portions of the subsidence profile curve is termed as *inflection point*. At the infection point, subsidence is half of the maximum subsidence value at the centre, the surface slope is the maximum, the curvature is zero and the horizontal strain is zero. The location of the inflection point depends on the size of opening or goaf or rock types. When the opening is very small as compared to depth, it is located beyond the edges of the opening on the solid side. As the opening grows in size, the inflection point moves inside the edges of the opening. For supercritical condition, the inflection will be inside the edges of the opening as shown in Figure 12.4. The distance from the inflection point to the nearest edge of the opening is the termed as *offset* of the inflection point. In case of critical width of the opening, point of inflection will be above the edge of the opening.

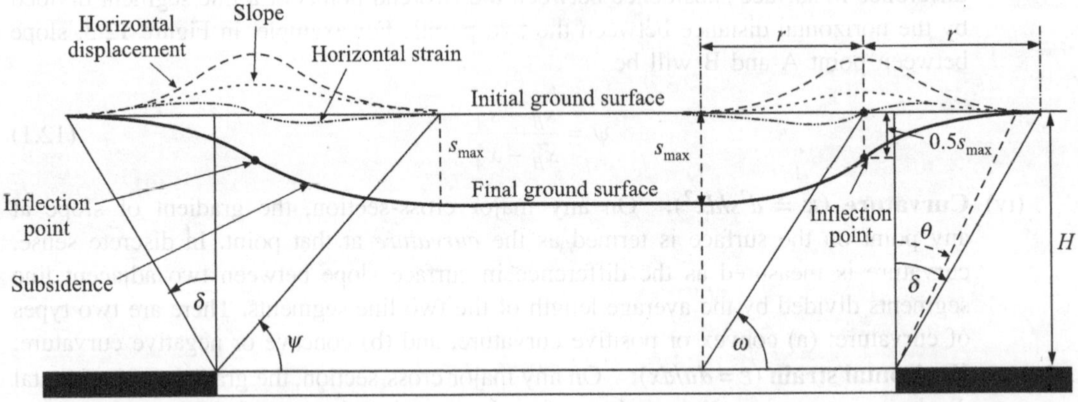

FIGURE 12.4 Terminologies used associated with supercritical subsidence profile.

(ii) Angle of draw (δ): Assuming a rectangular mined out panel, subsidence affected surface strata take the form of obtuse pyramid as mentioned in Figure 12.3. The angle subtended between the vertical line drawn at panel edge and a line joining the extraction edge or zero subsidence is termed as *angle of draw* or *angle of limit* of subsidence (Figure 12.3). Angle of draw generally ranges from 8° to 45° depending on the properties of overburden rock strata and their thickness. It is found that angle of draw is lower for stronger overburden rocks and higher for weaker rocks and soils. Pariseau (2007) mentioned that the presence of much sandstone is a likely cause of reduction in angle of draw. Sheorey et al. (2000) mentioned that the angle of draw for Indian coalfields varies from 9° to 25°. There are indication that angle of draw depend on:

- Depth
- Seam thickness
- Local geology especially major faults or fracture planes or the location of self-supporting strata above the coal seam.

(iii) Angle of critical deformation (θ): The angle between the vertical line at the opening edge and the line connecting the opening edge and the point of critical deformation on the surface is the *angle of critical deformation*. The point of critical deformation on the surface is defined based on the structural resistance to subsidence or horizontal strain. For example, if a structure on the surface is designed to withstand a certain amount of deformation or strain, it is obvious that angle of critical deformation of the structure will be less than angle of draw. The angle of critical deformation is on the average 5° to 10° less than the corresponding angle of draw.

(iv) Angle of break (η): On any major cross-section, the angle between the vertical line at the opening edge and the line connecting the opening edge and the point of maximum tensile strain on the surface is called the *angle of break*. The ground surface at the point of maximum tensile strain is the most likely place where tensile cracks occur.

(v) Radius (r) and angle of major influence (ω): When size of the opening or goaf reaches the critical size, the major surface deformations occur on both sides of the inflection point within a certain distance. This distance is called the *radius of major influence*. Beyond this distance, surface deformations are very small. For supercritical subsidence, infection point lies inside the panel. Then the radius of major influence is measured as the distance from the inflection to the zero subsidence (Figure 12.4). The *angle of major influence* is the angle between the horizontal and the line connecting the inflection point and the edge of the radius of major influence, mathematically, $\tan \omega = H/r$, where H is the mining depth.

(vi) Angle of full subsidence (ψ): On a major cross-section of the subsidence basin under supercritical width of mining the acute angle between the horizontal and the line connecting the edge of the flat bottom of the subsidence basin and the edge of the opening is called the *angle of full subsidence*. It indicates the degree of subsidence development and can be used to define the area within which subsidence has been fully developed.

12.3.3 Estimation of Maximum Subsidence

The maximum possible subsidence (s_{max}) at the middle of a trough is directly related to the height of the excavation or mining height (m), properties of overlaying rock strata and level of stowing or backfilling. For a longwall panel having a seam dip of α, the maximum possible subsidence can be approximated as:

$$s_{max} = am \cos \alpha \tag{12.3}$$

where, a is the subsidence factor. If the seam is nearly flat, the angle α is zero. It may be noted that the factor a mostly depends on the properties of overlaying strata and width to depth ratio of the panel. Subsidence factor is dimensionless and is dependent upon the (i) nature of the overburden, and (ii) type of goaf treatment such as caving and some form of stowing. In general, in France, Germany and in UK, it is 0.9 while in case of stowing pneumatically (0.4–0.5) and hydraulic sand stowing (0.12). Subsidence factor decreases with increasing depth.

The maximum subsidence and profile thereof also depend on the width to depth ratio of a longwall panel. As shown in Figure 12.5, the width and depth have substantial effect on the surface subsidence. The width to depth ratio for which there is no subsidence on the surface is called non-effective width (NEW). The non-effective width to depth ratio is found to vary approximately between 0.2 and 0.8 in Indian coalfields depending on strata competence. The figure shows that for particular W/H ratio the maximum subsidence is the same, although the area of influence will be more for greater depth for same angle of draw. Mathematically, it can be expressed as follows:

$$W_i' = W_i + 2H_i \tan \delta \tag{12.4}$$

In general, three types of surface subsidence can be expected based on length and width of longwall panel, overburden strata and depth of the seam. They are explained as follows:

(i) Critical area: An area of working just sufficient to cause the maximum possible subsidence (s_{max}) at one point on the surface. To get a maximum subsidence at the centre of the panel, both length and width of the opening are to be a size equal to the critical size. If any of them

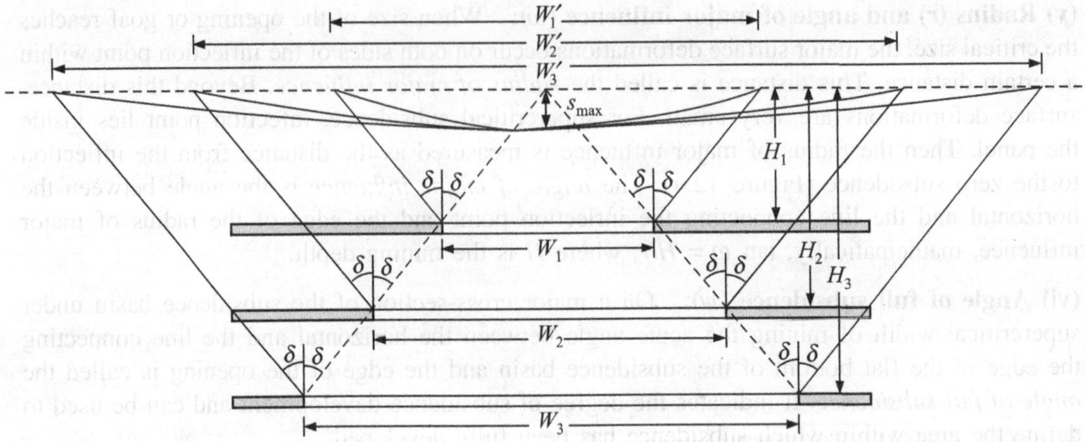

FIGURE 12.5 Effect of width to depth ratio on surface subsidence.

is less, then s_{max} cannot be achieved. Let us assume, δ denotes the abutment angle. Hence, critical width (W_c) [Figure 12.6(a)] equals to

$$W_c = 2H \tan \delta \tag{12.5}$$

From the data obtained from UK longwall panels, it is found that angle of draw or in this case, abutment angle is about 35°. This leads to $W_c = 1.4\ H$. The critical width for $\delta = 30°$, most common over longwall panels in former USSR and India is $W_c = 1.15\ H$.

(ii) Subcritical area: An area of working not sufficient to cause the maximum possible subsidence at a point on the surface. The maximum subsidence s_o occurred in a subcritical subsidence basin is less than s_{max}. In this case, width to height ratio is such that $W < W_c$ [Figure 12.6(b)].

(iii) Supercritical area: An area of working causing complete subsidence at a part of the surface. In other words, s_{max} occurs at more than one point on the surface. Figure 12.6(c) clearly shows that in this case, width to depth ratio will be high and hence, $W > W_c$.

Estimation of Factor (a)

Subsidence factor is strongly depended on the properties of overburden strata, depth of mining, panel dimension and time duration. This factor can be estimated for caving longwall panel as (Peng, 1992):

$$a = 0.35 + 0.5P \tag{12.6}$$

where P is the combined strata properties and are calculated as:

$$P = \sqrt{\frac{\sum_{i=1}^{N} (Q_i h_i)^2}{\sum_{i=1}^{N} (h_i)^2}} \tag{12.7}$$

where h_i is the thickness of ith strata and Q_i corresponds to the coefficient of stratum properties as given in Table 12.1. This table also shows the rock type and uniaxial compressive strength.

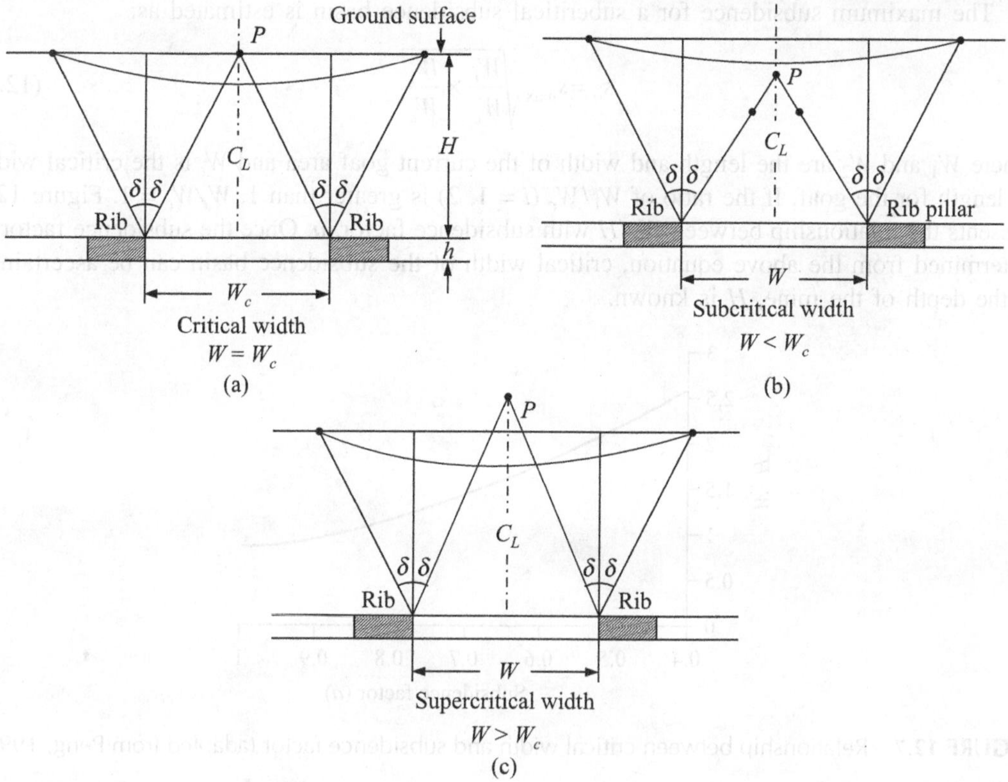

FIGURE 12.6 Concept of critical width in relation to rib-side subsidence profiles (Pariseau, 2007).

TABLE 12.1 Rock Factor Q for Different Rock Types (Peng, 1992)

Hardness		*Sandstone*	*Limestone*	*Shale*	*Coal*	*Clay*
Extremely hard	σ_c	>88	>86	>69	>27	>18
	Q	0.00	0.01	0.20	0.80	0.85
Very hard	σ_c	78 – 88	76 – 86	58 – 69	24 – 27	15 – 18
	Q	0.02	0.02	0.30	0.83	0.87
Hard	σ_c	69 – 78	68 – 86	44 – 58	20 – 24	13 – 16
	Q	0.05	0.06	0.40	0.87	0.89
Regular	σ_c	58 – 69	57 – 68	29 – 44	17 – 20	12 – 14
	Q	0.10	0.12	0.60	0.90	0.92
Soft	σ_c	48 – 58	48 – 57	19 – 29	10 – 17	7 – 12
	Q	0.2	0.22	0.7	0.93	0.95
Very soft	σ_c	44 – 48	39 – 48	10 – 19	7 – 10	3 – 7
	Q	0.30	0.40	0.80	0.97	1.00
Extremely soft	σ_c	<44	<39	<10	<7	<3 (400)
	Q	0.40	0.80	0.90	1.00	1.00

σ_c: Uniaxial compressive strength MPa; Q: Assigned rock factor.

The maximum subsidence for a subcritical subsidence basin is estimated as:

$$s_o = s_{\max}\sqrt{\frac{W_1}{W_c} \times \frac{W_2}{W_c}} \tag{12.8}$$

where W_1 and W_2 are the length and width of the current goaf area and W_c is the critical width or length for the goaf. If the ratio of W_i/W_c $(i = 1, 2)$ is greater than 1, $W_i/W_c = 1$. Figure 12.7 presents the relationship between W_c/H with subsidence factor, a. Once the subsidence factor is determined from the above equation, critical width of the subsidence basin can be ascertained if the depth of the mine, H is known.

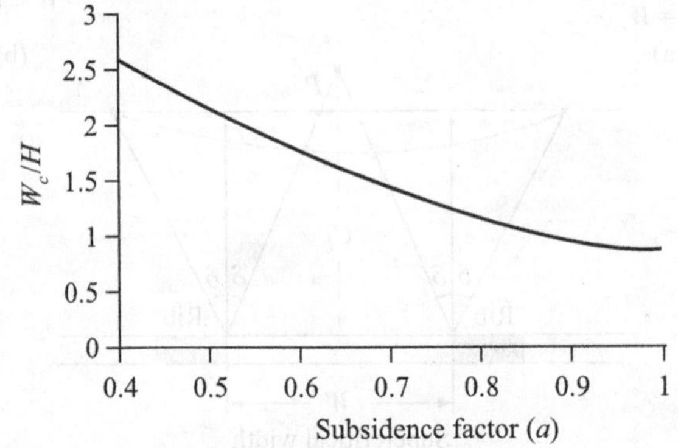

FIGURE 12.7 Relationship between critical width and subsidence factor (adapted from Peng, 1992).

Critical width can also be obtained by identifying "rock mass factor" as defined in Table 12.2 and using Eq. 12.9 (Saxena, 2003).

$$W_c = (2 - M)H \tag{12.9}$$

TABLE 12.2 Rock Mass Factor Based on Condition of Roof Strata

Sl. no.	*Nature of strata*	*Rock mass factor* (M)
1.	Totally disturbed roof strata (as in case of thick seam mining)	1.0
2.	Partially disturbed roof strata with thickness of intact strata being less than five times the seam thickness under extraction	0.9
3.	Partially disturbed roof strata with thickness of intact strata being more than five times the seam thickness under extraction	0.8
4.	Naturally weak and disturbed roof strata (having clay, shale or mudstone)	0.7
5.	Roof strata having interbedded layers of sandstones and shales	0.6
6.	Roof strata having predominantly sandstone beds	0.5

EXAMPLE 12.1 Given a longwall panel with following geological and mining data:

Rock type	*Hardness*	*Thickness* (m)
Clay	Regular	12
Shale	Hard	10
Sandstone	Very hard	35
Sandy shale	Hard	15
Shale	Regular	12
Sandstone	Soft	15
Shale	Very Soft	2
Coal	Hard	4

If the mined out length and width of the active panel are 250 m and 100 m, determine the maximum possible subsidence at current time. What will be the maximum subsidence over that panel?

Solution: The following table provides calculations of necessary parameters.

Rock type	*Thickness* (m)	*Hardness*	Q_i	$(h_i\, Q_i)^2$	$(h_r)^2$
Clay	12	Regular	0.92	121.882	144
Shale	10	Hard	0.40	16.000	100
Sandstone	35	Very hard	0.02	0.490	1225
Sandy shale	15	Hard	0.06	0.810	225
Shale	12	Regular	0.60	51.840	144
Sandstone	15	Soft	0.20	9.000	225
Shale	2	Very Soft	0.80	2.560	4
Coal	4	Hard	0.87		
				202.58	2067

From the above table, we estimate

$$P = \sqrt{\frac{202.58}{2067}} = 0.313,\ a = 0.35 + 0.5 \times 0.313 = 0.506 \text{ and } s_{\max} = 0.506 \times 4 = 2.024 \text{ m}$$

Depth of coal seam is 101 m. Using Figure 12.7, we find W_c = 222 m.

$$s_o = 2.024\sqrt{\frac{100}{222}} = 1.358 \text{ m}$$

Holla and Barclay, 2000 published a handbook on subsidence and proposed methods for estimating maximum subsidence based on mining induced subsidence data in the New South Wales, Australia. The subsidence data and resulting graphs in this method were obtained from collieries in the area between the Illawarra Escarpment and the Burragorang Valley

(Keilich, et al., 2006). These data were collected over a period of thirty years. The relationship between s_{max}/m and W/H for single panels is given in Figure 12.8. The figure shows that the largest longwall W/H ratio falls into the subcritical category ($W/H < 1.4$) since deep mining conditions exist in the Southern Coalfield.

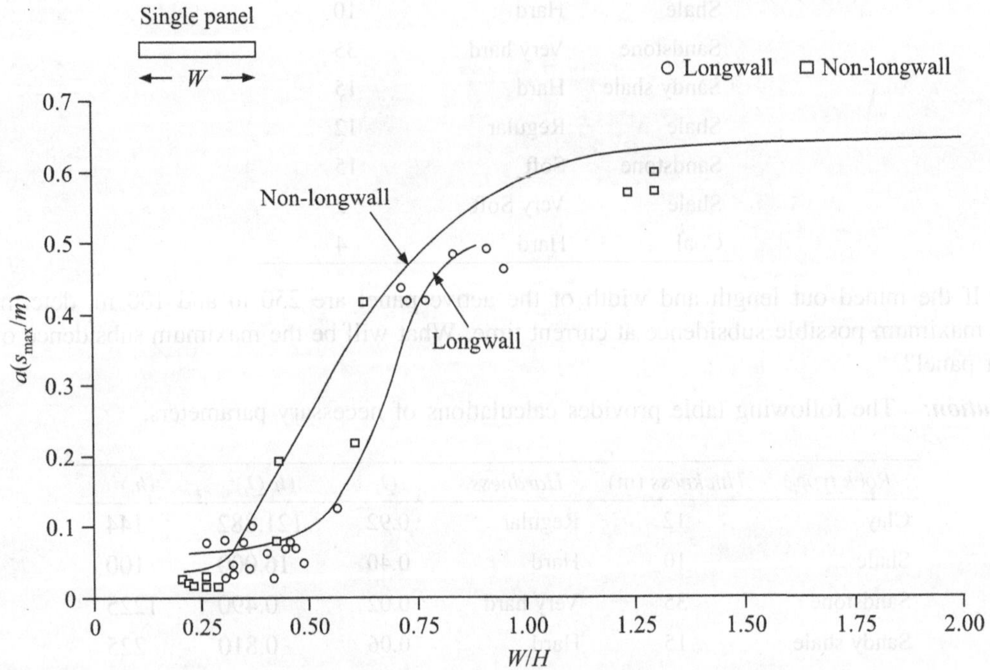

FIGURE 12.8 Relationship between s_{max}/m and W/H for single longwall panel (Holla and Barclay, 2000).

12.4 METHODS OF PREDICTING SUBSIDENCE

There are several techniques to predict mining induced surface subsidence. Each of this technique assumed that the maximum possible surface subsidence cannot exceed the height of extracted seam or mining height. Following are the six basic methods for prediction of mine subsidence:

(i) Empirical or graphical method
(ii) Analytical method
(iii) Profile function
(iv) Influence function
(v) Physical models
(vi) Numerical method

12.4.1 Empirical or Graphical Method

These relationships are established from observed surface subsidence behaviour. These relationships require sufficiently large number of observations and case studies in order to establish level of accuracy for prediction. As for example, the UK SEH (Subsidence Engineers' Handbook) empirical model is based on 200 sites in several coalfield of UK (Whittaker and Reddish, 1989). The subsidence data collected from these sites are rationalised by NCB (National Coal Board) and published in 1965 and 1975. The SEH empirical models are developed based on charts, graphs and tables and relate depth, mining height and width of the panel with maximum surface subsidence and also to generate subsidence profile, strain profiles and others. In general, the maximum possible subsidence, the maximum possible compressional and tensile strain and the maximum slope are obtained as function of depth, mining height and width of the panel as:

$$\frac{S_{max}}{m} = f\left(\frac{W}{H}\right) \tag{12.10a}$$

$$\varepsilon_{max} = g\left(\frac{S_{max}}{H}\right) \tag{12.10b}$$

$$\psi_{max} = h\left(\frac{S_{max}}{H}\right) \tag{12.10c}$$

The detail description of this method is beyond the scope of this chapter. Interested readers can refer to excellent references as Peng, 1992 and Whittaker and Reddish (1989).

12.4.2 Analytical Method

In this method, subsidence is considered as a problem which obeys laws of elasticity, plasticity and visco-elasticity. It is a mathematical expression that gives the value of the field variable at any location in the body. Their use is restricted to situations characterised by relatively simple geometry, boundary conditions and material behaviour.

For problems involving complex shapes, material properties and complicated boundary conditions, it is difficult and in many cases intractable (not easily controlled) to obtain analytical solution that satisfies the governing differential equations or gives extreme value to the governing functions. In order to overcome this limitation, numerical methods are nowadays frequently adopted for analyses related to complex design problems as mentioned below.

It is rarely found that analytical solution can be found for rock mechanics problems of practical concern. This may be because of any one reason or their combinations

(i) The boundary conditions cannot be described by simple mathematical functions
(ii) The governing partial differential equations are non-linear
(iii) The problem domain is inhomogeneous
(iv) The constitutive relations for the rock mass are non-linear
(v) Insufficiently simple mathematical calculation

12.4.3 Profile Function

The profile function $S = f(x)$, x being the distance either from infection point or from the centre of the panel, approaches to predict mining subsidence. This method is probably the most widely applied of all methods in view of its representation being conveniently expressed in mathematical terms or data-base form. This method predicts profile in set directions across an excavation. It uses the geometry of the excavation along with equations or tables of data to predict a longitudinal or transverse profile for subsidence. This empirical approach is extensively used because it is relatively easy to use and apply to new situations. Profile functions are divided into two types:

(i) Use of equations to define the profile in terms of mining geometry
(ii) Use of table of data to define the profile in terms of mining geometry

The constants for the equations are all derived empirically from local observations. The methods vary in complexity depending upon how many of the more significant factors are taken into account.

(i) Simple method can consider only depth
(ii) Advanced method can consider many parameters

A profile function simply defines the shape of the curve, and it requires a maximum value of subsidence (s_{max}) and a number of constants to calculate subsidence profile. The method is generally applied by adjusting the constants in the equation until a satisfactory shape is obtained which provides an acceptable match to the observed data (Whittaker and Reddish, 1989). Consequently, using this calibrated equation, it should then be possible to predict subsidence curve for that locality. As the profile functions are expressed in the form of equation, values of subsidence can be quickly and easily calculated by a computer program. When very little measured data are available, the method can still applied using limited data to calculate the constants. The predicted subsidence profile by this method may not match exactly the shape given by measured data consequently introduce an error. The shape of profile may vary depending upon various external parameters not taken into account by the function. Following are some of the most commonly used profile functions:

Functions based on point of inflection as origin as shown in Figure 12.9(a):

(i) Hyperbolic function (UK: King and Whetton, 1957, Wardel, 1965, Cherny, 1966, Peng and Chen, 1981)

For critical: $$S = \frac{s_{max}}{2}\left(1 - \tanh\left(\frac{cx}{R}\right)\right) \tag{12.11}$$

For subcritical: $$S = \frac{s_{max}}{2}\left(\tanh\left(\frac{n(x+W)}{R}\right) - \tanh\left[\frac{nx}{R}\right]\right) \tag{12.12}$$

where, c and n are arbitrary constants depending on the mine site. Generally, value of n is 2.

(ii) Hyperbolic function (Poland: Knothe, 1953)

$$S = \frac{s_{max}}{2}\left(1 - \left[\frac{2}{\sqrt{\pi}} \int_0^{\pi x R} \exp(-u^2)\, du\right]\right) \tag{12.13}$$

(iii) Exponential function (Hungary: Martos, 1958, Marr, 1958–59)

$$S = s_{\max} \exp\left(-\frac{1}{2}\left[\frac{x+R}{R}\right]^2\right) \tag{12.14}$$

(iv) Donets trigonometric function (USSR: Geological Institute of Mine Surveying, Anon, 1958)

For critical:
$$S = \frac{s_{\max}}{2}\left(1 - \frac{x}{R} - \frac{1}{\pi}\sin\left(\frac{\pi x}{R}\right)\right) \tag{12.15}$$

For subcritical:
$$S = a_{\max}(n_1, n_2)^{1/2}\left(n^2\left[1 - x + \frac{\sin 2\pi x}{2\pi}\right] + \frac{1-n^2}{4}[1+\cos \pi x]^2\right) \tag{12.16}$$

where, n_1, n_2 = coefficient related to width/depth ratio of either side of an inclined seam $n = n_1$ or n_2 depending on side of the panel.

Functions based on origin just above the middle of excavation as shown in Figure 12.9(b)

(i) Polish profile function (Kowalczyk, 1968)

$$S = s_{\max} \exp(-\bar{n}x^2) \tag{12.17}$$

where $\bar{n} = \frac{s_{\max}}{R^2 C}$; C = Average roof bed lowering.

(ii) Hungarian profile function (Martos, 1958)

$$S = s_{\max} \exp\left(-\frac{x^2}{2d^2}\right) \tag{12.18}$$

d = distance of transition point from panel centre

(iii) Niederhofer's method (Niederhofer, 1962)

$$S = s_{\max}\left(1 - \left[\frac{x}{P}\right]^2\right)^2 \tag{12.19}$$

(iv) Indian profile function (Kumar, Saxena and Singh, 1983)

For subcritical:
$$S = s_{\max} \exp\left(-\left[\frac{nx^2}{P^2 - x^2}\right]\right) \tag{12.20}$$

For critical:
$$S = s_{\max} \exp\left(-\left[\frac{nx^4}{P^4 - x^4}\right]\right) \tag{12.21}$$

where n = constant (1 to 5 generally), x = distance from extraction centre, R = radius of critical area of extraction and P = half width of subsidence profile, i.e., $W/2 + R$

(v) Exponential function (USA: Peng and Chen, 1981)

$$S = s_{max}\exp\left(-\left[\frac{cx}{R}\right]^d\right) \tag{12.22}$$

where, c and d are arbitrary constants depending on the mine site.

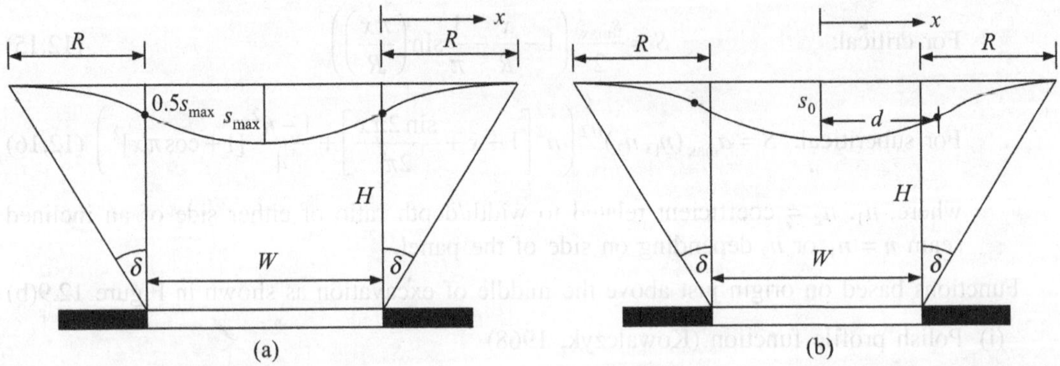

FIGURE 12.9 Parameters for profile function: (a) Critical subsidence, and (b) Subcritical subsidence.

EXAMPLE 12.2 Using data of Example 12.1, determine subsidence and slope profiles on the surface for angle of draw of 35°. Use hyperbolic function with $n = 2$. Also determine the location of inflection point.

Solution: Based on the data given in Example 12.1 the following plot is generated to describe subsidence, slope and inflection point. The data are obtained using Eq. 12.12. It can be noted that the maximum value of subcritical subsidence is found to be 1.798 m as opposed to 1.358 m as found by approximating using Eq. 12.8.

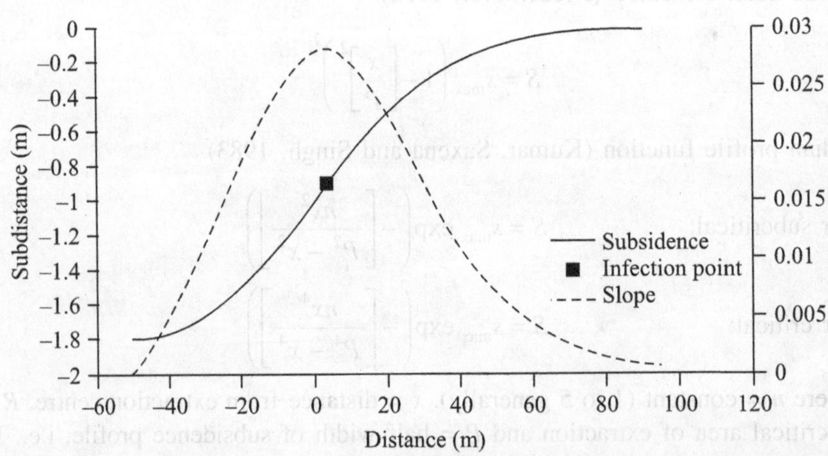

Assuming location of inflection point is d distance apart from the origin, we find

$$S(-W/2) = 2 \times S(d)$$

$$\Rightarrow \qquad \tanh\left(\frac{2(d+W)}{R}\right) - \tanh\left[\frac{2d}{R}\right] = \tanh\left[\frac{W}{R}\right]$$

Solving the above equation iteratively, we find d = 3.763 m.

12.4.4 Influence Function Method

Influence function method utilises the basic theory of superposition to determine the overall influence of an infinitesimal extraction area. Excavation opening is composed of finite number of small openings or elements. Each small opening produces an individual surface subsidence profile given by an influence function and its summation is the total subsidence. Since the overall subsidence is the summation of influence of each mined out elements, it is not necessary to have contiguous excavation like longwall panel. In other words, this method is well suited for determining surface subsidence caused due to bord and pillar panel.

It is based on the theory that an area of influence develops around a point on the surface due to extraction of a small element within the sphere of influence as shown in Figure 12.10. The influence is the maximum when element is directly below the surface point being considered. It becomes the minimum when the element is near the edge.

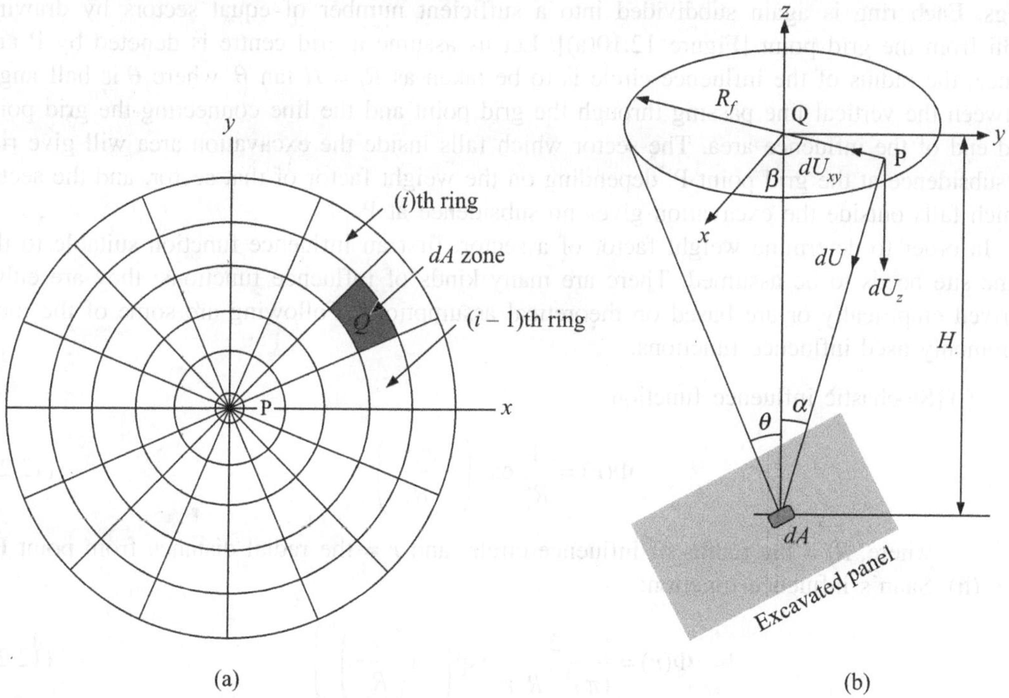

FIGURE 12.10 (a) Sectors of grid point P showing excavated zone dA, and (b) Displacement vector at point P due to excavation of element dA.

A typical influence function would look at annular zones around a point and use equations to relate the percentage extraction in each area to the subsidence of that point. The area of influence is generally assumed to be circular in shape, although hexagonal shapes have been employed occasionally. The functions commonly used can be subdivided as follows:

(i) Empirically derived functions which use arbitrary constants to produce the desired value and form of subsidence
(ii) Functions relating to the rock properties and mechanics of the problem rather than the arbitrary constants.

Second of these groupings is more popular but it is less flexible when carrying out calibration work.

The advantage of this method is the ability to predict subsidence movement at any point above extraction of any practical shape. Drawback of these method is that it is more difficult to apply than profile functions and much more difficult to check and calibrate.

Influence Functions and Weight Factors

The detailed procedure of influence function method is aptly available in various literatures (Peng, 1992; Singh, 1984; Whittaker and Reddish, 1989). The method consists of laying a suitable square or rectangular grid on the surface covering (i) the extraction panel area, and (ii) extending up to the subsidence draw limits (subsidence basin). An influence area of circular shape at each grid point (middle of the grid) is divided into a sufficient number of concentric rings. Each ring is again subdivided into a sufficient number of equal sectors by drawing radii from the grid point [Figure 12.10(a)]. Let us assume a grid centre is denoted by P and hence, the radius of the influence circle is to be taken as $R_f = H \tan \theta$, where θ is half angle between the vertical line passing through the grid point and the line connecting the grid point and end of the influence area. The sector which falls inside the excavation area will give rise to subsidence at the grid point P, depending on the weight factor of that sector, and the sector which falls outside the excavation gives no subsidence at P.

In order to determine weight factor of a sector, first an influence function suitable to the mine site needs to be assumed. There are many kinds of influence functions; they are either derived empirically or are based on theoretical assumptions. Following are some of the most commonly used influence functions:

(i) Stochastic influence function:

$$\Phi(r) = \frac{1}{R_f^2} \exp\left(-\frac{\pi r^2}{R_f^2}\right) \tag{12.23}$$

where, R_f = the radius of influence circle, and r = the radial distance from point P.

(ii) Sann's influence function:

$$\Phi(r) = \frac{2}{(\pi)^{3/2} R_f r} \exp\left(-4\left(\frac{r}{R_f}\right)^2\right) \tag{12.24}$$

(iii) Ehrhardt and Sauer influence function:

$$\Phi(r) = \frac{4.6}{\pi R_f^2} \exp\left(-4.6\left[\frac{r}{R_f}\right]^2\right) \tag{12.25}$$

(iv) USBM or Bal's influence function (1932/33):

$$\Phi(\zeta) = \frac{i}{H^2} \cos^2 \zeta \tag{12.26}$$

where, i = unit extraction, ζ = zone angle, which is in the range $0 \le \zeta \le \delta$ (angle of draw). For detailed description of this influence function, refer Peng (1992).

(v) *n-k-g* influence function for roof having strong floor (Nicieza et al., 2005)

$$\Phi(r) = \frac{Q}{(1+4n)\,(kr)^2}\left\{\exp\left(-\pi\left[\frac{H}{rk}\right]^2\right) + n\exp\left(-\pi\left[\frac{H}{rk}\right]^2 \Big/ 4\right)\right\} \tag{12.27}$$

where Q is a factor related to the volume of the excavated element and H is the depth of the excavated element, n and k are the two parameters that characterise the ground and that allow the shape and magnitude of the subsidence trough to be adjusted (Nicieza et al., 2005). The parameter n acts mainly on the magnitude of the subsidence, while k allows the shape of the trough to be fitted. The influence function can also be applied for inclined seam or ore body. For detailed study of this influence function and its applicability, refer to the above mentioned reference.

The weight factor for the ith ring, i.e., between radius r_{i-1} and r_i can be given by

$$S_i = \iint_{A_i} \Phi(r)\, r\,dr\,d\alpha \tag{12.28}$$

where A_i represents the area of ith annulus or ring. For example, S_i for Stochastic influence function can be derived as

$$S_i = \iint_{A_i}\left(\frac{e^{-(\pi r^2/R_f^2)}}{R_f^2}\right) r\,dr\,d\alpha = \int_0^{2\pi} d\alpha \int_{r_{i-1}}^{r_i}\left(\frac{e^{-(\pi r^2/R_f^2)}}{R_f^2}\right) r\,dr \tag{12.29}$$

$$= \exp\left(-\pi\left[\frac{r_{i-1}}{R_f}\right]^2\right) - \exp\left(-\pi\left[\frac{r_i}{R_f}\right]^2\right)$$

In general, 10 rings and 64 sectors or elements per ring or a total of 640 sectors provide reasonably good estimate of weight factor (Ren et al.,1987). In the calculation of weighting factors, the value of $S \neq 1$ in Eq. 12.30.

$$S = \iint_{\Omega} \Phi(r)\, r\,dr\,d\alpha \tag{12.30}$$

where, Ω = the whole area of the influence circle. However, the value of S must be unity and hence, it requires adjustment of weight factor of each ring as $S_i \geq S_i/S$. It may be noted that at *i*th ring, each sector has equal adjusted weight factor equals to $S_{in} = S_i/N_i$ where N_i denotes the number of sectors in i^{th} ring. Table 12.3 lists the weight factor of a sector in each ring for stochastic function. Note that if excavation is made in any sector, weight factor is added to grid P. The subsidence at point P is simply the sum of weight factors of all the element which fall in the extraction panel multiplied by the maximum possible subsidence, s_{max}. The weighting factor for the elements falling outside the extracted panel is zero. Hence, the total subsidence at grid point P will be

$$U_z = s_{max} \sum_{i=1}^{I} \sum_{n=1}^{N_i} S_{in} \tag{12.31}$$

where I signifies number of rings. Note that S_{in} is non-zero quantity even if a sector is extracted partially. For partial extraction, S_{in} can be multiplied by (A/A_s), where A is the extraction area inside the sector and A_s is the area of the sector.

TABLE 12.3 Value of S_i for *i*th Ring for Stochastic Function

Ring, i	1	2	3	4	5	6	7	8	9	10
S_i	0.03232	0.09109	0.13398	0.15551	0.15571	0.13923	0.11309	0.08424	0.05791	0.03687

Horizontal Displacement and Strain

Let us consider Figure 12.10 again. Assume that a vertical subsidence has occurred at grid point P due to extraction of an element dA in underground. The centroid of dA projected on the surface marks the point Q. Let us assume that the element dA lies inside any of the sectors of the influenced circle of grid point P. Then the displacement vector dU at grid point P has two components viz. (i) vertical component dU_z obtained by the weight factor, and (ii) horizontal component dU_{xy} in $x - y$ plane measured as given by Eq. 12.32.

$$dU_{xy} = dU_z \tan \alpha \tag{12.32}$$

where α is the angle between the vertical and the line joining the element centroid to the grid point P on the surface. It can be measured as:

$$\tan \alpha = PQ/H \tag{12.33}$$

It may be noted that if the circular area fully encompasses the extraction panel, then horizontal displacement at grid point P will be zero. The total horizontal movement at point P is again the sum of all extracted elements as $U_{xy} = \sum_N dU_{xy}$, where N = number of sectors in the influenced zone of grid point P. The resultant dU_{xy} can be resolved in x and y directions as:

$$dU_x = dU_{xy} \cos \beta \tag{12.34a}$$

$$dU_y = dU_{xy} \sin \beta \tag{12.34b}$$

where β is the angle between dU_{xy} and x-axis shown in Figure 12.10(b) and can be estimated as $\cos \beta = (x_Q - x_P)/PQ$. The total displacement at P is again a sum of all extracted elements. The slope and strain at point P are then readily calculated having known vertical and horizontal movements.

EXAMPLE 12.3 A longwall panel at a depth of 100 m has been excavated as shown in the following figure. Determine the vertical and horizontal displacements at point P for excavation of sectors A and B. Assume angle of draw of the area is 30°.

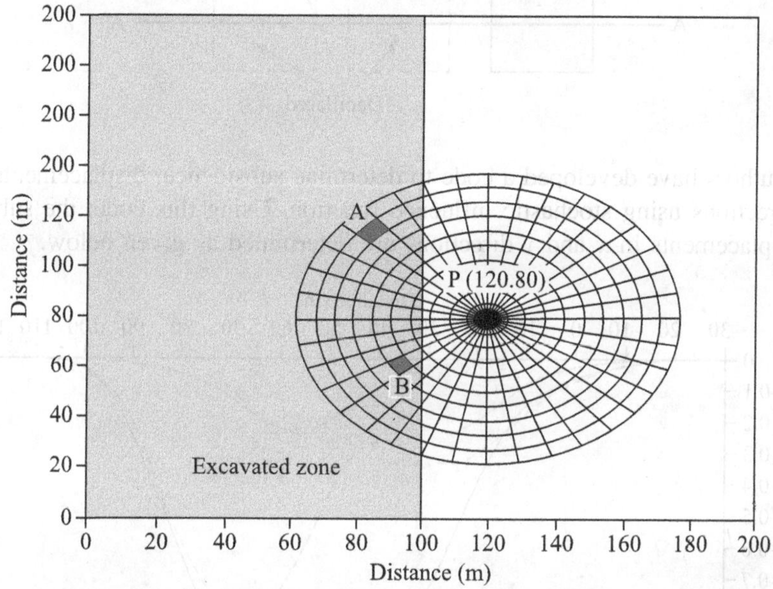

Solution: The influenced zone around point P is divided by 10 rings and 36 radials. The sectors A and B are located in 9th ring and 5th ring respectively. Based on Table 12.3, the contribution of vertical displacement at point P for extractions of these two sectors are 0.057911/36 = 0.00161 and 0.155713/36 = 0.00433 respectively. The table below summarises the other results considering s_{max} = 1.0 m.

Sector	U_z (m)	PQ/H	dU_{xy} (m)	x(m)	y(m)	$(x - x_P)$/PQ	$(y - y_P)$/PQ	dU_x (m)	dU_y (m)
A	0.00161	0.4907	0.000789	85.299	114.701	–0.70711	0.70711	–0.00056	0.00056
B	0.00433	0.2598	0.001124	98.718	65.098	–0.81915	–0.57357	–0.00092	–0.00065

EXAMPLE 12.4 A section of depillaring panel is shown in the figure given below. Pillar and gallery widths are 25 m and 5 m respectively. Assuming the depth of the seam is 50 m and angle of draw is 25°, plot the surface subsidence and horizontal displacement along the lines A–A′, B–B′ and C–C′.

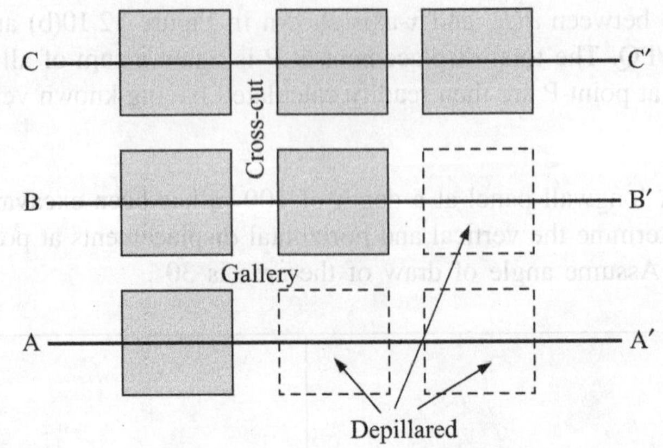

Solution: Authors have developed a code to determine subsidence, displacements and strains in x and y directions using stochastic influence function. Using this code, the subsidence and horizontal displacements in x and y directions are determined as given below.

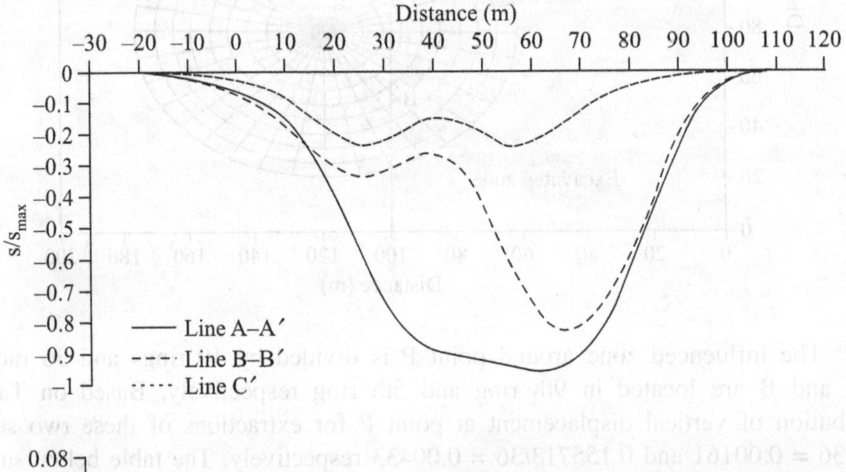

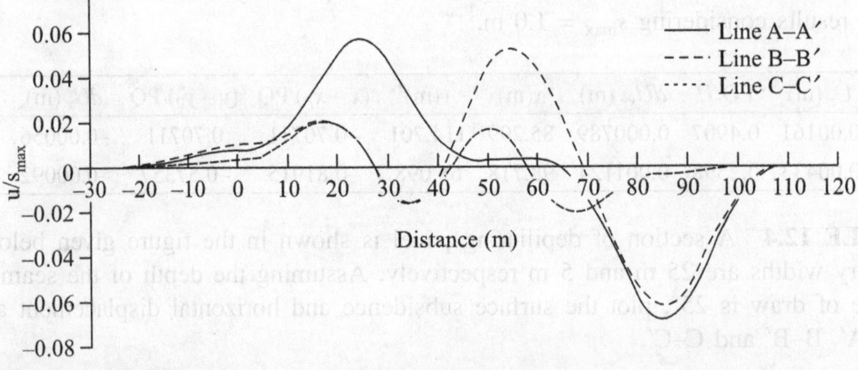

The overall subsidence is shown in the figure given below.

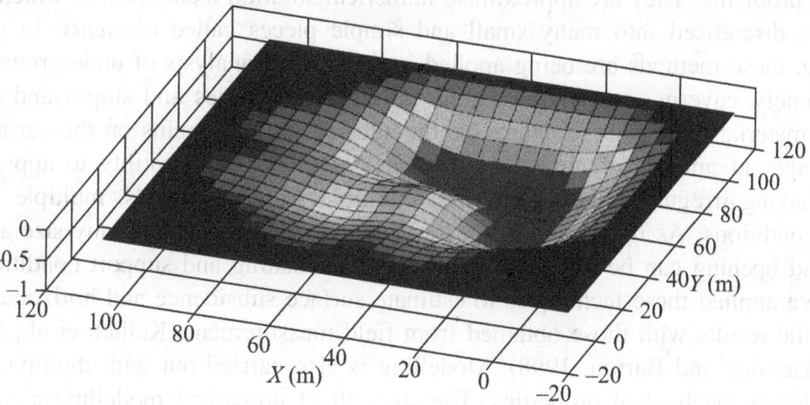

12.4.5 Physical Model

Physical models have played an important role in the development of knowledge pertaining to understanding ground behaviour in relation to mining subsidence. King and Whetton (1957) reported the finding of mining subsidence investigations using physical models constructed of solidified gelatin solution. Sand/plaster physical models are the most commonly used since other types do not readily provide the advantages given by such modelling materials. In particular, sand/plaster types easily allow the principle variations in geology to be built into the model. Additionally the propagation and growth of cracks caused by caving and subsidence can be observed and studied in relation to other factors. Surface subsidence can be studied as a separate feature, although it is usual to take account of sub-surface behaviour with such models. In this case, the rock strength can be readily adjusted by selecting appropriate sand to plaster ratio in the mixing stage of model construction.

The use of physical models, both two- and three-dimensional were popular from 1950s to the early 1980s using design of equivalent materials and equivalent load. Construction of the models have sometimes involved elaborate simulation of the sedimentary deposition process and the process is somewhat cumbersome. The use of physical models has declined greatly in recent years, due to:

(i) Their high cost and labour intensive nature, especially where several variations of the rock properties and loading conditions are required
(ii) The increasing capability of computer based numerical models to include the above-mentioned features, and to show these effects visually by computer graphics
(iii) The materials usually designed with scaled elastic properties may not realistically model and non-linear inelastic behaviour of actual rock mass

12.4.6 Numerical Method

Numerical methods especially finite element method (FEM), discrete element method (DEM), boundary element method (BEM) and others are alternative procedures to analytical methods, profile and influence function methods for analysing structures made in rock mass and associated subsidence problems. They are approximate numerical solution techniques in which continuous systems are discretised into many small and simple pieces called *elements*. In geotechnical engineering, these methods are being applied in design and analysis of underground structures such as tunnels, caverns, powerhouse, mine pillars and galleries and stopes and also utilises non-linear material behaviour to determine the subsidence and strains on the surface.

The major advantages of numerical methods reside on its versatility to apply in various problems having irregular shapes and sizes, multiple material models, multiple loading and boundary conditions. As for example, stresses and deformations around any size and shape of underground opening can be analysed under different loading and support conditions. Several authors have applied these techniques to estimate surface subsidence and horizontal strain and compared the results with those obtained from field measurement (Keilich et al., 2006; Guoet al., 2004, Heasley and Barton, 1998). Modelling is also carried out with multiple rock layers having different mechanical properties. The strength of numerical modelling also lies on the resemblance of the physical size of the problem with the model. There is no need to scale down or up the actual structure being analysed in order to obtain the stress and deformation behaviour. The actual engineering properties of rock or other material are used in finite element analysis. On the contrary, for physical modelling of structures, equivalent material properties are assumed with scale down geometries.

12.5 SINKHOLE SUBSIDENCE

Sinkhole subsidence occurs due to shallow depth abandoned mine workings, natural cavities, excessive withdrawal of ground water and others. Most of the old mines are worked with room and pillar method of mining and possesses great potential of surface subsidence by roof fall/pillar failure. In UK, more than 70,000 old mine workings are reported and some of them may be three centuries old (Whittaker et al., 1989). In the USA, 354 subsidence incidents were reported over Pittsburgh Coal bed, most of them were in the form of sinkholes (Gray et al., 1977). In 1985, Marino et al., reported that both trough and sinkhole type subsidence occurred in Illinois (Whittaker et al., 1989) over shallow depth room and pillar mines. After extensive study of surface subsidence in the USA, Gray et al., commented that the most prevalent subsidence features over abandoned mined land are sinkholes, with depth of sinkhole more than 3 ft, and trough or sags less than 3 ft (Peng, 1992). After studying subsidence incidents in Germany, Kratzsch commented that sudden cave-ins and irregular depressions in the form of sinkholes over near-surface abandoned mines possess a serious risk to the populated area nearby (Kratzsch, 1983). Kratzsch's (1983) study also suggests that size of the affected area must be established based on statistical investigations, taking into account type of workings, strength and thickness of roof, fluctuation of ground water and others. In Korea, more than 300 coal mines have been ceased their operations with the economic structural adjustments in early 1990s. With lots of abandoned coal mines, the surface subsidence problem has been coming up to the public, known as one of geohazards (Choi et al., 2004). Their study suggests that several parameters should be emphasised for evaluating the surface subsidence in coal mine area, and these parameters should be dealt with adding an extra weight for realistic analysis. In addition, the sinkhole type subsidence in urban area, whatever it caused by collapse of galleries in coal mines or in metal mines, can be a death-blow to many structures as well as public welfare.

All of the above mentioned literatures suggest that sinkhole development is a process of collapsed junction having weak/fractured roof and then progression of collapse chimney up to the surface at shallow depth of cover. In general, sinkhole is in the form of conical or cylindrical depression/cavity suddenly appears on the surface. Figure 12.11(a) shows plan of a 4-way junction with a cylindrical sinkhole of diameter *D*. Figures 12.11(b), (c) and (d) depict progressive failure of roof above the chimney or sinkhole. The major factors which contribute to sinkhole formation are width of mine opening/gallery (*w*), height of opening (*m*), depth of cover (*H*), rock type and thickness of roof, water condition, pillar strength, time elapsed after mining operation has been ceased, inclination of the seam, bulking factor (*K*) of roof rocks and others. Whittaker et al., (1989) have reported that width of gallery, depth of cover and roof conditions are the primary factors for the determination of sinkhole occurrences and higher chances are associated with lower *H*, higher *W* and weak roof conditions. Peng (1992) had also commented on the similar factors which contributed in sinkhole development in the Pittsburgh Coal bed.

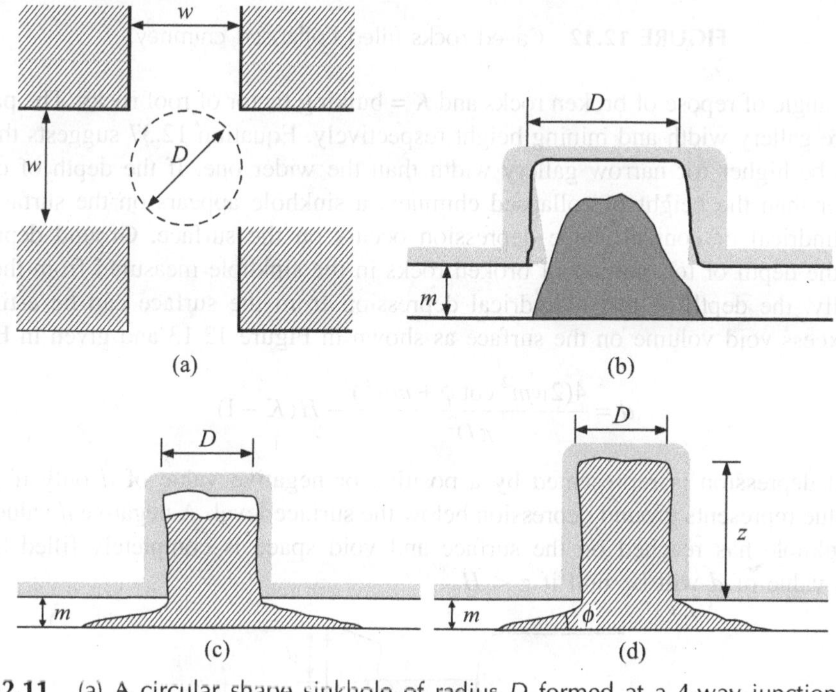

FIGURE 12.11 (a) A circular shape sinkhole of radius *D* formed at a 4-way junction, (b) Initial collapse of mine roof, (c) Progressive development of roof collapse, and (d) Caved rocks fills completely the void space (adapted from Whittaker and Reddish, 1989).

12.5.1 Volumetric Calculations

In this study, sinkhole is assumed to be cylindrical shape all through its length. Considering the diameter of the chimney or sinkhole as *D*, the height of collapsed chimney is estimated by equating the total void space and total volume of broken rock as (Figure 12.12) (Whittaker et al., 1989).

$$V_{\text{caved}} = \frac{Kz\pi D^2}{4} \tag{12.35a}$$

$$V_{\text{space}} = 4(1/2wm^2 \cot\phi) + mw^2 + z\pi D^2/4 \tag{12.35b}$$

$$z = \frac{4}{(K-1)\pi D^2}[2wm^2 \cot\phi + mw^2] \tag{12.36}$$

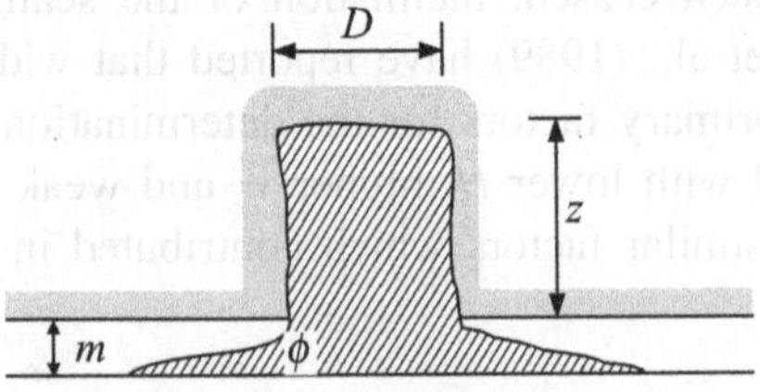

FIGURE 12.12 Caved rocks filled collapsed chimney.

where, ϕ is angle of repose of broken rocks and K = bulking factor of roof rocks. The parameters w and m are gallery width and mining height respectively. Equation 12.37 suggests that caving height will be higher for narrow gallery width than the wider one. If the depth of coal seam (H) is lower than the height of collapsed chimney, a sinkhole appears on the surface. In this case, a cylindrical or conical shape depression occurs on the surface. Ground depression is termed as the depth of top portion of broken rocks in the sinkhole measured from the surface. Theoretically, the depth of this cylindrical depression from the surface can be estimated by equating excess void volume on the surface as shown in Figure 12.13 and given in Eq. 12.37:

$$d = \frac{4(2wm^2 \cot\phi + mw^2)}{\pi D^2} - H(K-1) \tag{12.37}$$

Ground depression is represented by a positive or negative value of d only if $z > H$. A positive value represents ground depression below the surface level. A negative d value signifies that the sinkhole has reached on the surface and void space is completely filled by broken rocks. The value of d will be null if $z < H$.

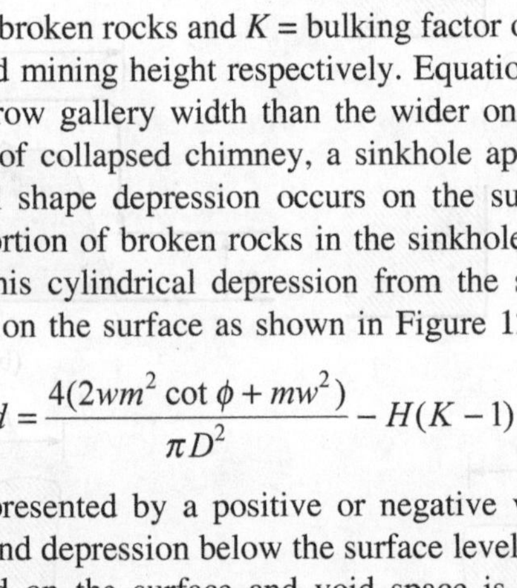

FIGURE 12.13 Depth of ground depression for $z > H$.

From earlier analysis, it is found that the volume occupied by the caved rocks in the adjoining rooms of a 4-way junction ($2wm^2 \cot \phi$) is significant as compared to volume of the junction itself (mw^2) and ratio of these two volumes is:

$$r = \frac{2m \cot \phi}{w} \tag{12.38}$$

Assuming similar area ratio in two-dimensional analyses, the total void area can be estimated as (Figure 12.14)

$$A_{space} = mw\,[1 + r] + A_p = mw\left[1 + \frac{2m \cot \phi}{w}\right] + zD \tag{12.39}$$

where A_p is the area released by the broken rocks in the roof. Hence, equating area of void space and area occupied by the broken rocks, the height of cavity will be

$$z = \frac{mw + 2m^2 \cot \phi}{D(K-1)} \tag{12.40}$$

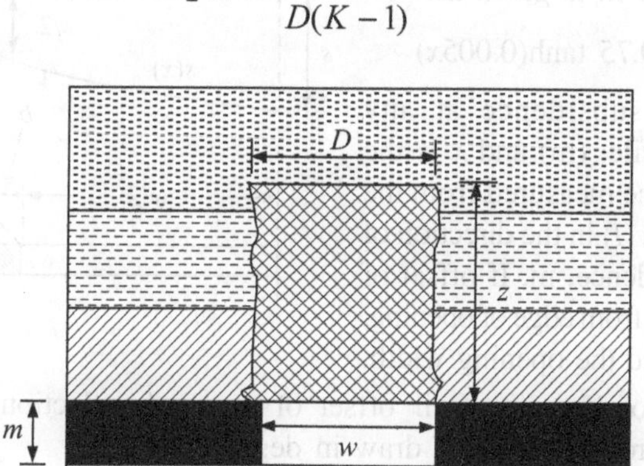

FIGURE 12.14 Two dimensional approximation of sinkhole formation.

As before, if the collapsed chimney reaches to the surface, a vertical ground depression is developed and it is estimated as:

$$d = \left[\frac{mw + 2m^2 \cot \phi}{D}\right] - H(K-1) \tag{12.41}$$

EXAMPLE 12.5 A chimney cave forms over a square excavation of side 6 m and height 3 m. The caved zone forms an inscribed circular cross-section. If the bulking factor is 1.2 and angle of repose of broken material is 40°, how high will cave zone extend above the excavation?

Solution: Based on Eq. 12.36, we estimate the height as:

$$z = \frac{4}{(K-1)\pi D^2}[2wm^2 \cot \phi + mw^2] = 251.94 \text{ m}$$

EXERCISES

1. If the ratio of critical width of the horizontal working to the mining depth of a longwall mine is 1.4. Find the angle of draw in degree for critical subsidence.
2. A 60 m wide longwall panel at a depth 60 m results in a subsidence trough of width 110 m. Find the angle of draw.
3. A flat longwall mine has panel width 200 m. Assuming angle of abutment as 30°, estimate the depth of mining for which supercritical subsidence occurs on the surface?
4. With neat sketches, draw and briefly explain the curves: (i) vertical subsidence, (ii) horizontal strain, and (iii) slope for subcritical and supercritical subsidence for a longwall panel with flat seam.
5. The subsidence profile over a longwall panel in a perfectly horizontal seam at a depth of 100 m is given as:

$$s(x) = 0.746 - 0.75 \tanh(0.005x)$$

where, $s(x)$ is subsidence at any point on the profile, (m) and x is the horizontal distance measured from the point of inflection (I) in the direction of decreasing subsidence, m. If offset of point of inflection from edge of opening, $b = 1$ m, then find the opening width.

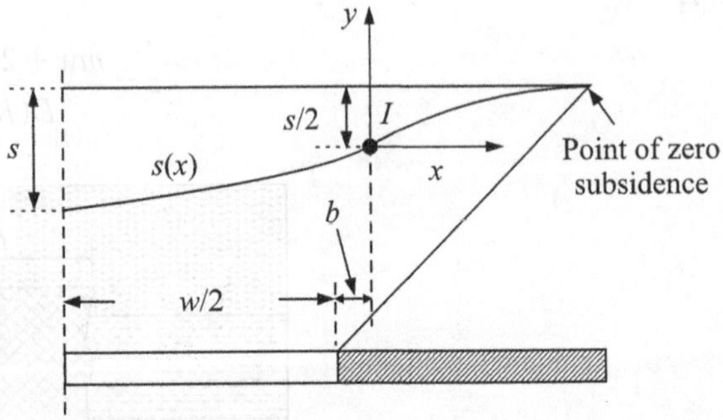

6. In continuation of Exercise 6, if offset of point of inflection from edge of opening, $b = 0$ m, then find the angle of draw in degree.
7. A rectangular excavation of height m, width w and length l induces a vertical chimney of elliptical cross section with semi axes $w/2$ and $l/2$ as shown in the Figure below. Develop a cave (chimney) height formula in terms of bulking factor (K), caving geometry and angle of repose (φ) of broken material.

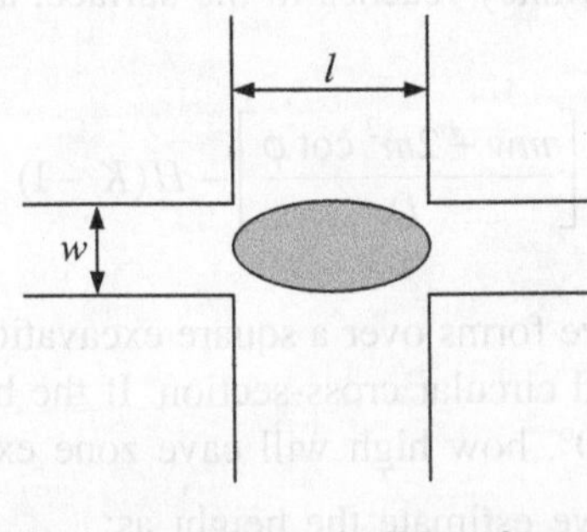

8. If the depth of the excavation is 100 m, bulking factor is 1.2, angle of repose of broken material is 30°, excavation height is 3 m, width is 4 m and length is 6 m, is there a danger on the surface due to sinkhole subsidence?

9. A subcritical subsidence profile function in m is given as:

$$S(x) = 0.74625 - 0.75\tanh\left(\frac{4x}{H}\right)$$

The inflection points are located 5 m inside the edges of the panel. Assuming the depth of the panel H is 200 m, draw the subsidence profile as s versus x. Here, x is the distance measured in m from the inflection point in the direction of decreasing subsidence. Calculate the radius of major influence and slope at inflection point and estimate the angle of draw.

10. Calculate the contribution of 7th ring in Example 12.3 in vertical and horizontal displacements at point P.

11. A mine in a flat seam at a depth of 500 m excavates panels with a face width of 260 m and a length of 2 km. Mining is full seam height at 4.8 m. The angle of draw and subsidence factor are estimated to be 25° and 0.7, respectively. Determine the maximum subsidence and trough width for a single panel.

12. Consider the data in Exercise 11 and suppose the seam dips 8° and panels are developed on strike so the face runs up and down dip. Estimate the shift in m, the location of the trough in relation to the panel using the rule of the normal. Assume the panel is at a depth of 500 m.

13. A mine in a flat seam at a depth of 500 m excavates panels with a face width of 250 m and a length of 2 km. Mining is full seam height at 4.8 m. The angle of draw and subsidence factor are estimated to be 35° and 0.9, respectively. Determine peak tensile and compressive strains after completion of a single panel. Also determine the secondary compressive strain maximum that may occur at the trough center.

14. Horizontal displacement, in m along a major section of a subcritical subsidence profile at a depth of 250 m is given as:

$$u(x) = \exp(-(x/50)^2)$$

where, x is the distance measured from the inflection point. Negative x denotes distance towards the centre and positive x signifies distance towards the panel edge. The panel width is 200 m. Determine the angle of critical deformation for a cut-off strain of 0.0005.

Chapter 13

Slope Stability

13.1 INTRODUCTION

Slope is an inclined surface, such as an open pit bench or cut slope in natural material and generally measured as a degree of inclination with respect to horizontal. Slope stability implies the resistance of that inclined surface to failure by sliding or toppling. Numerically, it is expressed same as factor of safety either as a ratio, decimal fraction or as a percentage. Figure 13.1 shows the basic configuration of four benches with terminologies.

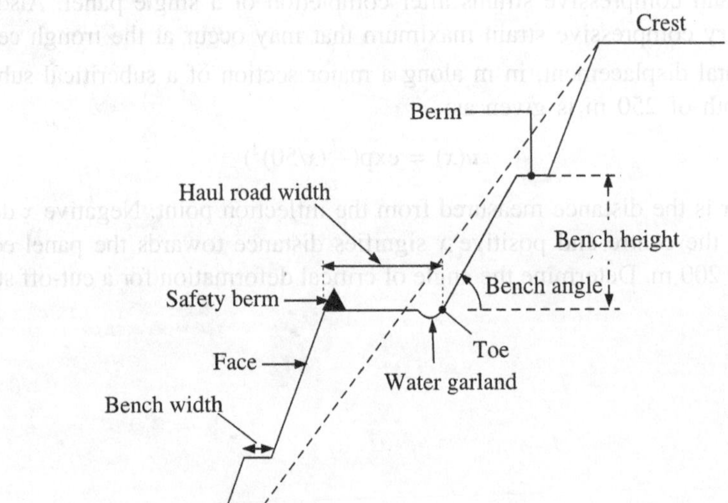

FIGURE 13.1 Slope configuration.

Bench: A horizontal shelf of sufficient width, which provides a working platform for men and machinery to excavate slope material. Generally, above or below, a wider bench isused for hauling excavated materials, called *haul road*. In Indian mining regulation, it is stated that haul road width should be at least three times the width of the widest dump truck.

Bench angle: It is an angle between the slope face with the horizontal.

Berm: A horizontal shelf built into the sloping wall to break the continuity of long slope by decreasing the overall slope angle. Primary objective of berm is to decrease the overall slope angle thereby increases the slope stability but also acts as an arrester for falling loose rock from higher levels. Berm width is generally larger than bench width.

Bench height: It is the vertical distance between the crest and toe of the bench.

Crest: The line of intersection of the slope with the unmined surface. This line defines the upper contour of the slope.

Overall slope angle: It is an angle between lines connecting the toe of the bottom most bench to the crest of the top most bench with the horizontal. Generally, slope stability study is conducted to determine whether overall slope of a cut is stable or not.

Toe: The line of intersection of the slope with the floor. This line defines the lower contour of the slope.

The causes of slope failure can be grouped into two major categories where the factors contribute to (a) increase shear stresses; and (b) reduces shear strength along failure plane. The shear stress increases due to (i) removal of lateral support, (ii) removal of underlying support, (iii) addition of surcharge to the slope, (iv) seismic activities, and (v) increase in overall slope. The shear strength reduces due to (i) presence of discontinuity, (ii) slope orientation, (iii) presence of weak rock layer below slope, (iv) weathering, (v) presence of water, (vi) loose material absorbing water, and (vii) change of structure. The failure of the slope depends on either one or combination of factors mentioned above.

13.2 MODES OF SLOPE FAILURE

The primary modes of slope failure depend on the rock type and interaction of geological discontinuities imbedded therein. Modes of slope failure are of four types

(i) Plane failure
(ii) Wedge failure
(iii) Toppling failure
(iv) Circular failure

Plane, wedge and toppling failure modes depends on orientation of discontinuity surface, slope face and shear strength of discontinuity in relatively hard rock. Hence in a rock slope, sliding blocks are identified based on geometry that is free to move along the discontinuity surface. Sliding block can be identified by stereonet projection of joint sand slope surfaces with their orientation of dip direction and dip. Circular mode of failure is very common in slope formed by weak/loose rock, waste dump and soil material.

13.3 PLANE FAILURE

In this failure mode, sliding block rests on a discontinuity plane and it slides once shear stress that acts on the discontinuity plane becomes greater than its shear strength. In this case, the discontinuity surface is more or less planar or gently undulating. The linear movement of the sliding block is controlled structurally by discontinuity, such as faults and joints, bedding planes, and properties of the contact surface. Figure 13.2 shows a picture of sliding plane along which the upper block (not shown in the figure) has failed. The following geometrical conditions control plane failure phenomena.

FIGURE 13.2 Natural plane failure (upper block not shown).

(i) The strike of the discontinuity plane is more or less within ± 20° of the strike of the crest of the slope (Figure 13.3).

(ii) The toe of the discontinuity plane should lie between the toe and crest of the slope.

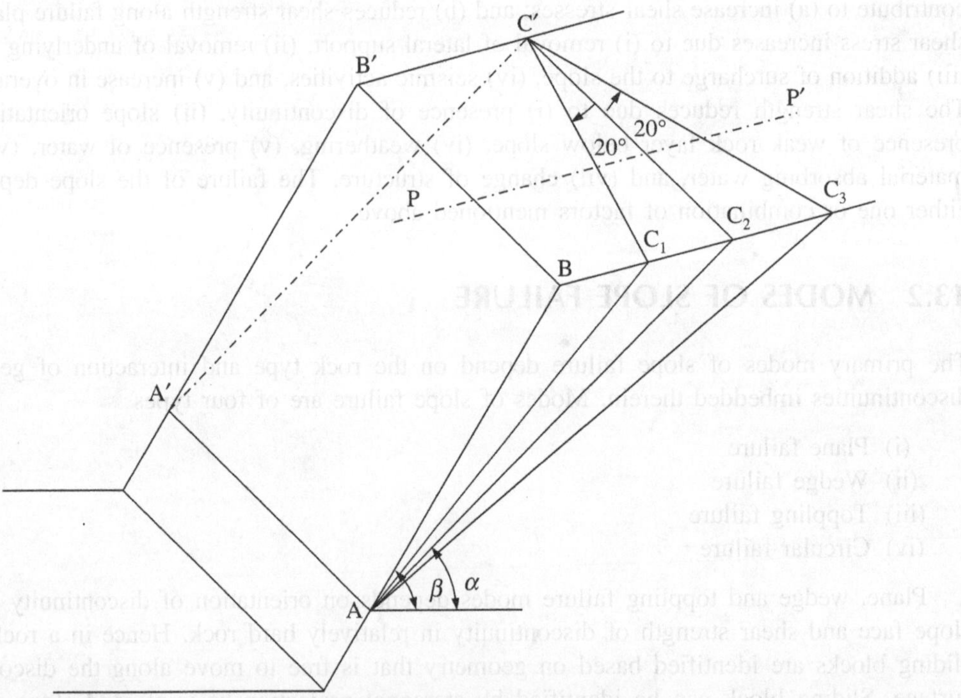

FIGURE 13.3 3D view of a slope geometrical orientation for plane failure.

(iii) The dip of the discontinuity plane must be less than the dip of the slope face, and the internal angle of friction for discontinuity surface must be less than the dip of the discontinuity plane.

For slope stability analysis, primary objective is to determine the factor of safety (FOS) of the sliding block. In Figure 13.3, a sliding block would mean the volume of rock enclosed by $AA'B'C'C_1B$ and so on. In this chapter, angle β would mean overall slope angle and α signifies the dip angle of discontinuity. FOS is defined as the ratio of resisting force (RF) to driving force (DF) acting parallel to the direction of sliding. Driving force is acting along the direction of movement whereas resisting force is acting against the movement. Thus, the FOS for plane failure is

$$\text{FOS} = \frac{\text{RF}}{\text{DF}} \tag{13.1}$$

A 3D slope shown in Figure 13.3 can be analysed in 2D plane strain condition by taking a vertical section along PP′. A 2D section is shown in Figure 13.4, where unit thickness of block is assumed. Considering a dry slope, the driving and resisting forces are shown in the figure below. In the above equation, resisting force is estimated along the discontinuity plane using Mohr–Coulomb equation.

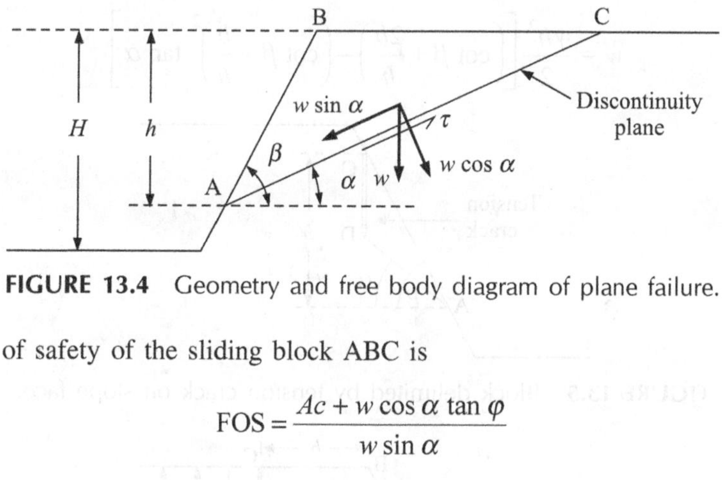

FIGURE 13.4 Geometry and free body diagram of plane failure.

The factor of safety of the sliding block ABC is

$$\text{FOS} = \frac{Ac + w\cos\alpha\tan\varphi}{w\sin\alpha} \tag{13.2}$$

where, weight of the sliding block (w) is $\gamma h^2(\cot\alpha - \cot\beta)/2$; γ is unit weight of sliding block; h is the vertical height of the sliding block, $A = L \times 1$ is area of contact of sliding block and equal to the length of contact of sliding block (AC) multiplied by unit thickness; and φ is angle of internal friction.

EXAMPLE 13.1 A planar block as shown in Figure 13.4 slides along the discontinuity plane AC. The geotechnical and dimensional data associated with the slope are $\alpha = 35°$; $\beta = 45°$; $c = 350$ kPa; $\gamma = 25$ kN/m^3; $\varphi = 29°$ and $h = 100$ m. Find the slope factor of safety.

Solution:

$$\text{Weight of sliding block} = \frac{\gamma h^2}{2}(\cot \alpha - \cot \beta) = 53.52 \text{ MN/m}$$

$$\text{Area of contact of sliding block (A)} = \frac{h}{\sin \alpha} \times 1 = 174.34 \text{ m}^2$$

$$\text{Resisting force} = cA + w \cos \alpha \tan \varphi = 85.53 \text{ MN/m}$$

$$\text{Driving force} = w \sin \alpha = 30.70 \text{ MN/m}$$

$$\text{FOS} = \frac{85.53}{30.70} = 2.78$$

Figures 13.5 and 13.6 shows two cases of plane failure where tension crack delimits the size of the sliding block. The delimitation of size of sliding block is either due to pre-existing discontinuity or tension crack. The tension cracks are created because of expansion of pre-existing crack either by blasting, water pressure or block sliding. Weight of sliding block shown in Figures 13.5 and 13.6 are given by Eq. 13.3(a) and 13.3(b) respectively as:

$$w = \frac{\gamma h^2}{2} \cot \beta \left(1 - \frac{\tan \alpha}{\tan \beta}\right) \tag{13.3a}$$

$$w = \frac{\gamma h^2}{2}\left[\left(\cot \beta + \frac{2b}{h}\right) - \left(\cot \beta + \frac{b}{h}\right)^2 \tan \alpha\right] \tag{13.3b}$$

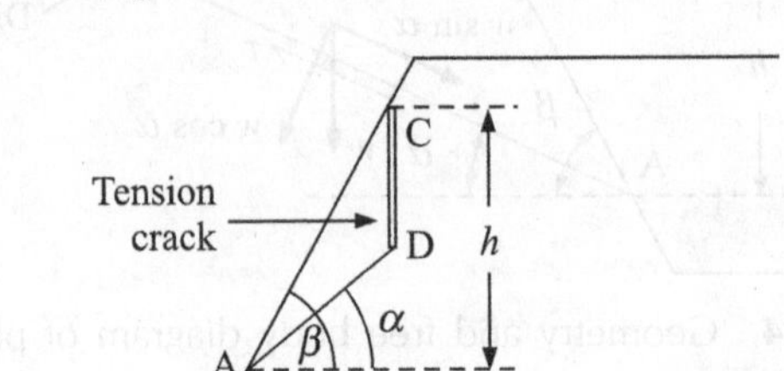

FIGURE 13.5 Block delimited by tension crack on slope face.

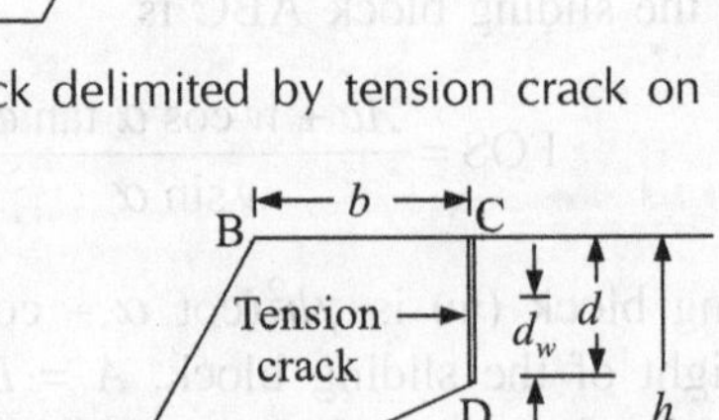

FIGURE 13.6 Block delimited by tension crack away from crest.

EXAMPLE 13.2 A planar block defined in Example 13.1 is delimited by tension crack CD of length 40 m shown in Figure 13.5. The vertical height of the sliding block is 80 m, other geotechnical and dimensional data are same. Find the slope factor of safety.

Solution:

$$\text{Weight of sliding block} = \frac{\gamma h^2}{2}\cot\beta\left(1 - \frac{\tan\alpha}{\tan\beta}\right) = 23.98 \text{ MN}$$

$$\text{Area of contact of sliding block (A)} = \frac{h-d}{\sin\alpha} \times 1 = 69.74 \text{ m}^2$$

$$\text{Resisting force} = c\text{A} + w\cos\alpha\ \tan\varphi = 33.93 \text{ MN}$$

$$\text{Driving force} = w\sin\alpha = 13.76 \text{ MN}$$

$$\text{FOS} = \frac{33.93}{13.76} = 2.57$$

EXAMPLE 13.3 If a planar block in Figure 13.4 is delimited by a tension crack shown in Figure 13.6 with b = 20 m. Find (a) length of tension crack, (b) area of contact of sliding surface, and (c) factor of safety of sliding block.

Solution:

$$\text{Weight} = \frac{\gamma h^2}{2}\left[\left(\cot\beta + \frac{2b}{h}\right) - \left(\cot\beta + \frac{b}{h}\right)^2 \tan\alpha\right]$$

$$= 48.95 \text{ MN}$$

$$\text{Length of tension crack (CD)} = h - (b + h\cot\beta)\tan\alpha = 16 \text{ m}$$

$$\text{Area of contact of sliding block (A)} = \frac{h-d}{\sin\alpha} \times 1 = 146.44 \text{ m}^2$$

$$\text{Resisting force} = c\text{A} + w\cos\alpha\ \tan\varphi = 73.48 \text{ MN}$$

$$\text{Driving force} = w\sin\alpha = 28.07 \text{ MN}$$

$$\text{FOS} = \frac{73.48}{28.07} = 2.62$$

13.3.1 Water in Tension Crack and/or Discontinuity Plane

In impervious rock mass, rain and/or ground water flows through the discontinuities. Water flowing through discontinuities imparts considerable change in resisting and driving forces because of buoyant force exerted by water on overlying rock mass. Eight conditions exist for water pressure distributions on a failure plane as illustrated in Figure 13.7. Water pressure distribution on discontinuity plane is equal to the area enclosed by water pressure triangle. The concentrated pressure acts at the centroid of pressure triangle directed perpendicular to the surface of contact. Water pressure is generated due to interconnected pore spaces from surface to below surface. Here, it is assumed that the distribution of water pressure varies linearly between two pressure points. But, the distribution depends on whether water seeps from slope face at atmospheric pressure or confined within the sliding surface. The condition of confinement exists when discontinuity surface is not fully opened on the slope face. If the sliding mass behaves like a rigid body, the condition of limiting equilibrium is reached when both the shear forces directed downward equal the shear strength directed upward along the sliding surface. Eight different conditions are discussed as follows:

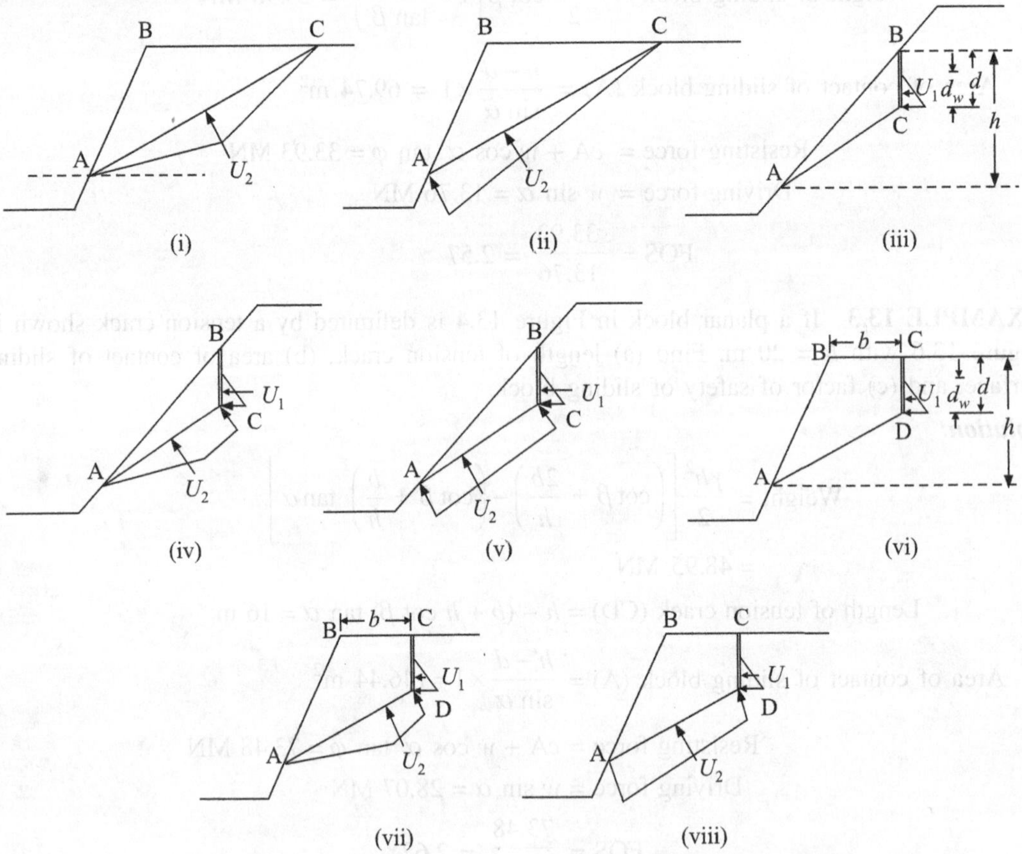

FIGURE 13.7 Water pressure distribution on sliding block.

(i) Water enters from crest and dripping away from slope face at atmospheric pressure. In this situation, maximum water pressure exists at the mid height of the water level.
(ii) This case is same as Case (i), with a difference that the drainage of water at slope face blocked. Hence, peak pressure occurs at the slope face. This is the worst case scenario, since it will result in the maximum hydrostatic force being applied perpendicular to the discontinuity surface.
(iii) Water filled tension crack on slope face delimits the discontinuity plane and water blocked at the bottom of tension crack. Hence, maximum water pressure occurs at the base of the tension crack.
(iv) This case is same as Case (iii) and water drains into the discontinuity and finally dripping away from slope face at atmospheric pressure. Maximum water pressure occurs at the mid height of the water level in the slope and the base pressure of tension crack is same at the point of its contact with the discontinuity.
(v) This is same as the previous condition with a difference that the drainage of water at slope face is blocked.

The water pressure distribution in Cases (vi), (vii), and (viii) are same with the corresponding Cases (iii), (iv), and (v) respectively with the exception that tension crack is located at the crest of the slope.

Equation 13.4 shows the generalized FOS relationships for the plane failure cases as shown in Figure 13.7. Table 13.1 shows the values of derived parameters for the eight cases of failure to calculate their FOS as per Eq. (13.4).

$$\text{FOS} = \frac{(c + \sigma_n \tan\varphi)A}{w\sin\alpha + U_1\cos\alpha} \tag{13.4}$$

TABLE 13.1 Factor of Safety of a Plane Slope with Water in Discontinuity

Cases	*Water pressure force* (U_1, U_2), *normal stress* (σ_n) *and Sliding block weight* (w)		
(i)	$U_1 = 0$; $U_2 = h\gamma_w A/4$; $A = h/\sin\alpha$; $\sigma_n = (w\cos\alpha - U_2)/A$		
(ii)	$U_1 = 0$; $U_2 = h\gamma_w A/4$; $A = h/\sin\alpha$; $\sigma_n = (w\cos\alpha - U_2)/A$		
(iii)	$U_1 = d_w^2\gamma_w/2$; $A = (h-d)/\sin\alpha$; $\sigma_n = (w\cos\alpha - U_1\sin\alpha)/A$		
(iv)	$d_w < \dfrac{h-d+d_w}{2}$	$U_1 = d_w^2\gamma_w/2$ $U_2 = \dfrac{\gamma_w}{2\sin\alpha}\left[\dfrac{3}{8}(h-d+d_w)^2 - d_w^2\right]$	$A = (h-d)/\sin\alpha$ $U_1 = d_w^2\gamma_w/2$ $\sigma_n = (w\cos\alpha - U_1\sin\alpha - U_2)/A$
	$d_w > \dfrac{h-d+d_w}{2}$	$U_1 = \gamma_w\left[\dfrac{d_w^2}{2} + (h-d)d_w - \dfrac{(h-d+d_w)^2}{4}\right]$ $U_2 = \gamma_w d_w A/2$	
(v)	$U_1 = d_w^2\gamma_w/2$; $U_2 = (h + 2d_w - d)\gamma_w A/2$; $A = (h-d)/\sin\alpha$; $\sigma_n = (w\cos\alpha - U_1\sin\alpha - U_2)/A$		
(vi)	$U_1 = d_w^2\gamma_w/2$; $A = (h-d)/\sin\alpha$; $\sigma_n = (w\cos\alpha - U_1\sin\alpha)/A$		
(vii)	$d_w < \dfrac{h-d+d_w}{2}$	$U_1 = d_w^2\gamma_w/2$; $U_2 = \dfrac{\gamma_w}{2\sin\alpha}\left[\dfrac{3}{8}(h-d+d_w)^2 - d_w^2\right]$	$A = (h-d)/\sin\alpha$ $\sigma_n = (w\cos\alpha - U_1\sin\alpha - U_2)/A$
	$d_w > \dfrac{h-d+d_w}{2}$	$U_1 = \gamma_w\left[\dfrac{d_w^2}{2} + (h-d)d_w - \dfrac{(h-d+d_w)^2}{4}\right]$ $U_2 = \gamma_w d_w A/2$	
(viii)	$U_1 = d_w^2\gamma_w/2$; $U_2 = (h + 3d_w - d)\gamma_w A/2$; $A = (h-d)/\sin\alpha$; $\sigma_n = (w\cos\alpha - U_1\sin\alpha - U_2)/A$		

In Table 13.1, γ_w is unit weight of water equal to 1000 kg/m^3 or 1 g/cm^3 at STP; γ is unit weight of sliding block; A is area of contact of sliding block; σ_n is normal stress on sliding surface.

Close examination of FOS formula shows that water effects the slope stability in two ways: (i) denominator increases with increasing water pressure, (ii) numerator decreases with increasing water pressure. Hence, FOS decreases drastically as water pressure increased along the discontinuity plane. It may also be noted that both c and φ of discontinuity surface may also decrease due to inrush of water through discontinuity. Hence, drainage of water is must to maintain stability of slope during rainy season.

EXAMPLE 13.4 Derive an expression for water pressure force exerted along a discontinuity plane AC as shown in figure. Assume that the water pressure developed along the discontinuity surface is due to the fully saturated slope upto the crest line BC.

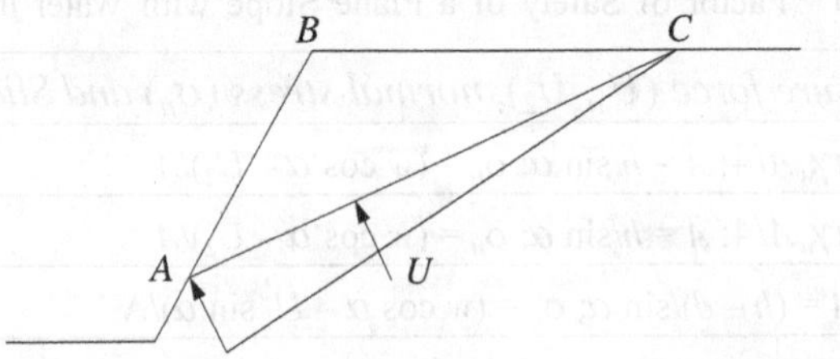

Solution: Assume that the water is blocked at point A, and hence the pressure developed along the sliding surface is shown by water pressure profile in Figure (b) below:

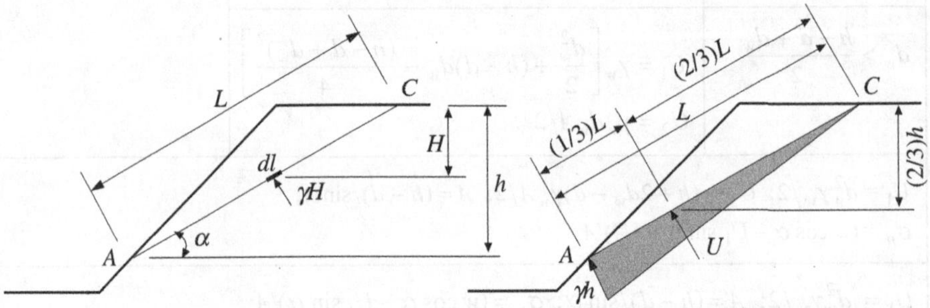

The water pressure at a depth of H meter from crest will be γH and will act perpendicular to the sliding surface. If pressure is acting on a very small area of length dl along discontinuity AC of unit width. Then, the water pressure force acting on area $dl \times 1$ will be γHdl. The resultant force acting on the discontinuity surface can be found by integrating the following expression:

$$U = \int_0^F dU = \int_0^h \gamma H dl = \int_0^h \gamma H\left(\frac{dH}{\sin\alpha}\right) = \frac{\gamma}{\sin\alpha}\int_0^h H dH$$

$$U = \frac{\gamma h^2}{2\sin\alpha}$$

The resultant uplifting force (U) on discontinuity surface is also equal to the area of the pressure triangle of height γh and base $L = h/\sin\alpha$. This force is acting at the centroid of the pressure triangle, i.e., 2/3rd length along AC as shown in Figure (b) from vertex C.

EXAMPLE 13.5 Consider Example 13.3 where slope is submerged with water and water table in tension crack is noticed at about 5 m below the crest. Find the factor of safety of the sliding block if water is dripping from toe of sliding block.

Solution: Water pressure profile along the tension and discontinuity plane is shown in Figure 13.7 (viii). Force body diagram is shown below. Table 13.1 shows the relationship to find the FOS as:

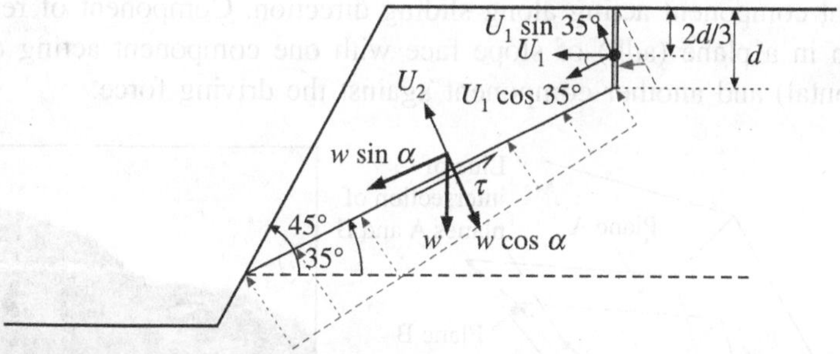

$$U_1 = \frac{\gamma_w d_w^2}{2} = 0.59 \text{ MN}; \quad U_2 = \frac{1}{2} \gamma_w d_w A = 7.90 \text{ MN}$$

$$\sigma_n = (w \cos \alpha - U_1 \sin \alpha - U_2)/A$$

$$\text{RF} = (c + \sigma_n \tan \varphi) A = 66.91 \text{ MN}$$

$$\text{DF} = w \sin \alpha + U_1 \cos \alpha = 28.56$$

$$\text{FOS} = \frac{\text{RF}}{\text{DF}} = 2.41$$

13.4 WEDGE FAILURE

Wedge failures are translational slides that occur when joint planes combine to form a rock block that may slide down along the line of intersection of the two joint planes. Wedge block may slide either in three-translational or three-rotational motion or their combinations. However, translational sliding along the line of intersection is the most important wedge failure mode in surface excavation. Possibility of wedge failure increases once the strike of two discontinuities cuts the slope face obliquely and their line of intersection lies on the slope face. A wedge block will slide down along the line of intersection if (i) inclination of the line of intersection is significantly greater than the angle of internal friction along the discontinuities, and (ii) plunge of the line of intersection lies between toe and crest of the slope. In geology, plunge is used to define tilt of lines, the term dip being used for planes. The plunge fully expresses the three dimensional orientation of line and has two parts (i) angle of plunge, and (ii) plunge directions.

The wedge block failure analysis is carried out using limit equilibrium method. Although, wedge failure is a three-dimensional (3D) problem, so 3D stability analysis is required. For simplicity of analysis and understanding, stability is popularly ascertained by principle of plane failure. Hence, the factor of safety of wedge block as shown in Figure 13.8 is defined as:

$$\text{FOS} = \frac{\text{RF}}{\text{DF}} = \frac{\text{RF}_A + \text{RF}_B}{\text{DF}}$$

where, $\text{RF}_A = c_A A_A + \text{R}_A \tan \varphi_A$ and $\text{RF}_B = c_B A_B + \text{R}_B \tan \varphi_B$, and R_A and R_B are the normal component of resisting forces; c_A and c_B are cohesion of discontinuity plane A and B respectively; φ_A and φ_B are angle of internal friction between sliding surfaces; A_A and A_B, area of contact of wedge. Subscripts A and B are used for plane A and plane B respectively. The driving force is equal to weight component acting along sliding direction. Component of reaction forces R_A and R_B are taken in a plane (adb) of slope face with one component acting opposite to each other (i.e. horizontal) and another component against the driving force.

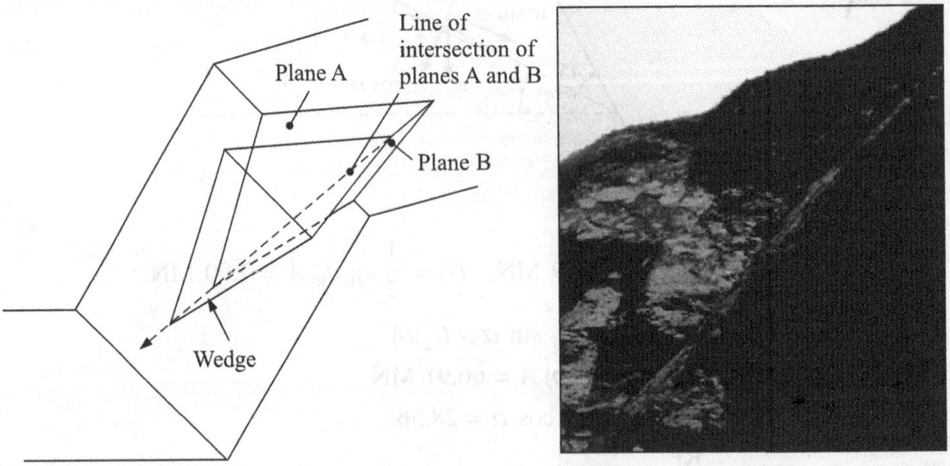

(a) (b)

FIGURE 13.8 (a) A typical wedge block, and (b) Natural wedge failure.

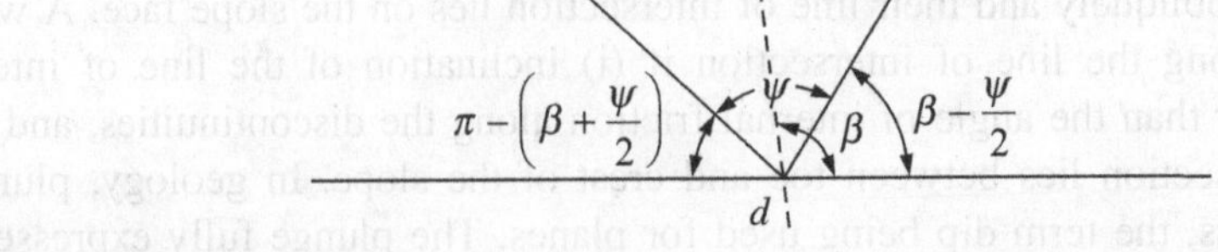

FIGURE 13.9 Front view of a typical wedge along the line of intersection.

Figure 13.9 shows the frontal view of the wedge along the line of intersection. Where, ψ is the angle between two planes or included angle of wedge; β is the angle of inclination of bisection line or angle of tilt; and α is the angle of plunge of bisection line. Force equilibrium produces

$$R_A \sin\left[\pi - \left(\beta + \frac{\psi}{2}\right)\right] = R_B \sin\left(\beta - \frac{\psi}{2}\right) \tag{13.5a}$$

$$R_A \cos\left(\beta + \frac{\psi}{2}\right) - R_B \cos\left[\pi - \left(\beta - \frac{\psi}{2}\right)\right] = w \sin\alpha \tag{13.5b}$$

Equating Eq. (13.5a) and (13.5b), gives

$$R_A = \frac{w \sin\left(\beta + \frac{\psi}{2}\right) \cos\alpha}{\sin\psi} \tag{13.6a}$$

$$R_B = \frac{w \sin\left(\beta - \frac{\psi}{2}\right) \cos\alpha}{\sin\psi} \tag{13.6b}$$

Using Eq. 13.4, the FOS for cohesionless ($c_A = c_A = 0$); and equal frictional coefficient ($\varphi_A = \varphi_B = 0$) for both the plane, i.e.,

$$\text{FOS} = \frac{\sin\beta \tan\varphi}{\sin\frac{\psi}{2} \tan\alpha} \tag{13.7}$$

EXAMPLE 13.6 Calculate FOS for a wedge block weight likely to fail. If the slope surface is cohesionless and the friction coefficient for both the planes are same. The wedge block is defined by angle between two planes or included angle of wedge ($\psi = 40°$); angle of inclination of bisection line or angle of tilt ($\beta = 40°$); and plunge angle of bisection line ($\alpha = 65°$). The friction angle of sliding surface is $\varphi = 40°$.

Solution: Use Eq. 13.7, the FOS is expressed as:

$$\text{FOS} = \frac{\sin\beta \tan\varphi}{\sin\frac{\psi}{2} \tan\alpha} = 1.68$$

13.4.1 Compass Co-ordinate

The orientation of a line in space (**l**, denoted by oc in Figure 13.10), can be defined by means of a Cartesian co-ordinate with axes $x(E)$, $y(N)$, $z(U)$, with origin at o as shown in Figure 13.10. The $z(U)$-axis is directed vertically up, $y(N)$-axis pointed towards magnetic north and axis $x(E)$ is perpendicular to $y(N)$-axis and directed towards east. From Figure 13.10, the orientation of the line in the given co-ordinate system is defined by unit vector (n_x, n_y, n_z) as:

$$\mathbf{l} = n_x\mathbf{i} + n_y\mathbf{j} + n_z\mathbf{k} \tag{13.8}$$

From the above definition, the plunge (α) and trend (β) of line oc. Similarly for plane (onmp), the dip direction is (β) and dip angle with horizontal plane is (α).

$$\alpha = \sin^{-1}\left(\frac{-n_z}{\sqrt{n_x^2 + n_y^2 + n_z^2}}\right); \ \beta = \sin^{-1}\left(\frac{n_x}{\sqrt{n_x^2 + n_y^2}}\right) \tag{13.9}$$

Then, the strike direction is (90 + β). Bedding or other geological layers and planes that are not horizontal are said to be dip. Dip is the slope of a geological surface. Two aspect to the dip of a plane are: (i) dip direction, which is compass direction towards which the plane slopes; and (ii) angle of dip, which is angle that the plane makes with the horizontal plane (Lisle, 2020). Hence, the dip of a plane is represented by two numbers as β/α, which represents a plane which dips at α angle in the direction of β angle in clockwise from magnetic north. In the field, the direction of dip is usually measured with a magnetic compass which incorporates Clinometer for measurement of dip angle.

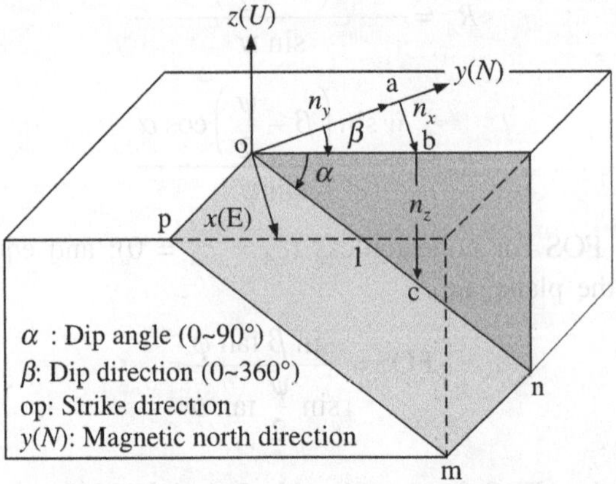

FIGURE 13.10 Dip direction (β) and dip angle (α) of plane mnop.

With reference to Figure 13.10, the trend (β) and plunge (α) of line (oc), is defined from compass survey then the unit vector of strike direction, dip direction, line oc and a line normal to plane is shown in Table 13.2.

TABLE 13.2 Direction cosine of a Joint Plane

Joint plane direction	*Compass direction*		
	x(E), (n_x) (East)	*y(N), (n_y) (Magnetic North)*	*Z(U), (n_z)*
op (strike direction)	cos (β)	−sin (β)	0
ob (dip direction)	sin (β)	cos (β)	0
oc (line on a plane onmp)	cos (α) sin (β)	cos (α) cos (β)	−sin (α)
Normal to plane (**op × oc**)	sin (α) sin (β)	sin (α) cos (β)	cos (α)

EXAMPLE 13.7 Data for the potential wedge failure by sliding down the line of intersection are shown in Figures (a) and (b). The average density of the sliding block is 2350 kg/m^3. From joint survey, it is found that the dip angle and dip direction (α, β) for plane A is (60°, 330°); for plane B is (60°, 120°); for slope face is (75°, 45°); for crest and toe is (0°, 45°). If the cohesion and angle of internal friction of joint plane A and joint plane B are 70 kPa, 30°; and 35 kPa, 25° respectively, then draw the stereographic projection with the given data to check whether wedge formation is possible or not. Finally, determine the following:

(a) plunge and trend of the line of intersection of joint planes A and B.
(b) the edge length of tetrahedron, if vertical height of sliding wedge is 20 m.
(c) the area of joint planes A and B
(d) volume of the sliding wedge
(d) the normal force on joint plane
(f) FOS of wedge

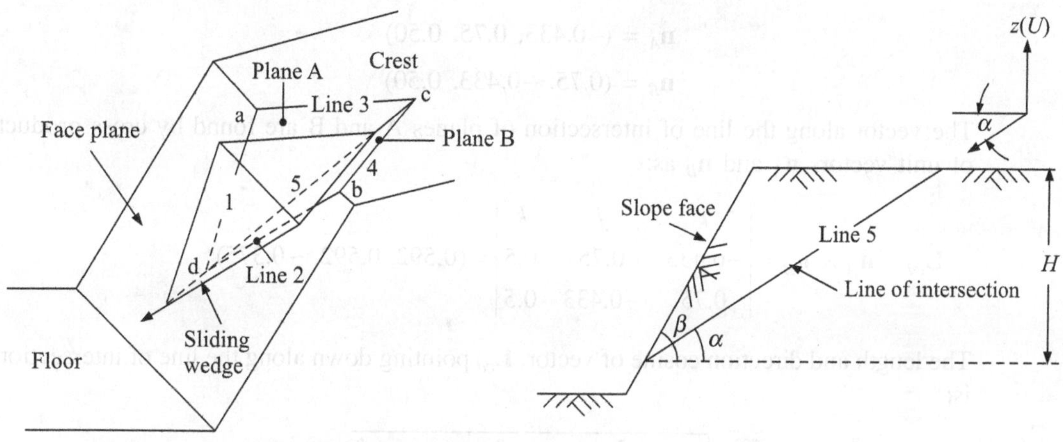

(a) A block failed as a wedge failure mode. (b) Sectional view normal to the line of intersection.

Solution: The stereographic projection is shown in Figure below using the procedure discussed in Chapter 1, Section 1.9 on streonets. Which shows the formation of wedge.

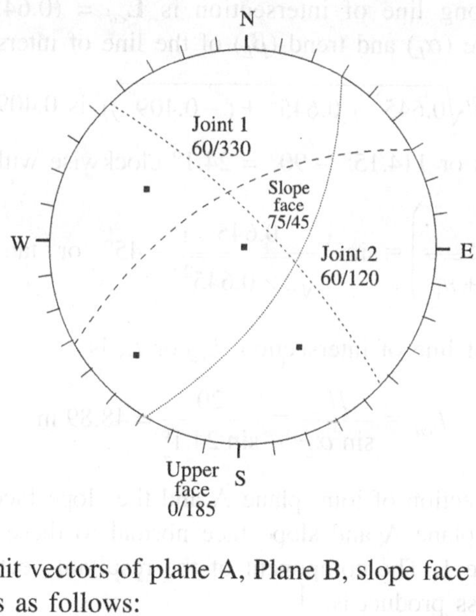

For further solution, unit vectors of plane A, Plane B, slope face and crest plane are defined using Table 13.2, which is as follows:

Plane A	(sin 60° sin –30°, sin 60° cos –30°, cos 60°)	$\mathbf{n}_A$ = (–0.433, 0.75, 0.50)
Plane B	(sin 60° sin 120°, sin 60° cos 120°, cos 60°)	$\mathbf{n}_B$ = (0.75, –0.433, 0.50)
Slope Face	(sin 75° sin 45°, sin 75° cos 45°, cos 75°)	$\mathbf{n}_F$ = (0.683, 0.683, 0.259)
Crest Plane	(sin 0° sin 45°, sin 0° cos 45°, cos 0°)	$\mathbf{n}_c$ = (0, 0, 1)

(a) the unit vectors normal to joint planes A and B are:

$$\mathbf{n}_A = (-0.433, 0.75, 0.50)$$

$$\mathbf{n}_B = (0.75, -0.433, 0.50)$$

The vector along the line of intersection of planes A and B are found by cross product of unit vectors $\mathbf{n}_A$ and $\mathbf{n}_B$ as:

$$\mathbf{L}_{cd} = \mathbf{n}_A \times \mathbf{n}_B = \begin{vmatrix} i & j & k \\ -0.433 & 0.75 & 0.5 \\ 0.75 & -0.433 & 0.5 \end{vmatrix} = (0.592, 0.592, -0.375)$$

The length and direction cosine of vector, $\mathbf{L}_{cd}$ pointing down along the line of intersection is:

$$|\mathbf{L}_{cd}| = \sqrt{(0.592)^2 + (0.592)^2 + (-0.375)^2} = 0.917$$

and
$$\left(\frac{n_x}{|\mathbf{L}_{cd}|}, \frac{n_y}{|\mathbf{L}_{cd}|}, \frac{n_z}{|\mathbf{L}_{cd}|}\right) = \left(\frac{0.592}{0.917}, \frac{0.592}{0.917}, \frac{0.375}{0.917}\right) = (0.645, 0.645, -0.409)$$

the unit vector along line of intersection is $\mathbf{L}_{cd}$ = (0.645, 0.645, –0.409), From Eq. 13.9, the plunge (α_L) and trend (β_L) of the line of intersection is:

$$\alpha_L = \sin^{-1}(0.409/\sqrt{0.645^2 + 0.645^2 + (-0.409^2)})$$ is 0.409. So, α_L = 114.15°

with positive *z*-axis or 114.15° – 90° = 24.1° clockwise with horizontal plane.

$$\beta_L = \sin^{-1}\left(\frac{n_x}{\sqrt{n_x^2 + n_y^2}}\right) = \sin^{-1}\frac{0.645}{\sqrt{2 \times 0.645^2}} = 45° \text{ or } \tan^{-1}\frac{n_x}{n_y} = 45°$$

Therefore, length of line of intersection, L_{cd} or L_5 is

$$L_{cd} = \frac{H}{\sin \alpha_L} = \frac{20}{\sin 24.1°} = 48.89 \text{ m}$$

(b) Line 1 is the intersection of joint plane A and the slope face (abd). The cross product of unit vectors of plane A and slope face normal to these planes results in a vector pointing along Line 1. The component of the resultant vector gives plunge and trend of Line 1. The cross product is:

Vectors normal to Plane A and slope face (abd) are

$$\mathbf{L}_1 = \mathbf{n}_A \times \mathbf{n}_F = \begin{vmatrix} i & j & k \\ -0.433 & 0.750 & 0.500 \\ 0.683 & 0.683 & 0.259 \end{vmatrix} = (-0.147), (0.454), (-0.808)$$

The length of vector ($\mathbf{L}_1$) are $|\mathbf{L}_1| = \sqrt{(-0.147)^2 + (0.454)^2 + (-0.808)^2} = 0.938$ and direction cosine is $\left(\frac{n_x}{|\mathbf{L}_1|}, \frac{n_y}{|\mathbf{L}_1|}, \frac{n_z}{|\mathbf{L}_1|}\right) = \left(\frac{-0.147}{0.938}, \frac{0.454}{0.938}, \frac{-0.808}{0.938}\right) = (-0.157, 0.483, -0.861)$. Hence, the unit vector of vector $\mathbf{L}_1$ is:

$$\mathbf{L}_1 = (-0.157, 0.483, -0.861)$$

The plunge (α) of Line 1 with Z-axis is $\sin^{-1}$ 0.861. So, α = 59.45° with horizontal plane. Hence length of line of intersection, i.e., line 1 is L_1 = H/sin α = 20/sin 59.45° = 23.22 m.

Similarly, the direction of line L_3 is along the direction of the cross product of unit vectors of plane A and crest plane, i.e., $\mathbf{L}_3 = \mathbf{n}_c \times \mathbf{n}_A = (-0.75, -0.433, 0)$ of length $|\mathbf{L}_3| = \sqrt{(-0.75)^2 + (-0.433)^2 + (0)^2} = 0.866$. The unit vector, $\mathbf{L}_3 = (-0.866, -0.5, 0)$ lies in $x - y$ plane in 3rd quadrant (as i and j components of unit vector is –ve and k component is zero) with trend, $\beta = \cos^{-1}(-0.866) = -150°$, i.e., oriented at 150° clockwise from magnetic north or x-axis or $\beta = \tan^{-1}(-0.5/-0.866) = 30°$, i.e., oriented 30° anti-clockwise from magnetic north.

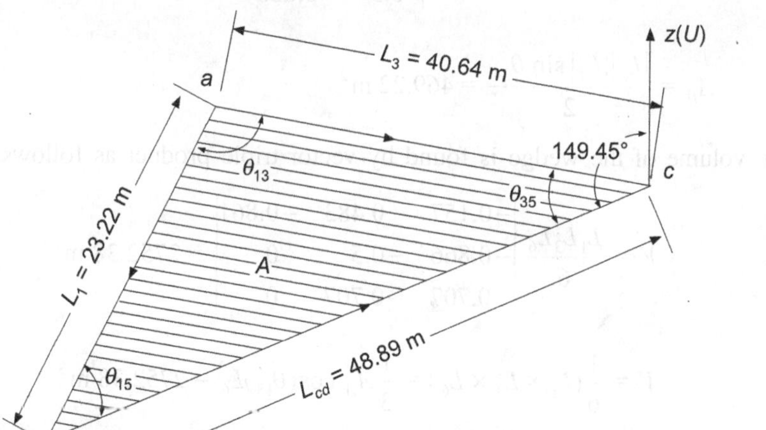

Above Figure shows plane A of unknown side (L_3) and included angles θ_{13}, θ_{15}, and θ_{35}. One of the included angle, i.e., θ_{15} and θ_{13} is determined by the dot product of unit vectors $\mathbf{L}_1$ and $\mathbf{L}_{cd}$ and unit vectors $\mathbf{L}_1$ and $\mathbf{L}_3$ respectively as:

$$\theta_{15} = \cos^{-1}\frac{\mathbf{L}_1 \cdot \mathbf{L}_{cd}}{|\mathbf{L}_1||\mathbf{L}_{cd}|} = 55.75°;\quad \theta_{13} = \cos^{-1}\frac{\mathbf{L}_1 \cdot \mathbf{L}_3}{|\mathbf{L}_1||\mathbf{L}_3|} = 96.1°$$

As for triangle acd, $\theta_{35} = 180 - \theta_{15} - 96.1° = 28.2°$

Now, using sine rule to find the side length L_3 as, $L_3 = \dfrac{\sin\theta_{15}}{\sin\theta_{35}} L_1 = 40.64$ m

Similarly for plane B, the length of the lines and included angles are determined. The unit vectors are $\mathbf{L}_2 = (0.483, -0.157, -0.861)$; $\mathbf{L}_4 = (0.5, -0.866, -0)$; and $\mathbf{L}_6 = (0.707, -0.707, 0)$; included angles are $\theta_{25} = 55.75°$; $\theta_{45} = 28.19°$; and $\theta_{24} = 83.93°$ and length of each sides are $L_2 = 23.22$ m; $L_4 = 40.64$ m and $L_6 = 21.04$ m respectively.

(c) Area of Plane A is the cross product of two sides (i.e., 1 and 3). Hence, area of plane A

$$A_A = \frac{1}{2}\left|\hat{\mathbf{L}}_1 \times \hat{\mathbf{L}}_3\right| = \frac{L_1L_3}{2}\begin{vmatrix} i & j & k \\ -0.157 & 0.483 & -0.861 \\ -0.866 & -0.5 & 0 \end{vmatrix} = 469.22 \text{ m}^2$$

or, $$A_A = \frac{|L_1||L_3|\sin\theta_{13}}{2} = 469.22 \text{ m}^2$$

Similarly, area of Plane B

$$A_B = \frac{1}{2}\left|\hat{\mathbf{L}}_2 \times \hat{\mathbf{L}}_4\right| = \frac{L_2L_4}{2}\begin{vmatrix} i & j & k \\ 0.483 & -0.157 & -0.861 \\ 0.5 & 0.866 & 0 \end{vmatrix} = 469.22 \text{ m}^2$$

or, $$A_B = \frac{|L_2||L_4|\sin\theta_{24}}{2} = 469.22 \text{ m}^2$$

(d) The volume of the wedge is found by vector triple product as follows:

$$V = \frac{L_1L_3L_6}{6}\begin{vmatrix} -0.157 & 0.483 & -0.861 \\ -0.866 & -0.5 & 0 \\ 0.707 & -0.707 & 0 \end{vmatrix} = 2752.38 \text{ m}^3$$

or, $$V = \frac{1}{6}(L_1 \times L_3 \times L_6) = \frac{1}{3}A_A\cos(\theta_{13})L_6 = 2752.38 \text{ m}^3$$

(e) Weight of the wedge ($W = V\rho g$) is 63.45 MN. The normal and tangent component of weight with respect to line of intersection (Line 5) are $W_n = W\cos\alpha_L = 57.9$ MN; $W_s = W\sin\alpha_L = 25.96$ MN. By equilibrium analysis of the normal force on joint planes and the weight of sliding block, the normal reaction to these surface are:

$$R_A = \frac{W_n\sin\alpha_B}{\sin(\alpha_A+\alpha_B)} = 52.83 \text{ MN};\quad R_B = \frac{W_n\sin\alpha_A}{\sin(\alpha_A+\alpha_B)} = 52.83 \text{ MN}$$

where, $\alpha_A = \tan^{-1}[\sin\alpha_L \tan(\beta_L - \beta_A)] = 56.77°$ and $\alpha_B = \tan^{-1}[\sin\alpha_L \tan(\beta_B - \beta_L)] = 56.77°$ are dip angle of plane A and plane B respectively. $\alpha_L = 45°$ is plunge of line of intersection and $\beta_A = -30°$ and $\beta_B = -60°$ are dip directions of plane A and plane B respectively.

(f) Using Eq. (13.3), the resisting force, $RF_A = c_A A_A + R_A \tan\varphi_A = 63.35$ MN and $RF_B = c_B A_B + R_B \tan\varphi_B = 41.06$ MN. The driving force is $W_s = W \sin\alpha_L = 25.96$ MN. Hence, the FOS is the sum of resisting force divided by driving force as 4.02. [Note: The above solution can also be found by other methods]

13.5 TOPPLING FAILURE

Toppling failure of rock slopes is associated with rock masses with a dominant regularly spaced discontinuity set (usually joint, bedding or foliation) with strike parallel, or nearly parallel to the slope face and dip at a steep angle into the face as shown in Figure 13.11. This geological structure forms a series of tall narrow slabs. If the dimensions of a slab are such that its centre of gravity acts outside the base of the slab, then there is a potential for the block to topple.

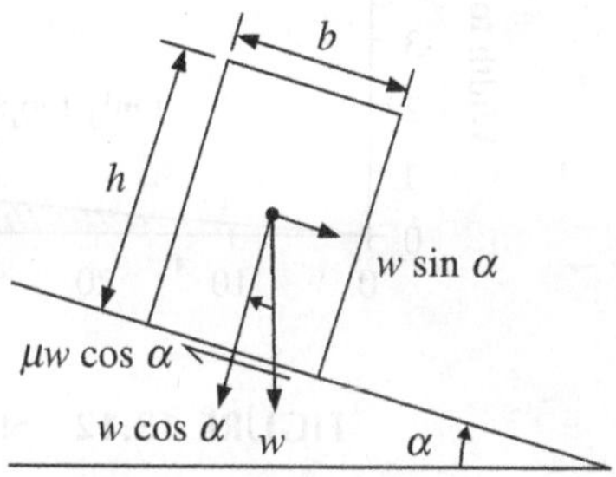

FIGURE 13.11 Toppling analysis of a block resting on inclined surface.

Before explaining the toppling failure, let us examine a block sitting on an inclined plane as shown in Figure 13.11.

The block will topple once the driving moment are less than the resisting moment about the outer edge of the block. Mathematically, by comparing moments about the right bottom corner of the block can be written as:

$$w(\sin\alpha)\frac{h}{2} > w(\cos\alpha)\frac{b}{2} \quad \text{or} \quad \tan\alpha > \frac{b}{h} \tag{13.10}$$

For sliding, the driving force should be more than the resisting force as:

$$w\sin\alpha > \mu w\cos\alpha \quad \text{or} \quad \alpha > \varphi \tag{13.11}$$

Based on Eqs. 13.10 and 13.11 the block fails in different modes and they are as follows:

(i) Topping only: $\frac{b}{h} < \tan\alpha$ and $\alpha < \varphi$

(ii) Toppling with sliding: $\frac{b}{h} < \tan\alpha$ and $\alpha > \varphi$

(iii) Sliding only: $\frac{b}{h} > \tan\alpha$ and $\alpha > \varphi$

(iv) No toppling and no sliding (stable): $\frac{b}{h} > \tan\alpha$ and $\alpha < \varphi$

From the above discussion, it is clear that geometry of the block (b/h) and friction angle between the block and resting surface are the key parameters which control the type of failure mode. To investigate this further, b/h versus α is plotted as shown in Figure 13.12. The figure shows the regions of four different failure conditions as mentioned above.

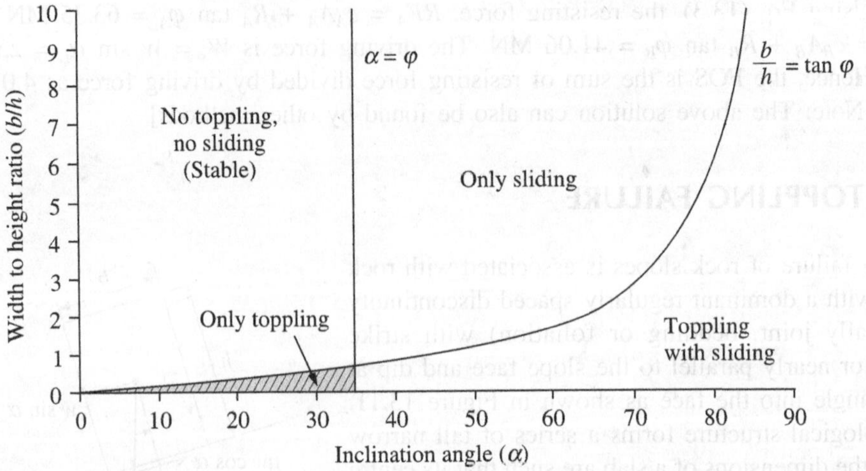

FIGURE 13.12 Stability diagram of a typical toppling failure.

EXAMPLE 13.8 A rock block of width 4 m and height 8 m is shown in Figure 13.11 rests on a plane inclined at an angle α = 48° with the horizontal. At what value of friction angle will the block (a) begin to slide and (b) begin to topple.

Solution:

(a) The limiting condition for sliding is, angle of internal friction should be more than equal to inclination of sliding surface, i.e.,

$$\varphi \geq 48°$$

Here, it should be noted that the angle of internal friction of rock reduces with moisture/water content. Hence, angle of internal friction changes due to sessional variation, i.e., dry or wet. In partially developed cross-joint blocks will be in tension.

(b) The limiting condition for only toppling is $\frac{b}{h} < \tan\alpha$ and sliding is $\alpha > \varphi$

Here, $\frac{b}{h}$ = 0.5 and tan 48° = 1.11.

As $\frac{b}{h}$ < tan 48°, so block will topple.

EXAMPLE 13.9 Two rock blocks as shown in given Figure rests on an inclined plane. Assume the friction angle is the same for sliding between the blocks and for sliding of a block along the surface. All the dimensions of smaller and larger blocks are same except height. Height of smaller block is half that of the larger block ($h_2 = 2h_1$). Discuss the conditions for equilibrium (sliding and/or toppling) of the two block system.

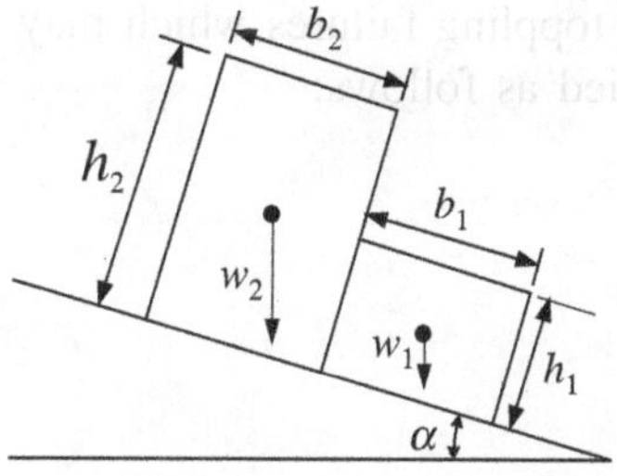

Solution:

(i) Both block will slide: if $\alpha > \varphi$.

(ii) Both block will stable, i.e., neither sliding nor rotation: if $b_1/h_1 > \tan\varphi$ and $b_2/h_2 > \tan\varphi$ and $\alpha < \varphi$.

(iii) No sliding in both blocks ($\alpha < \varphi$), and upper block stable ($b_2/h_2 > \tan\varphi$) while lower block rotates ($b_1/h_1 < \tan\varphi$). From drawing, however, $b_1/h_1 > b_2/h_2$ so this condition could not happen.

(iv) Rotation of upper block: If $\alpha < \varphi$ (i.e., no sliding of blocks), $b_1/h_1 > \tan\varphi$ (no rotation in lower block) and $b_2/h_1 < \tan\varphi$ (rotation of upper block). In this situation, due to interaction of both the blocks, the lower block may (i) slide or (ii) rotate as shown in Figure. Due to block interaction, let a force P transferred from upper block to the lower block.

$$P = \frac{w_2(h_2 \sin\alpha - b_2 \cos\alpha)}{2h_1}$$

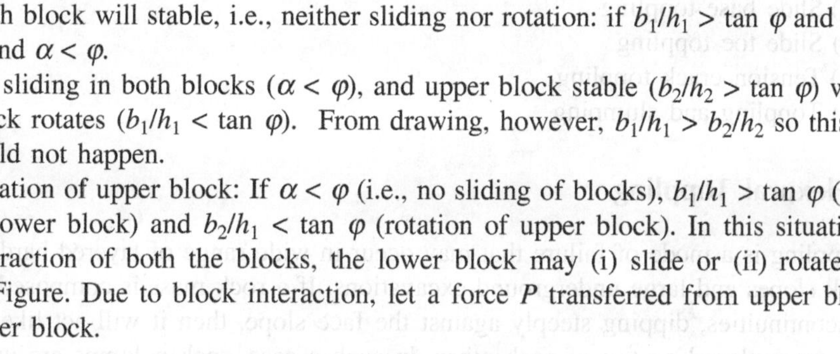

If the lower block slides, the liming condition is:

$$\frac{w_2(h_2 \sin\alpha - b_2 \cos\alpha)}{2h_1} = \frac{w_1(\cos\alpha \tan\varphi - \sin\alpha)}{1 - \tan^2\varphi}$$

If the lower block rotates, the limiting condition is:

$$\frac{w_2(h_2 \sin\alpha - b_2 \cos\alpha)}{2h_1} = \frac{w_1(b_1 \cos\alpha - h_1 \sin\alpha)}{2\,(h_1 - b_1 \tan\varphi)}$$

Toppling failure is one of the most serious and hazardous instability of rock slopes and underground openings. This mode of failure has been observed at different scales either in natural slopes or in open cuts for mining or civil engineering works. Goodman and Bray have described four principal types of toppling failures which may be encountered in the field. The toppling failure mode are classified as follows:

(i) Flexural toppling
(ii) Block toppling
   (a) Sliding toe
   (b) Tension toe
(iii) Block-flexural toppling
(iv) Secondary toppling
   (a) Slide head toppling
   (b) Slide base toppling
   (c) Slide toe toppling
   (d) Tension crack toppling
   (e) Toppling and slumping

13.5.1 Flexural Toppling

Flexural toppling is a mode of failure that may occur in wide range of layered hard rock strata in both rock slopes and large underground excavations. If a rock mass is composed of a set of parallel discontinuities, dipping steeply against the face slope, then it will act like some rock columns that are placed on top of each other. In such a case, rock columns are under tensile and compressive bending stresses due to their own weights. If the maximum tensile stress in each rock column exceeds its tensile strength, then it fails and topples. Such instability is categorised as flexural toppling failure (Figures 13.13 and 13.14).

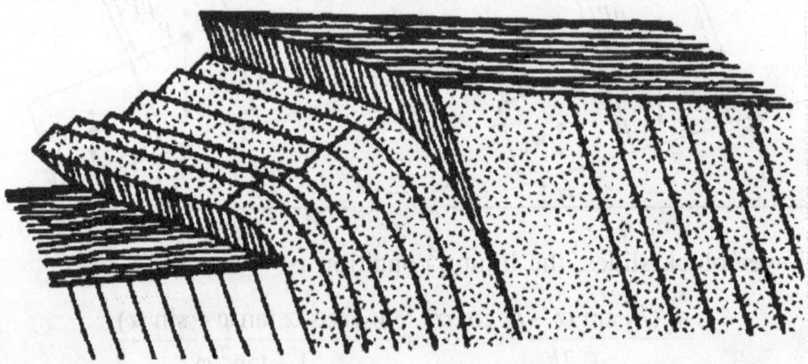

FIGURE 13.13 Schematic diagram of flexural toppling failure in rock slopes.

The body force of each rock column can be analysed as two components; normal to and parallel with the rock column longitudinal axis. The normal component causes the rock column to bend and transfers the load to the underlying one. Hence, to compute the factor of safety of the rock mass against flexural toppling failure, the magnitude of maximum tensile stress in rock columns must be determined (Amini et al., 2009).

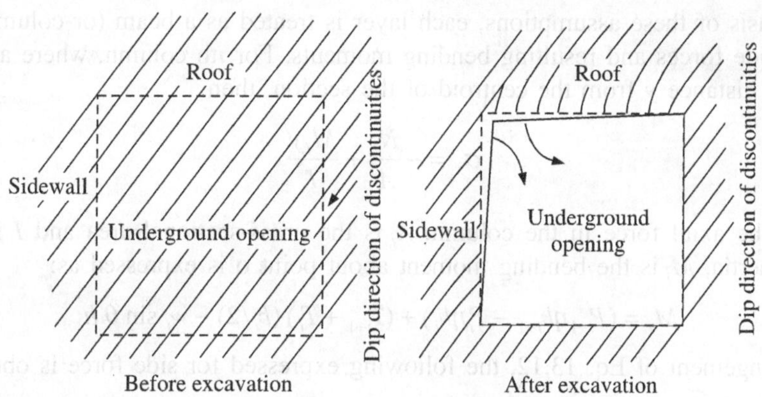

FIGURE 13.14 Flexural toppling failure in an underground opening (Amini et al., 2009).

This method is based on equations of limit equilibrium and the principle of compatibility. Aydan and Kawamoto (1992) derived expression for flexural toppling of foliated rock slope. The flexural toppling of foliated rock slope occurs due to bending induced tensile stresses initiating fractures of the rock columns at the location of maximum bending. The theoretical model by Aydan and Kawamoto (1992) is based on following assumptions:

(i) Each column or layer can be treated as a cantilever beam subjected to gravity forces and side forces as shown in Figure 13.15. For simplicity, the total side force is assumed to be acting at a point ηh_i, where h_i is the height of the ith layer, and η range between 0 and 1 is a parameter defining the inter-column force distribution relationship common to all columns.

(ii) A state of limiting equilibrium exists simultaneously across all the columns along the future fracture at least immediately prior to failure.

(iii) At the limiting state, the maximum tensile stress acting in each column along the future fracture plane ($\sigma_{x/y=b/2}$) is equal to the tensile strength of the material, σ_t.

(iv) The fracture will follow a plane inclined upward at an angle θ to the normal to the joint dip angle (Figure 13.15).

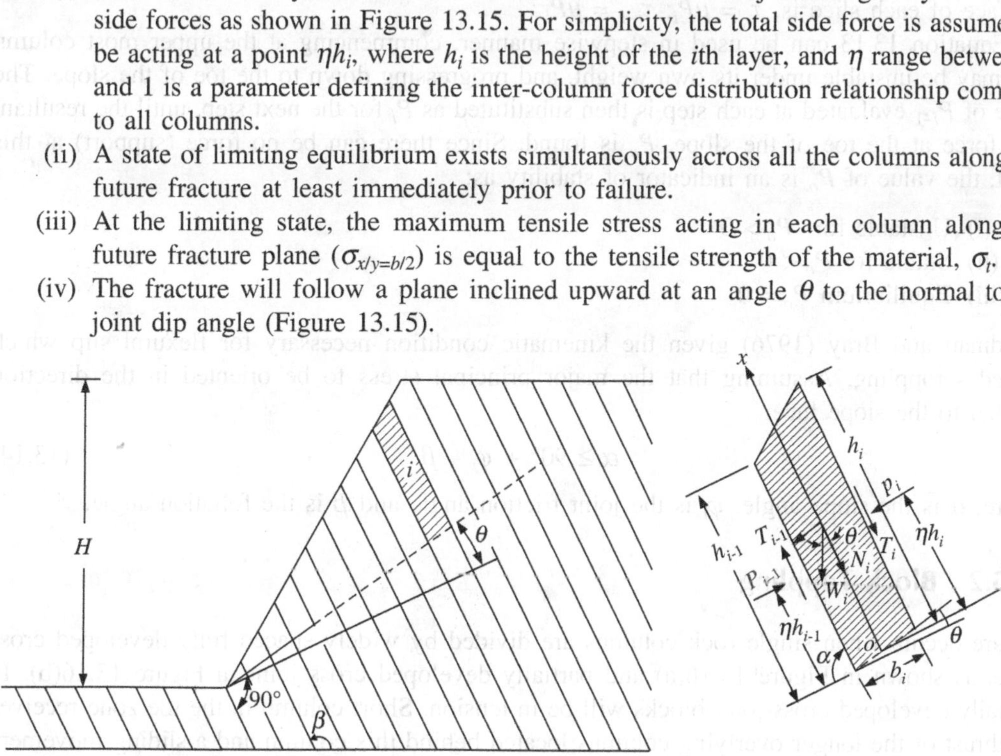

FIGURE 13.15 Force body diagram of flexural toppling failure.

On the basis of these assumptions, each layer is treated as a beam (or column) loaded by the gravity, side forces and resulting bending moments. For *i*th column, where axial stress σ_x is acting at a distance *y* from the centroid of the section, then

$$\sigma_x = -\frac{N_i}{A_i} + \frac{M_i y}{I} \tag{13.12}$$

where N_i is the axial force in the column, A_i is the cross-sectional area and I is the second moment of inertia, M_i is the bending moment about point o is expressed as:

$$M_i = (P_{i-1}\eta h_{i-1} - P_i \eta h_i) + (T_{i+1} + T_i)(b_i/2) - w_i \sin\theta\, h$$

By rearrangement of Eq. 13.12, the following expressed for side force is obtained:

$$P_{i-1} = \frac{\max(P_i,0)\left(\eta h_i - \dfrac{1}{2}\mu b_i\right) + w_i \sin\theta\, \bar{h}_i - \dfrac{2I_i}{b_i}\left(\sigma_{t_i} - \dfrac{N_i}{A_i}\right)}{\eta h_{i-1} + \dfrac{1}{2}\mu t_i} \tag{13.13}$$

where $I_i = b^3/12$; $w_i = \gamma_i b_i h_i$; $N_i = W_i \cos\theta$; $\bar{h}_i = (h_i + h_{i-1})/2$; $\mu = \tan\varphi_i$; φ_i is the layer friction angle; σ_{t_i} is the tensile strength, α is the column inclination angle; t_i is column thickness, and γ is the unit weight of the material. If P_i is the side force; the shear stress acting on right and left face of each slice is, $\tau_i = \mu P_i$; $\tau_{i-1} = \mu P_{i-1}$.

Equation 13.13 can be used in stepwise manner, commencing at the upper most column that may be unstable under its own weight, and progressing down to the toe of the slope. The value of P_{i-1} evaluated at each step is then substituted as P_i for the next step, until the resultant side force at the toe of the slope, P_o is found. Since there can be no force (support) at this point, the value of P_o is an indicator of stability as:

(i) Unstable for $P_o > 0$
(ii) Stable for $P_o < 0$
(iii) Equilibrium $P_o = 0$

Goodman and Bray (1976) given the kinematic condition necessary for flexural slip which precedes toppling. Assuming that the major principal stress to be oriented in the direction parallel to the slope face.

$$\alpha \geq 90° + \varphi_i - \beta \tag{13.14}$$

where, α is the slope angle, φ_i is the joint friction angle and β is the foliation angle.

13.5.2 Block Toppling

Failure occurs when single rock columns are divided by widely spaced fully developed cross joints as shown in Figure 13.16(a) and partially developed cross joint in Figure 13.16(b). In partially developed cross-joint blocks will be in tension. Short column in the toe zone receives the thrust of the longer overlying columns located behind this column and a sliding movement is induced at the slope foot. Sliding movement sets the toppling of the behind column free to develop higher up the slope. There are two modes of failure: (i) sliding toe; and (ii) tension toe.

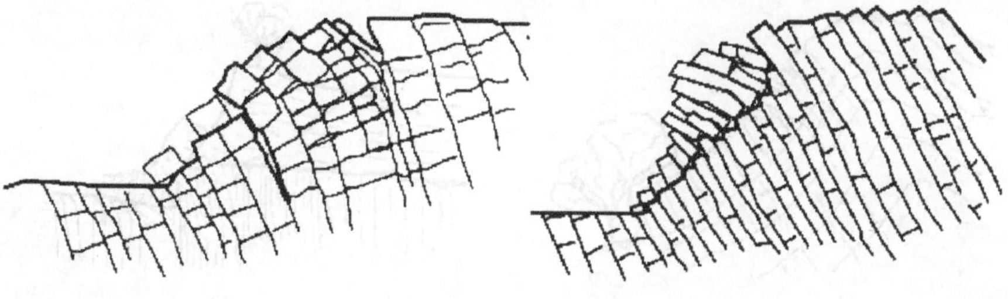

(a) (b)

FIGURE 13.16 (a) Block toppling, and (b) Block-flexural toppling (Goodman and Bray, 1976).

13.5.3 Block-Flexure Toppling

In this case, some rock blocks fail due to tensile bending stresses and some overturn under their own weights. Rock is a natural mass and its discontinuities are generally irregular and discontinuous; so, pure toppling failures (flexural and blocky) are rare and most of such failures occurring in nature are of the block-flexure type (Amini et al., 2009).

13.5.4 Secondary Toppling

Figure 13.17(a) shows a layered rock slope with cross joint system which isolates thin and high rock blocks. Toe erosion or excavation work induces a sliding movement in the upper layer. In the upper part of the slope, the rock blocks which are not involved in the sliding movement topples into the new void at the slide head. The toppling movement in the rock slab can also be induced by tensile stresses determined by the large difference in deformation characteristics between rock and soil deposits.

In slide base toppling [Figure 13.17(b)] the horizontal shear stress transmitted along the base of an incipient slide provokes toppling of steeply dipping layers forming the foundation for the slide toe. Figure shows a soil slope lying on a steeply dipping layered rock mass. Progressive rotational sliding and slumping movement induces a shear force along the toe of the rock mass able to provoke flexural toppling of the sub-vertical rock layer.

Slide toe toppling occurs when layers are blocked from sliding by a toppleable toe mass, as shown in Figure 13.17(c) of different geological formations. In effect, this is a two-block active-passive mechanism in which the passive region yields by toppling, rather than by sliding. In other words, the upper geological formation is layered and a layer sliding phenomenon is in progress whereas lower formation is like a blocky system and the thrust of the upper layers induces blocky toppling movement in the lower mass.

Figure 13.17(d) shows tension crack toppling, in which tension cracks of several metres high are formed. These tension cracks are generated above steep slope and may form potential blocks in soft rocks like chalk and damp sand for toppling.

Figure 13.17(e) shows rock slab lying on a soil deposit. The rock formation is made up of long column separated by a steeply dipping discontinuity. An instability phenomenon in the soil deposit, caused by rotational sliding may induce weathering of the rock slab toe with a consequent crack propagation of the non-perfect pre-existing rock discontinuity. The rock columns are completely free to move independently and are weathered at toe.

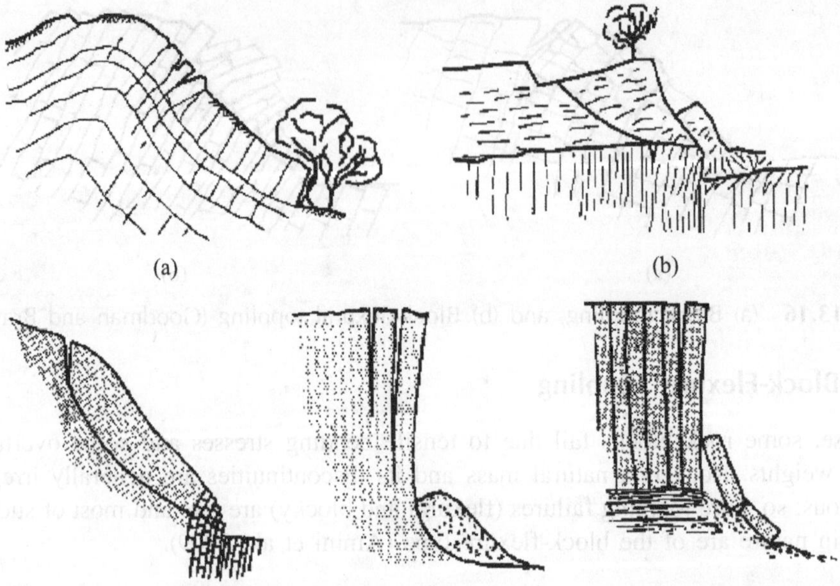

FIGURE 13.17 (a) Slide head toppling, (b) Slide base toppling, (c) Slide toe toppling, (d) Tension crack toppling, and (e) Toppling and slumping (Goodman and Bray, 1976).

13.6 CIRCULAR FAILURE

Soil, waste dump, Soil Rock Mixture (SRM) and severe fractured/weathered rock mass are likely to experience failure along a circular arc or generally non-linear path. The different types of circular failure includes (a) face failure, (b) toe failure, and (c) base failure as shown in Figure 13.18. The most critical failure surface are determined by trial and error method using limit equilibrium concept.

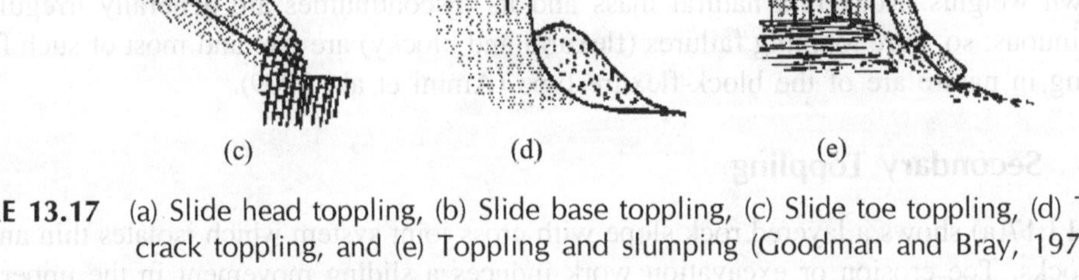

FIGURE 13.18 Modes of circular failure (a) face failure, (b) toe failure, and (c) base failure.

A soil or waste dump or SRM slope may start to fail along a non-linear path due to its own weight and/or presence of water. Waste dump or a heap of SRM is an uneconomical geomaterial removed before (or during) mining and civil operations and dumped at specific locations. Over the years several methods are developed to analyse this type of failure. The most common assumption is that slope fails in a circular path if shear stress at each point along this path exceeds its shear strength. Under this assumption, circular methods of slope failure are classified as:

(i) Circular arc method ($\varphi = 0$)
(ii) Friction circle method ($\varphi > 0$)
(iii) Slicing method
(iv) Design chart

13.6.1 Circular Arc Method

This is the simplest analysis method based on the assumption that a rigid block fails by rotation about its centre. The shear strength of failure surface is defined by its undrained strength, i.e., ($\varphi = 0$). The undrained strength is greater than the drained strength, because pore pressure decreases and effective stress increases during undrained shear. Example of drained materials are coarse grained such as sand and gravels with high permeability whereas undrained materials are fine grained such as silt and clay having low permeability. In case of drained material, both shear strength parameters, i.e. cohesion and angle of internal friction are mobilised whereas in undrained material only cohesion provide the shear strength. FOS of failure surface is defined as a ratio of resisting to driving moment about its centre as shown in Figure 13.19 and circular failure of waste dump of sand is shown in Figure 13.20.

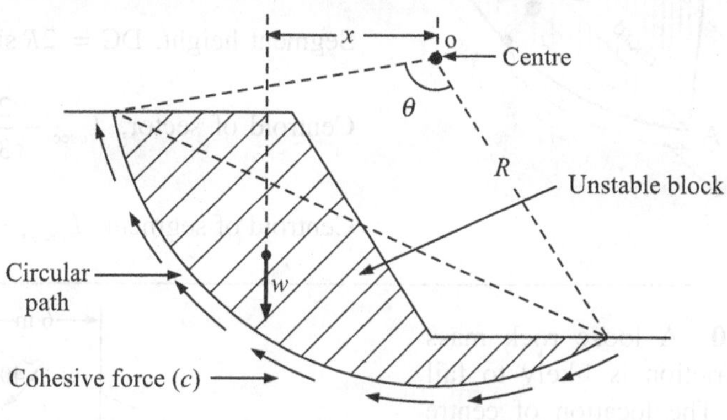

FIGURE 13.19 Force body diagram of a circular failure.

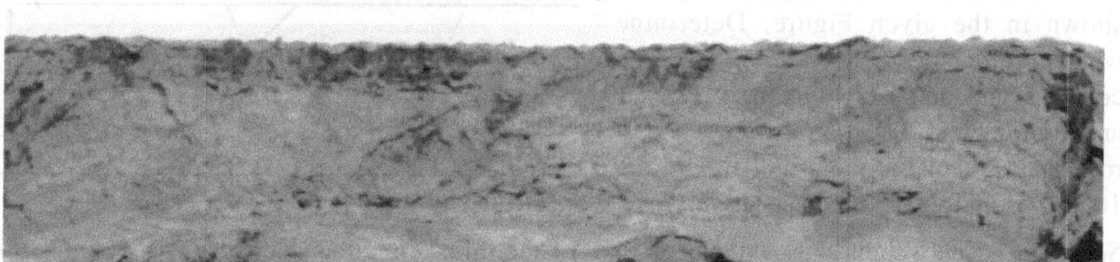

FIGURE 13.20 Circular failure of waste dump of sand.

Resisting moment (RM) = $cL_{arc} \times R = cR^2\theta$

Driving moment (DM) = wx

where, c is undrained cohesive strength; R is radius of failed circular path; w is weight of sliding or unstable block; x is horizontal distance between circular centre (o) and the centre of the unstable block. Hence, factor of safety can be expressed as

$$\text{FOS} = \frac{cR^2\theta}{wx} \tag{13.15}$$

This method can easily be evaluated for multiple layers as given in the following example.

Note: For solving the problems based on circular arc method, the following trigonometrical relations are useful consider θ in radian.

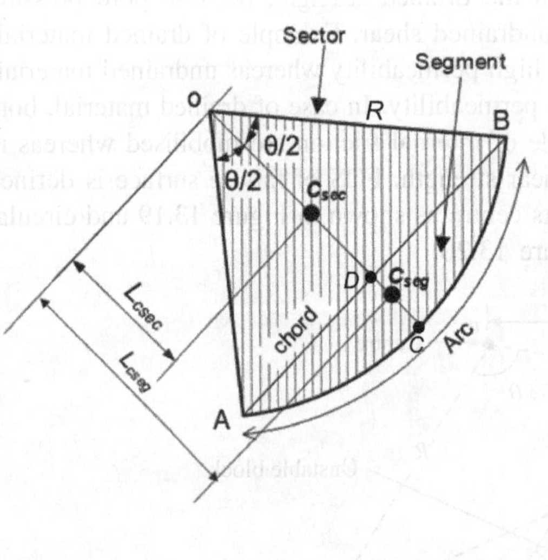

Arc length, $L_{arc} = R\theta$

Cord length, $L_c = 2R \sin (\theta/2)$

Area of segment, $A_{seg} = \frac{1}{2} R^2(\theta - \sin\theta)$

Area of sector, $A_{sec} = R^2\theta/2$

Area of triangle, $A_{tri} = \frac{1}{2} R^2 \sin\theta$

Segment height, DC $= 2R \sin^2 \frac{\theta}{4}$

Centroid of sector, $L_{c\sec} = \frac{2}{3} \frac{R \sin\theta}{\theta}$

Centroid of segment, $L_{cseg} = \frac{4R}{3} \left(\frac{\sin^3 (\theta/2)}{\theta - \sin\theta} \right)$

EXAMPLE 13.10 A loose rock mass (abcde) of zero friction is likely to fail in circular mode. The location of centre of the circle and other geometry are shown in the given Figure. Determine whether the block will fail or not. The cohesion and unit weight of rock layer 1 are 30 kN/m² and 20 kN/m³ respectively. For rock layer 2, cohesion and unit weight are 10 kN/m² and 20 kN/m³ respectively.

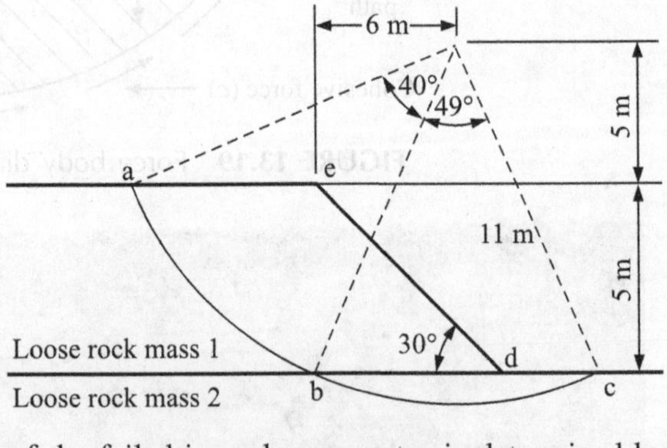

Solution: The weight and center of gravity of the failed irregular geometry is determined by splitting into blocks of regular geometry as shown in figure below. With the given geometry, the dimension of the regular geometry are determined as chord length L_{ab} = 7.52 m and L_{bc} = 9.12 m; length of the triangular faces are L_{ae} = 5.62 m; L_{bd} = 8.66 m. The moment arm of each regular geometry are determined and shown in figure below. For right angle triangle the centroid co-ordinates are 2/3rd away from the vertex and 1/3rd away from the right angle. For segment, centroid x co-ordinate will be mid-point of chord horizontal projection.

Block	*Area* (m^2)	*Weight* (kN)	*Moment arm* (m)	*Driving moment* (kNm)	*Resisting moment* (kNm)
1.	3.35	66.97	8.81	590.07	
2.	14.06	281.15	7.87	2213.83	2534.22
3.	21.65	433.01	2.89	1250.00	
4.	12.85	257.03	1.44	369.71	1034.81

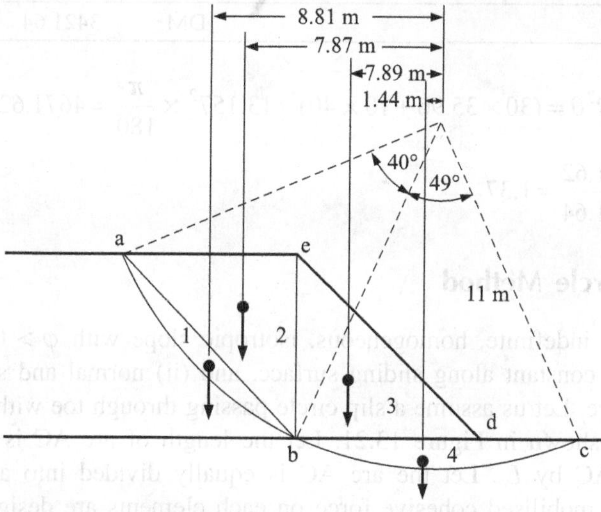

Total resisting moment = 3569.02 kNm
Total driving moment = 4423.61 kNm

$$\text{FOS} = \frac{3569.02}{4423.61} = 0.81$$

EXAMPLE 13.11 In continuation from Example 13.10, if the slip circle dimension as shown below, then determine the factor of safety.

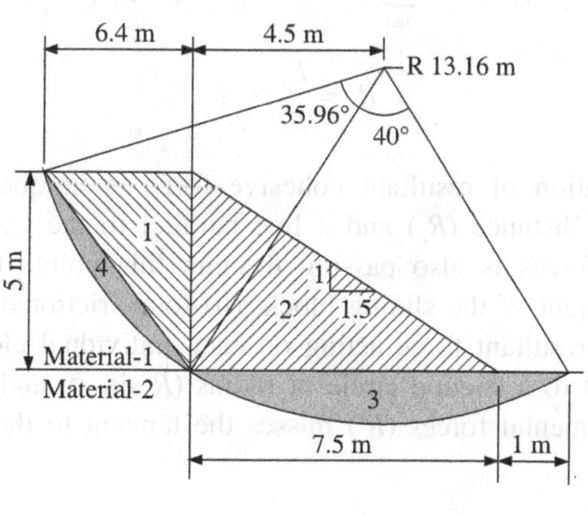

Solution: The weight of the failed surface is determined by spliting failed block into segments 1, 2, 3, and 4 of simple geometrical shape. The calculations are shown in table given below.

Segment	*Area* (m^2)	*Weight* (kN)	*Lever arm* (m)	*Driving moment* (kNm)
1	16.25	320	6.63	2121.60
2	18.75	375	2.00	750.0
3	4.79	47.89	0.00	0.0
4	3.50	70.27	7.836	550.64
			DM	3421.64

$$\text{RM} = \sum_{1}^{2} cR^2\theta = (30\times 35.96 + 10\times 40)\times 13.157^2 \times \frac{\pi}{180} = 4671.62 \text{ kNm}$$

$$\text{FOS} = \frac{4671.62}{3421.64} = 1.37$$

13.6.2 Friction Circle Method

This method considers indefinite, homogeneous, isotropic slope with $\varphi > 0$. The assumption is that (i) c and φ are constant along sliding surface, and (ii) normal and shear stresses vary along the sliding surface. Let us assume a slip circle passing through toe with centre of rotation (O) and radius (R) as shown in Figure 13.21. Let the length of arc AC is designated as L_a and length of chord AC by L_c. Let the arc AC is equally divided into a number of small elements (ΔL) and the mobilised cohesive force on each elements are designated as c_1, c_2, ... c_i, ... etc. The vector sum of the components of the cohesive force along and perpendicular to chord is such that the summation of cohesive force parallel to the chord length AC is equal to cL_c, whereas the summation of all normal components is zero as shown in Figure 13.21. The position of resultant cohesive force directed along the chord is determined by computing its moment to the sum of moments by elements cohesive force as:

$$(cL_c)R_c = R\sum_{i=1}^{n} c_i \Delta L \quad \text{or} \quad (cL_c)R_c = RcL_a \tag{13.16}$$

Hence,

$$R_c = \frac{L_a}{L_c} \times R \tag{13.17}$$

The point of application of resultant cohesive force is unique and determined by the intersection of the radial distance (R_c) and a line parallel to the chord. The resultant of the weight and pore water forces is also passing through this point. The forces acting on the sliding surface are (i) weight of the sliding block, (ii) total frictional resistance, and (iii) total cohesive resistance. The resultant force acting on each individual elements are normal to the elemental arc and tangent to a friction circle of radius ($R \sin \varphi$) and directed along PQ′. But, the vector sum of all elemental forces (R') misses the tangent to the friction circle as shown

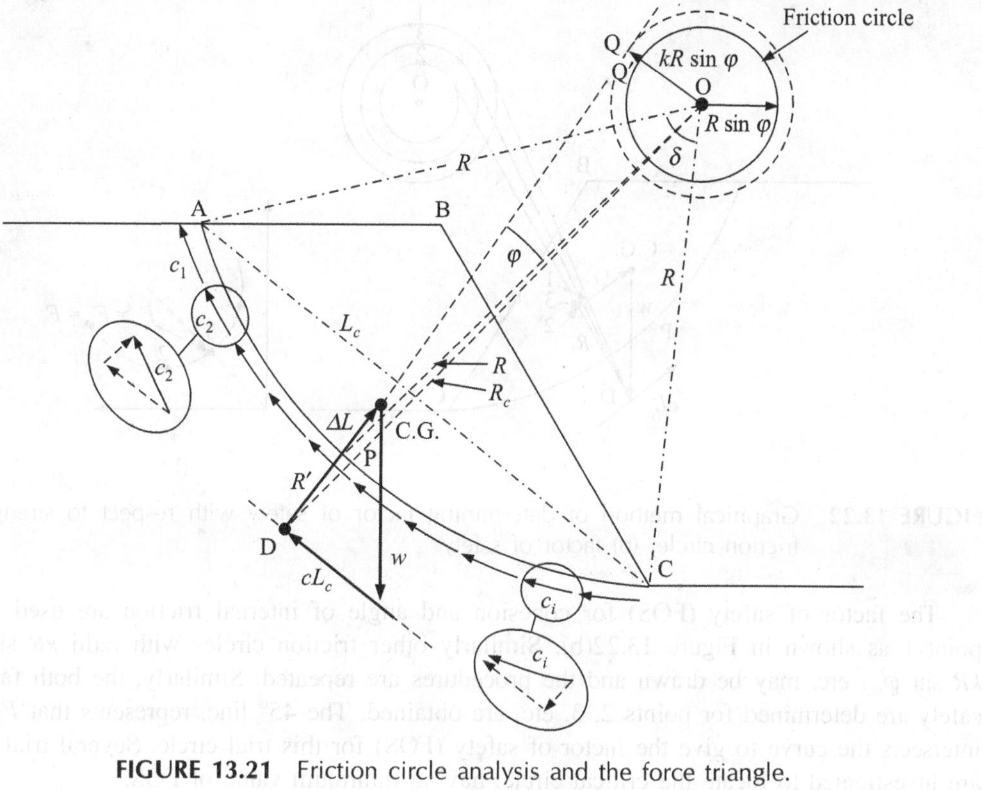

FIGURE 13.21 Friction circle analysis and the force triangle.

by line PQ, hence a modified friction circle is defined with radius ($kR\sin\varphi$). Where, k is a factor varies between 0 to 1 and depends upon the central angle (δ) and type of inter-granular pressure distribution along the sliding surface.

Let us take a trial circle of arc APC, as shown in Figure 13.22(a). The co-ordinate of points A, B, C and O are (x_A, y_A), (x_B, y_B), (x_C, y_C) and (x_O, y_O) respectively. The slope is made of geo-material having cohesion (c), angle of internal friction (φ) and unit weight (γ). Thus, the magnitude and direction of weight (w) and resultant cohesive force (cL_c) are known along with their meeting point (D). The trial circle arc makes an angle (δ) at the centre of the sliding circle. First trial using arbitrary friction angle designated by (φ'_{m1}) is shown by circle 1 of radius ($kR\varphi'_{m1}$). Subscript 1 is used for all other quantities of the first trial. The force (R_1) is then drawn through D and tangent to circle 1. Cohesive force c_1 is parallel to chord and point 1 is the intersection of force c_1 and R_1. The mobilised cohesion is equal to $c_1 = c'_{m1}L_c$. From this the mobilised cohesion c'_{m1} is evaluated. The factors of safety with respect to cohesion and friction are determined from the following expression.

$$F'_c = \frac{c'}{c'_{m1}} \tag{13.18a}$$

$$F'_\phi = \frac{\tan\varphi'}{\tan\varphi'_{m1}} \tag{13.18b}$$

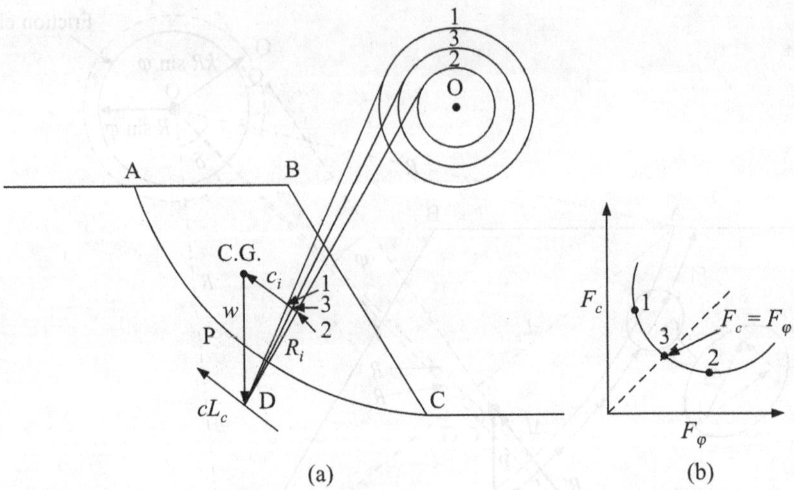

FIGURE 13.22 Graphical method of determining factor of safety with respect to strength: (a) friction circle, (b) factor of safety.

The factor of safety (FOS) for cohesion and angle of internal friction are used to plot point 1 as shown in Figure 13.22(b). Similarly other friction circles with radii $kR \sin \varphi'_{m2}$, $kR \sin \varphi'_{m3}$ etc. may be drawn and the procedures are repeated. Similarly, the both factor of safety are determined for points 2, 3, etc. are obtained. The 45° line, represents that $F_c = F_\varphi$, intersects the curve to give the factor of safety (FOS) for this trial circle. Several trial circles are investigated to locate the critical circle, having minimum value of FOS.

EXAMPLE 13.12 A 45 m high mine waste dump with an overall slope angle is 26° with horizontal. The properties of the dump material are cohesion of 40 kN/m^2, angle of internal friction of 5° and a unit weight of 20 kN/m^3. Consider any slip circle passing through the crest and toe as shown in the following Figure. Use the friction circle method to find the factor of safety with respect to cohesion.

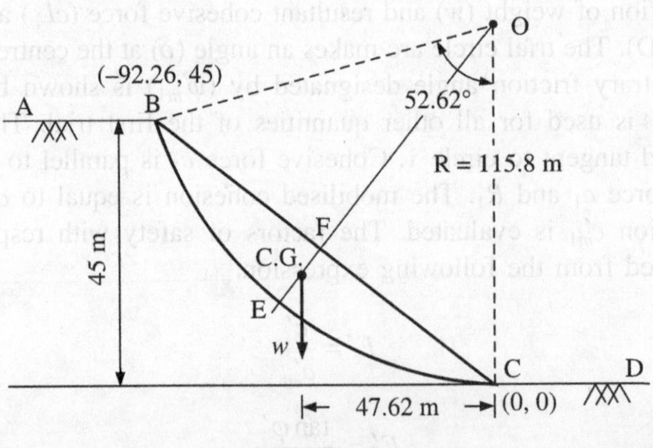

Solution: The first step of solution is to find the weight of the free body above the sliding/failure surface. With the given geometry, the area and weight of the segment (BECFB) of sliding circle is 829.86 m^2 and 16.60 MN respectively. The length of arc (L_a) BEC is 106.35 m and length of chord (L_c) BFC is 102.65 m. The line of action of the cohesive forces is:

$$R_c = \frac{L_a}{L_c} \cdot R = 119.97 \text{ m}$$

Equilibrium is satisfied by a closed polygon involving R_i, c_i, and w. A trial and error procedure is used to find the solution. Several friction circles are area assumed, those creating a closed polygon. For each assumed circle, two factor of safety are obtained one based on mobilised friction angle and another based on mobilised cohesion.

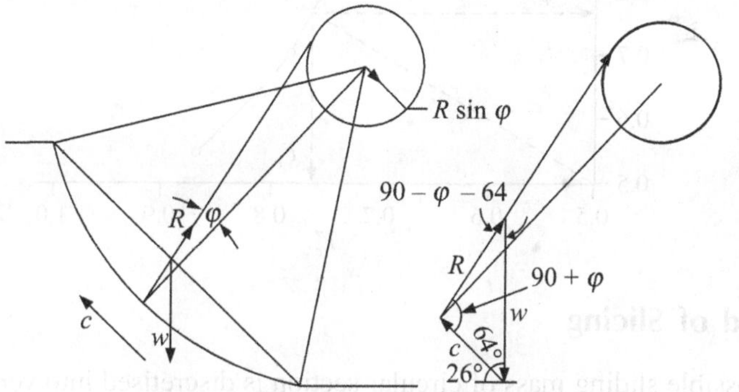

The force triangle of R_i, c_i, and w are formed as shown in the figure then sin rule is used to find the magnitude of mobilised cohesion $c_{m1} = c_1/L_c = 4.88/102.65 = 0.046$ MPa.

For first trial, let $F'_\varphi = 0.55$ as an arbitrary value, then mobilised friction angle will be:

$$\varphi'_{m1} = \tan^{-1}\left(\frac{\tan\varphi'}{F'_\varphi}\right) = \tan^{-1}\left(\frac{\tan 5^\circ}{0.55}\right) = 9.11^\circ$$

Using the sin rule to find the magnitude of resultant mobilised cohesive force, where, $a = 90 - \varphi'_{m1} - \gamma = 16.89$ and $\beta = 90 + \varphi'_{m1} = 99.11$.

$$c' = \left(\frac{\sin\alpha}{\sin\beta}\right) w = \left(\frac{\sin 16.89}{\sin 99.11}\right) \times 16.6 = 4.88 \text{ MN}$$

The corresponding cohesion is $c'/L_c = 4.88/102.65 = 45.91$ MPa

$$F'_c = \frac{c'}{c'_{m1}} = \frac{40}{45.91} = 0.87$$

Similar procedure is repeated for other trial friction circle. The calculations are shown in the following table and the final solution are determined by plotting the graphs as given below:

F'_φ	φ'_m (°)	c' (MN)	c'_m (kPa)	α (°)	β (°)	F'_c
0.55	9.11	4.88	45.91	16.89	99.11	0.87
0.70	7.16	5.40	50.79	18.84	97.16	0.79
0.85	5.90	5.73	53.92	20.10	95.90	0.74
1.00	5.01	5.97	56.11	20.99	95.01	0.71

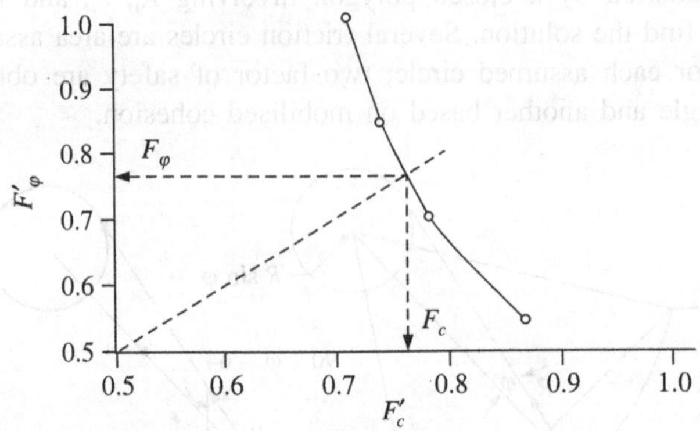

13.6.3 Method of Slicing

In this method, possible sliding mass of circular section is discretised into vertical slices of more or less equal width and treating individual slice as a unique sliding block. Figure 13.23(a) shows a slope having height H and inclination β, and a circular sliding path having a centre (X,Y) from the toe. The circular sliding path is assumed to be intersected a vertical tension crack at d m below the crest. It is also assumed that the tension crack is water filled up to d_w m height from the contact point of the sliding path and the tension crack. Phreatic surface is assumed to be passing through the toe of the slope. Water force $V = 1/2\gamma w d_w^2$ in the tension crack acts at height of $1/3d_w$ height from the contact point. It may be noted that there may not be any tension crack in the crest. In this case, the sliding path cuts the crest of the slope. In some cases, surcharge load due to heavy machinery working on the crest may also be applied as additional load.

In method of slicing, the entire volume of material bounded by the slope, circular sliding path and tension crack is divided into number of vertical slices as shown in Figure 13.23(a). If slice i is moving down or rotating, then $(i + 1)$th slice is resisting movement of ith slice. So, friction between i and $(i + 1)$th slice will be down and upward respectively. Similarly, the friction force due to $(i-1)$th slice will be upward. Consider a typical slice shown in Figure 13.23(b) has height h, inclination of the base α and height of water h_w. The forces acting on each slice are:

(i) Weight of the slice, w acting downward.
(ii) The effective normal force (N') and pore water force (U), In absence of interslice forces $N' + U = w \cos \alpha$ and in the absence of interslice forces and water $N' = w \cos \alpha$.
(iii) Shear force or resistive force T acts along the base

(iv) Interslice forces E_n and F_n act on the left vertical surface and E_{n+1} and F_{n+1} act on the right vertical surface. The locations of the forces are Z_n and Z_{n+1}.

(v) External surcharge (Q) may act on the crest of the slope if applicable.

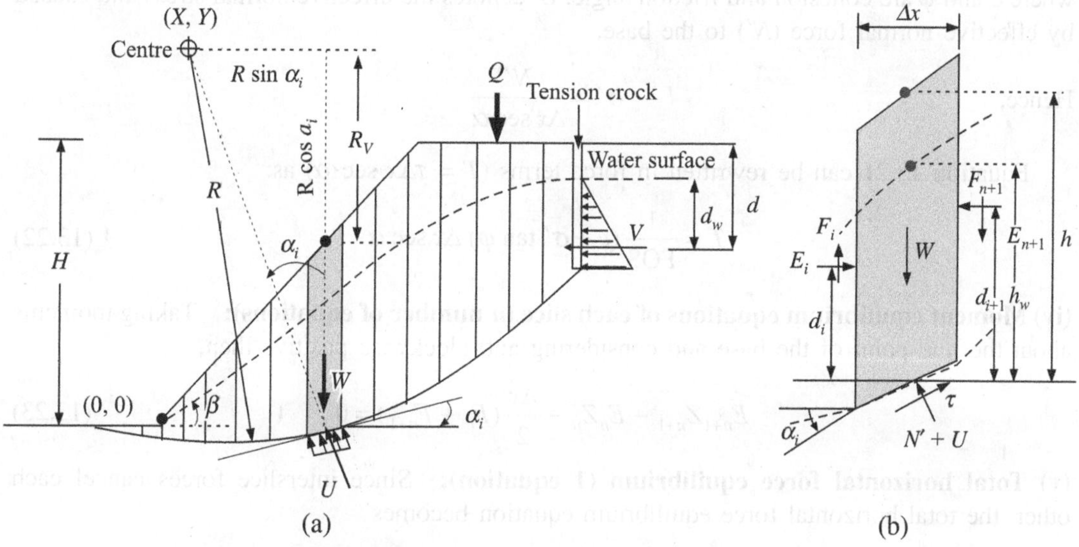

FIGURE 13.23 Forces acting on a typical slice.

The factor of safety of a slice is determined by (i) force equilibrium in vertical and horizontal direction at base for each slice, (ii) limit equilibrium condition of factor of safety (FOS) for each slice, (iii) moment equilibrium equation for each slice, (iv) total horizontal force equilibrium, and (v) total moment equilibrium. For rigorous analysis, force equilibrium in tangential direction of the base for each slice is also considered. Hence, four independent equations can be written for each slice to determine four unknown parameters. Similarly, for n number of slices, $4n$ equations can be written which give solutions for $4n$ unknown parameters. The limit equilibrium equations for the slice are as follows:

(i) Force equilibrium in vertical direction (*n* number of equations)

$$F_V = (N' + U)\cos\alpha + T\sin\alpha - w + (F_n - F_{n+1}) = 0 \tag{13.19}$$

where, N' is effective normal reactive force, U denotes pore water force and pore water pressure acting at the base of the slice is $u = U/\Delta x \sec\alpha$.

(ii) Force equilibrium in horizontal direction (*n* number of equations): For rigorous methods, equilibrium of horizontal forces is also satisfied for each slice as given below:

$$F_H = T\cos\alpha + (E_n - E_{n+1}) - (N' + U)\sin\alpha = 0 \tag{13.20}$$

(iii) Limit equilibrium relationship by assuming Mohr–Coulomb strength (*n* number of equations): It is assumed that at the time of base failure, the factor of safety (FOS) for all slices against the sliding are the same. The shear strength of the base is calculated based on Mohr–Coulomb failure criterion. So, the mobilised shear stress (τ) is:

$$\text{FOS} = \frac{\text{Shear strength}}{\text{Shear stress}} = \frac{c + \sigma' \tan \varphi}{\tau} \quad \text{or} \quad \tau = \frac{c + \sigma' \tan \varphi}{\text{FOS}} \tag{13.21}$$

where c and φ are cohesion and friction angle. σ' denotes the effective normal stress and caused by effective normal force (N') to the base.

Hence,
$$\sigma' = \frac{N'}{\Delta x \sec \alpha}$$

Equation 13.21 can be rewritten in force terms ($T = \tau \Delta x \sec \alpha$) as:

$$T = \frac{1}{\text{FOS}} (c + \sigma' \tan \varphi)\, \Delta x \sec \alpha \tag{13.22}$$

(iv) Moment equilibrium equations of each slice (*n* number of equations): Taking moments about the mid-point of the base and considering anti-clockwise positive then,

$$E_{n+1} Z_{n+1} - E_n Z_n - \frac{\Delta x}{2} (F_n + F_{n+1}) = 0 \tag{13.23}$$

(v) Total horizontal force equilibrium (1 equation): Since interslice forces cancel each other, the total horizontal force equilibrium equation becomes

$$T_H = \Sigma T \cos \alpha - \Sigma (N' + U) \sin \alpha - V = 0 \tag{13.24}$$

(vi) Total moment equilibrium about a centre of the sliding surface (1 equation): Taking moments about the centre of the circle, we can write

$$R \Sigma T_i - \Sigma w R_w - V R_V = 0 \quad \text{or} \quad R \Sigma T_i - R \Sigma w_i \sin \alpha_i - V R_V = 0 \tag{13.25}$$

where, R_w denotes the moment arm of weight w of each slice and defined as $R_w = R \sin \alpha_i$ and R_V is the moment arm of the water pressure in vertical tension crack as $R_V = Y - H + d - d_w/3$.

For n slices, total number of unknowns for factor of safety (1), normal force at base (n), location of normal force (n), shear force at base (n), interslice forces ($2n - 2$), and location of interslice force thrust ($n - 1$) are $6n - 2$, which is more than the number of possible equations ($4n + 2$). Hence, the solution is indeterminate. However, researchers have tried for a solution by reducing the number of variables by making some simplifying assumptions. One of the common assumptions is that the normal force on the base of the slice acts at the midpoint thus reducing the number of unknowns to ($5n - 2$). With ($5n - 2$) unknown, the problem is still indeterminate. So, to make the problem determinate further assumptions are incorporated into the generalised equations and number of different methods are conceived to determine factor of safety. In the following, ordinary (Fellenius) and simplified methods (Bishop's and Janbu methods) are described in details. The assumptions imposed for rigorous methods such as Spencer's and Morgenstern–Price methods are mentioned briefly. Detail discussions on these methods are beyond the scope of this book.

Ordinary or Fellenius Method (1927)

Fellenius method is one of the earliest and simplest analytical methods to estimate the stability of a slope by slicing method. It states that the resultant of interslice forces in both sides are zero meaning both normal and shear components (E and F) are zero. In other words, the resultant of interslice forces for all slices is inclined at an angle that is parallel to the base of the slice, this assumption fails to satisfy the interslice equilibrium where adjacent slices have different base inclination. Hence, number of unknowns becomes $2n + 1$, i.e. $[(5n - 2) - (3n - 3)]$ and the solution is overdetermined. As a result, this method does not satisfy the vertical and horizontal force equilibrium conditions given in Eqs. 13.19 and 13.20 respectively. Since there is no interslice forces, then FOS becomes

$$\text{FOS} = \frac{\sum_{i=1}^{n} RM_i}{\sum_{i=1}^{n} DM_i} = \frac{\sum_{i=1}^{n} [c\Delta x_i \sec\alpha_i + (w_i \cos\alpha_i - u_i)\tan\varphi]\, R}{\sum_{i=1}^{n} [w_i R \sin\alpha_i + (\gamma_w d_w^2/2) R_V]}$$

$$\text{FOS} = \frac{\sum_{i=1}^{n} [c \sec\alpha_i + \gamma_{ri} h_i\ (1 - r_u \sec\alpha_i)\tan\varphi]\ \Delta x_i}{\sum_{i=1}^{n} [w_i \sin\alpha_i + (\gamma_w d_w^2/2)\,(R_V/R)]} \tag{13.26}$$

The factor of safety derived from this method is over conservative and is lower than other methods of slices.

EXAMPLE 13.13 Determine the factor of safety for the slope given in Example 13.11 by Fellenius method for dry as well as for slope fully submerged in water.

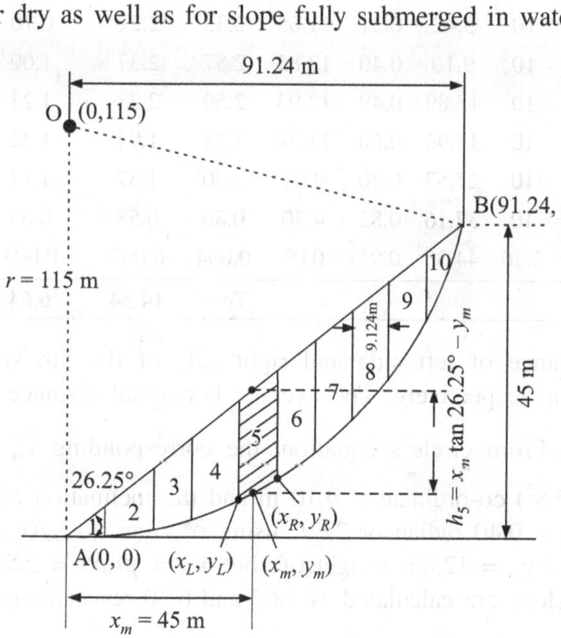

Solution: No tension crack is present so Eq. 13.26 can be modified and FOS relation is as follows:

$$\text{FOS} = \frac{\sum_{i=1}^{n}\left\{\frac{c\Delta x_i}{\cos\alpha_i} + \frac{(w_i \cos^2\alpha_i - u_i\Delta x_i)}{\Delta x_i}\tan\varphi\right\}}{\sum_{i=1}^{n}\{(w_i \sin\alpha_i)\}}$$

Now, divide the sector into slices of equal width, slices 1 to 9 is of 10 m and for 10th slice, width is 12.51 m. The equation of the sliding circle is written as:

$$x_i^2 + (y_i - 115)^2 = 115^2$$

Then, the slope of tangent at any point (x_i, y_i) on the circle is found by differentiating the circle equation w.r.to x

$$\tan\alpha_i = \frac{dy}{dx} = \left|\frac{-x_i}{y_i - 115.8}\right| \Rightarrow \alpha_i = \tan^{-1}\left|\frac{-x_i}{y_i - 115.8}\right|$$

Sample calculation for 5th slice is given below and complete results are shown in Table below.

Sl. no.	x_L (m)	x_R (m)	x_m (m)	Δx_i (m)	y_m (m)	α_i (rad)	h_i (m)	w_i (MN)	$w_i \cos\alpha_i$ (MN)	$w_i \sin\alpha_i$ (MN)	$\Delta x_i \sec\alpha_i$ (m)	$r_u \sec\alpha$
1	10	0	5	10	0.11	0.04	2.33	0.47	0.47	0.02	10.01	0.50
2	20	10	15	10	0.98	0.13	6.34	1.27	1.26	0.16	10.08	0.50
3	30	20	25	10	2.73	0.22	9.46	1.89	1.85	0.41	10.24	0.51
4	40	30	35	10	5.42	0.31	11.65	2.33	2.22	0.70	10.49	0.52
5	50	40	45	10	9.10	0.40	12.85	2.57	2.37	1.00	10.85	0.54
6	60	50	55	10	13.89	0.49	12.93	2.59	2.28	1.23	11.36	0.57
7	70	60	65	10	19.96	0.60	11.74	2.35	1.94	1.32	12.08	0.60
8	80	70	75	10	27.57	0.70	9.01	1.80	1.37	1.17	13.12	0.66
9	90	80	85	10	37.16	0.82	4.30	0.86	0.58	0.63	14.72	0.74
10	92.26	90	91.13	2.26	44.35	0.91	0.09	0.004	0.003	0.003	3.67	0.81
								Total	14.34	6.64	106.64	5.96

The horizontal distance of left side and right side of the 5th slice from toe of the circle is 50 m and 40 m respectively. The average horizontal distance of slice from toe is $x^2 = \frac{(x_{5L} + x_{5R})}{2} = 45\text{ m}$. From circle's equation, the corresponding y_m (use circle equation $x^2 + (y - 115.8)^2 = 115.8^2$) co-ordinate is 9.10 m and the inclination of the tangent at point (45, 10) on circle is $\alpha_5 = 0.40$ radian or 22.9° using $\alpha = \tan^{-1}[-x_m/(y_m - 115.8)]$; height of slice $h_5 = (x_m/\tan 26°) - y_m = 12.85$; weight of slice $w_5 = \gamma h_5\Delta x = 2.57$ MN. The factor of safety for dry and wet slope are calculated as 7.67 and 6.30 respectively.

Simplified Bishop Method (1955)

The simplified Bishop method neglects (or zero) interslice normal (i.e. $E_n = 0$ and $E_{n+1} = 0$) forces in Eqs. 13.19 to 13.25. Hence, number of variables become $4n - 1$ [$5n - 2 - (n - 1)$]. The method also ignores interslice normal forces (or can be assumed, $E_n - E_{n+1} = 0$) but ensures that the vertical stress given in Eq. 13.19 and overall moment equilibrium given by Eq. 13.23 is satisfied for each slice. Essentially, a number of variables reduce to $2n + 1$ (n number of effective normal forces, n number of shear forces and factor of safety). Due to this reason, this method is not suitable for high external horizontal stress applied to slope surface and mostly applies for circular sliding path problems.

Use Eq. 13.21 into Eq. 13.19 ($E_n = 0$; $E_{n+1} = 0$) by replacing σ' as given below:

$$(N' + U)\cos\alpha + \frac{1}{\text{FOS}}(c\Delta x \cos\alpha + N' \tan\varphi)\sin\alpha - w = 0$$

$$N' = \frac{w/\cos\alpha - U - \dfrac{c\tan\alpha}{\text{FOS}}\left(\dfrac{\Delta x}{\cos\alpha}\right)}{1 + \dfrac{1}{\text{FOS}}\tan\alpha\tan\varphi} \tag{13.27}$$

Now, N' can be estimated if FOS is known. Then, use Eq. 13.22 into Eq. 13.25 to find the FOS as:

$$R\Sigma T_i - R\Sigma w_i \sin\alpha_i - VR_V = \frac{R}{\text{FOS}}\sum\left[\frac{c + (w_i/\cos\alpha - u_i)\tan\varphi}{1 + \dfrac{1}{\text{FOS}}\tan\alpha_i\tan\varphi}\right] - R\Sigma w_i \sin\alpha_i - VR_V = 0 \tag{13.28}$$

The factor of safety is then obtained as:

$$\text{FOS} = \frac{\sum\left[\dfrac{(c + (\gamma_r h_i - \gamma_w h_{wi})\tan\varphi)(\Delta x \sec\alpha_i)}{1 + \dfrac{1}{\text{FOS}}\tan\alpha_i\tan\varphi}\right]}{\sum w_i \sin\alpha_i + V(R_V/R)} \tag{13.29}$$

where $w_i = \gamma_{r_i} h_i \Delta x$ and $u_i = \gamma_w h_{w_i} \Delta x \sec\alpha_i$. In the above equation, an initial guess of FOS is done based on the Fellenious result. Then iterative method is used to solve the equation until difference between two successive factor of safety is less than a predefined error.

The Bishop and Morgenstern method (1960) proposes the following equations for the evaluation of pore water pressure.

$$r_u = \frac{u}{\gamma_r h} = \frac{\gamma_w h_w}{\gamma_r h} \tag{13.30}$$

The pore pressure ratio r_u is assumed to be constant throughout the cross-section, which is called *homogenous pore pressure distribution*. In Eq. 13.30, the ratio γ_w/γ_r for soil type material is around 0.5. Therefore, the maximum value of r_u for 100% saturation ($h_w/h = 1$) would be around 0.5. Based on Eq. 13.30, Eq. 13.29 can be rewritten as:

$$\text{FOS} = \frac{\sum \left[\dfrac{(c + \gamma_r h_i (1 - r_u) \tan\varphi)(\Delta x \sec\alpha_i)}{1 + \dfrac{1}{\text{FOS}} \tan\alpha_i \tan\varphi} \right]}{\sum w_i \sin\alpha_i + V(R_V/R)} \tag{13.31}$$

Simplified Janbu's Method

This method is similar to Bishop's simplified method with the following exceptions:

(i) Janbu's simplified method uses a correction factor (f_0) to account for the effect of the interslice shear forces. The correction factor is related to cohesion, angle of internal friction, and the shape of the failure surface (Janbu et al., 1956).

(ii) No interslice shear forces ($F_n = 0$; $F_{n+1} = 0$) is considered but it satisfies vertical force equilibrium for each slice (Eq. 13.19), as well as overall horizontal force equilibrium (Eq. 13.20).

(iii) It does not satisfy overall moment equilibrium.

Equation 13.33 is derived considering vertical force equilibrium for each slice. Using this equation, the horizontal force parallel to the base of the slice equilibrium Eq. 13.24 is written as given below:

$$\sum \frac{1}{\text{FOS}} (c\Delta x + N' \tan\varphi) \cos\alpha_i - \sum (N' + U_i) \sin\alpha_i - V = 0 \tag{13.32}$$

Hence, factor of safety (FOS) is written as:

$$\text{FOS} = \frac{\sum (c\Delta x + N' \cos\alpha_i \tan\varphi)}{\sum (N' + U) \sin\alpha_i + V} \tag{13.33}$$

The solution of Eq. 13.33 is obtained using iterative method just as Bishop's method. A correction factor is applied to estimate the final factor of safety if $N' < 0$. The final factor of safety is obtained by multiplying a correction factor as given below:

$$\text{FOS}_{\text{Janbu}} = f_0 \times \text{FOS}_{\text{Calculated}} \tag{13.34}$$

The multiplying factor is a function of sliding geometry (d and l, see Figure 13.24) and the strength parameter of material. This factor is an attempt to compensate the assumption of negligible interslice force. This factor was established after more rigorous factor of safety analysis with the simplified one. There is no consensus concerning the selection of appropriate f_0. But for convenience the modification factor is calculated using the following formula:

$$f_0 = 1 + b_1 \left[\frac{d}{l} - 1.4 \left(\frac{d}{l} \right)^2 \right] \tag{13.35}$$

where, the parameter b_1 varies for soil type as follows:

(i) Cohesive soil only: $b_1 = 0.69$

(ii) Soil with friction only: $b_1 = 0.31$

(iii) Soil with cohesion as well as friction angle: $b_1 = 0.50$

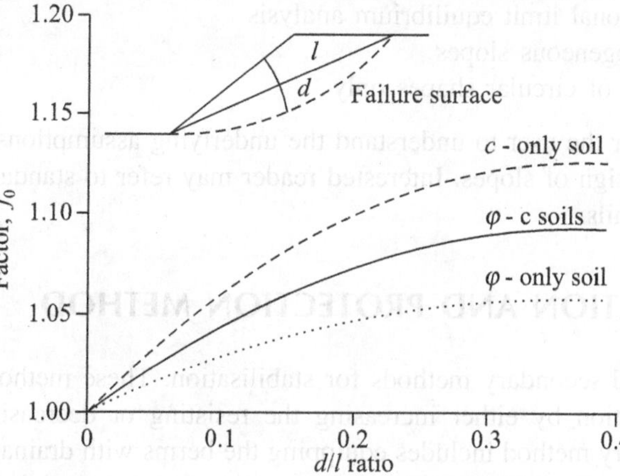

FIGURE 13.24 Janbu's correction factor for simplified method.

The appropriate b_1 value is selected for the use in Eq. 13.35 according to the type of soil encountered along the analysed failure surface.

Spencer's Method

Spencer's method assumes that ratio of shear interslice force to normal force (F/E) is constant, i.e., inclination of the resultant force is constant but unknown (Spencer, 1967). The number of variables becomes $4n - 1$ (n number variables for N', n number variables for T, 1 for factor of safety, $2n - 2$ for E and F). The unknown inclination adds one more variable and makes it $4n$. Considering the first 4 equilibrium equations (Eqs. 13.19–13.22), having $4n$ number of equations, factor of safety of slope is obtained in a rigorous manner.

Morgenstern-Price Method

It is similar to Spenser's method with only exception that the inclination of resultant interslice forces varies with a predefined arbitrary function. The variable of this function makes the total number of variable as $4n$ and then factor of safety is determined using iterative rigorous method.

13.6.4 Slope Stability Design Charts

Slope stability design charts are useful for preliminary analysis, to compare alternates that can later be examined by more detailed analysis. Chart solutions also provide a rapid means of checking the results of detailed analyses. It is a good practise to use charts before using a computer program to determine the approximate value of the FOS, as it allows quality control and checks for subsequent computer generated solutions. Another use is to back-calculate the strength for failed slopes to aid in planning for remedial measures. This is usually done by assuming an FOS of unit for field conditions and solving for the unknown shear strength.

The major shortcoming of design charts is that, it is developed for ideal rock/soil conditions, which are not encountered in practise. These charts have been devised using the following general assumptions:

(i) Two-dimensional limit equilibrium analysis
(ii) Simple homogeneous slopes
(iii) Slip surfaces of circular shapes only

It is imperative for the user to understand the underlying assumptions for the charts before using them for the design of slopes. Interested reader may refer to standard textbooks on slope stability for more details.

13.7 STABILISATION AND PROTECTION METHOD

There are primary and secondary methods for stabilisation. These methods can be used either singly or in combination by either increasing the resisting or decreasing the driving force/moment. The secondary method includes equipping the berms with drainage ditches to intercept surface runoff water, stepped benches in highly weathered rock materials to control erosion and to establish vegetation, the stepping helps in maintenance by scaling and trimming the bench slope. The primary methods like slope retarding, loading the toe, reducing height of slope are discussed in the following subsections.

13.7.1 Slope Regrading

Regrading helps in reducing active slope angle and also redistribute masses. Regrading of slope surface can be achieved in three ways: (i) flatter slope angle, (ii) loading the toe, and (iii) reducing the overall slope height.

Flatter Slope Angle

This can be done either by cutting material away from the slope, or adding (filling) material. This method is most effective where there are predominant shallow forms of instability (Figure 13.25).

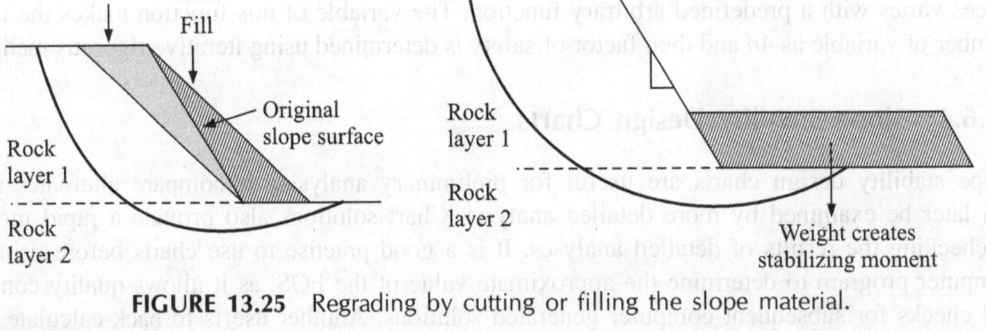

FIGURE 13.25 Regrading by cutting or filling the slope material.

Loading the Toe

The method is most effective for deep-seated instability. Slopes stabilised in this way may continue to suffer shallow forms of instability in near-surface layers, but the extent of the failure is well controlled. The material required can be used either from the crest of the slope (involving regrading of the slope), or brought to the site from elsewhere.

Reducing Height of the Slope

The method is not very practical as the height of the slope is difficult to change due to design. For a natural slope though, this may be considered. However, the reduction in instability achieved using this method is much lower than what achieved by loading the toe, unless the failure is deep-seated. It does not help much in the case of shallow failures. The material removed from crest and can be reused as a loading material for toe (Figure 13.26).

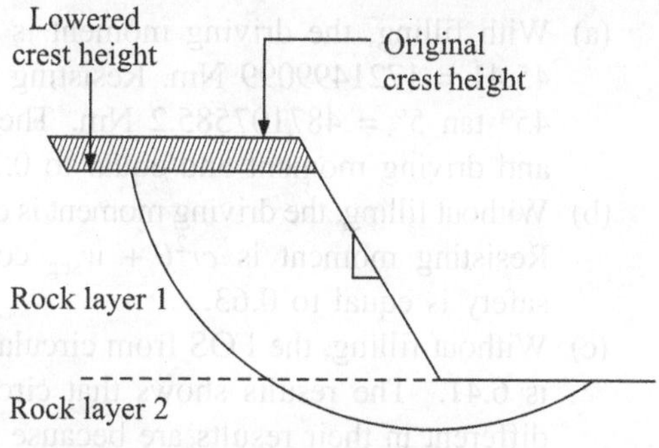

FIGURE 13.26 Regrading by reducing slope height.

EXAMPLE 13.14 In continuation with Example 13.13, if additional dump material is added on the slope face as shown in Figure by hatched area ABC. Determine the factor of safety (FOS) for the slope (a) with filling, (b) without filling, and (c) compare the results of without filling with other methods like circular arc method and ordinary slicing or Fellenius method.

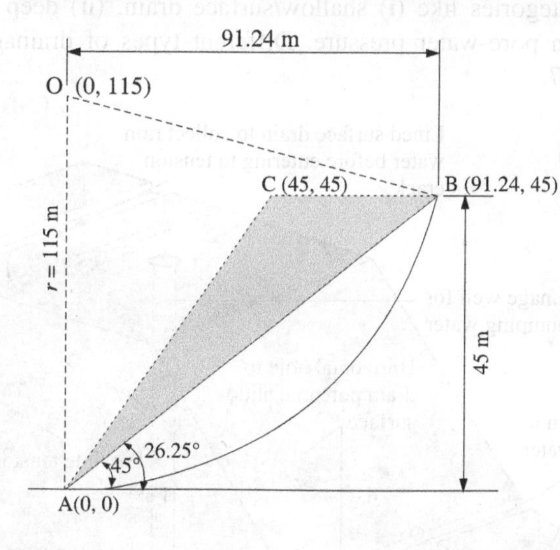

Solution: The centroid of the triangle ABC is (45.41, 30) and for segment it is (47.73, 103.14). The centroid of the triangle is calculated by taking average of the X and Y co-ordinates of the triangle vertex. Similarly, the segment, the distance of the centroid from center in m is calculated using relation $L_{c\,\text{seg}} = \dfrac{4r}{3}\left(\dfrac{\sin^3(\theta/2)}{\theta - \sin\theta}\right) = 107.90$, where θ is the angle subtended by arc at the origin and expressed in radian. Then the area in m^2 of the triangle and segment is calculated using relations $A_{\text{tri}} = \left|(1/2)[x_1(y_2 - y_3) + x_2(y_3 - y_1) + x_3(y_1 - y_2)]\right| = 1040.40$ and $A_{\text{seg}} = \dfrac{1}{2}r^2(\theta - \sin\theta) = 813.42$ respectively. Weight of segment and triangle are calculated by multiplying unit thickness and unit weight as 16268444.5 N and 20808000 N respectively.

(a) With filling, the driving moment is calculated as 16268444.5 × 47.73 + 20808000 × 45.41 = 1721499099 Nm. Resisting moment is cr^2 (52.51° π/180°) + (w_{seg} + w_{tri}) cos 45° tan 5° = 487107585.2 Nm. The factor of safety is the ratio of resisting moment and driving moment and equal to 0.28.

(b) Without filling, the driving moment is calculated as 16268444.5 × 47.73 = 776538458.9 Nm. Resisting moment is $cr^2\theta$ + w_{seg} cos 45° tan 5° = 485820322.7 Nm. The factor of safety is equal to 0.63.

(c) Without filling, the FOS from circular arc method is 0.63 whereas from ordinary slicing is 6.41. The results shows that circular arc method is underestimating the FOS. The different in their results are because of inherent assumption of the circular arc method.

13.7.2 Drainage Methods

It is difficult to quantify the effectiveness of most drainage methods. Drainage is best used as a short-term stabilising method, due to the fact that, in the long-term, the drains need much maintenance and repair, which is often difficult to perform, and expensive. Drainage methods are split into various categories like (i) shallow/surface drain, (ii) deep drain, and (iii) drain to eliminate construction pore-water pressure. Different types of drainage schemes for slope is shown in Figure 13.27.

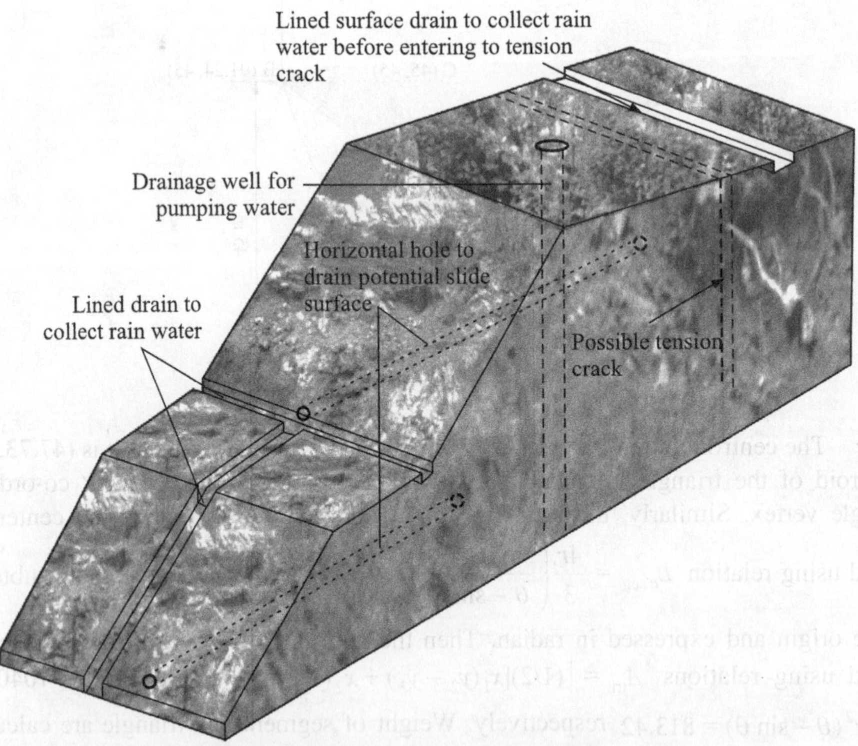

FIGURE 13.27 Different types of drainage schemes for slopes.

(i) Shallow/Surface Drain: It aims to reduce surface water and thereby reduce the pore water pressures at a much deeper level. These are the easiest drains to repair but get damaged easily. These drains are either (i) lined or unlined ditches, and (ii) shallow or gravel-filled trenches. Commonly a herringbone or chevron shape is used for the drains on the surface as this is most effective method to catch the run-off water.

(ii) Deep Drains: It performs the function of modifying the shape of the seepage flow in the slope material. It can be (i) deep trenches (although they are more commonly shallow), (ii) vertical bored drains—filled with sand or gravel, and (iii) horizontal bored drains—lined with perforated pipes.

(iii) Drains to eliminate construction porewater pressures: It is of two types: (i) drainage blankets which are installed either between layers of the fill material (which allows low permeability material to be used), or between the fill and foundation material, (ii) vertical drain wells are either sand filled or it may be constructed using geofabric materials, cardboard, or merely sand alone. Vertical drain wells are aimed to catch horizontal water flow.

13.7.3 Anchoring

The anchors are stressed against the slope, and the force they exert on the slope needs to be considered along with the other forces. The axial load on the anchor increases the effective stresses at depth, therefore, increasing the strength of the slope. Vector component of forces stabilises the slope against destabilising forces (Figure 13.28).

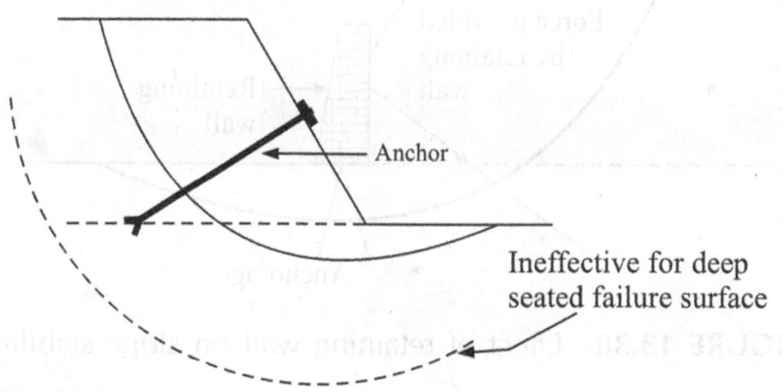

FIGURE 13.28 Anchoring by rock bolts or cable bolts.

13.7.4 Special Methods for Soil/Dump/Loose Rock

Sheet Piling

It is an earth retention and excavation support technique that retained loose soil, waste material, using sections with interlocking edges. It is installed in sequence upto the design depth as shown in Figure 13.29. This is an expensive remedial procedure, and is not commonly used unless the recovery scheme is very large. For installation of sheet piles, vibratory hammers are used but in case of hard or dense soil or dump material impact, hammer can complete the installation. If there is any problem with the vibration then the sheets can be hydraulically pushed into the ground.

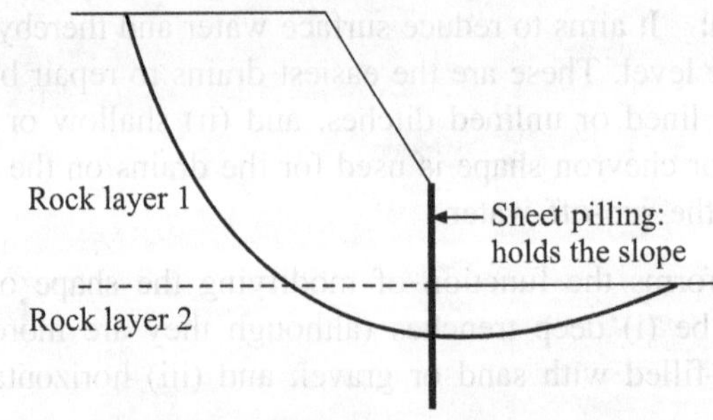

FIGURE 13.29 Sheet piling for slope stability.

Retaining Structures

As a whole, retaining structures are not particularly effective methods of remedy. They are very difficult to construct on an already moving slide. One use of these is to ensure complete stability of an existing (old) landslide, which may in future be reactivated. The force acting on a retaining wall can be estimated by using the interslice forces from stability analysis. The wall provides additional resistance which is only mobilised by further deformation of the slope. The force acts along the line of beneath the slope as shown in Figure 13.30.

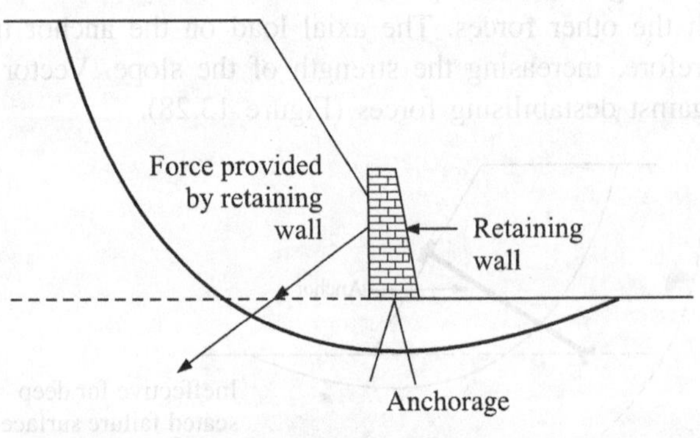

FIGURE 13.30 Effect of retaining wall on slope stability.

Gabion Structures

A gabion structure is prepared from pre-assembled rectangular cages made of hexagonal double-twist heavily galvanized mild steel woven wire mesh with or without polymer coating filled with rocks/boulders. All individual cages are laced together to act as a building block. Hence, the Gabion structure consist of three to four major components: (i) gabion cage filled with stone, (ii) filter media, and/or (iii) geotextile or geogrid reinforcement, and (iv) waste dump as backfill as shown in Figure 13.31.

A gabion wall is a gravity-retaining structure that retains soil by its weight. These structures are strengthened by reinforcing inside the soil/waste material using geotextile or geogrid. Filter media drains the water and does not allow to build water pressure behind the wall. The gabion structure is characterized by its monolithic and continuous construction process reinforced

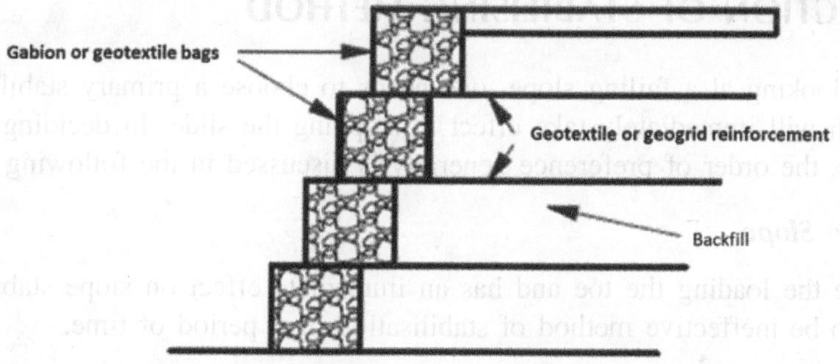

FIGURE 13.31 Basic component of Gabion wall.

structure, flexibility, permeability, durability, noise proofing, and beneficial environmental impact. Gabion structure is considered an ecological structure, for its merges into the natural environment. The stone filling and the layer of vegetation growing on the surface of the structure increase its landscape and durability. The Gabion facing prevents erosion on the front face, and the soil reinforcement increases the shear parameters of the waste dump material. Based on requirements, the gabion wall can be designed either as stepped outside, or stepped inside, or offset.

Geotextiles

Geotextiles are man-made (usually plastic based) reinforcement materials. In the area of slope stabilisation, geogrids are used in dump/embankment fill to reduce the amount of movement possible, keeping the fill in place. It is commonly used as an anchor which provides reaction against the disturbing moment. It is often used to repair small slides in soil/waste dumps (Figure 13.32).

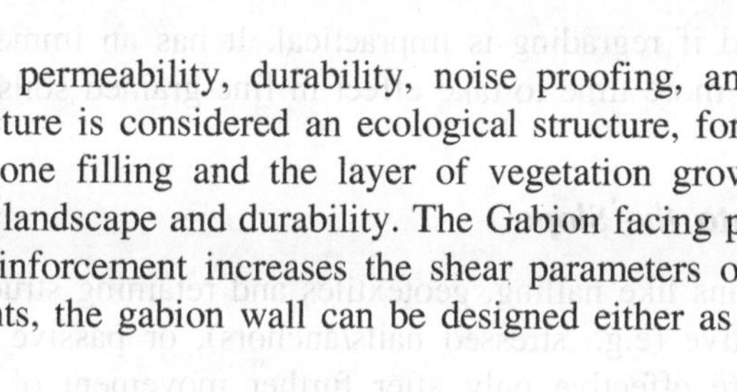

FIGURE 13.32 Geotextiles used in waste dump as layered material.

Grassing-over the Slope

By covering over a slope with either sand or grass, immediately reduces the amount of water infiltration. This method is often used in conjunction with more effective and long-term methods. Its benefits are being inexpensive and simple, whilst still performing a stabilising function, make it worth serious consideration.

13.8 SELECTION OF STABILISING METHOD

Generally, in looking at a failing slope, one needs to choose a primary stabilisation method, i.e., one which will immediately take effect in stopping the slide. In deciding which primary method to use, the order of preference generally is discussed in the following such sections.

Regrading the Slope

It also include the loading the toe and has an immediate effect on slope stability. Regrading are unlikely to be ineffective method of stabilisation over period of time.

Drainage

This method should be used if regrading is impractical. It has an immediate effect in high-permeability soils and takes more time to take effect in fine-grained soils. Surface drainage is very common.

Incorporating Structure into the Slope

It includes the artificial means like nailing, geotextiles and retaining structures into the slope. These structures can be active (e.g., stressed nails/anchors), or passive (e.g., walls or sheet piling). Passive structure are effective only after further movement of the slope, which of course may not be desirable.

13.9 INSTRUMENTATION

The instruments used in slope monitoring includes the measurement of displacement (surface and subsurface) and ground water measurement. There are variety of methods available, all these techniques have their own advantages and disadvantages. The traditional instruments generally have some disadvantages, such as low automation, low efficiency of monitoring, difficult to realize real-time monitoring, and easy to be subjected to electromagnetic interference. The effective slope monitoring methods and instruments can provide information on the current deformation and historical trends, which is important scientific information to analyze the mechanism of slope instability and provide a timely warning to the public (Gao et al., 2019). Surface monitoring systems are less expensive than subsurface systems, primarily due to requirement for drilling. Subsurface instrumentation gives additional, more reliable and more detailed information. In subsequent section, basic information are provided. Interested readers can extend their knowledge by going through the references.

13.9.1 Displacement Measuring Instruments

For surface measurement, the monitoring system includes cross-crack measurements (either manual or by wireline extensometer), survey instruments such as total station, global positioning system (GPS), digital photogrammetry, laser scanning, Radar (both ground-based and satellite-based InSAR; ground based LIDAR), tiltmeters and electrolevels (Read and Stacy, 2012). For sub-surface measurement, the monitoring system includes optical fibers, Inclinometers, Shear strips and time-domain reflectometer (TDR) cables, extensometers, thermistors, and Micro-seismic.

Surface Measurement

Different survey instruments such as level, theodolite, electronic distance measurement (EDM), total station and global positioning system (GPS) are used to measure the slope surface. Now a days, level and theodolite are rarely used. EDM only measures the distance. Total station without reflectors are the most common survey instrument used for pit slope monitoring. Basically a total station instrument incorporate the EDM and theodolite together into one unit. Conventional surveying methods are becoming more attractive with the increased availability and precision of global positioning system (GPS) receivers. GPS receiver requires an unobstructed view of at least four satellites to measure altitude and three satellites to determine horizontal position. These are the devices that determine position based on signals from satellites and measure position of any point once in 24 hours and can achieve high accuracies. For more accuracy differential GPS (DGPS) are used (Read and Stacey, 2012). Other survey methods of measurement are photogrammetry, Laser scanning using light detection and ranging (LiDAR), Radar, interferometric synthetic aperture radar analysis (InSAR) and their variations like differential InSAR, permanent scatter InSAR. Out which radar technology based slope stability radar is very popular now a days. **Slope stability radar (SSR)** is another standard alternative instrument for slope performance continuous monitoring of the active areas in real time without manual interference. The data generated by the device is continuously measured and communicated by wireless communication. This device is not useful to define the mode of instability, although it can provide a more accurate determination of the extent of the moving mass.

Subsurface Measurement

Here are some of the most commonly used devices are introduced such as OFS, TDRs inclinometers.

Optical fiber sensors (OFS): As a new type of monitoring technology, optical fiber sensors (OFS) have been developing rapidly in recent years because of its unique advantages, such as small size, light weight, corrosion resistance, low measurement cost, immunity to electromagnetic interference, remote transmission, high sensitivity and high resistance to hostile environments. Fiber Bragg grating (FBG) and Distributed optical fiber sensor (DOFS) are two important components of OFS for slope monitoring (Gao el al., 2019).

Time domain reflectometers (TDRs): A TDR sends a low voltage pulse into the cable anchored into the ground and at any impedance change within the cable a reflection will be seen. The TDR measures the time between release and return of the low voltage pulse from any reflections. By measuring the time and knowing the propagation velocity of the pulse, the distance to the reflection are calculated. This helps to give an information about location and type of impedance change or failure of the cable.

Inclinometers: An inclinometer is a instrument used to measure horizontal movements in the ground as a function of depth. These instruments are very helpful in slope stability studies, and can be used to locate shear surface and monitor the rate of shear displacement in slow-moving landslides. Inclinometers are installed in vertical borehole drilled below the potential zone of movement inside a plastic casing. The annular space between the casing and the borehole are filled with grouting material. Once the casing is installed, an initial set of reading is obtained by lowering the inclinometer probe inside. Which measures the inclination of the casing throughout its length in two perpendicular directions. Then, after sometime, the probe with the readout unit is inserted back into the casing and obtain a second set of readings. Comparison of these two or subsequent readings can determine the magnitude and direction of horizontal deformation (Coduto, 2002).

13.9.2 Groundwater Monitoring

Groundwater has a significant impact on slope stability, so information on the groundwater table position and pore water pressure is very important. For this purpose, apart from flow observation through horizontal drain and dewatering wells, piezometers are used to measure water level or pore pressure. Vibrating wire piezometers are more popular than the other alternative like hydraulic and electric type.

EXERCISES

1. Derive an expression for critical depth (d) and location of a vertical tension crack from the edge of the excavation (b) for a planer failure as a function of weight of the sliding block (w), cohesion (c) and angle of internal friction (φ) of the sliding surface, height of sliding block (h), inclination of bench slope (β) and direction of sliding (α).
2. Given the planer block slide data shown in the sketch. Where a uniformly distributed surcharge σ is applied to the slope crest over an area (bl). The cohesion, angle of internal friction and unit weight of the sliding surface are 345 kPa, 29° and 24.68 kN/m^3 respectively. It is given that the height of the slope is 152 m and b is 7.62 m.

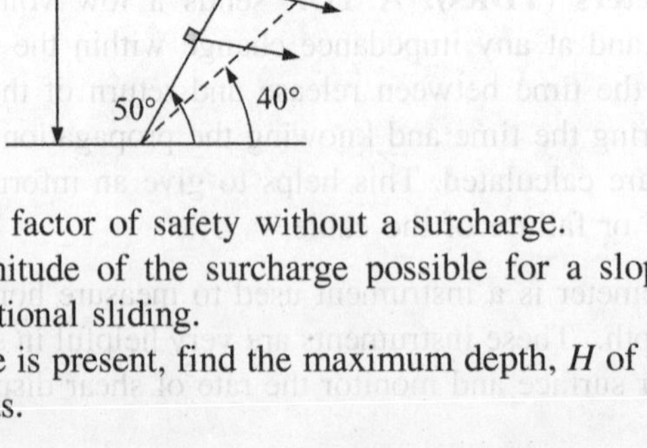

(a) Find the slope factor of safety without a surcharge.

(b) Find the magnitude of the surcharge possible for a slope factor of safety of 1.0 against translational sliding.

(c) If no surcharge is present, find the maximum depth, H of excavation possible before failure impends.

(d) Suppose the slope in the sketch is cable bolted. No surcharge is present. Bolt spacing is 15.2 m in the vertical direction and 7.6 m in the horizontal direction. Bolts ultimate strength are 2.22 MN and are installed in down holes (5°). Design tension is 60% of the ultimate strength. Find the factor of safety obtained by bolting.

(e) Consider the planer block slide in sketch without surcharge, seismic load and bolt reinforcement and suppose the water table rises to the top of the slide. Find the factor of safety of the flooded sliding rock mass, if continuous water seepage observed from slope toe.

3. A slope of height 20 m is shown below. The Cohesion and angle of internal friction of failure plane (AC) are 10 kPa and 30° respectively. The rock mass unit weight is 22 kN/m^3. Find the slope factor safety based on (a) Mohr-Coulomb failure criteria, (b) Barton's joint model, where JRC, JCS, and basic friction angle of the failure plane is 6.0, 44.7 MPa, and 20° respectively.

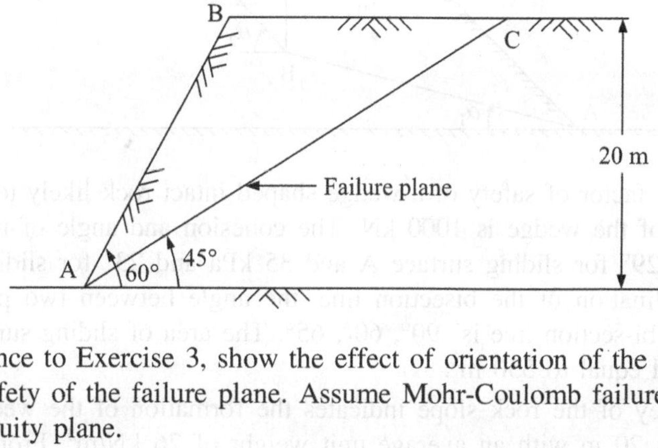

4. With reference to Exercise 3, show the effect of orientation of the failure plane on the factor of safety of the failure plane. Assume Mohr-Coulomb failure criteria for failure of discontinuity plane.

5. A metalliferous surface mine is operated with two benches, as shown in the Figure below. The upper bench is moving 10 m ahead of the lower bench. The properties of the failure plane (AA') are cohesion, angle of internal friction, basic friction angle, JRC and JCS are 10 kPa, 32.8°, 20°, 4.62, 44.7 MPa respectively. The unit weight of the rock mass is 22 kN/m^3. Find the slope factor safety based on (a) Mohr-Coulomb failure criteria, and (b) Barton's joint model.

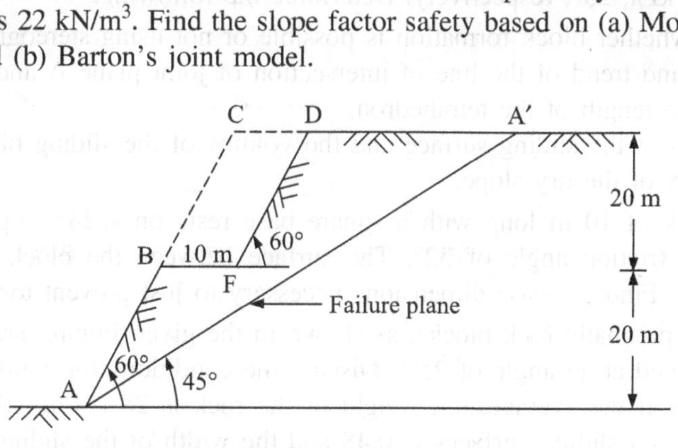

6. Figure shows a rock slope made of two blocks having weight W_1 and W_2. The blocks are separated by a vertical joint (BD), which lies on two inclined discontinuity surfaces. Both the blocks have the kinematic possibility of sliding. The contact surfaces AB, BC, and BD are cohesionless but not frictionless. The angle of internal friction to the contact surface is φ. Find the FOS of the blocks for which both these blocks slide together in terms of T, N_1 and N_2. Where, R and T are the normal force and frictional force respectively between two blocks on their contact; N_1 and N_2 are the normal reaction of the block 1 and 2 on sliding surface, i.e., AB and BC respectively.

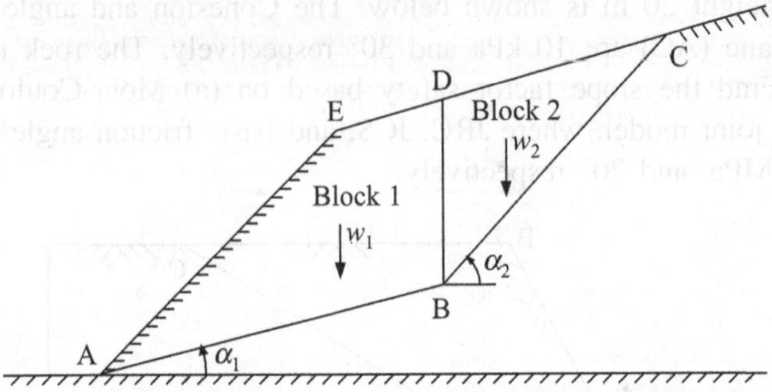

7. Calculate the factor of safety of a wedge shaped intact rock likely to fail from a slope. The weight of the wedge is 1000 kN. The cohesion and angle of internal friction are 70 kPa and 29° for sliding surface A and 35 kPa and 23° for sliding surface B. The angle of inclination of the bisection line, the angle between two planes, and the dip angle of the bi-section line is 90°, 60°, 65°. The area of sliding surfaces A and B are the same and equal to 550 m^2.

8. A joint survey of the rock slope indicates the formation of the wedge. The height of the wedge is 20 m with an average unit weight of 26 kN/m^3. From the joint survey, it is found that the dip angle and dip direction (α, β) for plane A is (45°, 360°); for plane B is (35°, 68°); for slope face is (75°, 45°); for crest and toe is (0°, 45°). If the cohesion and angle of internal friction of joint plane A and joint plane B are 70 kPa, 30°, and 35 kPa, 25°, respectively. Determine the following:

(a) Check whether block formation is possible or not using stereographic projection?
(b) Plunge and trend of the line of intersection of joint plane A and joint plane B.
(c) The edge length of the tetrahedron.
(d) The area of the sliding surface and the volume of the sliding block.
(e) The FOS of the dry slope.

9. A rock block of 10 m long with a square base rests on a 28° slope. The block base plane has a friction angle of 32°. The surface between the block and base plane is cohesionless. Find the base dimensions necessary to just prevent toppling.

10. Two square prismatic rock blocks, as shown in the given Figure, are resting on a rock surface inclined at an angle of 25°. Discuss the conditions for equilibrium of the two-block system if the average unit weight of the rock is 28 kN/m^3. If the coefficient of friction between sliding surfaces is 0.48 and the width of the sliding block, b is 4 m.

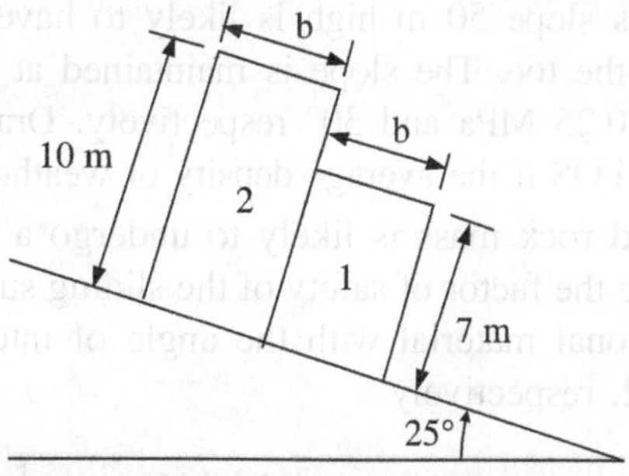

11. An unconsolidated rock slope is likely to undergo circular failure about a point 15 m above the toe of the bench. Calculate the FOS for slope assuming average density, Cohesion and angle of internal friction as 15 kN/m^3, 20 kPa and 30° respectively.

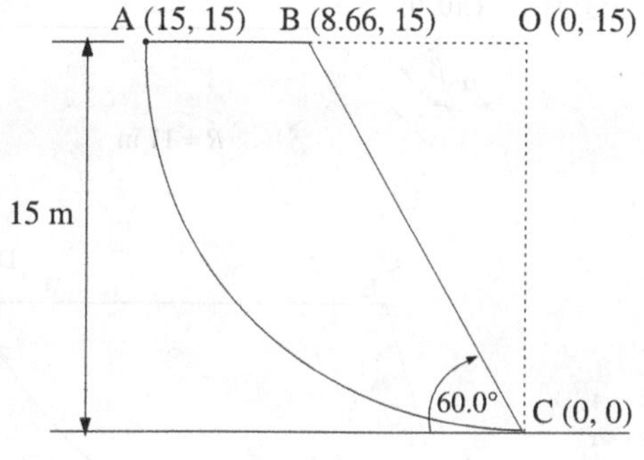

12. Calculate the factor of safety against shear failure along the slip circle shown in the Figure below. Assume Cohesion is 40 kN/m^2, and the total unit weight of the dump material is 20 kN/m^3. Find the factor of safety of the sliding volume if (a) the angle of internal friction, φ is zero, (b) φ = 25°, and (c) Comment on the effect of an increase in cohesion and friction angle on the slope stability.

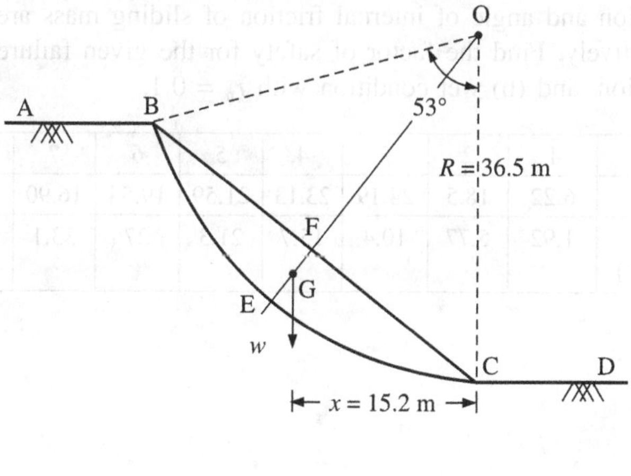

13. A weathered rock mass slope 50 m high is likely to have a toe failure about a center 50 m vertically above the toe. The slope is maintained at 45°. The cohesion and angle of friction is given as 0.25 MPa and 30° respectively. Draw the critical circle with the slope and calculate the FOS if the average density of weathered rock mass is 1250 kg/m^3.

14. A slope in a weathered rock mass is likely to undergo a base failure as shown in the Figure below. Calculate the factor of safety of the sliding surface ACD for (a) frictionless material, and (b) frictional material with the angle of internal friction of 30° and 25° for layer 1 and layer 2, respectively.

	Unit weight (kN/m^3)	*Cohesion* (kPa)
Rock layer 1	20	30
Rock layer 2	25	10

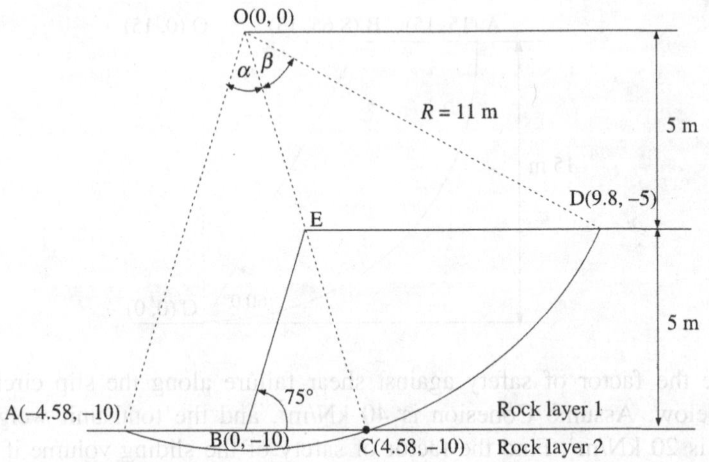

15. Toe failure of a 30 m high slope of slope angle 75°, is shown in the given Figure. The sliding mass is divided into ten slices. The radius of the sliding surface is 50 m. The unit weight, cohesion and angle of internal friction of sliding mass are 16 kN/m^3, 20° and 20 kPa respectively. Find the factor of safety for the given failure surface considering (a) dry condition, and (b) wet condition with r_u = 0.1.

Slice	1	2	3	4	5	6	7	8	9	10
Height (m)	6.22	18.5	24.19	23.13	21.59	19.54	16.90	13.53	9.24	3.60
Inclination of slice base, α (°)	1.92	5.77	10.4	15.7	21.3	27	33.1	39.6	46.8	55.1

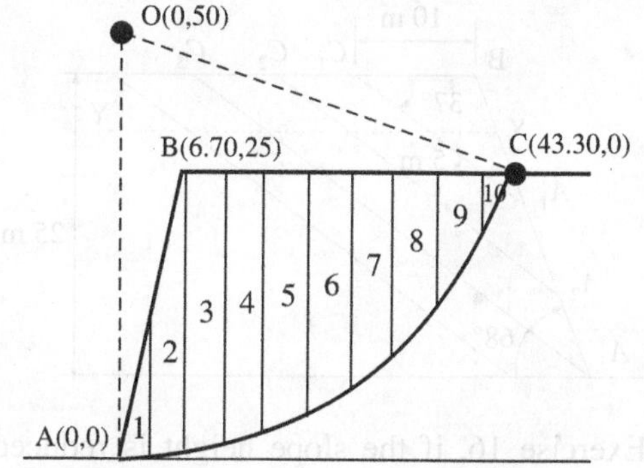

16. Face failure of a 30 m high slope of slope angle 50°, is shown in the Figure below. The sliding mass is divided into ten slices. The radius of the sliding surface is 45 m. The unit weight, cohesion, and angle of internal friction of sliding mass are 25 kN/m^3, 20° and 20 kPa respectively. Find the factor of safety for the given failure surface considering (a) dry condition, and (b) wet condition with $r_u = 0.1$.

Slice	1	2	3	4	5	6	7	8	9	10
Height (m)	6.35	13.61	29.96	29.38	28.13	26.15	22.34	19.49	14.18	6.36
Inclination of slice base, α(°)	19.9	24	28.1	32.5	37.4	43.4	52.3	76.3	10.9	42

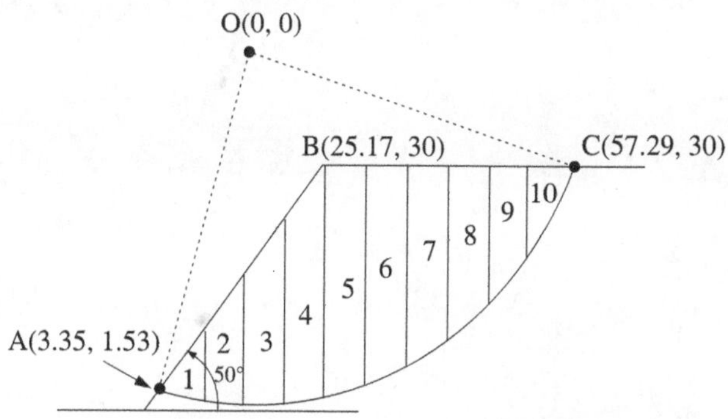

17. A hard rock slope shows one dominant joint set dipping at 37° with a joint spacing of 5 m. The Mohr-Coulomb properties of all joints are the same, with cohesion of 50 kPa and friction angle of 32°. The bulk density of the rock mass is 2600 kg/m^3. Estimate along which joint plane the failure is inevitable and determine the FOS of the slope.

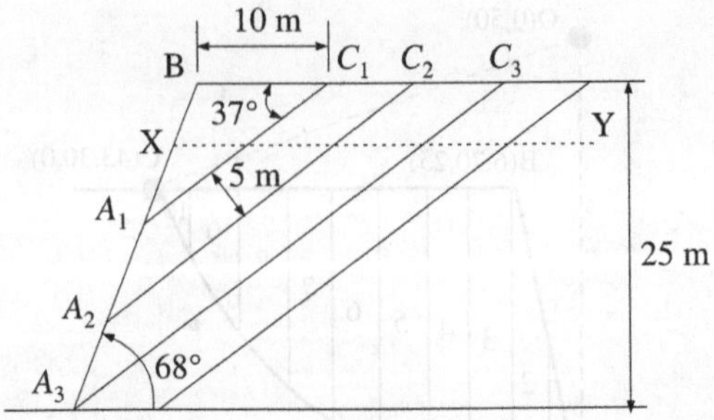

18. In continuation of Exercise 16, if the slope height is reduced to 20 m by removing a 5 m thick crest as shown by dotted line X-Y. Then, estimate the change in the factor of safety of each block.

19. In continuation of Exercise 8, if the wedge is 100% filled with water, then determine the FOS if peak pressure is developed at the (a) toe, and (b) mid height of the wedge.

20. In continuation of Exercise 8, if a 5 m vertical tension crack of dip 90° and dip direction 45°, intersect the wedge, then determine the wedge FOS in (a) dry condition, and (b) in 100% wet condition if peak water pressure is developed at the (i) toe, and (ii) at its mid height of the wedge.

Chapter 14

Numerical Methods, AI and ML Techniques in Rock Engineering

14.1 INTRODUCTION

Previous chapters have demonstrated that rock is a complicated natural material differing from other engineering materials due to the presence of discontinuities such as bedding planes, faults, fracture, joints and other mechanical defects. In this chapter, various numerical methods applied in rock engineering problems are introduced giving possible applications, their advantages and disadvantages. It is not the intent here to describe each method mathematically. There are excellent books and literatures written on each of the methods mentioned in the chapter. The idea is to expose the readers with various numerical tools those are available for solving complex rock engineering problems. The chapter also reviews a few artificial intelligence (AI) and machine learning (ML) techniques those are being applied in rock engineering.

Modelling excavations in isotropic and homogenous elastic rocks are in vogue since 1960s. Now, whether or not the discontinuity should be incorporated in design of rock structures either implicitly or explicitly is the primary decision to be made in selecting an analytical or computational method for application to a particular rock mechanics problem. With the rapid advancements in computer technology, numerical methods provide extremely powerful tools for the analysis and design of engineering systems with complex characteristics that are not possible or very difficult with the use of the conventional methods, like physical modelling, empirical solutions or closed form analytical solutions.

The use of physical models, both two- and three-dimensional were popular from 1950s to the early 1980s. Some of these, such as longwall coal mining models designed to study rock fractures and caving behaviour of roof, involved design of 'equivalent materials' of sufficiently low strength so that they fail at relatively low loads and experiments can be conducted in the laboratory. Construction of the models sometimes involved elaborate simulation of the sedimentary deposition process. Some models were several metres in height and 10 m or more in length. Similar 'equivalent materials' studies were made to understand slope stability in jointed rock mass, arch dam abutment load distributions in jointed rocks, tunnel support behaviour in over-stressed ground, and others. These studies were, of course, an attempt to compensate for the inability to carry out controlled full-scale tests and the inability of analytical tools available at that time to consider realistically the effects of heterogeneity, discontinuities, coupled effects, etc. The use of physical models has declined greatly in recent years, due to:

(i) their high cost and labour intensive nature, especially where several variations of the rock properties and loading conditions are required.
(ii) the increasing capability of computer numerical models to include the above-mentioned features, and to show these effects visually by computer graphics.

The visual observation of the actual effects of loading in physical models is an effective way of gaining understanding of the operative mechanics or physics, in a particular design problem. Questions may be raised, as whether the materials usually designed with scaled elastic properties, whether it can realistically model non-linear inelastic behaviour of the actual rock mass, whether the boundary loads applied in the model representative of those that obtain in reality, and others. The use of numerical models capable of simulating non-linear effects, various boundary conditions and others to analyse models could help in resolving some of those uncertainties.

The empirical procedure is based on the best fitting of the data obtained from a series of laboratories or field tests by means of function (polynomial, exponential, etc.), the characteristics of which are chosen just to meet the shape of the experimental design. This technique, that has been broadly used, for instance, for describing the time dependent bahaviour of rocks, leads to accurate results when applied to situations having boundary conditions, loading path, etc. similar to those characterising the tests on the basis of which the law has been formulated. However, the use of these laws for different conditions would in general require the introduction of some assumptions whose validity and influence on the final results is often difficult to assess a priori.

Analytical solution or closed form solutions or a mathematical expression gives the value of the field variable at any location in the body. Their use is restricted to situations characterised by relatively simple geometry, boundary conditions and material behaviour. Example: the Kirsch solution, in biaxial stress field explained in Chapter 7. For problems involving complex shapes, material properties and complicated boundary conditions, it is difficult and in many cases intractable (not easily controlled) to obtain analytical solution that satisfies the governing differential equations or gives extreme value to the governing functional. In order to overcome this limitation, the numerical methods are nowadays frequently adopted for analyses related to complex design problems.

It is rarely found that analytical solution can be found for rock mechanics problems of practical concern. This may be because of any one reason or their combinations

(i) The boundary conditions cannot be described by simple mathematical functions
(ii) The governing partial differential equations are non-linear
(iii) The problem domain is inhomogeneous and irregular
(iv) The constitutive relations for the rock mass are non-linear

There have been significant advances in the use of the computational methods in rock mechanics. Numerical methods and computing techniques have become tools for formulating conceptual models and mathematical theories integrating diverse information about geology, physics, construction techniques, complex geometry the environment and their interactions. This has greatly enhanced the development of modern rock mechanics from the traditional empirical art of rock deformability and strength estimation and support design to the rationalism of modern mechanics.

On the other hand, AI/ML techniques are now engulfing the entire gamut of engineering, social, management and medical disciplines. Several methods such as artificial neural network (ANN), support vector machine (SVM), fuzzy sets and logic, convolution neural network (CNN) and deep learning (DL) are being applied in solving rock engineering problems. However, it is to be noted here that these techniques are distinctly different from the numerical methods and mostly driven by data gathered from a specific problem.

14.2 COMPUTATIONAL METHODS IN ROCK ENGINEERING

In broad sense, numerical methods can be classified into continuum and discontinuum. The concepts of continuum and discontinuum are not absolute but relative and problem specific depending specially on geometric configuration. In Section 9.4, concept of continuum and discontinuum are described. Here, the similar concept is repeated for keeping the parity with the contents of the chapter. Figure 14.1(a) illustrates an underground opening in a medium without the presence of any discontinuity; in this case the displacement field is continuous and thus, continuum numerical methods are appropriate. Figure 14.1(b) shows a tunnel excavated in a medium in the presence of small number of discontinuities that divided the medium into a few number of continuous domains. In each of the domain, the displacement field is continuous but discontinuous across the domain. It is possible to model such medium with continuum description by considering special treatments on the specific discontinuities. Figure 14.1(c) depicts an opening in the presence of many discontinuities that divided the medium into blocks with spacing and continuity. In this case, displacements occur due to slip along the discontinuities and rotation of the blocks. Thus, a discontinuum numerical method is more appropriate.

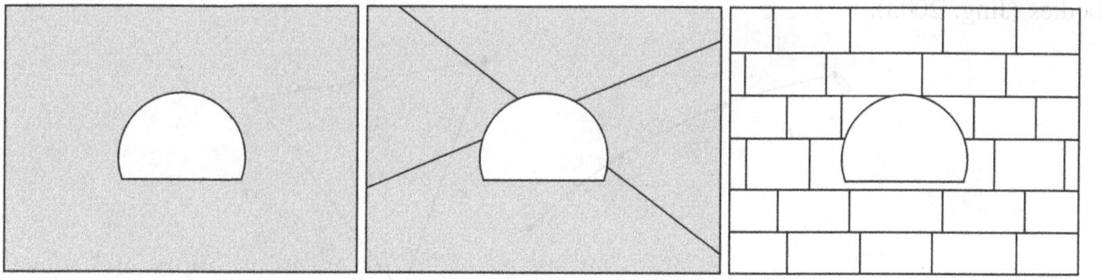

(a) Continuum (b) Discontinuum-continuum (c) Discontinuum

FIGURE 14.1 Continuum and discontinuum models.

Rock mass is largely discontinuous, anisotropic, inhomogeneous and inelastic in natural geological state. There are several difficulties arising in numerical modelling due to such complex geological conditions of rock mass. The complex combination of its constituents and its long history of formation make rock masses a difficult material for mathematical representation via numerical modelling. However, the advances in computer technology with rapid growth in algorithms make it possible for solving many complex problems in rock mechanics that do not have closed form analytical solutions. The most commonly applied continuum numerical methods in rock mechanics problems are finite difference method (FDM), finite element method (FEM), boundary element method (BEM) and meshfree methods (MMs). Discontinuum method such as discrete element method (DEM) and discontinuous displacement analysis (DDA) are commonly used to model jointed rock mass focusing mostly on applications in the fields of fractured or particulate geological media. There are some advantages and disadvantages of one method over others in a particular problem, as is explained later.

14.2.1 Boundary Element Method (BEM)

Boundary element method is an integral equation based numerical method. The main advantage of the BEM is the reduction of the computational model dimension by one (Figure 14.2), with much simpler mesh generation and therefore, input data preparation, compared with full domain discretisation methods such as the FEM and FDM is minimal. For example, area elements are used for 3D domain while line elements for 2D domain, hence, the number of element is less compared to other domain methods. Using the same level of discretisation, the BEM is often more accurate than the FEM and FDM, due to its direct integral formulation based on Betti's reciprocal theorem and Somigliana's identity. In addition, solutions inside the domain are continuous, unlike the pointwise discontinuous solutions obtained by the FEM and FDM. The solution domains of BEM can be divided into several sub-domains with different material properties, and this will often reduce the calculation time as well. The method is also suitable for considering infinite domains (full or half space/plane), due to its use of the fundamental solutions. In general, the BEM is not as efficient as the FEM in dealing with material heterogeneity, because it cannot have as many sub-domains as elements in the FEM. The BEM is also not as efficient as the FEM in simulating non-linear material behaviour, such as plasticity and damage evolution processes, because domain integrals are often presented in these problems. However, BEM is more suitable for solving problems of fracturing in homogeneous and linearly elastic bodies (Jing, 2003).

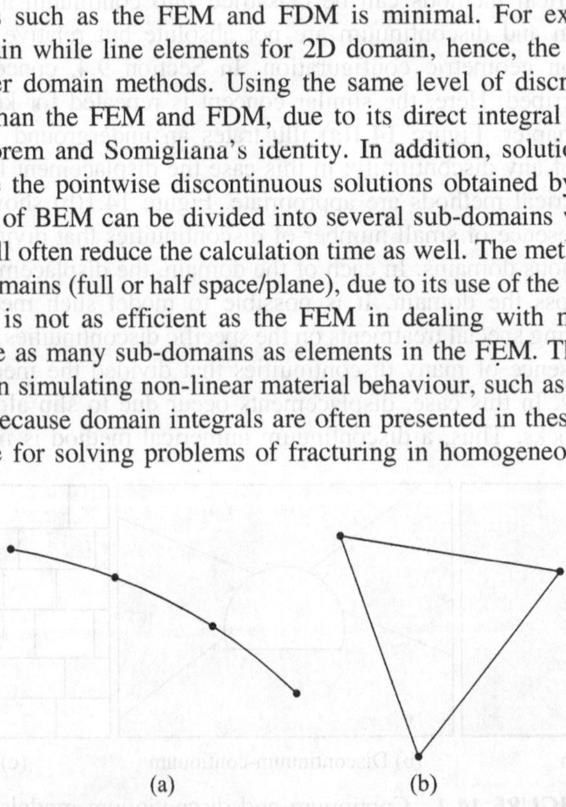

FIGURE 14.2 Boundary element used for (a) 2D domain, and (b) 3D domain.

The BEM formulation described above is called the direct formulation in which the displacements and tractions in the equations have clear physical meanings, are the basic unknowns of the boundary integral equations which are explicitly described on the problem boundary, and can be directly obtained by the solution of the integral equations. In the indirect formulation, on the other hand, the basic unknowns have no physical meanings and are just fictitious source densities related to the physical variables such as displacements and tractions.

The typical indirect BEMs are the displacement discontinuity method (DDM) by Crouch (1976) for 2D problems and Weaver (1977) for 3D problems and the fictitious stress method by Crouch and Starfield (1983). The basic concept of the indirect approach is to place the finite domain of interest into an imaginary infinitely large domain (full or halfplane or spaces) to derive the boundary integral equations relating the physical variables, such as displacements and tractions, to fictitious source densities, such as fictitious load (stress) or displacement.

14.2.2 Finite Difference Method (FDM)

This method is the oldest numerical technique to solve partial differential equations of engineering problems (Desai and Christian, 1977). The concept is to approximate differential equations into a system of algebraic simultaneous equations of field variables (e.g., displacement) at grids of nodes by using difference methods over certain spatial intervals Δx and Δy (Figure 14.3). Then the system of algebraic equations is solved by incorporating prescribed boundary conditions at the boundary nodes to evaluate the field variables at all nodes of the discretised domain. The conventional FDM employs rectangular grids as shown in Figure 14.3. The conventional FDM with regular grid systems does suffer from shortcomings, most of all in its inflexibility in dealing with fractures, complex boundary conditions and material heterogeneity. This makes the standard FDM generally unsuitable for modelling practical rock mechanics problems. However, several modification to the standard method have been made to use of irregular meshes (triangular grid or Voronoi grid systems), which leads to control volume or finite volume method (FVM) (Jing and Hudson, 2002).

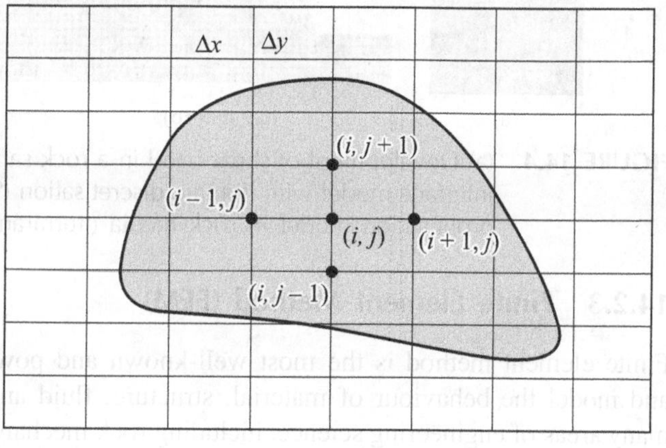

FIGURE 14.3 Spatial discretisation for finite difference method.

The combined FDM/FVM approach is much flexible in handling material behaviour such as inhomogeneity, plasticity and damage, without using any iterative scheme that may require in other numerical methods. Due to flexibility and unstructured mesh, it is still one of the most popular numerical method in rock mechanics and rock engineering, with application to the prediction of shear band formation in rock sample [Figure 14.4(a)], slope stability analysis [Figure 14.4(b)], wave propagation [Figure 14.4(c)], underground opening, thermo-hydro-mechanical (THM) process, rock mass characterisation, etc. Nowadays Fast Lagrangian Analysis of Continua, commonly known as FLAC is the most well-known computer package for stress analysis for engineering problems using FVM/FDM approach.

In spite of the simplicity and flexibility, there are many disadvantages associated with this method. It is very difficult to analyse the time history of field variables at a fix point on the material, because materials movements cannot be tracked using a spatially fixed mesh (Eulerian mesh). Since the FDM track the mass, momentum and energy flux across the grid cell boundaries, the position of the free surfaces, deformable boundaries, fracture initiation and moving material interfaces are difficult to be determined accurately.

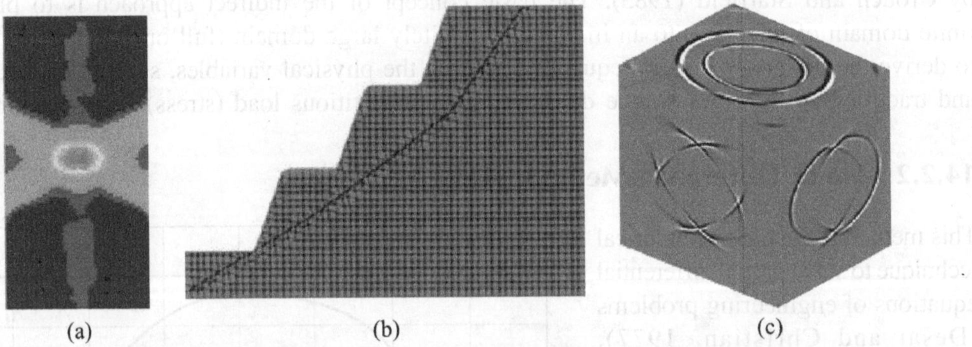

FIGURE 14.4 (a) Development of shear band in a rock sample (Zhao et al., 2013), (b) Slope single interface model with domain discretisation (Soren et al., 2014), and (c) Seismic wave propagation model in rock media (Iturrarán–Vivero sand Sánchez–Sesma, 2011).

14.2.3 Finite Element Method (FEM)

Finite element method is the most well-known and powerful numerical tool to study, predict and model the behaviour of material, structure, fluid and events. It is successfully applied in many areas of engineering science, including rock mechanics and rock engineering. Its popularity lies on the mathematical formulations in handling material complexities, complex boundary conditions, as well as its efficiency in dealing with complicated constitutive model and fractures.

In order to solve any problem using FEM, there are three primary steps to follow: domain discretisation, local approximations of field variables, and assemble of elemental matrices to a global matrix and its solution (Deb et al., 2010). The first task in a finite element solution is to accurately represent the physical domain of the problem by discretising the continuum into a finite number of non-overlapping elements or sub-regions (Figure 14.5). The finite element discretisation allows a variety number of shapes, for examples triangles (3-node or 6-node), quadrilaterals (4-node, 8-node or 9-node). Each element is formed on the connection with the nodes. The interpolation functions for approximation of field variables at elemental level depend on the type element associated with number of nodes. The interpolation functions must maintain the continuity of the primary variables across the inter-element boundaries and should be at least once differentiable since the governing equations are of second order. In the assembly process, the entire global stiffness matrix is zeroed initially. Then the assembly is done by adding up the contribution of

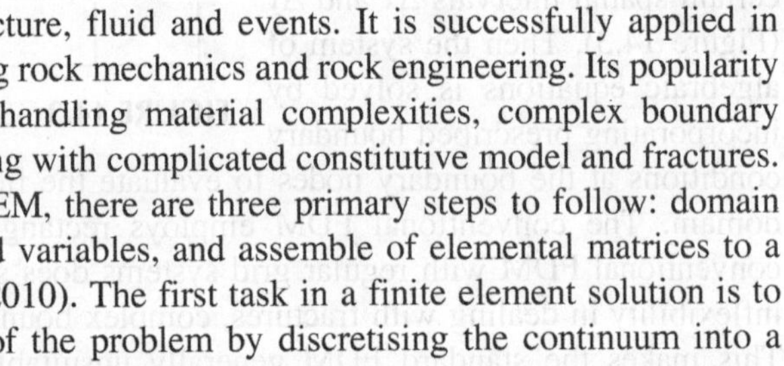

FIGURE 14.5 Finite element discretisation of circular opening in rock.

each node by the elements that share the node. This process leads to a global system of equations on a primary field variable (displacement) of all nodes. In order to have a unique solution, the determinant of the global stiffness matrix must be non-zero. This is done by incorporating prescribed boundary conditions on the corresponding boundary nodes in the stiffness matrix.

The modelling of discrete fractures in the FEM framework was put forwarded by several authors over the past years. FEM with interface elements is widely used approach in rock mechanics. In this approach, discontinuous function is approximated with a continuous function by representing discontinuity as a solid entity with zero thickness. The approach was first proposed as zero-thickness elements [Figure 14.6(a)] or Goodman's joint elements (Goodman et al., 1976). It has attracted attention to many researchers because of its simple implementation procedure into FEM framework. However, its formulation is based on the continuum assumption, therefore fracture opening, sliding and complete detachment of element cannot be simulated.

The easiest approach in FEM to deal with the discrete fractures is the element erosion technique (Belytschko and Lin, 1987). There are several algorithms for element erosion technique which do not need to prescribe fracture's topology. The main concept of this technique is that any element can be considered as either intact or fractured element. In the first case, element behaviour is elastic. Otherwise, the element is deleted if a fracture passes through an element [Figure 14.6(b)]. Most of the approaches do not really delete the element rather set the stress of the deleted element to zero. There are some approaches that indeed remove the element by replacing the element with a rigid mass (Song et al., 2008).

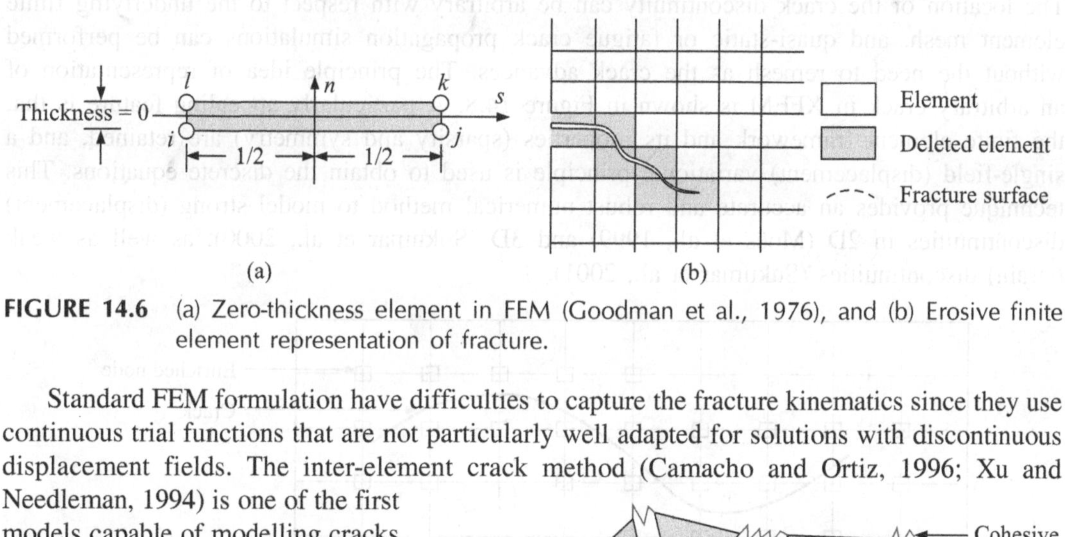

FIGURE 14.6 (a) Zero-thickness element in FEM (Goodman et al., 1976), and (b) Erosive finite element representation of fracture.

Standard FEM formulation have difficulties to capture the fracture kinematics since they use continuous trial functions that are not particularly well adapted for solutions with discontinuous displacement fields. The inter-element crack method (Camacho and Ortiz, 1996; Xu and Needleman, 1994) is one of the first models capable of modelling cracks or fractures kinematics. These were mostly applied to simulate dynamic fracture with cohesive cracks. The main concept is to model fracture by separation along the element edges (Figure 14.7). There are two approaches used in the inter-element crack method. In the formulation of

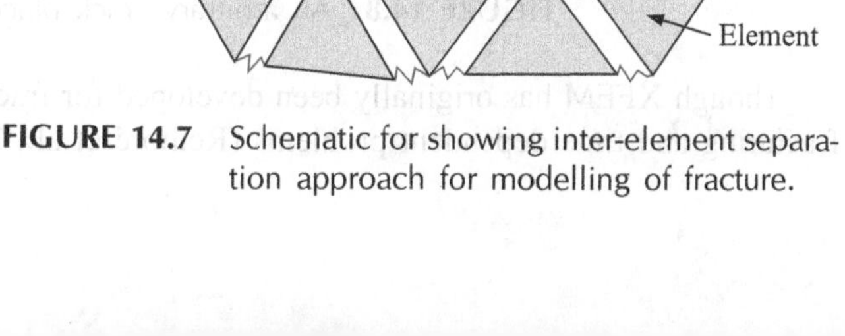

FIGURE 14.7 Schematic for showing inter-element separation approach for modelling of fracture.

Xu and Needleman (1994), all elements are separated from the beginning of the simulation. The edges are mechanically joined by cohesive laws. Camacho and Ortiz (1996) adaptively introduced cohesive surfaces at element edges when certain cracking criterion is met or the element edges are contiguous to a crack tip. In inter-element crack methods, cracks are only allowed to develop along existing inter-element edges. This method is comparatively simple, but can results in an overestimate of the fracture energy when the actual crack paths are not coincident with element edges. The results depend not only on the mesh size, but also on the mesh bias that can be compensated only by remeshing that is particularly cumbersome in 3D.

In general, the continuum assumption makes the FEM difficult in dealing with complete detachment and large-scale fracture opening problems (Jing 2003; Jing and Hudson 2002), which are the most concerned issues in micromechanical modelling of rock material failure. The locking effects, including element locking and numerical locking are others shortcoming of the FEM. They have more or less negative influences on the FEM application of rock fracturing.

A very flexible method that can handle linear and non-linear fractures initiation and their growth is the extended finite element method (XFEM) developed by Belytschko et al. (1999). The method employs special functions that are added to the finite element approximation using the framework of partition of unity (Duarte and Oden, 1996). For crack modelling, a discontinuous function (generalised Heaviside step function) and the two-dimensional linear elastic asymptotic crack-tip displacement fields are used to account for the crack. This enables the domain to be modelled by finite elements without explicitly meshing the crack surfaces. The location of the crack discontinuity can be arbitrary with respect to the underlying finite element mesh, and quasi-static or fatigue crack propagation simulations can be performed without the need to remesh as the crack advances. The principle idea of representation of an arbitrary crack in XFEM is shown in Figure 14.8. A particularly appealing feature is that the finite element framework and its properties (sparsity and symmetry) are retained, and a single-field (displacement) variational principle is used to obtain the discrete equations. This technique provides an accurate and robust numerical method to model strong (displacement) discontinuities in 2D (Moës et al., 1999) and 3D (Sukumar et al., 2000), as well as weak (strain) discontinuities (Sukumar et al., 2001).

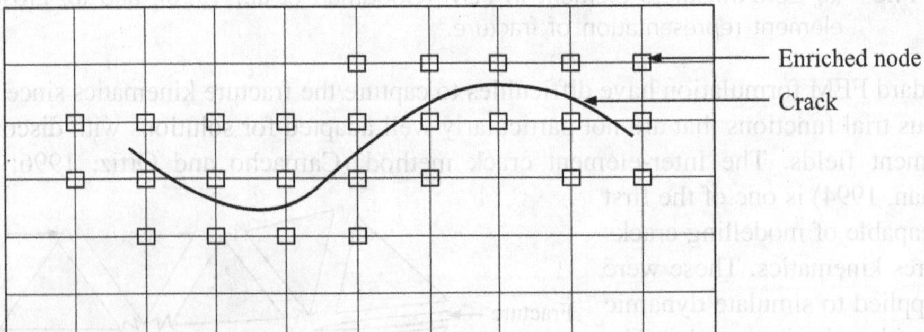

FIGURE 14.8 An arbitrary crack placed in XFEM mesh.

Though XFEM has originally been developed for fracture, it has been extended to dynamic fracturing and time dependent problems (Réthoré et al., 2005). Its recent development in rock mechanics include dealing with hydraulic fracturing (Remij et al., 2014), jointed rock mass (Deb and Das, 2010), fully grouted rock bolt (Deb and Das, 2011; Deb and Das, 2014), anisotropic problems (Asadpoure and Mohammadi, 2007) and hydro-mechanical modelling in multiphase porous media (Mohammadnejad and Khoei, 2013). Figure 14.9(a) and (b) shows a typical 3-noded element incorporating a rock bolt crossing a joint. It is termed as *bolt-crossing-joint* (BCJ) element. Stiffness matrix is determined in XFEM platform considering the effect of both bolt and joint. Figure 14.10 plots the displacement, axial stress and shear stress on a bolt crossing a joint at 0.5 m away from a circular tunnel boundary (Deb and Das, 2014).

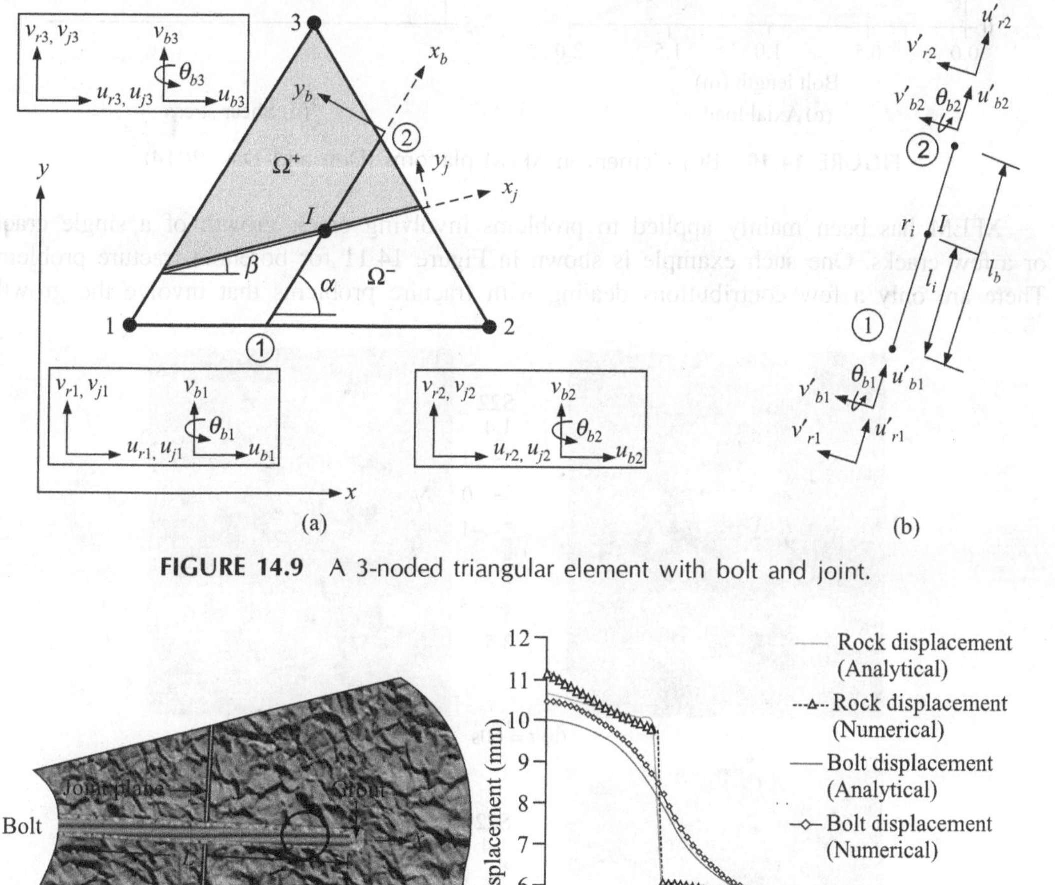

(a) (b)

FIGURE 14.9 A 3-noded triangular element with bolt and joint.

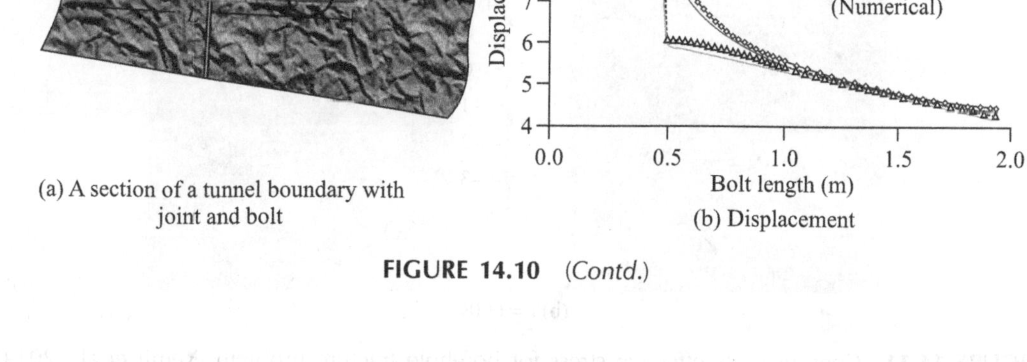

(a) A section of a tunnel boundary with joint and bolt

(b) Displacement

FIGURE 14.10 *(Contd.)*

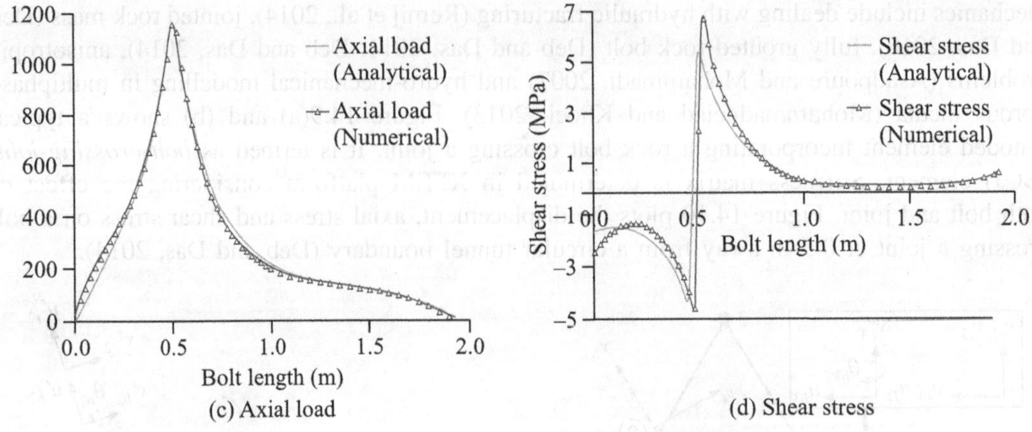

FIGURE 14.10 BCJ element in XFEM platform (Deb and Das, 2014).

XFEM has been mainly applied to problems involving crack growth of a single crack or a few cracks. One such example is shown in Figure 14.11 for borehole fracture problem. There are only a few contributions dealing with fracture problems that involve the growth

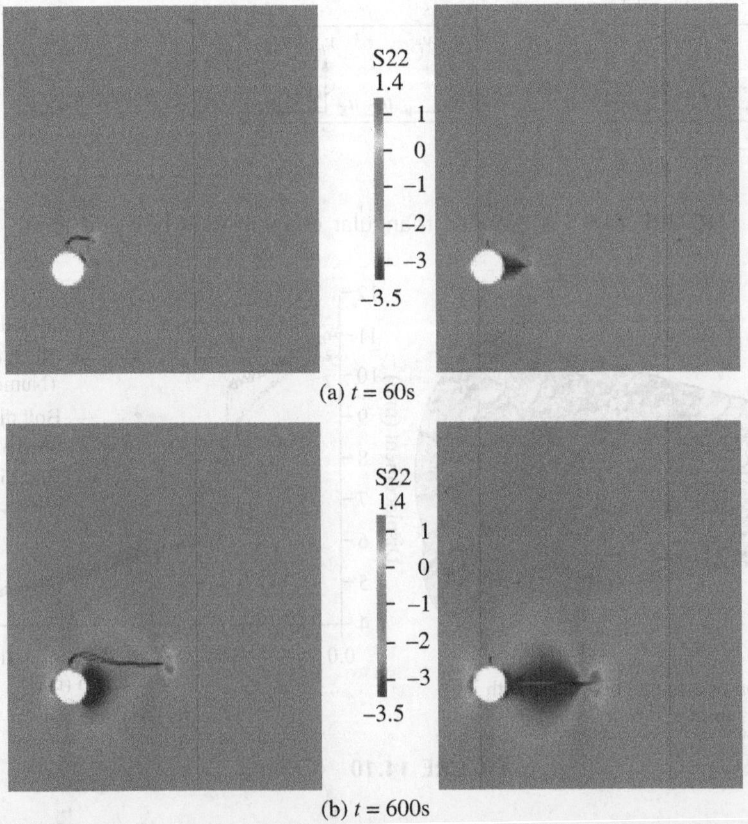

FIGURE 14.11 Contour plots effective stress for borehole fracture problem (Remij et al., 2014).

of numerous (several hundreds) of cracks. Methodologies have been developed in XFEM to account for complex crack patterns such as branching cracks and crack coalescence (within a single element). However, the formulation becomes cumbersome with increasing number of cracks and crack branches. Moreover, those methods suffer (in practise) from the absence of reliable crack branching criteria.

14.2.4 Meshfree Methods (MMs)

Meshfree methods, as indicated by its name, requires no mesh to discretise a problem domain (Figure 14.12). Ever since the beginning of their development, meshfree methods have been focused towards problems of changing initial geometry where a mesh-based method might struggle or not converge due to mesh distortion, or due to mesh sensitivity such as in fracture propagation, strain localisation and large deformation problems. Methods have particular advantages in dealing with large deformations, dynamic fracture initiation and propagation.

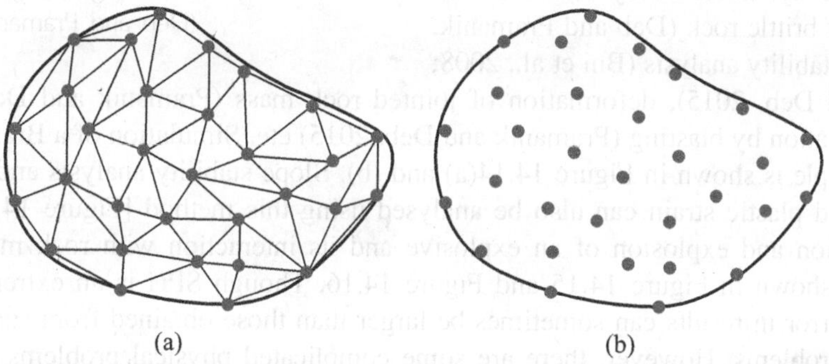

(a) (b)

FIGURE 14.12 Schematic of discretisation of a problem domain for (a) FEM, and (b) Meshfree analysis.

Meshfree methods form a class of numerical methods for solving PDEs in a discrete way where only nodal information is required. An ideal meshfree method would require no mesh throughout the analysis. The origin of meshfree methods dates from the 1970s with the smoothed partial hydrodynamics (SPH) method proposed by Gingold and Monaghan (1977). The SPH is a truly meshfree particle method based on a Lagrangian formulation where a set of moving particles represent a continuum where each particle represents a predefined mass or volume. The state of system is represented by particles, typically fixed mass, which possess material properties, follow the motion of the system, advancing contact discontinuities, preserving Galilean invariance, and reducing the computational diffusion of various material properties including momentum. The equation of governing evolution of the system can be formulated in such a way that mass, momentum and energy are conserved exactly.

Basically, there are two key steps for SPH formulations. The first step is the integral interpolation to represent a function and its derivative in the form of an integral equation, and the other one is the particle approximation of the integral representation. For particle approximation, the computational domain is represented by set distributed particles. The state of particles is represented by a set of points with fixed volume, which possess material properties interact with the all neighbouring particles by a weight function or smoothing function or smoothing kernel

(Monaghan 1992). This function is required to be continuous and differentiable. Each particle has a support domain specified by a smoothing length. The value of a function at a typical particle is obtained by interpolating the average of those values of that function at all particles in its support domain weighted by smoothing function (Figure 14.13). Gradients that appear in the flow equation are obtained via analytic differentiation of the smoothing kernel.

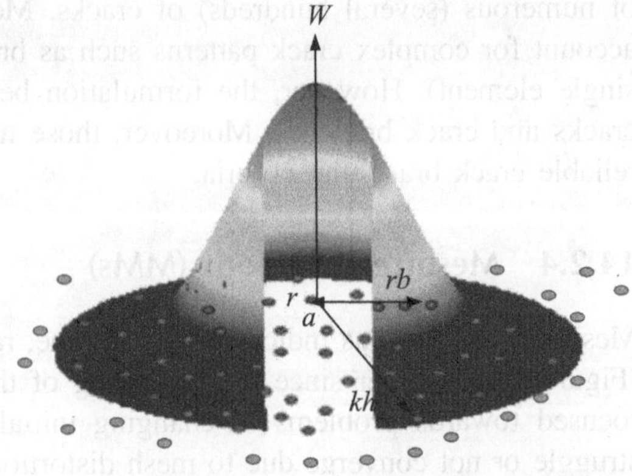

FIGURE 14.13 Diagram of the support domain of the particle at position r_a according to its weight function W (Deb and Pramanik, 2013).

In the SPH formulation, it is relatively easy to incorporate complicated physical effects and thus, it has been applied successfully to a vast range of rock mechanics problems. These include elastic dynamics (Gray et al., 2001), the fracture of brittle rock (Deb and Pramanik, 2013), slope stability analysis (Bui et al., 2008; Pramanik and Deb, 2015), deformation of jointed rock mass (Pramanik and Deb, 2015) and rock fragmentation by blasting (Pramanik and Deb, 2015) etc. Simulation of a Brazilian test of a disc rock sample is shown in Figure 14.14(a) and (b). Slope stability analysis and development of stresses and plastic strain can also be analysed using this method [Figure 14.14(c)]. Using SPH, detonation and explosion of an explosive and its interaction with rock medium can be simulated as shown in Figure 14.15 and Figure 14.16. Though SPH is an extremely versatile method, the error in results can sometimes be larger than those obtained from standard method for specific problems. However, there are some complicated physical problems which can be well simulated by SPH. Problems involving highly non-linear geometry and dynamic nature of loading can readily be solved by SPH. There are some special advantages of SPH over the traditional grid based numerical methods for which it can be exploited to give accurate results without requiring the use of complicated grid refinement techniques.

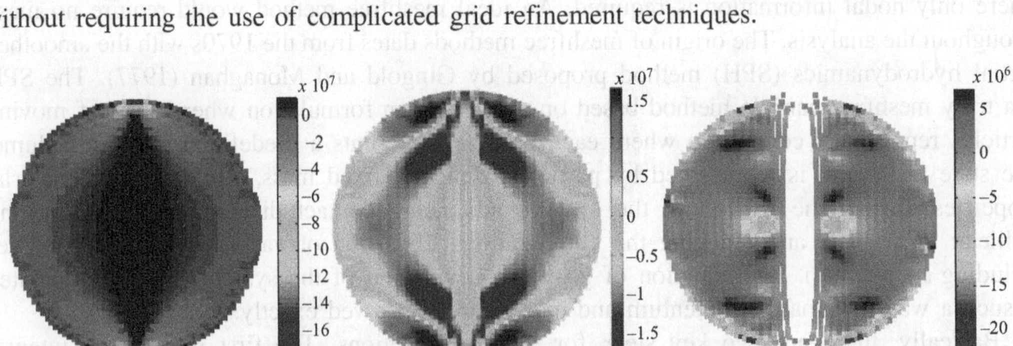

(a) Distribution of horizontal stress

FIGURE 14.14 (*Contd.*)

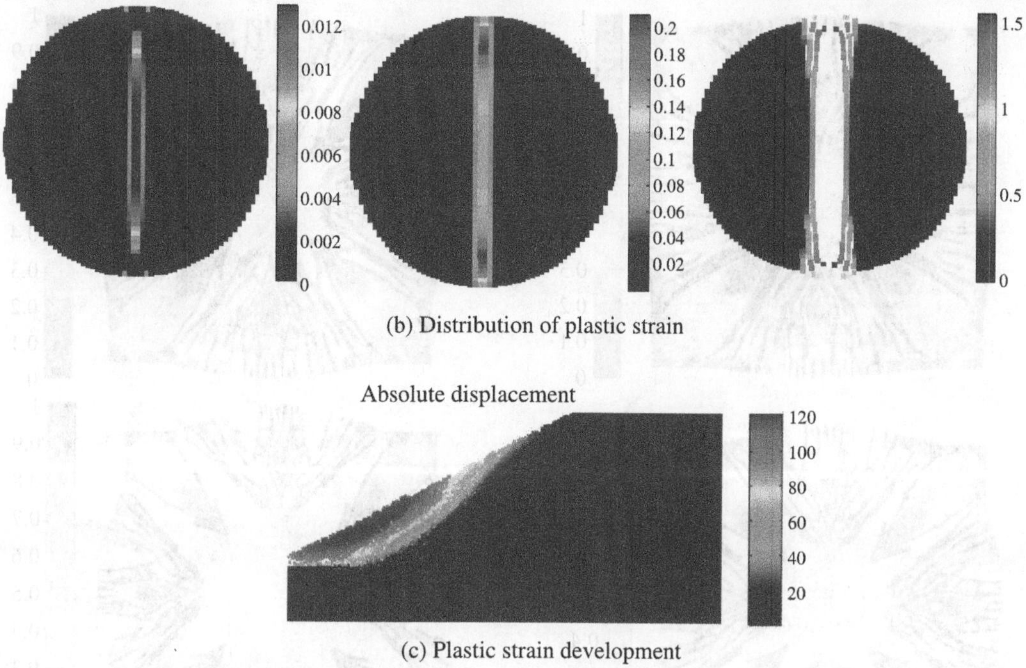

FIGURE 14.14 (a) and (b) Fracturing of a disc type rock sample in Brazilian test, (c) Plastic strain development in a slope (Deb and Pramanik 2013).

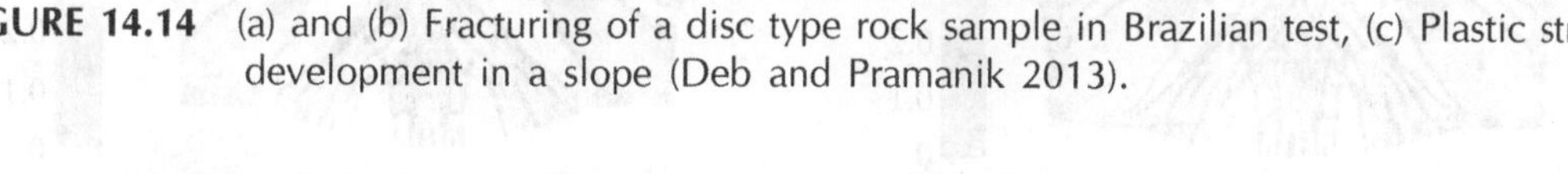

FIGURE 14.15 Detonation of an explosive column, explosion process and formation gas of an explosive (Pramanik and Deb, 2015).

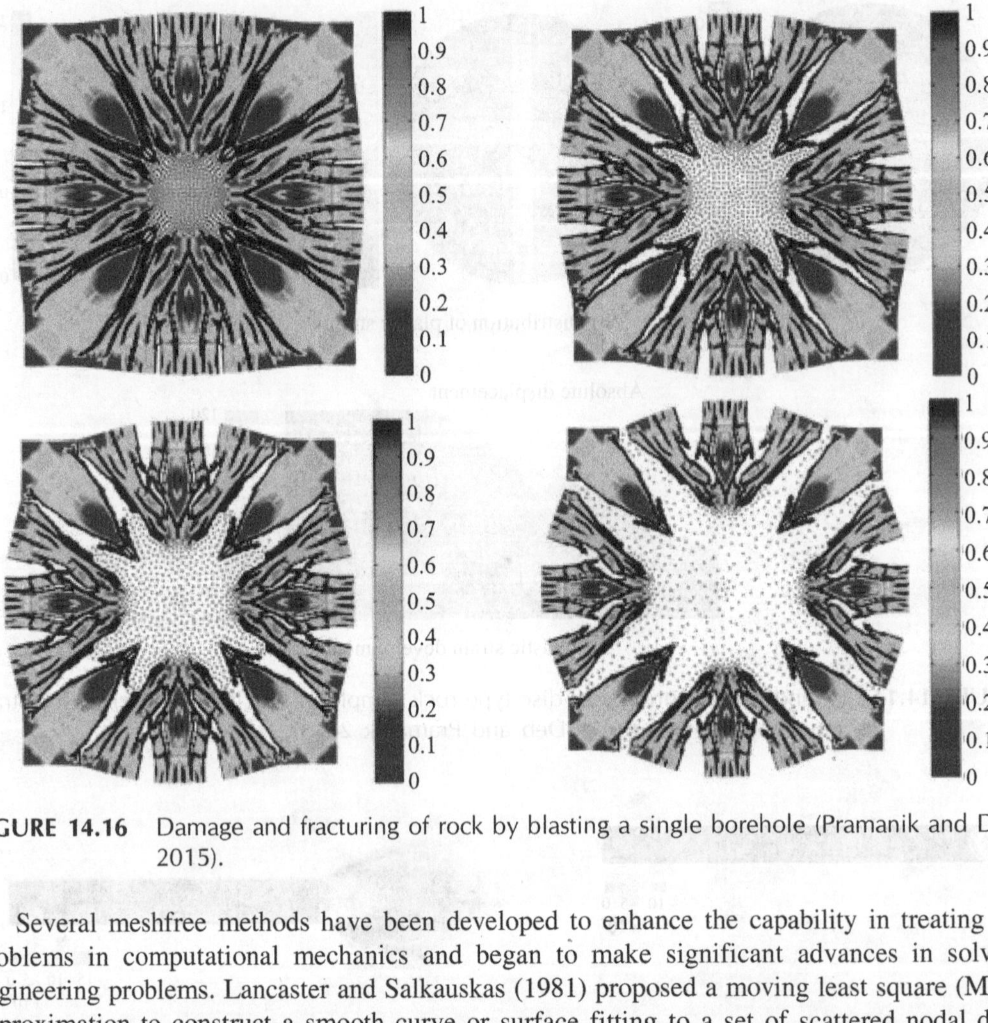

FIGURE 14.16 Damage and fracturing of rock by blasting a single borehole (Pramanik and Deb, 2015).

Several meshfree methods have been developed to enhance the capability in treating the problems in computational mechanics and began to make significant advances in solving engineering problems. Lancaster and Salkauskas (1981) proposed a moving least square (MLS) approximation to construct a smooth curve or surface fitting to a set of scattered nodal data. Based on this idea, Belytschko et al., (1994) developed a new method in which the MLS approximation was incorporated into a Galerkin weak form for solving solid mechanics problem. They named this method the element free Galerkin method (EFGM) and it is this method that is currently the most widely used meshfree method. Based on local Petrov-Galerkin weak forms and MLS approximation, a family of meshfree local Petrov-Galerkin methods (MLPG) (Atluri et al., 1999) has also been proposed.

14.2.5 Distinct Element Method (DEM)

DEM is one of the most rapidly developing areas of computational mechanics with a broad variety of applications in rock mechanics. DEM is relatively new and many think of it as "not yet proven" numerical technique for analysis and design in rock mechanics. Formulation and development of DEM have progressed over a long period of time since first studied by

Cundall (1988). This method has been extended to other areas of research even though it was originally developed for rock mechanics applications. This method was initially developed for 2D representation of jointed rock mass, but now has been extended to applications in particle flow research (Walton, 1980), studies of granular material (Cundall and Strack, 1983), and crack development in rock and concrete (Plesha and Aifantis, 1983). The universal distinct element code (UDEC) is the most popular commercial code used to perform static as well as dynamic analysis of jointed rock mass.

This method treats the domain of interest as a discontinuum medium rather than a continuum. In DEM, rock mass is discretised by assemblage of rigid or deformable discrete blocks/particles (Figure 14.17). The concept is to formulate the equations of motion at the block level for solution. The contact among the blocks needs to be determine at each time step during the deformation process in order to evaluate corresponding contact forces by means of constitutive models. The elements interact with one another through the forces developed at contact points. The equations of equilibrium are repeatedly solved until the laws of contacts and boundary conditions are satisfied. Generally, there are two formulations in DEM, namely explicit and implicit formulations.

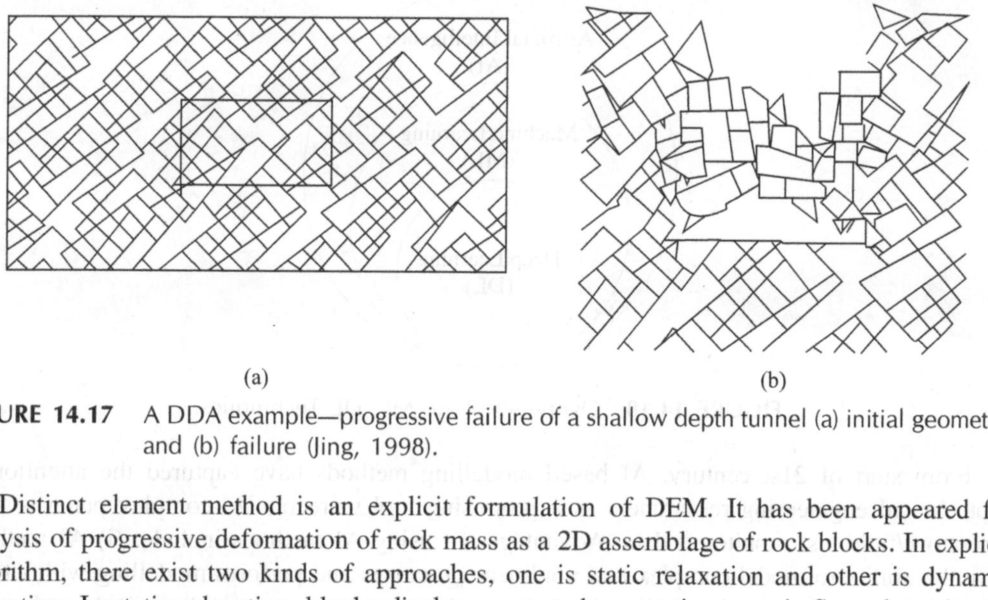

(a) (b)

FIGURE 14.17 A DDA example—progressive failure of a shallow depth tunnel (a) initial geometry, and (b) failure (Jing, 1998).

Distinct element method is an explicit formulation of DEM. It has been appeared for analysis of progressive deformation of rock mass as a 2D assemblage of rock blocks. In explicit algorithm, there exist two kinds of approaches, one is static relaxation and other is dynamic relaxation. In static relaxation, blocks displacement at the next time step is figured out by the equilibrium equations (Williams and Mustoe, 1987). Dynamic relaxation uses Newton's second law to determine the displacement of blocks at the next time steps (Cundall, 1988). Explicit DEM can be used to represent rock mass as a compact packing of irregularised particles interacting at their contact.

As the DEM implicit formulation, the discontinuous deformation analysis (DDA) is used to analyse mechanical behaviour of blocky systems. The DDA has some similarity with FEM, but accounts for the interaction of independent blocks along discontinuities in fractured and jointed rock masses. The DDA framework is typically formulated as a work-energy method.

It is also possible to derive using the principle of minimum potential energy or Hamilton's principle. The applications of the DDA are mainly on tunnelling, caverns, earthquake effects, and fracturing and fragmentation process of geological and structural materials. The DDA approach includes discretising the blocks with finite elements and handling the contacts as closed joints. The method does not include penetration criteria to improve the efficiency and to accelerate the convergence.

14.3 APPLICATIONS OF AI/ML IN ROCK ENGINEERING

Machine Learning techniques are subset of AI methods and mostly described by algorithms that learn patterns from huge data sets. If sufficient datasets are used for the learning process, it can then predict future outcomes, recognize unknown patterns or suggest different classes to the data. Deep learning is a subset of AI and ML that uses neural networks to learn from data or images. There may be overlap of techniques between AI/ML/DL. Figure 14.18 describes the differences pictorially.

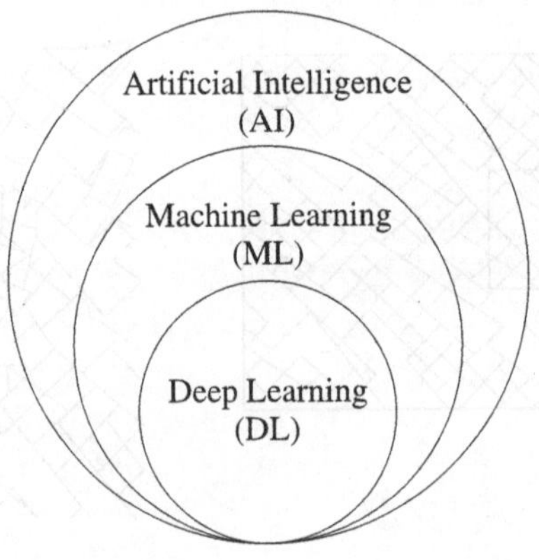

FIGURE 14.18 Overview of AI, ML, DL Techniques.

From start of 21st century, AI based modelling methods have captured the attention of geotechnical engineering researchers as a promising tool to recognize complex patterns from laboratory/field test monitored data. Advantages of using AI methods include: (i) AI methods have the ability to model complex and nonlinear processes and perform modelling without any initial assumptions about the relationship between input and output variables; (ii) AI techniques can be well used to predict, monitor, select, detect and identify in a variety of contexts; (iii) AI techniques can accurately predict even when physical relationships between parameters are unknown (Bagbhani, et al., 2022).

Some key topics that are relevant to the successful application of AI methods are useful for future research are still elusive. For example, the number of data sets needed for AI model training, is an important factor that affect the accuracy of the model. For any AI modelling, selection of input parameters is critical and can affect the performance of AI methods.

Brief Overview of Artificial Intelligence (AI) Techniques and Algorithms

AI methods most commonly used in geotechnical engineering research include Artificial Neural Network (ANN), Fuzzy Inference System (FIS), Adaptive Neuro-Fuzzy Inference System (ANFIS), Geotechnical Algorithm (GA), Particle Swarm Optimization (PSO), Support Vector Machine (SVM), and Deep learning.

14.3.1 Artificial Neural Network

The artificial neural network (ANN) is an intelligent data-modelling technique primarily applied for modelling complex non-linear relationships between parameters, without any prior assumptions or established mathematical relationships. The development of ANN takes inspiration from the structural, functional, and computational aspect of a Biological Neural Networks (BNNs). ANN replicates the parallelism of BNNs using a large number of operational non-linear computational elements, called artificial neurons (Haykin, 1999). This facilitates high computational rates produced by ANN and incorporates the variabilities and fluctuations related to the data sets. ANN methods developed in 1980s.

An artificial neural network consists of artificial neurons that receive and process a "signal" and then send it to neurons connected to it. The "signal" in an artificial neural network is a real number. The output of neurons is calculated based on nonlinear functions and the values of the inputs. Neurons have weights that are regulated by network learning. Each neuron with n inputs calculates its output a as:

$$a = f\left(\sum_{i=1}^{n} w_i p_i + b\right) \tag{14.1}$$

where, p_i and w_i are the values of ith input and weight, b is the bias, and f is the activation function of the neuron. Different types of activation functions such as, steep function, linear function, and sigmoid function can be used in ANNs.

The tasks performed by an ANN can be broadly classified into two categories: (i) pattern recognition/classification, and (ii) function approximation. The pattern recognition/classification based ANNs are trained recognise patterns in data or classify them into pre-defined categories. Function approximation ANNs are trained to learn and generalise the non-linear complex relationship between input and output parameters, enabling prediction of output parameters for a given unseen input condition.

Figure 14.19 shows the architecture of a ANN, in which the input parameters are uniaxial compressive strength, (σ_c) and minor principal stress, (σ_3) to predict the major principal stress (σ_1). Hence in the above Eq. 14.1, for two inputs $p_1 = \sigma_c$, and $p_2 = \sigma_3$; the output parameter $a = \sigma_1$ are used. In this study, Rafiai et al. (2013) used results of 254 polyaxial tests results on six different rock types, and for each rock type, the data are randomly divided into two subsets of training data and test data sets. Following the optimization of architecture and configuration of ANNs, the training data are used for training the networks to predict the value of major principal stress at failure from the values of minor and intermediate principal stresses. The same training data are also used to obtain the parameters of two failure criteria. Then, the test data are used to compare the accuracy of the different failure criteria. It is observed that in all cases, the ANN-based failure criterion leads to more accurate results, and the average value of root mean squared error for this criterion is about 50 and 20% less than those criteria used in this study.

The main disadvantage of using ANNs are that they do not show the relations between the input and output parameters.

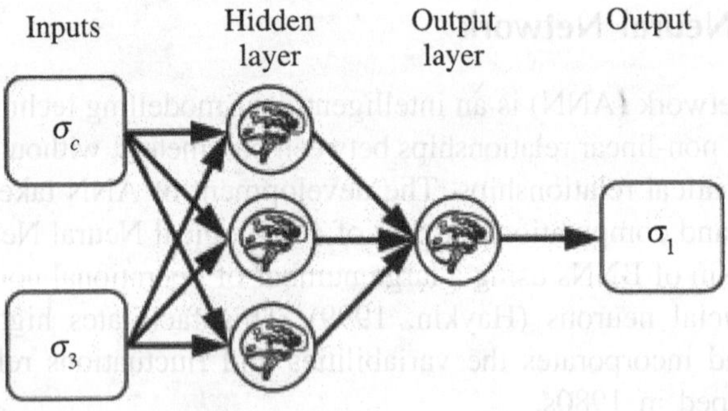

FIGURE 14.19 Architecture of ANNs used for prediction of major principal stress at failure (Rafiai et al., 2013).

14.3.2 Fuzzy Logic (FL) and Fuzzy Inference System (FIS)

Fuzzy logic can mimic human decision-making more quickly and can be used with neural networks. The Fuzzy Logic (FL) (Aliev and Tarassov, 2020) is a multivalued logic and is intended to facilitate defining of intermediate values between strict binary possibilities (Jouffe, 1998). Example, FL allows mathematical formulations of vague expressions like "very good". Such method of reasoning enables development of intelligent models that can simulate human reasoning and decision-making process.

FL architecture consists of four main parts: (i) fuzzification module, (ii) knowledge base (rules), (iii) inference engine, and (iv) defuzzification module. Rules include all the conditions that the user and the expert define for the system to control the decision system. In the fuzzification stage, inputs, including crisp numbers, are converted to fuzzy sets. In this stage, various functions, including sigmoid, triangle, gaussian and trapezium, are used. Then, the inference engine examines the degree of match between fuzzy inputs and rules to control measures. Finally, in the defuzzification step, the resulting fuzzy sets are converted to a crisp value. In this step, different de-fuzzifiers, including the center of gravity (COG) and middle of maxima (MoM), are used.

For example, Deb (2004) discussed the application of FL in rock mass classification to develop the membership grade of Coal Mine Roof Rating (for more details see Section 3.3.6 of Chapter 3) and shown in graphical form as shown in Figure 14.20.

The Fuzzy Inference System (FIS) uses Fuzzy Logic (FL) to map input to an output. FIS assigns output vectors through interpretation of input vectors based on pre-defined set of rules. One of the important advantages of FIS is that multivariate and complex problems as well as curves can be reasonably described in it.

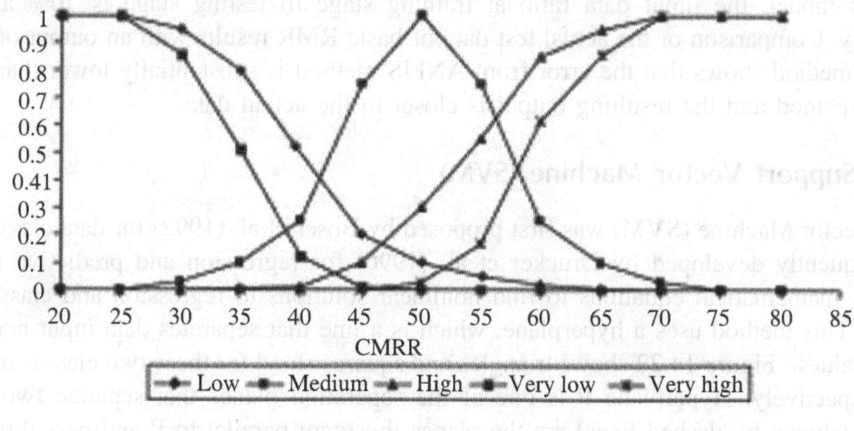

FIGURE 14.20 Fuzzy membership grades of coal mine roof rating (CMRR) for different linguistic hedges (Deb, 2003).

14.3.3 Adaptive Neuro-Fuzzy Inference System (ANFIS)

The Adaptive Neuro-Fuzzy Inference System (ANFIS) was introduced by Jang et al. (1997) to overcome difficult conditions about ANNs and FIS. ANFIS is a combination of the linguistic representation of FL and the generalization ability of ANN (Coley, 1999). In other words, ANFIS is an appropriate system to represent better and learn and describe the membership functions of the input data for the system (Rad et al., 2015).

For example, in rock mass rating (RMR) classification, the input parameters are divided into several classes, and each class has particular rating. In this classification system, because of uncertainties of the input parameters, determining the definite boundary between the classes and assigning a specified value to a particular class is difficult, so when the input parameters are close to the boundary between the classes, the class rating with certainty is not decided.

Figure 14.21 shows the ANFIS architecture on five parameters uniaxial compressive strength (UCS), rock quality designation (RQD), joint spacing (JS), joint condition (JC) and ground water (GW), the only output parameters of the basic RMR system.

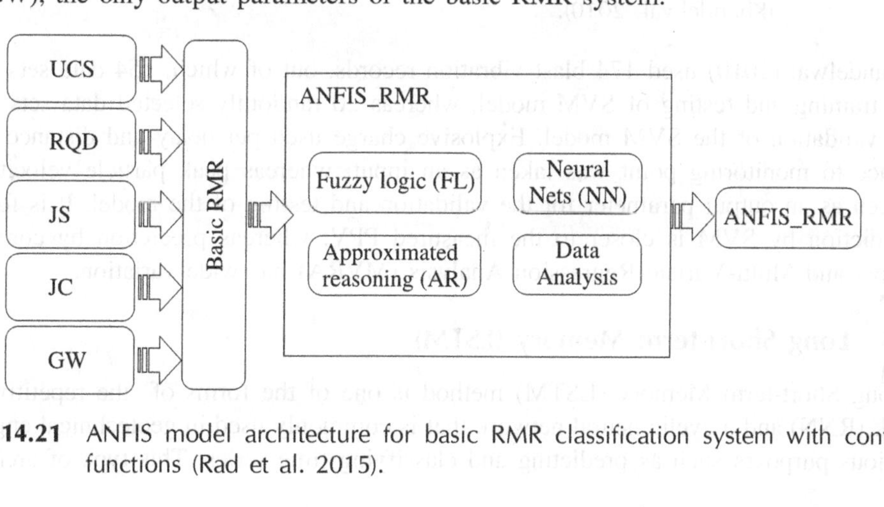

FIGURE 14.21 ANFIS model architecture for basic RMR classification system with continuous functions (Rad et al. 2015).

In this model, the input data ratio at training stage to testing stage is 70% and 30% respectively. Comparison of the actual test data of basic RMR results with an output of ANFIS and ANN method shows that the error from ANFIS method is substantially lower than that of the ANN method and the resulting output is closer to the actual data.

14.3.4 Support Vector Machine (SVM)

Support Vector Machine (SVM) was first proposed by Boser et al. (1992) for data classification and subsequently developed by Drucker et al. (1996) for regression and predictive analysis. SVM uses mathematical equations to find nonlinear solutions to regression and classification problems. This method uses a hyperplane, which is a line that separates data input nodes with different values. Figure 14.22 shows triangles and squares stand for these two classes of sample points, respectively. Hyperplane P is one of the separation planes that separate two classes. P_1 and P_2 (shown by dashed lines) are the planes those are parallel to P and pass through the sample points closest to P in these two classes. Margin is the distance between P_1 and P_2. The SVM tries to place a linear boundary between the two different classes, and orientate it in such way that the margin is maximized, which results in least generalization error. The nearest data points that used to define the margin are called support vectors.

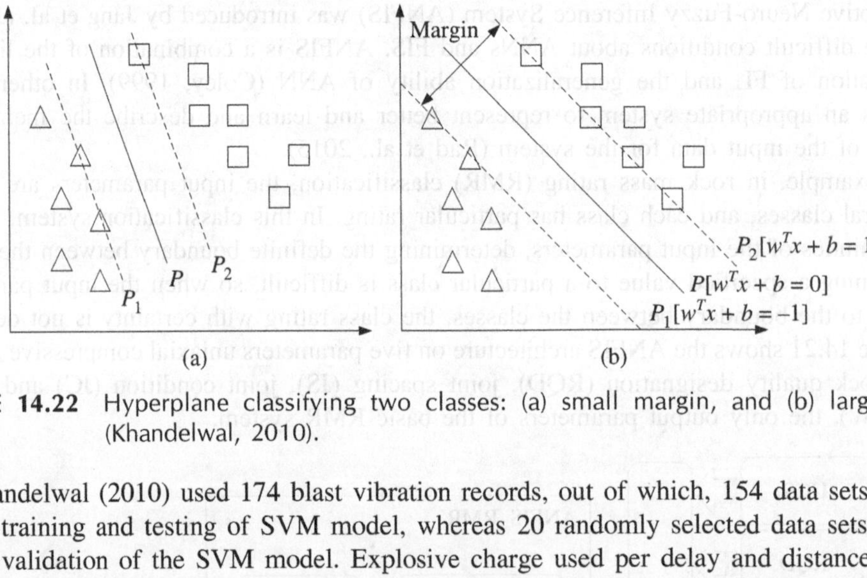

FIGURE 14.22 Hyperplane classifying two classes: (a) small margin, and (b) large margin (Khandelwal, 2010).

Khandelwal (2010) used 174 blast vibration records, out of which, 154 data sets are used for the training and testing of SVM model, whereas 20 randomly selected data sets are used for the validation of the SVM model. Explosive charge used per delay and distance between blast face to monitoring point was taken as an input, whereas peak particle velocity (PPV) was taken as an output parameter for the validation and testing of the model. It is found that the prediction by SVM is closer to the measured PPV, whereas prediction by conventional predictors and Multi-Variate Regression Analysis (MVRA) has wide variation.

14.3.5 Long Short-term Memory (LSTM)

The Long Short-term Memory (LSTM) method is one of the forms of the repetitive neural network (RNN) and a cyclic neural network that is commonly used in geotechnical engineering for various purposes such as predicting and classifying time series. This type of architecture,

LSTM, was introduced by Hochreiter and Schmidhuber (1997) as a new type of deep machine learning neural network with long-term and short-term memory capability. General recurrent neural networks have the problem of gradient explosion or gradient disappearance and only have the ability to learn well in information at short distances (Sahoo et al., 2019). In LSTM, this defect has been solved. LSTM consists of 4 main parts: memory cell and three types of gates, including input, forget and output. The data is received by the memory cell at separate times and three gateways is used to control and transmit data (Sundermeyer et al., 2012).

The structure of LSTM network is shown in Figure 14.23. A neural unit cell is mainly composed of four elements: forgetting gate f_t, input gate i_t, the state unit $\tilde{c}_t$ and output gate o_t. Input gate and forgetting gate are designed for controlling the state of memory cell. The output gate is used for controlling how much information is output for cell status (Zhang et al., 2021).

Zhang, et al. (2021) developed LSTM neural network with Synthetic Minority Oversampling Technique (SMOTE) and First difference transformation (FDT) for forecasting the frequency shift value of sensing fiber on distributed optical fiber sensor monitoring of overburden rock of a coal mine to predict the deformation monitored by optical fiber sensor. Accurately predicting the frequency shift value of fiber sensor is used to infer the deformation state of the rock. Ning, et al. (2021) used LSTM approach for modelling stress-strain behaviour of soil. The LSTM method includes three steps: data preparation, architecture determination, and optimisation. It is found that the results from LSTM deep-learning method outperforms the feed forward and feedback neural networks based on both accuracy and the convergence rate while reproducing the soil's stress–strain behaviour.

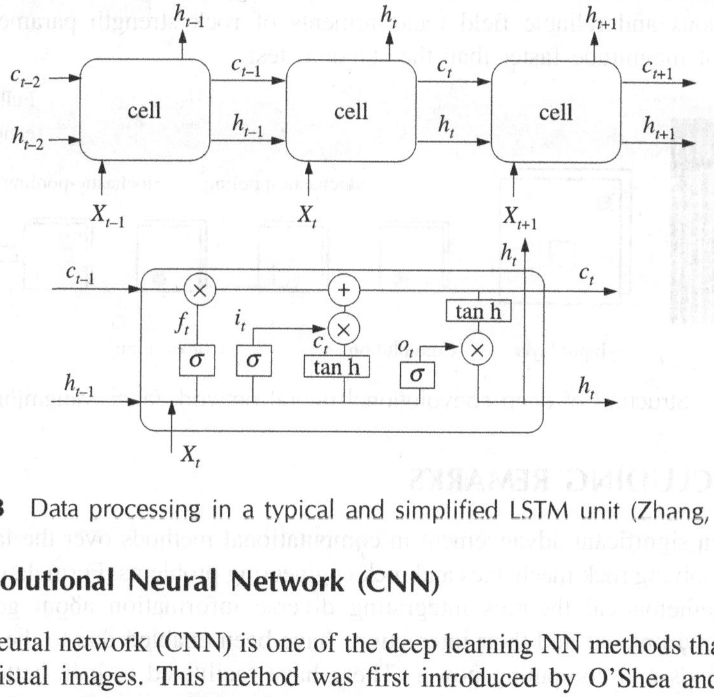

FIGURE 14.23 Data processing in a typical and simplified LSTM unit (Zhang, et al., 2021)

14.3.6 Convolutional Neural Network (CNN)

Convolutional neural network (CNN) is one of the deep learning NN methods that is commonly used to study visual images. This method was first introduced by O'Shea and Nash (2015). CNN, as a multilayer perceptron method, is a method that consists of networks of neurons, each neuron in each layer connected to all the neurons in the next layer. This type of connection is called a "full connectivity". CNN consists of a mathematical structure that has three types of layers: convolution, integration, and fully connected layers. Feature extraction is done by convolution and integration layers. A fully connected layer is responsible for mapping the extracted properties to the final output. One of the positive points of this method is its relatively low pre-processing, which causes high independence of the network from previous knowledge and human intervention in feature extraction. Generally, this method has different architectures, of which 4 important architectures include AlexNet, GoogLeNet, VGGNet, ResNet (Zhang et al., 2021a; Zhang et al., 2021c). ResNet method has been studied as a category due to the greater use by researchers in geotechnical engineering, whereas rest three CNN methods are considered as CNN.

The CNN consists of a convolution layer, stochastic pooling layer, and softmax loss, as shown in Figure 14.24. Mingming et al., (2019) proposed CNN model framework includes: (1) collecting drilling performance parameters of intact rock, (2) establishing a strength parameter database via standard laboratory tests, (3) training the proposed CNN model using images created from the data set of drilling performance parameters and rock strength parameters (unconfined compressive strength (UCS), cohesion, and internal friction angle), and (4) performing a new estimation to evaluate the reliability of the trained CNN model. The experimental results show that the predicted rock strength parameters are within the accepted error range of 10% compared to the results from the conventional standard test. The proposed CNN shows superior performance for UCS estimation of various rock types and achieves higher accuracy than the Mohr–Coulomb criterion. In particular, the proposed method can overcome major limitations of analytical models based on the force limit and energy equilibrium method. This method enables continuous and reliable field measurements of rock strength parameters at a speed several orders of magnitude faster than the standard test.

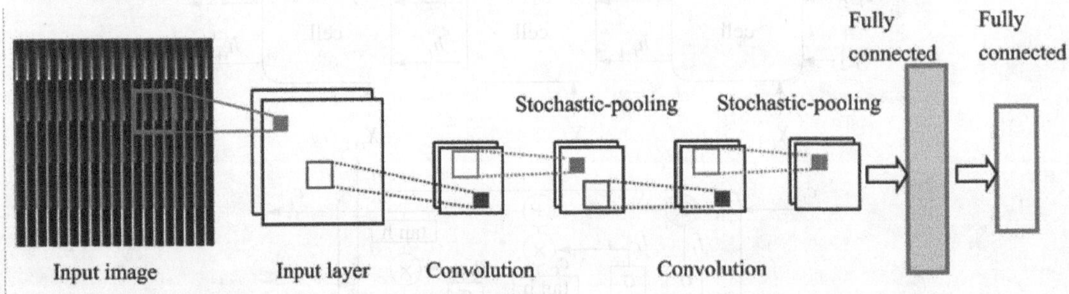

FIGURE 14.24 Structure of deep convolutional neural network (after Mingming, et al., 2019).

14.4 CONCLUDING REMARKS

There have been significant advancement in computational methods over the last five decades, specifically for solving rock mechanics and rock engineering problems. Formulation of conceptual models and mathematical theories integrating diverse information about geology, physics, economy, the environment and their interactions have been enabled due to the advancement of numerical methods and computing power. These have facilitated to gain better understanding of rock mass behaviour from the traditional empirical art of rock deformability and strength estimation, and support design to the rationalism of modern mechanics.

Applications of AI/ML techniques are fast increasing in the rock mechanics and rock engineering domain. These days, sophisticated instruments are installed to measure strata behaviour almost in real time. They collect huge sets of data in every hour. Analysis of these data sets require advanced AI/ML tools for understanding the patterns of displacements, stresses and loads that are being changed over the time or with excavation dimension.

Due to the inherent nature of rock mass containing discontinuity, fractures and inhomogeneity, numerical modelling sometimes becomes more challenging. Success of numerical modelling for rock mechanics can entirely depend upon the quality of the characterisation of the fracture system geometry, physical behaviour of the individual fractures and their interaction. Today's numerical modelling capability can handle very large scale and complex equation systems, but still there are limitations in the quantitative representation of the physics of fractured rocks.

It is very difficult to obtain very close numerical results comparing with experimental results due to the assumptions in mathematical formulations and inherent defects of naturally fractured rock mass. However, numerical models can be calibrated against laboratory and insitu experiments and the output of the results used to successfully analyse practical problems. This needs a combined scientific and engineering support for applying numerical methods to rock mechanics and rock engineering.

EXERCISES

1. Discuss the terminologies used in modelling:
   (a) physical (b) empirical (c) analytical (d) numerical
2. Distinguish between Continuum and Discontinuum methods applied to Rock Engineering.
3. What are the basic differences among FEM, FDM and BEM?
4. Differentiate between the discrete or distinct element method (DEM) and displacement discontinuity method (DDM).
5. A student would like to analyse the stability of a developed pillars in a Bord and Pillar mine. What numerical methods do you suggest and why? Draw the model boundaries in 2D considering typical coal mine rock strata and show the boundary conditions.
6. Define element, node and nodal shape function in the finite element method.
7. What is (i) Interior and exterior problems, and (ii) Explicit and implicit methods.
8. Define (i) Equivalent material modelling, (ii) Extended finite element method (XFEM), (iii) Smoothed partial hydrodynamics (SPH) method.
9. What types of data are required for analysing the performance of rock blasting with Machine Learning methods?
10. Discuss on the importance of AI/ML methods applied to rock engineering problems. Can these methods replace numerical methods?
11. If a student wants to analyse dump slope stability, what AI method do you suggest and why?
12. The performance of rock bolts is to be analysed using AI methods. Suggest a suitable AI method and describe the required data for the same.

Appendix A

Transformation of Displacements, Strain and Stress Tensors in 3D

Transformation or rotation matrices in 3D are straight forward extension of 2D. Illustration of 3D rotation matrices are given here based on the reference (Cook et al., 1989). Let us consider a displacement vector **OA** in 3D space represented by both $x - y - z$ and $x' - y' - z'$ coordinate systems as shown in the Figure A.1. Direction cosines are defined below:

$l_1 = \cos(x', x)$: cosine of angle between the axes x' and x.
$m_1 = \cos(x', y)$: cosine of angle between the axes x' and y.
$n_1 = \cos(x', z)$: cosine of angle between the axes x' and z.

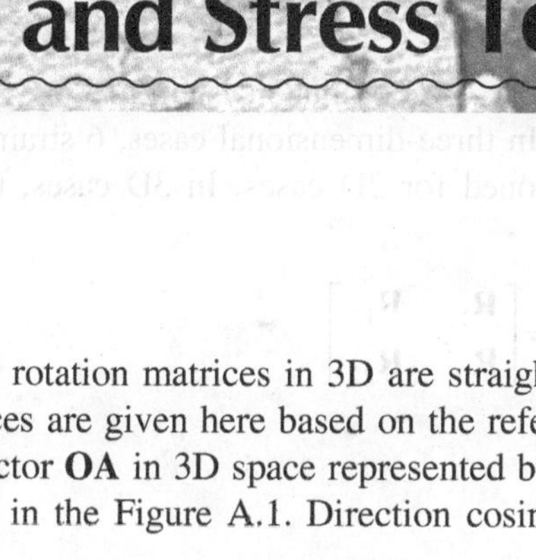

Direction cosines

	x	y	z
x'	l_1	m_1	n_1
y'	l_2	m_2	n_2
z'	l_3	m_3	n_3

FIGURE A.1 Transformation of displacement vector in 3D.

Similarly direction cosines of y' and z' axes with x, y and z axes can also defined as shown in Figure A.1. However, all 9 direction cosines are not independent to each other. It can be easily shown that there are relationships between the direction cosines as given in Eq. A.1.

$$l_1^2 + m_1^2 + n_1^2 = 1,\; l_2^2 + m_2^2 + n_2^2 = 1,\; l_3^2 + m_3^2 + n_3^2 = 1 \tag{A.1a}$$

$$l_1 l_2 + m_1 m_2 + n_1 n_2 = 0,\; l_2 l_3 + m_2 m_3 + n_2 n_3 = 0,\; l_3 l_1 + m_3 m_1 + n_3 n_1 = 0 \tag{A.1b}$$

Using similar approach as shown in Eq. 4.18, a three-dimensional displacement vector expressed in $x - y - z$ coordinate system can be transformed into a displacement vector in $x' - y' - z'$ coordinate system by transformation matrix **R** as given below:

$$\begin{Bmatrix} u' \\ v' \\ w' \end{Bmatrix} = \underbrace{\begin{bmatrix} l_1 & m_1 & n_1 \\ l_2 & m_2 & n_2 \\ l_3 & m_3 & n_2 \end{bmatrix}}_{\mathbf{R}} \begin{Bmatrix} u \\ v \\ w \end{Bmatrix} \tag{A.2}$$

Strain tensors are transformed from $x - y - z$ coordinate system to $x' - y' - z'$ system based on the concepts outlined in Section 4.3.3. In three-dimensional cases, 6 strain tensors are to be transformed instead of 3 tensors as mentioned for 2D cases. In 3D cases, the transformation matrix **R** is in the form of

$$\mathbf{R} = \begin{bmatrix} \mathbf{R}_{11} & \mathbf{R}_{12} \\ \mathbf{R}_{21} & \mathbf{R}_{22} \end{bmatrix} \tag{A.3}$$

where,

$$\mathbf{R}_{11} = \begin{bmatrix} l_1^2 & m_1^2 & n_1^2 \\ l_2^2 & m_2^2 & n_2^2 \\ l_3^2 & m_3^2 & n_3^2 \end{bmatrix}$$

$$\mathbf{R}_{12} = \begin{bmatrix} l_1 m_1 & m_1 n_1 & n_1 l_1 \\ l_2 m_2 & m_2 n_2 & n_2 l_2 \\ l_3 m_3 & m_3 n_3 & n_3 l_3 \end{bmatrix}$$

$$\mathbf{R}_{21} = \begin{bmatrix} 2l_1 l_2 & 2m_1 m_2 & 2n_1 n_2 \\ 2l_2 l_3 & 2m_2 m_3 & 2n_2 n_3 \\ 2l_3 l_1 & 2m_3 m_1 & 2n_3 n_1 \end{bmatrix}$$

$$\mathbf{R}_{22} = \begin{bmatrix} l_1 m_2 + l_2 m_1 & m_1 n_2 + m_2 n_1 & n_1 l_2 + n_2 l_1 \\ l_2 m_3 + l_3 m_2 & m_2 n_3 + m_3 n_2 & n_2 l_3 + n_3 l_2 \\ l_3 m_1 + l_1 m_3 & m_3 n_1 + m_1 n_3 & n_3 l_1 + n_1 l_3 \end{bmatrix}$$

The components of strain tensor are in order $\varepsilon^T = \{\varepsilon_{xx}\ \varepsilon_{yy}\ \varepsilon_{zz}\ \gamma_{xy}\ \gamma_{yz}\ \gamma_{xz}\}$ and must be followed while transforming from $x - y - z$ to $x' - y' - z'$ coordinate system.

The procedure for transforming stress tensors is also straight forward extension of two dimensional transformation matrix **T**. As before, the matrix **T** is expressed as:

$$\mathbf{T} = (\mathbf{R}^T)^{-1} \tag{A.4}$$

where the matrix **R** is given and thus, the matrix **T** can be obtained as:

$$\mathbf{T} = \begin{bmatrix} \mathbf{R}_{11} & 2\mathbf{R}_{12} \\ \frac{1}{2}\mathbf{R}_{21} & \mathbf{R}_{22} \end{bmatrix} \tag{A.5}$$

Appendix B

Closed Form Solution of Kirsch Equation

Stress functions around a circular tunnel having biaxial far fieldstressesof p_o and kp_o (Figure 7.1b) are solved by imposing suitable boundary conditions. Without body force, boundary conditions are:

(i) radial and shear stresses at the tunnel boundary is zero or mathematically,

$$\sigma_{rr} = \tau_{r\theta} = 0 \text{ at } r = a \tag{B.1}$$

(ii) radial and tangential stresses at the outer boundaries are estimated using stress transformation. Mathematically, in polar coordinates these boundary conditions at $r \to \infty$ are

$$\sigma_{rr} = \frac{p_o}{2}(1+k) - \frac{p_o}{2}(1-k)\cos 2\theta \quad \text{at } r = \infty \tag{B.2a}$$

$$\tau_{r\theta} = \frac{p_o}{2}(1-k)\sin 2\theta \quad \text{at } r = \infty \tag{B.2b}$$

Solution of Airy stress function (Eq. 4.79) can be decomposed into 2 parts. For the first case, axisymmetric condition of the problem is considered and secondly, 2θ term in equation B.2 is equated. In the following, solution procedures of both the cases are summarised. Note that the derived solution of Airy stress function in polar coordinates is restricted to the periodic case and is reproduced from the reference (Sadd, 2006) as given below:

$$\begin{aligned}\Phi = {} & a_0 + a_1 \log r + a_2 r^2 + a_3 r^2 \log r \\ & + (a_4 + a_5 \log r + a_6 r^2 + a_7 r^2 \log r)\,\theta \\ & + \left(a_{11} r + a_{12} r \log r + \frac{a_{13}}{r} + a_{14} r^3 + a_{15} r\theta + a_{16} r\theta \log r\right)\cos\theta \\ & + \left(b_{11} r + b_{12} r \log r + \frac{b_{13}}{r} + b_{14} r^3 + b_{15} r\theta + b_{16} r\theta \log r\right)\sin\theta\end{aligned} \tag{B.3}$$

$$+\sum_{n=2}^{\infty}(a_{n1}r^{n}+a_{n2}r^{2+n}+a_{n3}r^{-n}+a_{n4}r^{2-n})\cos n\theta$$

$$+\sum_{n=2}^{\infty}(b_{n1}r^{n}+b_{n2}r^{2+n}+b_{n3}r^{-n}+b_{n4}r^{2-n})\sin n\theta$$

where a_n, a_{nm} and b_{nm} are constants to be determined from the given problem and boundary conditions.

Case 1: Since the problem is axially symmetric about z-axis, the Airy stress function can be considered a function of r only. Equation B.3 reduces to:

$$\Phi(r)=a_0+a_1\log r+a_2r^2+a_3r^2\log r \tag{B.4}$$

The above equation can be further reduced by considering the radial and tangential stress functions mentioned in Eq. 4.79. It can be easily seen that $\log r$ term must be present in the equation to determine the variability of stresses in terms of radial distance. Since a constant term also exists in the boundary condition of radial stress (Eq. B.2a), r^2 term must also be present. Hence, the above equation can be reduced to:

$$\Phi(r)=a_1\log r+a_2r^2 \tag{B.5}$$

Case 2: Again from Eq. 4.79 and boundary conditions (Eq. B.2), it can be ascertained that all terms multiplied with cos 2θ must be present in the Airy stress function. Reader may notice that sin 2θ term in the shear stress function can be equated by taking the first derivative of Airy stress function with respect to θ. Hence, from Eq. B.3, all terms associated with cos 2θ is considered as given below.

$$\Phi(r,\theta)=(a_{21}r^2+a_{22}r^4+a_{23}r^{-2}+a_{24})\cos 2\theta \tag{B.6}$$

In infinite rock medium, a circular tunnel that is axially symmetric and satisfies far field stress conditions, the general solution of Airy stress function is the combination of Case 1 and Case 2, and it becomes

$$\Phi(r,\theta)=a_1\log r+a_2r^2+(a_{21}r^2+a_{22}r^4+a_{23}r^{-2}+a_{24})\cos 2\theta \tag{B.7}$$

The corresponding stress components, from Eq. 4.79, are:

$$\sigma_{rr}=\frac{1}{r}\frac{\partial\Phi}{\partial r}+\frac{1}{r}\frac{\partial^2\Phi}{\partial\theta^2}=\frac{a_1}{r^2}+2a_2+\left(-2a_{21}-\frac{6a_{23}}{r^4}-\frac{4a_{24}}{r^2}\right)\cos 2\theta \tag{B.8a}$$

$$\sigma_{\theta\theta}=\frac{\partial^2\Phi}{\partial r^2}=-\frac{a_1}{r^2}+2a_2+\left(2a_{21}+12a_{21}r^2-\frac{6a_{23}}{r^4}\right)\cos 2\theta \tag{B.8b}$$

$$\tau_{r\theta}=-\left(\frac{1}{r}\frac{\partial\Phi}{\partial\theta}\right)=\left(2a_{21}+6a_{22}-\frac{6a_{23}}{r^4}-\frac{2a_{24}}{r^2}\right)\sin 2\theta \tag{B.8c}$$

The constants of integration are now determined from the outer and inner boundary conditions as given above. These conditions give:

$$a_1 = -\frac{1}{2} p_0(1+k)a^2, \; a_2 = \frac{1}{4} p_0(1+k) \tag{B.9a}$$

$$a_{21} = \frac{1}{4} p_0(1-k), \; a_{22} = 0, \; a_{23} = \frac{1}{4} p_0(1-k)a^4, \; a_{24} = -\frac{1}{2} p_0(1-k)a^2 \tag{B.9b}$$

After rearranging the above equations, the stress functions around a circular tunnel excavated in infinite rock medium are obtained. These functions were first derived by Kirsch in 1898 (Brady and Brown, 2007). For the far field stresses of p_o and kp_o, the solutions of tangential ($\sigma_{\theta\theta}$), radial (σ_{rr}) and shear stress ($\tau_{r\theta}$) are reproduced in Eqs. 7.1(a–c) for tunnel radius of a.

$$\sigma_{rr} = \frac{p_o}{2}\left[(1+k)\left(1-\frac{a^2}{r^2}\right) - (1-k)\left(1-4\frac{a^2}{r^2}+\frac{3a^4}{r^4}\right)\cos 2\theta\right] \tag{B.10a}$$

$$\sigma_{\theta\theta} = \frac{p_o}{2}\left[(1+k)\left(1+\frac{a^2}{r^2}\right) + (1-k)\left(1+\frac{3a^4}{r^4}\right)\cos 2\theta\right] \tag{B.10b}$$

$$\tau_{r\theta} = \frac{p_o}{2}\left[(1-k)\left(1+\frac{2a^2}{r^2}-\frac{3a^4}{r^4}\right)\sin 2\theta\right] \tag{B.10c}$$

Displacement for Circular Hole

Strains are given using stresses as

$$\varepsilon_{rr} = \frac{\partial u_r}{\partial r} = \frac{1}{E}(\sigma_{rr} - \nu\sigma_{\theta\theta}); \; \varepsilon_{\theta\theta} = \frac{u_r}{r} + \frac{1}{r}\frac{\partial u_\theta}{\partial \theta} = \frac{1}{E}(\sigma_{\theta\theta} - \nu\sigma_{rr});$$
$$\gamma_{r\theta} = \frac{1}{r}\frac{\partial u_r}{\partial \theta} + \frac{\partial u_\theta}{\partial r} - \frac{u_\theta}{r} = \frac{2(1+\nu)}{E}\tau_{r\theta} \tag{B.11}$$

Replacing value of σ_{rr}, $\sigma_{\theta\theta}$ and $\tau_{r\theta}$ from Eq. B.10 into Eq. B.11, we get

$$\frac{\partial u_r}{\partial r} = \frac{p_0}{2E}\left[\begin{array}{l}(1+k)\left(1-\frac{a^2}{r^2}\right) - (1-k)\left(1+\frac{3a^4}{r^4}-\frac{4a^2}{r^2}\right)\cos 2\theta \\ -\nu\left((1+k)\left(1+\frac{a^2}{r^2}\right) + (1-k)\left(1+\frac{3a^4}{r^4}\right)\cos 2\theta\right)\end{array}\right] \tag{B.12a}$$

$$\frac{u_r}{r} + \frac{1}{r}\frac{\partial u_\theta}{\partial \theta} = \frac{p_0}{2E}\left[\begin{array}{l}(1+k)\left(1+\frac{a^2}{r^2}\right) + (1-k)\left(1+\frac{3a^4}{r^4}\right)\cos 2\theta \\ -\nu\left((1+k)\left(1-\frac{a^2}{r^2}\right) - (1-k)\left(1+\frac{3a^4}{r^4}-\frac{4a^2}{r^2}\right)\cos 2\theta\right)\end{array}\right] \tag{B.12b}$$

$$\frac{1}{r}\frac{\partial u_r}{\partial \theta}+\frac{\partial u_\theta}{\partial r}-\frac{u_\theta}{r}=\frac{p_0(1+\nu)}{E}(1-k)\left(1-\frac{3a^4}{r^4}+\frac{2a^2}{r^2}\right)\sin 2\theta \tag{B.12c}$$

Now, integrating Eq. B.12(a),

$$u_r=\frac{p_0}{2E}\left[\begin{array}{l}(1+k)\left(r+\dfrac{a^2}{r}\right)-(1-k)\left(r-\dfrac{a^4}{r^3}+\dfrac{4a^2}{r}\right)\cos 2\theta \\ -\nu\left((1+k)\left(r-\dfrac{a^2}{r}\right)+(1-k)\left(r-\dfrac{a^4}{r^3}\right)\cos 2\theta\right)\end{array}\right]+g_1(\theta) \tag{B.13}$$

where, $g_1(\theta)$ is a function of θ only. Putting the value of u_r from Eq. B.12(a) into Eq. B.12(b) and integrating it, we get

$$u_\theta=\frac{p_0}{2E}\left[\begin{array}{l}(1-k)\left(r+\dfrac{2a^2}{r}+\dfrac{a^4}{r^3}\right)\sin 2\theta \\ +\nu\left((1+k)\left(r-\dfrac{2a^2}{r}+\dfrac{a^4}{r^3}\right)\sin 2\theta\right)\end{array}\right]-\int g_1(\theta)\,d\theta+g_2(r) \tag{B.14}$$

where $g_2(r)$ is a function of r only. The differentiating Eq. (B.13) w.r.t. θ, and Eq. (B.14) w.r.t. r, we have

$$\frac{\partial u_r}{\partial \theta}=\frac{p_0}{2E}\left[2(1-k)\left(r-\frac{a^4}{r^3}+\frac{4a^2}{r}\right)\sin 2\theta+2\nu(1-k)\left(r-\frac{a^4}{r^3}\right)\sin 2\theta\right]+\frac{dg_1(\theta)}{d\theta} \tag{B.15a}$$

$$\frac{\partial u_\theta}{\partial r}=\frac{p_0}{2E}\left[(1-k)\left(1-\frac{2a^2}{r}-\frac{3a^4}{r^4}\right)\sin 2\theta+\nu(1-k)\left(1+\frac{2a^2}{r^2}-\frac{3a^4}{r^4}\right)\sin 2\theta\right]+\frac{dg_2(r)}{dr} \tag{B.15b}$$

Equation B.12(c), together with Eqs. (B.15a) and (B.15b) reduce to

$$\left[\frac{dg_1(\theta)}{d\theta}+\int g_1(\theta)\,d\theta\right]+\left[r\frac{dg_2(r)}{dr}-g_2(r)\right]=0 \tag{B.16}$$

As $g_1(\theta)$ is a function of θ only and $g_2(r)$ is a function of r only, each part of Eq. (B.16) must equal to some constant, thus

$$\left[\frac{dg_1(\theta)}{d\theta}+\int g_1(\theta)\,d\theta\right]=k_1;\quad \left[r\frac{dg_2(r)}{dr}-g_2(r)\right]=k_2 \tag{B.17}$$

The solution of ordinary differential can be

$$g_1(\theta)=A\sin\theta+B\cos\theta;\quad g_2(r)=Cr-k_2 \tag{B.18}$$

In the above either $k_1 + k_2 = 0$ or both $k_1 = 0$ and $k_2 = 0$. Substituting Eq. (B.18) into Eq. (B.13) and Eq. (B.14), we get

$$u_r = \frac{p_0}{2E}\left[\begin{array}{l}(1+k)\left(r+\frac{a^2}{r}\right)-(1-k)\left(r-\frac{a^4}{r^3}+\frac{4a^2}{r}\right)\cos 2\theta \\ -\nu\left((1+k)\left(r-\frac{a^2}{r}\right)+(1-k)\left(r-\frac{a^4}{r^3}\right)\cos 2\theta\right)\end{array}\right] + A\sin\theta + B\cos\theta \quad \text{(B.19a)}$$

$$u_\theta = \frac{p_0}{2E}\left[\begin{array}{l}(1-k)\left(r+\frac{2a^2}{r}+\frac{a^4}{r^3}\right)\sin 2\theta \\ +\nu\left((1-k)\left(r-\frac{2a^2}{r}+\frac{a^4}{r^3}\right)\sin 2\theta\right)\end{array}\right] + A\cos\theta - B\sin\theta + Cr \quad \text{(B.19b)}$$

The boundary condition for the displacements are $u_\theta = 0$ when $\theta = 0$ or $90°$ for all value of r

$$u_{\theta=0} = 0 = A + Cr; \quad u_{\theta=\pi/2} = 0 = -B + Cr$$

Solving these equations for A, B and C gives

$$A = B = C = 0$$

Again considering $\lim_{r\to\infty} u_r = 0$ and $\lim_{r\to\infty} u_\theta = 0$, the r term will be omitted and assuming $1/(1+\nu) \approx 1-\nu$ displacements are given by Eq. (7.2)

$$u_r = \frac{p_0 a^2}{4Gr}\left[(1+k)-(1-k)\left(4(1-\nu)-\frac{a^2}{r^2}\right)\cos 2\theta\right] \quad \text{(B.20a)}$$

$$u_\theta = \frac{p_0 a^2}{4Gr}\left[(1-k)\left(2(1-\nu)+\frac{a^2}{r^2}\right)\sin 2\theta\right] \quad \text{(B.20b)}$$

Appendix C

Plastic Flow

When the material is elastic, increase in stress corresponds to an increase in strain and then completely recover those strains by returning the stress state to its initial value. With plastic deformation (or yield) the strains will not be recovered fully called plastic flow. In order to describe the plastic flow a constitutive relationship was proposed by Huber-Von Mises in 1928. These relationship cannot be unique one-to-one relationship between plastic strain and stress since there may be unknown amount of plastic deformation in the material. Mises proposed that there existed a plastic potential function $Q(\sigma_{ij})$, and the plastic strain increments $d\varepsilon_{ij}^{p}$ can be derived as

$$d\varepsilon_{ij}^{p} = \frac{\partial Q}{\partial \sigma_{ij}} d\eta \tag{C.1}$$

where, $d\eta$ is a proportional positive scalar factor. To determine $d\eta$, the yield function should be used. The plastic flow rule shown in Eq. (C.1) is called plastic potential theory. The plastic potential $Q(\sigma_{ij})$ = constant represents a surface in the six-dimensional stress-space, and the plastic strain $d\varepsilon_{ij}^{p}$ can be represented to a vector which is perpendicular to the surface $Q(\sigma_{ij})$ = constant.

A common approach in plasticity theory is to assume that the plastic potential function $Q(\sigma_{ij})$ is the same as the yield function given in Chapter 9, as Eqns. (9.19), (9.22)–(9.26), and 9.29.

$$Q(\sigma_{ij}) = \Phi(\sigma_{ij}) \tag{C.2}$$

Equation (C.1) can then be rewritten as:

$$d\varepsilon_{ij}^{p} = \frac{\partial \Phi}{\partial \sigma_{ij}} d\eta \tag{C.3}$$

and the plastic flow vector is normal to the yield surface. This is called the *associated flow rule*. On the otherhand, if $Q \neq \Phi$, the flow rule is called *non-associative flow rule*.

Experimental observations shows that the plastic deformation of metals can be characterize quite well by associated flow rule, but for porous materials such as rocks, concrete, and soils, the non-associated flow rule provides a better representation of their plastic deformation (Yu, 2006).

Dilation Parameter

A dilation parameter can be defined for the material which behaves according to Mohr–Coulomb, Hoek–Brown or Drucker–Prager yield criteria. Dilatancy is a measure of how much volume changes when the material is sheared or yielded.

For a Mohr–Coulomb material, dilatancy is an angle that generally varies between zero (non-associative and dilation suppressed) and friction angle, φ (associative flow). For Hoek–Brown materials, dilatancy is defined using a dimensionless parameter m, that generally varies between 0 and m_b.

For very good quality hard rock masses, such as massive granites or quartzites, the analysis of spalling around highly stressed openings suggests that the rock mass behaves as an elastic brittle manner (Hoek and Brown, 1997). When the strength of the rock mass is exceeded, a sudden strength drop occurs. This is associated with significant dilation of the broken rock pieces. On the other hand, it is known that a very poor quality rock will behave like elastic–perfectly plastic material with practically no volume change. Hence, low dilation angles/parameters (close to 0) are generally associated with soft rocks while high dilation angles/parameters (i.e. φ or m_b) are associated with hard brittle rock masses. A good starting estimate is to use 1/8 m_b or 1/8φ for soft rocks and 1/4m_b or 1/4φ for hard rocks.

Appendix D

Beam Analysis

In underground excavations such as civil tunnels which is circular or in D-shape whereas in mining excavation the opening are mostly rectangular in shape called galleries or drive. These underground excavation are supported by artificial supports made of engineering or natural materials. These engineering or artificial supports are used as beam or pillars. In some of the condition steel arches are also a common support in loose ground condition in tunnels, decline, etc. The most common cross-section used are circular, square or rectangular. In this section, the various condition of loading on the rock structure are discussed to access their deformation and stresses.

Beams are classified based on the manner in which they are supported as shown in Figure D1. Different types of beams are: (i) cantilever, (ii) simply supported, (iii) continuous, (iv) overhanging, (v) fixed, and (vi) beam fixed at one end and simply supported at the other end. In any section of the beam three forces are acting, i.e., normal force, the shear force and the bending moment. The graphical representation of shear force and bending moment is called shear force diagram (SFD) and bending moment diagram (BMD). The calculation procedure to find normal, shear and bending moment is given for one beam is given which can be replicated to other beam as well.

A **cantilever beam** is supported at one end and free at the other, projecting horizontally into space. In underground excavation in bedded roof when only one end is supported as shown in Figure (D.1a). Example are from goaf or gob edges developed due to depillaring operation in Longwall and Bord and Pillar method of mining. Such type of beam are supported either by concentrated support like props or distributed support like cogs at goaf edge or hydraulic powered support at longwall mining by caving method.

Simply supported beam is the simplest and most commonly used type of beam. As the name suggests, it is supported at both ends and carries the load across its span. The weight is transferred to the supports, creating upward forces, and the beam resists bending due to these forces. The bedded roof of underground excavation at a shallow depth (less than 100 m) and the artificial supports like goalpost are some of the applications as shown in Figure (D.1b). Both ends of the beam are free to move because of smaller abutment load.

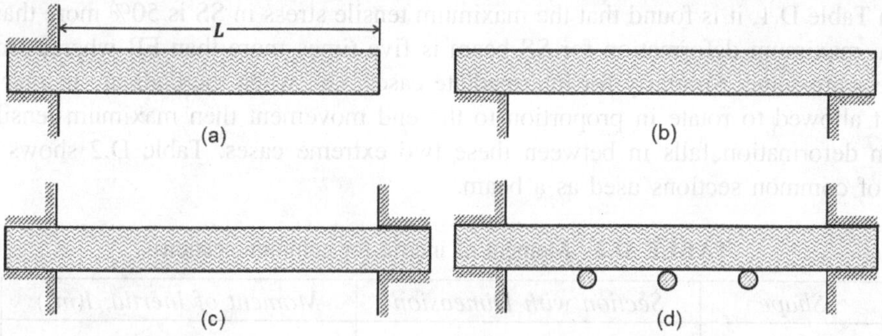

FIGURE D.1 Common beam types used in underground excavation.

A **fixed beam** is a beam that is firmly fixed and restrained at both ends. This type of beam prevents rotation at its supports and is highly effective at resisting bending moments. The example of fixed beam from underground excavation is the deformation of bedded roof at greater depth (> 100 m) and the use of artificial beam fixed into the wall to support the roof as shown in Figure (D.1(c)). This is also known as built-in or constraint beam in which ends are rigidly anchored without any movement and resembles deep excavation where larger abutment loads are induced.

Continuous beams are basically an assembly of multiple beams or a beam (either SS or FB) is supported between both ends. For example, in underground excavation roofs are not only supported as a FB or SS but also intermediate points. Figure (D.1d) shows that FB is supported at three points apart from the end support is an example of a continuous beam.

TABLE D.1 Various formula for uniformly distributed beam loaded transversely

	Simply Supported	*Fixed Beam*	*Remarks*
$\sigma_{t,\max}$	$\frac{3pL^2}{4h^2}$ (at center)	$\frac{pL^2}{2h^2}$ (at abutment)	$\sigma_{t,SS} = 1.5\ \sigma_{t,FB}$
$\tau_{\max}$	$\frac{3pL}{4}$ (at abutment)	$\frac{3pL}{4}$ (at abutment)	$\tau_{\max,SS} = \tau_{\max,FB}$
$\delta_{\max}$	$\frac{5pL^4}{384EI}$ (at center)	$\frac{pL^4}{384EI}$ (at center)	$\delta_{\max,SS} = 5\delta_{\max,FB}$
δ_y	$\frac{px}{24EI}(L^3 - 2Lx^2 + x^3)$	$\frac{px^2}{24EI}(L-x)^2$	—
V_x	$p\left(\frac{L}{2}-x\right)$	$p\left(\frac{L}{2}-x\right)$	equal
M_x	$\frac{px}{2}(L-x)$	$-\frac{p}{12}(L^2 - 6Lx + 6x^2)$	—

σ_t = is tensile stress; τ is shear stress; δ is deformation, p is vertical force per unit length, V is shear force, and M is bending moment.

From Table D.1, it is found that the maximum tensile stress in SS is 50% more than the FB condition, maximum deformation for SS beam is five times more than FB whereas maximum shear stress are same. Similarly for intermediate cases, i.e., when the ends are neither free nor fixed, but allowed to rotate in proportion to the end movement then maximum tensile stress, maximum deformation falls in between these two extreme cases. Table D.2 shows the area moment of common sections used as a beam.

TABLE D.2 Moment of inertia for common sections

Shape	*Section with Dimension*	*Moment of Inertia, I(m⁴)*
Circular	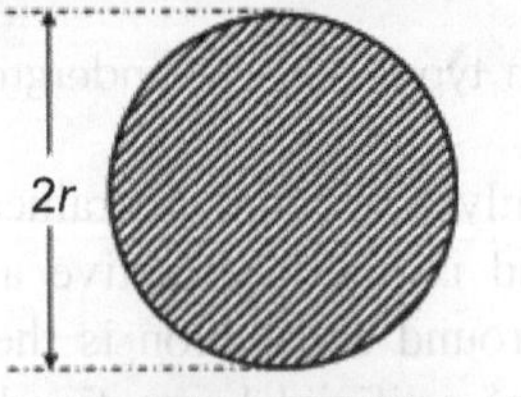	$\frac{\pi}{4}r^4$
Square	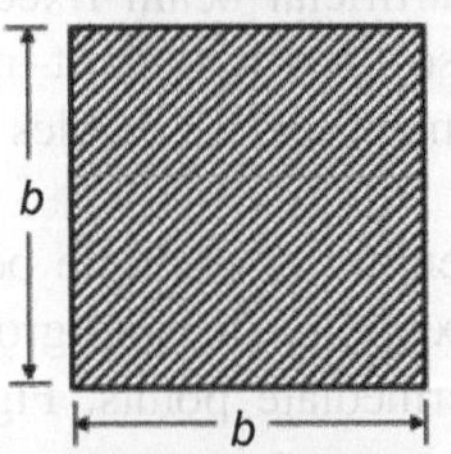	$b^4/12$
Rectangular	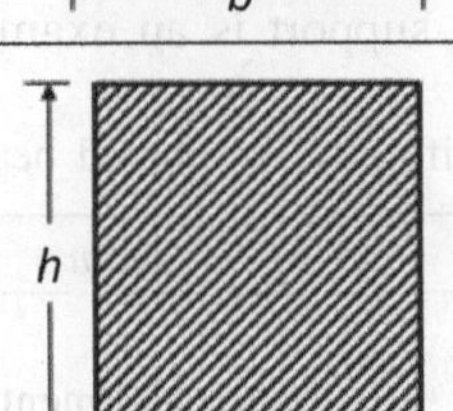	$bh^3/12$
I-section	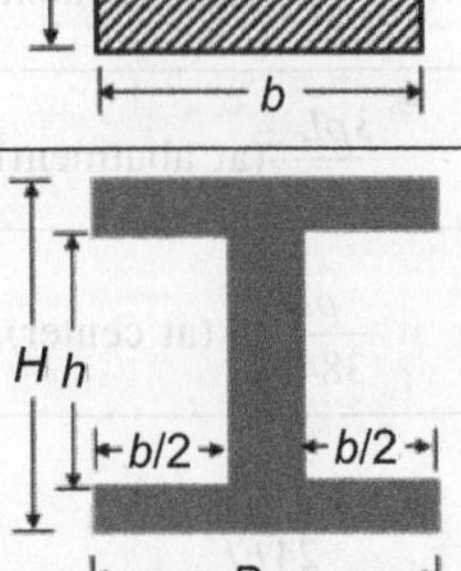	$\frac{BH^3 - bh^3}{12}$

Figure D.2 shows three loading conditions, i.e., (a) concentrated load at free end, (b) concentrated load at any point, (c) uniformly distributed load for a cantilever beam, resulting deformation are shown in Table D.3.

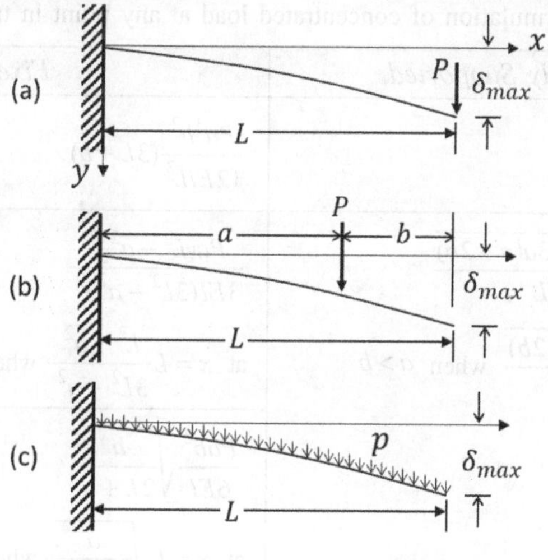

FIGURE D.2 Cantilever beam deflection.

TABLE D.3 Formulation of Cantilever beam for a rectangular beam loaded transversely

	Point load at free end	*Point load at any point*	*Uniformly distributed load*
Vertical displacement, y at any point x	$\frac{Px^2}{6EI}(3L-x)$	$\frac{Px^2}{6EI}(3a-x)$ for $0 < x < a$ $\frac{Px^2}{6EI}(3x-a)$ for $a < x < l$	$\frac{px^2}{24EI}(x^2+6L^2-4Lx)$
Maximum displacement, $\delta_{\max}$	$\frac{PL^3}{3EI}$	$\frac{Pa^2}{6EI}(3L-a)$	$\frac{pL^4}{8EI}$

Figure D.3 shows the point loading of (a) simply supported beam, and (b) fixed beam at any point from left side support and resulting deformation are shown in Table D.4.

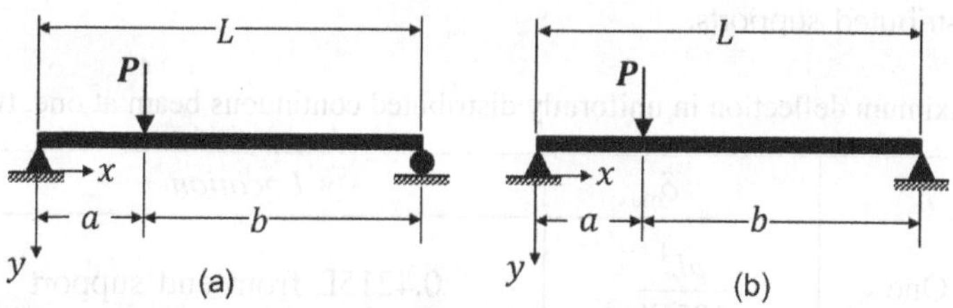

FIGURE D.3 Point load applied to (a) simply supported beam, and (b) fixed beam at any point.

TABLE D.4 Formulation of concentrated load at any point in transverse direction

	Simply Supported	*Fixed Beam*
δ_a	$\frac{Pa^2b^2}{3EIL}$	$\frac{Pa^3b^2}{12EIL^3}(3L+b)$
δ_{max}	$\frac{Pab(a+2b)\sqrt{3a(a+2b)}}{27EIL}$ at $x=\sqrt{\frac{a(a+2b)}{3}}$ when $a>b$	$\frac{Pa(L^2-a^2)^3}{3EI(3L^2-a^2)^2}$ at $x=L\frac{L^2+a^2}{3L^2-a^2}$ when $a<0.414L$
		$\frac{Pab^2}{6EI}\sqrt{\frac{a}{2L+a}}$ at $x=L\sqrt{\frac{a}{2L+a}}$ when $a>0.414L$
δ_x	$\frac{Pbx}{6EIL}(L^2-b^2-x^2)$ for $x<a$	$\frac{Pb^2x}{12EIL^3}(3aL^2-3Lx^2-ax^2)$ for $x<a$
		$\frac{Pa}{12EIL^3}(L-x)^2(3L^2x-a^2x-2a^2L)$ for $x>a$

Figure D.4 shows that a continuous beam is uniformly loaded and supported at one, two, and three points and resulting maximum deformation and their locations are shown in Table D.5.

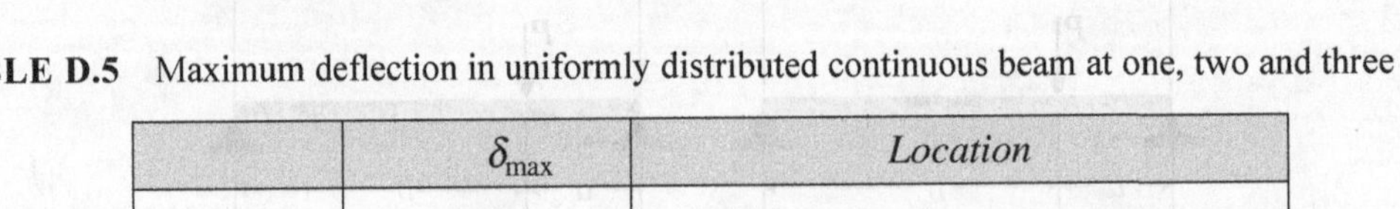

FIGURE D.4 A continuous beam simply supported by (a) one, (b) two, and (c) three equally distributed supports.

TABLE D.5 Maximum deflection in uniformly distributed continuous beam at one, two and three points

	δ_{max}	*Location*
One	$\frac{pL^4}{185EI}$	0.4215L from end support
Two	$\frac{0.0069pL^4}{EI}$	0.446L from end support
Three	$\frac{0.0065pL^4}{EI}$	0.440L from end support

References

Amadei, B. (1983), *Rock Anisotropy and the Theory of Stress Measurements*, Lectures Notes in Engineering, Springer, New York, 2, 478.

Amadei, B. and Stephansson, O. (1997), *Rock Stress and Its Measurement*, Chapman & Hall, UK, 490.

American Society of Testing Method, ASTM D2434-22, (2022), Standard Test Methods for Measurement of Hydraulic Conductivity of Coarse-Grained Soils.

American Society of Testing material, ASTM D6032/D6032M-17, 21st April 2017, Determining Rock Quality Designation (RQD) of Rock Core.

Amini Mehdi, Majdi Abbas and Aydan Omer (2009), "Stability analysis and the stabilization of flexural toppling failure", *Rock Mech. Rock Engg.*, 42, 751–782.

Asadpoure, A. and Mohammadi, S. (2007), "Developing new enrichment functions for crack simulation in orthotropic media by the extended finite element method", *Int. J. Num. Methods in Engg.*, **69**(10), 2150–2172.

ASTM, D4623-16, Jan 31, 2017, Standard Test Method for Determination of *In Situ* Stress in Rock Mass by Overcoring Method—Three Component Borehole Deformation Gauge.

ASTM (D4623–02), "Standard test method for determination of *In situ* stress in rock mass overcoring method", USBM borehole deformation gage, 13.

Atluri, S.N., Kim, H.G. and Cho, J.Y. (1999), "A Critical Assessment of the Truly Meshless Local Petrov-Galerkin (MLPG), and Local Boundary Integral Equation (LBIE), Methods", *Computational Mechanics*, **24**(5), 348–372.

Aydan, O. and Kawamoto, T. (1992), "Stability of slopes and underground openings against flexural toppling and their stabilization", *Rock Mech. Min. Engg.*, **25**(3), 143–165.

Baghbani, A., Choudhury, T., Costa S. and Reiner, J. (2022), Application of Artificial Intelligence in Geotechnical Engineering: A State-of-the-art review, *Earth-Science Reviews*, **228**, 103991.

Bandis, S.C., Barton, N.R. and Christianson, M. (1983), "Application of a new numerical model of joint behavior to rock mechanics problems", *Proc. Int. Symp. Fundamentals of Rock Joints*, Bjorkliden (Ed., O. Stephansson), 345–55.

Bandis, S.C., Lumsden, A.C. and Barton, N.R. (1981), "Experimental studies of scale effect on shear behaviour of rock joints", *Int. J. Rock Mech. Min. Sci. & Geomech. Abstr.*, **18**, 1–21.

Barton, N., By, T.L., Chryssanthakis, P., Tunbridge, L., Kristiansen, J., Loset, F., Bhasin, R.K., Westerdahl H. and Vik, G. (1992), Comparison of prediction and performance for a 62 m span sports hall in jointed gneiss, *4th Int. Rock Mech. Rock Engg. Conf.*, G. Barla (Ed.), Torino, Italy, 17.2–17.15.

Barton, N., Loset, F., Lien, R. and Lunde, J. (1980), Application of Q-system in design decisions concerning dimentions and appropriate support for underground installations, Subsurface space, Pergamon, 553–561.

Barton, N.R. (1973), "Review of a new shear strength criterion for rock joints", *Engineering Geology*, **8**(4), 287–332.

Barton, N.R. (1976), "The shear strength of rock and rock joints", *Int. J. Rock Mech. Min. Sci. & Geomech. Abstr.,* **13**, 255–279.

Barton, N.R. (2000), *TBM Tunneling in Jointed and Faulted Rock*, Balkema, Rotterdam, 173.

Barton, N.R. (2002), "Some new Q-value correlation to assist in site characterization and tunnel design.," *Int. J. Rock Mech. Min. Sci.*, **39**, 185–216.

Barton, N.R. (2013), "Shear strength for rock, rock joints, rockfill and rock masses: problems and some solutions", *J. of Rock Mech. and Geotechnical Engineering*, **5**, 249–261.

Barton, N.R. and Bandis, S.C. (1982), "Effects of block size on the shear behaviour of jointed rock", *23rd U.S. Symp. on Rock Mechanics*, Berkeley, 739–760.

Barton, N.R. and Choubey, V. (1977), "The shear strength of rock joints in theory and practice", *Rock Mech.,* **10**, 1–54.

Barton, N.R., Lien R. and Lunde, J. (1974), "Engineering classification of rockmasses for the design of tunnel support", *Rock Mech.*, 6(4), 189–236.

Basheer, I.A. and Hajmeer, M., (2000), Artificial Neural Networks: Fundamentals, Computing, Design, and Application. *Journal of microbiological methods*, **43**(1), pp. 3–31.

Bathe, K.J. (1982), *Numerical methods in finite element analysis*, Prentice-Hall.

Batugin, S.A. and Nirenburg, R.K. (1972), "Approximate relation between the elastic constants of anisotropic rocks and the anisotropy parameters", translated from Fizilo-Tekhnicheskie Problemy Razrabotki Poleznikh Iskopaemykh, **1**, 7–12.

Bauschinger, J. (1876), Mitteilungenausdem Mechanisch-Technischen Laboratorium der K, Technishen Hochschule in Munschen, 6.

Belytschko, T. and Black, T. (1999), "Elastic crack growth in finite elements with minimal remeshing", *Int. J. Numerical Methods in Engineering*, **45**(5), 601–620.

Belytschko, T. and Lin, J.I. (1987), "A three-dimensional impact-penetration algorithm with erosion", *Int. J. Impact Engg.*, **5**(1), 111–127.

Belytschko, T., Lu, Y.Y. and Gu, L. (1994), "Element-free Galerkin methods", *Int. J. Numerical Methods in Engineering*, **37**(2), 229–256.

Bieniawski, Z.T. (1968), "*In situ* strength and deformation characteristics of coal", *Engg. Geol.*, **2**, 325–340.

Bieniawski, Z.T. (1968), "The effect of specimen size on the compression strength of coal", *Int. J. Rock Mech. Mining Sci.*, **5**(4), 325–35.

Bieniawski, Z.T. (1973), "Engineering classification of joined rock masses", *The Civil Engineer in South Africa*, Dec., 335–343.

Bieniawski, Z.T. (1976), "Rock mass classifications in rock engineering", *Exploration for rock Engineering*, Bieniawski, Z.T., and Balkema, A.A. (Eds.), Johannesburg, 97–106.

Bieniawski, Z.T. (1978), "Determining rock mass deformability: Experience from case histories", *Int. J. Rock Mech. Min. Sci.*, **15**(5), 237–247.

Bieniawski, Z.T. (1989), "Engineering rock mass classifications: a complete manual for engineers and geologists in mining", *Civil and Petroleum Engineering*, Wiley, New York, 251.

Bieniawski, Z.T. and Van Heerden, W.L. (1975), "The significance of *in situ* tests on large rock specimens", *Int. J. Rock Mech. Mining Sci. & Geomech. Abstr.*, **12**(4), 101–13.

Bieniawski, Z.T. (1974), "Geomechanics classification of rock masses and its application in tunneling", *Proc. of the 3rd Int. Cong. Rock Mech.*, Denver, 27–32.

Biron, C. and Ergin, A. (1983), *Design of Supports in Mines*, John Wiley & Sons.

Bodonyi, J. (1970), "Laboratory tests of certain rocks under axially-symmetrical loading conditions", *Proceedings of the 2nd International Congress on Rock Mechanics*, Belgrade, **1**, 389–397.

Brady, B.H.G. (1987), "Boundary element and linked methods for underground excavation", E.T. Brown (Ed.), *Analytical and Computational Methods in Engineering Rock Mechanics*, Allen & Unwin, London.

Brady, B.H.G. and Brown, E.T. (1985), *Rock Mechanics for Underground Mining*, Allen & Unwin Publisher, London.

Brady, B.H.G. and Brown, E.T. (2007), *Rock Mechanics for Underground Mining*, 3rd ed., George Allen & Unwin, London.

Bray, J.W. (1986), "Some applications of elastic theory", *Analytical and Computational Methods in Engineering Rock Mechanics*, E.T. Brown (Ed.), 32–94, Allen & Unwin, London.

Brook, N. and Dharmaratne, P.G.R. (1985), "Support requirements for machine driven tunnels", *S. Afr. Tunneling*, **8**, 7.

Brown, E.T. and Hoek, E. (1978), "Trends in relationships between measured rock *in situ* stresses and depth, *Int. J. Rock Mech. Min. Sci. & Geomech. Abstr.*, **15**, 211–15.

Bui, H.H., Fukagawa R., Sako K. and Ohno S. (2008), "Lagrangianmeshfree particles method (SPH), for large deformation and failure flows of geomaterial using elastic–plastic soil constitutive model", *Int. J. Num. Anal. Methods in Geomech.*, **32**(12), 1537–1570.

Camacho, G.T. and Ortiz, M. (1996), "Computational modelling of impact damage in brittle materials", *Int. J. Solids and Structures*, **33**(20), 2899–2938.

Chakrabarty, J. (1987), *Theory of Plasticity*, Vol. 1, McGraw-Hill, New York.

Cherny, G.J. (1966), "Determination of surface subsidence and deformation by rock movements in the form of rheologic flow", *Izvestia VUZ Zhurnal*, **9**(7), 3–9.

Choi, S.O., Shin, H.S., Chung, S.K., Sunwoo, C., Song, W.K., Park, E.S., Jeon, Y.S., Jung, Y.B., Cheon, D.S. and Choi, B.H. (2004), "Development of the techniques for reducing the national disaster caused by ground subsidence", Project no. M1–0324–00–0013–03–B31–00–005–00, *Ministry of Science and Technology of Korea*, 232.

Chunlin, Li, C. (2006), "Rock support design based on the concept of pressure arch", *Int. J. Rock Mech. Mining Sci.*, **43**, 1083–1090.

Chunlin, Li, C. (2010), "A new energy-absorbing bolt for rock support in high stress rock masses", *Int. J. Rock Mech. Mining Sci.*, **40**, 396–404.

Coduto, D.P., (2002), *Geotechnical Engineering Principles and Practice*, Pearson Education, Asia, pp. 561–562.

Cook, R.D., Malkus, D.S. and Phesha, M.E. (1989), *Concepts and Applications of Finite Element Analysis*, John Wiley & Sons, US.

Crouch, S.L. (1976), "Solution of plane elasticity problems by the displacement discontinuity method", *Int. J. Numer Methods Engg.*, **10**, 301–43.

Crouch, S.L. and Starfield, A.M. (1983), *Boundary Element Methods in Solid Mechanics*, George Allen & Unwin, London.

Cummings, R.A., Kendorski, F.S. and Bieniawski, Z.T. (1982), "Caving rock mass classification and support estimation", *U.S. Bureau of Mines Contract Report* #J0100103, Chicago: Engineers International Inc.

Cundall, P.A. (1988), "Formulation of a three-dimensional distinct element model—Part I, A scheme to detect and represent contacts in a system composed of many polyhedral blocks", *Int. J. Rock Mech. Mining Sci. & Geomech. Abstr.*, **25**(3), 107–116.

Cundall, P.A. and Strack, O.D.L. (1983), "Modeling of microscopic mechanism in granular materials", Materials: New Model and Constitutive Relations, Jenkins, J.T. and Satake, M. (Eds.), *US-Japan Seminar on New Models and Constitutive Relations in the Mechanics of Granular Materials*, Elsevier, Amsterdam.

Das, K.C. (2012), Enriched Finite Element Method and Its Applications in Reinforced Jointed Rock, Ph.D. Thesis, IIT Kharagpur, India.

Das, M.N. (1986), "Influence of width/height ratio on post-failure behaviour of coal", *Int. J. Min. Geol. Engg.*, **4**, 79–87.

Das, B.M. (2002), *Principles of Geotechnical Engineering*, 5th ed., Thomson Brooks/Cole, Pacific Grove, CA.

Deb, D. (2002), "Coal Mine Roof Rating (CMRR)—A practical tool for evaluating coal mine ground control problems", *Journal of Mines, Metals and Fuels*, **50**(1 and 2), 15–19.

Deb, D. (2012), *Finite Element Methods: Concepts and Applications in Geomechanics*, 2nd ed., PHI learning, Delhi, 360.

Deb, D. and Choi, S.O. (2005a), "Comparison between direct and indirect implementation of generalized Hoek and Brown failure criterion in numerical analysis procedure, Tunnel and Underground Space", *J. Korean Society of Rock Mech.*, **15**(3), 142–149.

Deb, D. and Choi, S.O. (2005b), "Singularity adjustment on generalized Hoek and Brown yield surface for elastic-plastic analysis", *Proceeding of the 40th U.S. Rock Mech. Symp.*, June 25–29, Anchorage, Alaska.

Deb, D. and Das, K.C. (2010), "Extended finite element method for the analysis of discontinuities in rock masses", *Geotechnical and Geological Engineering*, **28**(5), 643–659.

Deb, D. and Das, K.C. (2011), "Enriched finite element procedures for analyzing decoupled bolts installed in rock mass", *Int. J. Num. Anal. Methods in Geomech.*, **35**(15), 1636–1655.

Deb, D. and Das, K.C. (2014), "A new doubly enriched finite element for modelling grouted bolt crossed by rock joint", *Int. J. Rock Mech. Min. Sci.*, **70**, 47–58.

Deb, D. and Pramanik, R. (2013), "Failure process of brittle rock using smoothed particle hydrodynamics", *J. Engg. Mech.*, **139**(11), 1551–1565.

Deere, D.U., Hendron, A.J., Patton F.D. and Cording, E.J. (1967), "Design of surface and near-surface construction in rock", Fairhurst, C. (Ed.), *Proc. of the US Rock Mech. Sysm., Failure and Breakage of Rock*, New York, Society of Mining Engineers of AIME, 237–302.

DGMS (1990), DGMS Circular No. DGMS (Tech)/5, *Directorate general of mines safety*, Dhanbad, India.

Dolbow, J.O.H.N. and Belytschko, T. (1999), "A finite element method for crack growth without remeshing", *Int. J. Numer. Meth. Engg.*, **46**(1), 131–150.

Duarte, C.A. and Oden, J.T. (1996), "An h-p adaptive method using clouds", *Computer Methods in Applied Mechanics and Engineering*, **139**(1), 237–262.

Dunnicliff, J. (1993), *Geotechnial Instrumentation for Monitoring Field Performance*, John Wiley & Sons, New York, 577.

Edelbro, C. (2004), "Evaluation of rock mass strength criterion," *Ph.D. Thesis, Lulea University of Technology*, Department of Civil and Environmental Engineering, Sweden, 152.

Edib, A.M. (2021), 35 Years of (AI) in Geotechnical Engineering: State of the Art, Geotech *Geol. Eng.,* **39**: 637–690, https://doi.org/10.1007/s10706-020-01536-7.

Evans, I., Pomeroy, C.D. and Berenbaum, R. (1961), "The compressive strength of coal", *Colliery Engg.*, March.

Fairhurst, C. (1964), "On the validity of the Brazilian test for brittle materials", *Int. J. Rock Mech. Min. Sci.*, **1**, 515–546.

Franklin, J.A. and Dusseault, M.B. (1989), *Rock Engineering*, McGraw-Hill, New York, 600.

Farmer, I.W. (1975), "Stress distribution along a resin grouted rock anchor", *Int. J. Rock Mech. Min. Sci. Geomech. Abstr.*, **11**, 347–351.

Franklin, J.A. (1971), "Triaxial strength of rock material", *Rock Mechanics,* **3**, 86–98.

Galvin, J.M., Hebblewhite, B.K. and Salmon, M.D.G. (1999), University of New South Wales Coal Pillar Strength Determinations for Australian and South African Mining Conditions, *Proc. Sec. Workshop on Coal Pillar Mech. Design*, NIOSH IC 9448, 63–72.

Gerarrd, C. (1985), "Formulations for the mechanical properties of rock joints", D. Stephansson (Ed.), *Proc. Int. Symp. Fundamentals of Rock Joints*, Bjorkliden, Lulea, Centek.

Gerrard, C.M. (1982a), "Elastic models of rock masses having one, two and three sets of joints", *Int. J. Rock Mech. Min. Sci. & Geomech. Abstr.*, **19**, 15–23.

Gerrard, C.M. (1982b), "Joint compliances as a basis for rock mass properties and the design of supports", *Int. J. Rock Mech. Min. Sci. & Geomech. Abstr.*, **19**, 285–305.

Ghabraie, B., Ren, G., Ghabraie, K. and Xie, Y.M. (2013), "A study on truss mechanism in controlling stability of underground excavation and cutter roof failure", *Geotech. Geol. Engg.*, **31**, 667–682.

Gingold, R.A. and Monaghan, J.J. (1977), "Smoothed particle hydrodynamics: theory and application to non-spherical stars", *Monthly Notices of the Royal Astronomical Society*, **181**(3), 375–389.

Goodman, R.E. (1970), "The deformability of joints", *ASTM Spec. Tech. Publ.* 477, 174–196.

Goodman, R.E. (1989), *Introduction to Rock Mechanics,* 2nd ed., John Wiley & Sons, New York.

Goodman, R.E. and Bray, J.W. (1976), "Toppling of rock slopes", *Proc. of the ASCE Specialty Conference on Rock Engineering for Foundations and Slopes*, Boulder, Colorado, August, **2**, 201–234.

Goodman, R.E. and Bray, J.W. (1977), "Toppling of rock slopes", *Proc. Specialty Conf. on Rock Engineering for Foundation and Slopes*, ASCE (Boulder, Colorado), **2**, 201–234.

Gramberg, J. (1989), "A non-conventional view on rock mechanics and fracture mechanics", A.A. Balkema, 250.

Gray, J.P., Monaghan, J.J. and Swift, R.P. (2001), "SPH elastic dynamics", *Computer Methods in Applied Mechanics and Engineering*, **190**(49), 6641–6662.

Gray, R.E., Bruhn, R.W. and Turka, R.J. (1977), "Study and analysis of surface subsidence over mined pittsburgh coalbed", *Final Report, Formally US Bureau of Mines*, Contract J0366047, NTIS PB 281 511, 362.

Greenwald, H.P., Howarth, H.C. and Hartmann, I. (1941), "Experiments on strength of small pillars of coal in the Pittsburgh bed", *Rpt. of Investigations* 3575, U.S. BuMines, 9.

Grimstad, E. and Barton, N.R. (1993), "Updating of the Q-system for NMT", Kompen, Opsahl, Berg (Eds.), *Proc. of the Int. Sym. on Sprayed Concrete—Modern Use of Wet Mix Sprayed Concrete for Underground Support*, Fagernes. Oslo: Norwegian Concrete Association, Hoek E., Strength of rock and rock masses, *ISRM News J.*, 2(2), 4–16.

Guo, H., Chen, S.G., Craig, M.S., Adhikary, D.P. and Poole, G. (2004), "Coal mine subsidence simulation using cosflow", *AusImm Bulletin*, **2**(47), March/April, 50–52.

Haimson, B.C. and Cornet, F.H. (2003), "ISRM suggested methods for rock stress estimation – Part 3: Hydraulic Fracturing (HF), and/or hydraulic testing of pre-existing fractures (HTPF), *Int. J. Rock Mech. Min. Sci.*, **40**, 1011–1020.

Haramy, K.Y. and Kneisley, R.O. (1990), "Yield pillars for stress control in longwall mines—case study", *Int. J. Min. Geol. Engg.*, **8**, 287–304.

Heasley, K.A. and Barton, T.M. (1998), "Coal mine subsidence prediction using boundary element method", *SME Annual Meet.*, March 9–11, 1998, Preprint Number 99–181.

Herget, G. (1988), "Stresses in rock", Rotterdam, Balkema.

He Mingming, Zhang Zhiqiang, Ren Jie, Huan Jiuyang, Li Guofeng, Chen Yunsheng, and Li Ning (2019), Deep convolutional neural network for fast determination of the rock strength parameters using drilling data, *International Journal of Rock Mechanics and Mining Sciences*, **123**, 104084.

Hobbs, D.W. (1964), "The Strength and Stress-Strain Characteristics of Coal in Triaxial Compression", *J. Geol.*, **72**(2), 214–231.

Hoek, E. (1968), "Brittle failure of rock", *Rock Mechanics in Engineering Practice*, K.G., Stagg and O.C., Zienkiewicz (Eds.), Wiley, London, 99–124.

Hoek, E. (2007), "Practical rock engineering", *Hoek's corner*, www.rockscience.com, 237.

Hoek, E. and Brown, E.T. (1980), "Underground excavation in rock", *London Instn. Mine Metallurgy.*

Hoek, E. and Brown, E.T. (1988), "The Hoek–Brown criterion—a 1988 update", *Proc. 15th Can. Rock Mech. Symp.*, Toronto, 31–38, Univ. Toronto Press.

Hoek, E. and Brown, E.T. (1990), "Underground excavations in rock", *The Institute of Mining and Metallurgy*, 329–365.

Hoek, E. and Brown, E.T. (1997), "Practical estimates of rock mass strength", *Int. J. Rock Mech. Min. Sci.*, **34**(8), 1165–86.

Hoek, E., Kaiser, P.K. and Bawden, W.F. (1995), "Support of underground excavation in hard rock", A.A. Balkema.

Hoek, E., Marinos, P. and Benissi, M. (1998), "Applicability of the geological strength index (GSI), classification for very weak and sheared rock masses: the case of the Athens Schist Formation", *Bull. Engg. Geol. Env.*, **57**(2), 151–60.

Hoek, E., Torres, C.C. and Corkum, B. (2002), "Hoek and Brown failure criterion", 5th ed., *North American Rock Mechanics Sym. and 17th Tunneling Association of Canada Conf. (NARMS-TAC)*, 267–271.

Hoffman, H. (1964), "The effects of direction of working and rate of advance on the scale—deformation of a self-loaded stratified model of a large body of ground", *Proc. Int. Con. on Strata Control*, New York, 397–411.

Holla, L. and Barclay, E. (2000), "Mine subsidence in the Southern coalfield", NSW, Australia, 1–16 (New South Wales Department of Mineral Resources).

Holland, C.T. and Gaddy, F.L. (1957), "Some aspects of permanent support of overburden on coal beds", *Proceedings of West Virginia Coal Mining Institute*, 43–66.

Holland, C.T. and Gaddy, F.L. (1964), "The strength of coal in mine pillars", *Proc. 6th US Symp. on Rock Mechanics* (University of Missouri, Rolla, MO), 450–466.

Hong, E.S., Lee, J.S. and Lee, I.M. (2008), "Underestimation of roughness in rough rock joints", *Int. J. Num. Anal. Methods in Geomech.*, **32**, 1385–1403.

Hudson, J.A. and Harrison, J.P. (1997), *Engineering Rock Mechanics*, Vol. 1, Chapter 12—Rock Mass Classification, Pragmons Publications, Elsevier Science, UK, 193–206.

Hudson, J.A. and Harrison, J.P. (1997), *Engineering Rock Mechanics*, Vol. 1, Chapter 4—*In situ* stress, Pragmons Publications, Elsevier Science, UK, 41–52.

Hustrulid, W.A. (1976), "A review of coal pillar strength formulas", *Rock Mechanics*, **8**, 115–45.

Hutchinson, D.J. and Diederichs, M.S. (1996), *Cable Bolting in Underground Mines*, BiTech Publishers, Vancouver, Canada.

Hyett, A.J., Moosavi, M., and Bawden, W.F. (1996), "Load distribution along fully grouted bolts, with emphasis on cable bolt reinforcement", *Num. Anly. Meth. in Geomech.*, **20**(7), Jult, 517–544.

Indian Standard, IS 130230: 1991: Method of test for laboratory determination of water content, porosity, density and related properties of rock material.

Indraratna, B. and Haque, A. (2000), "Shear behavior of rock joint", A.A. Balkema, Rotterdam, Brookfield.

ISRM (1978), "Suggested methods for the quantitative description of discontinuities in rock masses", *Int. J. Rock Mech. Min. Sci. & Geomech. Abstr.*, 15, 319–368.

ISRM (1981), "Basic geological description of rock masses", *Int. J. Rock Mech. Min. Sci. & Geomech. Abstr.*, **18**, 85–110.

ISRM (1987), "Suggested methods for rock stress determination", *Int. J. Rock Mech. Min. Sci. & Geomech. Abstr.*, **24**(1), 53–73.

ISRM Commission (2007), "The complete ISRM suggested methods for rock characterization, testing and monitoring: 1974–2006", R. Ulusay and J.A. Hudson (Eds.), Publisher: ISRM Turkish national Group, Ankara, Turkey, 628.

Iturrarán-Viveros, U. and Sánchez-Sesma, F.J. (2011), "Seismic wave propagation in real media: numerical modeling approaches", Encyclopedia of Solid Earth Geophysics, DOI 10.1007/978–90–481–8702–89, Gupta (Ed.), Springer Science + Business Media, 1200–1210.

Jaeger, J.C. (1960), "Shear fracture of anisotropic rocks", *Geol. Mag.*, **97**, 65–72.

Jaeger, J.C. and Cook, N.G.W. (1979), *Fundamentals of Rock Mechanics*, 3rd ed., Chapman and Hall, London, 593.

Jager, J.C., Cook, N.G.W., and Zimmerman, R.W. (2007), *Fundamentals of Rock Mechanics*, 4th ed., Chapter 13—State of stress underground, Blackwell publishing, USA, 399–418.

Janbu, N., Bjerrum, L. and Kijaernsli, B. (1956), "Soil mechanics applied to some engineering problems", *Norwegian geotechnical institute*, Pub. no., 16.

Janssen, H.A. (1895), Versucheuber getreidedruck in silozellen, Z. Vereines Deutscher Ingenieure, **39**, 1045–1049.

Jeremic, M.L. (1985), "Strata mechanics in coal mining", A.A. Balkema, Rotterdam, Boston, Neatherland.

Jing, L. (1998), "Formulation of discontinuous deformation analysis (DDA)—an implicit discrete element model for block systems", *Engineering Geology*, **49**(3), 371–381.

Jing, L. (2003), "A review of techniques, advances and outstanding issues in numerical modelling of rock mechanics and rock engineering", *Int. J. of Rock Mech. and Min. Sci.*, **40**, 283–353.

Jing, L. and Hudson J.A. (2002), "Numerical methods in rock mechanics", *Int. J. Rock Mech. Min. Sci. & Geomech. Abstr.*, **39**, 409–427.

Johnston, I.W. (1985), "Strength of intact geomechanical materials", *Journal of Geotechnical Engineering*, ASCE, **111**, 730–749.

Keilich, W., Seedsman, R.W. and Aziz, N. (2006), "Numerical modeling of mining induced subsidence", Aziz, N. (Ed.), Coal 2006, *Coal Operators' Conference*, University of Wollongong & the Australasian Institute of Mining and Metallurgy, 313–326.

Kendorski, F., Cummings, R., Bieniawski, Z.T. and Skinner, E. (1983), "Rock mass classification for block caving mine drift support", *Proc. 5th Congr. Int. Soc. Rock Mech.*, Melbourne, B51–B63, Rotterdam, Balkema.

KIGAM report (1993), "Study on the stability diagnosis and evaluation for the Boopyung mine area", 284.

King, H.J. and Whetton, J.T. (1957), "Mechanics of mine subsidence", *Proc. European Congress on Round Movement*, University of Leeds, UK, April, 27–38.

Knothe, S. (1953), "Rate of advance and ground deformation" (in German), Berhakademie, **5**(12), 513–518.

Kowalczyk, A. (1968), "Die Einwirkung des Bergbaus auf die erdoberflache in industrialisierten und besiedeltengebieten", Bergakad, **7**, 389–394.

Kratzsch, H. (1983), *Mining Subsidence Engineering*, Springer–Verlag Publication, 543, (translated from Germany by R.F.S. Fleming).

Kumar, B., Saxena, N.C. and Singh, B. (1983), "A new hypothesis for subsidence prediction", *J. Mines, Metals and Fuels*, **31**, 459–465.

Kwasniewski, M. (1984), "Anisotropy of elasticity of rocks and its effect on the magnitude and distribution of stresses around underground openings", Application of Rock Mechanics to Planning and Design Prior to Mining, A. Kidybinski and M. Kwasniewski (Eds.), A.A. Balkema, 5–38.

Ladanyi, B. and Archambault, G. (1969), "Simulation of shear behavior of a joined rock mass", *Proc. 11th U.S. Sump. On Rock Mechanics,* Berkeley, 102–125.

Lama, R.D. and Vutukuri, V.S. (1978), *Handbook on Mechanical Properties of Rocks—Testing Techniques and Results*, Vol. IV, Train Tech Publisher, USA, 515.

Lancaster, P. and Salkauskas, K. (1981), "Surfaces generated by moving least squares methods", *Mathematics of Computation*, **37**(155), 141–158.

Laubscher, D.H. (1975), "Class distinction in rock masses, coal, gold, base minerals", *S. Afr.*, **23**, August.

Laubscher, D.H. (1990), "A geomechanics classification system for the rating of rock mass in mine design", *J. South Afr. Inst. Miner. Metall.*, **90**(10), 257–273.

Lauffer, H. (1958), "Gebirgsklassifizierung fur den stollenbau", Geol. Bauwesen, **74**, 46–51.

Lawal, A.I., Kwon, S. (2021), Application of Artificial Intelligence to Rock Mechanics: An Overview, *Journal of Rock Mechanics and Geotechnical Engineering*, **13**, 248–266.

LeCun, Y., Bengio, Y., and Hinton, G. (2015), Deep learning. Nature, 521(7553), 436–444, https://doi.org/10.1038/nature14539.

Leeman, E.R. (1971), "The CSIR doorstopper and triaxial rock stress measuring instruments", *Rock Mechanics*, **3**, 25–50.

Lekhnitskii, S.G. (1968), *Anisotropic Plates*, Translated from the Second Russian Edition by S.W. Tsai and T. Cheron, Gordon and Breach, Science Publishers, Inc., New York, 157.

Li, C., and Stillborg, B. (1999), "Analytical models for rock bolt", *Int. J. Rock Mech. Min. Sci.*, **36**(8), 1013–1029.

Lo, K.Y., et al. (1975), "Stress relief and time dependent deformation of rocks", Final report to national research council of Canada, Special project S-7307.

Loset, F. (1992), "Support needs compared at the Svartisen road tunnel", Tunnel and tunnelling, June.

Mahmoodzadeh, A., Mohammadi, M., Ghafoor Salim, S., et.al., Machine Learning Techniques to Predict Rock Strength Parameters, Rock Mech. Rock Eng., **55**, 1721–1741 (2022), https://doi.org/10.1007/s00603-021-02747-x.

Mark, C. (1990), "Pillar design methods for longwall mining", Information Circular IC 9247, Washington, DC, U.S., Bureau of Mines.

Mark, C. (1997), "Analysis of retreat mining pillar stability", Proc. *New Technology for Ground Control in Retreat Mining*, IC9446, Washington, DC, NIOSH.

Mark, C. and Barczak, T.M. (2000), "Fundamentals of coal mine roof support", Mark, C., Dolinar, D.R., Tuchman, R.J., Barczak, T.M., Signer, S.P., and Wopat, P.F. (Eds.), Proceedings of *New Technology for Coal Mine Roof Support*, Pittsburgh, PA: U.S. Department of Health and Human Services, Public Health Service, Centers for Disease Control and Prevention, National Institute for Occupational Safety and Health, DHHS (NIOSH), Publication No. 2000–151, IC 9453, 23–42.

Mark, C. and Chase, F.E. (1997), "Analysis of retreat mining pillar stability", *Proceedings of New Technology for Ground Control in Retreat Mining*, IC 9446, NIOSH, 1–34.

Mark, C. and Molinda, G.M. (2005), "The coal mine roof rating (CMRR)—a decade of experience", *Int. J. Coal Geology*, **64**, 85–103.

Martin, C.D. and Maybee, W.G. (2000), "The strength of hard-rock pillars", *Int. J. Rock Mech. Mining Sci.*, **37**, 1239–1246.

Martos, F. (1958), "Concerning an approximate equation of the subsidence through and its time factor", *Proc. Int. Strata Control Congress*, Leipzig, Germany, 191–205.

Matula, M. and Holzer, R. (1978), "Engineering topology of rock masses", *In Proceedings of Felsmekanik Kolloquium, Grunlagen ung Andwendung der Felsmekanik*, 107–121, Karlsruhe, Germany.

Mazumder, B. (2012), *Coal Science and Engineering*, Woodhead, India, 459.

McCulloch, C.M., Maurice, Deul and Jeran, P.W. (1974), "Cleats in bituminous coalbeds", Report of Investigation 7910, U.S. Bureau of Mines.

Merrill, R.H. and Peterson, J.R. (1961), "Deformation of a borehole in rock", U.S. Bureau of Mines, Report RI 5881.

Moës, N., Dolbow, J. and Belytschko, T. (1999), "A finite element method for crack growth without remeshing", *Int. J. Num. Methods in Engg.*, **46**(1), 131–150.

Mohammadnejad, T. and Khoei, A.R. (2013), "Hydro-mechanical modeling of cohesive crack propagation in multiphase porous media using the extended finite element method", *Int. J. Num. and Anal. Methods in Geomech.*, **37**(10), 1247–1279.

Molinda, G.M. and Mark, C. (1994), "Coal mine roof rating (CMRR): A practical rock mass classification for coal mines", Pittsburgh, PA: U.S. Department of Interior, Bureau of Mines, IC 9387.

Molinda, G.M. and Mark, C. (1996), "Rating the strength of coal mine roof rocks", Pittsburgh, PA: U.S. Department of the Interior, Bureau of Mines, IC 9444, NTIS No. PB96–155072.

Molinda, G.M., Mark, C. and Deb, D. (2000), "Using the coal mine roof rating (cmrr), to assess roof stability in U.S. coal mines", *Journal of Mines, Metal and Fuels* (India), 25.

Molinda, G.M., Mark, C. and Dolinar, D.R. (2000), "Assessing coal mine roof stability through roof fall analysis, Mark et al. (Eds.), *Proc. New Technology for Coal Mine Roof Support*, Pittsburgh, PA, DHHS (NIOSH), Publication No. 2000–151, IC 9453, 53–71.

Monaghan, J.J. (1992), "Smoothed particle hydrodynamics", *Annual Review of Astronomy and Astrophysics*, **30**, 543–574.

Mortazavi, A., Hassani, F.P. and Shabani, M. (2009), "A numerical investigation of rock pillar failure mechanism in underground openings", *Computer and Geotechnics*, **36**, 691–697.

Murrell, S.A.F. (1965), "The effect of triaxial stress systems on strength of rocks at atmospheric temperature", *Geophys., J.R. Astron. Soc.*, **10**, 231-81.

Najjar, Y. and Zaman, M. (1993), "Numerical modeling of ground subsidence due to mining", *Int. J. Rock Mech. Min. Sci. & Geomech. Abstr.*, **30**(7), 1445–1448.

Nicieza, C.G., Fernandez, M.I.A., Diaz, A. and Vigil, A.E.A. (2005), "The new three-dimensional subsidence influence function denoted by n-k-g", *Int. J. Rock Mech. Min. Sci.*, **42**(3), 372–387.

Niederhofer, G. (1962), "A new method of precalculating surface subsidences, particularly over steep workings, Mitteilungen aus dem Markscheidewesen, **69**(2), 33–75.

NGI, 2022, Using the Q-system rock mass classification and support design, printed by Allkopi AS, Norway, 34–35.

Numerical methods in geotechnical engineering, C.S. Desai and J.T. Christian (Eds.), McGraw-Hill, 783, 1977.

Obert, L. and Duvall, W.I. (1967), *Rock Mechanics and the Design of Structures in Rock*, John Wiley & Sons, New York.

Osgoui Reja, R., Ulusay, Resat and Unal, Erdal (2010), "An assistant tool for the geological strength index to better characterize poor and very poor rock masses, *Int. J. Rock Mech. Min. Sci.*, **47**, 690–697.

Palmstrom, A. (1995), "RMi—A rock mass characterization system for rock engineering puroses," *Ph.D. Thesis*, University of Oslo.

Palmstrom, A., Stille, H., (2010), *Rock Engineering*, Thomas Telford Ltd., p. 408.

Pande, G.N, Beer, G. and Williams, J.R. (1990), *Numerical Methods in Rock Mechanics*, John Wiley & Sons, New York.

Pariseau, W.G. (2007), *Design Analysis in Rock Mechanics*, Chapter 8: Subsidence, Taylor & Francis, Canada, 397–467.

Patton, F.D. (1966), "Multiple modes of shear failure in rock", *Proc. 1st. Congr., Int. Soc. Rock Mech.*, Lisbon, **1**, 509–513.

Peng, S.S. (1978), *Coal Mine Ground Control*, John Wiley & Sons, New York, 450.

Peng, S.S. (1992), "Surface subsidence engineering", SME Publication, 161.

Peng, S.S. and Cheng, S.L. (1981), "Analysis of surface subsidence parameters due to underground longwall mining in the northern appalachian coalfield", Dept. of Mining Engg., West Verginia University, Morgantown, WV, TR81–1, 22.

Peng, S.S. and Chiang, H.S. (1984), *Longwall Mining*, John Wiley & Sons, New York, 17–70.

Plesha, M.E. and Aifantis, E.C. (1983), "On the modeling of rocks with microstructure", *Pme. 14lh U.S. SYlllp. Oil rock mecllmlic.v*, Texas A & M Univ., College Station, Tex., .Iun., 27–35.

Popov, E.P. (1978), *Mechanics of Materials*, 2nd ed., Prentice-Hall International, Inc., Englewood Cliffs, New Jersey.

Potvin, Y., Hudyma, M. and Miller, H. (1989), "Design guidelines for open stope support", *CIM Bulletin*, **82**(926), 53–62.

Poulos, H.G. and Davis, E.H. (1974), *Elastic Solutions for Soil and Rock Mechanics*, Wiley, New York.

Pramanik, R. and Deb, D. (2015), "Implementation of smoothed particle hydrodynamics for detonation of explosive with application to rock fragmentation", *Rock Mechanics and Rock Engineering* (Springer), DOI 10.1007/s00603–014–0657–y.

Pramanik, R. and Deb, D. (2014), "SPH procedures for modeling multiple intersecting discontinuities in geomaterial", *Int. J. Num. and Anal. Methods in Geomech.*, 39(4), 343–367.

Priest, S.D. and Hudson, J.A. (1976), "Discontinuity spacings in rock," *Int. J. Rock Mech. Min. Sci. & Geomech. Abstr.*, **13**, 135–148.

Rabcewicz, L. and Golser, T. (1972), "Application of the NATM to the underground works at Tarbela", *Water Power*, March, 88–93.

Ramamurthy, T. (1985), "Elastoplastic finite element analysis of slopes in rock mass", *Proc. Int. Conf. Finite Elements in Computational Mechanics*, Bombay, **1**, 291–300.

Ramamurthy, T. (1995), "Bearing capacity of rock foundations", Yoshinaka and Kikuchi (Eds.), *Rock foundation*, Balkema, Rotterdam.

Ramamurthy, T. (2014), "*Engineering in Rocks for Slopes Foundations and Tunnels*, 3rd ed., PHI Learning, Delhi.

Ramamurthy, T. and Arora, V.K. (1993), "A classification for intact and jointed rocks", *Geotechnical Engineering of Hard Soils and Soft Rocks*, Anagnostopoulos et al. (Eds.), Balkema, Rotterdam, ISBN 90 5410 344 2.

Ramamurthy, T., Sharma, K.G., and Ailawadi, O.P. (1995), "Bearing capacity of rock foundations", *Rock foundation*, Yoshinaka and Kikuchi (Eds.), Balkema, Rotterdam.

Read, J. and Stacey, P. (2012), *Guidelines for Open Pit Slope Design*, CRS Press, Australia, 342–356.

Remij, E.W., Remmers, J.J.C., Huyghe, J.M. and Smeulders, D.M.J. (2015), "The enhanced local pressure model for the accurate analysis of fluid pressure driven fracture in porous materials", *Computer Methods in Applied Mechanics and Engineering*, 286, 293–312.

Ren, G., Reddish, D.J. and Whittaker, B.N. (1987), "Mining subsidence and displacement prediction using influence function methods", *Min. Sci. Tech.*, **5**, 89–104.

Réthoré, J., Gravouil, A. and Combescure, A. (2005), "An energy-conserving scheme for dynamic crack growth using the extended finite element method", *Int. J. Num. Methods in Engg.*, **63**(5), 631–659.

Sadd, M.H. (2005), *Elasticity—Theory, Applications, and Numeric*, Academic Press, Elsevier, India.

Saeb, S. and Amadei, B. (1992), "Modelling rock joints under shear and normal loading", *Int. J. Rock Mech. Min. Sci. & Geomech. Abstr.*, **29**, 267–278.

Salamon, M.D.G. and Munro, A.H. (1967), "A study of the strength of coal pillars", *J.S. Afr. Inst. Min. Metal*, **68**, 55–67.

Sarkar S. and Kumar R. (2024), Behavior of Post-Yielding Parameters of Marble during Plastic Deformation, 58th U.S. Rock Mechanics/Geomechanics Symposium, Golden, Colorado, USA, June 2024, doi: https://doi.org/10.56952/ARMA-2024-0586.

Saxena, N.C. (2003), *Subsidence Management Handbook*, Scientific Publishers, Jodhpur, 205.

Serafim, J.L. and Pereira, J.P. (1983), "Consideration of the geomechanics classification of Bieniawski", *Proceedings of the International Symposium on Engineering Geology and Underground Constructions*, Lisbon, Portugal, II-33–II-43.

Sharan S.K. (2003), "Elastic-brittle-plastic analysis of circular openings in Hoek–Brown media", *Int. J. Rock Mech. Min. Sci. & Geomech. Abstr.*, **40**, 817–824.

Sheorey, P.R. (1988), "Estimation of gain in pillar strength due to bolting", *Int. J. Rock Mech. Mining Sci. & Geomech. Abstr.*, **25**, 321–326.

Sheorey, P.R. (1992), "Pillar strength considering *in situ* stresses", Iannacchione et al. (Eds.), *Proc. of the Workshop on Coal Pillar Mechanics and Design*, USBM IC 9315, Santa Fe, 122–127.

Sheorey, P.R. (1993), "Design of coal pillar arrays and chain pillars", *Comprehensive Rock Engineering*, Vol. 2, J.A. Hudson (Ed.), Pergamon, 631–670.

Sheorey, P.R., Biswas, A.K., and Choubey, V.D. (1989), "An empirical failure criterion for rocks and jointed rock masses", *Engg. Geol.*, **26**, 141–159.

Sheorey, P.R., Loui, J.P., Singh, K.B., and Singh, S.K. (2000), "Ground subsidence observations and a modified influence function method for complete subsidence prediction", *Int. J. Rock Mech. Mining Sci.*, **37**, 801–818.

Singh, B. (1993), "Norwegian method of tunneling workshop", New Delhi, CSMRS.

Singh, B. and Goel, R.K. (1999), *Rock Mass Classification*, Elsevier Science, Amsterdam, 267.

Singh, M.M. (1984), "Review of coal mine subsidence control measures", *Trans. SME-AIME*, **281**, 189–194.

Sjöberg, J., Christiansson, R., and Hudson, J.A. (2003), "ISRM suggested method for rock stress estimation: Part 2—overcoring methods", *Int. J. Rock Mech. Min. Sci.*, **40** (7–8), 999–1010.

Song, J.H., Wang, H. and Belytschko, T. (2008), "A comparative study on finite element methods for dynamic fracture", *Computational Mechanics*, **42**(2), 239–250.

Sonmez, H. and Ulusay, R. (1999), "Modifications to the geological strength index (GSI), and their applicability to stability of slopes", *Int. J. Rock Mech. Min. Sci.*, **36**, 743–760.

Sonmez, H. and Ulusay, R. (2002), "A discussion on the Hoek–Brown failure criterion and suggested modification to the criterion verified by slope stability case studies", Yerbilimleri (Earth Sciences), (26), 77–99.

Soren, K., Budi, G. and Sen, P. (2014), "Stability analysis of open pit slope by finite difference method", *Int. J. Research in Engg. Tech.*, **03**(05), 326–334.

Spencer, E. (1967), "A method of analysis of the stability of embankments assuming parallel inter-slice forces", *Geotechnique*, **17**(1), March, 11-26.

Stillborg, B. (1986), *Professional Users Handbook for Rock Bolting*, Trans. Tech Publications, Clausthal, 145.

Stille, H., Groth, T. and Fredriksson, A. (1982), "FEM-analysis of rock mechanical problems with JOBFEM", Stiftelsem Bergteknisk Forskning—BeFo, Stockholm, 307:1/82.

Su, W. and Peng, S. (1987), "Cutter roof and its causes", *Min. Sci. Technol.*, (4), 113–132.

Sukumar, N., Chopp, D.L., Moës, N. and Belytschko, T. (2001), "Modeling holes and inclusions by level sets in the extended finite-element method", *Computer Methods in Applied Mechanics and Engineering*, **190**(46), 6183–6200.

Sukumar, N., Moës, N., Moran, B. and Belytschko, T. (2000), "Extended finite element method for three-dimensional crack modeling", *Int. J. Num. Methods in Engg.*, **48**(11), 1549–1570.

Sun, Z., Gerrard, C.M. and Stephansson (1985), "Rock joint compliance tests for compression and shear loads", *Int. J. Rock Mech. Min. Sci. & Geomech. Abstr.*, **22**, 197–213.

Tao, Z.Y. and Chen, J.X. (1983), "Behaviour of rock bolts as tunnel support", *Int. Sym. Rock Bolting*, A.A. Balkema, 87–92.

Terzaghi, K. (1943), *Theoretical Soil Mechanics*, Wiley, New York.

Terzaghi, K. (1946), "Rock defects and loads on tunnel support", *Rock Tunneling with Steel Supports*, R.V. Proctor and T. White (Eds.), Commercial Shearing Co., Youngstone, OH, 15–99.

Terzaghi, K., Richart, F.E., (1952), Stress in Rock Around Cavities, Geotechnique, **3**, 57–99.

Twiss, R.J. and Moores, E.M. (2007), *Structural Geology*, W.H. Freeman, New York, 500.

Tzamos, S. and Sofianos, A. (2006), Extending the Q system's prediction of support in tunnels employing fuzzy logic and extra parameters, *International Journal of Rock Mechanics and Mining Sciences*, **43**(6), 938–949, https://doi.org/10.1016/j.ijrmms.2006.02.002.

Unal, E. (1996), "Modified rock mass classification: M-RMR system", *Milestone in Rock Engineering: A Jubilee Collection*, Z.T. Bieniawski (Ed.), Rotterdam, Balkema, 203–223.

Van Der Merwe, J.N. (2005), "Development and validation of a method to predict coal pillar life in South Africa", Post-Mining 2005, November 16–17, Nancy, France.

Varnes, D.J. (1970), "Model for simulation of residual stress in rock", *Proc. 11th US Symp. Rock Mech.*, Berkeley, SME/AIME, 415–26.

Vermeer, Pieter A. and R. De Borst (1984), Non-associated Plasticity for Soils, Concrete and Rock, 1–64.

Voight, B. (1966), "Interpretation of *in situ* stress measurements", Panel report on Theme IV, *Proc. 1st Cong. Int. Soc. Rock Mech.* (ISRM), Lisbon, Lab, Nac de Engg. Civil, Lisbon, Vol. III, 332–48.

Walton, O.R. (1980), "Particle dynamic modeling of geological materials", Lawrence Livermore National Laboratory, Livermore, CA, Report UCRL-52915.

Wang, H., Poulsen, B.A., Shen, B., Xue, S. and Jiang, Y. (2011), "The influence of roadway backfill on the coal pillar strength by numerical investigation, *Int. J. of Rock Mech. and Mining Sc.*, **48**, 443–450.

Wanless, H.R. (1964), "Local and regional factors in pennsylvanian cyclic sedimentation", D.F. Merriam (Ed.), Symposium on cyclic sedimentation: Kansas Geological Survey, *Bulletin*, **169**, 593–605.

Wardel, K. (1965), "Surface ground movements associated with the total and partial extraction of stratified mineral deposits", MSc Thesis, University of Nottingham, Nottingham, UK.

Weaver, J. (1977), "Three dimensional crack analysis", *Int J. Solids Struct.*, **13**, 321–30.

Wei Mingdong, Meng Wenzhao, Dai Feng, and Wu Wei, Application of machine learning in predicting the rate-dependent compressive strength of rocks, *Journal of Rock Mechanics and Geotechnical Engineering*, Volume 14, Issue 5, 2022, Pages 1356–1365.

Whittaker, B.N. and Reddish, D.J. (1989), *Subsidence Occurrence Prediction and Control*, Elsevier Publication, New York, 528.

Wickham, G.E., Tiedemann, H.R. and Skinner, E.F. (1972), "Support determination based on geological predictions", *Procd. Rapid Excav. Tunneling Conf.*, AIME, New York, 43–64.

Williams, J.R. and Mustoe, G.G. (1987), "Modal methods for the analysis of discrete systems", *Computers and Geotechnics*, **4**(1), 1–19.

Williams, T., Denton, D., Johnson, J., Maciosek, F. and Maslowski, M. (2004), "The use of instrumented split-set rock bolts to monitor strain in the walls of a deep underground mine stope", *SME annual meeting*, Feb. 23–25, Colorado.

Williamson, D.A. (1980), "Uniform rock classification for geotechnical engineering purposes", *Transp. Res.*, 783, 9–14.

Wilson, A.H. (1981), "Stress and stability is coal ribsides and pillars", *Proc. of First Conf. on Ground Control in Mining*, West Virginia Univ., Morgantown, 1–12.

Wu H, Guo Y, Xiong L, Liu W, Li G., and Zhou X., (1999), Optical Fiber-Based Sensing, Measuring, and Implementation Methods for Slope Deformation Monitoring: A Review, *IEEE Sensors Journals*, Vol. 19, No. 8, 2786–2800.

Xu, X.P. and Needleman, A. (1994), "Numerical simulations of fast crack growth in brittle solids", *J. of the Mechanics and Physics of Solids*, **42**(9), 1397–1434.

Yoshida, N., Morgenstern, N.R. and Chan, D.H. (1990), "Failure criterion for stiff soils and rocks exhibiting softening", *Canadian Geotechnical Journal*, **27**(2), 195–202.

Yu, Hai-Sai (2000), *Plasticity and Geotechnics*, Springer, 521.

Yu, M.H. and He, L.N. (1991), "A new model and theory on yield and failure of material under complex stress state", *Mechanics Behaviour of Materials*, VI, M. Jono and T. Inoue (Eds.)(ICM-6), Oxford, Pergamon, 841–846.

Yu, X. and Vayssad, B. (1991), "Joint profiles and their roughness parameters", *Int. J. of Rock Mech. and Mining Sci.*, **28**(4), 333–336.

Yudhbir, Lemanza W. and Pronzl, F. (1983), "An empirical failure criterion for rock masses", *Proc. 5th Int. Con. Rock Mech.*, Melbourne, Balkema, Rotterdam, **1**, B1–B8.

Zhang, W., Gu X., Hong L., Han L., and Wang L. (2023), Comprehensive review of machine learning in geotechnical reliability analysis: Algorithms, applications and further challenges, Applied Soft Computing, **136**, 110066.

Zhang Ning, Shen Shui-Long, Zhou Annan, and Jin Yin-Fu, Application of LSTM approach for modelling stress–strain behaviour of soil, Applied Soft Computing, Vol. 100, 2021, 106959,

Zhao, Z.H., Wang, W.M. and Yan, J.X. (2013), "Strain localization and failure evolution analysis of soft rock-coal-soft rock combination model", *Journal of Applied Sciences*, **13**, 1094–1099.

Zhao XC, Cai M., (2020), A mobilized dilation angle model for rocks, *Int. J. Rock Mech. Min. Sci.*, **47**(3), 368–384.

Zhu, W. and Wang, P. (1993), "Finite element analysis of jointed rock masses and engineering application", *Int. J. Rock Mech. Min. Sci. & Geomech. Abstr.*, **30**, 537–544.

Solutions to Chapter End Exercises

CHAPTER 1

1. Natural material is made by nature, whereas engineering materials are man-made using natural materials. In rock engineering, the purpose of natural materials are used for excavation purposes (like surface and underground), construction on top (used as the foundations for high-rise buildings), or structures that can be made with them (old monuments or houses made of building materials). Whereas the purpose of artificial materials is to enhance the strength of this natural material for engineering purposes. Examples of natural materials are rock and soil, whereas examples of engineering materials are metals [Aluminum, Iron, Steel, Brick, Reinforced Concrete Cement (RCC), Fibers, etc.].
2. Intact rock and rock mass are geological materials. Intact rock is defined as a rock that is free from geological discontinuities. The geological discontinuities include the joint, foliation, fracture, etc. Intact rock changes to rock mass once passing through the various phases of tectonic activities to infuse joints/foliation/fractures, etc. Hence, rock mass is defined as an intact rock with discontinuities. With the presence of geological discontinuities the physical and mechanical properties of rock reduces. The rock mass represents a rock material that is encountered during the excavation/construction of real-life projects, whereas the intact rock is generally a small rock sample tested in the laboratory to determine the intact rock properties.
3. The geological cycle is defined as the process of formation of rock from molten magma as an igneous rock, then passes through the phases of decomposition due to weathering by geological forces and then with the formation of sedimentary and metamorphic rocks and finally converted again to magma as shown in Figure 1.1. Based on formation, rock is classified as igneous rock, sedimentary rock, and metamorphic rock, and their most common examples are granite, basalt, serpentinite, gabbro, peridotite, etc. for igneous rock, for sedimentary rock common example is sandstone, limestone, coal, hematite, magnetite, limonite, etc., and for metamorphic rock the common example is quartzite, marble, slate, schist, gneiss, phyllite, etc. Slate, Phyllite, and Schist are formed from shale after its metamorphism.

4. Rocks are made of one or more minerals, i.e., an aggregate of minerals. The mineral is a naturally occurring substance having a fixed chemical composition and molecular structure formed from magma or naturally occurring chemical compounds making up a part of the earth's crust. Minerals may be gaseous (natural gas), liquid (petroleum, mercury, water) or solid (ore minerals, etc.). Almost 3000 minerals are known, out of which about 25 are rock-forming minerals. Most of the rock-forming minerals are classified as silicates, oxides, and carbonates. On the upper part of the earth's crust, Feldspars (sodium, potassium, and calcium aluminum silicates) consist of 60%; Amphiboles and Pyroxene 17%; Quartz 12%, and mica 3.8%. A mineral may be a single element, such as copper (Cu) or gold (Au), or it may be a compound made up of a number of elements, such as sodium chloride (NaCl), silica dioxide (SiO_2), etc. Mineralogy is a branch of geology that deals with various aspects related to minerals, such as their individual properties, their mode of formation, and their mode of occurrence.
5. The difference based on geology is that both rock and soil are natural materials formed from geological processes. Initially, rock formed from magma, and its weathering over millions of years decompose rock into soil. From an engineering perspective, rock is a cohesive material that is discontinuous, Inhomogeneous (or heterogeneous), Anisotropic, and Nonlinear Elasticity, whereas soil is a cohesionless and homogeneous material. As soil is formed from rock, hence the mineral composition will be the same (for the same type) but differ widely in physical and mechanical properties.
6. These are the very common structures found in the metamorphic rocks. Cleavage is defined as the capacity of the rock to split along certain directions, yielding parallel and subparallel smooth surfaces. The rock, which is made up of parallel or subparallel bands or layers of platy and flaky minerals like mica, chlorite, tourmaline, etc., is called Schistosity. Foliation is defined as a layered structure.
7. Joints are formed in the rock mass due to different reasons. Researchers have classified at least three principal processes of rock joint formation such as: (a) contraction during formation, (b) expansion and contraction, and (c) crustal disturbance. For example, in sedimentary rocks such as shale, limestone, clays, etc., which are plastic in nature and reach in moisture content, undergo some contraction on drying, resulting in the formation of regular joints. Similarly, igneous rocks, which form by cooling and differential crystallization from an originally hot and molten material (i.e., lava), leads to the formation of joint. Some rock expand with rise of temperature and contract on cooling. Such repeated and differential melting and cooling process leads to the joint formation. Crustal disturbance is responsible for raising mountains and deepening oceanic trenches, shifting and shaping continents, and triggering earthquakes. The strong force exerted by crustal disturbance develops joints in many rocks.
8. A fault is a fracture discontinuity along which the rocks on either side have moved past each other. The fault plane is identified by dip and strike. The trace of fault on the earth's surface is called a fault line. If the fault is not vertical, then the side above the fault plane is called the hanging wall or hanging wall, and the side below the fault plane is called the foot wall. The fault surface can be a plane or curve. Based on the relative translation movement of the side, faults are classified as: (a) dip-slip fault, (b) strike-slip fault, and (c) oblique-slip fault.

9. Stereographic project is a graphical method to represents a 3D geological features (say for fault) like fault line or fault plane on a 2D surface. This projection method is used to determine the included angle between two lines or two planes; length of a line; areá of a plane; and finally volume of a block created when two or more geological features are intersecting with each other (see these application in Chapter 12 on Slope Stability). For plane, strike and dip direction whereas for line trend and plunge. Pitch is defined as an angle within a plane with a horizontal line and line of intersection.
10. Figure shows the line of trend 245° and Plunge 35° plots as a point, in the south-west quadrant of the projection.

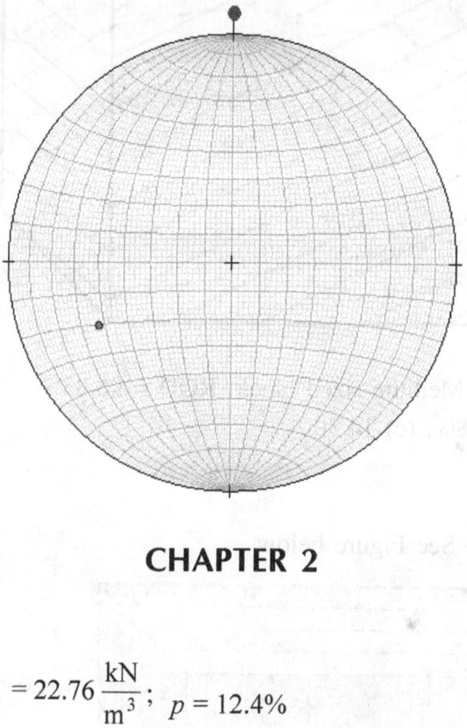

CHAPTER 2

1. $\sigma_V = 49.217$ MPa
2. $\gamma_{\text{wet}} = 24 \dfrac{\text{kN}}{\text{m}^3}$; $\gamma_{\text{dry}} = 22.76 \dfrac{\text{kN}}{\text{m}^3}$; $p = 12.4\%$
3. $p = 25.81\%$
4. $\kappa_{\text{eq}} = \dfrac{\sum_{i=1}^{n} (k_i T_i)}{\sum_{i=1}^{n} T_i}$
5. $p = 34\%$
6. $\Delta V = 22.22\%$
7. $w = 11\%$
8. $\kappa = 0.0002$ cm/s; $h = 32.54$ cm
9. $\kappa = 3.46 \times 10^{-6}$ m/s
10. $p = 33.28\%$; $v = 0.49$; $\rho_d = 1768$ kg/m^3; $s = 69.10\%$

CHAPTER 3

1. (a) 1.97 m, (b) 1.11 m, (c) 99.62
2. Figure below

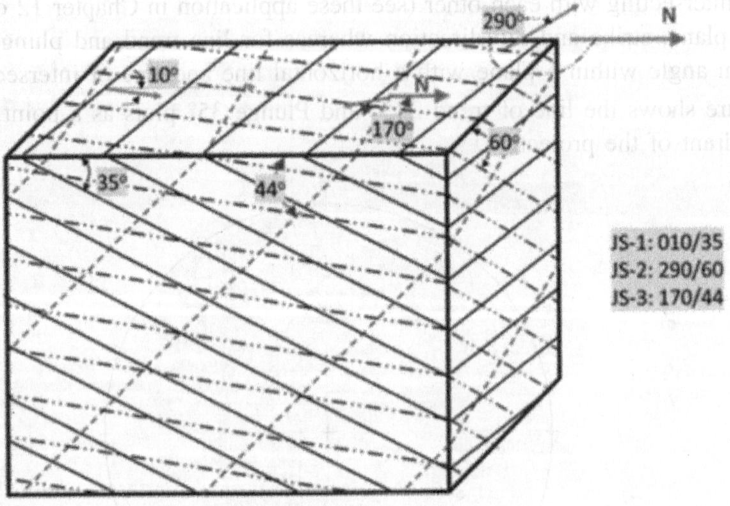

3. J_V = 4.9 Joints/m^3, Medium-sized block, RQD = 98.83%
4. (a) 67.8%, (b) 90.88%, (c) 14.30
5. Do it
6. 62%
7. (a) 53 and Fair, (b) See Figure below

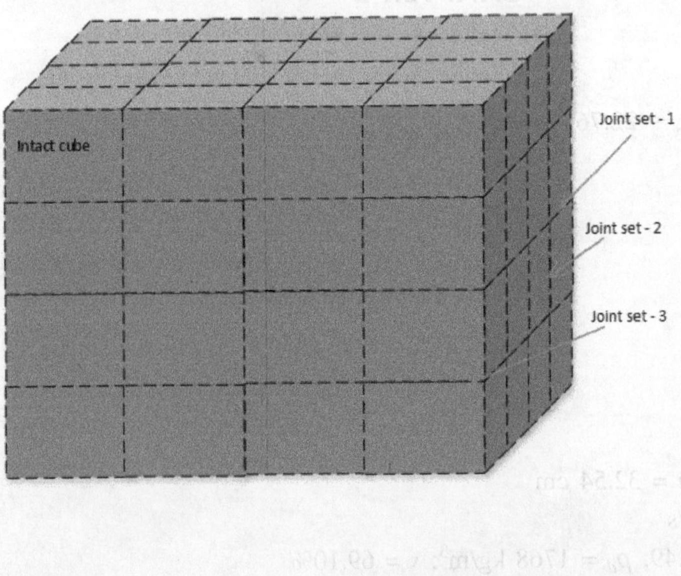

8. (a) Figure below, (b) RQD = 63% and Fair, (c) Stable with a standup time of 3000 hrs.

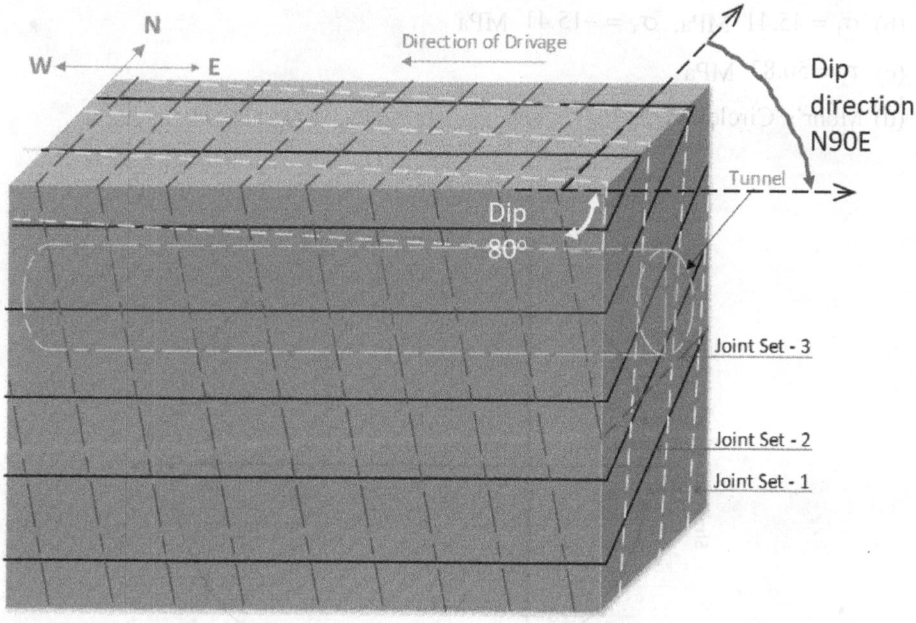

9. Q = 10, Fair
10. Q = 39.84 (Good); RMR = 77.16 (as per Bieniawski's 1989) and 74.00 (as per Barton's 1995) Good Quality Rock mass; GSI = 72.16 (as per Bieniawski's 1989) and 69.00 (as per Barton's 1995). Surface quality is good and Rock pieces are Blocky.
11. (i) UR for UCS = 1.61, DSR = 33, Shear Strength = 35, Moisture Rating = 0, Unit Rating for sandstone = 69.61; (ii) weak roof

CHAPTER 4

1. Proof is given in the book.
2. (a) $\sigma' = \begin{bmatrix} 2.85 \\ -5.85 \\ -1.53 \end{bmatrix}$ (b) $\begin{bmatrix} 0.00285 \\ -0.00585 \\ -0.00153 \end{bmatrix}$
3. Traction: $T = \begin{bmatrix} 4.92 \\ 1.79 \end{bmatrix}$ MPa, σ_n = 5.20 MPa, τ = –0.6 MPa
4. (a) N_1 = 16, N_2 = –68.75, and N_3 = 53.5

   σ_1 = 9.07 MPa, σ_2 = 5.93 MPa, σ_3 = 0.99 MPa; with direction cosine (0.20, 0.37, 0.96), (0.36, 0.88, 0.28), and (0.88, 0.27, –0.32) respectively

   (b) $\sigma_{x'x'}$ = 7.0 MPa, $\tau_{x'y'}$ = –3.18 MPa, $\tau_{x'z'}$ = –1.42 MPa

5. (a) $\sigma_n = -14.31$ MPa, $\tau_s = 8.10$ MPa

(b) $\sigma_1 = 45.41$ MPa, $\sigma_3 = -15.41$ MPa

(c) $\tau_s = 50.83$ MPa

(d) Mohr's Circle

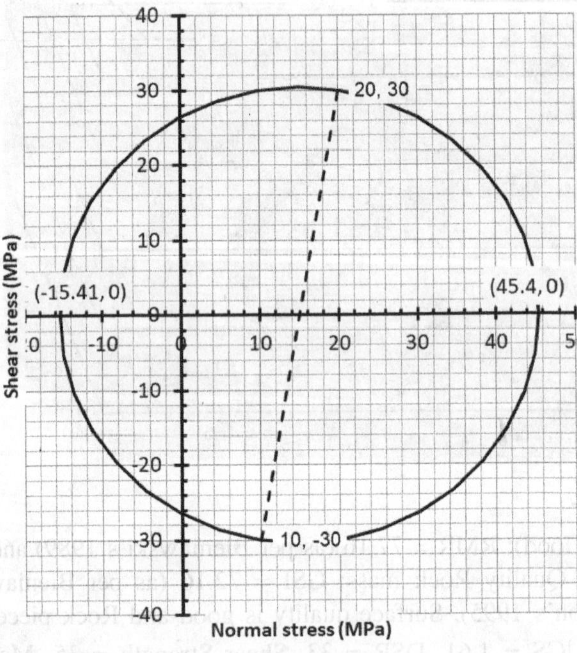

6. $\sigma_{x'x'} = 7.267$ MPa, $\tau_{x'y'} = -1.22, \tau_{x'z'} = 2.12$ MPa

7. $\sigma_1 = 9.22$ MPa, $\sigma_2 = 5.70$ MPa, $\sigma_3 = -0.72$ MPa; with direction cosine (0.92, −0.05, 0.39), (–0.125, 0.36, −0.33), and (0.318, 0.69, 0.64), respectively. $\tau_{max} = 4.97$ MPa

8. Mohr's Circle for (a) stress, and (b) strain transformation

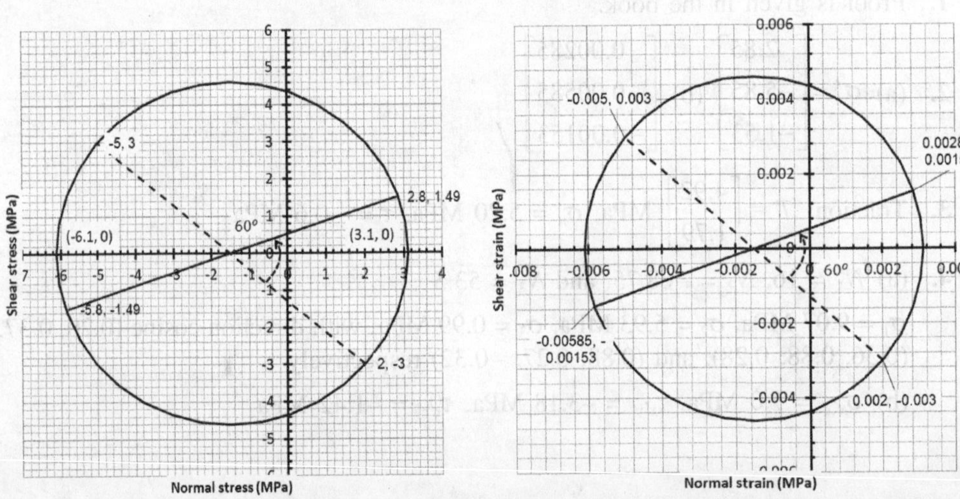

9. Exercise 4.3 using Mohr's circle of stress

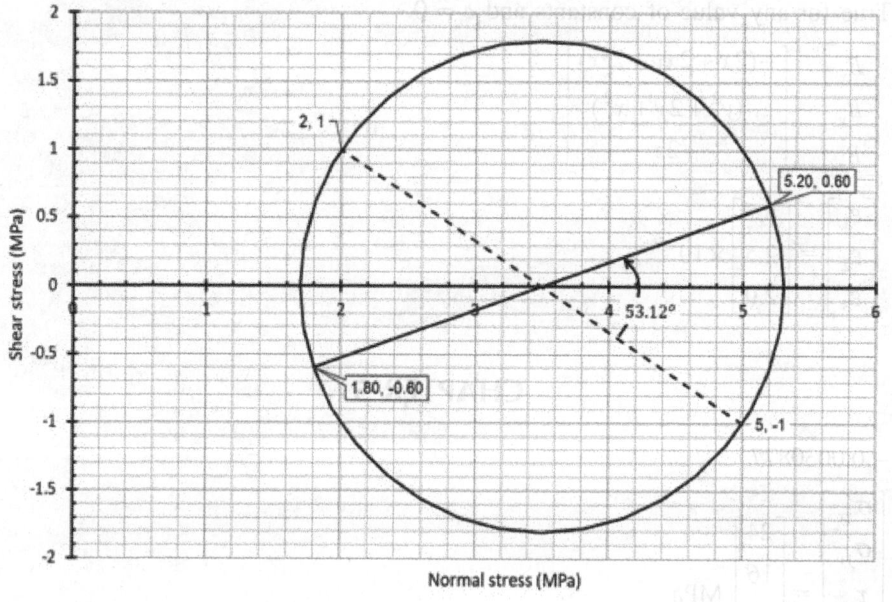

10. Draw Mohr circle (a) and (b)

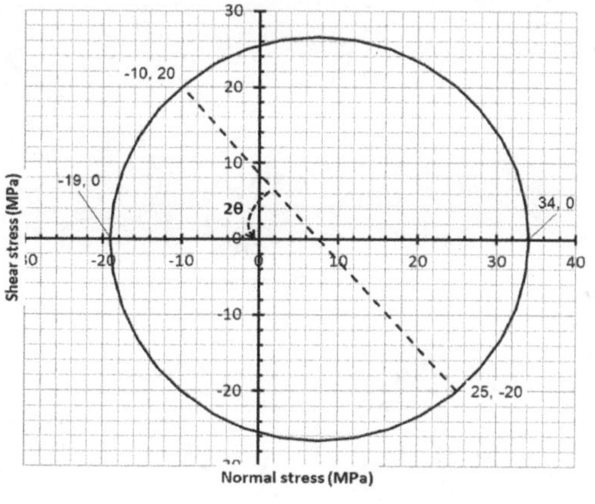

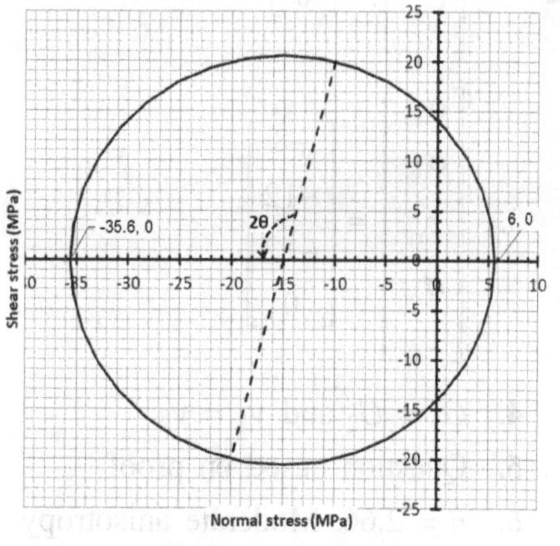

11. 0.74

12. $N_1 = 6$, $N_2 = -8$, $N_3 = 0$; $\sigma_1 = 4$ MPa, $\sigma_2 = 2$ MPa, $\sigma_3 = 0$ MPa

13. $$\begin{bmatrix} \varepsilon_{xx} \\ \varepsilon_{yy} \\ \gamma_{xy} \end{bmatrix} = \begin{bmatrix} 0.004 \\ 0.002 \\ 0.0 \end{bmatrix}, \begin{bmatrix} \varepsilon'_{xx} \\ \varepsilon'_{yy} \\ \gamma'_{xy} \end{bmatrix} = \begin{bmatrix} 0.0025 \\ 0.0035 \\ -0.00173 \end{bmatrix}$$

14. $$\begin{bmatrix} \varepsilon_{xx} \\ \varepsilon_{yy} \\ \gamma_{xy} \end{bmatrix} = \begin{bmatrix} -0.001 \\ -0.002 \\ -0.006 \end{bmatrix}, \quad \varepsilon_1 = 0.004, \ \varepsilon_3 = 0$$

15. Question based on Proof

16. True for any value of constants and $e = 0$.

17. $$\begin{bmatrix} B_x \\ B_y \\ B_z \end{bmatrix} = \begin{bmatrix} -(2x+2xy+2zx) \\ -(y^2+2y+x^2) \\ -2z \end{bmatrix}$$

18. $$\begin{bmatrix} \varepsilon_1 \\ \varepsilon_2 \\ \varepsilon_3 \end{bmatrix} = \begin{bmatrix} 5.5 \\ 2.5 \\ 2.0 \end{bmatrix} \times 10^{-5}$$

CHAPTER 5

1. – 0.0030877

2. $$\begin{bmatrix} \sigma_{xx} \\ \sigma_{yy} \\ \tau_{xy} \\ \sigma_{zz} \end{bmatrix} = \begin{bmatrix} 24 \\ 16 \\ 0 \\ 0 \end{bmatrix} \text{MPa}$$

3. $$\begin{bmatrix} \sigma_{xx} \\ \sigma_{yy} \\ \tau_{xy} \\ \sigma_{zz} \end{bmatrix} = \begin{bmatrix} -8 \\ -12 \\ -12 \\ 0 \end{bmatrix}$$

4. $G_1 = G_2$ and $\nu_1 = \nu_2$

5. Question based on proof

6. $n = 2.66$, Moderate anisotropy

7. $\sigma_1 = 4.58$ MPa, $\sigma_2 = 2.08$ MPa, $\theta_p = 8.05°$

8. $\varepsilon^e = 0.0825 \times 10^{-4}$, $\varepsilon^p = 0.517 \times 10^{-3}$

9. (a) 37.5 MPa; (b) 0.0002, 0.0008; (c) 6250 MPa

10. 2.30 mm

11. 0.34, 26.36 GPa, 0.26%

12. $0 \le \nu < 1/8$

CHAPTER 6

1. σ_1 = 8.31 MPa; σ_3 = 6.70 MPa and θ_p = 87.07°
2. σ_1 = 7.92 MPa; σ_3 = 7.02 MPa and θ_p = 82.17°
3. σ_{xx} = 9.24 MPa; σ_3 = 10.86 MPa and θ_p = 0.61°
4. (a) σ_{xx} = 22 MPa; σ_3 = 11.5 MPa and θ_{xy} = 0°; (b) $\begin{bmatrix} 22.75 & 0 & 0 \\ 0 & 22.00 & 0 \\ 0 & 0 & 11.50 \end{bmatrix}$
5. (a) σ_1 = 19.24 MPa; σ_3 = 6.76 MPa and θ_p = 41.78°
6. Rosette 1: σ_1 = 3.0 MPa; σ_3 = 2.1 MPa and θ_p = 39.49°;
   Rosette 2: σ_1 = 0.99 MPa; σ_3 = 0.65 MPa and θ_p = –24.12°;
   Rosette 3: σ_1 = 1.14 MPa; σ_3 = 0.11 MPa and θ_p = 21.73°.
7. Borehole 1: P = 14.22 MPa; Q = 7.85 MPa and θ_p = 15°;
   Borehole 2: P = 15.55 MPa; Q = 8.50 MPa and θ_p = 6.99°;
   Borehole 3: P = 14.86 MPa; Q = 1.65 MPa and θ_p = 28.29°.
8. P = 11.84 MPa; Q = 11.02 MPa and θ_p = 26.21°
9. P = 10.79 MPa; Q = 5.14 MPa and θ_p = –21.55°
10. σ_1 = 20.8 MPa; σ_2 = 17.0 MPa; σ_3 = 7.0 MPa
11. P_s = 71.6 MPa, P_B = 141.3 MPa
12. ΔR = 0.154 Ω
13. ΔR = 3.12 Ω
14. σ = 5.88 MPa
15.

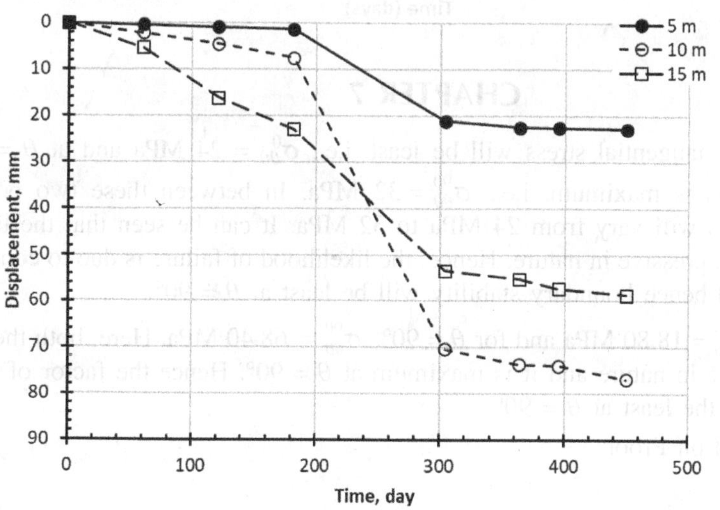

16. Table shows the calculation of developed pressure and plot between pressure versus time since start of reading

Sl. No.	*Day*	*Frequency* (Hz)	*Pressure* (kPa)
R_0	0	2873.8	0
R_1	50	2878.8	48
R_2	100	2886.8	125
R_3	250	2919.7	441
R_4	300	2927.7	518
R_5	350	2939.5	632
R_6	400	2941.6	652
R_7	450	2944.6	681
R_8	500	2944.6	681

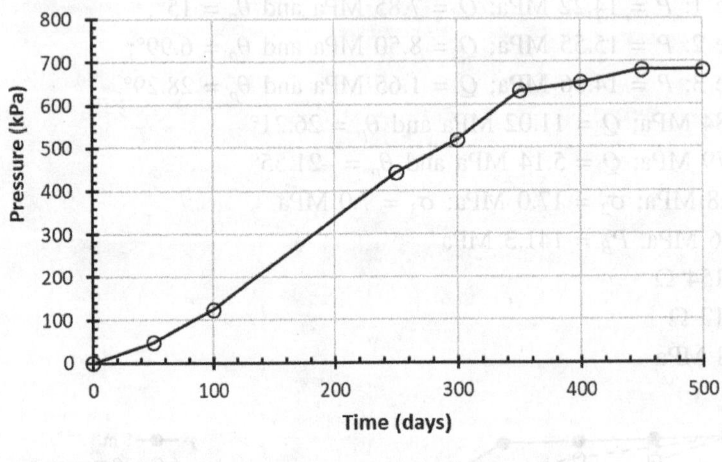

CHAPTER 7

1. For $\theta = 0°$ the tangential stress will be least, i.e., $\sigma_{\theta\theta}^{0}$ = 24 MPa and at $\theta = 90°$ the tangential stress is maximum, i.e., $\sigma_{\theta\theta}^{90} = 32$ MPa. In between these two points, the tangential stress will vary from 24 MPa to 32 MPa. It can be seen that the tangential stresses are compressive in nature. Hence, the likelihood of failure is due to compression stress only and hence boundary stability will be least at $\theta = 90°$.

2. For $\theta = 0°$, $\sigma_{\theta\theta}^{0}$ = 18.80 MPa and for $\theta = 90°$, $\sigma_{\theta\theta}^{90}$ = 68.40 MPa. Here, both the stresses are compressive in nature and it is maximum at $\theta = 90°$. Hence the factor of safety of the opening is the least at $\theta = 90°$.

3. Question based on Proof.

4. 2.08

5. Question based on Proof.

6. (a) $FOS_C = 2.67$, $FOS_T = 3.95$, (b) 3.61, (c) 0.80
7. $\sigma_{\theta\theta} = 3.75, 5, 6.25, 7.5, 11.25$ MPa and 15 MPa for $k = 0, 1/3, 2/3, 1, 2$ and 3, respectively.
8. For $\theta = 90°$ is $\sigma_{\theta\theta} = 75$ MPa, and for $\theta = 0°$, $\sigma_{\theta B} = 15$ MPa. The FOS at point A and B are 1.33 and 6.67, respectively.
9. (a) $FOS_A = 1.25$, $FOS_B = 0.59$; (b) major horizontal direction; (c) As per the question, the second shaft should be placed along the minor axis to achieve minimum distance between the shaft is 17.82 m; (d) $p_i = 14$ MPa.
10. Circumferences of both the tunnels are not stable.
11. Graph shows the shear distribution along line of discontinuity along with the table displaying the required calculations. As can be seen, (based on FOS) slippage occurs from 0 to 1 m on the left side and from 0.5 to 1.5 m on the right side.

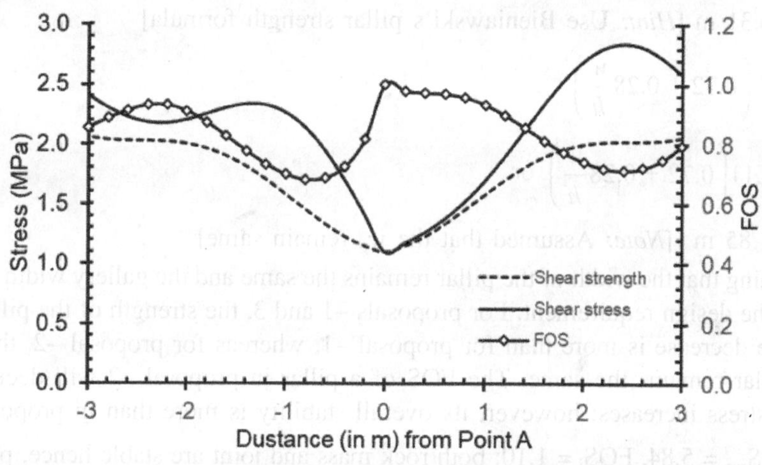

12. (a) 1.70 MPa on the fault at a distance of 4 m on both side of point A, (b) plot for normal and shear stress on fault surface is given in the figure.

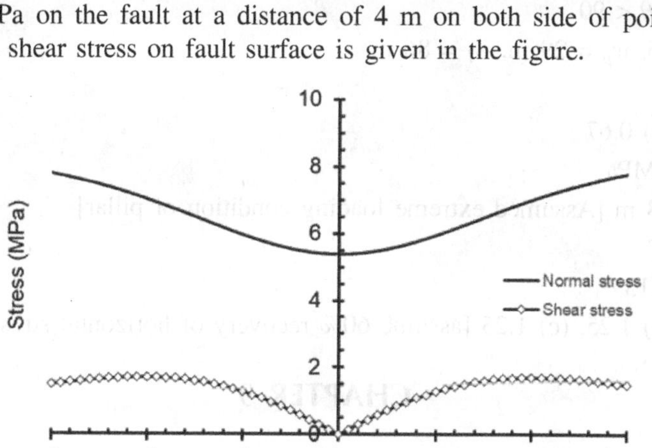

13. (a) $\sigma_\theta = 29$ MPa, $\sigma_r = 1$ MPa; $\tau = 0$; (b) $\sigma_{rr} = 1$ MPa; $\sigma_{\theta\theta} = 4.22$ MPa; and $\tau = 0$; (c) The reduction in tangential stress ($\sigma_{\theta\theta}$) is due to the fracture/yield zone created around the opening.

14. (a) p_i = 20 MPa; (b) p_i = 11.87 MPa
15. σ_1 = 22.08 MPa; σ_3 = 17.92; and θ = 43.03°
16. σ_1 = 33.88 MPa; σ_3 = 11.05; and θ = –23.41°
17. 58.2 m
18. FOS_A = 0.12, FOS_A = 0.24
19. p_i = 20.46 MPa
20. k = 1.37

CHAPTER 8

1. (a) 0.395, (b) $w_p \approx$ 21 m
2. w_g = 6.31 m [*Hint:* Use Bieniawski's pillar strength formula]
3. $S_p = S_l\left(0.72 + 0.28\dfrac{w}{h}\right)$
4. $S_p = 3.11\left(0.72 + 0.28\dfrac{w}{h}\right)$ or
5. w_g = 1.85 m [*Note:* Assumed that the w_g remain same]
6. Assuming that the width of the pillar remains the same and the gallery width is changed to fulfil the design requirement. For proposals –1 and 3, the strength of the pillar decreases, but the decrease is more than for proposal –1, whereas for proposal –2, the strength of the pillar remains the same. The FOS of a pillar in proposal –2 will decrease because pillar stress increases; however, its overall stability is more than in proposals 1 and 3.
7. (a) FOS_{rm} = 5.84, FOS_j = 1.10; both rock mass and joint are stable hence, pillar is stable, (b) $44.9 < \theta < 90$
8. R_{max} = 0.46, w_p = 29 m, w_l = 86 m
9. 10.5 m
10. (a) 1.01, (b) 0.67
11. 0.83, 1.48 MPa
12. w_{cp} = 79.08 m [Assumed extreme loading condition of pillar]
13. FOS = 1.17
14. 3.38 and 3.13
15. (a) 0.75, (b) 1.25, (c) 1.25 [assume 60% recovery of horizontal stress after backfilling]

CHAPTER 9

1. Question based on Proof
2. σ_1' = 16.33 MPa, σ_2' = 8.16 MPa, σ_3' = –4.08 MPa, θ = 6.59°

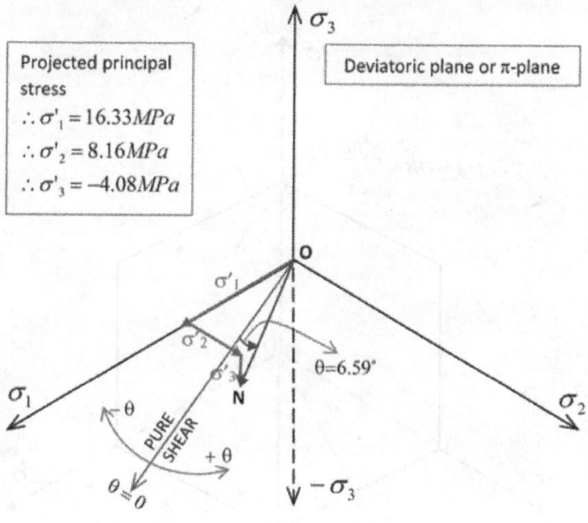

3. (a) 21.65 MPa, 30°; (b) Unsafe
4. Safe
5. Rock **fails** if we consider the yield surface which coincide inner apices and it is **safe** for the yield surface coinciding outer apices.
6. –26.09°

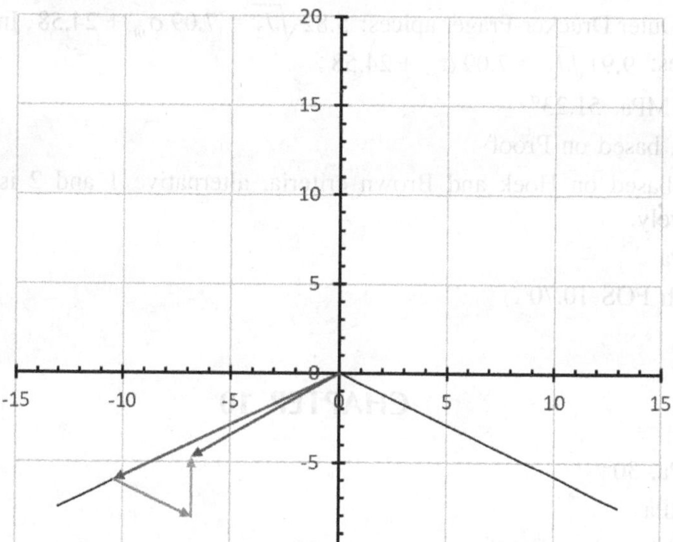

7. Safe

8. See plot

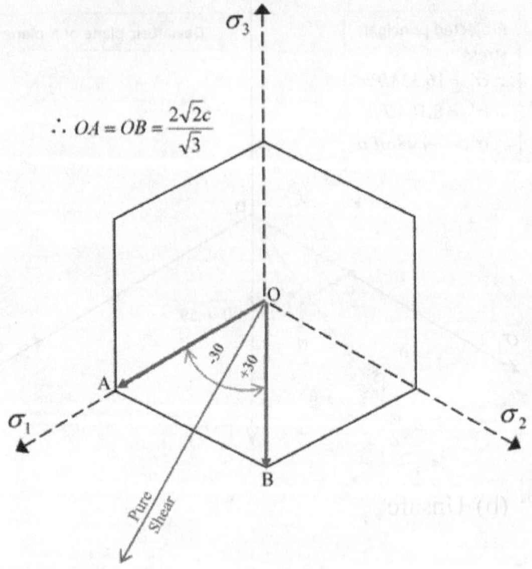

9. 1.70 MPa

10. (a) $\sigma_1 = 8.08$, $\sigma_3 = 24.58$,

(b) For Outer Drucker-Prager apices: $5.82\sqrt{J_2} = 7.09\,\sigma_m + 24.58$, Inner Drucker-Prager apices: $9.91\sqrt{J_2} = 7.09\,\sigma_m + 24.58$;

(c) 4.32 MPa, 51.23°

11. Question based on Proof

12. Safe as based on Hoek and Brown criteria, alternative 1 and 2 is –11.68 and –2.30 respectively.

13. 1.21 MPa

14. Safe with FOS 10.70

15. 4.53 m

CHAPTER 10

1. 0.43 MPa, 30°

2. 34.56 MPa

3. Question based on Proof

4. 2.47 MPa and slip will occur along joint plane 1

5. (a) 0.43 MPa, 35.37°; (b) $\tau = \sigma_n \tan\left(0.38 \log_{10}\left(\frac{30}{\sigma_n}\right) + 17.76\right)$; (c) See Table (d) See Plot

σ_n	τ_p	τ_r	$\tan^{-1}\left(\frac{\tau}{\sigma_n}\right)$	$\log\frac{30}{\sigma_n}$	$\tau_p(B)$	δ_p
0.50	0.8	0.25	1.012	1.78	0.755	19.36
1.00	1.2	0.54	0.876	1.48	1.188	16.08
1.50	1.4	0.90	0.751	1.30	1.558	14.16
2.00	1.9	1.20	0.760	1.18	1.889	12.80
2.50	2.1	1.41	0.699	1.08	2.193	11.75
3.00	2.5	1.70	0.695	1.00	2.476	10.89
3.50	3.0	2.00	0.709	0.93	2.742	10.16

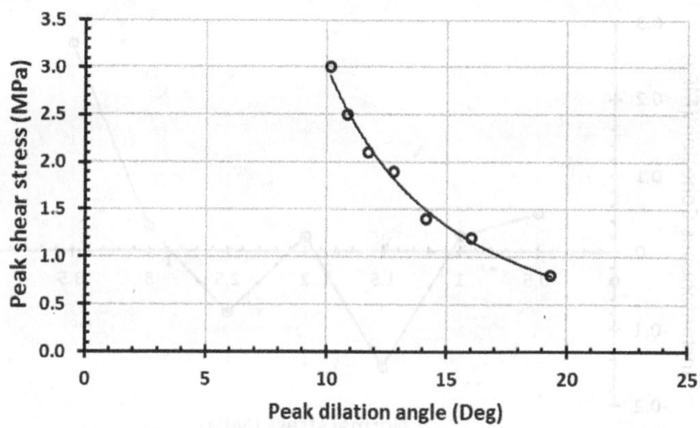

6. Table and Plots are

σ_n (MPa)	τ_p (MPa)	ϕ_i (Deg.)	C_i (MPa)
0.5	0.8	56.41	0.05
1	1.2	49.83	0.02
1.5	1.4	45.99	−0.15
2	1.9	43.26	0.02
2.5	2.1	41.14	−0.08
3	2.5	39.41	0.03
3.5	3	37.95	0.27

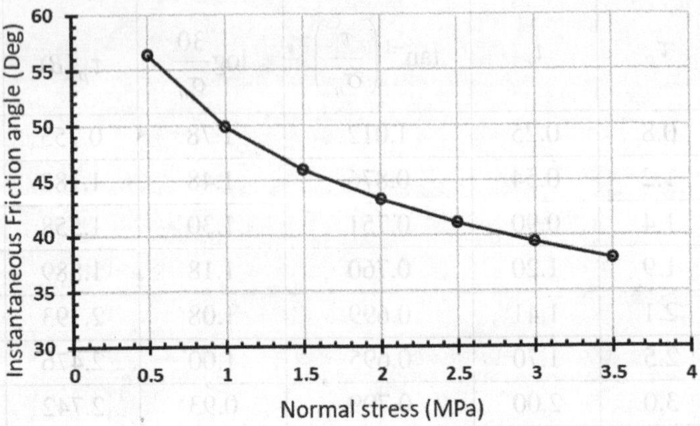

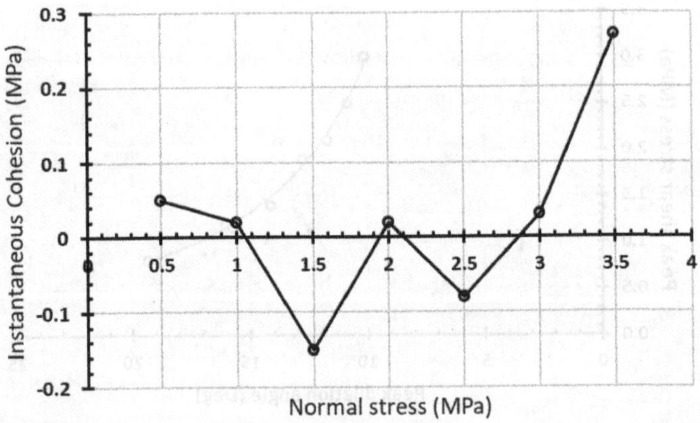

7. (a) $D_m = \begin{bmatrix} 0.1928 & -0.0599 & 0.0027 & -0.05 \\ -0.0599 & 0.1828 & -0.0175 & -0.05 \\ 0.0027 & -0.0175 & 0.4675 & 0 \\ -0.05 & -0.05 & 0 & 0.1667 \end{bmatrix} \times 10^{-6}$

(b) $\varepsilon = \begin{bmatrix} 0.599 \\ -1.343 \\ -0.140 \\ 0 \end{bmatrix} \times 10^{-3}$

8. $d_j = 0.02$ mm, $d_{eq} = 0.024$ mm (total)

9. No Slip: $\tau(Barton) = 5.7$ MPa, $\tau(estimated) = 4.9$ MPa

10. $C_m = \begin{bmatrix} 1.5/k & 0 & -0.5/k & 0 \\ 0 & 1.5/k & -0.5/k & 0 \\ -0.5/k & -0.5/k & 3.0/k & 0 \\ 0 & 0 & 0 & 1.0/k \end{bmatrix}, \ \varepsilon = \begin{bmatrix} 0 \\ 1.5 \\ -0.5 \\ 0 \end{bmatrix} \times 10^{-3}$

11. 1.16 MPa, 30°

12. $C_m = \begin{bmatrix} 0.2375 & -0.05 & -0.0717 & 0 \\ -0.05 & 0.2125 & -0.0717 & 0 \\ -0.0217 & -0.0217 & 0.55 & 0 \\ -0.05 & -0.05 & 0 & 0.2 \end{bmatrix} \times 10^{-9}$

CHAPTER 11

1. (a) 0 Pa, (b) 13.2 MPa
2. The pressure and deformation of active and passive supports are 5.67 MPa, 4.33 mm and 5.33 MPa, 4.67 mm respectively.

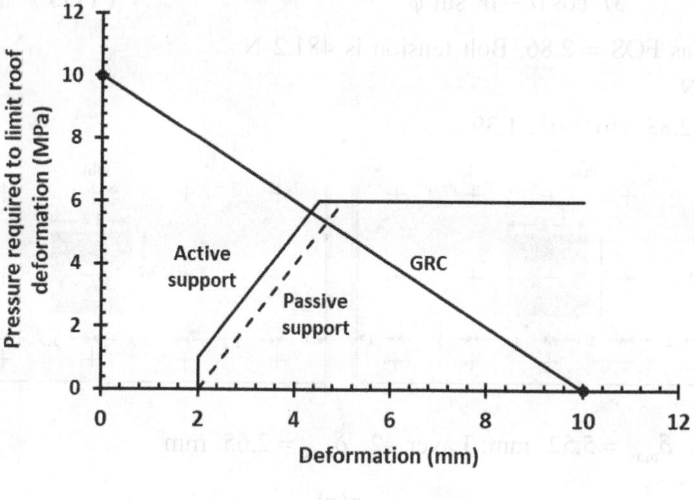

3. (a) 13.33 MPa, 8.33 mm, (b) 12.96 MPa, 10.19 mm
4. 3.96 MPa, 7.37 mm

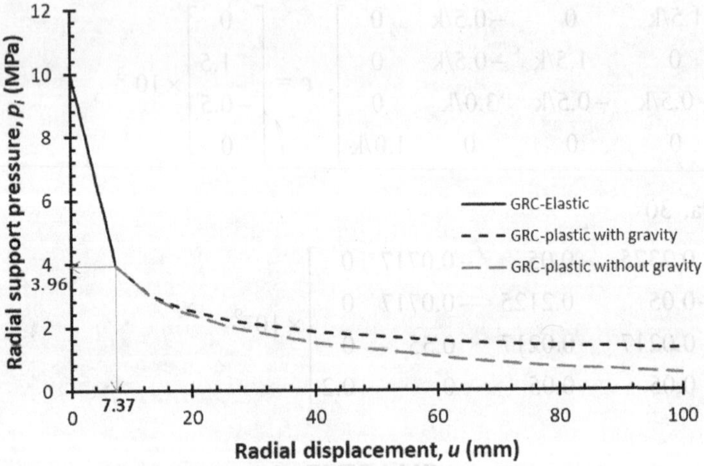

5. $T = \dfrac{\text{FOS}(W \sin \alpha) - W \cos \alpha \tan \varphi - cA}{\text{FOS} \cos(\alpha + \beta) + \sin(\alpha + \beta) \tan \varphi}$

6. $\text{FOS} = \dfrac{[c + (3T \sin \alpha + W \cos \psi) \tan \varphi] A}{3T \cos \alpha - W \sin \psi}, \beta = -\alpha + \tan^{-1}\left(\dfrac{\tan \varphi}{\text{FOS}}\right)$

7. No slide as FOS = 2.86, Bolt tension is 481.2 N

8. 112.10 kN

9. (a) 6.28, 2.88; (b) 3.03, 1.39

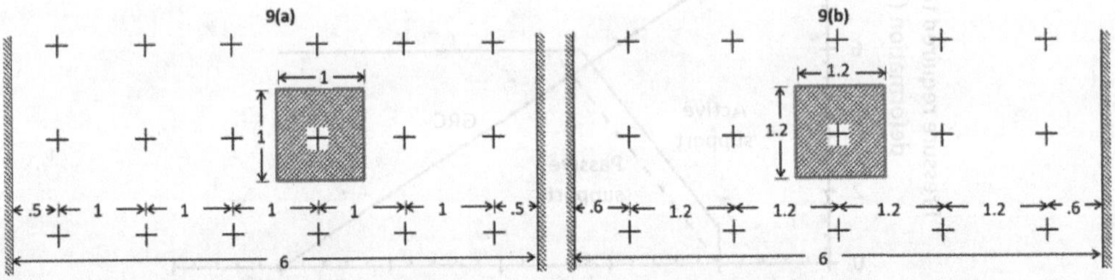

10. Layer –1, $\delta_{\max} = 5.52$ mm; Layer –2, $\delta_{\max} = 2.65$ mm

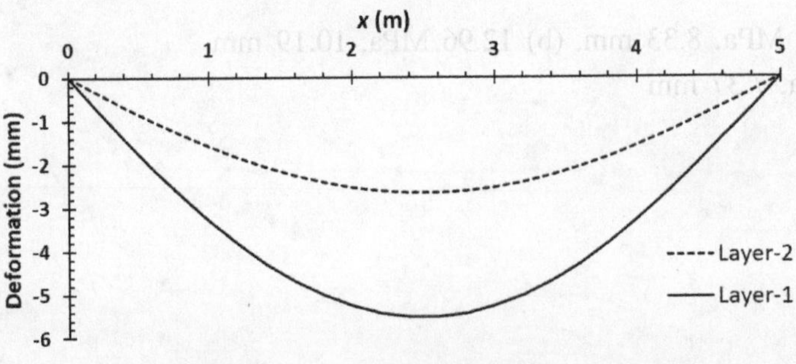

11. 0.79 m

12. Composite layer, $\delta_{max} = 0.76$ mm;

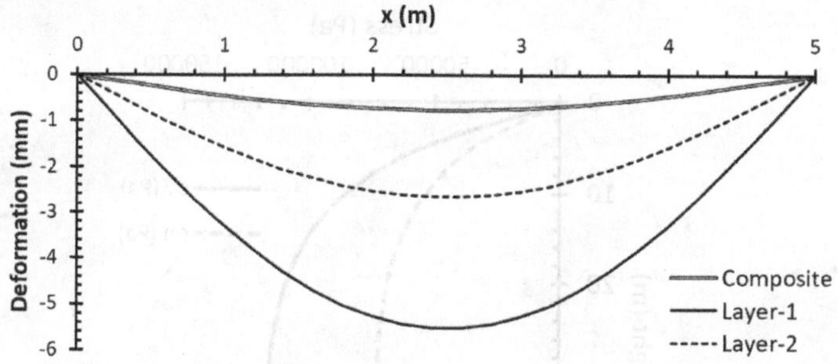

13.

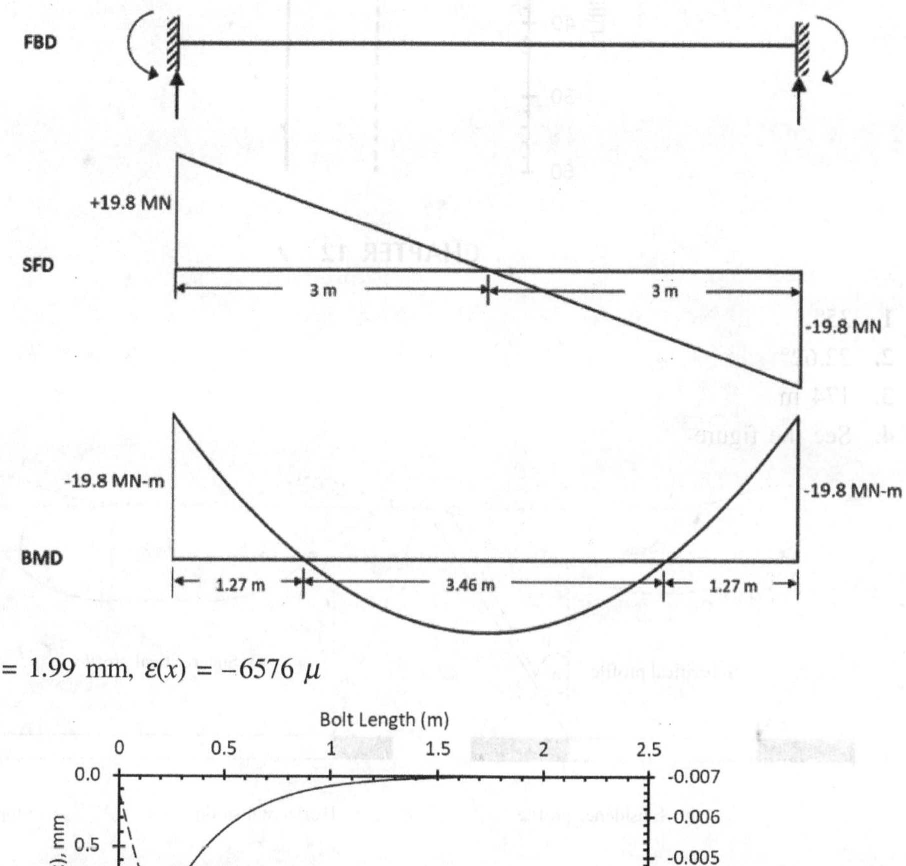

14. $\delta_{max} = 1.99$ mm, $\varepsilon(x) = -6576\ \mu$

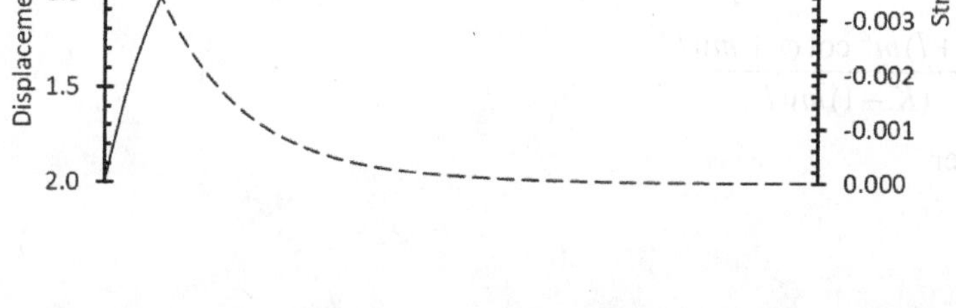

15. $\sigma_V = 1.84$ MPa, $\sigma_h = 0.75$ MPa

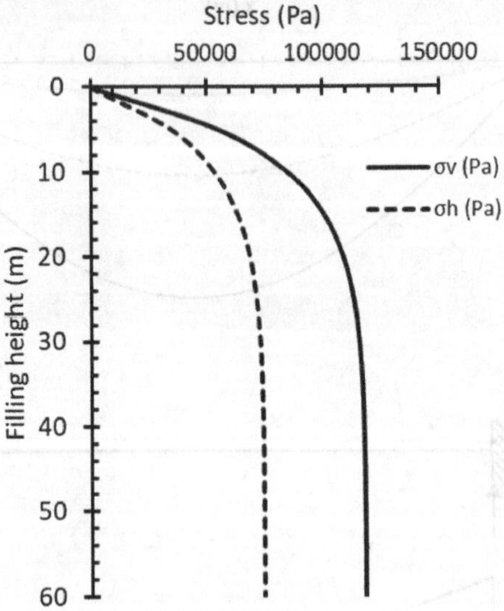

CHAPTER 12

1. 35°
2. 22.62°
3. 174 m
4. See the figure

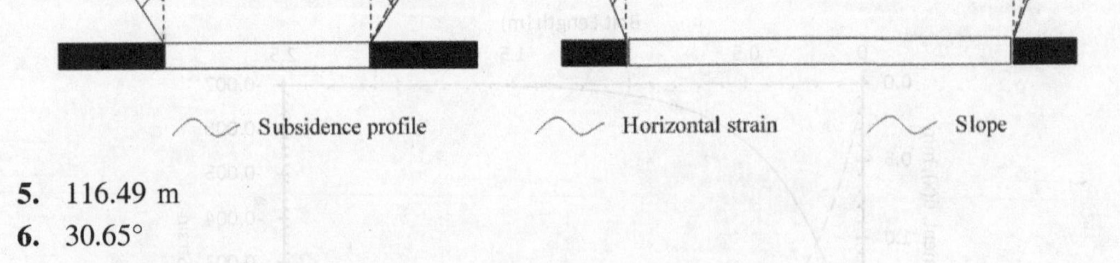

5. 116.49 m
6. 30.65°
7. $z = \dfrac{4(w+l)m^2 \cot\varphi + mwl}{(K-1)mwl}$
8. No danger

9. 149.72 m, – 0.015 radian, 37.73°
10. Horizontal and Vertical displacement is 14.09 mm and 37.7 mm
11. 2.71 m, 726.31 m
12. 70.27 m towards dip direction
13. $\varepsilon_t^p = 3.59 \times 10^{-3}$, $\varepsilon_c^p = 6.04 \times 10^{-3}$, Maximum secondary compressive strain is 6.05×10^{-3}
14. 27.1°

CHAPTER 13

1. $d = h - \frac{w}{c}\left[\sin^2\alpha - \frac{1}{2}\sin 2\alpha \tan\varphi\right]$, $b = \frac{w}{c}\left[\frac{1}{2}\sin 2\alpha - \cos^2\alpha \tan\varphi\right] - h\cot\beta$
2. (a) 1.92, (b) 2.08 GN, (c) 565.34 m, (d) 2.00, (e) 1.72
3. (a) 0.79, (b) 0.78
4. The factor of safety of the slope decreases with increase in orientation of discontinuity plane (α) and following a power law, $\text{FOS} = 34.89 \times \alpha^{0.99}$. Where α is in radian.

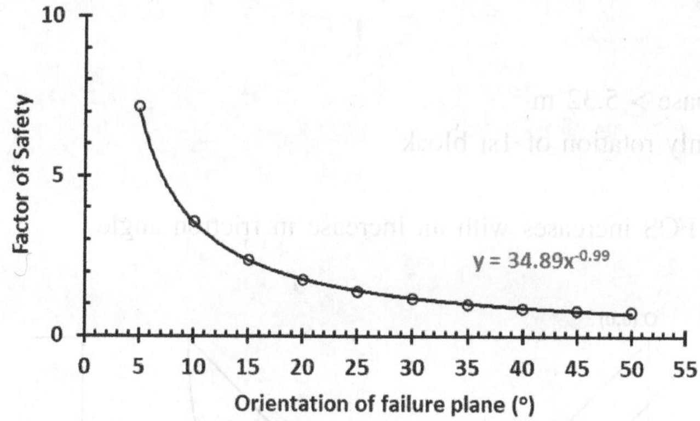

5. (a) 0.91, (b) 0.68
6. $\text{FOS}_1 = \text{FOS}_2$,

$$\text{FOS}_1 = \frac{N_1 \tan\varphi}{W_1 \sin\alpha_1 + R\cos\alpha_1 - T\sin\alpha_1}, \quad \text{FOS}_2 = \frac{N_2 \tan\varphi}{W_2 \sin\alpha_1 - R\cos\alpha_2 + T\sin\alpha_2}$$

7. FOS = 1.00
8. (a) Yes as shown in sterographic projection, (b) Plunge 33.45° in clockwise with horizontal plane, Trend is 311.35° in anti-clockwise (c) L_1 = 30.89 m, L_2 = 63.92 m, L_3 = 35.15 m, L_4 = 63.61 m, L_5 = 36.28 m (d) A_A = 497.05 m^2, A_B = 1108.93 m^2, V = 6909.16 m^3, (e) 1.54 [Note: Refer: Figure given in Example 13.6 for wedge nomenclature]

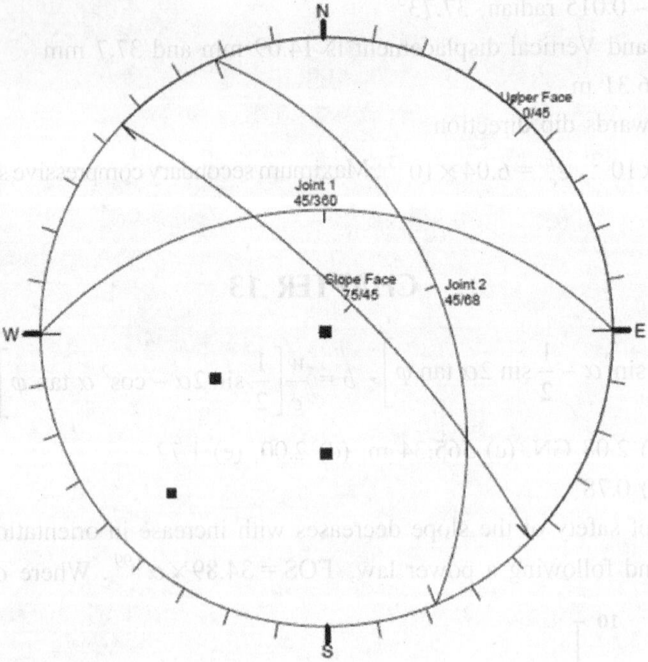

9. Side of the base > 5.32 m
10. No sliding only rotation of 1st block
11. 0.25
12. (a) 1.93, (b) FOS increases with an increase in friction angle
13. FOS = 3.84

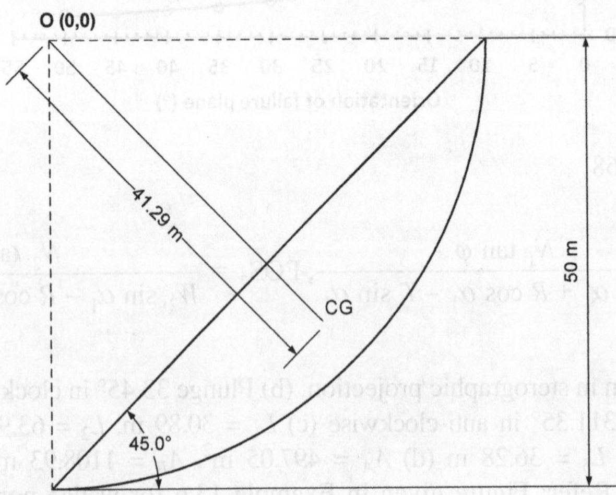

14. (a) FOS = 2.55, (b) FOS = 4.13
15. (a) FOS = 0.99, (b) FOS = 0.45
16. (a) FOS = 0.73, (b) FOS = 0.17

17. Least factor of safety is 1.24 for joint passing through toe

18. Least factor of safety is 1.32 for joint passing through toe

19. (a) 1.00, (b) 1.27

20. (a) 1.40, (b) (i) 0.25 and (ii) 0.73

CHAPTER 14

1. (i) In physical modelling, a field structure or an excavation is made in the laboratory by scaling down the dimensions and material properties using the dimensional analysis. Generally, density of field materials to those of physical model is kept fixed while modulus of elasticity and strengths are scaled down based on the dimensional proportionality constant. Preparation of equivalent materials in laboratory is a tedious process. The displacements and stresses those are observed during excavations in the physical model are then scaled up using the same proportionality constants. This way the physical behaviour of the structure are studied.

(ii) Empirical models are developed based on the curve fitting exercises of the data obtained from a series of laboratory or field tests. These data are fitted by means of function (polynomial, exponential, etc.) and generally a statistical regression analysis is employed. The characteristics of which are chosen just to meet the shape of the experimental design. For example, failure criteria like Mohr-Coulomb for intact rock (Eq. 9.15), Hoek and Brown for rock mass (Eq. 9.27 to 9.29), Barton's failure criteria for rock joints (Eqs. 10.16) are some of the popular empirical models. Other examples are also shown in Table 9.1 and Table 9.2.

(iii) Analytical models or closed form solution involves solving governing differential equations of a system or a body. As for example, the stress solutions given in Eqs 7.1 and 7.18 to 7.22 in Chapter 7 are derived based on the closed form solution from Airy stress function. In this case, understanding of proper boundary conditions is necessary. Analytical solutions may not be available or difficult to obtain for complex geometrical structures, multiple material layers and varying loading conditions.

(iv) Numerical models involve solving governing differential equations problems by discretizing a domain with proper boundary conditions. As for examples, FEM and DEM are continuum methods and UDEC is a discontinuum numerical method. Numerical methods are typically used when the problem cannot be solved analytically or when the analytical solution is too complex to compute. The application of numerical models for different problems are shown in Figures 14.10 to 14.11 and 14.14 to 14.16.

2. In rock engineering problems, mostly continuum methods are applied for evaluating deformations and induced stresses around excavations and pillars. Continuum methods are easy to apply if materials are considered to be isotropic and they provide quick results if elastic analysis is performed. On the contrary, if material or geometric nonlinearity are assumed, the analysis may take longer time depending on the dimensions of the "Region of Interest (ROI)". On the other hand, if the ROI contains multiple discontinuities then discontinuum models are best suited for simulating deformations and stresses. In this case, materials are anisotropic and hence, discontinum modelling approach is the best suited.

The equivalent material method approach can also be used to simulate discontinuities using continuum based methods such as finite element method (FEM), finite difference method (FDM), boundary element method (BEM), etc. Example of discontinuum method are discrete element method (DEM), particle flow code (PFC), Distinct element method (DEM), Discontinuous deformation analysis (DDA), Key block approach, Lattice models, Discrete fracture network method (DFN) and others.

3. The FEM, FDM, and BEM are numerical techniques used to solve partial differential equations. FEM is a numerical technique used to approximate solutions to differential equations by dividing the domain into smaller, simpler elements. Each element is represented by a set of equations, and the overall solution is obtained by combining the solutions of all elements. FDM is a numerical technique that approximates derivatives in differential equations by discretizing the domain into a grid of points. The differential equations are then approximated using finite difference formulas at each grid point.

BEM is a numerical technique that focuses on solving boundary value problems by discretizing only the boundary of the domain. The solution is obtained by representing the problem in terms of boundary integrals, which can be solved analytically or numerically. BEM is often used in problems involving potential flow, acoustics, and electromagnetic fields.

The drawback of these continuum method is that its formulation is based on appropriate stress-strain law for the material may not exist, or the law may be excessively complicated with many obscure parameters. The natural development of cracks is not handled properly by continuum methods. In that case, special or additional governing equations may be incorporated along with the momentum and mass balance equations. FEM is the most widely employed numerical method for rock mechanics and rock engineering followed by FDM.

4. Discrete or distinct element method (DEM) is used to describe the mechanical behaviour of discontinuous bodies introduced by Peter Cundall in 1971. Example of commercial codes based on this method are UDEC (Universal Distinct Element Code), 3DEC (Three-Dimensional Distinct Element Code), PFC (Particle flow code). The displacement discontinuity method (DDM) was first proposed by Couch in 1976 is one of the boundary element method (BEM) for modelling discontinuities in rock mass. In DDM, rock mass is represented as a distinct blocks each with their own displacement field. The block interaction results into sliding, closing and opening of the discontinuous plane.

5. Bord and pillar mining method is carried out in a sedimentary deposit. A sedimentary deposits are forming a bedding plane (a type of geological discontinuities) in which one rock types (such as coal, sandstone and shale) are lying on top of each other. This problem consists of two design problem: (i) design the Pillars, and (ii) design of Bord (i.e., Opening). For the solution of this problem both the continuum and discontinuum methods can be used. For the selection of the numerical model for a particular problem, we should know the physics, i.e., the force-displacement behaviour of rock structure.

The most popular method used to solve this problem is finite element method. There are three steps to develop the model: (i) pre-processing, which includes developing continuous geometry, define material properties, select of interpolation function and element type, discretize the geometry or domain of analysis, impose the boundary conditions like stresses and displacement, etc.; (ii) Solution, in this step required results are defined along points, line or area such as displacement, stresses, etc. and finally solving the system of equations; (iii) post-processing, in which the result output are collected for interpretation. Figure 8.1, shows the geometry and boundary conditions defined to solve the problem of a Bord and pillar mining method. The results are analysed for elastic and elasto-plastic properties of the rock mass, the vertical stress plots are shown in Figure 8.2.

6. Finite element is an approximate numerical solution technique in which continuous system are discretised into many small and simple pieces of finite size is called an element. Adjacent elements are connected on the boundaries by nodes. In addition to boundary nodes, an element may also have a few interior nodes. For each element it is necessary to make an assumption as to how the primary variable (such as Displacement, temperature, etc.) distributed in terms of geometric position. The nodal values of the field variable and the interpolation functions of the elements completely define the behaviour of the field variable within the element. These interpolation function are known as shape function. This function is estimated from the assumed polynomial (linear, quadratic, cubic, quartic, etc.) for element type. The strain tensors at a point inside an element is then related with spatial derivatives of the nodal shape functions and nodal displacements.

7. (i) In an interior problem, the boundaries enclose the problem domain, whereas in an exterior problem, the domain is an infinite or semi-infinite region outside the boundary of concern. Exterior problems are of greater interest in engineering rock mechanics and rock engineering.

(ii) Explicit and implicit methods are approaches used in numerical analysis for obtaining numerical approximations to the solutions of time-dependent ordinary and partial differential equations, as is required in computer simulations of physical processes. Explicit methods calculate the states (such as displacements and pressures) of a system at the next time step from the states of the system at the current time. For example, displacement at a point, u at time $t + 1$, can be found directly for a function involving variables such as force f, stiffness k and u at time t as:

$$u_{t+1} = h(u_t, f_t, k_t)$$

The implicit methods find a solution by solving an equation involving both the current states of the system and the later one. In this case, displacement u at $t + 1$ may also be dependent on force f at $t + 1$ as given as:

$$u_{t+1} = h(u_t, f_{t+1}, k_t)$$

8. (i) Equivalent material modelling is a technique in which an equivalent stiffness mostly in the form of elastic modulus or constitutive matrix is developed for a jointed rock mass. This equivalent rock material will then be considered the rock mass and is free from discontinuities. For example, the concept of equivalent material modelling is discussed as section 10.5 as a concept of joint compliance of rock mass with one, two, three or more joint sets.

(ii) XFEM is a very flexible method that can handle linear and non-linear fractures initiation and their growth developed by Belytschko, et al. (1999). The method employs special functions that are added to the finite element approximation using the framework of partition of unity. For crack modelling, a discontinuous function and the two-dimensional linear elastic asymptotic crack-tip displacement fields are used to account for the crack. This enables the domain to be modelled by finite elements without explicitly meshing the crack surfaces. The location of the crack discontinuity can be arbitrary with respect to the underlying finite element mesh, and quasi-static or fatigue crack propagation simulations can be performed without the need to re-mesh as the crack advances.

(iii) SPH is a meshfree method proposed by Gingold and Monagham in 1977. It is based on a Lagrangian formulation where a set of moving particles represent a continuum where each particle represents a predefined mass or volume. The state of system is represented by particles, typically fixed mass, which possess material properties, follow the motion of the system, advancing contact discontinuities, preserving Galilean invariance, and reducing the computational diffusion of various material properties including momentum. The equation of governing evolution of the system can be formulated in such a way that mass, momentum and energy are conserved exactly.

9. In rock blasting, generally two types of analysis are performed once the data are available. They are: (i) ground vibration analysis, and (ii) rock fragmentation analysis. Several parameters are considered to be responsible in ground vibration analysis. They are: (i) charge per delay, (ii) distance from the blast site, (iii) rock mass conditions, (iv) rock properties, and (v) blast design parameters such as burden spacing, diameter, length of the blast holes. Once all these data are available from several blasts, ANN can be used to forecast ground vibrations. If the classification of blast data is required, then SVM, tree classification, and random forest types of analysis can also be applied.

Rock fragmentation analysis may be required using CNN and DL methods to estimate the boulder size distributions and also to determine mean size. In this case, many images of blasted rocks are to be processed using the above methods.

10. Artificial intelligence (AI) refers to the field of research in computer science that focuses on developing methods and software enabling machines to exhibit intelligence. These intelligent machines, often computer systems, can perceive their environment, learn from it, and take actions to achieve specific goals. The field of Machine learning (ML) is a subset of artificial intelligence (AI). ML heavily relies on labeled data for training and making predictions. Deep learning is a captivating branch of machine learning that harnesses the power of artificial neural networks (ANNs) to learn from data.

Nowadays, numerous sensors are installed to obtain deformations, stresses and loads in any rock engineering project. These sensors are installed at multiple key locations of the excavations, pillars and supports. These data are captured and recorded every minutes if not in seconds. These huge set of time-series data contains information related to rock cutting/blasting and supporting of the walls and roofs. Therefore, AI/ML techniques are very useful for the analysis of these huge data sets and extract relevant information so that productivity and safety can be ensured or improved. This information will be vital for taking in-time decision.

No, numerical techniques analyze the mechanical responses of the structures based on the geometry, materials and load/boundary conditions. On the other hand, AI/ML are the sensor/data/image driven tools for evaluating the responses from key locations of the excavations. Therefore, the results of AI/ML techniques can be the complimentary inputs to the numerical analysis methods.

11. Dump slope stability analysis is generally performed by limit equilibrium method (LEM) or by numerical modelling methods. As mentioned before, deformations of benches can be monitored by installing tilt-meter, prisms, radar/laser based monitoring devices. AI/ML techniques will be useful to analyze those data and determine a trend of deformation patterns of different benches. These trend can be used to validate the results of LEM of numerical methods. Apart from that deep learning methods can be applied by taking images of the benches and determining the particle size distributions and then correlating them with material properties like cohesion, friction angle, and bulk density.

12. Rock bolts are regularly pulled out in the field and load-deformation data are recorded. These data are the direct manifestation of the anchoring capabilities with the given roof rock conditions, grout properties, material properties of bolt rod and its geometry. Apart from that dimension of the excavations and *in situ* stresses also play a vital role. If several such data are available, ANN or any classification method can be applied to evaluate/forecast the bond strength and stiffness based on all these geometric, material, load and spatial variables.

Index